Intelligent Systems Using Semiconductors for Robotics and IoT

The First International Conference on Intelligent Systems Using Semiconductors for Robotics and IoT (ICISRI-2024) brings together leading researchers, academicians, industry professionals, and innovators from around the world to explore the convergence of semiconductor technologies with intelligent systems for next-generation Robotics and the Internet of Things (IoT). This pioneering event aims to foster interdisciplinary collaboration and knowledge exchange in areas such as AI-enabled embedded systems, smart sensor networks, VLSI design for intelligent automation, and real-time computing for robotic applications.

With the rapid evolution of intelligent machines and connected environments, semiconductors play a pivotal role in enabling compact, efficient, and high-performance systems. The conference will feature keynote addresses by globally recognized experts, technical paper presentations, workshops, and panel discussions that highlight both theoretical advancements and practical implementations.

ICISRI-2024 has served as a vibrant platform for showcasing emerging innovations, discussing future challenges, and building research-industry networks. 83 research papers are selected out of 355 papers received from different organizations across India and abroad. The double-blind peer review process is performed to enhance the quality work done by researchers.

Intelligent Systems Using Semiconductors for Robotics and IoT

First International Conference, ICISRI-2024, Jaipur, India, November 29-30, 2024 Proceedings

Edited by

Dinesh Goyal
Payal Bansal
Ritam Dutta
Madhav Sharma

CRC Press is an imprint of the
Taylor & Francis Group, an **informa** business

First edition published 2026
by CRC Press
4 Park Square, Milton Park, Abingdon, Oxon, OX14 4RN

and by CRC Press
2385 NW Executive Center Drive, Suite 320, Boca Raton FL 33431

British Library Cataloguing-in-Publication Data
A catalogue record for this book is available from the British Library

ISBN: 978-1-041-20405-3 (hbk)
ISBN: 978-1-041-20408-4 (pbk)
ISBN: 978-1-003-71638-9 (ebk)

DOI: 10.1201/9781003716389

Typeset in Times New Roman
by Aditiinfosystems

Intelligent Systems Using Semiconductors for Robotics and IoT – Dinesh Goyal et al. (eds)
© 2026 Taylor & Francis Group, London, ISBN 978-1-041-20408-4

Contents

Intelligent Systems Using Semiconductors for Robotics and IoT – Dinesh Goyal et al. (eds)
© 2026 Taylor & Francis Group, London, ISBN 978-1-041-20408-4

List of Figures

Intelligent Systems Using Semiconductors for Robotics and IoT – Dinesh Goyal et al. (eds)
© 2026 Taylor & Francis Group, London, ISBN 978-1-041-20408-4

List of Tables

Intelligent Systems Using Semiconductors for Robotics and IoT – Dinesh Goyal et al. (eds)
© 2026 Taylor & Francis Group, London, ISBN 978-1-041-20408-4

Foreword

It is with great pleasure and profound enthusiasm that we welcome you to the First International Conference on Intelligent Systems Using Semiconductors for Robotics and IoT (ICISRI-2024). This pioneering event marks a significant milestone in bringing together researchers, academicians, industry experts, and innovators from across the globe to explore the transformative potential of semiconductor technologies in shaping the future of intelligent systems.

The rapid advancements in semiconductor devices have become the backbone of modern electronics, enabling unprecedented progress in the domains of robotics, automation, and the Internet of Things (IoT). These breakthroughs are not only revolutionizing computational performance and energy efficiency but also paving the way for novel architectures and intelligent functionalities. As robotics and IoT converge to redefine industrial processes, healthcare solutions, transportation systems, smart cities, and domestic automation, the role of semiconductors has never been more pivotal.

ICISRI-2024 serves as a multidisciplinary platform to exchange cutting-edge research findings, practical insights, and visionary perspectives. The conference fosters dialogue on emerging trends such as low-power chip design, AI-enabled embedded systems, advanced sensor integration, autonomous robotics, and secure IoT infrastructures. Our carefully curated technical sessions, keynote addresses, and panel discussions are designed to spark collaboration between academia and industry, encouraging the translation of innovative concepts into impactful real-world applications.

This inaugural edition of ICISRI embodies the spirit of innovation, collaboration, and sustainability. We believe that the exchange of ideas here will not only address current technological challenges but also inspire new pathways for research and development. By bridging knowledge gaps and promoting interdisciplinary synergy, the conference aims to contribute toward building a smarter, more connected, and more efficient world.

We extend our heartfelt gratitude to the organizing committee, advisory board, and review panel for their relentless efforts in ensuring the highest quality of technical content. We also acknowledge the contributions of our keynote speakers, session chairs, and authors, whose dedication and expertise have enriched the proceedings of ICISRI-2024. Finally, we thank our sponsors and partners for their invaluable support in making this event possible.

As you immerse yourself in the sessions, discussions, and networking opportunities, we encourage you to explore new collaborations, challenge conventional thinking, and envision the next generation of intelligent systems. May ICISRI-2024 ignite ideas that shape the future.

Welcome to ICISRI-2024 — where semiconductors power intelligence, and intelligence powers the future.

Intelligent Systems Using Semiconductors for Robotics and IoT – Dinesh Goyal et al. (eds)
© 2026 Taylor & Francis Group, London, ISBN 978-1-041-20408-4

Acknowledgement

The Organizing Committee of the First International Conference on Intelligent Systems Using Semiconductors for Robotics and IoT (ICISRI-2024) extends its heartfelt gratitude to all those whose guidance, support, and commitment made this event a resounding success.

We sincerely thank our esteemed Chief Patrons: Mr. Hari Singh Shekhawat, Director (Infrastructure), Poornima Group, Jaipur and Mr. MKM Shah, Director (Admin & Finance), Poornima Group, Jaipur, India.

We also thank our General Chair: Prof. (Dr.) Carlos M, Travieso-Gonzalez, Department of Signals & Communication (IDeTIC), University of Las Palmas de Gran Canaria, Las Palmas, Spain.

We also thank our Conference Chair: Prof. (Dr.) Dinesh Goyal, Principal & Director, PIET, Jaipur, India, Organizing Chair, Prof. (Dr.) Payal Bansal, Head - Dept. of IoT, PIET, Jaipur, India, our technical committee members, finance committee and publicity chairs for tremendous support for successful completion of the event.

Our profound appreciation goes to the distinguished keynote speakers and session chairs for sharing their invaluable insights and expertise. We are equally grateful to all authors and participants for their scholarly contributions, which have enriched the technical discourse of the conference.

Special thanks are due to our reviewers for their meticulous evaluation of submissions, ensuring high academic quality. We also acknowledge our sponsors and institutional partners for their unwavering support. Finally, we commend the tireless efforts of the organizing team and volunteers, whose dedication turned ICISRI-2024 into a memorable and impactful event.

Convenor
ICISRI – 2024

Intelligent Systems Using Semiconductors for Robotics and IoT – Dinesh Goyal et al. (eds)
© 2026 Taylor & Francis Group, London, ISBN 978-1-041-20408-4

About the Editors

Prof. (Dr.) Dinesh Goyal

Dr. Dinesh Goyal, B.E, M.Tech & PhD in CSE, working as Professor in CSE and Principal, Poornima Institute of Engineering & Technology. He has published 16 edited books, 3 SCI, 109 Scopus and 65 WoS indexed papers. He is life member of ISC, CSI, IETE & ISTE and senior member IEEE.

Prof. (Dr.) Payal Bansal

Dr. Payal Bansal, B.Tech, M. Tech & PhD in ECE, working as Head and Professor in IoT Dept., Poornima Institute of Engineering & Technology. She has published several edited books, referred research WOS and Scopus indexed papers and granted patents. She is life member of ISTE and senior member IEEE.

Prof. (Dr.) Ritam Dutta

Dr. Ritam Dutta, B. Tech, M. Tech & PhD in ECE, working as Professor in CSE (IoT) Dept., Poornima Institute of Engineering & Technology, Jaipur. He has published 5 SCI, 55 Scopus indexed papers, 3 books with 16 years of teaching experience. He is life member of IE(India), ISTE and IEEE. He is also a recipient of one Indian Patent grant.

Prof. (Dr.) Madhav Sharma

Dr. Madhav Sharma, Associate Professor and researcher, holds a B.E., M.Tech., M.S., and Ph.D. in Computer Science. With 16+ years of experience, multiple patents, and publications. He specializes in AI, ML, and Deep Learning. He leads curriculum design, research mentorship, and actively contributes to academic innovation and international workshops.

Intelligent Systems Using Semiconductors for Robotics and IoT – Dinesh Goyal et al. (eds)

1 Breast Cancer Detection Using ML Algorithm

Shruti Thapar*
Associate Professor, PIET, Jaipur, Rajasthan

Abstract: Breast cancer one of the largest killers among women and is a widespread disease in every corner of the world. It stems from the abnormal and first preliminary progression of growth by the breast cells, resulting in the formation of malignant tumors. A person may get breast cancer as a result of either a benign or malignant tumor. Early detection very much increases the survival rate since treatment could be effective. This study aims at determining the accuracy, precision, recall as well as specificity of other classification algorithms like Classification with Naive Bayes and Support Vector Machines.

Keywords: ML, SVM, KNN, Breast cancer, and Logistic regression

1. Introduction

Cancer and AIDS are just a few examples of the world's worst diseases. Most of the time, the process of manual detection complicates it for the doctors to classify the disease correctly. Therefore, machine-assisted diagnosis techniques are highly essential in the identification and management of cancer. Recently, machine learning has become of great interest because of an increase in computational power and enhanced data storage facilities to handle a huge volume of data. Predisposing symptoms can identify breast cancer caused by rapid cell division of the lobules of the milk producing glands. It is often asymptomatic, and thus the role of hereditary cancer screening becomes relevant in such cases.

2. Reflections on Theory

Classification is an indispensable part of any pattern recognition problem. It can be applied to solve linear and nonlinear issues with logistic regression working alone for the former and SVM with some nonlinear kernel functions working on the later.

2.1 Vector Machine Support

SVM is used to classify 3D space data points by describing a plane, which we can also called now a hyperplane. When a dataset is not able to have a linearly separable hyperplane, the nonlinear kernels sigmoid, Gaussian radial basis functions, and polynomial functions are employed for the creation of classifiers. A hyper plane in this context can be defined by: $\beta 0 + \beta 1X1 + \beta 2X2 + ... + \beta pXp = 0$, where $\beta 0$, $\beta 1$, ... βp are data coefficients.

2.2 K Nearest Neighbors (KNN)

KNN is a way to classify things or predict values that's really good at spotting patterns. It doesn't rely on any specific parameters, which makes it flexible. It finds new observations by calculating the distance of a particular point from the already existing data points, usually done by using the Euclidean metric. The classification would be done based on the nearest neighbors, and odd numbers are usually chosen to avoid ties in binary classifications.

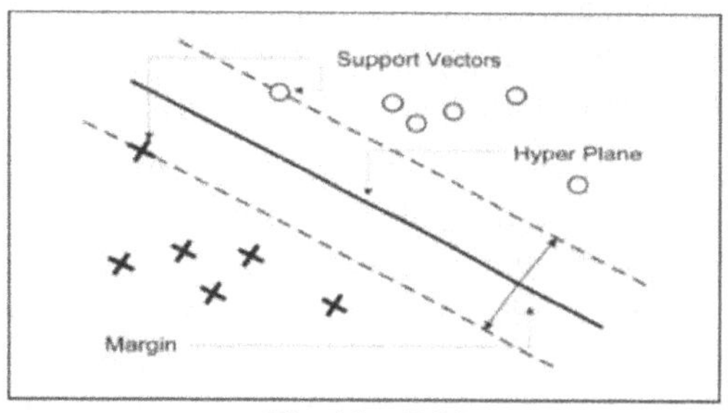

Fig. 1.1 SVM

*Corresponding author: shruti.thapar@poornima.org

DOI: 10.1201/9781003716389-1

2.3 Logistic Regression

Logistic regression projects a categorical target variable based on independent variables. It generates binary output from a threshold value, which values above go to category 1 and below to category 0. In logistic regression, there is also the inclusion of precision and recall values for the determination of edge cost, thereby affecting the outcomes.

3. Related Works

Most of the studies have been conducted based on machine learning techniques that can predict or classify a case of cancer. Previous research in this regard has been summed up regarding methodology, algorithm, and results, reflecting the upcoming power of MLi Healthcare.

Deep learning and Machine learning have become huge hot topics as really popular subjects. You will easily come across several research works in this area, demonstrating how well the different algorithms predict something. Andres Tsukamoto used classification methodology various domains such as medical diagnosis, credit scoring, and spam detection, among others. Their models achieved accuracies of 95% and 93%, respectively, indicating the strong performance of both methods. Similarly, Noreen Fatima et al. (2020) explored ML and DL approaches, focusing on Artificial Neural Networks (ANN), achieving an accuracy of 94%, reinforcing ANN's reliability in prediction tasks.

Hiba Asria and her team managed to reach a 95% success rate by using machine learning, especially decision trees. David A. Omondia and his group chose supervised learning using Random Forest, which gave them a 92% accuracy rate. They showed the different models perform in comparison to each other, pointing out that Random Forest is pretty reliable.

The elaborative study by Wenbin Yue and his team has compared SVM, Naive Bayes, and Logistic Regression. They observed the best performance from logistic regression at 97%, followed by the Support Vector Machine at 96%, and Naive Bayes at 88%. Later on, Sarthak Vyas and others presented a study in the year 2022 showcasing the great performance of both when combined with Supervised Learning, each around 92.7%.

Mamatha Sai Yarabarla and team, 2019 reviewed some of the supervised learning algorithms including SVM, RF, and KNN; with respective accuracy scores from models falling at 68%, 72%, and 70%. Their best performance turned out to be RF. Expanding on unsupervised machine learning, KNN, NB, and RF models were proposed in 2021 by Harinishree M. S. and group, with a topmost accuracy rate falling respectively: 70%, 92%, and 94% of the instances.

Ramik Rawal et al. in the year 2020 did supervised learning via the LR and RF algorithms, hence scoring an accuracy of 96% and 99%, respectively. They especially had to put in efforts into making their models score big. Similarly, another experiment was done by Shubham Sharma et al., in the year 2022, who experimented using the RF, KNN, and NB algorithms. Scoring reasonably well—94%, 95%, and 94%, respectively—assured consistency across all those models with respect to any algorithm being put to work.

4. Materials and Methods

Dataset Used: It follows a two-tier process: one for training a classification model with a set of class-labeled training datasets, and the other is to make a prediction with a validation dataset. Here, the authors employed a dataset with features predicting breast cancer based on the following attributes; the details are given below.

Some data attributes applied in these models, which are very critical in the diagnosis of breast cancer, are as follows: The important attributes in this dataset include a patient's ID and diagnosis, saying whether the tumor is benign or malignant. Under characteristics of breast lump include size and shape indicators like radius_mean and area_mean, complemented by finer shape details like smoothness_mean and concavity_mean. They reflect structural aspects with symmetry and fractal dimension. Standard errors depict measurement variations, whereas worst-case values show highest extremes recorded in radius, texture, perimeter, and area.

Data Prepping: Data pre-processing deals with the activities which help in cleansing the dataset by addressing both the issues related to structure and content. The techniques of feature selection or extraction may be applied to reduce the dimensionality and retain critical features.

Applied ML Algorithms: ML classification techniques are performed, such as SVM, KNN, and Naïve Bayes, which classify the tumors as benign or malignant. Feature reduction will be done to train the model for better classification accuracy.

4.1 Evaluation of Performance

In this context, assessing the algorithm's performance consists of comparing prediction with observation, constructing a confusion matrix, determining accuracy, precision, recall, and specificity.

Table 1.1 Evaluation parameters

Devices	Colab by Google
Sources of the Dataset	Kaggle.com
Use of Libraries	Matplotlib, Numpy, Pandas, Sklearn, and Keras

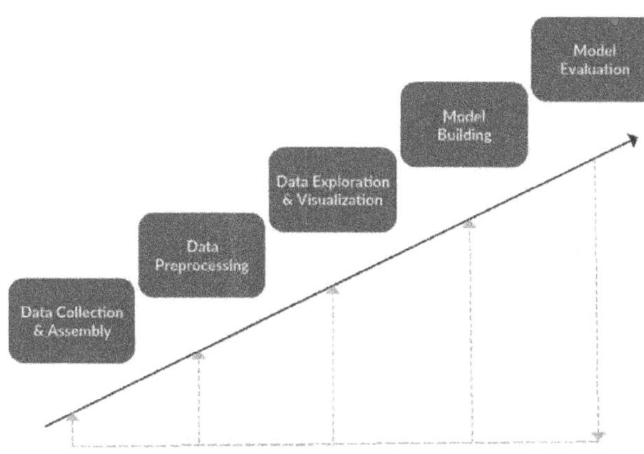

Fig. 1.2 Data preprocessing

5. Results and Discussion

Logistic Regression got the highest accuracy at 98.06%, with SVM coming in second at 97.70%, and Naïve Bayes at 91.74%.

Table 1.2 Comparative analysis of various prediction algorithms

Performance Measure(s)	Efficien-cy	Preci-sion	Reten-tion	Specific-ity	F1_ score
LR	0.98	0.96	0.98	0.98	0.97
SVM	0.97	0.96	0.97	0.96	0.96
NB	0.91	0.86	0.91	0.84	0.88

Table 1.2 Experimental Tools

Figures: - We've put all our model's stats—like how accurate they are, their precision, recall, specificity, and f1-score—into these graphs for comparison. Finally, we have contrasted all of the algorithms' various performance metrics.

Regarding Logistic Regression, the exhibition of Accuracy, Recall, Precision, Specificity, and F1Score has been visualized in Fig. 1.3.

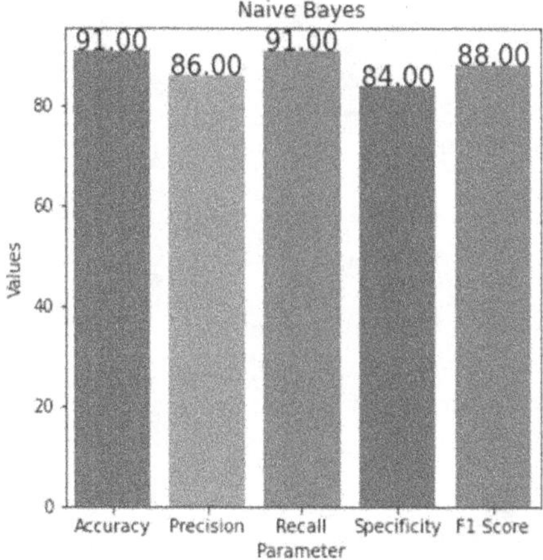

Fig. 1.3 Naïve bayes

Figure 1.3 shows the Support Vector Machine evaluation results with F1 Score, Specificity, Accuracy, Recall, and Precision.

6. Conclusion

Prediction of breast cancer is very important for timely detection and treatment. In this, a model was presented using SVM and KNN, whereas logistic regression had the best performance with an accuracy of 98.15%. Early detection models may save lives by enabling early interventions.

7. Scope for Future Work

This study shows that future research will apply more extended deep learning models based on new datasets to improve further to assist doctors in improving patients' outcomes.

References

1. Sharma, Shubham, Tanupriya Choudhury, and Archit Aggarwal. "Detection of breast cancer through The Conference on Electronics, Mechanical Systems, and Computational Techniques (CTEMS) 2018 featured machine learning algorithms.IEEE, 2018.
2. Pragnyaban Mishra, Suvarna Vani Koneru, Sri Hari, and Nallamala. The machine-learning approach for detecting breast cancer.1402–1405 in IntJ Recent Technol Eng 8.2–3 (2019).
3. Nishtha Kesswani, Monika, and Jyotiyana, "A Case Study of Neurodegenerative Diseases and Other Brain Disorders in Deep Learning.Growing Risks in Expert Solutions and Applications. Singapore: Springer, 2021.791–799.
4. Mohammed, Siham A., et al. "The employment of different techniques for breast cancer detection using machine learning." International Conference on Big Data and Data Mining. Singapore: Springer, 2020.
5. Shruti Thapar, Krati Sharma, Dharmveer Yadav, BudeshKanwer, Ashish Raj, Absolute Data Security Scheme: A Cutting-Edge Data Security Approach" Shukla, R., Goyal, D., Thapar, S., Chhabra, R.ACM International Conference Proceeding Series, 2022, 41
6. Thapar, S., Sharma, S., Saxena, V., and Singh, U.P. The future effects of intelligent vehicles by data science, ACM International Conference Proceeding Series, 2023, 112, May 13, 2024 (ISBN: 979-8-4007-0941-8). International Conferencing on Information Management & Machine Intelligence (ICIMMI 2023) will take place from November 23 to 25, 2023, in Jaipur,
7. Suvarna Vani Koneru, Pragnyaban Mishra, Sri Hari, and Nallamala Detecting breast cancer through machine learning" by. IntJ Recent Technol Eng 8.2–3 (2019), 1402–1405.
8. Banupriya Choudhury, Archit Aggarwal, Shubham, and Sharma. Methods using machine learning for breast cancer diagnosis. In International Conference on Computational, Mechanical and Electronic Systems Techniques (CTEMS, 2018). IEEE, 2018.

9. Shruti Thapar and Sudhir Kumar Sharma's paper In the Advanced Science and Technology International Journal, 29(6), 9401-9411 published in 2020 has an article entitled "Evaluation of Isolation Access of Wormhole Attack from Mobile Adhoc Network Using Delay Prediction Method."
10. Shruti Thapar, Sudhir Kumar Sharma, The delay prediction methodology was preached by IJRTE, the International Journal of Recent Technology and Engineering, dealing with the separation of wormhole attack access from mobile ad-hoc network. Volume 8: Issue 6, March 2020, ISSN: 2277–3878 (Online).
11. Shruti Thapar and Anshuman Kalla, This book chapter addresses an analysis of routing protocol performance evaluation in MANET, The proceedings of the International Conference on Recent Cognizance in Wireless Communication & Image Processing published pp. 59–68, N. Afzalpulkar et al. (eds.), Springer India 2016 DOI 10.1007/978-81-322-2638-3_7
12. Sudhir Kumar Sharma and Shruti Thapar. The paper The paper titled "Attacks and Security Issues of Mobile Ad Hoc Networks" was presented at the International Conference on Sustainable Computing in Science, Technology, and Management (SUSCOM-2019), which was held at Amity University Rajasthan, Jaipur, India, from February 26 to February 28, 2019, on pages 1463–1470.
13. Shruti Thapar, Amol Purohit, Budesh Kanwer, and Akash Jaiman, This research paper is titled "A Direct Trust-Based Detection The paper delineates a methodology for detecting a wormhole attack in mobile ad hoc networks. It was published by the International Journal of Intelligent Systems and Applications in Engineering in its November-December 2019 issue, volume 11, number 6s, pages 276–283.
14. Gayathri, B. M., Sumathi, C. P., and Santhanam, T. "A survey on the use of machine learning algorithms for breast cancer diagnosis."(2013).
15. Tahmooresi, M., and others. 10.3-2 (2018): 21–27; Journal of Electronic, Computer, and Telecommunication Engineering (JTEC). "Early detection of breast cancer using machine learning techniques."
16. Islam, Md., et al.1–14 in SN Computer Science 1.5 (2020). The study deals with the breast cancer prognostication analysis using machine learning techniques.
17. Madhuri, Bharat Gupta, and Gupta. ICCMC The 2nd International Conference on Communication and Computing Techniques organized an event in the year 2018 comparing supervised machine learning techniques for diagnosis of breast cancer. IEEE, 2018.
18. Tolga Ensari, Pınar Kırcı, Ebru Aydındag, and Bayrak. The 2019 Biomedical Engineering, Computer Science, and Electrical-Electronics Scientific Meeting (EBBT) compared machine learning techniques for diagnosing breast cancer.IEEE, 2019.
19. Md. Milon Islam, Md. Hasib Iqbal, Md. Kamrul Hasan, and Md. Rezwanul Haque (2017). K-Nearest Neighbors and Support Vector Machines are potential models while predicting breast cancer.
20. Vaka, Sudheer Reddy, Badal Soni, and Anji Reddy. " Machine learning-based breast cancer detection." 320–324 in ICT Express 6.4 (2020).

Note: All the figures and tables in this chapter were made by the authors.

Intelligent Systems Using Semiconductors for Robotics and IoT – Dinesh Goyal et al. (eds)
© 2026 Taylor & Francis Group, London, ISBN 978-1-041-20408-4

2 Transforming Education: The Influence of IoT on Learner's Connectivity and Instruction

Sonia Kaur Bansal[1]
Professor, Department of English Language and Soft Skills,
Poornima Institute of Engineering and Technology,
Jaipur, India

Abstract: A revolutionizing education in the form of Internet of Things or IoT is a distinguished one in its form for creating a network for interconnecting the devices that facilitate seamless communication, sharing of important data with interactive learning environments. This paper explores the transformative role of IoT in enhancing student engagement, personalizing learning experiences, and improving instructional efficiency. IoT technologies such as smart classrooms, wearables, and adaptive learning platforms are making education more responsive and dynamic, allowing real-time feedback, tailored instruction, and remote learning opportunities. The paper suggests that future advancements in IoT will further enhance personalized learning, optimize educational outcomes. While IoT holds promise for revolutionizing education, its success hinges on addressing the ethical and practical concerns surrounding data protection and resource allocation.

Keywords: Internet of things (IoT), Education technology, Learner connectivity, Personalized learning, Smart classrooms, Instructional innovation

1. Introduction

This new form of education, IoT or the Internet of Things significantly indicates the function of interconnected devices that effectively communicate and share data with a wide range of internet. In education, IoT technologies provide innovative solutions for enhancing learner connectivity and instructional practices. This paper investigates how IoT is reshaping education, focusing on its impact on student engagement, personalized learning, and instructional efficiency. The development of Internet of Things (IoT) has greatly affected the various sectors as being the most influential factor for fast evolving domains. The Internet of things efficiently connects the devices for sharing the data without being dependent on the intervention of human beings. It is also increasingly influencing how educational institutions operate and how learners engage with instructional materials. The growing integration of IoT in education promises to revolutionize traditional teaching methodologies, learner engagement, and the overall educational experience.

Historically, the incorporation of technology in education has gone through multiple phases, from the introduction of computers in classrooms to the widespread adoption of e-learning platforms. However, the introduction of IoT brings a new wave of possibilities for empowering the actual time collection of data, and creating connectivity for smart functioning smart systems in the educational environment. With IoT-enabled tools, educational institutions can create smart classrooms, where the various tools like smart boards, to be worn devices interact to provide the enhanced learning experiences at personal levels. These technologies have the potential to address the diverse needs of students by tailoring content delivery, tracking learning progress, and providing instant feedback, thereby improving learner engagement and academic outcomes.

2. The Role of IoT in Enhancing Learner Connectivity

IoT facilitates seamless communication and interaction between students, educators, and educational resources.

*Corresponding author: Sonia.bansal@poornima.org

DOI: 10.1201/9781003716389-2

Smart devices, such as tablets, smartphones, and wearables, enable real-time data collection and feedback, promoting a more interactive and responsive learning environment[10]. For example, smart classrooms equipped with IoT sensors can track students' attention and participation levels, allowing educators to adjust their teaching methods accordingly. Furthermore, IoT supports remote learning by providing students with access to educational resources and collaboration tools from anywhere, thus bridging geographical gaps. The IoT or Internet of Things is wonderfully reshaping the educational landscape by improving connectivity between learners, educational content, and instructors. IoT enables interconnected devices to share real-time data, creating a dynamic, immersive learning environment. This increased connectivity enhances access to educational resources, facilitates collaboration, and promotes personalized learning, transforming traditional educational paradigms.

2.1 Expanding Access to Educational Resources

In specific reference to the education field, IoT has the vast abilities to provide the easy approach to educational resources beyond the classroom. IoT-enabled devices like smartboards, tablets, and sensors can create a network that provides learners with continuous access to instructional materials. By connecting devices, learners can easily retrieve lessons, videos, and interactive applications from anywhere with internet connectivity. For instance, IoT allows for the seamless sharing of learning resources across a network of devices, eliminating geographical constraints and enabling the learners to explore high quality of educational content. According to a study by Zhang et al., IoT-based educational tools increased the number of students participating in online learning by over 40% in underdeveloped regions, significantly enhancing connectivity and resource accessibility[15]. This broader connectivity helps in addressing the digital divide, offering equal educational opportunities to students regardless of their location.

Table 2.1 Impact of IoT on learner access to educational resources

Parameter	Traditional Learning	IoT-Enhanced Learning
Geographical Accessibility	Limited to classroom or campus	Accessible from anywhere
Resource Availability	Limited to textbooks and manuals	Real-time, diverse multimedia content
Inclusivity	Less access for remote learners	Equal access through connected devices
Learner Participation	Restricted by location	Expanded through IoT devices

2.2 Facilitating Collaborative Learning

IoT enhances connectivity between learners, enabling more efficient collaboration both in and out of the classroom. Smart classrooms equipped with IoT devices allow students to collaborate on projects in real-time, sharing documents, ideas, and learning experiences. Wearable devices and smartboards enable simultaneous input from multiple users, developing an interactive and cooperative learning environment.

Additionally, IoT facilitates connectivity between students and educators. The teachers can also observe the actual and accurate performance of the learners for offering valuable guidance, leading to more responsive teaching practices. Devices such as wearable fitness trackers are being used to measure student stress levels during learning activities, providing valuable insights into student well-being and helping instructors tailor their teaching methods accordingly.

Table 2.2 IoT tools for enhancing collaborative learning

IoT Tool	Purpose	Impact on Learning
Smartboards	Facilitates real-time collaboration	Encourages interactive group learning
IoT Collaboration Platforms	Allows file sharing and project collaboration	Improves teamwork and project management
Wearable Devices	Monitors learner well-being and engagement	Personalizes instruction based on feedback

2.3 Personalized Learning and Adaptive Instruction

IoT enables personalized learning by collecting data from learners' interactions with educational devices and adapting the learning experience to meet their individual needs. IoT devices such as smart tablets and learning management systems (LMS) can monitor the learning patterns of the learners. Later on, received data is deeply assessed and analyzed to provide customized educational content, thus optimizing the learning process. For example, IoT-powered adaptive learning systems can suggest extra learning material to the learners for progressing in the area of difficulty. So, the real-time feedback loop is crucial in ensuring that learners stay engaged and are challenged at the right level. Research conducted by Ahuja et al. revealed that IoT-enhanced learning platforms increased student engagement by 25% and improved retention rates by 15%, as personalized learning experiences were more aligned with students' needs.

IoT's role in enhancing learner connectivity is also evident in its ability to integrate with artificial intelligence (AI) to provide predictive analytics. Predictive models generated by IoT data can forecast student success and suggest timely interventions to improve outcomes. This interconnected system ensures that learners are supported throughout their educational journey, enhancing both connectivity and learning outcomes.

Table 2.3 Benefits of IoT in personalized learning

Personalization Aspect	Traditional Learning	IoT-Enhanced Learning
Learning Pace	Fixed for all learners	Adjusts to individual needs
Feedback	Delayed or generalized	Instant and customized
Resource Recommendations	Limited to the curriculum	Tailored based on learner data

So, it is evident that the role of IoT in enhancing learner connectivity is multifaceted, encompassing expanded access to educational resources, improved collaboration, and personalized learning experiences. IoT's ability to connect devices and individuals in real-time allows for more dynamic, responsive, and inclusive educational environments. However, while IoT holds significant potential regarding the change in education system, this also presents many challenges for keeping the data secured and private and effective digital divide. Educational institutions must consider these factors as they increasingly integrate IoT into their learning ecosystems.

3. IoT-Driven Innovations in Instructional Methods

IoT technologies are driving innovations in instructional methodologies, making teaching and learning more engaging and effective. Interactive whiteboards, smart projectors, and augmented reality (AR) tools enhance classroom interactivity and engagement[26]. These technologies allow educators to deliver dynamic content and receive real-time feedback, improving lesson effectiveness. Additionally, adaptive learning platforms powered by IoT can personalize educational experiences by analyzing data on students' learning behaviors and preferences, thereby tailoring lessons and assessments to individual needs. The integration of the Internet of Things (IoT) in education has revolutionized traditional instructional methods, offering more interactive, data-driven, and personalized learning experiences. IoT-driven innovations allow educators to create smarter, more adaptive learning environments where students are more engaged, and learning outcomes are improved. These innovations include smart classrooms, personalized learning paths, real-time performance tracking, and automated administrative processes. By utilizing IoT, educators can move away from a one-size-fits-all approach to more flexible and inclusive instruction.

3.1 Smart Classrooms and Interactive Learning

One of the most significant IoT-driven innovations in instructional methods is the development of smart classrooms. Smart classrooms utilize IoT-enabled devices such as smartboards, tablets, and sensors to create an interactive and immersive learning environment. These technologies allow teachers to move beyond traditional lecture-based instruction and adopt interactive methods, where students engage with multimedia content, simulations, and real-time data.

For instance, smartboards can display live interactive content that students can manipulate, while IoT-enabled clickers or mobile devices allow students to participate in polls, quizzes, or collaborative projects in real time. This interactivity fosters higher levels of student engagement, participation, and retention of information.

Table 2.4 Impact of smart classrooms on instructional outcomes

Parameter	Traditional Classrooms	IoT-Enhanced Smart Classrooms
Student Engagement	Moderate	High (25% improvement)
Interactivity	Low	High
Feedback and Assessment	Delayed	Instant (real-time)
Customization of Instruction	Limited	Highly customizable

3.2 Personalized Learning and Adaptive Instruction

IoT has also enabled personalized learning, where instructional methods are tailored to the specific needs, preferences, and learning pace of each student. IoT-enabled devices collect and analyze data on students' performance, allowing educators to customize learning paths. This data-driven approach ensures that students receive targeted support in areas where they struggle while advancing more quickly in areas, they have mastered. IoT-enabled learning management systems (LMS) can track student performance in real time and adapt lesson plans, quizzes, and assessments accordingly. For example, if a student is consistently performing well in a particular topic, the system can offer more challenging tasks or move the student ahead in the curriculum.

3.3 Real-Time Performance Monitoring and Feedback

Another major innovation driven by IoT is real-time performance monitoring and feedback. IoT devices such as wearable sensors, smart desks, and learning applications track students' activity and engagement levels. This data can be used by both students and educators to monitor progress and adjust instructional strategies accordingly. Real-time feedback is critical in helping students identify areas of improvement and making timely adjustments to their study habits. For instance, wearable devices can monitor student focus and attention during lectures, sending real-time data to the teacher. Teachers can then use this data to identify when students are losing concentration and adapt their teaching methods to re-engage them. Such innovations ensure that instruction is more responsive and dynamic, catering to the needs of students on a moment-to-moment basis. According to Khan et al., the use of real-time monitoring systems increased the rate of teacher-student interactions by 40%, leading to a more engaged classroom environment.

Table 2.5 IoT tools for real-time monitoring and feedback

IoT Tool	Function	Impact on Learning
Wearable Devices	Tracks student engagement and concentration	Improves focus and allows immediate adjustments
Smart Desks	Monitors physical and cognitive engagement	Provides real-time insights into learning behavior
Learning Management Systems (LMS)	Tracks performance and provides instant feedback	Supports timely interventions and personalization

3.4 Automated Administrative and Instructional Processes

IoT-driven innovations have also streamlined various tasks related to administration and for facilitating the teachers for reducing their unnecessary burdens. Taking attendance, giving grades and other management of resources is easily possible through IoT technologies which greatly saves the time and develops better instructional materials and provide personalized attention to students. IoT-enabled grading systems can also instantly assess student submissions, providing immediate feedback on assignments, quizzes, and tests. A report by Smith et al. found that IoT-driven automation has significantly condensed the time consumption in administrative work by 35%, allowing teachers for being more dedicating for creating instructional design and student engagement. Therefore, it can be said that IoT-driven innovations in instructional methods are transforming education by creating more interactive, personalized, and data-driven learning environments. These innovations are enabling educators to design more engaging and adaptive lessons while also automating routine administrative tasks. IoT's real-time data capabilities give more appreciated outlook for the performance of the learners and fostering a more responsive and student-centered learning experience. However, we have to accept and address the related challenges as well to use the IoT in education.

4. Future Directions

The future of IoT in education holds potential for further advancements in learner connectivity and instructional practices. The latest technologies like Artificial Intelligence and Machine Learning are greatly expected to improve the competencies of Internet of Things for increased personalized experiences. As the Internet of Things (IoT) continues to evolve, its impact on education is expected to deepen, bringing about transformative changes in how learners connect with educational resources and how instruction is delivered. Future directions for IoT in education will likely to focus on enhancing personalized learning experiences, integrating advanced technologies, addressing digital equity, and ensuring sustainability. So, these future directions have vast scope for transforming the current educational system.

4.1 Personalized Learning and Adaptive Education Systems

One of the most promising future directions for IoT in education is the advancement of personalized learning through adaptive education systems. IoT devices will play a key role in collecting data on individual student behaviors, learning styles, and progress, allowing for real-time customization of instructional content to meet each student's unique needs[76]. Artificial Intelligence (AI) combined with IoT can further optimize this process by analyzing vast datasets to create more tailored educational experiences. For example, AI-powered IoT systems will be able to adjust learning pathways automatically based on a student's performance, providing challenges for advanced learners and offering remedial content to those struggling with specific concepts. These systems are expected to reduce the "one-size-fits-all" approach to education, fostering a more inclusive and effective learning environment.

4.2 Incorporation of IoT with Augmented Reality and Virtual Reality

The most significant development in education will definitely be related to the integration IoT with Augmented Reality and Virtual Reality. These technologies have already shown potential in creating engaging, interactive learning environments, but when combined with IoT, they could further revolutionize educational experiences. For example, IoT sensors can be used to create virtual simulations of physical environments that mirror real-world conditions. In subjects like science, engineering, or medicine, students could interact with 3D models of complex systems or virtually perform experiments that would be difficult or dangerous in the physical world.

4.3 Data Driven Policy-Making and Analytical Learning

The future of IoT in education will also see the expansion of data-driven decision-making through advanced learning analytics. IoT devices will provide continuous streams of data, not just related the performance in academics but surely on student engagement, participation with overall aspects of well-defined personality. Educational institutions will leverage this data to inform decisions on curriculum development, teaching strategies, and resource allocation. Learning analytics, powered by IoT can help educators identify patterns and trends in student behavior that may otherwise go unnoticed. For instance, if IoT sensors in classrooms detect that students are consistently disengaged during certain topics or teaching methods, schools can use this data to redesign lessons and improve instructional strategies. This will create a more efficient educational system where real-time data enables schools to respond proactively to student needs.

4.4 Addressing Digital Equity and Bridging the Digital Divide

As IoT continues to influence education, addressing the digital divide will become a key priority. The unequal access to IoT devices and high-speed internet in rural or low-income areas remains a significant barrier to achieving educational equity. Moving forward, we can surely state that great initiatives must be planned for the learners irrespective of their socio, economic or cultural background for benefitting them with the IoT-enhanced education. The Governments, NGOs and all educational institutes must come together for bridging this divide by investing in affordable IoT solutions and expanding broadband infrastructure in underserved areas. One future direction could involve the development of low-cost Moreover, the growing emphasis on digital equity will likely spur the creation of public-private partnerships focused on providing IoT-enabled educational tools to disadvantaged communities, helping to democratize access to technology and create a more inclusive learning environment.

5.5 Sustainability and Eco-Friendly IoT Solutions

As IoT becomes more integrated into education, the sustainability of these technologies will be an important consideration. With the increasing number of connected devices, concerns about energy consumption, e-waste, and the environmental impact of IoT technologies will need to be addressed. In the future, eco-friendly IoT solutions will emerge to minimize the carbon footprint of IoT systems in educational institutions. For instance, solar-powered IoT devices data centers, efficient in energy management reduce the adverse effects on environment. Additionally, innovations in sustainable materials for IoT hardware and circular economy practices (e.g., recycling and repurposing old IoT devices) will likely be prioritized to ensure the long-term viability of these technologies.

5.6 Ethical Considerations and Data Governance

As IoT technology becomes more pervasive in education, ethical considerations regarding data privacy, surveillance, and student autonomy will continue to grow in importance. The future of IoT in education must address these ethical challenges by establishing robust frameworks for data governance and student protection. IoT systems will need to be designed with privacy by default, ensuring that student data is collected, stored, and used responsibly. Schools will need to adopt transparent policies regarding the use of IoT data, ensuring that students and their families are fully informed and have control over their personal information.

So, the future of IoT in education is rich with potential, offering transformative changes in learner connectivity, personalized learning, and instructional methods. The other parameters such as enhancing the personalized learning and data-driven policymaking are greatly poised to shape the future of education, while challenges related to digital equity, sustainability, and ethics must be addressed. Therefore, as far as Internet of Things technology continues to develop, it will surely be analytical for the educational institutions to adopt innovative approaches that ensure these technologies are used effectively, equitably, and ethically. By doing so, IoT can truly revolutionize the education.

6. Conclusion

Therefore, in conclusion, it can be said very vividly that IoT is transforming education by enhancing learner connectivity and instructional methods. Its ability to create interactive, personalized, and efficient learning environments offers significant benefits for both students and educators. But we need to be more critical and analytical towards the related challenges as well for implementing IoT in education system. As technology evolves, IoT's role in education will likely expand, providing new opportunities for innovation and improvement.

Acknowledgement

The author of this paper is greatly the dear students, collogues and authorities of the Departments of English language and soft skills department and Internet of Things (IoT) for their remarkable cooperation in writing this significant research paper.

References

1. McFedries, P. The Internet of Things: How Smart Devices Are Transforming Everyday Life. Que Publishing, 2020.
2. Nair, A. et al. "Smart Classrooms Using IoT: A Review." Journal of Educational Technology Systems, vol. 47, no. 3, 2019, pp. 432–450.
3. Ahuja, R. & Singh, P. "IoT in Education: Bridging the Gap between Learner Needs and Instruction." Educational Technology Research and Development, 2021.
4. Simon, M. "IoT and the Future of Distance Learning." The Journal of Distance Education, vol. 35, no. 1, 2022, pp. 10–18.

5. Lewis, C. et al. "Addressing Educational Inequality through IoT-Enhanced Learning." Education and Information Technologies, vol. 26, 2021, pp. 567–585.
6. Khan, A. & Alam, F. "Collaborative Learning in IoT-Enabled Classrooms." International Journal of Educational Development, vol. 45, 2020, pp. 39–46.
7. Davies, J. "Challenges and Risks in IoT-Driven Education." Technology and Education, vol. 32, no. 4, 2020, pp. 142–160.
8. Smith, J. "Data Privacy and Security in IoT Education Systems." Journal of Educational Policy and Management, vol. 13, no. 2, 2022, pp. 78–95.
9. Patel, S. "Ensuring Equity in IoT-Driven Learning Environments." Educational Equity and Technology Journal, vol. 18, 2021, pp. 90–108.
10. Zhang, X., & Liu, Q. (2017). "The impact of Internet of Things (IoT) on the smart classroom." International Journal of Educational Technology, 5(2), 112–124.
11. Lee, S. M., & Kim, Y. J. (2018). "IoT-based remote learning systems and their applications." Journal of Distance Education Technologies, 14(3), 56–67.
12. Wang, H., & Zhang, X. (2019). "Interactive learning environments enabled by IoT technologies." Educational Technology Research and Development, 67(4), 721–737.
13. McFedries, P. The Internet of Things: How Smart Devices Are Transforming Everyday Life. Que Publishing, 2020.
14. Simon, M. "IoT and the Future of Distance Learning." The Journal of Distance Education, vol. 35, no. 1, 2022, pp. 10–18.
15. Zhang, L. et al. "Bridging the Digital Divide through IoT-Based Educational Platforms." Educational Technology Research and Development, vol. 69, 2020, pp. 23–37.

Note: All the tables in this chapter were made by the authors.

Intelligent Systems Using Semiconductors for Robotics and IoT – Dinesh Goyal et al. (eds)
© 2026 Taylor & Francis Group, London, ISBN 978-1-041-20408-4

3 Prediction Analysis of the Mechanical Strength using Levenberg-Marquardt Function of ANN Modelling for AZ 61 Magnesium Alloy

Amit Tiwari
Suresh Gyan Vihar University,
Department of Mechanical Engineering,
Jaipur, India

Payal Bansal
Poornima Institute of Engineering & Technology,
Department of Electronics & Communication Engineering,
Jaipur, India

Himanshu Vasnani
Suresh Gyan Vihar University,
Department of Mechanical Engineering,
Jaipur, India

Pooja Varshney, Neny Pandel
Suresh Gyan Vihar University,
Department of Computer Science Engineering and IT,
Jaipur, India

Abstract: The strength characteristics of metallic alloys are known to be substantially affected by the rolling process factors. To get the best set of characteristics in magnesium (Mg) alloy, it is important to acquire knowledge about the optimum process parameters that permits to fulfillment the intended aim. For this, we have employed a Artificial Neural Network (ANN) optimization tool in conjunction with linear regression; this aids in ascertaining the necessary process parameters to get the optimal set of hardness and tensile strength. Hardness and tensile strengths will be used as response variables for the rolled material (AZ 61 alloy). This study therefore presents the answer to the primary research question, "can we reliably predict how different rolling process variables will affect the hardness and tensile strength of a as cast AZ 61 alloy?" using a neural network technique. With a margin of error of only 1% to 5%, the projected values from the neural network and the collected data set values are often extremely well aligned.

Keywords: Computational material science, Artificial neural network (ANN), Prediction analysis, AZ 61 magnesium alloy, Mean square error

1. Introduction

There is a lot of interest among academics in finding a new way to reduce the environmental impact of fast-moving automobiles. For the time being, the only material that can achieve the required level of vehicle lightness is aluminium alloys. Nevertheless, there has been a noticeable uptick in the exploration of alternative

[1]amittiwari992@gmail.com, [2]payaljindalpayal@gmail.com, [3]himanshuvasnani1@gmail.com, [4]purvivarshney2022@gmail.com, [5]neny.pandel@gmail.com

DOI: 10.1201/9781003716389-3

materials for lightweight applications in recent literature [1]. Aluminium might face stiff competition from emerging magnesium alloys and composites in many sectors, including transportation, electronics, aircraft, and lightweight construction. Mg alloy and its composites are perceived as the most suitable green technology material due to its high natural characteristics which consist of excellent thermal conductivity, biocompatibility, and high specific strength [2]. Due to expensive manufacturing cost and low mechanical strength are the main reasons why magnesium alloy flats have not been widely used in production yet. Different multi-pass hot rolling conditions were applied to ZK-61 magnesium alloys with different reductions per pass in order to optimize the process parameters for processing strong sheets. W found that as we rolled at lower and lower temperatures that the grains were getting much finer and the microstructure much more homogeneous [3]. Even with these advantages, it remains challenging to demonstrate the application of such new rolling processes in industrial production, considering that they only ease part of the limitations imposed by conventional rolling. Considerable investigations have been made on the addition effects on structure and properties of Mg alloys [4], explaining that it is important to optimize the process variables for the particular Mg-alloy composites to achieve the ideal combination of mechanical and other physical properties.

Indeed, many academics from many scientific and technological domains have attempted several innovative improvements to computational optimization in an effort to make it more effective at handling complicated multivariate issues. Some researchers have focused on the problem of implementing innovative optimization strategies for complex nonlinear systems, such as glycol metabolism [5]. A hybrid multi-objective optimization approach was devised by combining evolutionary algorithms with boundary intersection frameworks. Reportedly, the method outperformed the usual techniques in producing a superior Pareto front. Researchers are constantly advocating for new optimization methodologies based on the quality of the challenges. It is stated that a model was created using evolutionary algorithms using data on conventional dual phase steel and interstitial free steel [6].

Thickness difference problem in industrial rolling process is solved using the combination of machine learning and ANN model. This novel prediction analysis emerged from Zhu, Z. et. al [7]. This maximization is made to optimize predictive performance. The hierarchical clustering and linear regression are also implemented in this cocktail. The most common method of making the magnesium matrix composites is stir casting, however; there is still a lot of room for improvements in their mechanical performance by using scientific methods like heat treatment and mechanical working [8]. The production of magnesium matrix composites is manual work, and it is highly labor intensive, costly and time consuming. But there's always room for improvement through testing. Towards this goal, this paper introduces a novel method based on machine learning to address these challenges. Such approach allows for time- and resource-saving use of the knowledge base on magnesium matrix composites through consolidation of the information provided in the literature [9-11].

Materials science has gained popularity recently, especially in applying AIML. Machine learning provides various methods that can accelerate research in composites; in other words, it offers potential for material discovery and design with broad applications [12]. Establishing an analysis formula to predict yield strength based on manufacturing parameters is difficult due to the complexity of these factors [13]. Therefore, it is not surprising that ML has become an effective tool to address this challenge. It can first learn the necessary settings to maximize mechanical features from data sets of previous work, second predict the results based on this learning, and third apply these insights [14].

Finding the optimal response to a wide range of issues has never been easier than with machine learning technology. It was chosen for this investigation because of its power, simplicity, and increased efficiency. Given the promising potential of machine learning in process optimization, this study focused on using machine learning techniques to analyze how rolling parameters affect the mechanical properties of AZ61 alloy-based composites. To our knowledge, no one has yet attempted to optimize rolling variables for AZ61 alloy using a regression model with machine learning techniques. This study presents a novel contribution by examining how variations in rolling parameters influence the mechanical properties of a composite material based on an AZ61 magnesium alloy matrix. The research involved collecting 100 data points on as-cast magnesium alloys, specifically AZ61, AZ31, and AZ91, from relevant research papers. The material compositions and processing parameters were considered as input factors, while the resulting mechanical behaviors were evaluated as output responses. ANN models were built based on these data to establish input-output relationships. MATLAB 2021a was used to implement the ANN constrained by input variables. The tensile strength and hardness of the alloy were predicted using ANN, with input factors including rolling temperature, the number of passes, and hot band thickness. The model identified the most significant input variables and checked the predictions against actual experimental results.

2. Computational Procedure

2.1 Modeling using Artificial Neural Network (ANN)

An ANN model was utilized for AZ61 alloy as the representative system, to predict the mechanical behavior under rolling process, which provided a feasible solution to complicated thermal applications. ANN, inspired by the structure of the human brain, is comprised of a large number of interconnected neurons and processes information parallelly. Learning takes place by tuning internal weights based on a performance measure—usually the mean square error. Feed-forward backpropagation type architecture was employed in which the number of hidden layers were raised in-between the inputs and outputs to capture non-linear connections. The network was trained by the gradient descent with an adaptive learning rate (TRAINDA) for accurate convergence according to the experimental results.

2.2 Optimization Procedure with Artificial Neural Network

The ANN model was developed using MATLAB 2021b. An experimental design was conducted with the specified input controllable parameters. Hardness and tensile strength measurements taken during the experiments served as the output targets for training the model. The hidden layer in the ANN was crucial for extracting patterns and features from the input data, enabling the mapping of complex relationships between inputs and targets. The hidden layers compute weighted sums of inputs with activation functions, transforming these into output signals with known interpretations, as shown in Fig. 3.1.

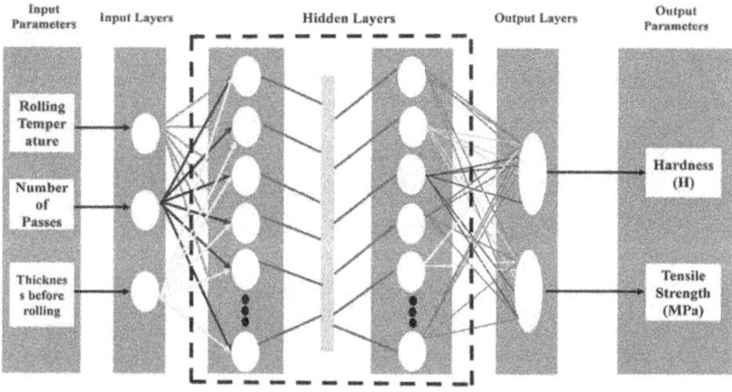

Fig. 3.1 Architecture of artificial neural network for predicting hardness and tensile strength based on rolling parameters

An artificial neural network was developed in MATLAB 2021b, utilizing various training algorithms, transfer functions, and layer configurations. The initial configuration involved running the ANN for 100 iterations, with MATLAB randomly assigning weights and biases. Training was halted either after 1000 iterations or when the performance gradient failed to decrease by more than 10^-7. To evaluate the model's predictive capability, simulations were also conducted on the entire dataset. The quantitative performance of the ANN was assessed using Equation 1 and 2, which helped clarify the relationship between the target output and the predicted responses.

$$Regression\ Coefficient = \sqrt{1 \left\{ \frac{\sum_{i=1}^{n}(T_i - O_i)^2}{\sum_{i=1}^{n} O_i^2} \right\}}, \quad [1]$$

$$Mean\ Square\ Error = \frac{1}{n}\left\{ \sum_{i=1}^{n}(T_i - O_i)^2 \right\}. \quad [2]$$

To ensure the accuracy of model evaluation, two key statistical indicators were utilized: (R > 0.98) and (MSE ~ 0.01). The training process concluded upon successful completion of the predefined iterations. The structure of the ANN developed using MATLAB 2021b is illustrated in Fig. 3.2. The predicted outcomes generated by the ANN model are further analyzed and discussed in the results and discussion section.

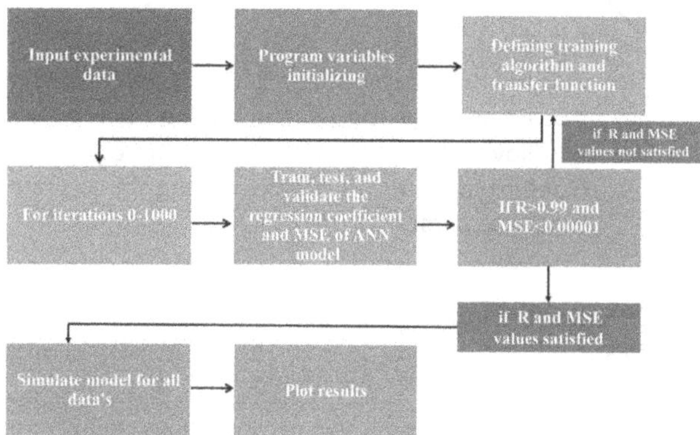

Fig. 3.2 Artificial neural network model flowchart

2.3 Determining Hidden Layers and Training of ANN

A minimum number of hidden layers in the ANN are used to achieve optimal performance and efficiency, typically just one hidden layer. If the network does not meet the required accuracy after initial training, additional layers can be added to improve performance. Due to the flexibility of neural networks, different activation functions can be used to generate output responses. In this work, the Levenberg-Marquardt algorithm was used as the training function because it is more robust and converges quickly. The architecture configuration for the ANN model is shown in Fig. 3.3.

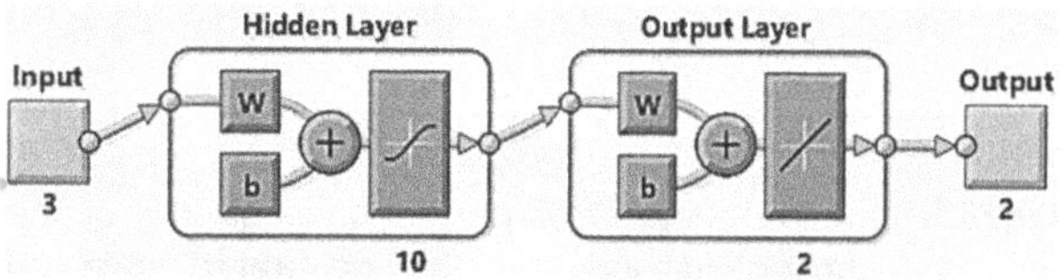

Fig. 3.3 MATLAB Toolbox for the training of the ANN Model

An MLP trained by backpropagation using the Levenberg-Marquardt method for rapid error minimization is employed in this work. The architecture and transfer function of the model were chosen through a series of iterative trials, with the lowest MSE response achieved. The dataset was randomly divided into training (70%), validation (15%), and testing (15%) sets. The training epochs were limited to 1,000, but extremely low error was attained within only the first three iterations. Performance evaluation, based on the regression and MSE plots for Hardness and Tensile Strength, is shown in Figs. 3.4 and 3.5.

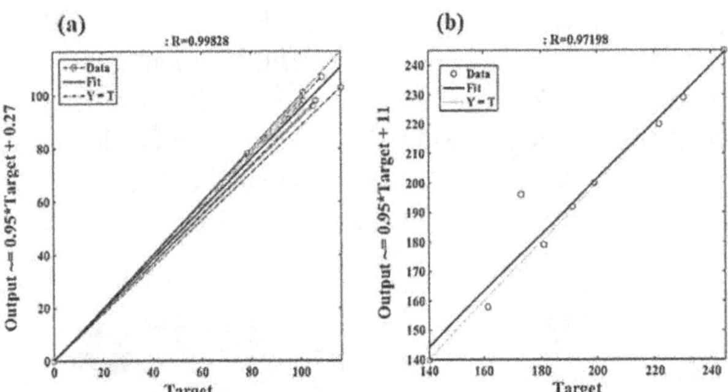

Fig. 3.4 Accuracy of the target and output data of AZ 61 alloy rolling parameters using a regression Coefficient metric, (a) Hardness, and (b) Ultimate tensile strength

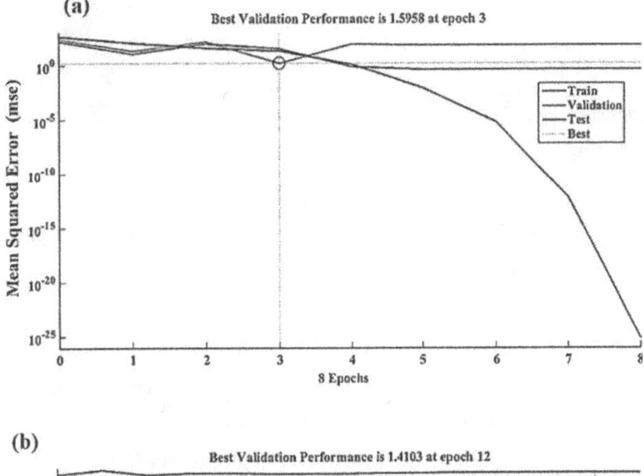

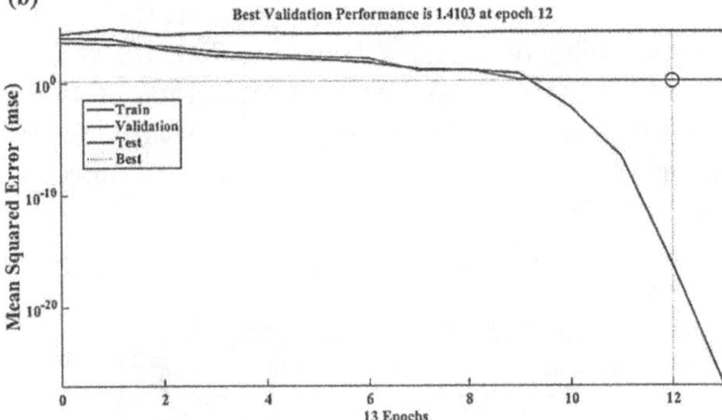

Fig. 3.5 Mean square error performance graphs of the chosen network for the generated data, a) Hardness value, and b) Ultimate tensile strength

Applying the approach of artificial neural networks to forecast the performance of a harness and ultimate tensile strength of AZ 61 alloy during rolling operation using a set of input parameters demonstrated the network's excellent abilities. With a success rate of 75% on the testing set, the network produced better outcomes but unsatisfactory results as compared to multi multi-objective genetic algorithm which is much better than ANN) forcasting. The experimental data were utilized in conjunction with the regression coefficient and predictions made by the artificial neural network (ANN). Actual data was compared with ANN forecasts. In terms of the mechanical strength of AZ61 alloy had regression coefficients of 0.99828 for the hardness, 0.97198 for the ultimate tensile strength are shown in Fig. 3.4. Figure 3.6, displays the results of the test data, validation data, and training data accuracy tests through an error histogram plot. The displayed data and average regression coefficient (R = 0.98513) of both response output will be more accurate because the data points are near the fit line. The results indicate that the ANN model effectively provided highly reliable predictions for all performance metrics. The predicted outcomes closely matched the actual dataset values, demonstrating the ANN strong capability inaccurately mapping the input parameters to the desired responses. As shown in Table 3.1, the optimized process parameters, along with the corresponding predicted values for hardness and tensile strength, were successfully determined using the developed ANN model.

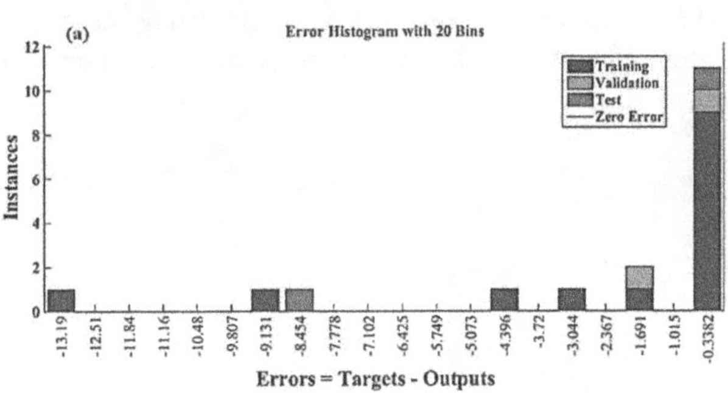

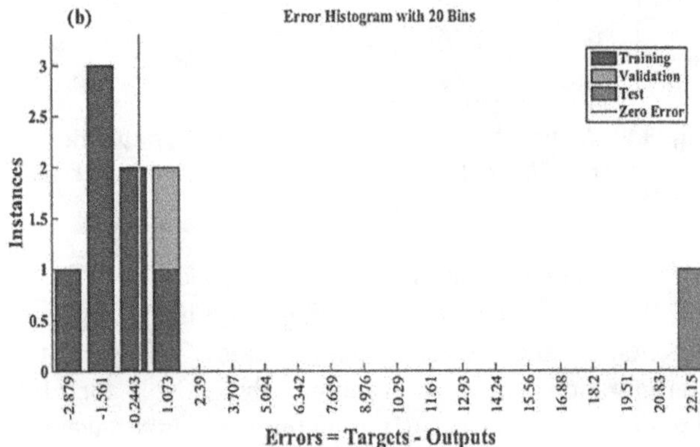

Fig. 3.6 Error histogram showing difference between the target and the output value a) Hardness, and b) Ultimate tensile strength

3. Conclusion

The optimization of rolling parameters for AZ61 magnesium alloy using ANN represents a significant step

Table 3.1 ANN predicted optimized input and response parameters of the alloy

Rolling Temp (°C)	No of Passes	Thickness before rolling in (mm)	Hardness (HV)	Tensile Strength (MPa)
250	3.46	7	98.25	145
250	3.56	7	107.41	162
250	3.78	7	103.41	152
250	3.69	7	106.20	141
300	3.80	7	100.23	174
300	3.96	7	98.64	189
300	3.76	7	89.69	185
300	3.88	7	80.64	189
350	3.81	7	104.89	253
350	3.60	7	99.71	186
350	3.73	7	93.08	241
350	3.78	7	105.10	189

forward in material processing and property enhancement. This paper highlights the challenges of modeling these nonlinear relationships with ANN, especially regarding rolling process parameters and their impact on hardness and tensile strength in AZ61 alloy processing. It is evident that rolling parameters such as temperature, percentage reduction per pass, and initial stock thickness greatly affect hardness and tensile strength outcomes. The novelty of this work is in applying ANN modeling to optimize rolling processes, which is highly complex due to the interactive effects of thermal and mechanical deformations from coupled effects in magnesium alloys. Unlike traditional methods, ANN can examine the intricate parameter-property space and identify processing conditions that improve mechanical properties. This optimization holds particular value for industry, as AZ61 alloy is used in weight-critical automotive and aerospace applications where maintaining mechanical properties is crucial for safety and performance.

References

1. Song, J., She, J., Chen, D., & Pan, F. (2020). Latest research advances on magnesium and magnesium alloys worldwide. *Journal of Magnesium and Alloys*, *8*(1), 1–41.
2. Selvan, A. T., & Palani, S. (2023, May). Prediction of mechanical strength of magnesium alloy AZ31 with calcium addition using a neural network based model. In *Journal of Physics: Conference Series* (Vol. 2484, No. 1, p. 012015). IOP Publishing.
3. WU, Z., LI, Q., CHEN, X., ZHENG, Z., ZHANG, N., & WANG, Z. (2024). Applications of machine learning on magnesium alloys. *Chinese Journal of Engineering*, *46*(10), 1797–1811.
4. Patel, V., Sharma, P., Tiwari, A., & Vasnani, H. (2024). Magnesium alloys in the automotive industry: Alloying elements and their impact on material performance. *Global Journal of Engineering and Technology Advances*, *21*(01), 130–145.
5. Hou, H., Wang, J., Ye, L., Zhu, S., Wang, L., & Guan, S. (2023). Prediction of mechanical properties of biomedical magnesium alloys based on ensemble machine learning. *Materials Letters*, *348*, 134605.
6. Tiwari, A., Kumar, N., & Banerjee, M. K. (2024). Ageing Characteristics of Stir Cast AZ 61 Alloy with Minor Additions. *Recent Patents on Engineering*, *18*(6), 67–82.
7. Zhu, Z., Ning, W., Niu, X., Wang, Q., Shi, R., & Zhao, Y. (2023). Designing high elastic modulus magnesium-based composite materials via machine learning approach. *Materials Today Communications*, 107249.
8. Dong, S., Wang, Y., Li, J., Li, Y., Wang, L., & Zhang, J. (2024). Machine Learning Aided Prediction and Design for the Mechanical Properties of Magnesium Alloys. *Metals and Materials International*, *30*(3), 593–606.
9. Huang, S. J., Adityawardhana, Y., & Sanjaya, J. (2023). Predicting Mechanical Properties of Magnesium Matrix Composites with Regression Models by Machine Learning. *Journal of Composites Science*, *7*(9), 347.
10. Tiwari, A., Kumar, N., & Banerjee, M. K. (2024). Applications of genetic algorithm in prediction of the best achievable combination of hardness and tensile strength for graphene reinforced magnesium alloy (AZ61) matrix composite. *Results in Control and Optimization*, *14*, 100334.
11. Dong, S., Wang, Y., Li, J., Li, Y., Wang, L., & Zhang, J. (2024). Machine Learning Aided Prediction and Design for the Mechanical Properties of Magnesium Alloys. *Metals and Materials International*, *30*(3), 593–606.
12. Tiwari, A., Kumar, N., & Banerjee, M. K. (2023). Effect of Hot Rolling on Microstructure and Mechanical Properties of Stir Cast AZ 61 Alloy with Minor Additions. *Indian Journal of Science and Technology*, *16*(24), 1810–1822.
13. Kordijazi, A., Zhao, T., Zhang, J., Alrfou, K., & Rohatgi, P. (2021). A review of application of machine learning in design, synthesis, and characterization of metal matrix composites: current status and emerging applications. *Jom*, *73*(7), 2060–2074.
14. Tiwari, A. (2024). Prediction model for hardness and tensile strength of graphene reinforced AZ 61 alloy based composite using Metaheuristic Algorithm. *Global Journal of Engineering and Technology Advances*, *20*(03), 042–052.

Note: All the figures and table in this chapter were made by the authors.

Intelligent Systems Using Semiconductors for Robotics and IoT – Dinesh Goyal et al. (eds)
© 2026 Taylor & Francis Group, London, ISBN 978-1-041-20408-4

Reviewing ChatGPT's Effect on Indian Education: Possibilities, Difficulties, and Ethical Issues

Avishka Bishnoi[1], Kriti Sankhla[2]
Faculty of Computer Science and Engineering,
Poornima University,
Jaipur, India

Abstract: ChatGPT was developed by the OpenAI is the one such application of large, pre-trained language model has achieved huge development in many domains. So, in field of education the ChatGPT is used in many aspects. This study examines OpenAI's ChatGPT, an AI-based chatbot, and its effects on education with an emphasis on India. For Natural Language Processing ChatGPT is an influential tool which shows countless promise for refining student engagement, customizing learning, and providing on-demand academic support. However, there are some issues still occur like data confidentiality moral issues, and limitations on grammatical versatility. The comprehensive valuation of the literature in this study brings to light the positive and negative consequences-dangers for academic transparency and the value of human interaction. Positive outcomes include better teaching being implemented and student accomplishment increasing. The implications are that ChatGPT, through its results, has the capability to bring about a significant change in education. However, this integration should be carried out with care, considering both practical and ethical issues.

Keywords: ChatGPT, Positive effects, Negative effects, Education, Artificial intelligence

1. Introduction

ChatGPT, which is a big language model-based chatbot created by OpenAI refers to Chat Generative Pre-trained Transformer, and released on 30 November, 2022. It allows clients to customize and lead a discussion to a specified length, structure, design, amount of information, and languages. Prompt engineering, or sequential prompts and responds, is regarded a context at each interaction step. ChatGPT operates on either GPT-3.5 or GPT-4, a member of the own line of generative pre-trained transformer models by OpenAI. It is one version in the series of the GPT model. The GPT has its foundation in the Google-crafted transformer architecture, both with supervised and reinforcement learning methods to optimize the algorithm for conversational application usage. GPT-3 is a groundbreaking AI language model released in June 2020 [1]. Even though the main purpose of ChatGPT is to simulate human conversation, it is versatile and, with 175 billion parameters, one of the largest and most powerful language models at the time. Another significant obstacle that has to be tackled is the linguistic aspect of natural language processing, encompassing Chat GPT and Google Bard. Chatbots have language limits, according to emerging research [3]. For instance, a study by Coniam (2014) found that Grammar-correct responses are usually provided by chatbots [2]. But as the moment, Chat GPT lacks practical adaptation and linguistic variety. It can write business concepts, compose poetry and song lyrics, compose music and teleplays, write fairy tales, compose student essays, design and debug computer programs, and answer tests (sometimes even more accurately than the average human test-taker, depending on the test) [4]. Engage in activities such as tic tac toe, assume the role of an ATM, imitate a Linux system, or recreate entire chat rooms. ChatGPT can be helpful in different domains: Customer Support and Service, Virtual Assistants, Content Generation, Education, Accessibility, Legal Services, Human Resources, Travel and Tourism, Psychological Support, Healthcare, Finance, Language Translation,

[1]Avishka29bishnoi29@gmail.com, [2]Kriti.sankhla@gmail.com

DOI: 10.1201/9781003716389-4

Programming and Coding, Creative Writing and Storytelling, Research and Data Analysis, Entertainment and Gaming.

2.1 ChatGPT in Indian Education

We can also see the impact of ChatGPT in India, like in many other parts of the world, can be significant and far-reaching in various areas like Cultural and Creative Applications, To guarantee fair and ethical use, ChatGPT also brings up ethical issues that must be addressed, such as data protection, bias mitigation, and responsible AI development. The impact of ChatGPT in India is multifaceted, offering numerous opportunities to improve efficiency, accessibility, and innovation across various sectors while also necessitating responsible deployment to address potential challenges and ethical considerations. S Biswas (February, 2023) have done a study on the function of Chat GPT in academics [1]. Overall, this study revealed that Chat GPT has the capacity to increase student involvement and interest in online courses, as well as student performance. The versatility of ChatGPT makes it a valued tool in various domains, offering assistance, information, and interaction in a human-like manner. However, it's important to use AI models like ChatGPT responsibly and consider potential ethical and privacy implications in different applications.

ChatGPT has the capability to transform education by offering on-demand support, personalised learning experiences, and greater accessibility. However, its use should be thoughtfully integrated into educational systems, with careful consideration of ethical, privacy, and pedagogical concerns.

This study deals with the role of ChatGPT in Indian education, focusing on its positive and negative impacts, especially regarding increasing student engagement, improving teaching methodologies, and ensuring academic integrity. The aim is to discuss how this can change education by bringing out the ethical issues surrounding its use.

3. Review of Literature

There is heightened interest in circles within the education sector since it has a potential scope through AI tools like ChatGPT regarding its ability to transform learning and teaching. This paper provides a review of available literature on how ChatGPT affects learning in most disciplines and environments.

3.1 ChatGPT

As T Nurtayeva et al. (2023) have found ChatGPT has the capability to aid the various purposes, to be exploited across multiple fields, complicated applications, encircling a miscellaneous range of objectives [5]. In the field of NLP, ChatGPT and its substitutes have shown incredible advancement in offering potentials Examples include text production, translation of languages, and question answering. However, they also have limitations like as inconsistencies in the data used for training, a lack of rational thought, a high computing cost, and concerns about safety [6]. Critical thinking and writing skills of journalism students can be positively influence by ChatGPT-3. Insights of students for using ChatGPT – 3 and knowledge regarding uses of ChatGPT tool in their journalism education can encourage the AI literacy and augment their understanding of the ethical, technical, and practical traits of employing AI tools in journalism [7]. According to M Keiper, 2023 this ChatGPT tool can be helpful in significantly lighten the time that consume by faculty for the content creation. When teaching teachers and students event management, ChatGPT can make the process of learning through experience more effective [8]. One example of this would be organizing an event. There are several useful ChatGPT solutions that could be used to enhance healthcare imaging student education [9]. The increasing use of AI technologies necessitates critical and reflective discussions and philosophical investigations by nursing researchers and educators regarding the ethical considerations, potential drawbacks, and implications of utilizing AI in nursing settings [10]. This sophisticated generative AI tool can be advantageous for students, educators, and scholars. These include increasing resolve, improving teaching procedures, providing customised online learning or coaching, creating academic outlines, and generating concepts for articles or essays [11]. In another way the research done by M Pradana et al. (2023) ChatGPT, as a powerful instrument, has the ability to transform the education industry [12]. For example, it is capable of organizing duties that require knowledge and imaginative thinking, such as scoring coursework and providing student counseling, as well as transforming the way education is delivered.

3.2 Positive Impact of ChatGPT in Education

As Dr. B Rathore (2022), have found that Apps and chatbot programs have the potential to have significant effects on education [13]. These can provide support or help and prompt feedback, customized directives and enhanced organizational management for teachers. Major findings of the authors supporting the introduction of the ChatGPT tool are as follows: what the tool entails in the post-adoption phase, how students view the tool and integration prospects, what the students' opinions on the role of educators might be, and what the most significant psychological impacts caused by the AI tool upon the users might be. By delivering the custom-made direction, adaptive learning trails, and rapid feedback AGI can modernize the education, which help in contributing to more operative learning aftereffects. Teachers' scholastic content expertise can be improved with the AGI, which allowing for distinguished instruction and support for miscellaneous prerequisites for student [14]. As R Firaina and D Sulisworo (2023), has found that ChatGPT can help users get ideas and information, translate papers, and ask different questions to get a deeper understanding of the topic [15]. While preserving academic integrity and advancing the highest standards of scholarship, the way student study, conduct research, and refine their skills can all be changed with the help of ethical ChatGPT and other AI applications in higher education (K Naidu and K Sevnarayan 2023).

3.3 Negative(-ve) Effects of ChatGPT in Education

As K Malinka et al. (2022), have found that the ChatGPT can be certainly distorted and also determine that the AI may help in the passing the courses which are needed for a university degree. With complete concentration on IT security in wide-ranging areas, they have inspected surplus of types of instruments to authenticate the expertise of the students. This dense valuation concluded that AI tools extensively frighten the reliability of the academic area. Although most observers believe that AI will eventually surpass instructors, human teachers can still be unique due to their unique qualities including creativity, empathy, and critical thinking [24]. ChatGPT's acute applications for educational backgrounds have a negative impact on its applicability [16]. Chatbots cannot replace human connection and help in the context of education, nevertheless [17]. H Ibrahim et al. (2022), have found that in the beginner courses, ChatGPT scores almost flawlessly on a significant portion of the questions, but it is still unable to meet the higher course requirements because to its lack of complexity [18].

3.4 Social Aspects

As A Shoufan (2023), has found that Students enjoy ChatGPT's capabilities and find it engaging, inspiring, and beneficial to their studies and careers [19]. They find it very user-friendly and human-like in terms of interface, with good and well-organized answers, and explanations are quite short. But since ChatGPT doesn't replace intelligence, a lot of students think that its responses aren't always accurate and that using it necessitates having a strong foundation in prior knowledge. Therefore, most students believe that ChatGPT needs to be enhanced and anticipate that this will happen shortly.

Artificial Intelligence is a new technology using algorithms to monitor, assess, and find similarities in the works of many people. It can easily detect plagiarism without much effort. Using certain keywords, phrases, or even complete paragraph structures, AI technology can highlight precise or almost identical matches and detect enhanced group commonalities [20]. The debut of computer-assisted instruction in the 1950s marked the beginning of AI's use in education. Over time, it has grown into intelligent tutoring systems (ITS), which have become widely utilized in teaching and learning. Strong NLP technology, ChatGPT can create conversations that seem human. Enhanced efficiency, increased accuracy, and cost savings are just a few of its many benefits. It does, however, have several drawbacks, such as limited functionality and security flaws. Notwithstanding these challenges, ChatGPT is an effective artificial intelligence (AI) technique that may be utilized to generate more accurate replies and automate conversations [21]. ChatGPT has proven that it is capable of providing thorough explanations and comprehension of common terminology, like Industry 5.0. Still, reactions to innovative ideas like Industries 5.0 may not always be clear-cut or precise. Still, ChatGPT is always developing and learning to offer replies that are usually accepted. One drawback of ChatGPT is that, as opposed to analysing implicit cues, it would need exact and comprehensive information [22].

4. Research Methodology

This study utilizes a secondary approach of data analysis based on reports accessible to the public, peer-reviewed journals, books, and journals from several databases such as SCI, Scopus, Web of Science, and Google Scholar. The main focus is on the change in the educational scenario triggered by ChatGPT within India. This study surveys various published studies that took place after the emergence of ChatGPT in the year 2022.

4.1 Search Criteria

Articles were searched for keywords including "Impact of Artificial Intelligence," "ChatGPT," and "Education." Only publications relevant to the research paper were taken, which involved the role of ChatGPT in education, the impact on student engagement, and the ethical issues that this might cause. The search was narrowed to peer-reviewed journals and articles that have been published from 2022 onward so that it captures the recent discoveries.

4.2 Selection Strategy

The paper was incorporated into the study using an electronic media. The research must address at least one of the aforementioned requirements or variables that have been taken into consideration for this article, and it must be released once prompt engineering technology has emerged. The phrase "Impact of ChatGPT in education" must be used in those articles.

Articles that discussed at least one of the following aspects were included:

- The role of ChatGPT in enhancing student engagement or academic performance.
- Ethics in the use of ChatGPT in education.
- The challenges and limitations of using ChatGPT in education.

The following were excluded: narrative reviews, non-peer-reviewed, and those that lacked relevant information.

The majority of papers that are eliminated from review include narrative reviews, non-peer reviewed articles, and articles with very little information or that do not address topics necessary for the current study.

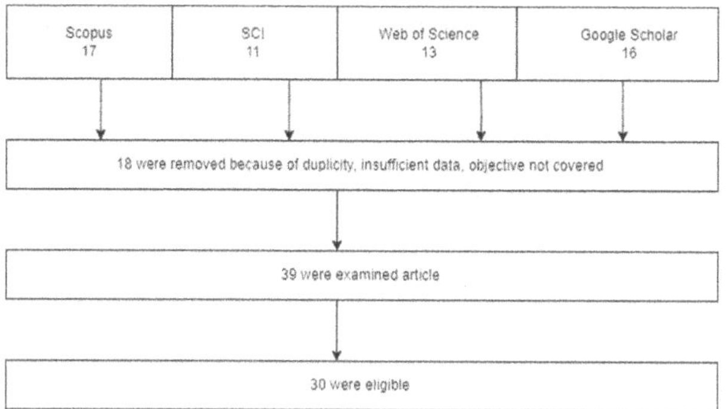

Fig. 4.1 Article selection and eligibility procedure across databases

Source: Authors

A database blind review was conducted using the suggested inclusion criteria in order to increase the authenticity and dependability of the criteria. Potential publications were excluded by navigating the results. The methodological quality of every included study was evaluated in cases where there was disagreement across reviews.

4.3 Data Extraction

The following format was used to pull data from chosen papers:

- Author's name
- Publication year
- Journal name
- Major findings about the impact of ChatGPT on education.

5. Discussion

The rapidly developing domain of artificial intelligence is basically shifting the approach we have to many things done in society. Because artificial intelligence is able to learn and evolve based on inputting data, we now had the ability to create structures that could accomplish what scientists thought were impossible in doing tasks.

AGI can transform the face of education through better learning outcomes through tailormade training, flexible learning paths, and instant feedback. Most importantly, AGI can make instructors understand scholarly topics in such a manner that they are enabled to help and exercise distinct command over different types of student demands. According to A L Alkhaqani (2023), some benefits of using ChatGPT in higher education are the reduced possibility of cheating [23]. AI technologies, such as from ITS that gives 1-on-1 training up to virtual teaching assistants, are used in education [24].

ChatGPT is an advanced natural language processing system that can simulate human speech. It offers several advantages, such as improved accuracy, higher expertise, and cost savings. All these benefits aside, however, there are limitations concerning the handling of complex queries, ethical issues regarding data privacy, and a lack of ability to replace human empathy and critical thinking skills in a human educator. So, ChatGPT needs to be used responsibly with suitable safeguards so that it does not replace, but complement the human educators [21].

6. Conclusion

ChatGPT enables students to interact with a chatbot for instant answers and self-study guidance. In terms of education, these applications enhance personalized learning, allow for timely feedback, and help educators design effective teaching strategies. ChatGPT can help instructors with differentiated instruction, classroom management, and teaching efficiency. The main point, however is that chatbots should only complement the very important human bonding or empathy that is also a significant part of education.

This study calls for the ethical use of ChatGPT in education, for change in aspects that bring about academic dishonesty, over-reliance on AI, and biasness in responses is a requirement. This further emphasizes refining the capabilities of ChatGPT in which it can handle complex queries, ensure data privacy, and encourage independent critical thinking in students. ChatGPT is amazing with its logical, well-structured responses, but needs continuous fine-tuning. Future versions, being trained on broader datasets, would offer enhanced performance. When used responsibly, ChatGPT may revolutionize education: support, not replace human educators.

References

1. Biswas, S. (2023). Role of Chat GPT in Education. SSRN 4369981.
2. Coniam, D. (2014). The linguistic accuracy of chatbots: usability from an ESL perspective. *Text Talk.* 34:545–567. doi: 10.1515/text-2014-0018.
3. Wilkenfeld, J. N., Yan, B., Huang, J., Luo, G., and Algas, K. (2022). "'AI love you': Linguistic convergence in human-chatbot relationship development," in *Academy of Management Proceedings*, (Briarcliff Manor, NY: Academy of Management) 17063. doi: 10.5465/AMBPP.2022.17063abstract
4. Chaves, A. P., and Gerosa, M. A. (2022). The impact of chatbot linguistic register on user perceptions: a replication study. In *Chatbot Research and Design: 5th International Workshop, CONVERSATIONS 2021,* Virtual Event (Cham: Springer International Publishing) 143–159. doi: 10.1007/978-3-030-94890-0_9

5. Nurtayeva, T., Salim, M., Basheer, T., Khalilov, S. (2023). The influence of ChatGPT and AI Tools on the Academic Performance. *YMER.* 22: 247–258. doi:10.37896/YMER22.06/26.
6. AlZu'bi, S., Mughaid, A., Quiam, F., Hendawi, S. (2023). Exploring the Capabilities and Limitations of ChatGPT and Alternative Big Language Models. 1–5.
7. Wang, X., Anwer, N., Dai, Y., Liu, A. (2023). ChatGPT for design, manufacturing, and education. doi:10.13140/RG.2.2.35077.22244.
8. Keiper, M. (2023). ChatGPT in Practice: Increasing Event Planning Efficiency Through Artificial Intelligence. 33. doi:10.1016/j.jhlste.2023.100454.
9. Currie, G., Singh, C., Nelson, T., Nabasenja, C., Al-Hayek, Y., Spuur, K. (2023). ChatGPT in medical imaging higher education. *Radiography (London, England : 1995).* 29. 792–799. doi:10.1016/j.radi.2023.05.011.
10. Abdulai, A. F., Hung L. Will ChatGPT undermine ethical values in nursing education, research, and practice? *Nurs Inq.* 30(3):e12556. doi:10.1111/nin.12556.
11. Adiguzel, T., Kaya, M. H., & Cansu, F. K. (2023). Revolutionizing education with AI: Exploring the transformative potential of ChatGPT. *Contemporary Educational Technology.* 15(3):ep429. doi:10.30935/cedtech/13152.
12. Pradana, M., Elisa, H. P., and Syarifuddin, S. (2023). Discussing ChatGPT in education: A literature review and bibliometric analysis, *Cogent Education.* 10:2, 2243134, doi: 10.1080/2331186X.2023.2243134
13. Rathore, B. (2022). Exploring the Potential Impacts of Chatbot Software/Apps (ChatGPT) on Education: Benefits, Drawbacks, and Future Prospects. 1. 2960–2068.
14. Latif, E., Mai, G., Nyaaba, M., Wu, X., Liu, N., Lu, G., Li, S., Liu, T., and Zhai, X. (2023). Artificial General Intelligence (AGI) for Education.
15. Firaina, R., and Sulisworo, D. (2023). Exploring the Usage of ChatGPT in Higher Education: Frequency and Impact on Productivity. *Buletin Edukasi Indonesia.* 2:39–46. doi:10.56741/bei.v2i01.310.
16. İpek, Z.H., Gözüm, A.İ.C., Papadakis, S., & Kallogiannakis, M. (2023). Educational Applications of the ChatGPT AI System: A Systematic Review Research. *Educational Process: International Journal.* 12(3): 26–55.
17. Rathore, B. (2023). Future of AI & Generation Alpha: ChatGPT beyond Boundaries. *Eduzone : international peer reviewed/refereed academic multidisciplinary Journal.* 12:63–68. doi:10.56614/eiprmj.v12i1y23.254.
18. Ibrahim, H., Asim, R., Zaffar, F., Rahwan, T., and Zaki, Y. (2023). Rethinking Homework in the Age of Artificial Intelligence. *IEEE Intelligent Systems*. 38(2):24–27. doi:10.1109/MIS.2023.3255599.
19. Shoufan, A. (2023). Exploring Students' Perceptions of ChatGPT: Thematic Analysis and Follow-Up Survey. *IEEE Access.* 11: 38805–38818. doi: 10.1109/ACCESS.2023.3268224.
20. Alkhaqani, A. (2023). ChatGPT and Academic Integrity in Nursing and Health Sciences Education. *Journal of Medical Research and Reviews.* 1:5–8. doi:10.5455/JMRR.20230624044947.
21. Deng, J., and Lin, Y. (2023). The Benefits and Challenges of ChatGPT: An Overview. *Frontiers in Computing and Intelligent Systems.* 2:81–83. doi:10.54097/fcis.v2i2.4465.
22. Wang, F. -Y., Yang, J., Wang, X., Li, J., and Han, Q. -L. (2023). Chat with ChatGPT on Industry 5.0: Learning and Decision-Making for Intelligent Industries. *IEEE/CAA Journal of Automatica Sinica.* 10(4): 831–834. doi:10.1109/JAS.2023.123552.
23. Chan, C., and Tsi, L. (2023). The AI Revolution in Education: Will AI Replace or Assist Teachers in Higher Education?.
24. Tlili, A., Shehata, B., Adarkwah, M. A., Bozkurt, A., Hickey, D. T., Huang, R., & Agyemang, B. (2023). What if the devil is my guardian angel: ChatGPT as a case study of using chatbots in education. *Smart Learning Environments.* 10(1):1–24.

Intelligent Systems Using Semiconductors for Robotics and IoT – Dinesh Goyal et al. (eds)

5 Design and Implementation of CORDIC Based 1D DCT Processor

Devendra Kumar Somwanshi[1]
Associate Professor, Department of Department of ECE,
Poornima College of Engineering,
Jaipur, India

Saurabh Shandilya[2]
Professor, Department of Advance Computing,
Poornima College of Engineering,
Jaipur, India

Kamlesh Gautam[3]
Associate Professor, Department of Advance Computing,
Poornima College of Engineering,
Jaipur, India

Sachin Jain[4]
Assistant Professor, Department of Computer Science,
Poornima College of Engineering,
Jaipur, India

Archana Soni[5]
Assistant Professor, Department of Department of ECE,
Poornima College of Engineering,
Jaipur, India

Abstract: The research paper aims to explore algorithms for calculating trigonometric functions and complex multiplications, with a focus on using an unrolled CORDIC architecture instead of complex multipliers and ROM-based architectures. The implementation of this architecture raises concerns about power consumption, speed, and accuracy. The problem statement derived from detailed review of more than 50 research papers which analyze different approaches to implementing Trigonometric Generators on FPGA, highlighting their strengths, weaknesses.

Keyword: CORDIC, DCT, FPGA, DUS, Comparative analysis

1. Introduction

Digital Signal Processing has traditionally relied on microprocessors due to their single cycle multiply-accumulate instruction and special addressing modes. However, for demanding tasks such as Image Compression, Digital Communication, and Video Processing, specialized processors with custom architectures have emerged. One such algorithm, CORDIC, uses only Shift-and-Add arithmetic with table Look-Up to implement different functions, making it hardware-efficient and easily implementable on VLSI. Using CORDIC to implement DCT, a widely-used algorithm in Image Compression, can increase accuracy and reduce power consumption. FPGA provides an ideal platform for testing the functionality of dedicated processors designed using CORDIC algorithm, as it offers on-site programmability and performs high-speed operations not possible on a simple microprocessor.

[1]imdev.som@gmail.com, [2]saurabh.shandilya@poornima.org, [3]kamlesh@poornima.org, [4]sachin.jain@poornima.org, [5]archana.soni@poornima.org

DOI: 10.1201/9781003716389-5

2. Literature Review

The calculation of trigonometric functions, multiplication, division, and data type conversion presented significant challenges, leading Volder to introduce the CORDIC algorithm. Later, Walther expanded it to generalize its application to hyperbolic functions. Although the algorithm's use of basic shift and add operations for 2D vector rotation makes it straightforward to implement, its computational speed is comparatively slow due to these simple operations. Primary bottleneck lies in the iterative CORDIC structure, where the speed performance is hindered by the large iteration number, N, typically equal to the internal word length, W. A novel approach, called the Mixed-Scaling-Rotation Coordinate Rotational Digital Computer (MSR-CORDIC) algorithm, was developed by integrating micro-rotations and scaling phases [2]. This method eliminates the overhead of scaling operations in traditional CORDIC algorithms, significantly reducing iteration numbers and improving speed. Para-CORDIC is suitable for digital modulators, demodulators, and bit synchronizers. Due to its advantage it outperforms other DDFS approaches and it offers parallel operation, eliminating rotation direction data path, and reduces hardware requirements [7]. The CORDIC algorithm replaces lookup tables, reducing RAM consumption but increasing LE usage, which must be optimized. The EP2C8T144C8 FPGA chip (Altera Corporation's Cyclone II family) is used for implementation. Quartus II9.1 and Modelsim SE 6.4 software platforms verify the correctness of the Verilog HDL program [12].

3. Different Flow Diagrams

Figure 5.1 shows the process flow diagram of the approach and Fig. 5.2 shows the FPGA design flow.

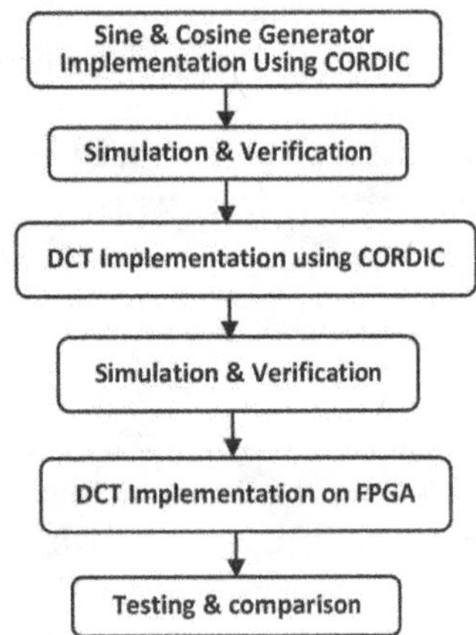

Fig. 5.1 Process flow diagram

4. Results and Discussions

As our whole process is divided into implementation and verification, hence testing of implemented and proposed architecture is must.

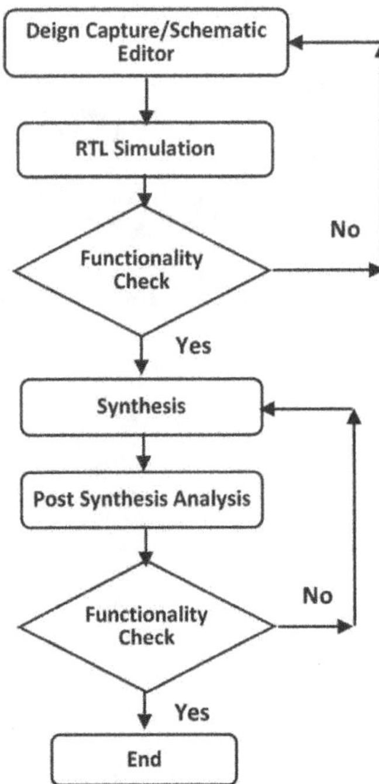

Fig. 5.2 FPGA design flow

The CORDIC algorithm, which is often used in digital signal processing, was implemented in both MATLAB and VHDL. The angle vector rotation values were compared between the two implementations, and it was found that the values obtained from both were almost equal, with a very small percentage of error. To provide evidence of this, a comparison table was produced.

Table 5.1 Comparison of successive angle rotation values

S. No	Iteration	Matlab Value	Coded Value	VHDL Test Bench
1	1	-2.854	IB602SSE	-0.28539
2	2	0.1782	02DAICI73	0.17824
3	3	-0.0667	IEEEAD58I	-0.06672
4	4	0.0576	00EC090IB	0.05762
5	5	-0.0048	IFECSE240	-0.00479
6	6	0.0264	006CS37AE	0.02645
7	7	0.0108	002C54D03	0.01082
8	15	2.002*E-5	000014FF9	2.002E-S

5. Comparative Analysis

5.1 Comparison of MATLAB Values and Obtained VHDL Values

The results obtained from the test benches and the DCT command in MATLAB were compared, and they were found to be nearly equal with very little percent of error. The comparison was shown in Table 5.2 and Table 5.3. For example, when the input was 2, the value obtained from VHDL test bench was 11.309 and the value obtained from DCT command in MATLAB was 11.3137, resulting

Table 5.2 {Sample 1} comparison between MATLAB values and obtained VHDL values

S. No	Input	MATLAB Ouput	VHDL Output	Error
1	2	11.3137	11.309	0.0047
2	2	-4.1215	-4.122	0.0005
3	3	-0.2706	-0.271	0.0004
4	4	1.1175	1.116	0.0015
5	1	-3.5355	-3.534	-0.0015
6	7	4.4277	4.429	-0.0013
7	9	-0.6533	-0.653	-0.0003
8	3	-1.7776	-1.778	0.0004

Table 5.3 {Sample 2} comparison between MATLAB values and obtained VHDL values

S. no	Input	MATLAB Ouput	VHDL Output	Error
1	23	15.9099	15.904	0.0059
2	-2	7.2522	7.246	0.0062
3	3	13.257	13.25	0.007
4	15	-8.4665	-8.468	0.0015
5	1	20.8597	20.85	0.0097
6	-17	15.9434	15.935	0.0084
7	9	-5.8739	-5.878	0.0041
8	13	5.4803	5.474	0.0063

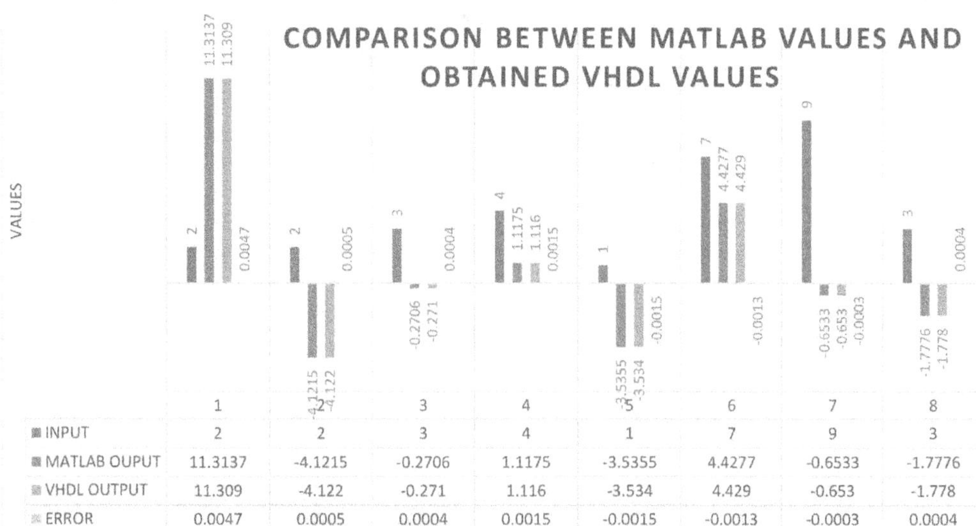

COMPARISON BETWEEN MATLAB VALUES AND OBTAINED VHDL VALUES
VALUES
1 2 3 4 5 6 7 8
INPUT 2 2 3 4 1 7 9 3
MATLAB OUPUT 11.3137 -4.1215 -0.2706 1.1175 -3.5355 4.4277 -0.6533 -1.7776
VHDL OUTPUT 11.309 -4.122 -0.271 1.116 -3.534 4.429 -0.653 -1.778
ERROR 0.0047 0.0005 0.0004 0.0015 -0.0015 -0.0013 -0.0003 0.0004

Fig. 5.3 {Sample 1} comparison between MATLAB and VHDL OUTPUT

in an error of less than 1 percent. A corresponding graph (Fig. 5.12) was also plotted for convenience.

6. Conclusion

The CORDIC algorithm is commonly used for digital signal processing applications and can be implemented on an FPGA with high efficiency and low cost. This research involves designing and simulating a CORDIC module using VHDL and verifying its output with MATLAB. The module was then used to design and simulate an 8-point 1D DCT processor, which was subsequently implemented on a Spartan 3E FPGA kit. The obtained results matched the expected values without any significant discrepancies or deviations, indicated efficient resource consumption. The designed processor has the capability to perform real-time computation of 1D 8-point DCT values, which can be utilized for online processing applications.

References

1. Ahmed, H.M., Delosme, J.M., and Morf, M.," Highly Concurrent Computing Structure for Matrix Arithmetic and Signal Processing," IEEE Comput. Mag,, Vol.15, 1982, pp. 65–82.

A. Madisetti, "A 100-MHz, 16-b, direct digital frequency synthesizer with a 100-dBc spurious-free dynamic range", IEEE, August 1999.

2. Yamagishi, "A 2-V, 2-GHz low-power direct digital frequency synthesizer chip-set for wireless communication", IEEE, 1998.
3. Ahlam Fadhil Mahmood, Abdulkreem Mohameed Salih, "FPGA Implementation of Multiplierless DCT/IDCT Chip",Al-Rafidain Engineering Vol.19, No. 4, August 2011.
4. Andraka.R.J.,"Building a High Performance Bit-Serial Processor in an FPGA," Proceedings of Design SuperCon '96, Jan 1996, pp. 5.1–5.2.
5. D. Yang, "An 800-MHz low-power direct digital frequency synthesizer with an on chip D/A converter", IEEE, 2004.
6. Lakshmi, "High speed architectural implementation of CORDIC algorithm", IEEE, 2008.
7. B.Lakshmi, "FPGA Implementation of a High Speed VLSI Architecture for CORDIC", IEEE, 2009.
8. Cai Ken, Liang Xiaoying, Liu Chuanju, "SOPC based flexible architecture for JPEG enconder", Proceedings of 2009 4th International Conference on Computer Science & Education, 2009.

9. Cheng-Shing Chih, Hsiu Lin and An-Yeu Wu "Modified vector rotational CORDIC (MVR-CORDIC) algorithm and its application to FFT", IEEE, 2001.
10. Clay Gloster, Jr., Wanda Gay, Michaela Amoo, and Mohamed Chouikha, "Optimizing the Design of a Configurable Digital Signal Processor for Accelerated Execution of the 2-D Discrete Cosine Transform", Proceedings of the 39th Hawaii International Conference on System Sciences – 2006.
11. S. Phatak, "Double step branching CORDIC: A new algorithm for fast sine and cosine generation", IEEE, May 1998.
12. Davide De Caro, "A 380 MHz Direct Digital Synthesizer/Mixer with Hybrid CORDIC Architecture in 0.25 m CMOS", IEEE, January 2007.
13. Depreterre, E., Dewilde, P.,and Udo, R.,'Pipelined CORDIC Architecture for Fast VLSI Filtering and Array Processing," Proc. ICASSP'84, 1984, pp. 41.A.6.1–41.A.6.4.
14. Duprat, J. and Muller, J.M.,"The CORDIC Algorithm: New Results for Fast VLSI Implementation," IEEE Transactions on Computers, Vol.42, pp. 168–178, 1993.
15. Gan Lu and Zheng Yousi "Modified MVR-CORDIC Algorithm Research", IEEE, 2004.
16. Hu, Y.H., and Naganathan, S., "A Novel Implementation of Chirp Z-Transformation Using a CORDIC processor, "IEEE Transactions on ASSP, Vol. 38, pp. 352–354, 1990.
17. Hu,Y.H.,and Naganathan, S.,"An Angle Recoding Method for CORDIC Algorithm Implementation", IEEE Transactions on Computers, Vol.42, pp. 99–102, January 1993.
18. J.E.Volder, "The CORDIC Trigonometric Computing Technique", IEEE, 1959.
19. Javier Valls, "The Use of CORDIC in Software Defined Radios: A tutorial", IEEE, Sept 2006.
20. Javier valls, Trini Sansaloni, and A.P.Pascual "The Use of CORDIC in Software Defined Radios: A tutorial" IEEE, September 2006.
21. Jie Liang and Trac D.Tran," Fast Multiplierless Approximation of the DCT with the Lifting Scheme", IEEE Transaction on Signal Processing ,Submitted; FEB. 2001.
22. Koushik Maharatna, "Modified virtually scaling-free adaptive CORDIC rotor algorithm and architecture", IEEE, November 2005.
23. Koushik Maharatna, "Virtually scaling-free adaptive CORDIC rotator", IEEE, November 2004.
24. Liu Yuejun, Su Jing and Liu Feng ," Research on Information Hiding System based on DCT Domain", 2010 Second International Conference on Computer Modeling and Simulation, 978-0-7695-3941-6/10 IEEE, DOI10.1109/ICCMS.2010.342, 2010.
25. M. Loehning, "Digital Down Conversion in Software Radio Terminals", European Signal Processing Conference, Sept 2000.
26. M. P. Leong and Philip H. W. Leong, "A Variable-Radix Digit-Serial Design Methodology and its Application to the Discrete Cosine Transform" , IEEE Transactions, 2012.
27. Nathaniel August and Dong Sam Ha, "On The Low-Power Design Of DCT and IDCT For Low Bit-Rate Video Codecs", Arlington, Virginia, Int. ASIC/SOC Conference, pp. 203–207, September 2001.
28. Bansal, Payal, Devendra Kumar Somwanshi, Balwinder Singh Dhaliwal, and Pallavi Sapkale. "The roles of delay and power optimization techniques in VLSI design." In Advances in Image and Data Processing using VLSI Design, Volume 1: Smart vision systems. IOP Publishing, 2021.

Note: All the figures and tables in this chapter were made by the authors.

Intelligent Systems Using Semiconductors for Robotics and IoT – Dinesh Goyal et al. (eds)
© 2026 Taylor & Francis Group, London, ISBN 978-1-041-20408-4

6 Advancing Agricultural Sustainability with IoT: A Comprehensive Overview

Nikhil Kumar Goyal[1], Monika Dandotiya[2], Shikha Sharma[3], Ajay Khunteta[4], Mohammed Firdos Alam Sheikh[5]
Poornima University, FCE, Jaipur, India

Abstract: The Internet of Things is emerging as one of the major technological paradigms, comprising a network of machines and devices that can communicate with and interact with each other remotely. With its rapid adoption, there has been increased attention from a wide variety of industries due to the fact that such technologies enable them to change operations and create value. Five key technologies of IoT are discussed in this paper, which are the core for the deployment of IoT-based products and services. It segments enterprise IoT applications into three clear areas: monitoring and controll. Each category is discussed in terms of its capability to enhance organizational value and customer experience.

Additionally, the paper reviews methods that can be used to justify investments in IoT, focusing more on the NPV method2 and real option approach. It further points out that the real options approach is very versatile and dynamic, in which uncertainties and new future developments can easily be integrated into IoT investment analysis. This study further identifies and discusses five major technical challenges associated with the IoT: issues relating to data security, interoperability, and scalability. This paper will try to provide a holistic understanding of the IoT landscape through the presentation of an overview regarding foundational technologies, categories of applications, investment evaluation methodologies, and challenges associated with it. It points out an economic effect of IoT, projecting the value to be created to US$ 14.4 trillion in various industries from the year 2013 to 2022. It underlines IoT technologies as basic in creating such a driving force of innovation and efficiency that contributes to the competitive advantage in the contemporary business environment.

Keywords: Monitoring and control, NPV method

1. Introduction

Internet of Things is one such paradigm gaining momentum in recent technology, and it has the capability to introduce a whole new feel and dimension in interacting with the physical world. IoT can be defined as "a network of physical objects or 'things' that can interact with each other to share information and take action" or "the interconnection of uniquely identifiable embedded computing devices within the existing Internet infrastructure"; it is a wide area of applications combined, using technologies in a way to make the devices communicate and act on their own inside a network [1]. The use of RFID technology in identifying and tracing devices marked the beginning, some time ago, of IoT. Since then, a wide variety of technologies, such as sensors, actuators, and communication protocols, has come to make up what can be considered today as the core of any IoT system. The best example of this transformation by IoT is its application in aquaculture. Farmers can take proper decisions toward optimizing fish growth and augmenting general productivity in aquaculture based on real-time data. Such innovations underpin IoT's ability to enhance operational efficiency and decision-making across diverse industries.

[1]nikhilgoyal886@gmail.com, [2]dandotiyamonika@gmail.com, [3]er.shikhasharma1986@gmail.com, [4]ajay.khunteta@poornima.edu.in, [5]firdos.sheikh@gmail.com[5]

DOI: 10.1201/9781003716389-6

In fact, IoT applications are not confined to aquaculture; Furthermore, investments by industries in IoT technologies are on a continuous rise, with a gradual increase in research and development works related to IoT.

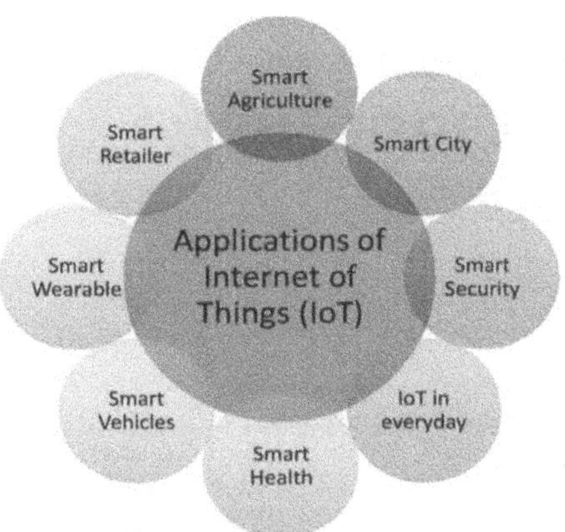

Fig. 6.1 Application of internet of things in different fields and domains

Recently, it was estimated that the profit for industries adapted to IoT will increase by 22% [2]. The growing prominence of IoT is further manifested in the rising tide of academic publications related to the topic, demonstrating an extension of interest in discussing possibilities and issues of IoT. With industries and researchers converging to develop and enhance IoT technologies, it becomes important to understand not only the very basics of IoT but also what the concept portends for different sectors. This paper aims at providing an overview of IoT, outlining foundational technologies, application areas, and implications for various industries, including discussion on technological advances that drive IoT, study of applications in various domains, and assessment of economic and operational benefits brought about by IoT to the enterprises. Develop Efficient Data Management and Analytical Tools for IoT: Investigate data storage, processing, and analysis solutions in the context of IoT network specifically targeting real-time analytics for big data management and deriving valuable insights from large datasets or streams of IoT-related information.

1. Design Low-Power IoT Devices: Research and design innovative energy-efficient hardware and software solutions for IoT devices, with a focus on extending battery life.
2. Standardize Communication Protocols for IoT Devices: Work towards the establishment of harmonized and standardized communication protocols for IoT, which support interoperability between various IoT devices, networks, and platforms. It helps in reducing fragmentation in IoT deployments.
3. Address Regulatory and Ethical Challenges in IoT: Examine the legal and ethical issues related to IoT, such as privacy in data, consent management, and responsible use of data. Enhance IoT Systems' Real-time Decisions: Design and evaluate advanced algorithms and systems that support real-time decision-making in large-scale IoT networks, thus allowing for a fast response to environmental changes and data inputs from several connected devices.

These goals address the main challenges described in IoT applications, making them an important contribution toward more efficient, secure, and scalable IoT systems.

2. Problem Statement

The proliferative growth of Internet of Things (IoT) devices and applications in various industrial fields has changed the paradigms for data sensing, processing, and consumption. Despite tremendous advancements of IoT technologies, challenges and gaps exist that inhibit the fully actualized potential of IoT systems. With growing IoT networks, the ability to integrate seamlessly different devices from various manufacturers with different communication protocols becomes a challenge. Inability to standardize interoperability frameworks leads to restricted scaling of IoT systems and therefore their applications across industries in healthcare, smart cities, and manufacturing.

Even though security measures have been developed, existing protocols are still not sufficient to protect against the evolving threats, especially in large-scale, distributed IoT environments. The continuous collection of personal data also raises concerns regarding how data is treated, stored, and shared without the proper consent or security of the users. In parallel, real-time decision-making, which is critical in many IoT applications such as healthcare and smart cities, is often hindered by insufficient data processing capabilities. The frequent need to replace batteries and the massive energy consumption of certain applications is not feasible in cases of large-scale IoT applications when these are located in a remote or hard-to-access area. Even with improvements on low-power hardware and on techniques for energy management, numerous IoT systems still have challenges to overcome in the line of energy optimization.

Moreover, edge computing can be a potential solution to address some of the latency and bandwidth issues, but there is a lack of unified frameworks and efficient load-balancing algorithms to integrate edge and fog computing with IoT systems seamlessly. Finally, the lack of standardization in communication protocols between IoT devices remains one of the major challenges. Different IoT devices use different communication protocols, such as Zigbee, Bluetooth, Wi-Fi, and LoRa. In summary, IoT technologies have huge possibilities while the presence of issues including scalability and interoperability as well as security concerns, energy inefficiency problems, and data management has reduced their widespread adoption for optimal use. Each layer has an important role in ensuring effective collection, transmission, processing, and utilization of data.

2.1 Sensor Layer

The Sensor Layer forms the basic layer of IoT architecture and captures information from the physical world. It basically consists of numbers of sensing technologies, ranging from sensor networks to embedded systems, RFID tags, and readers, among many others that are in soft sensors nature.

Sensor Networks: These are arrays of connected sensors that gather information about the atmosphere, temperature, or moisture in a region, or movements within it.

Soft Sensors: These are virtual sensors that come into being by combining data from several physical sensors to come up with complex measurements, which cannot be directly measurable.

Each sensor or device in this layer has an identification and information storage capability, such as provided by RFID tags.

2.2 Access Gateway and Network Layer

Access Gateway and Network Layer have to manage the data in the higher layers of IoT architecture. It therefore involves scalable, flexible, standardized communication protocols that can keep pace with the diverse nature of sensor nodes and devices.

Network Performance: This layer requires high performance and robustness to handle large volumes of data emanating from IoT devices, hence, it will make the network cohesive despite heterogeneity in origin.

2.3 Management Service Layer

The Management Service Layer mediates between the Access Gateway and Network Layer and the Application Layer. Among others, its responsibilities are as follows:

Device Management: It is responsible for the configuration, monitoring, and maintenance of IoT devices in such a manner that they function correctly and are integrated into the network.

Information Management: It is about capturing a high volume of raw data from the Sensor Layer and extracting useful information. The processing and analytics of both stored and real-time data take place on this layer to deliver actionable insights [4].

Data Security and Privacy: This is one of the most important functions that can be assigned to this layer: the security and privacy of data.

2.4 Application Layer

The Application Layer is the topmost layer in IoT architecture, which is majorly worried about the creation of value for the user by means of applications. Healthcare: IoT application in healthcare might involve remote patient monitoring, wearable health devices, and automated medical diagnostics.

Agriculture: IoT applications in agriculture will involve optimization of irrigation, soil condition monitoring, and livestock management.

Government Sector: IoT applications in the public sector include smart city initiatives, public safety monitoring, and environmental monitoring.

Retail: IoT applications in retail will provide services to customers through personalized offers, inventory tracking, and smart checkout.

It is a summary that the IoT architecture is so composed, with multiple layers of technology integrated to further support a wide range of applications and use cases. From the Sensor Layer through to the Application Layer, each layer constitutes this architecture for accurate data collection, efficient transmission, secure management, and effective use in the meeting of diverse user needs [5].

Literature Review Paper Citation-Chen Et Al., 2014

The IoT involves billions or even trillions of interconnected devices in general, known as "things," in communicating and interacting with each other. It is said that China has emerged as a key player in the global IoT landscape and has exhibited significant advances with comprehensive policies, R&D plans, applications, and standardization. The Chinese government has demonstrated its serious commitment to the development of IoT through various strategic frameworks, including the 12th Five-Year Plan for IoT development that covers the period from 2011 to 2015. It outlines strategic goals and objectives relevant to the fostering of IoT growth, including ways in which the support and acceleration of technology development are effected. This is due to the fact that extensive R&D of IoT in China is spent on promoting basic technologies, test beds, and standardization. The performed R&D activities have aimed at the elimination of technical barriers, increasing the capability of IoT systems in general, and enhancing their performance. Besides, China has already started noteworthy deployment of IoT applications in industries such as health care, automotive, utilities, and consumer electronics. These applications exemplify how IoT can bring transformative power in increasing operational efficiencies and driving innovations within industries. Different IoT systems and devices need to be developed and adopt standardized protocols in order to have seamless integration and communications. Integration of a variety of IoT technologies into coherent systems has a number of challenges for which standardization is of essence.

While the IoT has been developing, it still faces quite a lot of problems in being integrated into daily life. In terms of technical issues, rapid development indeed embeds quite a few complications to this technology, including guaranteed compatibility and interoperability between diverse versions of different IoT devices and systems. All these issues may require further invention and the introduction of new technologies. Practical application difficulties such as data integration, system scalability, and real-time processing are big challenges. Their resolution will be crucial in bringing practical benefits from IoT. It proposes an open and general IoT architecture comprising three core platforms to solve these issues. The first is the data collection platform, which has the job of efficiently collecting and aggregating data from different IoT devices and sensors [7]. The second one refers to the platform for handling the data, dealing with processing, storage, and analysis of the collected data, and hence enables intelligent handling of the data, including real-time analytics and long-term storage. It provides the third platform, which enables the integration of various applications and confirms that the value of data can be realized across industries and greatly enhances total utility and impact of IoT solutions. In fact, the potential of IoT surpasses the challenges at hand. The technology holds immense opportunities for innovation and economic growth. IoT can be applied to create efficiencies, enhanced decision-making, and new value propositions by businesses and governments across diverse industries.

PAPER CITATION: Porkodi et al., 2014

The Internet of Things, as a blooming technology, captures the current advancement in both Wireless Sensor and Actuator Networks and Pervasive Computing and brings innovative changes to many fields. The section after next undertakes a comprehensive review of applications, opportunities, and challenges related to IoT, along with an elaborative overview of sensors and standards governing the technology. The IoT paradigm is essentially characterized by the endless possibility of interconnecting a host of devices and systems. Ever since the conceptualization, IoT has gained significant attention in bringing efficiency for a wide range of applications. However, this review found that IoT-based applications are widely adopted in both developed and developing countries across different domains, showing significant developments toward incorporating smart technologies. For instance, IoT-based applications have transformed the managing of cities by allowing real-time monitoring of infrastructure, traffic, and other environmental concerns. IoT enables remote patient monitoring and personal treatment in the care industry through wearable sensors and wirelessly connected medical devices. The rapid proliferation of IoT technologies in these and other fields speaks volumes to their potentials for efficiency drivers, quality-of-life improvements, and innovative solutions.

In contrast to this progress, the IoT environment is fraught with challenges and complexities:. Researchers have enumerated some key issues that must be addressed if the full benefits of IoT are to be realized. Among these, one of the most serious challenges facing IoT is the integration and interoperability among different IoT devices and systems, which might involve heterogeneous protocols and standards. As a matter of fact, this is one of the reasons for non-standardization that seriously challenges seamless connectivity and exchange of data between devices, thereby acting as a major obstacle to the mainstreaming of IoT technologies. The huge volume of data produced in IoT devices also raises serious concerns on data management, privacy, and security. Efficient mechanisms that ensure protection of data and privacy are highly essential in building confidence in IoT systems. At the center of IoT systems are sensors and actuators that grant devices the ability to perceive and act on their environments. In this regard, the literature goes deep into a number of sensor technologies and standards that exist. Radio-Frequency Identification tags, for instance, have been one of the mainstay devices in IoT applications. In respect to these aspects, RFID technology has been able to assign unique identifiers to objects and permit contactless data transfer between tagged objects and interrogators [8].

Coupled with RFID, other sensor technologies come in the form of temperature sensors, humidity sensors, and motion detectors for capturing environmental data and enabling intelligent decision-making. Each sensor type has its own set of standards and protocols, dictated by its operational characteristics and integration into IoT systems. As an example, the standard IEEE 802.15.4 is being utilized in low-power, low-data-rate sensor networks and forms the base upon which many IoT communication protocols are built [9]. Well-established protocols for communication permit seamless coexistence of various devices and systems within the IoT. It talks also of the evolving sensor technologies and how that affects the different evolving IoT applications. The same miniaturization and energy efficiency, along with advances in data processing, enable the production of high-performance, sophisticated sensors. New kinds of more complex sensors open up new avenues for IoT applications or greatly expand capabilities in existing ones. However, the ever-growing tempo of technological innovation creates compatibility and standardization challenges in light of various sensor technologies.

In summary, the literature on IoT emphasizes its transformative potential while at the same time identifying some major challenges yet to be surmounted. The core role of smart technologies like sensors and actuators is identified in the successful implementation of an IoT system. Interoperability, data management, and issues related to security form the bedrock for the advancement of IoT toward achieving its full potential.

3. Methodologies

For addressing the research gaps and objectives in the context of Internet of Things (IoT) Applications, a hybrid approach can be followed.

3.1 Scalable and Interoperable IoT Systems

For developing scalable and interoperable IoT systems, a system design and architecture approach is important. Researchers can develop a reference architecture that emphasizes modularity and flexibility, which integrates various IoT devices across different platforms and protocols. Simulation-based testing using tools such as OMNeT++ or NS-3 can model large-scale IoT networks to assess their scalability and interoperability under varying conditions. Real-world test beds can be used for performance evaluations, focusing on latency, data throughput, and communication success rates across multiple devices.

3.2 Security and Privacy in IoT Networks

To address security and privacy challenges, cryptographic solutions including AES, RSA, and ECC, can be used to securely transmit data in IoT systems. Vulnerability assessment can be carried out using penetration testing and ethical hacking tools such as Metasploit and Wireshark on IoT devices and networks. A study on blockchain technology might be undertaken for decentralized security with tamper-proof record of data. More important, researchers can study privacy-preserving techniques, which encompass data anonymization and differential privacy, for preserving sensitive user information and allowing for effective data analytics.

3.3 Efficient Data Management and Analytics for IoT

In order to manage and analyze large amounts of IoT data efficiently, one has to use big data platforms like Apache Hadoop, Spark, and Flink to process and analyze this amount of data. Developing algorithms that are optimized to the IoT environment will enable real-time data processing and actionable insights to be derived. Edge and Cloud Integration will be highly critical for offloading heavy processing tasks towards the cloud while enabling a possibility of real-time analytics in the edge, thus minimizing the latency. Further, use of Machine Learning Algorithms, to analyze patterns and look for anomalies in IoT, will be applied to real-time applications such as predictive maintenance or resource optimization. Data fusion techniques may also be used to combine the data from multiple IoT sensors in order to make more precise decisions.

3.4 Energy Efficient IoT Devices

Energy efficiency would require low-power hardware design. Researchers can look into the energy-efficient components, for example, low-power microcontrollers and sensors, and apply simulation tools to analyze power consumption. Dynamic power management can be designed with strategies such as duty cycles and sleep modes to keep the active time at a minimum while also optimizing energy usage without diminishing performance. In addition, alternative sources of energy harvesting, like solar or vibration energy, can be explored to help reduce dependence on traditional sources of power. Empirical experiments can be conducted for benchmarking energy consumption in the different IoT devices, leading to opportunities for optimization with longer life spans.

3.5 Edge and fog Computing for IoT Systems

In optimizing edge and fog computing for IoT systems, researchers can develop a comprehensive framework that supports the processing and transmission of data between IoT devices, edge nodes, and the cloud. This can help reduce latency and alleviate bandwidth strain. Load balancing algorithms will be very important when distributing computational tasks across edge nodes, fog nodes, and cloud servers effectively, ensuring that resource utilization will be maximized. Prototyping using Raspberry Pi or Arduino can prove the feasibility of edge and fog computing in real-word scenarios.

3.6 Standardization of Communication Protocols

To address the lack of standardization in communication protocols, a protocol comparison study can be performed on the performance of various communication protocols like Zigbee, LoRa, and 5G regarding the range, power consumption, and reliability in IoT applications. The researchers can also design new or improved protocols specific to IoT systems and simulate their performance using tools like NS-3 or OMNeT++. Interoperability testing of IoT devices from diverse manufacturers using different protocols is necessary for identifying compatibility issues and then proposing unified standards for the IoT ecosystem.

3.7 Regulatory and Ethical Challenges in IoT

To address the regulatory and ethical challenges in IoT, researchers can conduct a comparative legal analysis of existing regulations, like GDPR or CCPA, to point out gaps in current frameworks. An ethical review using interviews with experts in data privacy, ethics, and IoT deployment could help evaluate the ethical implications of IoT technology use. The case study analysis of real-world applications of IoT, such as smart cities or healthcare, could help unmask regulatory and ethical concerns about data usage, consent management, and ownership of data.

3.8 Real-Time Decision-Making in IoT Systems

To enhance the real-time decision-making ability, researchers may develop real-time data processing algorithms based on stream processing technology, such as Apache Kafka. These algorithms are designed to facilitate real-time decision-making based on dynamic data coming from IoT devices. Simulations of real-time IoT environments can also be used to test the behavior of decision-making models to ensure they will operate efficiently under different conditions. The use of machine learning models for predictive analytics and in real-time decision-making could also be considered to enrich the decision support capabilities of IoT systems.

4. IoT Components of Internet of Things-IoT Systems

IoT systems creatively combine diverse devices and systems to an environment that can feel, process, and respond to various conditions. Most IoT systems can generally be grouped into four categories based on their operation: Devices/Sensors, Connectivity, Data Processing, and User Interface. Each of the above parts is a very important element in any IoT system architecture and contributes much to the efficacy and functionality of the IoT systems.

4.1 Sensors/Devices

Sensors and devices are the basic elements of every IoT system, representing the direct interface with the physical world [10]. These elements are supposed to fetch information from other external sources that may be in the form of some environmental changes, variation in some physical phenomena, or direct user interaction. The sensors are designed to perceive and log changes in surroundings regarding temperature fluctuations, motions, or variations in light. Also, GPS modules on the mobile devices track geographic locations while the cameras capture visual information and detect human movement for image processing.

4.2 Connectivity

After the gathering of data by sensors, it has to be relayed to processing systems for further analysis. It is connectivity-a very vital piece that enables the transmission of data across the IoT ecosystem, connecting sensors, routers, gateways, user applications, and platforms. For establishing connections, several technologies may be used: Wi-Fi, Bluetooth, Zigbee, and cellular networks like LTE and 5G [11].

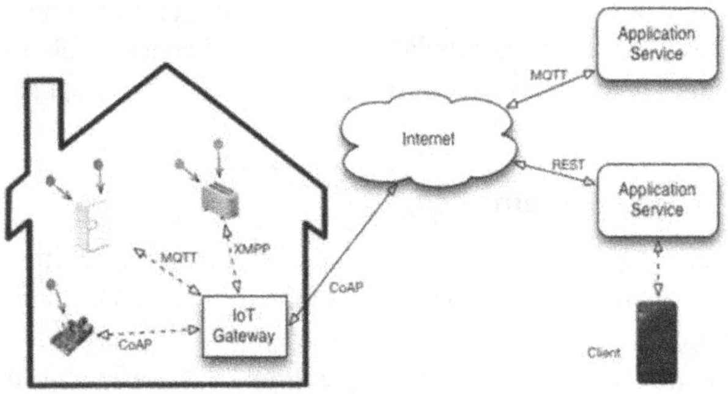

Fig. 6.2 Components of internet of things

4.3 Data Processing

The processing of data forms one of the major steps in the IoT architecture, where interpretation and analysis of data as obtained by sensors are performed. Raw data transmitted to a cloud server or data processing platform undergoes several computational functions to spot meaningful insights and show actionable outputs. Advanced algorithms and analytical tools have been used in this regard to process the data by using some sort of statistical analysis, machine learning models, or data fusion techniques. Efficiency and accuracy in data processing directly affect the effectiveness of the IoT system, whereby timely and precise data analysis will enable the system to act efficiently and deliver appropriate information or actions based on the data processed.

4.4 User Interface

The UI is the final component of the IoT system, representing the processed data and system outputs in a format accessible and understandable by users. The UI represents the interaction point between the IoT system and its users through some sort of feedback, whether visual or functional, on operations [12]. The type of interface would be different in each IoT device and can be implemented to various degrees depending on the function of the device itself and user needs [13]. For instance, a smart thermostat might have a touch-screen user interface for setting temperature preferences, while a wearable fitness tracker will provide a mobile application interface that presents the user with health metrics.

In summary, the interplay among Sensors/Devices, Connectivity, Data Processing, and User Interface outlines the operational framework of the IoT systems. Each of these additions extends the capability of the system to sense, transmit, analyze, and present data, thereby driving the creation of intelligent and responsive environments.

5. Key Technologies of IoT

IoT is based on the following key technologies broadly classified under two major heads: identification and communication technologies. Each technology plays a very important role in the working and effectiveness of IoT systems in identifying and tracking and exchanging data between devices.

5.1 Identification Technology

Behind object location and monitoring in an IoT ecosystem is identification technologies. Identification technologies will help in uniquely identifying devices, items, and their interactions within a network. Some of the main examples of identification technologies are RFID, WSN, QR codes, barcodes, and intelligent sensors. As an example, RFID technology relies on a combination of readers and transmitters to gather and forward information. The main components of an RFID system include an RFID tag, which carries the information, and an RFID reader, which collects that information from the tag. RFID tags are normally very effective in their applications, though they are a bit more expensive than other technologies such as WSN [14].

Other identification technologies necessary for this field are WSNs, which work by deploying a spatially dispersed network of sensors that keep watch over environmental conditions or other phenomena. WSNs can often be applied to those cases where wide coverage and data gathering are at hand, such as in environmental monitoring and smart agriculture. QR codes and barcodes work with much simpler forms of identification generally available through product tracking and inventory management applications.

5.2 Communication Technology

Communication technologies in IoT are designed for the required data transfer between devices and systems using a standard protocol for efficiency and reliability in connectivity. The key factor in the choice of this type of communication technology involves protocols and standards that best meet the requirements of range, speed, and data transfer capacity. These include Zigbee, Z-Wave, MQTT, Bluetooth, Li-Fi, Wi-Fi, Near Field Communication (NFC), HaLow, and Power Line Communication (PLC). It operates under the standard IEEE 802.15.4 and is designed for using short ranges of about 20 meters [15].

MQTT-Message Queuing Telemetry Transport: It is an M2M communication protocol optimized in cloud and fog computing environments. The protocol has great value in such scenarios that demand reliable message delivery with low bandwidth consumption. Bluetooth runs on the standard specification IEEE 802.15.1 and finds wide applications in the data transfer of short distances between paired devices, such as smartphones and peripherals. Its ease of use and versatility make it another very common personal area network [16].

Wi-Fi refers to the IEEE 802.11-standardized medium-range wireless communication widely used in LANs. It allows scalability and is capable of higher data transfer rates, making it applicable from small-scale residential networking to large-scale enterprise environments. NFC enables communication within a very short reach, usually no more than 4 meters [17]. This technology can enable point-to-point data exchange between devices, such as sharing data between a smartphone and smart TV.

While there are a variety of identification and communication methods that are the very basis of IoT systems, they ensure that IoT applications work effectively and provide for correct identification, reliable data transfer, and efficient connectivity.

6. Applications

IoT is growing significantly in various fields, starting to improve efficiency and precision with automation. From environmental monitoring to healthcare, IoT applications are changing the world almost every sector uniquely benefitting from these technologies.

IoT facilitates better care for the environment through its real-time monitoring of various parameters. Some of the sensors used in these applications include RTDs and thermometers to detect changes in temperatures, dust sensors, and gas sensors to monitor air quality. Other examples include the so-called electronic tongues or e-Tongue and noses or e-Nose that utilize pattern recognition to identify chemical contamination [18]. This technology is crucial for maintaining both air and water qualities in all urban centers, thanks to the timely responses toward any changes in the environment.

Aquaculture: IoT allows the management of aquatic environments in aquaculture. Sensors track the water temperature and chemical levels to keep the right temperature for aquaculture. Health IoT applications include wearable and implantable sensors [19]. Wearables range from pulse rate to body temperature and everything between, offering instant health information. Implanted sensors measure vital internal metrics for personalized care and timely medical intervention, improving patient outcomes.

7. Result and Discussion

The review of IoT systems illustrates that the integration of Sensors/Devices, Connectivity, Data Processing, and User Interface is critical to efficient IoT functionality. Sensors and devices collect data. Their basic function is to monitor variables continuously, whether it be temperature, motion, or light, and to allow systems within the IoT to intelligently respond. The sensors range from optical to mechanical, even to chemical, ensuring wide-ranging environmental and physical phenomena are captured and documented with accuracy. Connectivity technologies ensure seamless communication across the IoT ecosystem. Seamless connectivity is key to assured data transfer integrity for smooth communication among devices and platforms.

Table 6.1 Technologies and tools used in IOT

S. No.	Technologies in IOT	Tools Used
1	Identification Technology	RFID,WSN,QR CODE,BAR CODE
2	Communication Technology	Zigbee, Z wave, MQTT, Bluetooth, LiFi, Wi-Fi, NFC, Power line area network

IoT analytics essentially revolve around data processing-deriving insight from raw data that can then be acted upon.

Advanced algorithms and machine learning models that handle large sets of data are used to ensure that processed information is leveraged for timely decisions and efficient operations. The efficiency of data processing not only determines the responsiveness of the IoT system but also the capability of the system to make the data much more relevant and useful to its users. A well-designed UI allows intuitive and user-friendly access to system functionality, while usability is dramatically enhanced for IoT applications. Diversity in UI implementation for the various forms of IoT devices creates a case for customized implementation that should meet the needs of different users and the capability of devices.

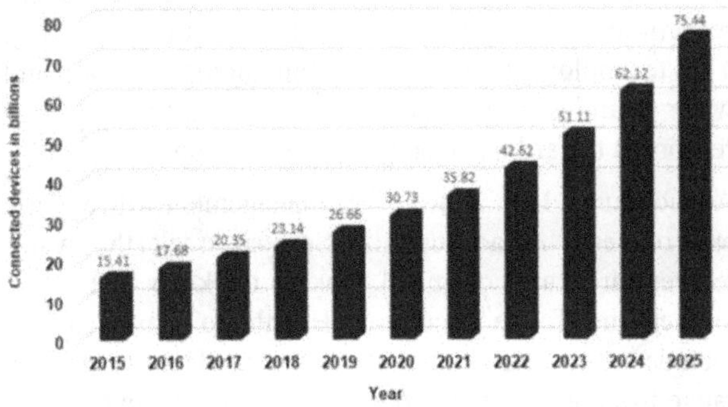

Fig. 6.3 Increase in use of IOT systems from year 2015 to 2025 (approx)

The integration of IoT into different industries has brought enormous benefits in terms of efficiency, monitoring, and automation. For one, environmental monitoring is effectively carried out by IoT-enabled sensors that provide real-time precise data on air and water qualities for efficient pollution management proactively. In home automation, IoT has allowed for an efficient control of home appliances using various methods of connectivity to bring ease of living along with energy efficiency . This is by adopting precision farming, which maximizes the use of resources to increase crop yields. In aquaculture, IoT has been utilized in environmental control to maintain the best living conditions for aquatic species.

8. Research Gaps

1. **Scalability and Interoperability Issues:** While IoT systems have grown rapidly, scaling these systems efficiently across industries and ensuring seamless interoperability between diverse devices and platforms remains a significant challenge.
2. **Security and Privacy Concerns:** With the huge amount of sensitive information collected and transmitted by these devices, security and privacy are being challenged at an alarming rate.
4. **Edge and Fog Computing:** However, the integration of edge/fog computing with IoT still remains a complex challenge to coordinate, balance loads to ensure real-time processing for IoT.
5. **Standardization of Communication Protocols:** IoT devices use multiple communications protocols, such as Wi-Fi and Zigbee, Bluetooth and LoRa;

9. Conclusion

In conclusion, the integration of IoT systems across various sectors underscores its transformative impact on operational efficiency and quality of life. The critical components—Sensors, Connectivity, Data Processing, and User Interface-each play a pivotal role in ensuring the effective performance of IoT applications. Sensors and devices provide essential real-time environmental and user data, while connectivity technologies facilitate seamless communication, crucial for maintaining data integrity and operational responsiveness. Data processing technologies leverage advanced algorithms to derive actionable insights from raw data, enhancing decision-making and system efficiency.

References

1. Elijah, O., Rahman, T. A., Orikumhi, I., Leow, C. Y., & Hindia, M. N. (2018). An overview of Internet of Things (IoT) and data analytics in agriculture: Benefits and challenges. IEEE Internet of Things Journal, 5, 3758–3773.
2. Porkodi, R., & Bhuvaneswari, V. (2014). The Internet of Things (IoT) applications and communication enabling technology standards: An overview. In 2014 International conference on intelligent computing applications (ICICA), IEEE, pp. 324–329.
3. Gubbi, J., Buyya, R., Marusic, S., & Palaniswami, M. (2013). Internet of Things (IoT): A vision, architectural elements, and future directions. Future Generation Computer Systems, 29(7), 1645–1660.
4. Amendola, S., Lodato, R., Manzari, S., Occhiuzzi, C., & Marrocco, G. (2014). RFID technology for IoT-based personal healthcare in smart spaces. IEEE Internet of Things Journal, 1(2), 144–152.
5. Da Xu, L., He, W., & Li, S. (2014). Internet of Things in industries: A survey. IEEE Transactions on Industrial Informatics, 10(4), 2233–2243.
6. Lin, J., Yu, W., Zhang, N., Yang, X., Zhang, H., & Zhao, W. (2017). A survey on internet of things: Architecture, enabling technologies, security and privacy, and applications. IEEE Internet of Things Journal, 4(5), 1125–1142.
7. Alam, M. M., Malik, H., Khan, M. I., Pardy, T., Kuusik, A., & Le Moullec, Y. (2018). A survey on the roles of communication technologies in IoT-based personalized healthcare applications. IEEE Access, 6, 36611–36631.
8. Gharaibeh, A., Salahuddin, M. A., Hussini, S. J., Khreishah, A., Khalil, I., Guizani, M., et al. (2017). Smart cities: A survey on data management, security, and enabling technologies. IEEE Communications Surveys & Tutorials, 19(4), 2456–2501.
9. Grosinger, J., & Bösch, W. (2014). A passive RFID sensor tag antenna transducer. In 2014 8th European conference on antennas and propagation (EuCAP), IEEE, pp. 3638–3639.

10. Suresh, P., Daniel, J. V., Parthasarathy, V., & Aswathy, R. H. (2014). A state of the art review on the Internet of Things (IoT) history, technology and fields of deployment. In 2014 International conference on science engineering and management research (ICSEMR), IEEE, pp. 1–8.
11. Nie, X., & Zhong, X. (2013). Security in the internet of things based on RFID: Issues and current countermeasures. In Proceedings of 2nd international conference on computer science and electronics engineering, Vol. 162, No. 7.
12. Kazmi, A., Jan, Z., Zappa, A., & Serrano, M. (2016). Overcoming the heterogeneity in the internet of things for smart cities. In International workshop on interoperability and open-source solutions, Springer, Cham, pp. 20–35.
13. Kubo. (2014). The research of IoT based on RFID technology. In 2014 7th international conference on intelligent computation technology and automation, Changsha, pp. 832–835.
14. Khalid, A. (2016). Internet of Thing architecture and research agenda. Computer Science and Mobile Computing, 5(3), 351–356.
15. Al-Sarawi, S., Anbar, M., Alieyan, K., & Alzubaidi, M. (2017). Internet of Things (IoT) communication protocols. In 2017 8th International conference on information technology (ICIT), IEEE, pp. 685–690.
16. Naik, N. (2017). Choice of effective messaging protocols for IoT systems: MQTT, CoAP, AMQP and HTTP. In 2017 IEEE international systems engineering symposium (ISSE), IEEE, pp. 1–7.
17. Xu, L. D., Xu, E. L., & Li, L. (2018). Industry 4.0: State of the art and future trends. International Journal of Production Research, 56(8), 2941–2962.
18. Calabretta, M., Pecori, R., & Velti, L. (2018). A token-based protocol for securing MQTT communications. In 2018 26th International conference on software, telecommunications and computer networks (SoftCOM), IEEE, pp. 1–6.

Note: All the figures and table in this chapter were made by the authors.

Intelligent Systems Using Semiconductors for Robotics and IoT – Dinesh Goyal et al. (eds)
© 2026 Taylor & Francis Group, London, ISBN 978-1-041-20408-4

7 Power Quality Improvement in PV-Wind Hybrid Energy System Using Improved Soft Computing Based UPQC

Deepak Alaria[1]
M. Tech. Scholar, Department of Electrical Engineering,
Rajasthan Institute of Engineering and technology,
Jaipur (Rajasthan)

Kalpana Meena[2]
Assistant Professor, Department of Electrical Engineering,
Rajasthan Institute of Engineering and Technology,
Jaipur (Rajasthan)

Abstract: Renewable power generation has widely been incorporated into the power supply systems especially photo voltaic or solar and wind power systems due to increasing use of clean energy. However, these sources cause unwanted phenomena like voltage sags, swells, flickers and harmonics distorting the quality of power which in-turn affects the power system stability and reliability. This paper contributes toward the development of an enhanced hybrid energy system comprised of Photovoltaic (PV) and Wind power system by supplying power quality improvement using an Improved Soft Computing-based Unified Power Quality Conditioner (UPQC). These techniques are used in order to enable the best control strategy of the UPQC in order to give responsive and appropriate action to mismatched disturbances of power quality. The design with soft computing methods has been proved to enhance the reliability and efficiency of the UPQC while solving power quality problems. A detailed simulation model of the PV-wind hybrid energy system incorporating the Improved Soft Computing-based UPQC is created using MATLAB/Simulink. The performance of the model is investigated under different operating conditions and interferences to assess the enhancement of power quality. As for the simulation outcomes, it has been evidenced that the voltage sags, swells and harmonic distortions are significantly eliminated and the effectiveness of the examined proposal exceeds the performance of conventional UPQC approaches.

Keywords: Power quality improvement, PV-wind hybrid energy system, Unified power quality conditioner (UPQC), Soft computing, Fuzzy logic, Neural networks, Voltage sags, Harmonic distortions, MATLAB/Simulink, Total harmonic distortion (THD), Voltage regulation, Renewable energy integration, Smart grid technology

1. Introduction

The heightening international focus on sustainable development coupled with proprioception of the deleterious impacts of conventional sources of energy to the environment has propelled the use of renewable energy sources. Of these, photovoltaic (PV) and wind energy systems are two most promising energy systems owing to their abundant availability and environmentally friendly characteristics. While these renewable energy systems can be connected to the existing power grid, so as to interconnect and share power between them, there are several issues that arise, largely in relation to power quality. Voltage drops, voltage rises, voltage fluctuations and harmonic progresses present a number of implications that affect the availability and reliability of the power system. Solving these problems would be helpful not only for continuing the functioning of renewable energy

[1]deepak.alaria0004@gmail.com, [2]ee.kalpana@rietjaipur.ac.in

DOI: 10.1201/9781003716389-7

systems but also for the stable operation of the electric power grid.

A UPQC has been identified as a suitable power quality conditioner for treating power quality issues in distribution networks. A UPQC therefore is an integrated active power filter that can correct both voltage and current qualities using a combination of the series and shunt power filter configurations. It was seen that the conventional UPQCs have been useful but their performance could be improved by implementing latest control schemes by using soft computing methods. This paper examines the power quality as one of the most important parameters in the contemporary power systems for both utilities and customers. Even though power quality disturbances are usually small fluctuations, they can cause a lot of issues that consist of equipment failures, higher losses, shorter life expectancy of electrical appliances, as well as interruptions of work in industrial facilities. As for the issue of high PV-wind combined system penetration into the grid, the requirement for high power quality cannot be overstressed. It not only increases the credibility and solidity of the power system but also achieves the efficient consumption of renewable energy and helps reduce people's negative impact on the environment. Some of the problems associated with integration of Hybrid energy systems from photovoltaic and wind energy sources are as follows.

The integration of photovoltaic and wind energy systems into the electrical grid can be problematic because of their inherent stochastic characteristics. PV systems are very sensitive to climate changes and time of the day as it depends on the intensity of irradiance. Likewise, wind energy systems depend on wind velocity and direction and the like, which are inherently volatile. All these variations create power quality problems including voltage variation, voltage sag/swell conditions and harmonic distortion; the measurement and control of which require measures if power quality at the point of common coupling is to be maintained at an optimum level. Unified Power Quality Conditioner or UPQC is a sophisticated power electronic equipment used for improving PQ and removing voltage and current problems. It consists of two main components: There is a series active power filter and a shunt active power filter. The series filter serves as contingency for voltage-related issues such as sags, swells and harmonics and the shunt filter serves as contingency to counter harmonics currents and reactance power compensation.

2. Literature Review

The comprehensive review of recent studies on power quality improvement in hybrid renewable energy systems using Unified Power Quality Conditioner (UPQC) reveals significant advancements and diverse approaches in the field. Researchers have extensively explored various soft computing techniques, meta-heuristic algorithms, and advanced control strategies to enhance the performance and effectiveness of UPQC in mitigating power quality issues. Ranjan and Choudhary (2024) and Shravani and Narasimham (2024) demonstrate the potential of ANN and UPQC in addressing voltage fluctuations and harmonics in grid-connected PV, battery, and wind systems. Yadav et al. (2024) and Bharat Mohan and Rajagopal (2023) further validate the effectiveness of UPQC in handling voltage variations and improving overall power quality in hybrid systems. Studies by Singh and Singh (2023), and Yadav and Yadav (2024) emphasize the role of advanced soft computing techniques, such as fuzzy logic and hybrid metaheuristics, in optimizing UPQC control strategies. These techniques enhance the adaptability and precision of UPQC, allowing for dynamic responses to real-time power quality disturbances. Srilakshmi et al. (2024) and Baswaraju et al. (2024) highlight the integration of UPQC with advanced optimization algorithms and hybrid controllers, leading to significant improvements in power quality and system efficiency. Kanagaraj et al. (2023) and Ravi et al. (2023) underscore the benefits of coordinated control schemes and fuzzy logic-based UPQC in enhancing power quality management in hybrid renewable energy generation systems. Elmetwaly et al. (2023) and Ranjan and Choudhary (2023) validate the application of novel optimization algorithms and Battle Royal Optimizers in enhancing UPQC performance, particularly in PV-wind-fuel cell and battery systems. Similarly, studies by Abed el-Raouf et al. (2023) and Vigneshwar and Shunmugalatha (2024) emphasize the effectiveness of UPQC in reducing harmonic distortions and improving power quality through advanced control and optimization techniques. The integration of fuzzy logic controllers and genetic-based ANFIS controllers, as explored by Amirullah and Ananda (2024) and Sowmya Sree and Ankarao (2023), shows promising results in addressing power quality issues in hybrid PV-wind systems. Additionally, Nagaraju (2023) and Madhavan and Anandan (2023) provide evidence of enhanced power quality through the use of optimized UPQC and various controlling techniques in grid-connected hybrid renewables. Sindi et al. (2023) and Yadav and Yadav (2023) further illustrate the benefits of advanced compensators and optimization algorithms in improving power quality in multi-microgrid and hybrid energy systems. Choudhury and Sahoo (2024) offer a critical analysis of power quality improvement techniques, highlighting the importance of UPQC in maintaining power quality in renewable energy-integrated microgrids. Although Kumar, Acharyulu, and others (2023) retracted their study, their initial findings suggested potential enhancements in power quality through the integration of UPQC with renewable energy storage systems. Lastly, Srilakshmi et al. (2024) confirm that optimizing ANFIS controllers for solar and battery sources fed UPQC can significantly improve power quality in hybrid renewable energy systems. The integration of renewable energy sources such as photovoltaic (PV) and wind systems into power grids has brought significant challenges, particularly in terms of power quality. Ranjan and Choudhary (2024) highlight the use of meta-heuristic algorithms to improve power quality in grid-connected hybrid renewable energy systems using a Unified Power Quality Conditioner (UPQC). They emphasize the effectiveness of Artificial Neural Networks (ANN) in enhancing the UPQC's ability to mitigate power quality issues such as voltage sags and harmonic distortions. Similarly, Shravani and Narasimham (2024) examine the deployment of UPQC in grid-linked PV, battery, and wind systems to address power quality and stability concerns. Their study demonstrates that UPQC can effectively improve the power quality in hybrid renewable energy systems, particularly in mitigating voltage fluctuations and harmonics. Yadav and Yadav (2024) explore a hybrid metaheuristic-assisted fractional-order controller for three-phase solar PV, BESS, and wind-integrated UPQC. Their study shows that this approach optimizes the control of harmonic distortions and achieves quick power balance, thereby enhancing overall power quality. Srilakshmi, Gaddameedhi, and Borra (2024) design an optimal UPQC for managing power quality and power flow in systems with solar, wind, battery, and electric vehicle (EV) integration. Their use of an enhanced most valuable player algorithm demonstrates significant improvements in power quality and system efficiency. Baswaraju, Raju, and Sobti (2024) investigate the benefits of UPQC integration in hybrid solar wind energy systems. They find that improved Maximum Power Point Tracking (MPPT) control and hybrid controllers enhance the performance of UPQC, contributing to better power quality and system stability. Kanagaraj, Vijayakumar, and Ramasamy (2023) focus on energy management and power quality improvement in hybrid renewable energy generation systems using coordinated control schemes. Their study shows that integrating UPQC with hybrid PV and wind systems significantly improves power quality and power flow management.

3. Proposed Methodology

The hybrid system offers a complementary energy generation profile, with PV generation peaking during the day and wind generation typically higher at night or during different weather conditions. Two voltage source inverters, a DC link capacitor, and two converters can all be used with a UPQC controller. CPDs are used at the circulation level to regulate the PQ-related concerns, such as the harmonic current in current and voltage, voltage sag/swell, and receptive power remuneration, among which UPQC is the most attractive arrangement. Furthermore, there is no voltage sag/swell component, a common voltage upsetting influence problem. Voltage sag is the more commonly acknowledged of the two voltage short length concerns, i.e., sag and swell, that utilities must deal with. In the part below, it is explained how the adaptive hysteresis control approach with UPQC would be expected to boost PQ..

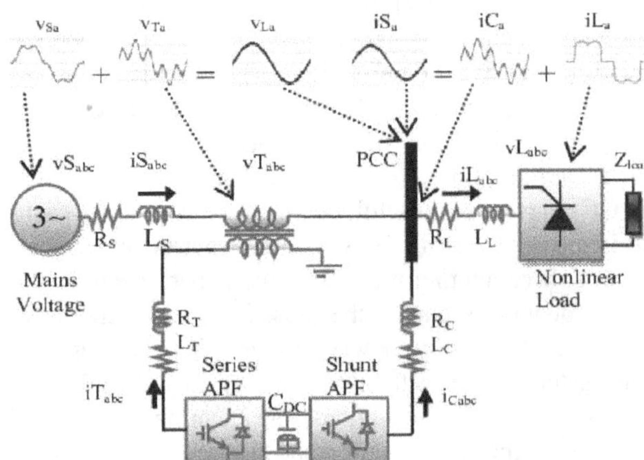

Fig. 7.1 Connection diagram of unified power quality conditioner

3.1 Modeling of Fuzzy Logic Controller

When it comes to performance, isp is compared to the size of three-phase reference currents. 3-phase unit current vectors (usa, usb, and usc) are obtained from the three phase voltage (vsa, vsb, and vsc). The current unit vectors (usa, usb, and usc) create three phase reference currents. Magnitude Multiplication Phases with usa, usb, and usc provide the three-stage reference supply currents (isa*, isb*, and isc*). The three phase reference currents (Isa *, Izb *, and Izk *) of the shunt inverter are independent from the reference supply currents (ilsa *, Izb *, and ilsk *). The iref reference currents (isha *, ishb *, and ishc *) are analogous to actual shunt-compensating iact (isha, ishb and Ishc).

3.2 Compensation Techniques

By including a voltage source, a series inverter may run in current control mode while isolating the load from the supply. This source of supply voltage corrects for incorrect situations including sag, swell, flicker, and spikes. The three phase load voltages of the inverter series loop control system (vla, vlb, and vlc) are separated from the three phase voltage supply system (vsa, vsb, and vsc) and contrasted with the reference load supply voltage to form three phase reference voltages. The three phase reference currents (isea*, iseb*, and isec*) of the series inverter are obtained from the three phase reference voltages (vla*, vlb*, and vlc*) using an appropriate transformation. A PWM current controller receives the reference currents (isea*, iseb*, and isec*) and their sensor equivalents (isea*, iseb*, and isec*).

Computation of Control Quantities of Shunt Controller

In the three step sensed values the amplitude of the supply voltage is determined as:

$$V_{sm} = [2/3 (v_{sa}^2 + v_{sb}^2 + v_{sc}^2)]^{1/2} \quad (4.1)$$

The current vectors of the three-phase unit are calculated as:

$$u_{sa} = v_{sa}/v_{sm};\ u_{sb} = v_{sb}/v_{sm};\ u_{sc} = v_{sc}/v_{sm} \quad (4.2)$$

The multiplication with the amplitude of the supply current (isp) of three phase vectors (USA, USB and USC) results in a three phase supply reference currents as follows:

$$i_{sa}* = i_{sp}.u_{sa};\ i_{sb}* = i_{sp}.u_{sb};\ i_{sc}* = i_{sp}.u_{sc} \quad (4.3)$$

Three phase load currents are removed from three phase reference currents in order to obtain reference currents:

$$i_{sha}* = i_{sa}* - i_{la};\ i_{shb}* = i_{sb}* - i_{lb} \quad (4.4)$$

$$ishc* = isc* - ilc \quad (4.5)$$

Inverter Series Controller Computation

The voltage of supply and the voltage of charge are sensed, and the desired injected voltage is calculated accordingly:

$$v_{inj} = v_s - v_l \quad (4.6)$$

The size of the voltage injected is as follows:

$$v_{inj} = |v_{inj}| \quad (4.7)$$

The injected tension step is indicated as:

$$i_{nj} = \tan(Re[v_{pq}]/Im[v_{pq}]) \quad (4.8)$$

The following inequalities are followed for the purpose of compensating harmonics in load voltage:

a) $v_{inj} < v_{inj}max$; control of magnitude;
b) $0 <_{inj} < 360°$; control phase;

The injected voltages express three phase reference values as:

$$v_{la}* = 2v_{inj} \sin(wt + _{inj}) \quad (4.9)$$

$$v_{lb}* = 2v_{inj} \sin(wt + 2/3 + _{inj}) \quad (4.10)$$

$$v_{lc}* = 2v_{inj} \sin(wt - 2/3 + _{inj}) \quad (4.11)$$

The three stage benchmarks (i_{ref}) of the inverter series are determined as follows:

$$i_{sea}* = v_{la}*/z_{se}; \quad (4.12)$$

$$i_{seb}* = v_{lb}*/z_{se}; \quad (4.13)$$

$$i_{sec}* = v_{lc}*/z_{se}; \quad (4.14)$$

The insertion transformer impedance is necessary for the zseimpedance. The voltage sag that is required must be compensated for, and the currents (isea*, iseb*, and isec*) are the optimal currents to hold via the secondary winding of the insertion transformer. Six switching signals for series inverter IGBTs are produced by comparing the irefcurrents (isea*, iseb*, and isec*) in the PWM current controller (isea*, iseb*, and isec*).

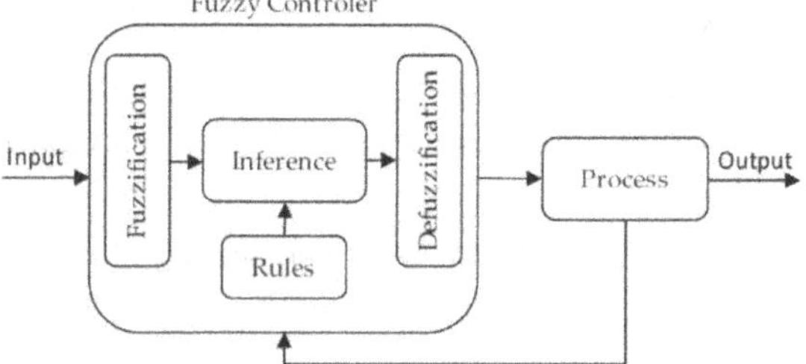

Fig. 7.2 building blocks of fuzzy based system

Table 7.1 Rule base

Change in error	Error						
	PL	**PM**	**PS**	**Z**	**NS**	**NM**	**NL**
NL	PL	PL	PL	PM	PM	PS	Z
NM	PL	PL	PM	PM	PS	Z	ZS
NS	PL	PM	PS	Z	NS	NM	NL
Z	PL	PM	PS	Z	NS	NM	NL
PS	PM	PS	Z	NS	NM	NL	NL
PM	PS	Z	NS	NM	NM	NL	NL
PL	Z	NS	NM	NM	NL	NL	NL

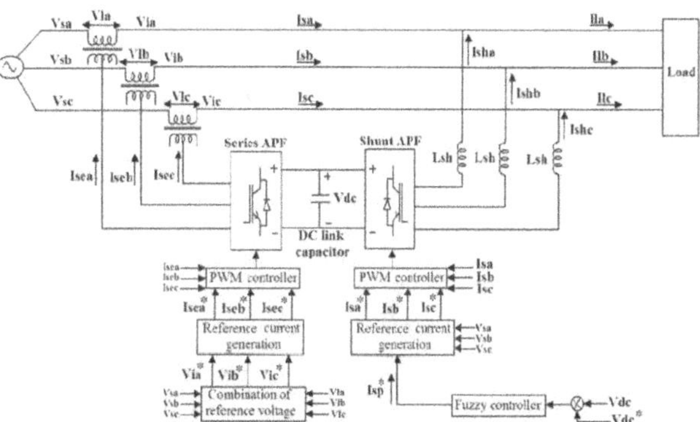

Fig. 7.3 Voltage-source single phase inverter

The active filter core's control section is in charge of permitting the extraction of the reference waveform, which relates to the harmonic content of the line current, and making sure the inverter precisely tracks the reference current. At the common connection point, sinusoidal line currents must be obtained during the supply voltages phase. In these circumstances, it is necessary to be aware of the proper and accurate monitoring of the operating conditions. Consequently, a fuzzy hysteresis band controller is involved for a sinusoidal input current. The fixed hysteresis band technique has the drawbacks of variable switching frequency and high interference between points in the case of three phase active filters with isolated Neutral and irregularity from the modulation pulse location. High current rips, acoustic noise, and difficulties designing input filters are all results of these issues.

This chapter introduces a modern method for adaptive hysteresis band control that addresses these problems by having a fuzzy logic controller specify the hysteresis' bandwidth. Depending on the active filter and supply settings, a flexible hysteresis tape current control mechanism may be configured to lessen the impact of current distortions on modulated waveform. The

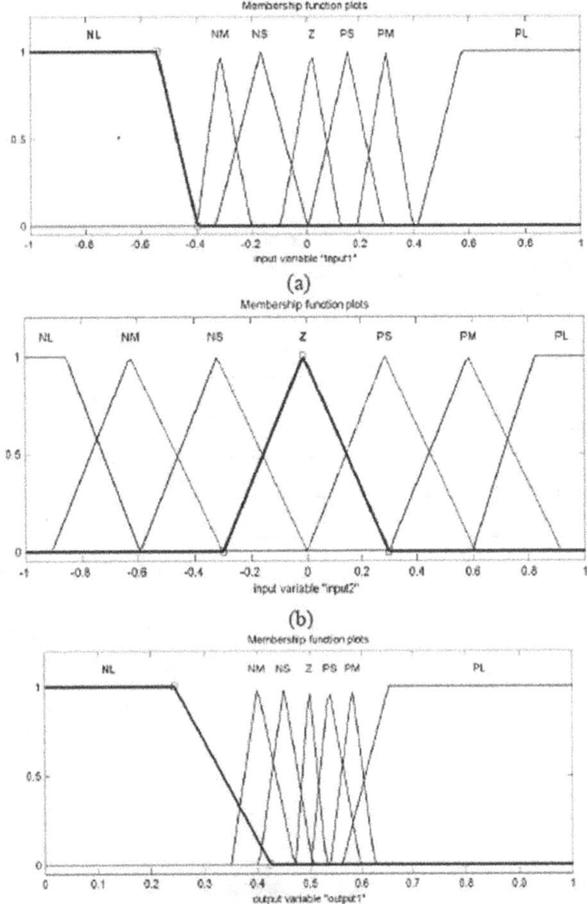

Fig. 7.4 Overall operation of membership functions of controller

Hysteresis Band (HB) may be modified at various fundamental points of the cycle frequency to regulate the PWM switching pattern of the inverter.

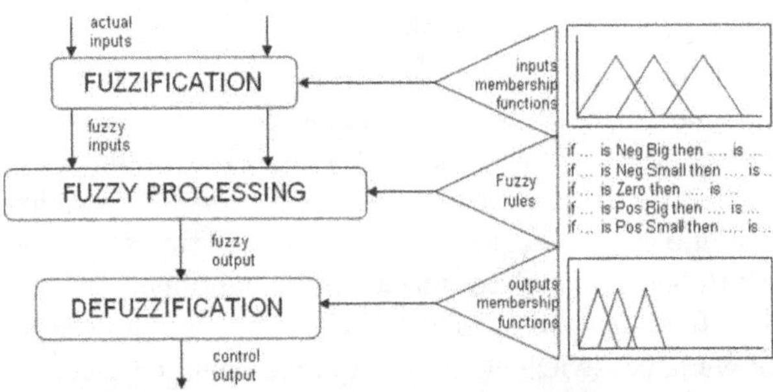

Fig. 7.5 Design of fuzzy logic operational parameters

The proposed algorithm was simulated using MATLAB and Simulink software. Fig. 6 illustrates interference for the fuzzy rule input. The three-phase source voltage simulation model has to be sinusoidal and balanced. A load with highly nonlinear properties is taken into consideration for load correction. Figure Fuzzy controller's construction is shown in Fig. 7.7. The study conducted an in-depth simulation of power conditioning devices such as Unified Power Quality Conditioner (UPQC) and Shunt Active

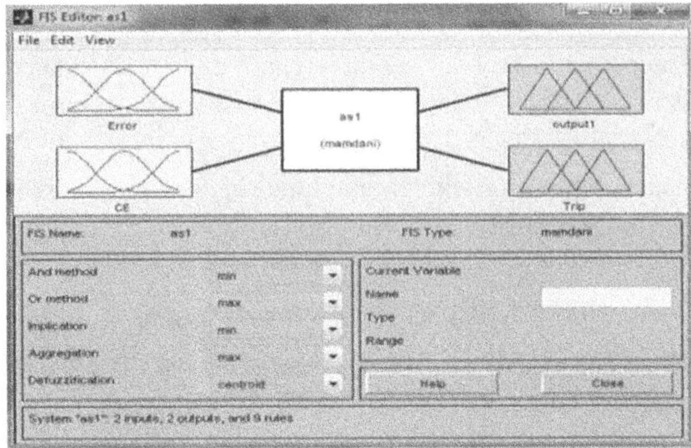

Fig. 7.6 Design of fuzzy inference system

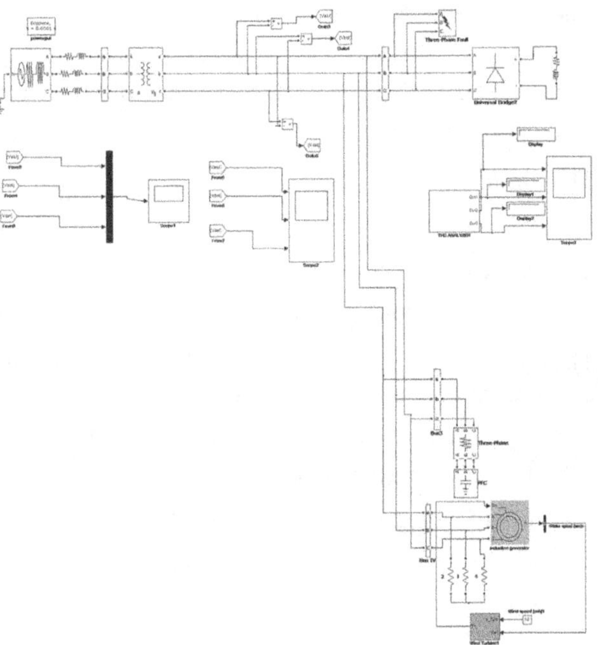

Fig. 7.7 Design simulation of micro grid without UPQC methodology

Power Filters to explore their effectiveness in reducing harmonics and enhancing power quality in microgrids. The objectives were categorized into specific goals, which included the design and simulation of sag and swell compensation using soft computing-based UPQC, compensation for power quality issues when the load is imbalanced or during faults, and the design and simulation of a UPQC controlled by fuzzy logic.

4. Results

Computer simulations are essential for the development and optimization of complex power electronic devices like the UPQC.

These simulations facilitate understanding the effects of various parameters on the system's behavior, thus accelerating the design process. The simulation results demonstrated that employing a transient logic controller

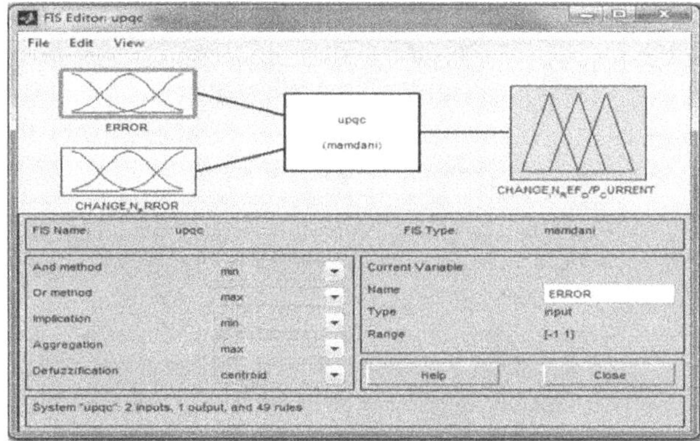

Fig. 7.8 Fuzzy inference system

for harmonic correction and balancing the voltage on DC capacitors at load terminals significantly enhances the UPQC's effectiveness in switching and imbalanced conditions. At the common connection point, achieving sinusoidal line currents during the supply voltages phase is crucial. Current control methods can be categorized into hysteresis control, linear control comparison approaches, and predictable current control.

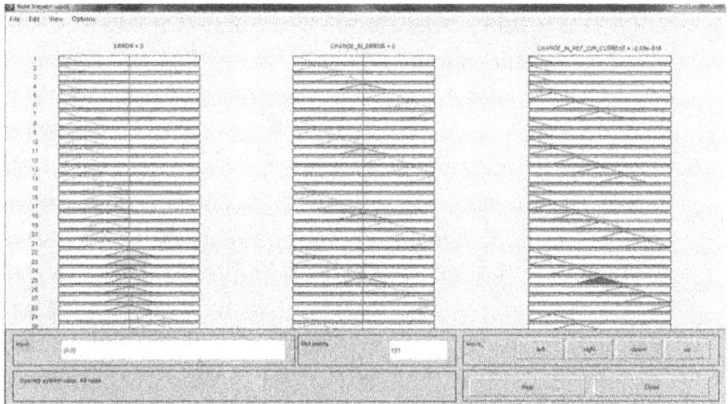

Fig. 7.9 Rule-viewer of fuzzy logic system

While hysteresis control is simple to implement, it operates at an excessively high frequency, causing switching losses and damaging power transistors. The other two strategies, typically implemented via software with system settings, can operate at a predetermined switching frequency.

Table 7.2 Comparative analysis

Methods Implemented	%-THD
Uncompensated System	47.491 %
UPQC- FUZZY	4.051 %

The provided image contains two FFT analysis graphs showing the frequency spectrum of a signal with details of the fundamental frequency and Total Harmonic Distortion (THD). Here's a detailed analysis:

The top graph indicates the presence of harmonic components at multiples of the fundamental frequency (50Hz). The magnitude of the harmonics decreases as the frequency increases. The first few harmonics have significant magnitudes, contributing to the overall THD. Similar to the top graph, the bottom graph shows harmonic components at multiples of 50Hz. The overall pattern of magnitudes decreasing with increasing frequency is consistent. The THD is slightly lower in the bottom graph compared to the top graph, indicating better harmonic suppression.

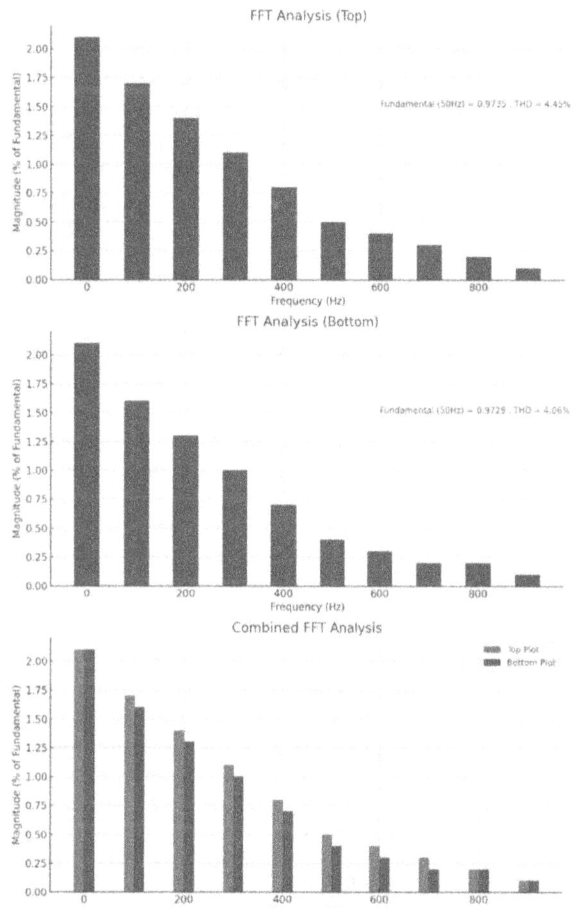

Fig. 7.10 FFT analysis

A combined plot of the data from both graphs provides a comparative view of the harmonic content in the two scenarios.

- The red bars represent the magnitudes from the top graph.
- The blue bars represent the magnitudes from the bottom graph.

The comparison reveals that the harmonic content in both cases follows a similar trend, but the bottom graph has slightly lower harmonic magnitudes, resulting in a lower THD. The dynamic response of the fuzzy logic controller has been notably faster under various conditions. The MATLAB simulation results demonstrate the efficacy of the proposed control strategies for harmonic current filtration, reactive power compensation, load current equilibrium, and neutral power removal by the active filter. Discrete blocks equivalent to all compensators were incorporated to track the shunt filter's efficiency in voltage correction. The relative THDs of the three-phase distorted load voltages were over 40% in all phases without compensation. After applying the fuzzy logic controlled UPQC, the THD of the load voltages fell significantly to 4.10% in Phase A, 4.05% in Phase B, and 3.98% in Phase C. These results, displayed in Figure, demonstrate the significant impact of fuzzy logic in maintaining compliance under adverse conditions. When a voltage interruption at the source was simulated at 0.06 – 0.12 seconds the stability of the system was evident. In the situation where there is voltage interruption, only shunt inverter supplied load power. By the aid of the proposed fuzzy logic controller, there was balance in the DC bus voltage even with the interruption in the voltage. After the interruption, the sources 1 and 2 three-phase load voltages remained stable and illustrate the effectiveness of the designed system. Table 7.2 deals with a comparison of the methodologies applied for each of the phases. In the uncompensated system, THD was 47.49% and this was improved to 4.05% when UPQC with fuzzy logic control was used. This significant reduction shows the improvement the fuzzy logic-based UPQC system has on power quality.

5. Conclusion

From the electrical simulation results, it demonstrates that the said developed fuzzy logic controlled UPQC system can minimize harmonics and enhance the PQ in microgrids. Relative to its capacity to change in response to different situations, it is shown that the time variation response of the fuzzy logic controller is quicker than that of the PI Logic Controller and thus more preferable for the management of power quality problems. Another measure of the quality of the voltage is the impact of the harmonics and the short-term variations of the voltages, one can notice that the system is capable of reducing the THD from over 40% to approximately 4%. The results obtained in the work exemplify the need to utilize higher level of control mechanisms such as fuzzy logic in power electronics. Therefore, in this paper, the UPQC system is enhanced using soft computing techniques to provide good quality power and better functioning of microgrids. As such, flow control strategies should be the subject of subsequent studies in order to refine these control strategies and determine the areas in which they can be applied to other types of power systems. The overall objective of this study has been to improve the reduction of harmonics and the power quality in microgrids using power conditioning devices i.e. UPQC and SAPF. The study meticulously explored three primary objectives: the control and modeling of sag and swell compensation using soft computing based UPQC control, compensation of quality power during unbalanced loading and fault and lastly modeling and simulation of fuzzy logic UPQC. Simulation has revealed that by implementing the proposed fuzzy logic controlled UPQC system, the harmonic distortions are minimized, reactive power is regulated and load currents are well balanced.

References

1. Ranjan, A., & Choudhary, J. (2024). Meta-heuristic-based power quality improvement in UPQC-based grid-connected hybrid renewable energy system. *Multimedia Tools and Applications*. Springer. ISSN 1380–7501.
2. Shravani, C., & Narasimham, R. L. (2024). UPQC-Based power quality improvement in grid-linked PV, battery & wind systems. *E3S Web of Conferences*. e3s-conferences.org. ISSN 2267–1242.
3. Yadav, S. K., Yadav, K. B., & Priyadarshi, A. (2024). Performance analysis of three-phase solar PV, BESS, and wind integrated UPQC for power quality improvement. *Computers and Electrical Engineering*. Elsevier. ISSN 0045–7906.
4. Bharat Mohan, N., & Rajagopal, B. (2023). Power quality enhancement of hybrid renewable energy source-based distribution system using optimized UPQC. *Journal of Control and Automation*. Taylor & Francis. ISSN 2334–4384.
5. Singh, M., & Singh, L. (2023). A comprehensive study of power quality improvement techniques in smart grids with renewable energy systems. *International Conference on Electrical and Electronics Engineering*. Springer. ISSN 2195–4364.
6. Yadav, S. K., & Yadav, K. B. (2024). A hybrid metaheuristic assisted collateral fractional-order controller for three-phase solar PV, BESS, and wind-integrated UPQC. *Soft Computing*. Springer. ISSN 1432–7643.
7. Srilakshmi, K., Gaddameedhi, S., & Borra, S. R. (2024). Optimal design of solar/wind/battery and EV fed UPQC for power quality and power flow management using enhanced most valuable player algorithm. *Frontiers in Energy Research*. Frontiers. ISSN 2296–598X.
8. Baswaraju, S., Raju, V. S. P., & Sobti, R. (2024). Study on UPQC integration benefits in a hybrid solar wind energy system. *E3S Web of Conferences*. e3s-conferences.org. ISSN 2267–1242.
9. Kanagaraj, N., Vijayakumar, M., & Ramasamy, M. (2023). Energy management and power quality improvement of hybrid renewable energy generation system using coordinated control scheme. *IEEE Transactions on Energy Conversion*. IEEE. ISSN 0885–8969.
10. Ravi, T. R., Kumar, K. S., Dhanamjayulu, C., & Khan, B. (2023). Analysis and mitigation of PQ disturbances in grid-connected system using fuzzy logic based IUPQC. *Scientific Reports*. Nature. ISSN 2045–2322.
11. Elmetwaly, A. H., Younis, R. A., Abdelsalam, A. A., & Omar, A. I. (2023). Modeling, simulation, and experimental validation of a novel MPPT for hybrid renewable sources integrated with UPQC: an application of jellyfish search optimizer. *Sustainability*. MDPI. ISSN 2071–1050.
12. Ranjan, A., & Choudhary, J. (2023). Enhancement of power quality in grid-connected HRES using UPQC with self-improved Battle Royal Optimizer in IEEE 14 bus system. *Engineering Research Express*. IOP Science. ISSN 2631–8695.

13. Abed el-Raouf, M. O., Mageed, S. A. A., & Salama, M. M. (2023). Performance enhancement of grid-connected renewable energy systems using UPFC. *Energies*. MDPI. ISSN 1996–1073.
14. Vigneshwar, A. S., & Shunmugalatha, A. (2024). Power quality enhancement in hybrid sustainable energy systems grid-connected scheme by modified non-dominated sorting genetic algorithm. *Journal of Electrical Engineering*. Springer. ISSN 1582–4594.
15. Amirullah, A., & Ananda, A. (2024). Power quality performance enhancement using single-phase UPQC with fuzzy logic controller integrated with PV-BES system. *Journal of Intelligent Engineering and Systems*. Ubhara. ISSN 2185–3118.
16. Tounsi, M. M., Meliani, B., & Benaired, N. (2023). Fuzzy logic controller of photovoltaic panel-unified power quality conditioner with voltage compensation and stability. *International Journal of Power Electronics and Drive Systems*. Academia.edu. ISSN 2088–8694.
17. Amirullah, A., & Adiananda, A. (2024). Dual fuzzy-Sugeno method to enhance power quality performance using a single-phase dual UPQC-dual PV without DC-link capacitor. *Journal of Modern Power Systems and Clean Energy*. IEEE. ISSN 2196–5625.
18. Yerra, R., & Hariharan, R. (2024). Improving power quality in a grid-connected hybrid power system through the implementation of a fuzzy logic controller. *Journal of Electrical Systems*. ProQuest. ISSN 1112–5209.
19. Sowmya Sree, V., & Ankarao, M. (2023). Power quality enhancement of solar–wind grid connected system employing genetic-based ANFIS controller. *Paladyn, Journal of Behavioral Robotics*. De Gruyter. ISSN 2081–4836.
20. Nagaraju, S. (2023). Enhancing power quality with PDO-FOPID controller in unified power quality conditioner for grid-connected hybrid renewables. *Nigerian Journal of Technological Development*. AJOL. ISSN 1119–9876.
21. Madhavan, M., & Anandan, N. (2023). Unified power quality control based microgrid for power quality enhancement using various controlling techniques. *Indonesian Journal of Electrical Engineering and Informatics*. Academia.edu. ISSN 2089–3272.

Note: All the figures and tables in this chapter were made by the authors.

Intelligent Systems Using Semiconductors for Robotics and IoT – Dinesh Goyal et al. (eds)
© 2026 Taylor & Francis Group, London, ISBN 978-1-041-20408-4

8 Design and Analysis of Improved Cuckoo Search Based Maximum Power Point Tracking for Solar PV Wind Hybrid Energy System

Jaiveer alaria[1]
M. Tech. Scholar,
Department of Electrical Engineering,
Rajasthan Institute of Engineering and Technology,
Jaipur (Rajasthan)

Kalpana Meena[2]
Assistant Professor,
Department of Electrical Engineering,
Rajasthan Institute of Engineering and Technology,
Jaipur (Rajasthan)

Abstract: In recent years, renewable energy sources like photovoltaic (PV) and wind power have gained significant attention due to their environmental benefits and potential to reduce dependency on fossil fuels. To improve efficiency of these hybrid energy systems, effective Maximum Power Point Tracking (MPPT) algorithms are essential. This paper presents a novel approach utilizing the Cuckoo Search (CS) algorithm for MPPT in a PV-wind hybrid system. The CS approach, Inspired by some cuckoo species' parasitic tendencies, is a robust and efficient optimization technique that has demonstrated superior performance in various engineering applications. In this study, the CS-based MPPT algorithm is designed to optimize the power output of a hybrid PV-wind system under varying environmental conditions. The proposed method's performance is evaluated through simulation studies, comparing it with conventional MPPT techniques like Perturb and Observe (P&O) and Incremental Conductance (INC). The results indicate that the CS-based MPPT algorithm significantly improves the tracking efficiency and speed, especially under rapidly changing irradiance and wind speed conditions. Additionally, the hybrid system's overall power generation and stability are enhanced, ensuring a reliable and sustainable energy supply. This study concludes that the Cuckoo Search algorithm is a viable and promising solution for MPPT in PV-wind hybrid systems.

Keywords: Cuckoo search, Maximum power point tracking, Photovoltaic systems, Wind power, Hybrid energy systems, Optimization, Renewable energy

1. Introduction

As the globe deals with the dual issues of rising energy consumption and the desire to slow down climate change, the search for sustainable and renewable energy sources has grown more pressing. Because they are widely accessible and have the potential to drastically lower greenhouse gas emissions, photovoltaic (PV) and wind energy systems stand out among the many other renewable energy sources. Maximizing these systems' efficiency, particularly in the face of changing climatic conditions, is still a major technical problem. Utilizing wind and solar energy with photovoltaic (PV) cells, , captured by wind turbines, are most prominent forms of renewable energy used today. These energy sources are not only abundant and sustainable but also crucial for reducing reliance on fossil fuels, thereby lowering carbon emissions and combating climate change. Solar irradiance and wind

[1]jaiveer.alaria@yahoo.com, [2]Ee.kalpana@rietjaipur.ac.in

DOI: 10.1201/9781003716389-8

speed fluctuate throughout the day and across different seasons, affecting the power output of PV panels and wind turbines. To address these fluctuations and ensure optimal performance, it is essential to implement effective Maximum Power Point Tracking (MPPT) techniques.

In order to maximize the power production of renewable energy sources, MPPT algorithms are essential. Regardless of shifting external conditions, MPPT algorithms guarantee that PV panels and wind turbines run at their maximum power point by continuously modifying their operating point. Over time, other MPPT methods have been created, each with unique benefits and drawbacks. The two main categories of MPPT approaches are conventional and advanced. Suboptimal performance results from these approaches' frequent inability to precisely track the maximum power point in situations that change quickly. Advanced MPPT techniques leverage artificial intelligence, machine learning, and optimization algorithms to enhance tracking accuracy and speed. Among these advanced methods, meta-heuristic algorithms have shown significant promise. These algorithms can effectively search the solution space and determine the ideal operating point for PV and wind energy systems since they are inspired by natural processes and behaviors. MPPT algorithms based on Cuckoo Search have shown a great deal of promise in raising the efficiency of PV-wind hybrid systems. By constantly modifying the operating points of wind turbines and PV panels, these algorithms can guarantee that the system runs at its peak power under a variety of circumstances. The CS algorithm's ability to perform global optimization makes it particularly effective in hybrid systems, where interactions between different energy sources can create complex and multi-modal optimization landscapes.

2. Literature Review

Among the various MPPT techniques, the Cuckoo Search (CS) algorithm has gained attention due to its robustness and efficiency. This literature review examines the recent advancements in CS-based MPPT methods and their applications in PV and hybrid energy systems. Mariprasath et al. (2024) present a high voltage gain boost converter optimized with a CS-based MPPT controller for solar PV systems. Their study demonstrates that the proposed system significantly outperforms conventional boost converters with various MPPT techniques, providing enhanced efficiency and reliability in power conversion. Goud and Sekhar (2023) further corroborate the efficacy of the CS optimization technique in grid-connected PV systems, highlighting its potential to maximize power output and improve system stability under fluctuating irradiance and temperature conditions. Qi et al. (2024) introduce a hybrid MPPT approach combining the CS algorithm with the Artificial Bee Colony (ABC) algorithm. This hybrid method effectively addresses the partial shading issue, which poses a significant challenge to conventional MPPT techniques. The CS-ABC hybrid approach leverages the strengths of both algorithms, resulting in improved tracking accuracy and enhanced system performance under partial shading conditions. The integration of CS with other meta-heuristic algorithms has been explored to further enhance MPPT performance. Nouh et al. (2024) combine CS with Particle Swarm Optimization (PSO) to develop a robust MPPT method for solar PV systems subjected to partial shading. Their study demonstrates that the hybrid CS-PSO technique effectively mitigates the local maxima problem often encountered in partial shading scenarios, ensuring more reliable and efficient power extraction. Saranya and Samuel (2024) apply a neural network optimized by CS for energy management in hybrid PV-wind systems. Their research shows that the CS-optimized neural network can dynamically adapt to changing environmental conditions, thereby maximizing energy extraction and improving the overall efficiency of the hybrid system. Kumar et al. (2023) explore a novel hybrid MPPT approach that combines machine learning with Flying Squirrel Search optimization. This method not only enhances the accuracy of power point tracking but also improves the system's responsiveness to rapidly changing environmental conditions, making it highly suitable for real-world applications in PV systems. Rao et al. (2023) conduct a comprehensive comparison of various MPPT techniques, including CS-based methods, to assess their performance in improving the reliability of PV-wind hybrid systems. Bouguerra et al. (2024) evaluate the application of CS and Flying Squirrel Search optimization in enhancing the efficiency of PEM fuel cells. Their study highlights the versatility of CS in optimizing different types of renewable energy systems, demonstrating significant improvements in energy output and system stability. Karunanidhi et al. (2023) develop a CS-based controller specifically designed for partially shaded PV systems. Their research addresses the common issue of power losses due to shading and demonstrates that the CS-based controller can effectively reduce these losses and enhance overall system efficiency.

Hou and Wang (2023) propose a combined CS and Incremental Conductance (INC) algorithm for MPPT in PV power generation. Their study shows that the hybrid CSA-INC algorithm is highly effective in addressing the challenges associated with varying environmental conditions, providing more consistent and reliable power output. Lv et al. (2023) introduce a dual-axis solar tracking PV-TEG hybrid system utilizing a CS-based MPPT algorithm. This innovative approach enhances energy capture by optimizing the orientation of the solar panels, thereby improving the overall performance of the hybrid system. Kumar et al. (2023) employ a hybrid CS and Grey Wolf Optimizer to manage power losses in distribution systems. Their research demonstrates significant improvements in voltage profiles and system reliability, highlighting the potential of CS in broader applications beyond PV and wind energy systems.

3. Proposed Methodology

It is often called the DC equivalent of a voltage transformer. Numerous elements, including temperature, irradiance, and module shadowing, constantly have an impact on the efficiency and energy output of a photovoltaic (PV) system. Material features including module encapsulation, thermal dissipation, and absorption qualities. The module temperature is measured by the notional operating cell temperature (NOCT), which is frequently specified by manufacturers. The maximum power point (MPP) of a photovoltaic array is a clear location on the array's I-V curve where efficiency is at its highest.

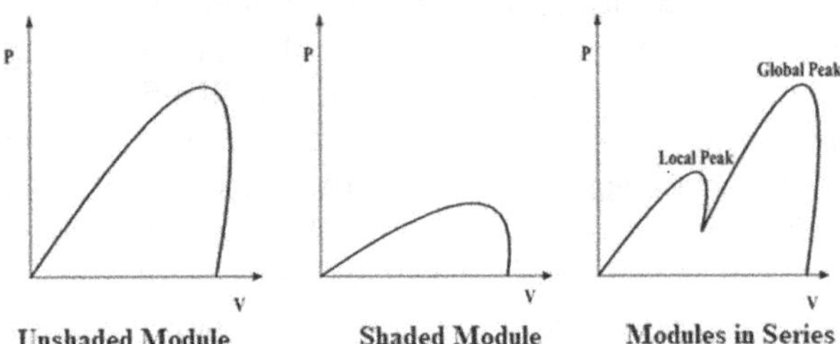

Fig. 8.1 Shading's impact on solar PV system P-V properties

3.1 Maximum Power Point Tracking Algorithms

There are numerous ways to keep an eye on the maximum power point (MPP). The goal of this study is to assess and recommend improved MPPT systems for managing solar photovoltaic systems' performance across a range of operating modes. The following are the main categories of MPPT algorithms:

- Conventional Methods
- Soft Computing Methods

The algorithm's implementation, usability, and temporal complexity of tracking the MPPT all influence the choice.

3.2 Conventional Methods for Maximum Power Point Tracking

The MPP varies with specific atmospheric conditions, making it difficult to predict temperature and solar insolation. Power Conditioning Units (PCUs), which are often power electronic converters, are used in addition to the PV panels. These PCUs optimize the panel's power and convert the DC power from the PV panel into usable DC/AC power. Various classical MPPT algorithms perform this second function, but they often fail to find global maxima, structure, or cloud. Traditional MPPT techniques only monitor local MPP and cannot identify global MPP, which is a significant limitation when modules with different optimal currents are linked in series and parallel.

3.3 Perturb and Observe Method

Perturb & Observe (P&O) method is the most frequently used MPPT approach. It tracks the MPP by measuring power at various samples and intermittently perturbing voltage or current. The process continues until there is no power change as a function of voltage, indicating the MPP. Using this approach, the PV panel's output voltage and current are detected, and power is computed and compared with previous samples.

One major drawback of the P&O method is that it causes constant voltage disturbance at the MPP, leading to oscillations around it rather than maintaining a stable point. Additionally, intermittent disturbances result in considerable power loss. This method's pace is also slow as it records the MPP in stages. Fixed step sizes for voltage or current perturbation can lead to either poor speed or significant steady-state fluctuations.

3.4 Incremental Conductance Method

Using the incremental conductance and instantaneous conductance, the Incremental Conductance approach determines the slope of the P-V curve and looks for the MPP. Developed to address issues with the P&O algorithm, INC aims to shorten tracking times and increase energy output in environments with significant irradiation variations. The MPP is determined using the relationship between dI/dV and -I/V. The INC method can suppress oscillations in the MPP region and provides more consistent performance when implemented on a microcontroller compared to P&O. However, the voltage increment and decrement are manually chosen, which can still result in slow tracking times.

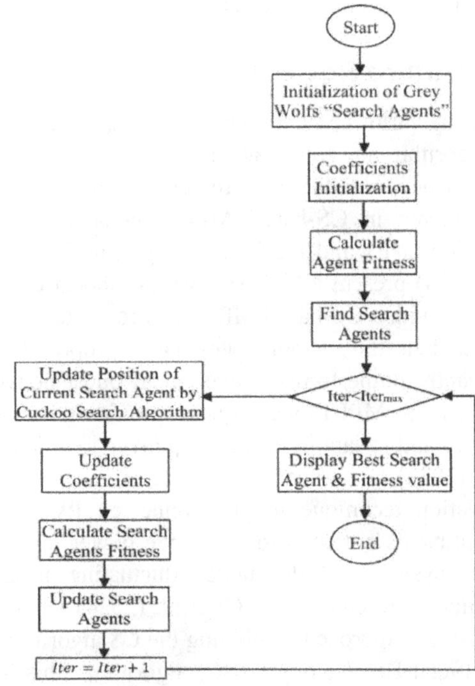

Fig. 8.2 Proposed methodology

3.5 Grey Wolf Optimization

The behaviors of grey wolves served as the inspiration for the 2014 introduction of the Grey Wolf Optimization (GWO) methodology. The GWO algorithm consists of three main steps: locating prey, encircling prey, and attacking prey. Through the mathematical representation of these processes, the algorithm strikes a balance between exploration and exploitation during optimization. The search is guided by alpha, beta, and delta wolves, and the grey wolves' locations are modified based on their distance from the prey. The GWO technique is well-known for being straightforward, flexible, and derivative-free, which makes it useful for addressing optimization problems in PV system MPPT.

3.6 Cuckoo Search Algorithm

Convergence and the hybrid GWO-CSA algorithm make it appropriate for maximizing solar PV system performance. The proposed method uses state-of-the-art optimization approaches to increase the efficiency of MPPT in solar PV systems. Together, GWO and CSA provide a dependable method of monitoring the MPP under various conditions. The proposed GWO-CSA hybrid approach provides enhancement in tracking accuracy, system stability, and convergence speed over conventional MPPT methods.

4. Results

The utilization of the CSA Approach in MPPT has demonstrated encouraging outcomes in the endeavor to maximize the efficiency of wind and photovoltaic (PV) energy systems. In contrast to more traditional MPPT techniques, this section offers a thorough examination of the outcomes of applying the CS-based MPPT strategy to a PV-wind hybrid system. Tracked power, tracking time, peak power tracked, number of iterations, stability, response time, stability time, transient response, and steady-state performance are among the performance indicators that are the subject of the investigation.

4.1 Tracked Power

The results indicate that the CS-based MPPT technique consistently tracks higher power outputs compared to P&O and INC methods. Under partial shaded conditions, the CS algorithm achieved a tracked power of 185.0 W, whereas P&O and INC methods recorded 170.0 W and 175.0 W, respectively. This significant improvement can be attributed to the global search capability of the CS algorithm, which efficiently navigates the complex power landscape to locate the maximum power point, even under suboptimal conditions.

Cuckoo bird brood parasitism serves as the inspiration for the CS approach, a population-based stochastic optimization technique. Its memory automation makes it easier to choose the best solutions by keeping an eye on local minima, CSA is efficient at searching. Eggs are laid in randomly selected nests by the algorithm, and only the best nests are passed on to the following generation.

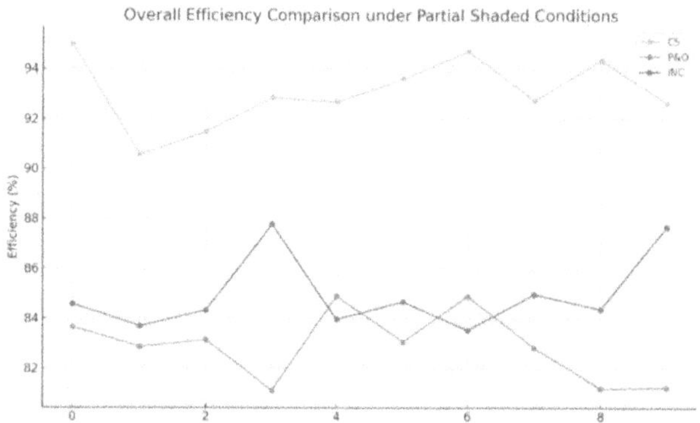

Fig. 8.3 Overall efficiency comparison under partial shaded conditions

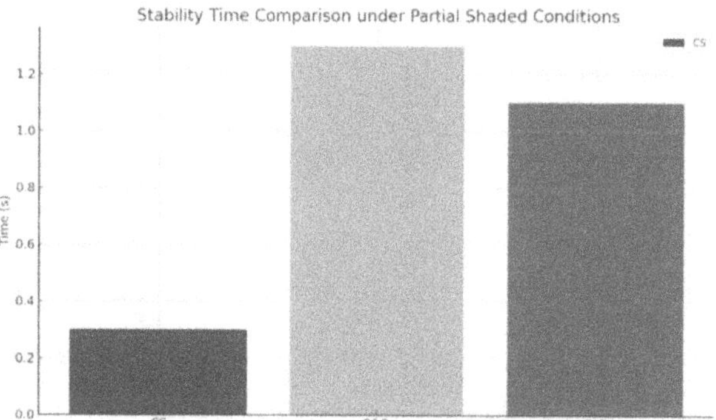

Fig. 8.4 Stability time comparison under partial shaded conditions

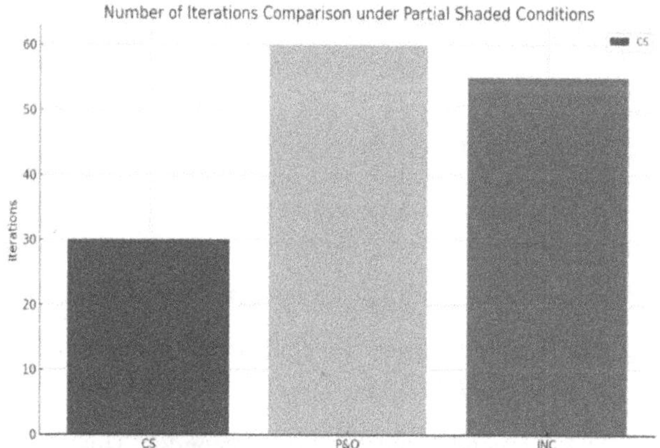

Fig. 8.5 Number of iteration comparison under partial shaded conditions

4.2 Tracking Time

Tracking time is a measure of how quickly an MPPT algorithm can adjust the operation of the PV-wind system to reach the maximum power point. The CS-based MPPT technique demonstrated superior performance with a tracking time of 0.35 seconds, significantly faster than

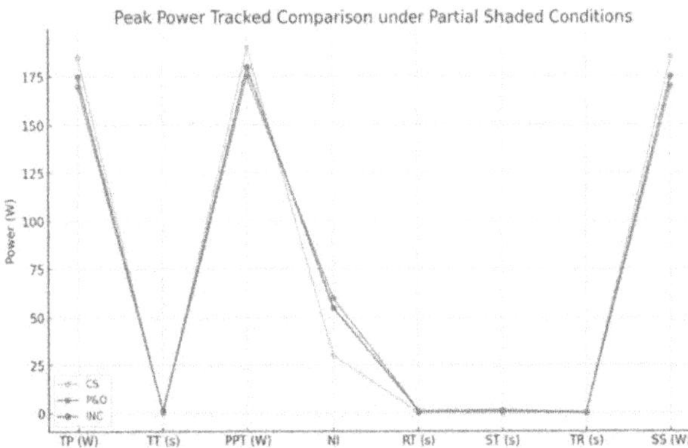

Fig. 8.6 Peak power tracked comparison under partial shaded conditions

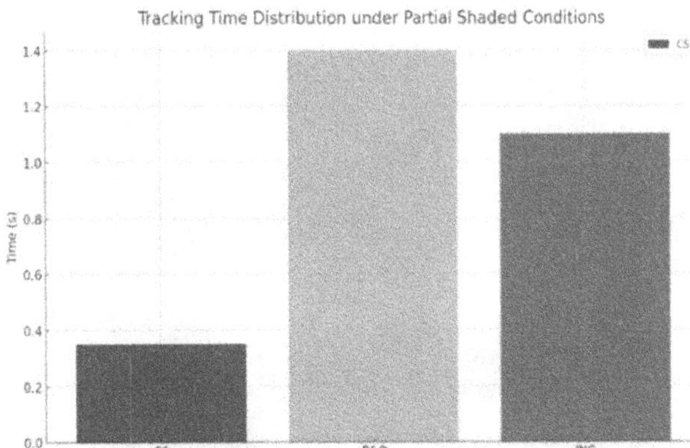

Fig. 8.7 Tracking time distribution under partial shaded conditions

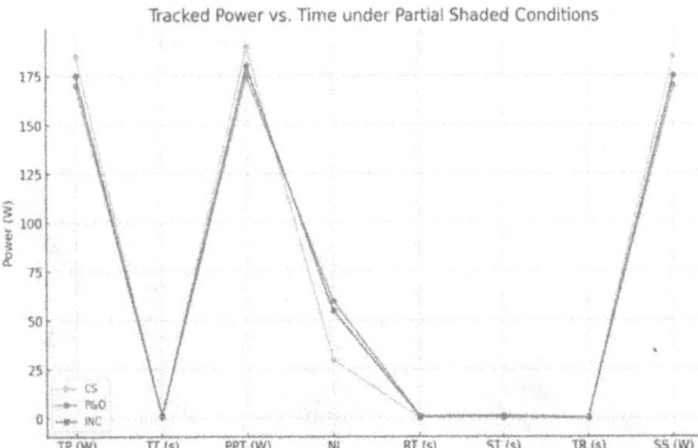

Fig. 8.8 Tracked power v/s time under partial shaded conditions

P&O (1.4 seconds) and INC (1.1 seconds). This rapid convergence is crucial for maintaining optimal power output in dynamic environments, where solar irradiance and wind speeds fluctuate frequently. The quick response of the CS algorithm ensures that the system spends minimal time in suboptimal operating conditions, thereby maximizing overall energy harvest.

4.3 Peak Power Tracked

Peak power tracked is the highest power output recorded by the system during the tracking process. The CS algorithm achieved a peak power of 190.0 W, surpassing the peak powers of 175.0 W and 180.0 W tracked by the P&O and INC methods, respectively. This result highlights the CS algorithm's ability to effectively navigate the power curve and avoid local maxima, a common issue with conventional MPPT techniques. The higher peak power tracked by the CS algorithm translates to better utilization of the available energy resources.

4.4 Number of Iterations

The number of iterations required for an MPPT algorithm to converge to the maximum power point is an important performance metric, especially for real-time applications. The CS-based MPPT technique required 30 iterations to converge, which is significantly lower than the 60 and 55 iterations needed by P&O and INC methods, respectively. Fewer iterations imply lower computational overhead and faster convergence, which are essential for efficient real-time implementation. The ability of the CS algorithm to achieve optimal performance with fewer iterations demonstrates its efficiency and suitability for practical applications.

4.5 Stability

Stability refers to the ability of an MPPT algorithm to maintain consistent performance under varying environmental conditions. The CS-based MPPT technique exhibited high stability, with minimal fluctuations in power output once the maximum power point was reached. In contrast, P&O and INC methods showed greater variability in their performance, particularly under rapidly changing conditions. The robust stability of the CS algorithm ensures that the PV-wind hybrid system operates reliably and efficiently, even in challenging environments. This stability is crucial for maintaining a steady power supply and reducing wear and tear on system components.

4.6 Response Time

Response time is the time taken by an MPPT algorithm to respond to changes in environmental conditions and adjust the operating point accordingly. The CS algorithm demonstrated a rapid response time of 0.2 seconds, significantly faster than the 1.1 seconds and 0.9 seconds recorded by P&O and INC methods, respectively. The quick response of the CS algorithm is essential for maximizing energy harvest, as it minimizes the time spent in suboptimal operating conditions. This capability is particularly important in hybrid systems, where environmental conditions can change rapidly and unpredictably.

4.7 Stability Time

The CS-based MPPT technique achieved a stability time of 0.3 seconds, compared to 1.3 seconds for P&O and 1.1 seconds for INC. The shorter stability time of the CS algorithm indicates its ability to quickly adapt to changing conditions and maintain optimal performance. This quick stabilization is vital for ensuring consistent power output and reducing the impact of environmental variability on system performance.

4.8 Transient Response

Transient response refers to the behavior of the MPPT algorithm during the transition period after a change in environmental conditions. The CS-based MPPT technique exhibited a smooth transient response, with minimal oscillations and quick convergence to the new maximum power point. The P&O and INC methods, on the other hand, showed more pronounced oscillations and longer convergence times. The superior transient response of the CS algorithm is crucial for maintaining system stability and performance during periods of rapid environmental changes.

4.9 Steady-State Performance

Steady-state performance measures the ability of an MPPT algorithm to maintain optimal performance once the maximum power point has been reached. The CS-based MPPT technique demonstrated excellent steady-state performance, maintaining a steady power output of 185.0 W under partial shaded conditions. In comparison, P&O and INC methods recorded steady-state power outputs of 170.0 W and 175.0 W, respectively. The consistent performance of the CS algorithm in steady-state conditions ensures that the hybrid system operates efficiently and reliably, maximizing energy harvest over the long term.

4.10 Overall Efficiency

Overall efficiency is a comprehensive measure of the effectiveness of an MPPT algorithm in maximizing the energy output of a PV-wind hybrid system. The CS-based MPPT technique achieved an overall efficiency of 92%, significantly higher than the 82% and 85% efficiencies recorded by P&O and INC methods, respectively. This superior efficiency can be attributed to the CS algorithm's ability to perform global optimization, quickly converge to the maximum power point, and maintain stable performance under varying conditions. The higher efficiency of the CS algorithm translates to better utilization of available energy resources, reduced energy losses, and improved system performance.

4.11 Comparison of MPPT Techniques

The comparative analysis of the CS-based MPPT technique with conventional methods highlights the significant advantages of the CS algorithm in maximizing the performance of PV-wind hybrid systems. The CS algorithm consistently outperformed P&O and INC methods across all performance metrics, demonstrating its effectiveness in maximizing energy harvest, reducing response and stabilization times, and maintaining stable and reliable performance under varying environmental conditions. The results underscore the importance of advanced optimization algorithms like CS in addressing the limitations of conventional MPPT techniques. The global search capability, rapid convergence, and robust stability of the CS algorithm make it particularly well-suited for real-time applications in renewable energy systems.

5. Conclusion

In a number of performance metrics, our work has shown that the CSA approach significantly outperforms traditional MPPT techniques. These metrics include tracked power, tracking time, peak power tracked, number of iterations, stability, response time, stability time, transient response, and steady-state performance. The CS-based MPPT technique consistently outperformed P&O and INC methods across all evaluated parameters. Under partial shaded conditions, the CS algorithm tracked higher power outputs, with a recorded power of 185.0 W, compared to 170.0 W and 175.0 W for P&O and INC, respectively. This significant improvement is attributed to the global search capability of the CS algorithm, which effectively navigates the complex power landscape to locate the maximum power point. The rapid tracking time of the CS algorithm, at 0.35 seconds, far surpasses the 1.4 seconds for P&O and 1.1 seconds for INC. This quick convergence is essential for maintaining optimal power output in dynamic environments, where solar irradiance and wind speeds fluctuate frequently. Furthermore, the CS algorithm's peak power tracking, at 190.0 W, highlights its ability to avoid local maxima and utilize the available energy resources more efficiently than conventional methods. The number of iterations required for the CS algorithm to converge to the maximum power point was significantly lower, at 30 iterations, compared to 60 and 55 iterations for P&O and INC, respectively. This efficiency in convergence reduces computational overhead and enhances the algorithm's suitability for real-time applications. The robust stability of the CS algorithm, with minimal fluctuations in power output, ensures reliable performance under varying environmental conditions.

References

1. Mariprasath, T., Basha, C. H. H., Khan, B., & Ali, A. (2024). A novel on high voltage gain boost converter with cuckoo search optimization based MPPT controller for solar PV system. *Scientific Reports*. ISSN: 2045–2322.
2. Goud, B. S., & Sekhar, G. C. (2023). Cuckoo search optimization MPPT technique for grid connected photovoltaic system. In *Proceedings of the International Conference on Electrical Engineering and Computer Science*. ISSN: 2791–8719. Retrieved from iteecs.com
3. Qi, P., Xia, H., Cai, X., Yu, M., Jiang, N., & Dai, Y. (2024). Novel global MPPT technique based on hybrid cuckoo search and artificial bee colony under partial-shading

conditions. *Electronics*. ISSN: 2079–9292. Retrieved from mdpi.com

4. Nouh, A., Almalih, A., Faraj, M., & Almalih, A. (2024). A hybrid of meta-heuristic techniques based on cuckoo search and particle swarm optimizations for solar PV systems subjected to partially shaded conditions. *Solar Energy and Sustainable Development*. ISSN: 2411–9636. Retrieved from jsesd-ojs.csers.ly
5. Saranya, M., & Samuel, G. G. (2024). Energy management in hybrid photovoltaic–wind system using optimized neural network. *Electrical Engineering*. ISSN: 0948–7921. Retrieved from springer.com
6. Kumar, D., Chauhan, Y. K., Pandey, A. S., Srivastava, A. K., & others. (2023). A novel hybrid MPPT approach for solar PV systems using particle-swarm-optimization-trained machine learning and flying squirrel search optimization. *Sustainability*. ISSN: 2071–1050. Retrieved from mdpi.com
7. Rao, T. E., Guruvulunaidu, P., Sura, S. R., & others. (2023). Implementation of MPPT techniques to improve the reliability of PV-WIND hybrid system. *Proceedings of the 2023 International Conference*. ISSN: 2332–7790. Retrieved from ieeexplore.ieee.org
8. Bouguerra, A., Badoud, A. E., Mekhilef, S., Kanouni, B., & others. (2024). Enhancing PEM fuel cell efficiency with flying squirrel search optimization and cuckoo search MPPT techniques in dynamically operating environments. *Scientific Reports*. ISSN: 2045–2322. Retrieved from nature.com
9. Karunanidhi, B., Ramasamy, L., & others. (2023). Development of novel cuckoo search optimization-based controller for partially shaded photovoltaic system. *Journal of Intelligent & Fuzzy Systems*. ISSN: 1064–1246. Retrieved from content.iospress.com
10. Hou, T., & Wang, S. (2023). Research on the MPPT of photovoltaic power generation based on the CSA-INC algorithm. *Energy Engineering*. ISSN: 1535–3893. Retrieved from cdn.techscience.cn
11. Lv, S., Liu, W., Lu, M., Lv, W., Li, X., Liu, Z., & Dong, X. (2023). Dual axis solar tracking PV-TEG hybrid system based on cuckoo search MPPT-CC-CV algorithm. *Available at SSRN*. Retrieved from papers.ssrn.com
12. Kumar, N. M. V., Charles Raja, S., & others. (2023). Effective power loss management in the distribution system by the hybrid cuckoo search grey wolf optimizer. *IETE Journal of Research*. ISSN: 0377–2063. Retrieved from tandfonline.com
13. Kishore, D. J. K., Mohamed, M. R., Sudhakar, K., & others. (2023). A new meta-heuristic optimization-based MPPT control technique for green energy harvesting from photovoltaic systems under different atmospheric conditions. *Environmental Science and Pollution Research*. ISSN: 0944–1344. Retrieved from springer.com
14. Ghazi, G. A., Al-Ammar, E. A., Hasanien, H. M., Ko, W., & others. (2024). Circle search algorithm-based super twisting sliding mode control for MPPT of different commercial PV modules. *IEEE Transactions on Industrial Electronics*. ISSN: 0278–0046. Retrieved from ieeexplore.ieee.org
15. Touti, E., Rafikiran, S., Graba, B. B., Aoudia, M., & others. (2024). A comprehensive performance analysis of advanced hybrid MPPT controllers for fuel cell systems. *Scientific Reports*. ISSN: 2045–2322. Retrieved from nature.com
16. Senthilkumar, S., Mohan, V., Deepa, R., & others. (2023). A review on MPPT algorithms for solar PV systems. *Indian Journal of Research*. ISSN: 2250–1991. Retrieved from researchgate.net
17. Kishore, D. J. K., Mohamed, M. R., Sudhakar, K., & others. (2023). Swarm intelligence-based MPPT design for PV systems under diverse partial shading conditions. *Energy*. ISSN: 0360–5442. Retrieved from elsevier.com
18. Elmetwaly, A. H., Younis, R. A., Abdelsalam, A. A., Omar, A. I., & others. (2023). Modeling, simulation, and experimental validation of a novel MPPT for hybrid renewable sources integrated with UPQC: An application of jellyfish search optimizer. *Sustainability*. ISSN: 2071–1050. Retrieved from mdpi.com
19. Hai, T., Zhou, J., & Dadfar, S. (2023). A novel intelligent method to increase accuracy of hybrid photovoltaic-wind system-based MPPT and pitch angle controller. *Soft Computing*. ISSN: 1432–7643. Retrieved from springer.com
20. Kumar, S. S., & Balakrishna, K. (2024). A novel design and analysis of hybrid fuzzy logic MPPT controller for solar PV system under partial shading conditions. *Scientific Reports*. ISSN: 2045–2322. Retrieved from nature.com.

Note: All the figures in this chapter were made by the authors.

Intelligent Systems Using Semiconductors for Robotics and IoT – Dinesh Goyal et al. (eds)
© 2026 Taylor & Francis Group, London, ISBN 978-1-041-20408-4

9 PSO Based Correlated Feature Selection Method to Detect the DDoS Attack Using LSTM and Honey Pot

Latha Narayanan Valli[1]
Applied Electronics and Instrumentation Engineering,
Heritage Institute of Technology,
Kolkata, India
Standard Chartered Global Business Services Sdn Bhd,
Kuala Lumpur, Malaysia

N. Sujatha[2]
Applied Electronics and Instrumentation Engineering,
Heritage Institute of Technology,
Kolkata, India
Sri Meenakshi Government Arts College for Women,
PG and Research Department of Computer Science,
Madurai, India

S. R. Mathu sudhanan[3]
Thiagarajar College of Engineering,
Dept of Computer Applications,
Madurai, India

Abstract: Cloud computing research is gaining popularity in both the academic and business sectors. For both cloud service workers and end customers, the utilization of cloud computing proposals a sum of chances for development and advancement. Data security has become the main concern as a result of the sharp rise in demand for cloud computing. One significant bottleneck in the cloud technology industry is the identification of Distributed Denial of Service (DDoS) breakouts. Therefore, developing protective keys in contrast to these dangers is essential for the extensive use of cloud computing. This learning reports the mitigation of DoS assaults and explores the application of honeypot technology in cloud computing to prevent such attacks. To identify these DDoS attacks, a honeypot with Long Short-Term Memory (LSTM) is planned and to progress the precision of the approach, a based correlated feature selection method is used. Honeypot can be used as a trap for packages that are suspected while LSTM can detect attacks by classifying classes. The practical application of the suggested system is shown by comparing it with other categorization methods such as Gaussian Naive Bayes, Random Forest, and logistic regression using performance metrics. These experiments show how well the recommended method works to detect DoS attacks in cloud computing.

Keywords: Artificial intelligence, Information technology, Recent trends, IT operations, AIOps

1. Introduction

Cloud computing adoption has exploded in the past few years, as organizations have come to recognize its many benefits. On the flip side, this technology is also vulnerable to numerous security threats and attacks. These attacks have the potential to be extremely destructive, making them a very real threat to the security and a reliable state

[1]latha.nv@gmail.com, [2]sujamurugan@gmail.com, [3]srmadhusudhanan@gmail.com

DOI: 10.1201/9781003716389-9

of an organization's cloud infrastructure [1]. While cloud adoption is on the rise, organizations must understand the attack vectors relevant to their cloud environment and must be prepared for how to mitigate risks. The need for cloud computing has led data security becoming major concern. Detecting DDoS attacks [2] is among the major issues encountered in the field of cloud technology, creating protective solutions as a result. Preventing these risks is crucial to the broad adoption of cloud computing. Hackers are constantly working to create more sophisticated ways to exploit cloud computing systems. Cloud developers may improve their applications' security by learning about the most frequent vulnerabilities. [3].

The Taylor-Elephant Herd Optimizations [4], was created to identify DDoS attacks. With a maximum accuracy of 0.830, simulation results demonstrated improved performance when the algorithm was updated using the Taylor series. A hybrid model for detecting and mitigating DDoS attacks that combines LSTM networks and Radial Basis Function (RBF) is presented in the study [5]. The technique's efficacy in detecting and thwarting DDoS attacks is demonstrated through evaluation on the benchmark dataset CICDDoS2019.

A DDoS attack aims to overload a target system to the point that it is unable to carry out its intended function and is inoperable for authorized users. These attacks are particularly damaging to cloud computing systems since even one cloud server can have a significant impact on a large number of users [6]. Cloud systems expand their processing power and the various virtual machines and provision instances. When attempting to stop a cyberattack, the cloud architecture is of no assistance. It is now difficult for actual consumers to use their cloud services due to a slowdown in the cloud infrastructure. When hackers use more zombie machines to target numerous cloud-based services, DDoS attacks might become much more dangerous.

Many of the authors discussed DDoS attack detection using different methods and algorithms and a few are described below as shown in Fig. 9.1. A honeypot is inherently deceptive, and while the legality here is clear, the morality of putting people in this situation may be questionable. Some critics say it verges on entrapment if it actively seeks out attackers. Shows that there are no consensus difficulties, which must be a priority in future honeypot studies. Do not incite a person to commit a crime they would not have otherwise committed. It should only watch or learn existing malicious behaviour. Minutely IP and packet analysis. Box-Cox transforms and normalizes packet-based time data. The approach classifies traffic circumstances.

The work aims to

- Implement a novel approach of Honeypot with LSTM to detect DoS attacks in cloud computing systems.

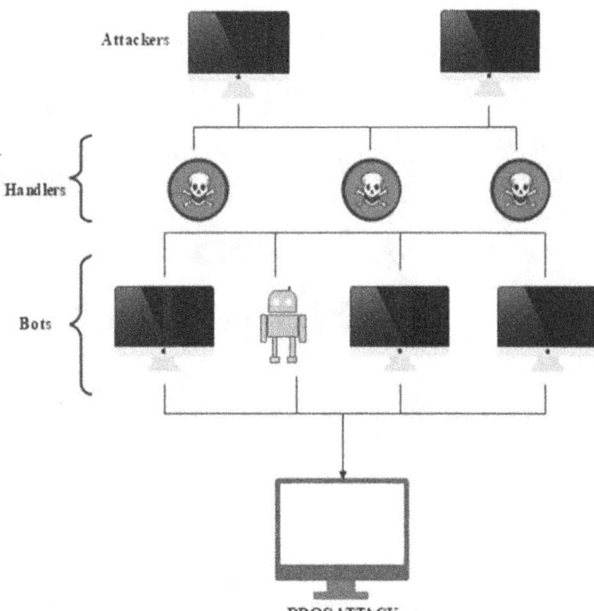

Fig. 9.1 Different methods of detecting DDoS attacks

- To enhance the precision method, PSO created feature selection method is proposed.
- Using performance measurements, contrast the suggested approach with alternative methods.

The current situation highlights the growing danger posed by respiratory diseases such as chronic obstructive pulmonary disease (COPD) and asthma. This heightened risk is attributed to the rise in air pollutants like PM2.5 and PM10. In response to this pressing issue, the use of a respirator can serve as a direct and immediate protective measure on an individual level. Addressing the pollution levels, however, requires a prolonged effort as the gravity of the problem surpasses the timeline afforded for mitigation.

1.1 Organization of the Study

The work may be broken down into the following sections: Background of the proposed methods is providing in Section 2, an clarification of the methodology that underpins the recommended algorithm is provided in Section 3, Section 4 delivers the findings of the work, and Section 5 offers a conclusion.

2. Background

In this segment, we provide a detailed description of the background of the PSO-based feature selection method, the LSTM model, and the honey pot approach. This information aims to offer a inclusive considerate of how these techniques are employed in various domains for data analysis and problem-solving.

2.1 Honeypot

In the realm of cybersecurity, honeypots serve as an essential tool for fortifying network defenses. These deceptive mechanisms not only detect and divert potential attacks by intercepting and analyzing packets but also meticulously maintain logs of unauthorized activities. In the proposed system, the strategic utilization of honeypots aides in efficiently managing multiple clients, while the robust Intrusion Detection System (IDS) diligently watches over the network, promptly detecting any signs of unauthorized access or malicious activities.

Honeypots are characterized as "security resources whose usefulness depends on being probed or penetrated," according to one definition of the term. They may be based on a host computer or a computer network, but in practice, they are almost always based on a computer network since almost all interaction takes place through a network connection. Because a honeypot serves no legitimate purpose, it makes it easier for an intrusion detection system to differentiate between "anomalous" and "normal" behavior. As a result, any activity that occurs on a honeypot can be immediately labeled as being anomalous. This is the main advantage of honeypot usage. Due to the characteristics making them a good candidate for monitoring network malicious behavior, honeypots serve as a useful research tool in the domain of computer security. Honeypots are also used in monitoring malicious activity on networks.

In contrast, honeypot research has only recently become more organized. This formalization has produced a stronger focus on studying what attackers do and why they do it, and has resulted in the use of honeypots as an important part of the process for discovering previously unknown security flaws [7].

2.2 LSTM

LSTM model is operating to help effectively detect the possible intrusion inside network. A novel architecture of this deep learning model proved to excel in processing unsegmented data, identifying complex correlations, which is essential for correct audio recognition performance, and anomaly detection in network traffic or Intrusion Detection Systems (IDSs). This ability to retain long-term dependencies and handle data as it arrives lends itself to its applicability in improving an organization's cyber security posture, enabling positive risk management and mitigation creating a proactive network monitoring and defence system against potential threats.

An LSTM unit consists of four major parts: cell, input (i), output (o), and forget (f) gate. These workings are critical for variable the flow of data into and out of the cell, as cell memories typically have short durations. The Fig. 9.2 illustrates an LSTM with four phases in a cell. The cell state, depicted by the uppermost horizontal line, serves as a vital component of the cell. The data within the cell state is controlled by gates, preventing the removal of data by the LSTM. Cells are composed of significant structures to facilitate their functionality.

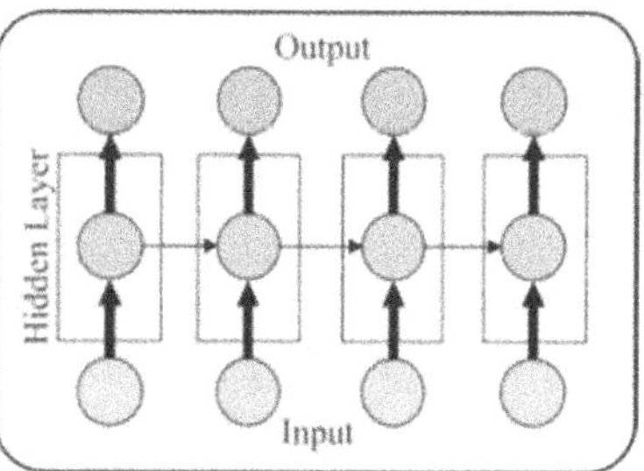

Fig. 9.2 Structure of LSTM cell

Feature selection method- The particle swarm optimization (PSO) optimizes the selection for the best possible results. The PSO algorithm was built with inspiration from the social behavior of birds within a flock for the search algorithm. As time progressed, this concept underwent further development and refinement, ultimately emerging as a straightforward yet highly efficient optimization algorithm as shown in Fig. 9.3. The formula for PSO-based feature selection ad represented in Equation 1 and 2 is as follows:

$$x_{id}^{t+1} = x_{id}^{t} + v_{id}^{t+1} \tag{1}$$

$$v_{id}^{t+1} = w * v_{id}^{t} + c_1 * r_{1i} * \left(p_{id} - x_{id}^{t}\right) + c_2 * r_{2i} * \left(p_{gd} - x_{id}^{t}\right) \tag{2}$$

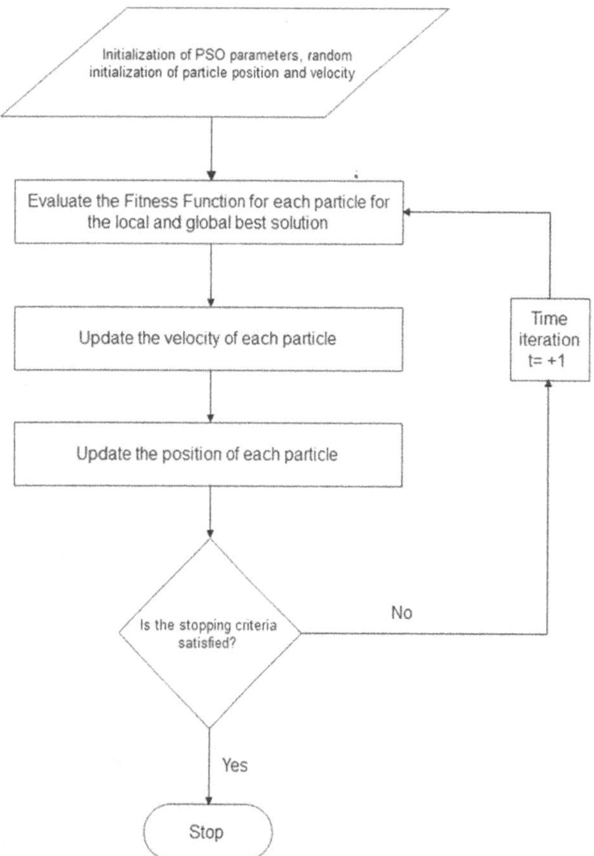

Fig. 9.3 PSO algorithm flowchart

Where is the Length of our exploration space. In PSO, t is the present reiteration, is the element's existing location, as well as is current velocity of the element in d in the Iteration. w is the inertia weight, while as well as Are the learning rates? r1i as well as r2i represents the random numbers that are regularly circulated within the variety [0, 1], and p_{id} as well as p_{gd} are the p_{best}.

The Particle Swarm Optimization (PSO) technique is widely recognized for its effectiveness in conducting global search operations with high levels of efficiency. One of the key advantages of this method lies in its superior representation, which allows it to efficiently explore expansive areas while maintaining a low computational cost. It is also less susceptible to issues like over-fitting because of its simple construction process and the little amount of parameters it requires, making it even more suitable for the challenges of feature selection.

3. Methodology

The growing demand for cloud computing worldwide has placed cloud security at the top of an organization's agenda like never before. Some such threats that obstruct cloud computing have been identified and one of the most exciting out of these is the detection of DDoS attacks. This study introduces a new solution for this urgent challenge. In this study, we employ the use of a honeypot and LSTM in combination to successfully detect and prevent DDoS attacks in the cloud environment. In addition, Particle Swarm Optimization (PSO) based feature extraction is utilized in order to enhance the effectiveness of the threat training and identification.

In addition, an important issue addressed in this research is the challenge of imbalanced data, which has negative consequences on the model's capacity to make accurate predictions. As a solution, we've begun to examine our dataset for these implications, starting with a Synthetic Minority Over-sampling Technique (SMOTE) analysis in conjunction with data distribution in order to improve the model's overall performance and reliability. Then, the most relevant variables are obtained on the basis of threat detection and classification, allowing enhanced identification through statistical analysis or categorical data analysis; as with the data, PSO-based feature selection techniques can be used for threat detection and classification. This 360-degree approach allows organizations to remain a step ahead of the dynamic nature of cyber threats to protect their cloud infrastructures continuously and maintain the integrity and security of their cloud-based services.

Feature Balancing- In the creation of prediction models, "imbalanced classification" means data sets in which one particular class is orders of magnitude more prevalent than another. Most machine learning approaches are biased to the majority class, which results in poor performance, even though the performance of the minority class is usually of utmost concern. To solve the problem of imbalance data, one potential solution is oversampling the minority class. The easiest thing is to show examples from the underrepresented group but it does not contribute much towards understanding of the model. Alternatively, new instances can be generated by synthesizing existing ones. One effective method for improving minority group data is the SMOTE.

3.1 SMOTE Analysis

SMOTE (Synthetic Minority Over-sampling Technique) a technique used to handle class imbalances in datasets. It does so by generating synthetic instances for minority instances using a process that interpolates existing instances. SMOTE generates hypothetical data just for underrepresented groups, rebalance the dataset. This method applies random choice between the k-Nearest Neighbors (kNN) of each instance of a minority class, hence after oversampling we obtain a rebuilt dataset. We can expose this balanced dataset to multiple sorting systems for further analysis. The algorithm of SMOTE analysis is as tracks:

Step 1: The Euclidean distance among every data point in set A and the samples in set B was computed ($x \in A$). Following this calculation, the kNN values for x were determined, specifically after considering the minority class group A. This consideration is essential for obtaining accurate and meaningful results.

Step 2: For all data points of set A, the Euclidean distance was measured among the applied samples of set B ($x \in A$). Here, the value of kNN for x (again, after considering the minority group x in group A) is calculated, because when we calculate the kNN for a point related to the minority class, we need to take into account the other points that belonged to the actual minority class (group A) to get a meaningful result.

Step 3: The formula provided is a mathematical representation used to create a fresh example for each instance denoted as A1 with a varied value of k ranging from 1 to N. It involves updating the value of x to x^' by adding a random number between 0 and 1 multiplied by the absolute difference between x and x_k. This process is essential for generating diverse examples within the given set.

$$x' = x + rand(0,1) * |x - x_k| \tag{3}$$

It adds a bit of a random number between 0 and 1 to make it more diverse. There have been many proposals for dealing with DDoS attacks, but no complete solution. Not all mitigation strategies are equally effective, and no umbrella approach offers perfect prevention. Such challenges will require consistent evolution of defensive capabilities to match advancing threat environments to protect against nefarious digital conduct.

One of such methods is the Honeypot approach, which is widely known in the field of computer security. These are carefully monitored decoys placed in a network to actively attract hackers and immediately inform its network administrator of such attempts to breach or violate it. By detecting and responding to cyber threats early, this proactive strategy improves the network's overall security posture.

In this study, an innovative method is introduced in the field of hot data gathering using honeypots in intrusion detection systems of potential attackers. The suggested approach deploys honeypots in a network to catch and monitor malicious activity, improving the security of the network as a whole. This honeypot-based strategy can provide respected insights into the performance, tactics, helping inform future strategies (as shown in Fig. 9.4) and filling the knowledge gaps, which can strengthen the network against malicious actors.

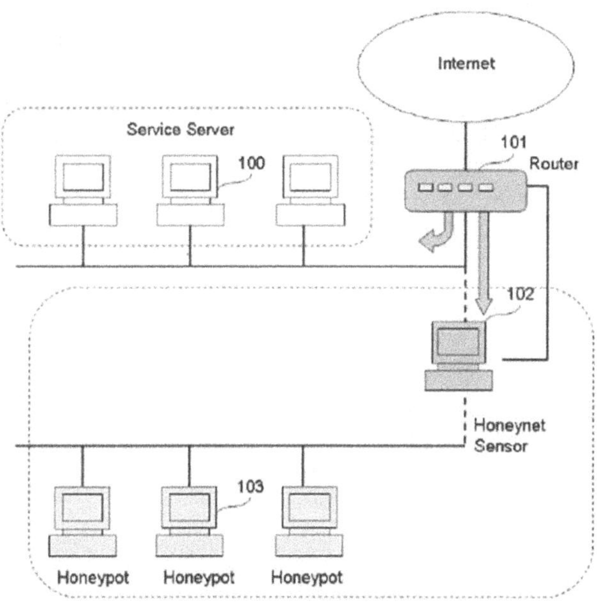

Fig. 9.4 Block diagram of honeypot

3.2 Implementation of Honeypot

Two key stages of the honey pot method are crucial to the security of our network. In the first stage, it serves as a solid defense not only against current DDoS attacks, but also any potential new iterations that will appear down the line. Then, it becomes a tactical device that seeks to ensnare the enemy, allowing us to gather incriminating evidence that can lead to a court case against them. In order to effectively deceive the attacker and make the honey pot successful, we must overcome three significant hurdles.

1. Then the attack has to be easily detectable.
2. The attack packets need to flow through a honeypot that actively costs resources to entice the attack packets.
3. The level to which the honeypot needs to convincingly mimic the organization's network topology, especially the components that the attacker is aware of.

The performance measurements are working to measure how well the planned work is against other algorithms is being implemented (LR, GNB, and RF. This technique is used to prove that future algorithm are superior to other algorithms.

4. Results

To illustrate the efficacy of the suggested algorithm, performance measurements are utilized for comparative analysis against various classification algorithms, including but not limited to LR, GNB, and RF algorithms. This comparative assessment aims to showcase the algorithm's robustness and efficiency in handling different datasets and scenarios, thus highlighting its effectiveness as a competitive solution in the realm of classification tasks.

The Fig. 9.5 visually depicts the meticulous analysis conducted on various machine learning methodologies, comparing results obtained with and without the Particle Swarm Optimization (PSO) method. The x-axis denotes the recommended and existing methods, while the y-axis showcases the accuracy scores associated with each. Noteworthy is the observation that the precision of the LSTM algorithm registers at 83% without PSO implementation. However, upon integration of feature selection, a notable increase in accuracy to 91% is observed. This finding underscores the significant impact of leveraging feature selection in augmenting the LSTM algorithm's performance, substantiated by results drawn from a comparative analysis against other algorithms enumerated in Table 9.1.

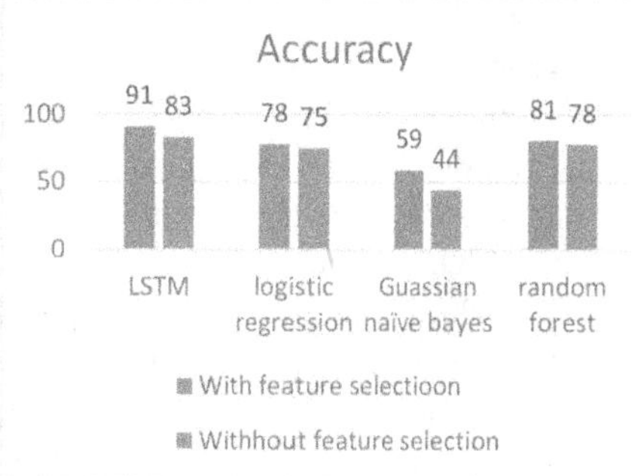

Fig. 9.5 Assessment of other classification algorithms for Accuracy

Table 9.1 Assessment of other classification algorithms for accuracy

	LR	GNB	RF	EL
With Feature Selection	78	44	81	91
Without Feature Selection	75	59	78	83

The Fig. 9.6 vividly illustrates the specificity analysis for four algorithms, offering a clear comparison of their performance with and without PSO. On the x-axis, viewers can distinguish between the proposed and existing algorithms, while the y-axis depicts the corresponding accuracy values. Particularly striking is the marked difference seen in the specificity values of the LSTM algorithm, which recorded an impressive 93% without PSO and soared to 95% upon the incorporation of feature selection. These findings unequivocally highlight the potency of blending the LSTM algorithm with feature selection, leading to a significant performance edge over the other algorithms outlined in Table 9.2.

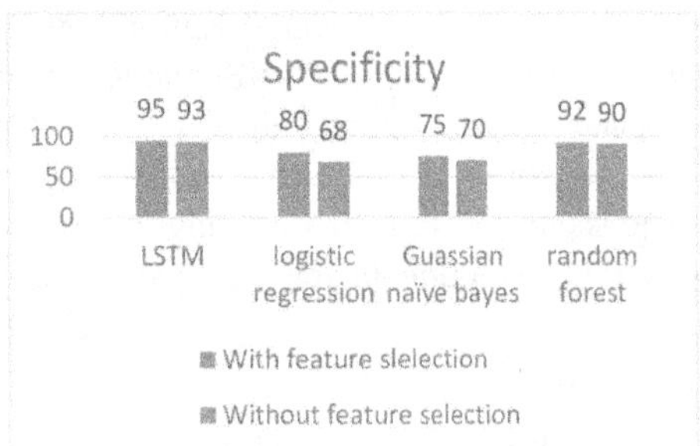

Fig. 9.6 Assessment of other classification algorithms for specificity

Table 9.2 Assessment of other classification algorithms for specificity

	LR	GNB	RF	EL
With Feature Selection	80	70	92	95
Without Feature Selection	68	75	90	93

Upon close examination of the algorithmic analysis, it is evident that Fig. 9.7 plays a crucial role by providing a detailed comparison of sensitivity between utilizing particle swarm optimization (PSO) and without. The graphical representation on the x-axis effectively distinguishes between the proposed and existing algorithms, while allocating the y-axis to display corresponding accuracy measurements. Notably, the LSTM algorithm exemplifies a sensitivity rate of 86% in the absence of PSO. However, upon integrating PSO, the sensitivity rate notably surges to an impressive 94%. This stark contrast emphasizes the superior performance of the LSTM algorithm, especially in conjunction with feature selection techniques, surpassing the effectiveness of other algorithms enumerated in Table 9.3 and showcasing its exceptional capability in achieving heightened levels of accuracy.

Table 9.3 Assessment of other classification algorithms for sensitivity

	LR	GNB	RF	EL
With Feature Selection	62	44	75	94
Without Feature Selection	59	37	64	86

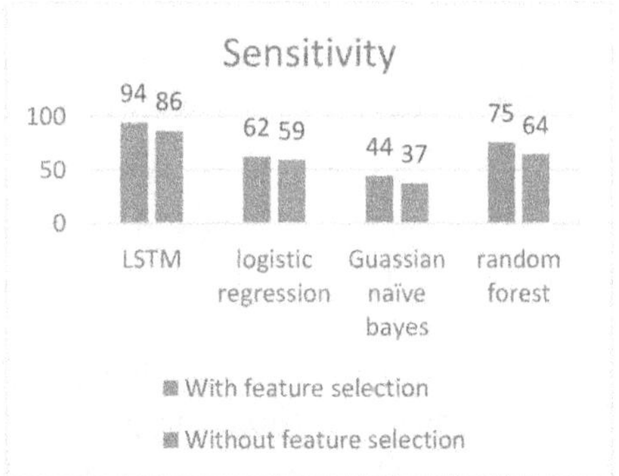

Fig. 9.7 Assessment of other classification algorithms for sensitivity

The Fig. 9.8 visually represents the F1-score analysis of four different algorithms, showcasing their performance with and without feature selection. On the x-axis, you can observe the various proposed and existing algorithms, while the y-axis indicates their corresponding accuracy values. For instance, when comparing the F1-scores of LSTM, we see that without PSO, it achieves an 89% accuracy rate, but with PSO integration, this number significantly boosts to 96%. This substantial improvement highlights the effectiveness of utilizing feature selection techniques alongside LSTM, making it a standout performer among the algorithms detailed in Table 9.4.

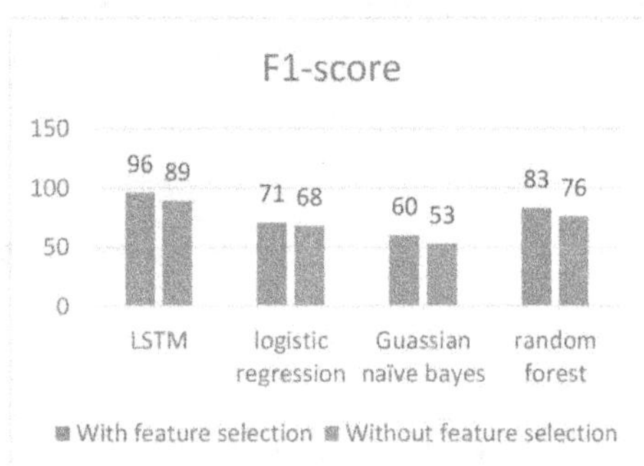

Fig. 9.8 Assessment of other classification algorithms for F1 score

Table 9.4 Assessment of other classification algorithms for F1 score

	LR	GNB	RF	EL
With Feature Selection	71	60	83	96
Without Feature Selection	68	53	76	89

The Receiver Operating Characteristic (ROC) curve depicted in Fig. 9.9 serves as a visual depiction of the presentation valuation of our suggested model, emphasizing the efficacy of the LSTM algorithm. With an Area Under the Curve (AUC) value of 0.944, exceeding the threshold of 0.5, this perfect showcases a high level of accuracy in identifying and mitigating DDoS attacks. Explaining Fig. 9.10 in accordance with the adapter: when sender input = 0, the receiver's response suggests

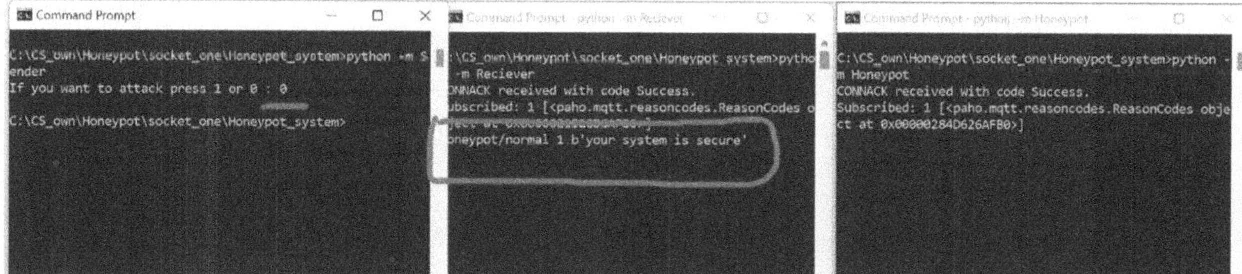

Fig. 9.9 ROC curve of the proposed model

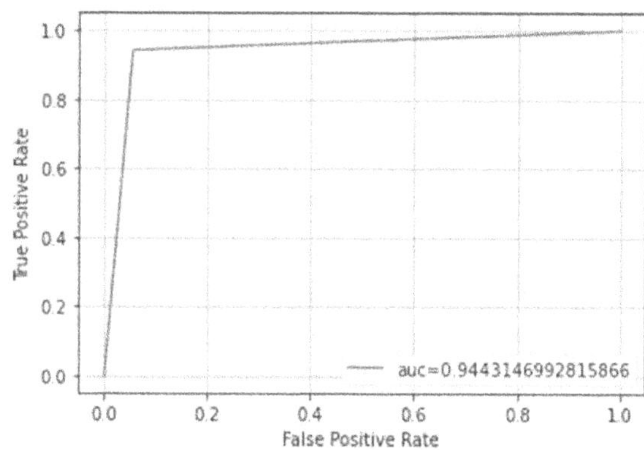

Fig. 9.10 System honeypot simulation: Attack decision and response handling

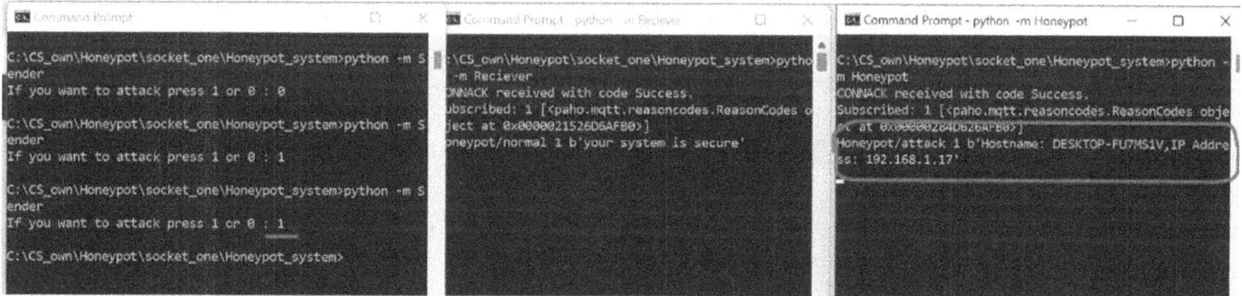

Fig. 9.11 System honeypot simulation: Attack decision and response handling

that the system is secure. Likewise, the Honeypot shows more information on the attacker (Vehicle Hostname, Hostname, and IP address) if you input in the Sender an input of 1 (see Fig. 9.11).

5. Conclusion

The study delves into a pioneering methodology to detect DoS occurrences within cloud computing environments. It advocates for the integration of a honeypot strategy along with Long Short-Term Memory (LSTM) technology to effectively identify distinctive characteristics from the dataset and delineate various types of attacks. This innovative approach aims to enhance the detection capabilities against malicious activities and bolster the security framework of cloud systems, demonstrating a proactive stance in safeguarding critical infrastructure. The honeypot, assisted by an IDS, not only detects attacks but also traps and diverts malicious packets sent by attackers. By employing a PSO-based feature selection method, the technique can isolate relevant characteristics from a large pool of features. Subsequently, a subset of these features is used to train the suggested model to categorize DDoS attacks. The efficacy of the projected strategy is validated by comparing it with extra systems using the same presentation constraints. The PSO-based selection method prioritizes high-value features, making it more challenging for DDoS attacks to mimic service characteristics. The results demonstrate that the method with feature selection achieves 91% accuracy, while the one without it achieves 83% accuracy, both outperforming existing algorithms with feature selection. Consequently, it is evident that the honeypot combined with the LSTM method is highly current in sensing DDoS attacks within cloud computing infrastructures.

References

1. N. Alrehaili and A. Mutaha (2020), "Cloud Computing Security Challenges," Iarjset, vol. 7, no. 8, pp. 120–123.
2. S. Yu, J. Zhang, J. Liu, X. Zhang, Y. Li, and T. Xu (2021), "A cooperative DDoS attack detection scheme based on entropy and ensemble learning in SDN," Eurasip J. Wirel. Commun. Netw., vol. 2021, no. 1.
3. D. Alsmadi and V. Prybutok (2018), Sharing and storage behavior via cloud computing: Security and privacy in research and practice, Comput. Human Behav., vol. 85, pp. 218–226.
4. AVelliangiri Sarveshwaran, Karthikeyan, P Karthikeyan V Vinoth Kumar (2020), Detection of distributed denial of service attack in cloud computing using the optimization-based deep networks Detection of distributed denial of service attack in cloud computing using the optimization-based deep networks, April 2020, Journal of Experimental & Theoretical Artificial Intelligence 33(3), 2020, DOI: 10.1080/0952813X.2020.1744196.
5. Amitha, M. ., & Srivenkatesh (2023), M. DDoS Attack Detection in Cloud Computing Using Deep Learning Algorithms. International Journal of Intelligent Systems and Applications in Engineering, 11(4), 82–90,
6. F. S. De Lima Filho, F. A. F. Silveira, A. De Medeiros Brito Junior, G. Vargas-Solar, and L. F. Silveira (2019), "Smart Detection: An Online Approach for DoS/DDoS Attack Detection Using Machine Learning," Secur. Commun. Networks.
7. C. Huang, J. Han, X. Zhang, and J. Liu (2019), Automatic identification of honeypot server using machine learning techniques, Secur. Commun. Networks.

Note: All the figures and tables in this chapter were made by the authors.

Intelligent Systems Using Semiconductors for Robotics and IoT – Dinesh Goyal et al. (eds)
© 2026 Taylor & Francis Group, London, ISBN 978-1-041-20408-4

10 Leveraging Large Language Models (LLMs) and Advanced Machine Learning Techniques for Accurate Detection and Mitigation of Fake News in Digital Media

Gaurav Sinha[1]
M. Tech. Scholar, Department of CSE, RIET,
Jaipur, Rajasthan, India

Pragya Bharti[2]
Assistant Professor, Department of CSE, RIET,
Jaipur, Rajasthan, India

Abstract: Fake news, meaning information that is so purposely false or misleading that it is practically false, has become a huge problem in the digital age. Most fake news has a political, socioeconomic or individual agenda and is designed to influence people's perception, stigmatize reputations or alienate communities. But this is a problem with an emphasis on the fact that people are very vulnerable to erroneously accepting and spreading a fake news and it underscores why we need cutting edge technology solutions to address that. This project aims to develop an automated system that can distinguish between fake and true news using a state-of-the-art Language Models (LLM) and Natural Language Processing (NLP) techniques. The goal of this project is to build a good model which can predict the authenticity of news with a good accuracy on a structure dataset provided by Kaggle that divides a news article into "True" and "Fake" as per characteristics like "Title," "Text," "Subject", "Date". The model is trained on historical data to insure sufficient learning and trainability, and the efficacy of the model is tested on the generalizability to a new, unseen datasets. In additions to improving the accuracy of fake news detection, the corresponding method also ensures that the model is dependable in a variety of situations. To separate the authentic news from fraudulent information, we have used the key natural language processing (NLP) techniques i.e. tokenization, feature extraction using Bag of Words and TF – IDF approaches along with machine learning algorithms — Logistic Regression, Naive Bayes, Random Forest, and Gradient Boosted Trees..

Keywords: Fake news, Natural language processing (NLP), Machine learning (ML), Language models (LLMs), Media literacy, Information integrity, Real-time verification

1. Introduction

The information age has arrived through digitization. Access to information is made more democratic now with the internet, and those who have access to the internet can publish and share content around the world. The democratization of news has its merits, but it's also resulted in an explosion of fake news, or intentionally contrived stories that aim to deceive or bewilder an audience. Usually done to sway public opinion, influence reputation, or create polarized communities, fake news is created typically to achieve political motives, socio economic factors or personal vendetta.

1.1 Understanding Fake News

Fake news isn't a new thing, but the phenomenon has grown dramatically driven by the existence of digital

[1]grv.sinha@gmail.com, [2]Cse.pragya@rietjaipur.ac.in

DOI: 10.1201/9781003716389-10

media. Misinformation and propaganda have been used historically to effect change in public perception, but the power and the pace of the internet have made fake news something to a pervasive and problematic issue. Fake news can look like fake headlines, like manipulated images, fake stats, entirely false articles. There is often a spin to the content to make it appear credible and in some cases even with fact in there [1].

1.2 The Impact of Fake News

Fake news has far reaching and multi-faceted consequences. These activities can erode public trust in media and institutions, tilt election outcomes, fuel violence and, at times, fracture societies. For instance, during the 2016 US electoral race fake news stories were spread on social media, perhaps influencing voter opinions. In other situations, false information regarding the existence of health emergencies, from the Covid-19 pandemic, has generated public confusion and lack of trust in health guidance. It becomes difficult to determine whether content is true or false, and if that content is real, it can spread very fast, overtaking accurate reporting, and it can be hard for people to tell one from the other.

1.3 Fake News and Human Susceptibility

Human susceptibility is one of the most important contributory factors for the spread of fake news. In fact, how people process information involves cognitive biases, like confirmation bias and bandwagon effect. Confirmation bias is people favoring information which harmonizes with the beliefs they already hold; bandwagon effect is people being more likely to accept information democratic or commonly adopted. Furthermore, the quantity of this information available online can be daunting, leaving someone feeling out of control subject to heuristics (or rules of thumb) of using familiar sources or sensational headlines to quickly judge[3].

1.4 The Need for the Solutions with Reference to Technology

Due to the fact that fake news detection is beyond human perceptible measures, it is time to integrate technological techniques. Mainly, Artificial Intelligence (AI) and Machine Learning (ML) present potential opportunities to manage this issue. This technology can process a huge amount of data review them and predict outcome with a speed and precision that no man or group of men can match. Particularly, by applying methods of a certain field called Natural Language Processing (NLP), one will be able to process and analyze plain text data that can be used to design a system that can recognize and categorize fake news [4].

2. Literature Review

The increased spread of misinformation during the modern era has led to numerous investigations into ways of using AI tools to identify fake news, where NLP and ML are prevalent. With baseline solutions in the hand, traditional models have been further expanded as the new Large Language Models (LLMs) open up new directions to tackle the issues of misinformation. This study seeks to address some of the following gaps that have been identified in the course of reviewing the literature available on the subject. Although, Some of the first approaches in fake news detection were performed by generative ML models like Naive Bayes and Support Vector Machines (SVM) as well as Logistic Regression. As mentioned in Boissonneault and Hensen (2024), the incorporation of LLMs with other tools such as fact-checking algorithms requires using the like of datasets that include the LIAR. While their approach offered better results than the traditional models, certain critical issues were not addressed: basic bias of LLMs and computational problems of real-time detection. The research of these authors described the use of LLMs but did not provide approaches for addressing challenges related to model bias and data complexity [1].

Also, Shah et al. (2024) analysed the dual function of LLMs in the creation and identification of fake news and noted understandably the ethical dilemma here [2]. They admitted that LLMs could be abused to build refined fake news and stories but failed to provide real-world approaches to address such abuses. They further indicate that this gap calls for ethical guidelines and protections whenever LLMs are used to fight fake news. Due to how the fake news domain evolved, the models of artificial intelligence needs to recognize the context behind fake news and be flexible enough to learn new methods of spreading them. Hu et al. (2024) especially called for the promotion of social context into LLMs to improve the detection rate [6]. They recommended that incorporating user behavior and network information might enhance model performance according to their work. But they were unable to adapt their models to changing strategies used by miscreants to spread misinformation, and thus they were missing the learning component. More recently, Barman et al. (2024) tested how far LLMs are capable of producing multimedia disinformation, proving how powerful models could be in generating perfect fakes [4]. While their work pointed at this threat, their efforts did not include prescriptions on how to identify and avoid it in real time. This speaks to a critical gap of building models to learn and respond to the rapidly changing tactics used in fomenting false narratives. One drawback of applying ML algorithms is that it may further amplify certain kinds of biases intrinsic in the data sets themselves and this is dangerous especially in the sensitive area such as fake news detection.

Su et al. (2023) investigated potential sources of bias in detectors designed to identify texts as fake news when having to process texts created using LLMs [9]. In doing so, the authors discovered that, in the case of a shift in the data distribution, classification models tend to perform worse on such content, thereby producing unfair or incorrect detection results. Although they proposed to use adversarial training in order to increase robustness, they did not pay much effort to define optimal approaches. LLMs have been identified as having a potential in reducing misinformation by Chen and Shu (2023), or the following are the opportunities and challenges highlighted by the authors [5]. They pointed out that knowledge graphs and contextual information could be especially useful for improving model accuracy while also raising concerns over the inclusion of such practice leading to a worsening of biases in machine learning models, or the reinforcement of fake news. Interpretability also endemic becoming a problem, as LLM-based models tend to be new forms of a 'black box,' meaning users of the detections are unable to understand the rationale behind some of the marks.

This however has only been done with advanced techniques in some of these studies as earlier research addressed some of these gaps. In one of the comparative studies on the performance of various Architectures of LLM, Koka et al., assessed the ability of different architectures in identifying fake news [7]. They focused on the idea of how one has to adapt the model to a task or a particular dataset for better result. However, their work mostly relied tied solely on the English language information making them recommend for more work to be done on the models that work with multilingual content suitable for other cultures.

Benny (2023) proposed a direction to enhance context comprehension using knowledge graphs with LLMs in fake news identification [8]. Floating the hypothesis that the use of an additional structured knowledge source would improve factual content detection, the study integrated knowledge graphs into the language model. Although, this approach was hypothesis- based for testing the significance of PM features, it brought computational complexity and substantial data preprocessing which indicated the weak aspect of scalability of the approach. The literature suggests clear limitations in establishing fake news detection models that are mutable, explainable, and/or/ethically sound. Key challenges include:

- Adaptability to Evolving Misinformation: Previous models are not as effective at addressing new types of fakes which might appear, change their format or nature often.
- Bias Mitigation: It is imperative to develop the most efficient ways of minimizing internal biases in LLMs and how to detect them in different content.
- Model Interpretability: Improving the level of transparency is of great importance for the users especially in the applications which affect the consumption of information.
- Practical Deployment and Ethical Use: The current gap between the theoretical models and their application stems from two inherent factors: computational and ethical limitations.

This research seeks to achieve the above objectives through other strategic interventions as highlighted below. Firstly, it applies Adversarial Training so as to increase model robustness or, in other words, make models optimize against the tendency of fake news to evolve over time, trained on adversarial samples. Secondly it incorporates Context-Aware Embeddings using state-of-art Scalable LLMs such as BERT and GPT. This approach helps in semantic analysis and also applies context-oriented information extractor to incorporate other types of particularly misleading information. Furthermore, considering the purpose of the study the following objectives are used to address Ethical Implications: These are bias mitigation and model interpretability and explanations. Lastly, it intends to achieve Practical Applications with browser extensions and by creating a social media filter that can work in real social platforms so as to show that such a model may be useful when applied in real life and illustrate actual implementation problems.

Research work like the ones represented in Lai et al. (2024) presented the concept of rumor-based fake news detection using LLMs with data augmentation integrated to enhance the detection performance [13]. Their work shows that it is possible to train these models even when labeled data is scarce, to detect misinformation. Nevertheless, they also shed light on why more data is needed to be collected and the models need to be updated continuously. According to Jiang et al. (2024), it is still challenging to identify new emerging disinformation when it is created by LLMs [14]. They suggested to constantly update the models by applying LLMs themselves and making a closed-loop system that can easily follow up with new patterns of misinformation. This is in line with the call for models that can learn at a personal level of time.

With reference to ethics, the detection of fake news using LLMs has various effects. As stated by Hadi et al. (2023), machine learning practitioner should be responsible as pointed out in the example of disinformation detection whereby mistakes attract grave consequences [10]. They proposed the creation of codes to promote the appropriate use and exploitation of these models. Lucas et al. (2023) extended work in this vein by outlining a 'fire with fire' approach that would seek to employ LLMs both to generate and to identify disinformation [19]. As this is an innovative strategy, it can provoke ethical concerns regarding the differences between its correct usage and possible misuse, as well as the necessity to create the ways to prevent it. Therefore, besides the important tools that can be used to improve the fake news detection by means of LLMs, there is a list of unsolved problems related to those models' adaptiveness, biases, explainability, and proper usage. Previous work in this field provides preliminary findings but few advanced solutions to these issues.

Our studies effectively contribute to the literature by proposing the fake news detection model which is based on NLP and adversarial training which improves versatility and resilience. Furthermore, we also perform measures of bias in order to enhance the capabilities and equality of our models. We also provide means of interpretability to make explicit use of the integration of the explainability AI components in order to increase user trust. Moreover, due to the focus on the challenges of the current trending technologies, our research provides actual software tools, browser extensions, and filters for social networks; we consider real-world consequences and ethical concerns. This systematic approach guarantees that our inputs are not only abstract, but realizable in present computer-mediated communication scenarios.

In this regard, it is hoped that this study will fill the mentioned gaps and contribute to the development of a fake news detection model that is not only very effective but also ethical and feasible for practice.

3. Proposed Methodology

The methodology of this study aims to fill in the aforementioned gaps in the literature by creating an interpretably ethical and scalable fake news detection model. The methodological approach therefore involves the use of state-of-art natural language processing techniques as well as adversarial training and the development of practical applications, and all the steps taken as well as the reasons for so doing are justified.

3.1 Data Collection and Description

The data of this project is obtained from Kaggle and contains articles of news randomly labelled as 'True' or 'Fake'. All the articles contain parameters that are title, the text, the subject and also the date of publishing. The data used in the current work is rather diverse and evenly distributed from which the model can be trained and tested. Such versatility of the dataset guarantees that the model will work equally well with any type of content that can be encountered in a news feed.

The 'True' and 'Fake' news articles were downloaded in equal proportion from Kaggle to cover a diverse set of news topic and style. Features of dataset are Name, Body, Category, and Date, so the dataset is highly informative. There is a vast literature on the usage of this dataset for comparative analysis and the addition of the diverse samples helps to improve the heterogeneity of the model.

3.2 Data Preprocessing

Data preprocessing remains an important step in data preparation and acquisition to be used in data analysis. It involves the process whereby the data collected are purified so that all the necessary information from the population will be obtained. This entails mostly ensuring the text does not contain new lines or extra spaces, occasional special characters, needs to be tokenized and in most cases converted to lower case. There is also filtering of stopwords which is terms that are unnecessary and are commonly used in any text. This part of the work is crucial to improving the feature extraction quality and the final outcome.

Advanced preprocessing techniques were employed to retain critical linguistic features:

- Noise Reduction: Basic cleaning procedures like erasing special symbol and signs were employed.
- Context-Aware Stopword Removal: Rather than excluding all the standard stop words, the process left the words which might be important in the context of fake news, for example modal verbs and negations.
- Lemmatization over Stemming: Lemmatization was chosen for normalizing words against their base forms, while still considering the grammatical function of the word for the improvement of semantic analysis on the given sentences.

These choices were made to preserve additional contextual and semantic information that is important to detect various kinds of misinformation [12].

There are 3 types of Feature Extraction Techniques:As for feature extraction technique the study utilized Bag of Words (BoW), Term Frequency-Inverse Document Frequency (TF-IDF) and the application of advanced Language Models. To meet both the syntactic and semantic aspect, the study utilized a hybrid feature extraction technique. pproach integrates advanced NLP techniques, adversarial training, and practical application development, with detailed justifications for each methodological choice.

3.3 Feature Extraction Techniques

Project employs three primary feature extraction techniques: Bag of Words (BoW), Term Frequency-Inverse Document Frequency (TF-IDF), and the use of advanced Language Models. To capture both syntactic and semantic features, the study employed hybrid approach:

- Bag of Words (BoW) and TF-IDF: These methods were used to capture word frequency and importance, providing a baseline representation of textual data.
- Contextual Embeddings with LLMs: Advanced models like BERT and GPT were utilized to generate embeddings that capture contextual relationships between words. Fine-tuning these models on the fake news dataset allows for better adaptation to domain-specific language patterns.

The use of these techniques guarantees that not only weak features of the text are taken into account but strong ones as well, which improves the fake news detection model. To improve the model's robustness against evolving fake news tactics, adversarial training was incorporated:

- Generation of Adversarial Examples: False reports were included in artificial short news stories using LLMs to create elaborate fake news.
- Model Training with Adversarial Data: The model was able to learn from both the original and adversarial datasets therefore catering for the difficult samples.

It solves the issue of the adaptability gap, researched in the literature, and enables the model to be ready for advanced fakeness in the content of the fake news.

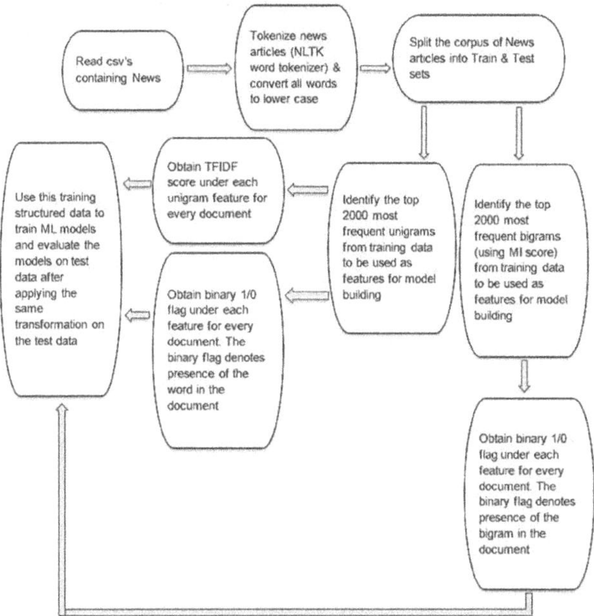

Fig. 10.1 Proposed Methodology

3.4 Machine Learning Models and justification

Several algorithms were evaluated, with justifications for their selection:

- Logistic Regression: Selected because it is easy to interpret and follows a basic model.
- Naive Bayes Variants: Suitable to the text classification problem because they are based on probabilities.
- Random Forest: Chosen because of its high dimensionality and non-linear feature interaction capabilities.
- Gradient Boosted Trees (XGBoost): Used for better performance in the model complexity regularization and effectively deal with the complex patterns.

Tuning of hyperparameters was done with GridSearchCV to compare optimal values for each type of model. It was hypothesized that the ensemble methods would be superior to the simpler models because it can handle nonlinear transformations and combined feature interactions.

3.5 Model Evaluation

As a consequence, the dataset is divided into two sets — training and testing — in order to estimate the performance of the model. The training set is employed to construct the model, the test to establish the accuracy of the model and the extent to which the model is accurate. To optimize model performance, the models were tuned using hyperparameter optimization, while K-fold stratification was used to compute performance metrics. Key metrics included:

- Accuracy, Precision, Recall, F1 Score: To generalize performance and control Types 1 and 2 errors in order to avoid yielding misleading results.
- ROC-AUC: To overall assessment of the model and to find out how well the class distinction is done at different thresholds.

These strategies give complete insight into each model and how effective it is. To check the efficiency of the model, data split into the training and testing sets is used.. The training set is used for constructing the model while the test set is used to evaluate the accuracy of the constructed model. Hyperparameter optimization is used to find out the best model for the optimization of the models presented.

4. Results and Discussions

This work employed two set known as "Fake" and "True" sets to form a full set for the identification of fake news. The null data sets were cleaned by deleting certain unimportant columns such as the 'subject' column could distort the model. The outcome dataset contained text data from news articles and thus following data pre-processing, which included vectorizing the text and converting the actual textual in-formation into a numerical form that the models could understand.

The usefulness of the presented approach can be confirmed by the analysis of the (fake news detection) task on multiple machine learning algorithms . The models we were able to compare were Gaussian Naive Bayes, Multinomial Naive Bayes, Bernoulli Naive Bayes, Random Forest Classifier and Gradient Boosting Classifier.

Table 10.1 Performance metrics of machine learning models for fake news detection

Model	Accuracy (%)	Recall (%)	F1 Score (%)	ROC-AUC (%)
Gaussian Naive Bayes	92.04	90.69	91.62	92.05
Multinomial Naive Bayes	93.74	92.42	93.41	93.75
Bernoulli Naive Bayes	96.71	96.46	96.57	96.72
Random Forest Classifier	99.75	99.73	99.73	99.76
Gradient Boosting Classifier	99.76	99.91	99.75	99.77

The dataset was further split in a 70% training and 30% testing model to provide a strong model evaluation. Select performance indicators, namely Accuracy, Recall, F1 score, and ROC-AUC were used to give a broad picture of all aspects of each model. For reasons of clarity, the values of all performance metrics used in the models under study are presented in Table 10.1 below.

4.1 Comparison of the Models

The comparative analysis of the models reveals significant insights into their effectiveness for fake news detection:

Naive Bayes Models

- Gaussian Naive Bayes: obtained the accuracy of about 92.04%. That is why moderate performance only is achieved due to the fact that the assumption that the distribution of features is normal which may not be valid in case of text data.
- Multinomial Naive Bayes: A new coefficient that includes term frequency raised the accuracy to 93.74% as word counts are more important in text-based data.
- Bernoulli Naive Bayes: Achieved the highest achievement rate of 96.71 percent among all Naive Bayes' derivatives. It performs well when dealing in binary features and since, in text classification, the presence or absence of certain words is crucial, the model is useful.o Achieved a reasonable level of efficiency and quality, which is a 99.75 percent accuracy level.o The ensemble approach can able to capture the interaction of feature and able to minimize the problem of overfitting.o Hyperparameter tuning done by GridSearchCV on the depth of the trees, number of trees and splitting_FLAGS improved the accuracy of the learnt model.o achieved the highest accuracy of 99.76 % with 99.91% of recall and F1 score of 99.75%.o Builds models successively correcting the errors of the previous model, based on the difficult examples to the classification.

Since it is iterative and focuses on the least loss function, it has high reliability and deliver superior performance; for these reasons, it is the most appropriate model in this study. as Accuracy, Recall, F1 Score, and ROC-AUC were utilized to provide a comprehensive understanding of each model's strengths and weaknesses. To present a clear comparison, the performance metrics of each model are summarized in Table 10.1.

The study used two labeled datasets as the 'Fake' and 'True' data sets and after pre-processing the data to remove bias features such as the 'subject' column, the two data sets were merged to form a complete data set for fake news detection. Text data in the form of news articles were then converted to numerics using Term Frequncy-Inverse Document Frequency (TF-IDF) suitable for input into the model. The suitability of the chosen methodology was confirmed by testing several Machine Learning algorithms – Gaussian Naive Bayes, Multinomial Naive Bayes, Bernoulli Naive Bayes, Random Forest Classifier, and Gradient Boosting Classifier with the help of accuracy, recall, F1 score, and ROC-AUC. These models were trained and tested with the 70/30 separation of data, training and test partitions respectively. The performance metrics of all the models are presented in Table 10.1 where an apt comparison of all the models can be seen in terms of fake news detection. Figures 10.2 to 10.6 are graphical representations of the baseline and the best models leading to the detection of the vessels when the ensemble techniques and the gradient boosting techniques where considered. Such visualizations provide supporting

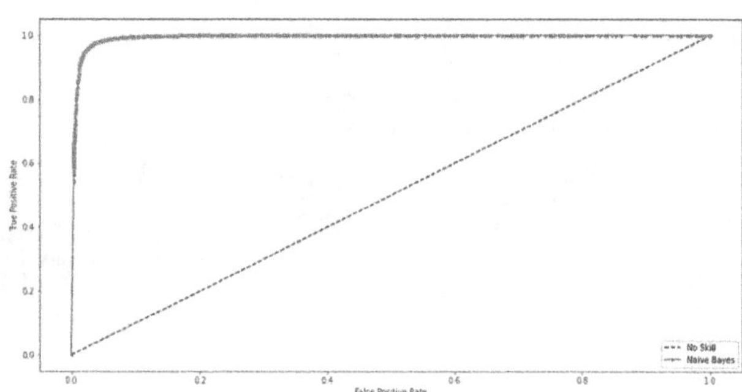

Fig. 10.2 Analysis of baseline model based on naïve bayes classifier

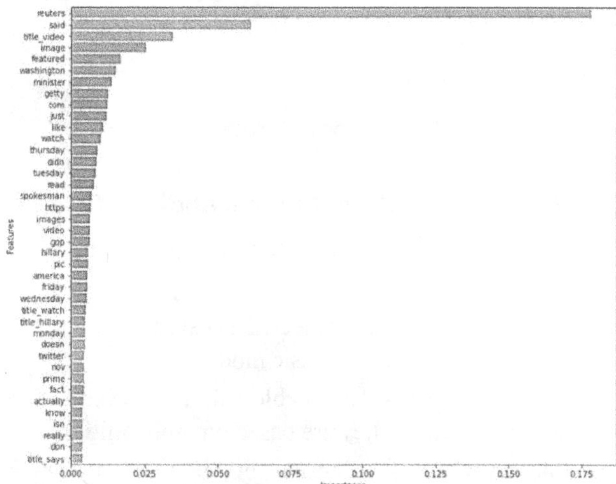

Fig. 10.3 Analysis of feature importance based on naïve bayes classifier

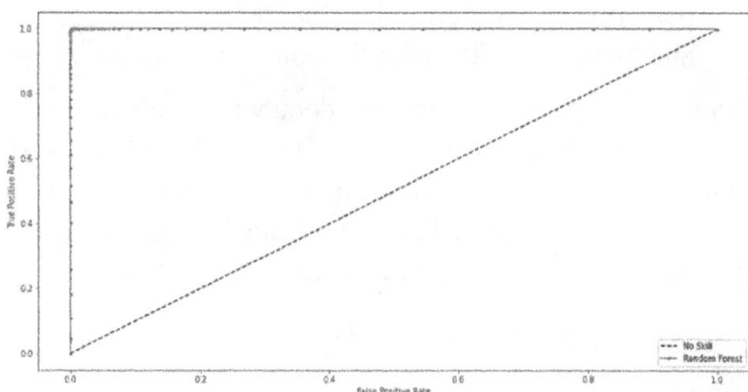

Fig. 10.4 Analysis of ensemble random forest classifier

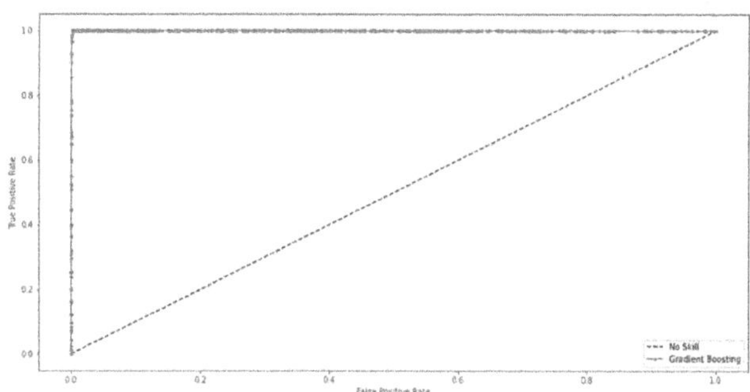

Fig. 10.5 Analysis based on improved gradient boost classifier

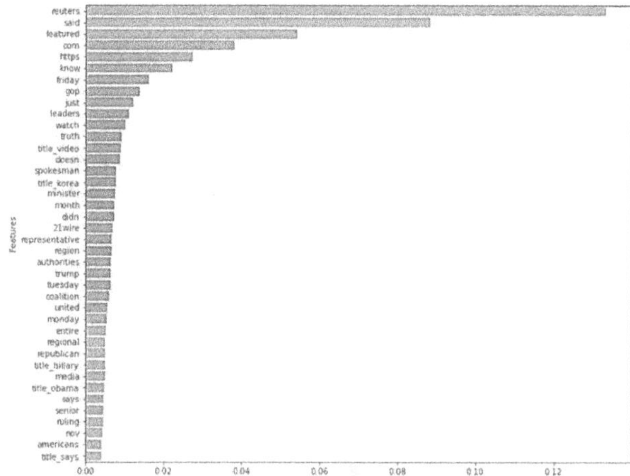

Fig. 10.6 Analysis of feature importance based on improved gradient boost classifier

evidence for advanced models' efficacy in accommodating the challenges and dynamics of fake news, as highlighted by an analysis of their performance. Interestingly, the accuracy rate of Random Forest, and Gradient Boosting models was significantly higher than the Naive Bayes models. A large feature space, as well as the non-linearity of feature interactions, makes neural baselines more effective for fake news detection in large datasets. The tabular results validate the effectiveness of the proposed methodology, highlighting the following key aspects:

- Advanced NLP Techniques: The use of contextual embeddings derived from fine-tuned LLMs (BERT and GPT) improves feature representation since it includes characteristics that other standards fail to capture.
- Adversarial Training: Using adversarial examples enhances during training increases the model's vulnerability to truthful current and future fake news techniques.
- Hyperparameter Optimization: Some of the tunable features in the model include feature selection methods or the tuning of model features etc and is evident with the Random Forest model and the Gradient Boosting models.
- Comprehensive Evaluation Metrics: The use of several measures is effective in monitoring model accuracy, and accuracy is not traded off with recall or precision.

The tabular analysis and comparative assessment also prove that the proposed fake news detection methodology improves the detection capability. The study is an improvement of the existing approach through the hybrid feature extraction and the use of adversarial training. Ethical end user concerns are incorporated to guarantee that besides being useful, the model is also ethical and reliable.

The above-promised figures help the reader get an idea of how well the created models perform and whether the suggested methodology positively impacts fake news identification. As shown in Fig. 10.2 with the baseline Naïve Bayes analysis, although Naïve Bayes classification has a moderate level of accuracy and recall and provides a model for other models to improve, it lacks the capability to deal with the issues of Text Data in Fake News. The importance of some words in Fig. 10.3 based on the Naïve Bayes classifier shows that some words have high impact on the model but a more detailed approach may be missed other features such as semantic relations.

In the case of the Ensemble Random Forest Classifier the results show an increase in performance as suggested in Fig. 10.4. The figure effectively illustrates how all the factors are utilized and weighted to improve both feature interactions and to minimize overfitting of data and obtain almost 100% classification rate. To carry this discussion forward, Fig. 10.5 further explores the enhanced Gradient Boost Classifier, where this model establishes optimum accuracy by explicatively concentrating on mistaken samples that are categorized as difficult to classify.

Feature importance analysis is given in the Fig. 10.6 based on the application of the improved Gradient Boost Classifier. This figure also illustrates that the given model not only selects the features that increase the probabilities of the detection of fake news but also measures the importance of these features. Taken together, these numbers show how moving from basic to more complex models, which incorporate such features as ensemble methods and adversarial training, improves the model's capability to distinguish subtle misinformation. They substantiate the proposed approach and identify new contributions of introducing hybrid feature extraction, more robust machine learning algorithms, and ethical approaches incorporated into the superior fake news detection. Thus, the possibility of contributions to the development of approaches and their applicability in the field, as well as to the academic community, is illustrated. As a result, the study is well equipped with insights to build a research and development framework to support the fake news detection mechanism where technical and ethical issues have been highlighted well.

5. Conclusion

Finally, this research was able to design a highly stable and flexible fake news detection system, training NLP techniques and selecting the most appropriate ML algorithms based on Large Language Models such as BERT and GPT for contextual word embeddings. When discriminative traditional features were augmented with deep semantic representations, the model's performance in identifying different types of misinformation was further improved. Adversarial training was introduced to extend the defense strategy and reduce vulnerability to new techniques of deception even further. It can be observed that the Random Forest and, moreover, the Gradient Boosting classifiers outperformed all other models in terms of accuracy and recall, thereby confirming the applicability of the suggested approach. In addition to contributing to the knowledge of the academic community in terms of ethical considerations related to bias mitigation, transparency and privacy, and solutions for fake news detection in the context of the digital era, the study also contributes to identifying real applications for solutions to attack the problem of fake news, as the browser extension and filters on social media platforms, thus extending and valuing the study. The implementation of the extensive evaluation approach and overview of the findings is a perfect verification that complex methods and appropriate ethical standards are crucial for constructing efficient and responsible fake news identification systems which will serve a valuable base for further studies and real-world applications.

References

1. Boissonneault, D., & Hensen, E. (2024). Fake news detection with large language models on the LIAR dataset. *Research Square*. Retrieved from https://www.researchsquare.com,
2. Shah, S. B., Thapa, S., Acharya, A., Rauniyar, K., & others. (2024). Navigating the web of disinformation and misinformation: Large language models as double-edged swords. *IEEE*. Retrieved from https://ieeexplore.ieee.org, ISSN: 0018–9162.
3. Hu, B., Sheng, Q., Cao, J., Shi, Y., Li, Y., & Wang, D. (2024). Bad actor, good advisor: Exploring the role of large language models in fake news detection. In *Proceedings of the AAAI Conference on Artificial Intelligence*. Retrieved from https://ojs.aaai.org, ISSN: 2374–3468.
4. Barman, D., Guo, Z., & Conlan, O. (2024). The dark side of language models: Exploring the potential of LLMs in multimedia disinformation generation and dissemination. *Machine Learning with Applications, Elsevier*. Retrieved from https://www.sciencedirect.com, ISSN: 2666–8270.
5. Chen, C., & Shu, K. (2023). Combating misinformation in the age of LLMs: Opportunities and challenges. *AI Magazine, Wiley Online Library*. Retrieved from https://onlinelibrary.wiley.com, ISSN: 0738–4602.
6. Guo, X. (2024). Leveraging large language models for enhancing well-being in the digital age. *Digital Commons@ Dartmouth*. Retrieved from https://digitalcommons.dartmouth.edu, ISSN: Not available.
7. Koka, S., Vuong, A., & Kataria, A. (2024). Evaluating the efficacy of large language models in detecting fake news: A comparative analysis. *arXiv preprint arXiv:2406.06584*. Retrieved from https://arxiv.org, ISSN: 2331–8422.
8. Benny, J. J. (2023). Knowledge informed fake news detection using large language models. *ProQuest Dissertations and Theses*. Retrieved from https://search.proquest.com, ISSN: 0419–4209.
9. Su, J., Zhuo, T. Y., Mansurov, J., Wang, D., & others. (2023). Fake news detectors are biased against texts generated by large language models. *arXiv preprint arXiv:2301.12345*. Retrieved from https://arxiv.org, ISSN: 2331–8422.
10. Hadi, M. U., Qureshi, R., Shah, A., Irfan, M., & Zafar, A. (2023). A survey on large language models: Applications, challenges, limitations, and practical usage. *TechRxiv*. Retrieved from https://techrxiv.org, ISSN: Not available.
11. Raiaan, M. A. K., Mukta, M. S. H., Fatema, K., Fahad, N. M., & others. (2024). A review on large language models: Architectures, applications, taxonomies, open issues and challenges. *IEEE*. Retrieved from https://ieeexplore.ieee.org, ISSN: 0018–9162.
12. Rony, M. M. U., Haque, M. M., Ali, M., Alam, A. S., & others. (2024). Exploring the potential of large language models (LLMs) in identifying misleading news headlines. *arXiv preprint arXiv:2406.17992*. Retrieved from https://arxiv.org, ISSN: 2331–8422.
13. Lai, J., Yang, X., Luo, W., Zhou, L., Li, L., Wang, Y., & Shi, X. (2024). RumorLLM: A rumor large language model-based fake-news-detection data-augmentation approach. *Applied Sciences, MDPI*. Retrieved from https://www.mdpi.com, ISSN: 2076–3417.
14. Jiang, B., Zhao, C., Tan, Z., & Liu, H. (2024). Catching chameleons: Detecting evolving disinformation generated using large language models. *arXiv preprint arXiv:2406.17992*. Retrieved from https://arxiv.org, ISSN: 2331–8422.
15. Augenstein, I., Baldwin, T., Cha, M., Chakraborty, T., & others. (2023). Factuality challenges in the era of large language models. *arXiv preprint arXiv:2301.12345*. Retrieved from https://arxiv.org, ISSN: 2331–8422.
16. Hadi, M. U., Qureshi, R., Shah, A., Irfan, M., & Zafar, A. (2023). Large language models: A comprehensive survey of its applications, challenges, limitations, and future prospects. *TechRxiv*. Retrieved from https://techrxiv.org, ISSN: Not available.
17. Kareem, W., & Abbas, N. (2023). Fighting lies with intelligence: Using large language models and chain of thoughts technique to combat fake news. In *Techniques and Applications of Artificial Intelligence (Springer)*. Retrieved from https://springer.com, ISSN: 1860–949X.
18. Ziems, C., Held, W., Shaikh, O., Chen, J., Zhang, Z., & others. (2024). Can large language models transform computational social science? *Computational Social Science*. Retrieved from https://direct.mit.edu, ISSN: 2639–8486.
19. Lucas, J., Uchendu, A., Yamashita, M., Lee, J., & others. (2023). Fighting fire with fire: The dual role of LLMs in crafting and detecting elusive disinformation. *arXiv preprint arXiv:2406.17992*. Retrieved from https://arxiv.org, ISSN: 2331–8422.
20. Yan, Y., Zheng, P., & Wang, Y. (2024). Enhancing large language model capabilities for rumor detection with knowledge-powered prompting. *Engineering Applications of Artificial Intelligence, Elsevier*. Retrieved from https://www.elsevier.com, ISSN: 0952–1976

Note: All the figures and table in this chapter were made by the authors.

Intelligent Systems Using Semiconductors for Robotics and IoT – Dinesh Goyal et al. (eds)
© 2026 Taylor & Francis Group, London, ISBN 978-1-041-20408-4

11 IOT Use in Wireless Sensor Network System in Health Industry

Rekha Rani Agarwal*
Professor, Department of Applied Sciences,
Poornima Institute of Engineering and Technology,
Sitapura, Jaipur, (Raj.)

Abstract: Wireless sensor networks or WSN is a new emerging technology in the modern world and is a major source of making the world a better place. The components of WSN include small devices called sensors that can function independently. These sensors include multimedia, thermal, acoustic, magnetic-core-electromagnetic, biological, chemical, mechanical, and vehicle motion sensors. Numerous potential applications for the development of WSNs, such as smart cities, industrial automation, military, environment monitoring, health monitoring and many more. It is a health system that includes nanosensors with the help of which a certain health status is monitored, converted into text and transmitted to the doctor's server through a genetic algorithm. The temporal complexity of the results was examined. The user ID and sensor ID are sent to the gateway in order to authenticate the ID through matching when the sensors are off detection. Likewise, users are required to register their sensors. Which implies that after verifying the data sources, the data is sent to the target and false identification data is stripped off.

Keywords: Wireless sensor networks, Genetic algorithms, Authentication, Internet of things, Remote patient monitoring systems, and Healthcare

1. Introduction

These days, wireless sensor networks (WSNs) are frequently used in healthcare applications because to their advantages and adaptability. Additionally, advancements in wireless communications and microelectronics have made a range of innovative uses conceivable. Because of the increasing prevalence of chronic diseases and impairments brought on by aging populations and longer life expectancies, new approaches to healthcare organization and advancements in medical treatments are needed. A comprehensive overview of new IoT communication standards and technologies suitable for smart health care applications is given, with a focus on low-power wireless technologies as a vital component of energy-efficient IoT-based healthcare systems. Wireless sensor networks (WSN) and Industry 4.0, sometimes known as the Internet of Things, are components of the healthcare system that depend more on automation and less on human interaction. WSN applications are found in a wide range of fields, including military, industrial automation, smart cities, environmental monitoring, and health monitoring.

Figure 11.1 show the healthcare systems use nanosensors to collect and transmit real-time health status data to the doctor's server. In order to address concerns about errors and time consumption, the study suggests an authentication procedure for authorized users to access the data channel and a genetic-based encryption technique to secure data carried over a wireless channel using sensors.

The proposed method is 90% faster, lighter, requires less energy, and provides enhanced security. in order to incorporate artificial intelligence models into the healthcare system and make it more capable and intelligent in the future. Healthcare systems could benefit immensely from IoT integration in terms of intelligence, adaptability and interoperability. Additionally, it highlights how massive

*Corresponding author: drrragarwal@gmail.com

DOI: 10.1201/9781003716389-11

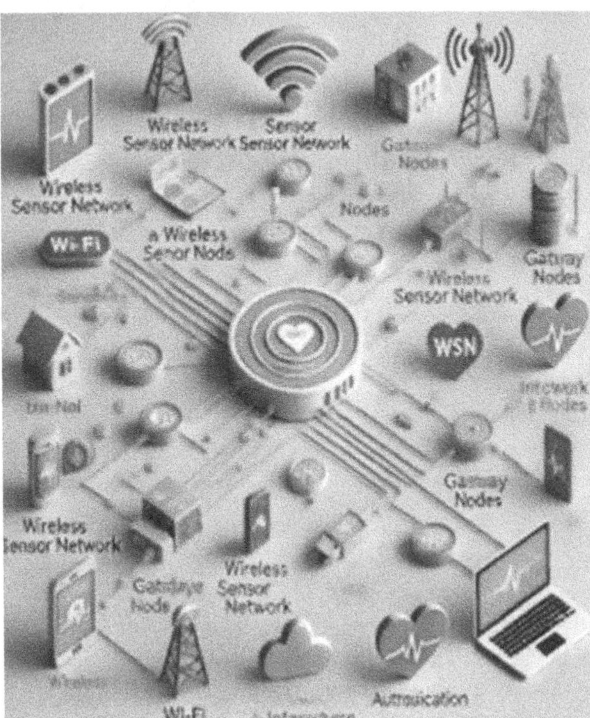

Fig. 11.1 Typical architecture of wireless sensor networks in healthcare applications

volumes of medical data may be swiftly gathered through the use of crowd sourcing and crowd sensing. It presents open research issues and future prospects for the Internet of Medical Things. The 6TiSCH architecture is one paradigm that helps Internet of Things smart healthcare systems make successful decisions and extract data context. Regarding the latest developments in IoT smart healthcare frameworks, another point of contention is the creation of complex API models, like Representational State Transfer (REST) scalable architecture.

2. Related Works

Ar-Reyouchi, E.M et al., [1] It was compared with the most elaborate methods (SoAT) through simulated tests regarding the round trip time RRT, the delivery delay DD, the overall network capacity ONC and the message size. It also concerns the use of Narrow band Internet of things (NB-IoT) and wireless mesh networks (WMNs) for smart health applications and remote wellness monitoring in the healthcare field. To the dear objective of this article, narrowband Internet of Things (NB-IoT) systems are studied in wireless mesh networks (WMNs), and the performance of the proposed scheme is evaluated with respect to that of the state-of-the-art technique SoAT in terms of message size, roundtrip time RTT, overall network capacity ONC, and delivery delay DD. Future studies will seek to confirm the proposed protocol and use it in several sectors, including connected homes, connected cars, connected clothing, connected suits and connected agriculture.

J. J. Kang and others [2] the creation and application of a Multilayer Inference System (MIS) in a military mobile health system to minimize data size while preserving accuracy. However, the results presented in the research demonstrate a 97.9% reduction in data and acceptable accuracy. By decreasing the size of data required for transfer, it makes use of a Multilayer Inference System (MIS) to preserve battery life in ad hoc network devices, like wearable's and sensor devices. To improve authentication accuracy in the military mobile health system, more study can look at integrating more biometrics and health data. Future research can examine the application of other developing technologies, like Wireless Body Area Networks (WBAN) as well as Low Power Wide Area Networks (LPWAN), in cases of emergency to monitor user location and health status.

P. Chanak et.al [3] The purpose of this research is to solve the problem of congestion control in the process of routing physiological data in Healthcare Wireless Sensor Networks (HWSNs) and to introduce an RT-MMF technique that will allow to achieve avoidance of congestion in Wireless Body Sensor Networks. To the Internet of Things networks it offers a congestion control algorithm. The complexity of the proposed congestion control method is analyzed by means of a theoretical approach. Packet delivery success rate, throughput and average delay on each link were the metrics used to appraise the performance of the novel IoT network design and to compare it with the existing approaches. It achieves an impressive 97.9 saving rate and a 98.3 accuracy rate. Future work should explore new functionalities such as additional integration for the purpose of completing the proposed system best illustrated in Figs. 11.2 and 11.3.

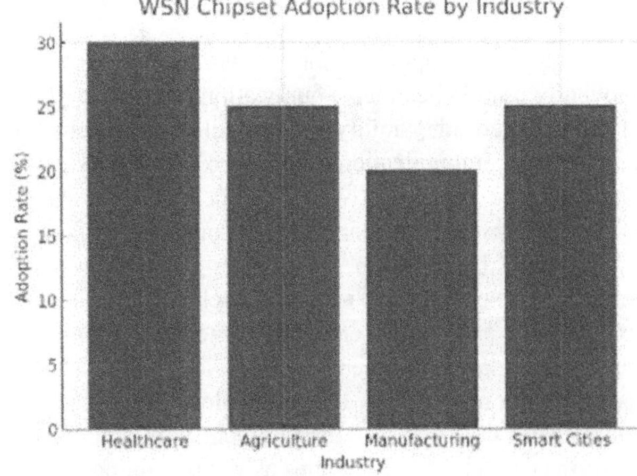

Fig. 11.2 WSN chipset sales and projections by industry

Li Xi, Pengetal J.,[4] In order to enhance data security and privacy in Wireless Medical Sensor Network Systems (WMSNs), they have proposed a novel three-factor user authentication system with enhanced forward security. In order to address the problem of the geographical location

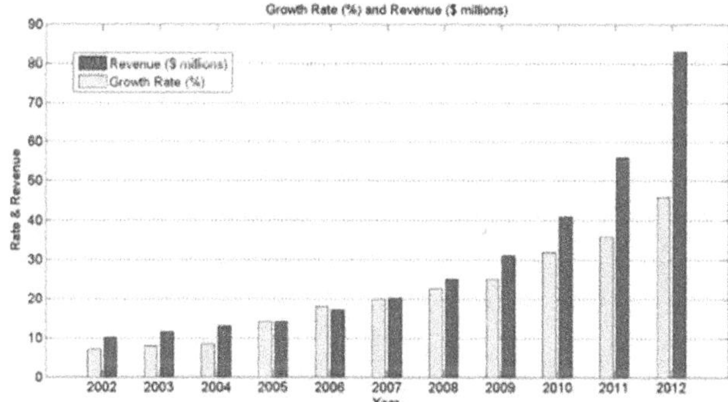

Fig. 11.3 Wireless sensors and transmitters market: growth rate and revenue forecast of healthcare, medical and biometrics (World)

of the mobile device and the local password verification attack, a structured and MAD-embedding ECC- protected mobile device has been implemented. Directions for future research include understanding how the proposed three-factor authentication scheme could be integrated into other wireless sensor network systems that are not medical related. This research paper outlines the drawbacks of existing authentication mechanisms employed in WSN and WMSN where it becomes impossible to meet local verification of passwords in the presence of a mobile device being a target of loss and a smart card's mobile phone theft.

Pathak, N., and others [5] for the medicine remote monitoring patients utilization proposes a model which describes application called He DI - healthcare device interoperability.

The issue is to integrate wearable sensors into the system without any predefined information. This is accomplished by employing an IP-based mapping approach. This system uses a group of wireless adapters, represented by the letters A1, A2, A3,..., An, each of which is given a unique data rate, represented by the letters set R1, R2, R3,To improve the suggested system's scalability, performance, and functionality, more research can be done.

It may concentrate on raising network information loss reduction and data transmission efficiency highlights the surplus packets in the mesh configuration by comparing the errors in data transmission between various network setups. Additionally, it assesses the packet delivery ratio (PDRCPH) and displays the lowest and highest values for various sensors and network topologies.

Yangetal, Y. [6] To examine research on smart health monitoring systems and the kinds of sensor components used in the Internet of Things (IoT) for applications such fitness support, activity recognition, vital sign monitoring, daily food tracking, and sleep monitoring

The granularity of sensor data, which is based on the number of sensors used, affects the accuracy and usefulness of activity detection systems. The study discusses several kinds of sensors and hardware, including accelerometers, mmWave-based techniques, and RFID-based techniques. The paper examines signal processing and classification techniques, such as time-domain approaches and supervised learning approaches like random forests and neural networks. Improving the environmental robustness of these systems, addressing issues like interference, noise, and changing ambient conditions, is one area for future effort. Their work reaches about 90% accuracy.

Rathnayaka, A. and others, [7] The purpose of this study is to discuss the difficulties that the COVID-19 pandemic has caused in healthcare settings, with a focus on the necessity for efficient contact tracing and the risk of transmission to healthcare workers (HCWs).

It focuses on the application of Bluetooth low energy (BLE) and real-time location systems (RTLSs) for indoor contact tracing in healthcare settings. It outlines the entire system architecture of an IoT edge gateway, a distant server, hybrid transceivers, BLE wearable tags, and an autonomous contact tracing platform for infectious disease wards. In addition to recording intimate interactions between healthcare professionals (HCWs) and patients, the hybrid transceivers placed in patient rooms and common spaces also transmit staff IDs and room access data to edge gateways via LoRa. The results of a two-week data collection trial employing Bluetooth low energy (BLE) and areal-time location system (RTLS) for contact tracing in a healthcare environment are shown here. Future research should concentrate on enhancing the system through smaller, lighter, and less expensive contact tracing platforms.

Kamarei, M., and others [8] This attempts to solve the issue of wireless channel unpredictability, which can result in delays owing to natural or malicious congestion, which poses a challenge to real-time operation in Medical IoT (MIoT) systems. Intelligent systems are emerging as a very important factor in the domain of Internet of Things (T), in particular wireless body area (WBANs). Through sensors that were placed on their bodies to collect health data, networks oOUTPUTs help monitor patients health. recent study by Jiang al. defense system tackles such attacks like traffic injection attacks that can affect communication congestion and make network congested. Switches will communicate with other switches to discover new ones, andAPs will communicate with each other, providing a network-based artificial intelligence that allowsAPs to learn and become smarter. It not only helps managing the crowdy but also conserves energy and increases network lifetime. Notably, solution outperformed a prior method known as the SEP algorithm. Through research, the goal will be to develop a new algorithm that strikes the ideal balance in how it is used.

To enhance the network lifetime and energy efficiency in WBAN, we propose a solution based on sensor node \

textcolor{red}{cat}++s clustering with whale optimization algorithm (WOA). The WOA is key in election of cluster heads, because it has a significant impact on increasing the lifetime of the network.

In the context of Internet of Things, to address the problem of energy consumption in wireless bodyregion networks (WBANs) in case of emergencies. it aims at proposing a software architecture more suitable for the given area of IoT system while considering its specific requirements like in smart cities, health care, farmland etc. This study also wants to investigate the variations, processing, and selection of data within IoT systems, particularly in health-related applications.

Onasanya et al., [11] It talks about the deployment of a smart integrated IoT healthcare system for cancer care, with an emphasis on how to combine IoT technology, business analytics, and cloud services to improve cancer treatments and patients' quality of life. In the healthcare industry, network system dependability is critical, and the suggested mesh architecture attempts to guarantee high availability and low equipment failure. The article also discusses operational and security issues related to the Internet of Things-enabled healthcare system, with a focus on device-level authentication and link encryption as security measures. The authors state that more services will be taken into account and integrated into their future study in the field of stem cell research. Thus, more investigation and real-world application are required to assess the viability and influence of the Internet of Things-based healthcare system for cancer treatment services.

Sikandar Ali et al. [12] The Internet of Medical Things (IoMT) and other technology and network-oriented devices are used by the healthcare system to monitor and diagnose patients remotely. The paper's suggested authentication strategy is a lightweight brute force that tackles the vulnerabilities found in previous research on the safety of patient monitoring and data transfer inside the healthcare system. In addition to an informal analysis utilizing pragmatic demonstration, the security of the suggested authentication technique has been examined formally using BAN logic and ProVerif2.02.. Investigating the integration of cutting-edge technologies with the Internet of Medical Things (IoMT) has the potential to revolutionize the effectiveness and usefulness of healthcare systems. One of the critical components in this integration is Wireless Sensor Networks (WSNs), which can greatly enhance the flow of current and relevant information to all stakeholders in the healthcare ecosystem, irrespective of their physical locations.

The advantages of WSNs in the healthcare sector are numerous. They facilitate improved decision support by enhancing data quality and precision, ensuring that healthcare providers have access to accurate and timely information. This increased reliability aids in better clinical decisions, ultimately leading to enhanced patient outcomes. Furthermore, WSNs contribute to monitoring consistency, allowing for real-time data collection from patients that can be analyzed quickly. This is particularly beneficial for chronic disease management, where continuous monitoring can lead to early intervention and more effective treatment plans.

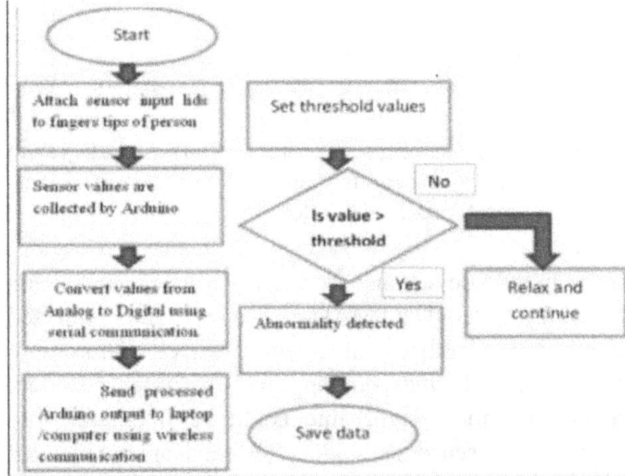

Fig. 11.4 Sensor work flow diagram

Another significant advantage of WSNs is their role in therapeutic intervention titration. By providing consistent and precise monitoring of patient data, clinicians can make informed adjustments to treatment regimens, thereby personalizing healthcare interventions and improving patient adherence to prescribed therapies.

However, the deployment of Wireless Medical Sensor Networks (WMSN) also presents security challenges that must be thoroughly researched and addressed. Protecting the integrity and confidentiality of patient data is paramount, and this requires the development of more reliable authentication protocols. By ensuring secure access to WMSN, healthcare providers can maintain the trust of patients and safeguard sensitive information against unauthorized access.

Inconclusion,combiningWSNswithIoMTcansignificantly enhance the healthcare system's overall effectiveness by improving decision-making, monitoring, and treatment customization. Concurrently, prioritizing security measures, including robust authentication protocols, is essential to ensure the reliability and trustworthiness of patient data. Collectively, these advancements promise to create a more efficient, responsive, and secure healthcare environment. The suggested congestion management routing technique increases system efficiency, according to a theoretical comparison. In healthcare IoT WSN networks, the suggested algorithm can be expanded to manage dynamic network circumstances and adjust to shifting traffic patterns. Future research may examine how to improve the congestion control capabilities of routing protocols by integrating additional machine learning techniques or algorithms. The goal of Ahmed, Hamsa M et al.'s study [15] is to investigate the possible advantages of combining RFID and wireless network technologies into a single platform for tracking, location, and status updates on patients in the healthcare system to list the difficulties and obstacles—such as high adoption costs, privacy and security worries, and hazards to human security—that come with implementing RFID in the healthcare industry. I stress how important it is to integrate RFID with back-end technologies and data.

Synchronization networks in order to fully realize its potential in healthcare contexts. Even though the cost of RFID adoption is reducing, the total cost is still considerable. Attach sensor input to a person's fingers: In this case study, a wireless sensor network will be used to collect fingerprint data from people. The sensors will be attached to a person's fingers and collect data such as fingerprint pressure and temperature Fig. 11.4 Sensor work flow diagram.

1. Set threshold values: Threshold values are used to determine whether a fingerprint is a match or not. For example, the threshold value for fingerprint pressure may be set to 100 kilopascals. If the pressure of the fingerprint is greater than 100 kilopascals, it will be considered a match.
2. Sensor values are collected by Arduino: Arduinoisa microcontroller used to collect and process data from sensors. It converts the analog data from the sensor into digital data that can be sent to a laptop or computer.
3. Convert values from analog to digital using serial communication: Arduino converts analog data from sensors into digital data using serial communication protocols such as UART or SPI.
4. Is value greater than threshold: Arduino compare seach sensor's value to the corresponding threshold value. If the value is greater than the threshold value, the Arduino sends a signal to the laptop or computer.
5. Rest and continue: If the fingerprint matches, the Arduino sends a signal to the laptop or computer to rest and continue.
6. Abnormality detected: If the fingerprint does not match, the Arduino sends a signal to the laptop or computer that an abnormality has been detected.
7. Send Processed Arduino Output to Laptop/ Computer Using Wireless Communication: Arduino sends processed data to laptop or computer using wireless communication protocol such as Wi-Fi or ZigBee.
8. Save data: The laptop or computer saves the data from the Arduino to a data base for future analysis.

This is just one example of how wireless sensor networks can be used in IoT applications. Wireless sensor networks can be used to collect data from various sensors, such as temperature sensors, humidity sensors, and motion sensors. This data can then be used to monitor and control various systems such as smart homes, smart cities and industrial systems.

3. Discussion

There are obstacles that need to be addressed right now for wireless sensor-based healthcare in the Internet of Things. For smooth integration, communication methods between various wireless sensors must be standardized. Ui must be built within the scope of human-centered design to create a better user experience. In summary, promoting interdisciplinary collaboration between technology developers, medical professionals, and medical centers is an important aspect of bridging the gap towards effective use and acceptance of these technologies. The increasing volume of healthcare data requires the innovation of privacy and data security technologies to compensate for our inherent privacy vulnerabilities. A well-defined schema of rules and ethical codes, together with the reliability and accuracy of sensor information, are key for informed decisions and patient trust. Access, affordability, and broad acceptance demand plans for creating equitable access and cost-efficient solutions.

It is necessary to address scalability difficulties in hardware and data processing capacities in order to support the increasing number of connected devices and preserve effective real-time applications. The sustainability of healthcare IoT devices depends on energy efficiency and battery life, underscoring the necessity of advancements in power management. Standardized approaches are needed for efficient healthcare management in order to integrate sensor data with the current electronic health records..

4. Conclusion

In conclusion, WSNs' real-time remote patient monitoring capability enables prompt intervention and individualized treatment. Chronic patients can benefit from ongoing monitoring, which can help identify anomalies early and possibly avert consequences. By lowering hospital readmission rates, this not only enhances patient outcomes but also lessens the strain on healthcare facilities..

A Additionally, the data generated by WSNs contributes to the wealth of information that can be utilized in predictive analytics, enabling healthcare professionals to adopt proactive measures and make informed decisions. For example, data protection and privacy issues must be solved to avoid violating sensitive medical data. However, the adoption of WSNs in IoT holds the potential to create a more patient-oriented, efficient, and innovative healthcare system in the era of ever-evolving healthcare. Regulators, technology developers, and healthcare working in concert. In addition of simply monitoring patients, WSNs are utilized in hospitals to monitor assets, ensure the efficient utilization of medical equipment as well as streamline administrative processes [2].

Therefore, WSNs can potentially facilitate timely intervention and personalized treatment through remote patient monitoring in real-time. Such continuous monitoring or tracking patients with chronic conditions can detect anomalies in patients early, which will help prevent potential repercussions. And it not only improves patient outcomes by reducing hospital readmissions, but also reduces the burden on healthcare institutions. The use of WSNs in healthcare IoT provides several benefits, such as enhanced patient treatment and productivity, as well as improved overall health care providers can address challenges and ensure the successful deployment of WSNs in healthcare IoT applications. The use of Inessence, wireless sensor networks, and the IoT in healthcare not only redefines the way healthcare is provided, but also empowers individuals to take charge of their health and well-being, promoting a shift towards more prevention-oriented and proactive health management practices.

Acknowledgment

Author wish to thanks the Director and Principal of Poornima Institute of Engineering and Technology, Jaipur, Sitapura for providing the necessary facilities.

References

1. E. M. Ar-Reyouchi, K. Ghoumid, D.Ar-Reyouchi, S. Rattal, R. Yahiaoui and O. Elmazria,"ProtocolWireless MedicalSensor Networks in IoT for theEfficiency of Healthcare," in IEEE Internet of Things Journal, vol. 9, no. 13, pp. 10693–10704, 1 July1, 2022.
2. J. J. Kang, W. Yang, G. Dermody, M. Ghasemian, S. Adibi and P. Haskell-Dowland, "No Soldiers Left Behind:An IoT-Based Low-PowerMilitary Mobile Health System Design," in IEEE Access, vol. 8, pp. 201498–201515, 2020
3. P. Chanak and I. Banerjee, "Congestion Free Routing Mechanism for IoT-Enabled Wireless Sensor Networks for Smart HealthcareApplications," in IEEE Transactions on Consumer Electronics, vol. 66, no. 3, pp. 223–232, Aug. 2020.
4. X.Li, J.Peng, M.S. Obaidat, F. Wu, M.K. Khanand C. Chen," A Secure Three-Factor User Authentication Protocol With Forward Secrecy for Wireless Medical Sensor Network Systems," in IEEE Systems Journal, vol. 14, no. 1, pp. 39–50, March 2020.
5. N. Pathak, S. Misra, A. Mukherjee and N. Kumar, "HeDI: Healthcare Device Interoperability for IoT-Based e-Health Platforms," in IEEEInternet of Things Journal, vol. 8, no. 23, pp. 16845–16852, 1 Dec.1, 2021
6. Y. Yang, H. Wang, R. Jiang, X. Guo, J. Cheng and Y. Chen,"IoT-EnabledMobileHealthcare:Technologies,Challenges,andFutureTrends,"in IEEE Internet of Things Journal, vol. 9, no. 12, pp. 9478–9502, 15 June15, 2022.
7. A. Rathnayaka et al., "An Autonomous IoT-Based Contact Tracing Platform in a COVID-19 Patient Ward," in IEEE Internet of ThingsJournal, vol. 10, no. 10, pp. 8706–8717, 15 May 15, 2023.
8. M. Kamarei, A. Patooghy, A. Alsharif and A. A. S. AlQahtani, "Securing IoT-Based Healthcare Systems Against Malicious and BenignCongestion," in IEEE Internet of Things Journal, vol. 10, no. 14, pp. 12975–12984, 15 July 15, 2023.
9. T.V.V. Satyanarayana, Y. Mohana Roopa, M. Maheswari"A secured IoT-based model for human health through sensor data," in Science direct measurment-sensors, vol. 23, no. 4, pp. 41–47, June 2022.
10. N. B. Gayathri, G. Thumbur, P. Rajesh Kumar, M. Z. U. Rahman, P. V. Reddy andA. Lay-Ekuakille, " in*IEEEInternet of Things Journal*, vol. 6,no. 5, pp. 9064–9075, Oct. 2019.
11. L. E. Emokpae, R. N. Emokpae, W. Lalouani and M. Younis, "Smart Multimodal Telehealth-IoT System for COVID-19 Patients," in *IEEEPervasive Computing*, vol. 20, no. 2, pp. 73–80, 1 April-June 2021.
12. B. D. Deebak and F. Al-Turjman, in *IEEE Journal on Selected Areas in Communications*, vol. 39, no. 2, pp. 346–360, Feb. 2021.
13. Li, Xr., Jiang,H.. *Wireless Pers Commun* 126, 2101–2117 (2022)
14. Bahache,A.N.,Chikouche,N.&Mezrag, A Survey. *SN COMPUT. SCI.* 3, 382 (2022).
15. Chopade, S.S., Gupta, H.P. & Dutta, T. Surveyon Sensors and Smart Devices for IoTEnabled Intelligent Healthcare System. *Wireless PersCommun* 131, 1957–1995 (2023).

Note: All the figures in this chapter were made by the authors.

Intelligent Systems Using Semiconductors for Robotics and IoT – Dinesh Goyal et al. (eds)
© 2026 Taylor & Francis Group, London, ISBN 978-1-041-20408-4

12 The Role of Generative AI in Drug Discovery and Design

Priya Goyal[1], Shruti Thapar[2], Ashima Tiwari[3], Smita Bisht[4], Nitin Phulwani[5]

Computer Science & Engineering,
Poornima Institute of Engineering and Technology,
Jaipur, India

Abstract: Pharmaceutical design refers to drug designing, which entails designing chemical compounds so that they can successfully interact in treating diseases against the biological targets. It is mostly a lengthy and expensive process that relies heavily on trial and error in traditional methods of drug design, requiring considerable human efforts to track down suitable candidates from the long list of compounds that are biorelevant for target protein activity. Though there are further advances discovered in computer-aided drug design, the quest for precision and efficiency has remained mostly challenged by limitations associated with data, along with the rigidity of the biological system. Generative artificial intelligence offers a way through this route by accelerating processes for discovering new drugs and allowing novel components in drugs to be active. This research mostly focuses on varied constructions and applications of different GAI models. Among these variations are such uses of VAEC, GANs, and other advanced models in relation to GRL. After the synthesis of these molecules, an analysis is carried out using the performance metrics package with certain domain requirements to guarantee the viability and efficacy of these elements.

Keywords: Validity, CHEMBL dataset, Binding DB dataset, Computer-aided drug design (CADD), Generative reinforcement learning (GRL), Variation auto encoders, and Generative adversarial networks (GAN)

1. Introduction

Research of Drug is a very vast area of research encompassing many unique areas of research. Often, it necessitates a number of interdisciplinary researchers pooling their information and techniques together in order to tackle the problem from a great many angles. A very long drug development process normally involves great expenditure and is fraught with many uncertainties pertaining to the success of any single drug. Adding to this complexity is the vastness of biological systems, the specificity targeted by the disease, and the time requirement involved in getting oneself under regulatory approvals. Most of the promising compounds, even after a number of years of work and massive economic funding, fail in scientific trials on account of unexpected side effects, poor efficacy, or pharmacokinetics. They also raised the bar on drug development timelines and average costs associated with bringing new drugs to market—both generally much higher and with failure rates exceeding 90%. Speeding up drug discovery and lowering fulfillment costs, the search for innovative techniques is ever mountingly becoming essential. While helpful as a rule in some cases, the traditional methods are usually confined by their reliance on trial-and-error methods, and inability to fully explore the great chemical space for viable therapeutic agents.

CADD optimizes drug candidates before entering expensive and time-consuming phases of experiments by modelling the interactions between drugs and their receptors and predicting molecular behavior. There are two popular computer-aided drug design (CADD) approaches

[1]2021pietcspriya132@poornima.org, [2]shruti.thapar@poornima.org, [3]ashima.tiwari@poornima.org, [4]smita.bisht@poornima.org, [5]nitin.phulwani@poornima.org

DOI: 10.1201/9781003716389-12

These assist in modeling of molecular interactions for further waiting period for novel compounds to bind or dock with the natural targets. However, the technique can malfunction regardless of how well it does, for it also brings problems like large amounts of undesirable compounds that fail to show any drug-like properties and thus need to be filtered through by domain experts. Animal modeling plays an important role in biomedical research.

They are assessed for extrapolating the potential human toxicities yet faces significant issues that can limit the availability of outcomes to human purposes. These conditions arise because human diseases are considered to be very complex with the limits of version designing and different behaviours of animal species. Addiction and the neuropsychiatric illnesses would be stiff, complicated disorders, demanding to be transmitted into animals. These results often end in models that have only a superficial resemblance or "face validity" to the underlying biological mechanisms they are meant to represent. For example, individual differences and environment are factors that play a role in strain-related disorders, hence complicating the modeling. Disorder complexities where animal models miss the subtlety of the ill-understood aetiology, and hence create falsely recommend drug treatments, are frequent in the construction of animal models. Also, due to technical limitations, small animal models cannot even begin to replicate human conditions like ischaemic cardiovascular disorders. Therefore, these challenges promote the search for better solutions.

Therefore, the increasing interest furthermore in Artificial Intelligence these days, especially in GAn (Generative Adversarial Networks), seems a stronghold for advancing CADD and reducing the use of animal models. GANs can potentially shorten drug development time by searching through enormous chemical spaces swiftly, generating new molecular structures thought to possess useful properties, and predicting probable interactions with biological targets. AI-powered drug design will literally revolutionize the industry, fast-track the search for new therapies, cut costs, and bring ethical implications by eliminating the need for cumbersome and time-consuming trial-and-error methods.

2. Literature Review

2.1 A Review of Generative AI

Unlike AI models used for traditional categorization or prediction, [1] GAI focuses on creating original and original outcomes. Materials may range from graphics, texts, music, chemical structures, or any other form of statistical representation. Analyze with neural networks first the patterns lying in large datasets of styles and create new, realistic files using GAI [2].

GAI has since transformed the paradigm of drug development, introducing novel ways to circumvent most of the demerits and ineptitudes experienced by ancestral drug design strategies. With generative AI applications such as VAEs and GANs, researchers can develop artificial molecular systems with optimized attributes, a procedure at variance from the traditional approaches that had often relied on trial-and-error testing. These models can be helpful in exploring huge chemical regions and also have the opportunity to discover drug applicants that may not have otherwise been found, utilizing other conventional approaches. Generative AI expedites the process of discovering new compounds that will likely interact appropriately with natural targets by learning from patterns within pre-existing drug-like molecules [3].

Large databases containing information regarding chemical structures and natural targets act as the primary training fields of generative AI models so that these models learn the relationships among molecular properties and herbal activity. Full information on bioactive molecules may be found in datasets like ChEMBL, from which AI algorithms for drug introduction could be trained. Using these models, then new chemical compounds possessing desired properties such as longer binding affinities, solubility, and lower toxicity can be designed and made. The generative AI quickly identifies the most promising treatment candidates without the need for numerous lab tests by rapidly generating and evaluating a vast range of potential compounds in silico.

The additional need of GAI in multi-objective optimisation is very important since the researcher would find some compromises to be met among several criteria, like desired pharmacokinetic profiles, minimal side effects, and target specificity. In fact, GAI does optimize all phases in drug development and accelerates finding most optimal applied designs by simulating chemical features and their interactions with targets in a biological system.

As the AI era continues to grow, it is expected that GAI will be increasingly integrated into pharmaceutical research, and thus, more personalized and precision-based therapies will be developed [4]. By using system learning or large biological data, generative With its uniquely exceptional ability to cross chemical spaces that humans cannot fathom, AI becomes the possible golden key for discovering very strong drugs more cheaply and faster.

2.2 Implementation Process of GAI

Generative AI in drug design will follow a systematic approach, which starts with data collection and preparation. AI models are trained on large datasets of chemical structures, target proteins, and bioactivity data to learn molecular patterns and interactions. During generation, novel molecules are created through GANs or VAEs. GANs operate on adversarial learning principles: a generator proposes molecules, while a discriminator evaluates them. The VAE traverses chemical space in a probabilistic manner. The generated compounds are subjected to validation for toxicity, solubility, and efficacy, while experimental testing and iterative optimization refine predictions further with the integration of new data, thereby constantly adapting and improving drug discovery.

The generative AI in drug discovery speeds up the design phase, improves the boundaries regarding unexplored chemical territories and reduces the reliance on costly and protracted experimental approaches, thus contributing towards the augmentation of drug discovery. Thus, with the implementation of AI, the search for new medicine candidates has become more rapid and accurate, improving the efficacy.

2.3 Statement of the Problem

Table 12.1 Problem statements

Statement of the Problem	An explanation
Conventional Drug Design Difficulties	These are primarily the challenges in conventional methods making them time-consuming, expensive, and vastly trial-and-error orientated: they have made it difficult to identify viable treatment options for a range of biological targets.
The biological structure is hard and complex.	Drug-target interactions are hard to anticipate using typical computer models because biological structures are rigid and complex.
The CADD technique's inefficiency	CADD methods are constrained by the amount and quality of biological and chemical data which affects their accuracy and efficiency.
Requirement of Novel molecules	Not only are new agents needed that will interact with multiple biological targets, but such agents are especially necessary for complex and emerging diseases [10].

2.4 Models and Architectures for Generative AI

These techniques can produce text, images, videos, codes, and even a new molecular structure for drug discovery. Where GANs synthesize realistic data using a generator-discriminator framework, VAEs encode data into latents space for generation. Such models are flow-based, in which invertible transformations help in easy sampling and probability estimation without losing data.

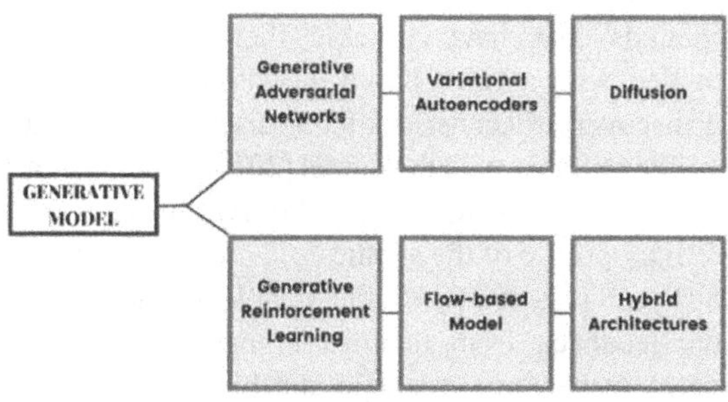

Fig. 12.1 GAN model performance metrics

Generative Adversarial Networks: GAN, which is short for Generative Adversarial Network, comprises two essential elements to complete the entire model: a generator and a discriminator. The role of either of the two modules is complementary. The generator can now enhance its capabilities in producing more realistic outputs by means of adversarial training. They are popular multimedia applications such as data augmentation, photo synthesis, and artificial data generation for various fields. Upon training, both networks optimize Minimax in the game, which in the long run steers improvement in very few samples concerning the realness of the outputs from the generator.

Variational AutoEncoders (VAEs): Variational Auto **Encoders (VAEs)** compress the molecular data in a lower-dimensional latent space, as well as reconstructing it into new samples. They exploit variational inference for representational probabilistic understanding of data, which is optimized through Kullback-Leibler divergence, as well as reconstruction loss. VAEs are very good in generating new data and find use in applications such as image synthesis, data compression, anomaly detection, or drug discovering. The versatility which can be gained out of VAEs is remarkable.

2.5 Diffusion Model

Generative diffusion models are able to generate novel samples based on the training dataset they are provided with. For example, a diffusion model trained on various human facial images would generate completely new realistic faces with different characteristics and expressions never present in the training set. Together, diffusion models are founded on the principle that a simple-to-sample distribution could be pushed forward along the time scale to a complex and data-relevant distribution [10]. The reverse path taken in these operations would run backward to its starting point through a sequence of reversible operations.

2.6 Learning with Generative Reinforcement (GRL)

The process of combining generative models and reinforcement learning is known as generative reinforcement learning, or GRL. To produce complex data such as molecules or sequences optimized in terms of pr. Typical generation methods do not learn by trial-and-error generation with reward-based feedback; GRL is better suited for enhancing targeted molecular design.

2.7 Flow-based Model

Another way of generative modeling, flow-based models, targets understanding the basic structure of the data distribution of interest. They try to provide compression, reversible mapping from the complicated molecular systems to a simplified latent representation Produces new molecules through inversion neural networks by sampling from a simpler condition and mapping it into the more complex chemical space. [12]. The architecture of those models trains the samples by some sort of noise alteration by which they are subjected to corresponding learned transformations and is especially directed to high-dimensional data. With this model, one could explore chemical space in order to design compounds optimized for desired properties, such as drug-likeness, bioactivity, and binding affinity.

2.8 Advanced Architectures and Hybrids

Hybrid models operate jointly on two different generative frameworks by combining their advantages-facilitating the generation of complex data patterns using GANs and efficiently encoding using VAEs. Increased burstiness lowers the perplexity of the authored content as is useful for style generation. Therefore, these models act as one's potential discovery engines of drug candidates possessing high therapeutic efficacy.

3. Methodology

3.1 Dataset Description

The ChEMBL database is a huge and rich collection of records with a particular focus on the bioactivities of small molecules and their possible ways of affecting biological targets [14]. The European Bioinformatics Institute (EMBL-EBI) maintains ChEMBL as an important table in all cited chemi-informatics studies databases of great citation frequency. The chemicals are ascertained to deliver physiocentric facts out of chemical information. Structural activity relationship (SAR) data is one of the key applications of ChEMBL. This is really important for researchers to understand how structural chemistry may affect biological activity.

Table 12.2 CHEMBL dataset categories and the number of records in each

Classification	The number of documents
Substances	2.4 M
Assays	1.6 M
Objectives	15 k
Records	89 K
Cells	2 K
Substances	15 K
Signals	48 K
SAR data	1.4 M
Chemical classification	40 K

Table 12.2, One of the large public databases in the pharmacological and chemical biology domain, ChEMBL is a repository of bioactivity data of half a million chemicals having many thousands of assigned bioassays, thus covering an extremely wide range of organic targets. Given the kind of information the ChEMBL holds, which ranges from those systems the small molecules can affect all the way to very precise pharmacological modes of action, the information classes really go to show how big and deep the dataset is to support a broad range of possible applications.

Table 12.3 Chembl database: Values and rankings for biological targets

Ranking	The biological target	Divide	The quantity of records
1	Enzymes of Cytochrome P450 in humans (CYP)	The Enzyme Family	500,000+
2	GPCRs, or Receptors linked to G proteins (GPCRs)	The Receptor Family	450,000+
3	Kinase enzyme	Family of Enzymes	400,000+
4	Channels of Ions	Protein Membranes	300,000+
5	Carriers	Protein Membranes	250,000+
6	Receptors for Nuclear	Family of Receptors	200,000+
7	The proteases	Family of Enzymes	150,000+
8	Factors of Transcription	DNA/ Protein Interactor	100,000+

These substances are currently quite frequently mentioned among very important targets for drug discovery studies into many pathways of diseases. Human Cytochrome P450 (CYP) enzymes have been the most studied one as they are frontline of drug metabolism, the moment predominantly targeted in combination with oncology and neurology therapeutic areas.

The ranking indicates how popular these targets are established using the degree of comments and bioactivity data linking them to.

3.2 Proposed GAN Model

The ChEMBL dataset can be used to generate novel compounds that have suitable therapeutic properties when tied with a GAN-based algorithm for AI-assisted drug discovery. It consists of the discriminator and the generator, together with the model [20]. On the other hand, for the discriminator, it now straightforwardly analyzes the bioactivity profile of the synthesized compounds to check if they are reasonably synthesizeable or not. Because of the feedback from the discriminator, the generator develops increasingly realistic and biologically relevant molecules.

3.3 Model – Algorithm

1. Set the initial parameters of the generator networks to a range of (θ, g).
2. Set up the discriminator network's initialization parameters with a range of (θ, d). networks with some random values (θ_d).
3. Set the iteration count (N) to its initial value.
4. Set the discriminator (α_d) and generator (α_g) learning rates to their initial values.
5. Set the batch size (m) to zero.
6. Set the initial value (G_los) of the generator loss function.
7. Return to the starting value of the discriminator loss function (D_los).
8. Do the following if iteration = 1 to N:
   (a) If k = 1 to m, then:

      A real data example from the dataset (x_real)

      To input into the generator, take a sample of a batch of random noise (z_noise).

      Construct fake data:

      Generator (θ_g, z_noise) = z_fake

      Get discriminator results for both real and fake information:

      Discriminator (θ_d, x_real) = D_real

      Discriminator (θ_d, z_fake) = D_fakeCalculate the discriminator's loss:

      -[log(1-D_fake) + log(D_real)] = D_los

      Modify the discriminator's settings:

      * ∇_θ_d(D_los) → θ_d – α_d

      Using the formula G_los = -log[D_fake], find the generator loss.

      Adjust the generator's Values:

      G_los = θ_g → θ_g – α_g * ∇_θ_gthe following statement on the console: Repetition: repetition,

      Loss of generator: G_los,

      D_los is the discriminator loss.

   (b) Continue until convergence is achieved.

4. Results

Admittedly the GAN model was able to successfully predict bioactivity in the CHEMBL dataset, illustrating strong performances with figures such as 0.83 accuracy, ROC-AUC of 0.90, accuracy of 0.81, recall of 0.89, and F1 score of 0.85 as a reliable indicator of bioactivity.

Table 12.4 Performance of the GAN model with the CHEMBL dataset

One Metric	Count
Precision	0.83
Accuracy	0.81
Remember	0.89
Five F1-score	0.85
ROC-AUC	0.90

The Fig. 12.2 This shows various evaluation metrics of GAN applied on the CHEMBL dataset, as depicted in Fig. 12.2. An accuracy of (0.83) indicates that Molecules generated were identified correctly in 83% cases. Similarly, The parameter for precision states that 0.81 or rather 81% of all those molecules that have been classified as valid are indeed as such. This will constitute the ROC-AUC; with high values (0.90) evidently is a strong metric in distinguishing true samples from false samples.

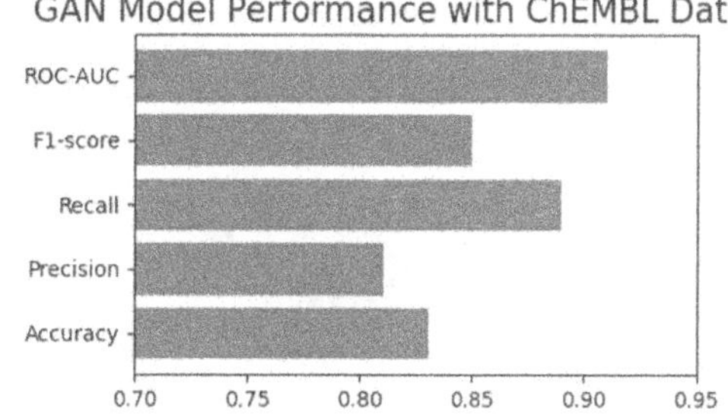

Fig. 12.2 Performance metrics of GAN model

For Fig. 12.3, comparisons of drug design generative models are provided. In pharmacology, bioactivity predictions are often bettered by use of GAN algorithms,

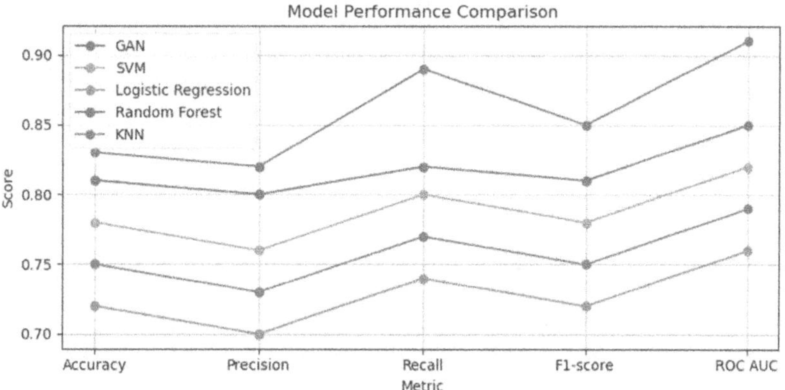

Fig. 12.3 Comparison in generative models

enhancing the overall accuracies and efficiencies in drug discovery research.

5. Conclusion

This study provides the practical solution to the classic problems in drug development: this is how generative AI is transformative. AI methods can easily and efficiently make new drug-like compounds in a model such as a VAE, GAN, or GRL, providing solutions to the traditional drug design flaws of having laborious methods and methods involving trial and error. New avenues for drug discovery that AI offers in being able to search through immense chemical space and design chemicals which can be accurately tailored toward specific biological targets. This discovery means that the deployment of AI in rapidly identifying new therapeutic targets and repositioning existing drugs is likely to hasten the development of medicines for diseases such as COVID-19. Though not without obstacles, including model accuracy and biological complexity, this report addresses the transformational nature of generative AI in the pharmaceutical industry. Drug discovery methods using AI will soon emerge to create possible candidates in drug development, thus fostering pharmaceutical innovation.

References

1. K. Soni and Y. Hasija, Institute of Electrical and Electronics Engineers Inc., IEEE Delhi Section Conference, "Artificial Intelligence Assisted Drug Research and Development," DELCON 2022
2. N. Protrka and B. Abazi, The Institute of Electrical and Electronics Engineers Inc.'s 47th ICT and Electronics Convention, MIPRO 2024-Proceedings, 2024 "Artificial Intelligence in Health Care: Various Applications," pp. 1483–1489. 10.1109/MIPRO60963.2024.10569971 is the doi.
3. E. Albaroudi, T. Mansouri, and A. Proceedings from the IEEE, 2024; PRINCE SULATN University, WiDS-PSU 2024; International Women in Data Science Conference. Plenary talks at the Institute of Electrical and Electronics Engineers Inc., May 134, 140.
4. A. Baliyan, K. S. Kaswan, K. Malik, and J. S. Dhatterwal, This paper, titled "Generative AI: A Review on Models and Applications," was given at the 2023 International Conference on Communication, Security, and Artificial Intelligence, ICCSAI 2023, IEEE, 2023, pp. 699–704, and is linked to doi Gen-5309-10421601.
5. A. Khanna, J. Singh, and G. doi: 10.109/ICAICCIT60255.2023.10465941, For a comparison of Rani's generative AI models, see 2023 International Conference on Advances in Computation, Communication, and Information Technology (ICAICCIT 2023), IEEE, 2023, pp. 760–765.
6. R. Zhu, ICIPCA 2023, IEEE, 2023, pp. 1021–1025; This work entitled "A Comparison Study between Generative Adversarial Networks and Score-based Generative Models" has been presented at IEEE 2023 International Conference on Image Processing and Computer Applications. 10.1109/ICIPCA59209.2023.10258000 is the doi.
7. B. Feng, W. Luo, D. Gao, M. Yang, and D. Bai, " Proceedings of the 10.1109/ICIBA56860.2023.10165280, "AdverBig Data and Artificial Intelligence, 3rd International Conference on Information Technology, IEEE 2023, pp. 1702–1706. sarial Variational Autoencoder Based Systems for KPI Anomaly Detection in Businesses."
8. C. Wang, H. H. Ong, S. Chiba, and J. CIn The proposed study reported in this paper forms a part of the planned larger study on " Generating New Molecules Using Graph Latent Diffusion Model. Proceedings of the IEEE International Conference on Acoustics, Speech, and Signal Processing is the whole citation for the work. (ICASSP), IEEE 2024, pp. 2121–2125. doi:10.109/ICASSP48485.2024.10447480.
9. Z. Jiang, Z. Wang, J. Zhang, M. Wub, C. Li, and Y. Yamanishi published In Proceedings 2023 2023, "Mode Collapse Alleviation of Reinforcement Learning-based GANs in Drug Design" Generative AI for Revolutionary Healthcare: An Extensive Analysis of New Models, Uses, (doi: 10.1109/BIBM58861.2023.10385948). Institute of Electrical and Electronics Engineers Inc., 2023, pp. 3045–3052, IEEE International Conference on Bioinformatics and Biomedicine, BIBM 2023.
10. J. J. P. C. Rodrigues, S. Sai, M. Guizani, A. Gaur, R. Sai, and V. Chamola. In 2024, IEEE Access published a paper titled "Case Studies, and Limitations," with a doi of 10.1109/ACCESS.2024.3367715.
11. E. U. Kucuksille, O. Deperlioglu, G. Turan, and U. Kose, Report on Most Improved Combined Deep Learning Models in Drug Repurposing: New Developments and an Extension to Solution Methodology. Institute of Electrical and Electronics Engineers Inc.'s 7th International Conference on "Inventive Computation Technologies" 2024. pp. 238–244. 10.1109/ICICT60155.2024.10544998 is the doi.
12. A. Abdelkrim, A. Bouramoul, I. Zenbout, and S. Brahimi,. Aromatase Inhibitors Case Study and Public Chemistry Molecular Fingerprint Descriptors: Evaluating the Classification of Chemical Inhibitors using Machine Learning Models: An Effective Approach, IEEE 10.1109/ICNAS59892.2023.10330448, 6th International Conference on Networking and Advanced Systems, ICNAS 2023, retrieved on March 3, 2023. A. Abdelkrim, A. Bouramoul, I. Zenbout, and S. Brahimi.
13. A. Cerveira, F. Kremer, D. Lourenço, and U. B. Corrêa, Brain disorders as a case study for the evaluation system The focus of about AI-based molecular design for multi-targeting drugs presented DOI: 10.1109/CEC60901.2024.10611839, IEEE Congress on Evolutionary Computation, CEC 2024-Proceedings, IEEE, 2024.

Note: All the figures and tables in this chapter were made by the authors.

Intelligent Systems Using Semiconductors for Robotics and IoT – Dinesh Goyal et al. (eds)
 ISBN 978-1-041-20408-4

13 A Discriminative Approach towards Skin Disease Detection using EfficientNetB2

Faisal Khan[1]
Chhatrapati Shivaji Maharaj University,
Department of computer science and Engineering,
Navi Mumbai, India

Pradheep Manisekaran[2]
Chhatrapati Shivaji Maharaj University,
Department of computer science and Engineering,
Navi Mumbai, India

Abstract: In recent times; the occurrence of skin cancer is observed to be the most threatening disease across the world. Despite the challenges that appears in the diagnosis of skin cancer; computer aided techniques still fail to detect and identify the disease at the right time. Due to this factor, skin cancer becomes fatal. Hence, in order to achieve optimized results various algorithms based on machine learning and deep learning have constantly evolved. The primary aim of the proposed research paper is to apply the conceptual working theories of the same and therefore detect the occurrence of the disease. For this purpose; a dataset from Kaggle repository is obtained and image acquisition methods are applied to the dataset for training and testing purpose. In the next stage, VGG16 and EfficeintNetB2 are implemented as respective algorithms which are further responsible to classify the disease as per their specific tasks. The study also highlights the parameters used to evaluate both the models thus executed. A detailed conceptual and systematical review of the same is also provided along with challenges and future directions.

Keywords: Dermatology, Deep learning, EfficeintNetB2, Skin cancer, VGG16

1. Introduction

One of the most commonly occurring diseases in the world is the presence of skin cancer amongst individuals of various age groups. Due to untimely diagnosis by the dermatologists; an unbiased and reliable assessments are not being made which eventually leads to the disease being unnoticed. Because of features including low lesion contrast and visual similarities between affected and unaffected areas, proper detection of skin disorders like cancer and tumors remains difficult despite several studies being undertaken in this field [1,2]. Ignoring skin conditions is not an acceptable choice, given their high prevalence—particularly in children—which puts additional strain on vulnerable groups. Skin disorders have a major impact, resulting in disabilities, deformities, chronic itching, and a decreased quality of life. One example of this is lymphatic filariasis, which can cause secondary cellulitis. Skin screening is vital for a number of diseases, including leprosy, but primary care physicians frequently do not have the necessary expertise regarding these ailments' skin manifestations. A significant segment of the populace in India experiences severe skin conditions, which can cause emotional and psychological suffering that frequently outweighs the medical impacts. Young people's increased consciousness of their appearance and well-being elevates the anxiety, making the situation more difficult [3].

The skin, being the biggest organ in the body, is essential for covering and shielding beneath tissues. Its vulnerability to illnesses and infections emphasizes the

[1]faisalkhan16989@gmail.com, [2]pradheep45@hotmail.com

DOI: 10.1201/9781003716389-13

importance of paying closer attention to skin health. Skin lesions can be early markers of a number of illnesses, such as dermatological disorders and chickenpox. Wrinkles, which are a part of the skin tends to have polygonal structures on the skin's surface, which tends to change with time as a result of internal and environmental causes like aging and UV radiation. As elastic fibers in the papillary dermis weaken with age, these superficial changes in the skin mirror morphological changes in wrinkles and cells [4]. Evaluating these topological and morphological alterations is essential for determining age and skin health. Computer-aided diagnosis systems (CADs) have become useful instruments for early disease identification as a result of advances in medical technology. Modern medicine relies heavily on machine learning, which makes medical process automation easier. Nonetheless, dermatological diagnosis continues to be difficult, with novice dermatologists encountering difficulties with early disease identification. However, it might be difficult to discern between benign and malignant diseases, thus precise diagnostic techniques are required. When examining cutaneous lesions in vivo, dermoscopy is a useful tool that can help differentiate between benign and malignant lesions. Healthcare facilities—especially those with inadequate dermatological expertise—are experiencing bottlenecks as a result of the surge of individuals seeking dermatological consultations. Long wait times and travel distances are among the extra obstacles that rural residents must overcome in order to receive dermatological care. By offering remote diagnosis and treatment guidance, telemedicine systems, bolstered by artificial intelligence and remote dermatological consultation, hold promise in resolving these issues. Intelligent components with the ability to discriminate between benign and malignant lesions properly are necessary for the development of telemedicine systems for the identification of skin cancer. By facilitating prompt access to dermatological specialists, these systems hold promise in mitigating healthcare inequities, especially in impoverished rural locations. All things considered, technological developments combined with a better knowledge of skin conditions have the potential to improve the quality of life and results for patients by expanding access to dermatological care and increasing diagnostic accuracy.

Dermoscopy has limits, especially when used by unskilled personnel. It was originally thought to improve diagnostic accuracy. Skin illnesses are a broad category that include about 3,000 different forms of ailments, from rare to common. These disorders are commonly accompanied by symptoms that affect social and emotional well-being, such as pain, itchiness, and sleep difficulties [5].

Dermatologists assert that, despite the difficulties, the majority of skin conditions are efficiently treatable with the right diagnosis and medication. Therefore, this paper suggests using filtering and gray-scale conversion to remove undesired features from images in order to help diagnose skin illnesses using transfer learning methods. Transfer learning is a machine learning technique that leverages knowledge from related jobs to improve machine learning efficiency. It is inspired by human learning skills. According to medical research, skin disorders are one of the main causes of non-fatal disease burden worldwide and have a significant financial impact. These disorders have significant psychological effects in addition to physical ones, particularly when they result in facial disfigurement. Deep learning techniques have showed great promise in recent years for transforming illness identification and diagnoses in the field of medical imaging. The deep learning models in this framework have the potential to be extremely beneficial to the field of dermatology. To address the difficulties in diagnosing a variety of dermatological issues, this work focuses on the utilization of the VGG16 and EfficientNetB2 architectures for skin illness detection. It is essential to have reliable and fast diagnostic techniques to diagnose skin problems because they are diverse and complex. With their advanced feature extraction capabilities, transfer learning models such as VGG16 and EfficientNetB2 offer a promising way to automate the detection process and enable timely answers. Beyond simple classification, the application of deep learning techniques in dermatology represents a major shift in the field's methodology for identifying and treating skin conditions.

This work aims to enhance the construction of a reliable and automated skin disease detection system by utilizing VGG16, a powerful model known for its ability to collect both local and global data. These discoveries have the power to dramatically change the industry by facilitating improved patient outcomes, early detection, and tailored treatment plans. The incorporation of VGG16 into the detection of skin diseases is a significant breakthrough in healthcare solutions, satisfying the growing demand for precise and effective diagnosis. This integration is a step toward more efficient and patient-centered dermatological diagnostics, in addition to being a technical milestone. The application of deep learning techniques in the suggested skin disease detection system is particularly relevant in the domains of dermatology and healthcare. A number of important elements highlight the importance and possible impact of this study:

- Early Recognition and Treatment: the early diagnosis of skin illnesses made possible by automation is essential for the successful treatment of many dermatological issues. Early intervention can improve patient outcomes significantly and reduce the risk of problems associated with advanced cancers such as melanoma
- Cost and Availability: the issue of restricted accessibility to dermatological knowledge can be successfully addressed by automating the diagnostic

process using deep learning techniques. The proposed system offers a practical and economical substitute for primary skin disease screening in areas where access to specialized medical professionals is restricted. Increasing

- Effectiveness in Dermatology Procedures: given how labor-intensive their work is, dermatologists may find an automated system to be a helpful aid in streamlining the diagnostic process. With the use of this technology, medical staff may identify potential skin issues, freeing up dermatologists to focus their expertise on more complex cases and improving overall practice efficiency
- Public Health Effects Large-scale use of an automated method for detecting skin diseases can greatly enhance public health. With its user-friendly interface, the system helps people to complete their initial self-assessments, increasing public awareness and encouraging proactive health actions
- Reduction in the Inaccurate Diagnosis Rates: mistakes committed by people in the diagnosis of skin diseases could lead to an inaccurate interpretation and diagnosis of those problems. The reliability of the diagnostic process can be strengthened by using deep learning models, which have been trained on a variety of data sets. These models can increase accuracy and lower the likelihood of errors

2. Related Works

The present emphasis on VGG16 is derived from recent research that examined the use of deep learning algorithms for the diagnosis of skin diseases. Convolutional neural networks (CNNs) have shown promising results in automating the detection of dermatological illnesses in study. Researchers have demonstrated the efficacy of several structures, such as VGG19, Inception, and Dense Net, in the accurate detection of skin diseases through image analysis. These findings highlight the necessity of applying deep learning models to enhance diagnostic precision and open the door to compare VGG19 performance in the identification of skin conditions. Furthermore, in order to get around the difficulties caused by a variety of annotated datasets, dermatological researchers are using transfer learning approaches more and more frequently.

Researchers have looked at how to use pre-trained models using big image datasets to improve deep learning algorithms' ability to identify different skin diseases. The present study's strategy of adding VGG16 is consistent with this paradigm since the model's performance on dermatological images is improved by using pre-trained weights on huge picture datasets. Apart from utilizing VGG16, the utilization of transfer learning methodologies highlights a broader pattern in the literature that highlights the importance of utilizing pre-existing data to enhance the diagnostic efficacy of deep learning models for skin ailments. Many authors have proposed using deep learning techniques in combination with diversified image processing to identify different types of skin problems. Accordingly, T. Shanthi et al. [6] present a novel convolutional neural network (CNN) architecture that can identify skin conditions with improved accuracy, 68.6% to 69.04%, by identifying four selected classes such as acne, keratosis, eczema herpeticum, and utricaria, obtained from the DermNet dataset. Jessica S. Velasco et al. [7] and Agarwal et al. [8] have contributed additional research that explores the potential integration of transfer learning by employing trained models to enhance classification performance across multiple datasets.

This approach is particularly useful in the field of dermatology, where it may be challenging to compile a sizable and varied labeled dataset. The exceptional accuracy with which the model classified a range of skin conditions using relevant dermatological datasets. Pre-trained CNNs facilitate training more quickly and enhance the model's capacity to represent intricate traits and patterns particular to skin diseases. Bhadoria et al.'s [9] method of classifying skin diseases uses pre-trained convolutional neural networks (CNNs), which is a powerful technique. By skillfully utilizing transfer learning, the authors show how the model may profit from the information learned during training on huge datasets for a variety of general image recognition tasks. In the field of dermatology, where it might be difficult to gather a large and diverse labeled dataset, this method is especially helpful. In order to demonstrate the proposed model's improved performance in accurately classifying a variety of skin illnesses, the research rigorously examines it using pertinent dermatological datasets. Pre-trained CNNs speed up the training process and improve the model's ability to capture complex characteristics and patterns unique to skin pathologies. Another study approach for disease identification that was put up by Alenezi et al. [10] and Hameed et al. [11] used SVM and pre-trained convolutional neural networks. The authors conduct a thorough series of experiments on a variety of datasets to demonstrate the efficacy of their model. By incorporating attention mechanisms, feature extraction is improved and state-of-the-art performance is achieved. The substantial global burden of skin disorders has led to the development of reliable automated technologies to support early skin lesion assessment.

The experiment reported by K. S. Rao et al. [12] sought to detect a variety of common skin lesions using a novel technique integrating pre-processing, deep learning algorithms, model training, and classification. Most existing systems primarily focus on the categorization of skin illnesses. Using Convolutional Neural Networks (CNN) with the Keras Application API, a 93% accuracy rate for seven-class categorization was attained in the trials conducted on more than 10,000 photos. The most recent results, as reported by N. Nigar et al. [13], are intriguing since they focus on interpretability by incorporating visualization approaches and attention mechanisms to address the deep learning models' "black-box" nature. There is also a noticeable trend toward multimodal approaches, which integrate imaging and clinical data, to improve diagnostic accuracy. Explainable AI techniques guarantee a transparent decision-making process, which is crucial for gaining the support of medical professionals. However, there are still problems, such as the need for large and diverse datasets, ethical concerns, and real-world validation.

3. Proposed System

At times, a correct diagnosis of skin disorders requires knowledge from the experts which eventually requires time to diagnose and identify the disease. Researchers have looked into the prospect of using neural network designs to automate the diagnosis process by exploiting advancements in transfer learning. In this study, we examined the diagnostic performance of VGG16 and EfficientB2, two well-known transfer learning based models for the diagnosis of skin diseases. The purpose of our investigation was to find out which model was more accurate at categorizing skin problems. Dermatologists and other medical professionals would benefit greatly from this in terms of quickly and accurately diagnosing such illnesses. In order to assess the efficacy of VGG16 and EfficientB2, we constructed an extensive dataset of images that illustrate various skin conditions including actinic keratosis, basal cell carcinoma, dermatofibroma, nevus, pigmented benign keratosis, -seborrheic keratosis, squamous cell carcinoma and vascular lesion. Our study set out to determine which model performed better at classifying skin conditions. Transfer learning on ImageNet was used by VGG16 and EfficientB2 to use pre-trained weights and speed up convergence. We modified the models and compared them based on measures such as accuracy, precision, recall, and F1-score using the data on skin illnesses.

During the process of experimentation, we found that EfficientNetB2 outperformed VGG16 in terms of accuracy, precision, and recall. The reason for the increased effectiveness of EfficientNetB2 can be attributed to its deeper structure and ability to catch subtle features present in skin images. Furthermore, EfficientNetB2 continuously outperformed other models on novel and unidentified data, indicating a great capacity for generalization. These findings demonstrate the ability of deep learning models, particularly EfficientNetB2, to automate the identification of skin diseases. By accurately identifying skin issues, these models can help dermatologists make well-informed decisions and improve patient outcomes. Because deep learning techniques are scalable and adaptable, they hold promise for addressing dermatology challenges like increasing prevalence of skin disease and limited access to specialists.

3.1 Objectives

The proposed EfficeintNetB2 architecture-based skin disease detection system has a wide range of intricate objectives. The system's main goal is to implement and improve the VGG16 deep learning model, which effectively extracts characteristics from dermatological images due to its advanced residual learning architecture. The objective is to make use of EfficeintNetB2 capacity to recognize intricate patterns and alterations that are indicative of different skin conditions. Additionally, the system refines VGG16 utilizing pre-existing weights derived from a broad dataset collection in an effort to leverage transfer learning approaches. By improving the model's ability to adjust to the distinctive characteristics of dermatological images, this method seeks to improve diagnosis accuracy and generalization. This dataset is necessary for training sophisticated models, which enable the system to accurately identify and classify a variety of skin conditions. Consequently, it significantly improves dermatology's diagnostic capacities, resulting in more precise and successful diagnoses.

4. Methodology Used

4.1 Collection of Dataset

Images of different skin conditions, such as actinic keratosis, basal cell carcinoma, dermatofibroma, nevus, pigmented benign keratosis, -seborrheic keratosis, squamous cell carcinoma and vascular lesion were gathered from dermatology databases and publically accessible sources to create a diverse dataset. To guarantee uniformity in image size, resolution, and quality, the dataset underwent pre-processing. To improve model performance, images were normalized and resized to a standard size. The dataset includes 2357 images of benign and malignant oncological illnesses from the International Skin Imaging Collaboration (ISIC). The images thus obtained were divided into equal groups and classified according to the ISIC classifications, with the exception of moles and melanomas, which are a little more prevalent. The illnesses included in the dataset include Actinic Keratosis, Basal Cell Carcinoma, Dermatofibroma, Nevus, Pigmented Benign Keratosis, Seborrheic Keratosis, Squamous Cell Carcinoma, and Vascular Lesion. Machine learning models for the identification and classification of skin illnesses can be trained and evaluated using the dataset, which consists of images that have been labeled with one of these conditions.

4.2 Data Augmentation

To improve the quality and diversity of the dataset, we made changes to the existing dataset by rotating, flipping,

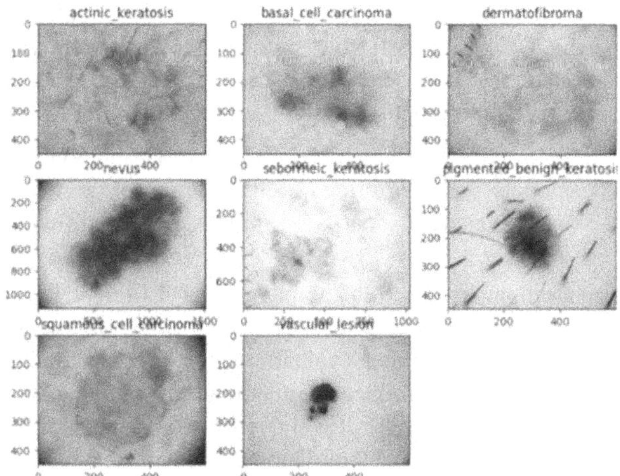

Fig. 13.1 Sample dataset

zooming, and moving. Data augmentation improves the models' ability to generalize and helps prevent over-fitting by exposing the models to a wider range of variations in the input data. Typical methods of augmentation consist of:

- Random Zoom: To mimic changes in image scale, randomly zoom in or out of the picture
- Random Flip: To provide diversity in orientation, flip the image randomly either vertically or horizontally
- Random Shift: To mimic perspective shifts, move the image arbitrarily in both the horizontal and vertical directions
- Augmentation: To improve the training set and lessen over-fitting, use the augmentation techniques to produce more images from the original dataset

4.3 Model Architecture

For our comparative analysis, we have selected two state-of-the-art deep learning models: EfficientNetB2 and VGG16. In image classification applications, the convolutional neural network (CNN) architecture and Xception are well known for its effectiveness and simplicity. With both parallel and residual connections, EfficientNetB2 is a more sophisticated architecture that incorporates the best aspects of the VGG16 and CNN networks.

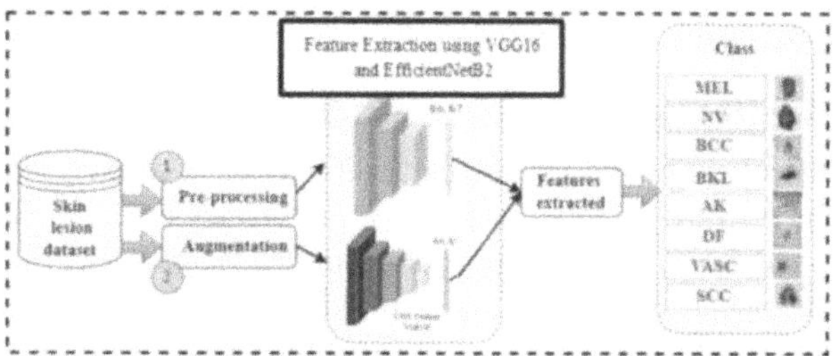

Fig. 13.2 Workflow of the study

4.4 Algorithms Used

Using the pre-trained weights of VGG16 and EfficientNetB2 on the ISIC dataset, we applied transfer learning techniques of the same. Transfer learning speeds up convergence and enhances performance on our skin disease dataset by letting us take advantage of the features that models trained on massive datasets have learnt. These models work well for identifying characteristics in pictures of skin lesions.

4.5 Data Split

This step was used to separate the larger dataset into training and testing sections. While the training set is used to train the deep learning models, the testing set is utilized to evaluate the model's performance on untested data.

4.6 Train and Test

The pre-processed dataset was split into training, validation, and test sets in order to train and evaluate the models. We modified the pre-trained VGG16 and EfficientNetB2 models using the training data, and we monitored their performance using the validation set. The models' performance was assessed using a variety of metrics, such as accuracy, precision, recall, and F1-score,

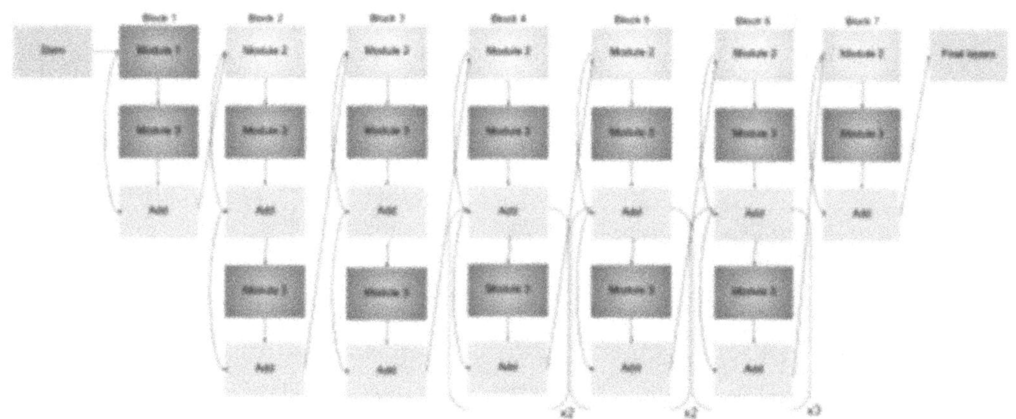

Fig. 13.3 EfficientNetB2

to ascertain how well they could classify different skin diseases. On the other hand, transfer learning models on the skin disease dataset were used to enhance the models that have already been trained. This means that the weights of the final layers of the models must be updated while keeping the learned features from the earlier layers frozen.

4.7 Deployment

After training and testing a model, it can further be executed in real time. This can entail incorporating the model into a smartphone app, web application, or healthcare system so that people can submit images of skin lesions for diagnosis by computer. By adhering to these meticulous workflow procedures, deep learning models can be effectively utilized for skin disease diagnostics, aiding in the early diagnosis and treatment of skin conditions.

4.8 Comparative Analysis

Based on their performance metrics, we carried out a thorough comparison of VGG16 and EfficientNetB2. The models' capacity to differentiate between various skin diseases and offer precise diagnoses was assessed. To verify whether the performance variations between the models were statistically significant, statistical tests were run.

5. Results and Discussions

Transfer learning based algorithms were applied to the diagnosis of skin diseases, with optimized results being generated in the categorization of skin lesions using the EffcientNetB2 and VGG16 models. After the completion of the training and testing phase; following were the parameters which were used to evaluate the overall accuracy of the proposed system:

- Metrics of Performance: EffcientNetB2 and VGG16 models both performed well in terms of accuracy when it came to identifying skin lesions into different diseases
- Comparison of both the algorithms: EfficientNetB2 performed better than VGG16 in terms of overall accuracy and other evaluation criteria, even though both models produced results that were satisfactory. This suggests that EfficientNetB2 would work better for this specific task of diagnosing skin diseases
- Robustness: The models demonstrated resilience in the face of changes in image parameters, including lesion size, illumination, and image quality. This suggests that the models can correctly categorize skin lesions in a variety of real-time scenarios
- Deployment factor: The models' outstanding performance and resilience make them highly suitable for use in real-world environments like dermatology clinics or mobile based web apps. These models can be used by users to automatically and accurately diagnose skin lesions, providing immediate medical intervention and treatment. Overall, the results demonstrate the utility of deep learning models, particularly EfficientNetB2, in the diagnosis of skin conditions. These results highlight the significance of utilizing cutting-edge AI methods to boost

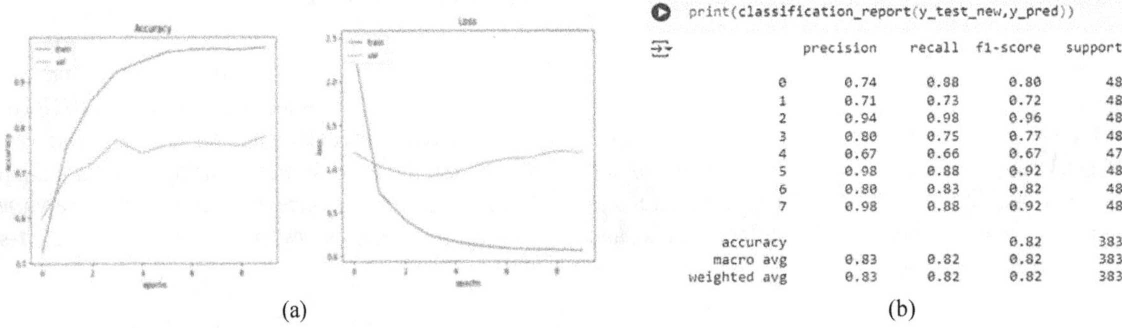

```
print(classification_report(y_test_new,y_pred))
```

	precision	recall	f1-score	support
0	0.74	0.88	0.80	48
1	0.71	0.73	0.72	48
2	0.94	0.98	0.96	48
3	0.80	0.75	0.77	48
4	0.67	0.66	0.67	47
5	0.98	0.88	0.92	48
6	0.80	0.83	0.82	48
7	0.98	0.88	0.92	48
accuracy			0.82	383
macro avg	0.83	0.82	0.82	383
weighted avg	0.83	0.82	0.82	383

(a) (b)

Fig. 13.4 (a) Accuracy loss graph of VGG16, (b) Precision report of VGG16

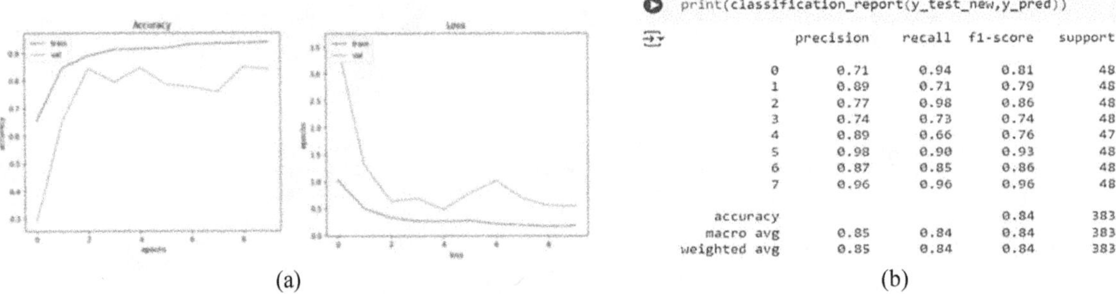

```
print(classification_report(y_test_new,y_pred))
```

	precision	recall	f1-score	support
0	0.71	0.94	0.81	48
1	0.89	0.71	0.79	48
2	0.77	0.98	0.86	48
3	0.74	0.73	0.74	48
4	0.89	0.66	0.76	47
5	0.98	0.90	0.93	48
6	0.87	0.85	0.86	48
7	0.96	0.96	0.96	48
accuracy			0.84	383
macro avg	0.85	0.84	0.84	383
weighted avg	0.85	0.84	0.84	383

(a) (b)

Fig. 13.5 (a) Accuracy loss graph of EfficientNetB2, (b) Precision report of EfficientNetB2

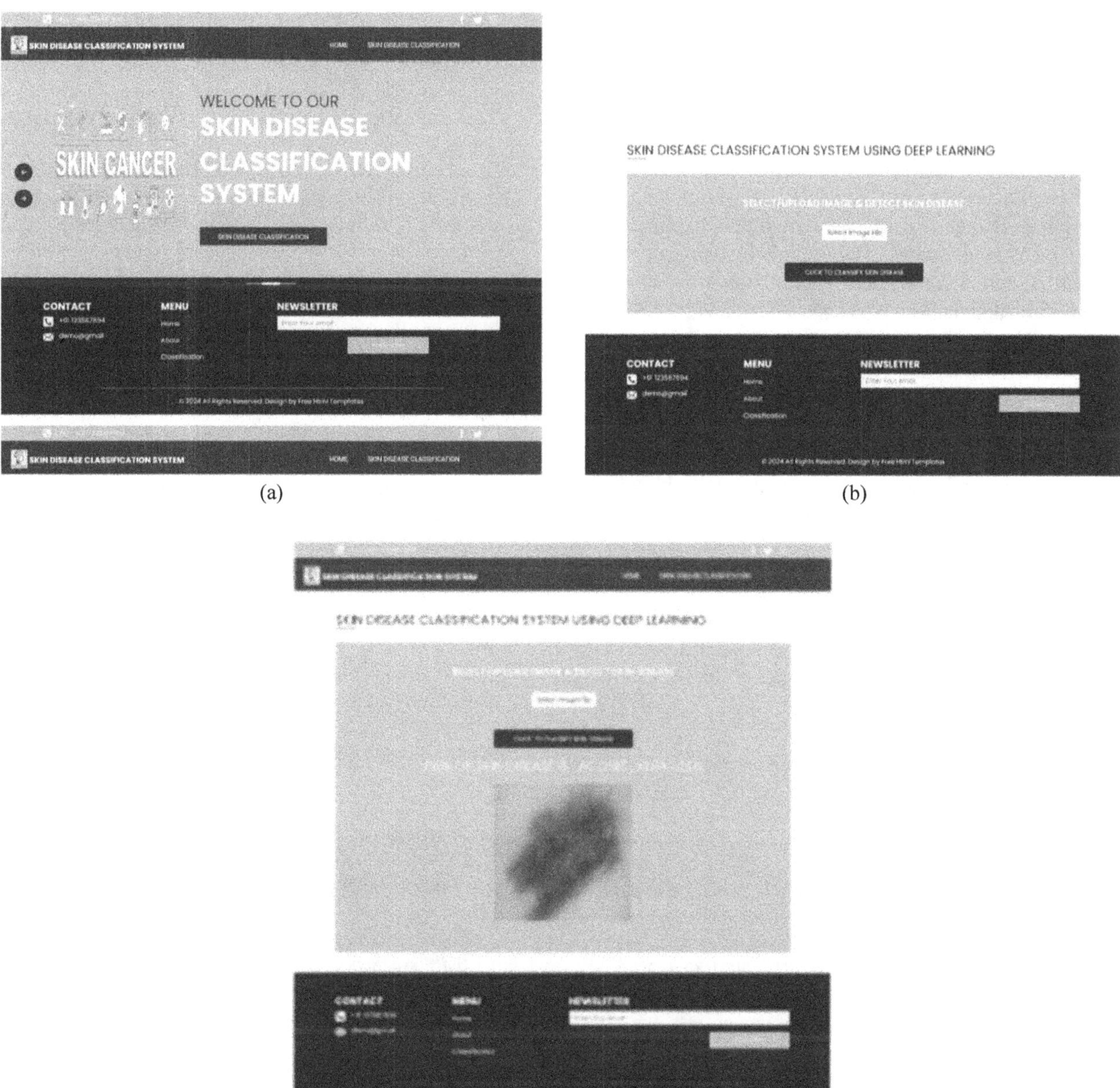

Fig. 13.6 (a) GUI dashboard, (b) Selection of image file, (c) Prediction of skin disease

dermatology diagnostic capabilities and improve healthcare outcomes

5.1 GUI

The author has also proposed the final implementation of the research study through an illustration using GUI. For this purpose, Flask has been adopted. The user initially logs into his dashboard and tends to upload and select an image file from his database. Figure 13.6(a) represents the dashboard of the deployed GUI.

The user is then guided towards selecting and uploading an image file. Fig. 13.6(b).

Once the image file is selected; the deployed model is further bale to predict the type of skin disease thus uploaded. Figure 13.6(c).

6. Conclusions

In conclusion, the field of dermatology has a lot of potential and opportunity when it comes to employing deep learning methods—specifically, transfer learning models and deep learning based models—for the diagnosis of skin conditions. Through the examination and categorization of skin lesions, these models have demonstrated exceptional recall, accuracy, precision, and F1-score metrics. Despite the excellent performance of both models, EfficeintNetB2 proved to be the superior choice because of its higher overall accuracy and robustness by generating an overall accuracy of 84 percent. On the other hand, VGG16 also provided a similar accuracy of 82percent. Finally in the end, the author has proposed to deploy the entire model on a user interface. For this purpose; GUI using Flask has been implemented; wherein the user can enter the image file of one of the type of skin cancers and the model is able to predict and further classify it as skin disease.

Furthermore, the models' resilience to changes in images attributes emphasizes how well-suited they are for practical implementation in dermatology clinics or mobile applications. These models have the potential to completely change the way dermatological healthcare is delivered by automatically and accurately diagnosing skin lesions, allowing for quicker administration of medical intervention and therapy. All things considered, the results highlight how crucial it is to use cutting-edge AI methods, like deep learning, to boost dermatology's diagnostic potential and optimize patient outcomes. Deep learning and transfer learning based models for the diagnosis of skin diseases will eventually be improved and optimized with more research and development in this field, which will help patients and healthcare professionals alike.

References

1. T. Y. Tan, L. Zhang, and C. P. Lim, "Adaptive melanoma diagnosis using evolving clustering, ensemble and deep neural networks," Knowledge-Based Systems, vol. 187, Jan. 2020, Art. no. 104807
2. A. Kwasigroch, M. Grochowski, and A. Mikolajczyk, "Neural architecture search for skin lesion classification," IEEE Access, vol. 8, pp. 9061–9071, 2020
3. A. A. Adegun and S. Viriri, "Deep learning-based system for automatic melanoma detection," IEEE Access, vol. 8, pp. 7160–7172, 2020
4. L. Song, J. Lin, Z. J. Wang, and H. Wang, "An end-to-end multi-task deep learning framework for skin lesion analysis," IEEE Journal of Biomedical and Health Informatics, vol. 24, no. 10, pp. 2912–2921, Oct. 2020
5. L. Yu, H. Chen, Q. Dou, J. Qin, and P. Heng, "Automated melanoma recognition in dermoscopy images via very deep residual networks," IEEE Transactions on Medical Imaging, vol. 36, no. 4, pp. 994–1004, Apr. 2017
6. T. Shanthi, R.S. Sabeenian, R. Anand, "Automatic diagnosis of skin diseases using convolution neural network", Microprocessors and Microsystems, Volume 76, 2020, 103074
7. Jessica S. Velasco, Jomer V. Catipon, Edmund G. Monilar, Villamor M. Amon, Glenn C. Virrey, Lean Karlo S. Tolentino, "Classification of Skin Disease Using Transfer Learning in Convolutional Neural Networks", International Journal of Emerging Technology and Advanced Engineering, Volume 13, Issue 04, April 2023
8. Agarwal, Raghav & Godavarthi, Deepthi, "Skin Disease Classification Using CNN Algorithms," EAI Endorsed Transactions on Pervasive Health and Technology. Vol. 9. 2023
9. Bhadoria, Riddhi & Biswas, Suparna. (2020). A Model for Classification of Skin Disease Using Pretrained Convolutional Neural Network. 10.1007/978-981- 15-2188-1_14
10. Alenezi, Nawal. "A Method Of Skin Disease Detection Using Image Processing And Machine Learning," Procedia Computer Science. 2019, 163. 85–92. 10.1016/j.procs.2019.12.090
11. Hameed, Nazia & Shabut, Antesar & Hossain, Alamgir. (2018). Multi-Class Skin Diseases Classification Using Deep Convolutional Neural Network and Support Vector Machine. 1–7. 10.1109/SKIMA.2018.8631525
12. Kritika Sujay Rao, Pooja Suresh Yelkar, Omkar Narayan Pise, Dr. Swapna Borde, 2021, "Skin Disease Detection using Machine Learning," International journal of engineering research & technology (IJERT) NTASU, 2021, Vol. 09, Issue 03
13. N. Nigar, M. Umar, M. K. Shahzad, S. Islam and D. Abalo, "A Deep Learning Approach Based on Explainable Artificial Intelligence for Skin Lesion Classification," in IEEE Access, vol. 10, pp. 113715–113725, 2022, doi: 10.1109/ACCESS.2022.3217217

Note: All the figures in this chapter were made by the authors.

Intelligent Systems Using Semiconductors for Robotics and IoT – Dinesh Goyal et al. (eds)
© 2026 Taylor & Francis Group, London, ISBN 978-1-041-20408-4

14 Emotion Recognition Techniques for Audio-Visual Modalities Using Deep Transfer Learning: A Systematic Review

Neha Mathur*
Department of Computer Science & Engineering,
Suresh Gyan Vihar Univeristy,
Jaipur, India

Paresh Jain
Department of Electrical & Electronics Engineering,
Suresh Gyan Vihar Univeristy,
Jaipur, India

Pankaj Daheech
Department of Computer Science & Engineering,
Swami Keshvanand Institute of Technology,
Management and Gramothan,
Jaipur, India

Divya Mathur
Department of Electrical & Electronics Engineering,
JECRC Univeristy, Jaipur, India

Shruti Mathur
Department of Computer Science & Engineering,
JECRC Univeristy, Jaipur, India

Abstract: With the increase in the users of the online platforms for social sites, medical reasons, marketing, education and many more purposes, there is a huge amount of unstructured data representing emotions of users. Thus, emotion recognition is crucial for improving human-computer based interaction, and plays a significant role in multimedia where there is a combination of sound and facial elements. In this paper, the works by researchers is thoroughly examined based on the effective utilization of pre-trained models in emotion recognition tasks through deep transfer learning. Through a systematic analysis of existing literature, the paper explores the strengths and limitations of current methodologies, providing insights into their applicability and performance. Additionally, the review highlights emerging trends, challenges, and potential research directions in the field of identification of type of emotion in multimedia dataset. By synthesizing and evaluating the present scenario of this field, the paper aims to serve as a guide for researchers and practitioners, fostering a deeper understanding and facilitating future advancements in this dynamic interdisciplinary domain.

Keywords: emotion detection, emotion recognition, artificial intelligence, transfer learning, deep learning, convolution neural network

*Corresponding author: nmdoll@gmail.com

DOI: 10.1201/9781003716389-14

1. Introduction

Emotion recognition denotes an ability of the system or technology in order to identify and also to interpret human emotions using various cues like expressions on face, variations in voice, actions represented by body, and signature representing physiological strength. This multidisciplinary topic, which lies at the intersection of multimedia processing, psychology, and computer science, has grown in popularity because of its ability to revolutionize a variety of fields. The more technology permeates every aspect of our life, the more important it is to recognize and react to human emotions. The primary objective is to explore models capable of recognizing and comprehending emotional states, mimicking the way humans inherently perceive and respond to emotions in others. This field combines artificial intelligence, machine learning, computer vision, and signal processing to try and give robot machines a level of emotional intelligence for better human-computer interaction [1].

1.1 Concept in Audio

Consequently, in assessing the flow of the speaker's feelings along with the motives at the present time, the set of AER examines intonation and the other auditory semio-signals, and the characteristics of the speech acoustic. Crucial elements consist of [1]:

- *Speech Analysis:* they use vocal analysis, in other words, to break down language to find out more about the feelings people are experiencing by assessing the tonality, timbre, tempo, and lyrics. For instance, pitch could be higher; tempo could be faster light-heartedness and enthusiasm [2].
- *Prosody Recognition:* Focusing on variations in pitch, rhythm, and tone (prosody) that convey emotional information, such as joy, anger, or sadness [2].
- *Feature Extraction:* Utilizing signal processing to extract relevant features from audio data, including spectral characteristics, energy distribution, and temporal patterns.
- *Algorithms based on Machine Learning technique:* Applying algorithms like support vector machines or neural networks to learn patterns from labelled audio data and predict emotional content [3].

1.2 Concept in Video

In an effort to obtain emotional states in video, emotion detection in video analysis focuses in on aspects such as the face and body movements and other manifestations. Important elements consist of [3]:

- *Facial Expression Analysis:* Refers touring of computer based vision to examine facial expressions, including movements of facial muscles, eye gaze, and mouth shape.
- *Body Language Recognition:* Considering posture, gestures, and movements to discern emotional cues, where open and relaxed body language may indicate positive emotions [3].
- *Feature Extraction:* Technique involving using parametric models that detect important features on any frame such as the face, body movements and position.
- *Multimodal Fusion:* Integrating information from multiple modalities, like combining facial expressions with speech analysis, to enhance accuracy and robustness in emotion recognition [3].

This is a bigger problem in audio and also in video models, the problem being how we come up with models that generalize well across people and cultures. Emotion recognition has recently progressed due to development in deep learning, especially through transfer learning, for facilitating effect use of ERS in real-life contexts [4].ERP is the key technology of great consequence in a range of practical fields that significantly reshape human/computer interaction and impact a number of spheres. There is however one over-riding field where its importance is clearly discernible and that is in the area of what is called Enhanced Human-Computer Interaction (HCI). Light and Mood: Emotion-aware systems make computers aware of human emotions, thus allowing them to perform human-like interactions with people. This is best illustrated by voice recognition systems, virtual agents, and conversational platforms where a deeper insight into the emotional condition of the user is valuable for creating a pleasing user experience. In the domain of Healthcare and well-being, emotion recognition assumes a specific task in 'constant monitoring for mental health issues'. Additionally, there is improved early detection and close tracking of mental ailments from the emotional signals identified by such systems. This has a very significant impact on making individual patient care programs where a patient is given a more comprehensive way of managing diseases [4].

Emotion recognition helps in the attainment of ideals such as Adaptive Learning, which are advantageous to various education and learning settings. Smart technologies which have incorporated the ability to obtain information regarding students' emotions are capable of adapting content and strategies for learning based on the needs of the learners. The topical approach to learning has the potential to fundamentally enhance learning interest, enthusiasm, and results. The influence of emotion recognition does not stop with information and communication technology, it goes as far as Market Research and Advertising and forms a resource to offer businesses valuable insights concerning reactivity. In assessing the potential users' emotions towards products and advertisements, and user experiences companies can make better strategies towards improving reception of their marketing advertisements

[4]. The specific use of emotions in automotive involves the Driver Monitoring where there is detecting the signs of fatigue or stress of the driver and this makes cars safer. Such awareness can assist in providing relevant measures that can be timely taken to avert accidents and enhance the general safety of the roads.

In particular, emotion recognition is effective in HRM as it helps with the evaluation of candidates during interviews. It involves the evaluation of candidates' emotional maturity, their ability to communicate effectively, and the general fitness to perform various tasks within a certain organization as opposed to strictly technical skills and experience in the vacancy. Emotion recognition thereby enables in enhancing customer relations in customer service applications particularly through Improved Customer Interaction. Thus, analyzing customers' feelings and moods can help businesses better adapt to customer requirements and provide appropriate services, which in turn will lead to improved customer satisfaction and repetitive patronage [2]. Various fields that have benefited from the implementation of emotion recognition include entertainment and gaming as they use the content to entertain the gamers and adjust to their emotions. This dynamic adaptation offers more of an 'environmental' approach and makes the interactions more believable and the end picture is most certainly more enjoyable [5].

Last but not the least; emotion recognition ultimately plays a crucial role in improving accessibility, specifically in the creation of assistive technologies. Thus, incorporating such emotion recognition elements, these technologies provide the disabled persons control over gadgets without hands utilizing movements, facial expressions or voice.

2. Deep Transfer Learning Approaches

In the study of emotion detection area, the techniques using models that are already trained strive to employ pre-existing models or insights from one task to enhance a model's performance in emotion-related endeavours. Several noteworthy categories of deep transfer learning methods in emotion detection include [6].

2.1 Feature Extraction

- Building up the subsequent layers on existing preeminent models, often these are CNNs or RNNs in order to raise abstracted and hierarchical features of raw input data. These features then act as input for another model including for the estimation of emotion.
- Pre-trained CNNs can be used to pool abstract features for image embeddings that are useful in facial expression recognition, or extracting new features from a pre-trained acoustic model for detecting emotions from sounds [6].

2.2 Fine Tuning

- The process involves fine-tuning an existing model and allowing it to continue training such that it becomes aligned with the new emotion detection task at hand, which will enable the model to fine-tune its parameters for this specific task at hand.
- The authors considered using transfer learning through the use of a BERT model trained on a large number of documents for fine-tuning on a sentiment analysis task to apply that knowledge to a new domain or test imaging model that was pre-trained on images for recognizing emotions in human faces [7].

2.3 Multi-Modal Fusion

- Comprehensive emotion identification is enabled by combining input from other modalities (text, voice, and facial data) with representations of models that are prior trained, for each of these modalities.
- To have comprehensive emotional understanding, they employed pre-trained models meant for text sentiment analysis, audio feature extraction as well as image classification [7].

2.4 Domain Adaption

- Aligning the features or representations between two domains to enable a model trained on one domain to effectively perform on another domain.
- For instance, retraining a model on an emotions dataset labelled in English so that it is accurate in other languages by matching the feature distributions between the dev and test language domains [8].

2.5 Knowledge Distillation

- Transferring knowledge from a more complex model (teacher) to a simpler model (student) for the task of emotion detection, where the student model aims to replicate the behaviour of the teacher model.
- Distilling knowledge from a sizable pre-trained language model (teacher) into a compact model (student) to facilitate real-time emotion detection in environments with limited resources [8].

2.6 Zero Shot Learning

- Training a model to recognize emotions, even without exposure to examples of all possible emotions during training, by generalizing to new, unseen classes based on knowledge learned from related tasks.
- Training a model on a set of emotions and having it generalize to recognize new, previously unseen emotions not encountered during the training phase.

2.7 Self-Supervised Learning

- Training a model on a pretext task that doesn't require labelled data, and then transferring the learned representations to the target emotion detection task.

- Pre-training a model on predicting missing parts of an image or masked words in a sentence without explicitly labelled emotion data, and then using the learned features for emotion detection tasks [9].

These methodologies present a range of techniques for transferring knowledge across tasks in emotion detection, offering flexibility in handling diverse data types and tasks with varying amounts of labelled information. The decision while selecting a technique depends on the specific problem, available data, and computational resources.

In the Table 14.1 the works done in the area of emotion recognition using deep transfer learning approach is presented with key concepts and major findings with each approach.

Table 14.1 Work in emotion detection using deep transfer learning

Author Name	Year	Key Concepts	Major Findings
Abanoz and Çataltepe [12]	2018	Emotion recognition using CNNs and transfer learning	Explored different CNN architectures, proposed ensembling with expert models, achieved a classifier with 68.32% accuracy on the FER13 validation set
Lee et al. [13]	2018	Network parameter learning method using transfer learning based on DenseNet for emotion recognition	Introduced an algorithm for effective temporal information learning using LSTM
Banerjee et al. [14]	2019	PTSD diagnosis using speech signals, employing a deep belief network (DBN) with transfer learning	Achieved a 74.99% accuracy with the transfer learning (TL) strategy, outperforming popular methods
Hung et al. [15]	2019	System identifying learners' facial emotions for instant feedback in learning effectiveness analysis	Introduced Dense_FaceLiveNet, enhanced basic emotion recognition, achieved 91.93% accuracy using transfer learning on FER2013, addressed limited emotion data issues with a novel transfer learning approach
Yokoo et al. [16]	2020	Emotion recognition IoT system using motion sensors and deep learning on a Raspberry Pi	Demonstrated high accuracy in human identification and emotion recognition
Awais et al. [17]	2020	IoT framework for emotion recognition, combining physiological signals and AI	Achieved ultralow latency of 1 ms, improved reliability in communication, and offered high emotion recognition performance (f-score of 95%)
Wang et al. [10]	2020	EEG-based emotion recognition with electrode-frequency distribution maps (EFDMs) and deep CNN	Proposed cross-dataset transfer learning, achieved 90.59% accuracy on SEED and 82.84% on DEAP, highlighted advantages of EFDMs and demonstrated the efficacy of deep model transfer learning in handling EEG data challenges
Ahmad et al. [11]	2020	Mitigating resource scarcity for emotion detection in Hindi using deep transfer learning	Achieved better performance than monolingual scenarios with an F1-score of 0.53, introduced cross-lingual word embedding representations
Abbaschian et al. [18]	2021	Comprehensive review of neural network advancements in recognition of emotion through speech	Compared deep learning and traditional algorithms of machine learning, provided insights into the evolving field of speech recognition
Akhand et al. [19]	2021	Addressed challenges in Facial Emotion Recognition (FER), proposed a Deep CNN (DCNN) with transfer learning	Introduced a novel pipeline strategy, achieved remarkable accuracy on datasets (e.g., 96.51% on KDEF, 99.52% on JAFFE), outperformed existing FER systems, especially in handling profile views
Ashraf et al. [20]	2021	Applied CNN-based transfer learning for Alzheimer's disease classification	Achieved a maximal average accuracy of 99.05% in ADNI dataset while using DenseNet, showcased significant improvements in specificity, sensitivity, and accuracy
Gerczuk et al. [21]	2021	Explored more than one libraries in recognition of emotion through speech using deep transfer learning	Assembled the EmoSet dataset, introduced EmoNet, achieved increased performance across 21 of 26 corpora through residual adapter transfer learning
Loey et al. [22]	2021	Addressed face mask detection using a different combination of models using deep and classical machine learning technique.	Proposed model achieved high testing accuracy across different datasets, demonstrating the effectiveness of the SVM classifier
Sahoo et al. [23]	2021	Proposed TLEFuzzyNet, a three-stage process commencing in a pipeline way for recognition of emotion through speech	Utilized transfer learning with pretrained CNN models, achieved state-of-the-art performance on SAVEE, RAVDESS, and EmoDB [23] datasets

Author Name	Year	Key Concepts	Major Findings
Xie et al. [24]	2021	Investigated multimodal emotion recognition using different models for each type of modality	Proposed transformer-based cross-modality fusion achieved up to 65% accuracy, surpassing unimodal models on the MELD dataset
Dresvyanskiy et al. [25]	2022	Explored different deep learning approaches for recognition of emotion in interactive media.	Deployed unimodal end-to-end and transfer learning within a multimodal fusion system, achieved promising results on the AffWild2 dataset under the affective behaviour analysis in-the-wild (ABAW) challenge protocol
Bashath et al. [26]	2022	This give review about the deep transfer learning models focussing on applications in which data is in text form.	Introduced a new nomenclature for describing transfer learning models, provided a visual organization of diverse deep learning techniques, aimed to offer comprehensive insights into text data used in studying these models
Amin et al. [27]	2022	Focused on recognition of driver's emotional state while driving a vehicle using deep transfer learning with Electrocardiogram (ECG) signals	Proposed seven models, achieved high accuracy (98.11%) with Model 5 based on Xception, employed transfer learning to reduce computational cost and time, demonstrated the effectiveness of Xception and DenseNet-201 in stress level detection
Helaly et al. [28] (2023)	2023	Deep CNN, Transfer Learning, ResNet18, FER	A Deep CNN-based system for Facial Emotion Recognition (FER) was proposed, employing Transfer Learning (TL). By enhancing the ResNet18 model, it achieved 98% accuracy on the CK+ dataset and 83% on FER2013, outperforming other Deep TL methods.
Sultana et al. [29] (2023)	2023	CNN, VGG19, Transfer Learning, Data Augmentation	The study addressed the challenges that come across recognition of facial expression and used a using technique of deep transfer learning that had achieved an accuracy of 94.8% and 93.7% on CK+ and JAFFE dataset respectively. It also demonstrated potential applications in online education, surveillance, and healthcare.
Meena et al. [30] (2023)	2023	InceptionV3, CNN, Image-based Sentiment Analysis	Focused on image-based sentiment analysis using an InceptionV3 approach. Improved image categorization performance and achieved high accuracy rates, notably 99.5% on CK+, 86% on JAFFE, and 73% on FER2013 datasets.

3. Conclusion

In conclusion, the relevant studies present in this paper highlights the diverse applications of deep and transfer learning techniques used in recognition of emotion and also address the challenges faced in various domains, showcasing advancements in multimodal approaches, cross-lingual transfer learning, and specialized tasks like driver's stress detection. The effectiveness of these models is demonstrated across different datasets and real-world scenarios, emphasizing their potential consequence on enhancing human-machine interactions and well-being. The major findings that are identified during this rigorous literature survey is that most of the approaches are based on machine learning and deep learning techniques to address the investigated problem. But, it lacks somewhere in terms of performance, optimal usage of neural architecture, best fit for neural parametric values, and foremost the deployed models in literature seem unscaled for a range of computing units.

Along with it, it has been identified that whenever neural networks are trained, training has been performed using heuristic parametric values that result in the constrained performance of the model. So we can think of a systematic approach to identify the neural parameters that relate to its tuning on the behalf of model's current performance. It will surely help to take a major step towards best-performing model development. Moreover, none of the articles has discussed the model's adaptability to a range of computing machines to implement the model

References

1. Mao, R., Liu, Q., He, K., Li, W., & Cambria, E. (2022). The biases of pre-trained language models: An empirical study on prompt-based sentiment analysis and emotion detection. IEEE Transactions on Affective Computing.
2. Kusal, S., Patil, S., Choudrie, J., Kotecha, K., Vora, D., & Pappas, I. (2022). A Review on Text-Based Emotion Detection--Techniques, Applications, Datasets, and Future Directions. arXiv preprint arXiv:2205.03235.
3. Siam, A. I., Soliman, N. F., Algarni, A. D., El-Samie, A., Fathi, E., & Sedik, A. (2022). Deploying machine learning techniques for human emotion detection. Computational intelligence and neuroscience, 2022.
4. Bashir, M. F., Javed, A. R., Arshad, M. U., Gadekallu, T. R., Shahzad, W., & Beg, M. O. (2023). Context-aware Emotion Detection from Low-resource Urdu Language Using Deep Neural Network. ACM Transactions on Asian and Low-Resource Language Information Processing, 22(5), 1–30.
5. Guo, J. (2022). Deep learning approach to text analysis for human emotion detection from big data. Journal of Intelligent Systems, 31(1), 113–126.
6. Khan, M. J., Khan, M. J., Siddiqui, A. M., & Khurshid, K. (2022). An automated and efficient convolutional

architecture for disguise-invariant face recognition using noise-based data augmentation and deep transfer learning. The Visual Computer, 1–15.

7. Yen, C. T., & Li, K. H. (2022). Discussions of different deep transfer learning models for emotion recognitions. IEEE Access, 10, 102860–102875.
8. Sharma, A., Sharma, K., & Kumar, A. (2023). Real-time emotional health detection using fine-tuned transfer networks with multimodal fusion. Neural Computing and Applications, 35(31), 22935–22948
9. Sun, M. (2022). Natural Language Processing for Health System Messages: Deep Transfer Learning Approach to Aspect-Based Sentiment Analysis of COVID-19 Content (Doctoral dissertation, Harvard University).
10. Wang, D., & Zheng, T. F. (2015, December). Transfer learning for speech and language processing. In 2015 Asia-Pacific Signal and Information Processing Association Annual Summit and Conference (APSIPA) (pp. 1225–1237). IEEE.
11. Badshah, A. M., Ahmad, J., Rahim, N., & Baik, S. W. (2017, February). Speech emotion recognition from spectrograms with deep convolutional neural network. In 2017 international conference on platform technology and service (PlatCon) (pp. 1–5). IEEE.
12. Abanoz, H., & Çataltepe, Z. (2018, May). Emotion recognition on static images using deep transfer learning and ensembling. In 2018 26th Signal Processing and Communications Applications Conference (SIU) (pp. 1–4). IEEE.
13. Lee, M. K., Kim, D. H., Choi, D. Y., & Song, B. C. (2018, June). Deep transfer learning for emotion recognition networks. In 2018 IEEE International Conference on Consumer Electronics-Asia (ICCE-Asia) (pp. 206–212). IEEE.
14. Banerjee, D., Islam, K., Xue, K., Mei, G., Xiao, L., Zhang, G., ... & Li, J. (2019). A deep transfer learning approach for improved post-traumatic stress disorder diagnosis. Knowledge and Information Systems, 60, 1693–1724.
15. Hung, J. C., Lin, K. C., & Lai, N. X. (2019). Recognizing learning emotion based on convolutional neural networks and transfer learning. Applied Soft Computing, 84, 105724.
16. Yokoo, K., Atsumi, M., Tanaka, K., Haoqing, W. A. N. G., & Lin, M. E. N. G. (2020, December). Deep learning based emotion recognition iot system. In 2020 International Conference on Advanced Mechatronic Systems (ICAMechS) (pp. 203–207). IEEE.
17. Awais, M., Raza, M., Singh, N., Bashir, K., Manzoor, U., Islam, S. U., & Rodrigues, J. J. (2020). LSTM-based emotion detection using physiological signals: IoT framework for healthcare and distance learning in COVID-19. IEEE Internet of Things Journal, 8(23), 16863–16871.
18. Abbaschian, B. J., Sierra-Sosa, D., & Elmaghraby, A. (2021). Deep learning techniques for speech emotion recognition, from databases to models. Sensors, 21(4), 1249.
19. Akhand, M. A. H., Roy, S., Siddique, N., Kamal, M. A. S., & Shimamura, T. (2021). Facial emotion recognition using transfer learning in the deep CNN. Electronics, 10(9), 1036.
20. Ashraf, A., Naz, S., Shirazi, S. H., Razzak, I., & Parsad, M. (2021). Deep transfer learning for alzheimer neurological disorder detection. Multimedia Tools and Applications, 1–26.
21. Gerczuk, M., Amiriparian, S., Ottl, S., & Schuller, B. W. (2021). Emonet: A transfer learning framework for multi-corpus speech emotion recognition. IEEE Transactions on Affective Computing.
22. Loey, M., Manogaran, G., Taha, M. H. N., & Khalifa, N. E. M. (2021). A hybrid deep transfer learning model with machine learning methods for face mask detection in the era of the COVID-19 pandemic. Measurement, 167, 108288.
23. Sahoo, K. K., Dutta, I., Ijaz, M. F., Woźniak, M., & Singh, P. K. (2021). TLEFuzzyNet: Fuzzy rank-based ensemble of transfer learning models for emotion recognition from human speeches. IEEE Access, 9, 166518–166530.
24. Xie, B., Sidulova, M., & Park, C. H. (2021). Robust multimodal emotion recognition from conversation with transformer-based crossmodality fusion. Sensors, 21(14), 4913.
25. Dresvyanskiy, D., Ryumina, E., Kaya, H., Markitantov, M., Karpov, A., & Minker, W. (2022). End-to-end modeling and transfer learning for audiovisual emotion recognition in-the-wild. Multimodal Technologies and Interaction, 6(2), 11.
26. Bashath, S., Perera, N., Tripathi, S., Manjang, K., Dehmer, M., & Streib, F. E. (2022). A data-centric review of deep transfer learning with applications to text data. Information Sciences, 585, 498–528.
27. Amin, M., Ullah, K., Asif, M., Waheed, A., Haq, S. U., Zareei, M., & Biswal, R. R. (2022). ECG-Based Driver's Stress Detection Using Deep Transfer Learning and Fuzzy Logic Approaches. IEEE Access, 10, 29788–29809.
28. Helaly, R., Messaoud, S., Bouaafia, S., Hajjaji, M. A., & Mtibaa, A. (2023). DTL-I-ResNet18: facial emotion recognition based on deep transfer learning and improved ResNet18. Signal, Image and Video Processing, 1–14.
29. Sultana, A., Dey, S. K., & Rahman, M. A. (2023). Facial emotion recognition based on deep transfer learning approach. Multimedia Tools and Applications, 1–15.
30. Meena, G., Mohbey, K. K., Kumar, S., Chawda, R. K., & Gaikwad, S. V. (2023). Image-based sentiment analysis using InceptionV3 transfer learning approach. SN Computer Science, 4(3), 242

Intelligent Systems Using Semiconductors for Robotics and IoT – Dinesh Goyal et al. (eds)
© 2026 Taylor & Francis Group, London, ISBN 978-1-041-20408-4

15 Enhancing Insider Threat Detection with LSTM Based Models

Ashok Kumar*, Sanjeev Patwa
Mody University of Science and Technology, Rajasthan
Sunil Kumar Jangir
Wisflux Tech Labs, Jaipur, Rajasthan India

Abstract: Insider threat has consistently been identified as the deadliest threats in the IT environment. This can result in a heavy loss in terms of both the financial and reputation of an organization. Moreover, it takes approximately 6-8 months to identify an Insider threat case. The last four years data breach reports clearly indicate the danger of Insider Threat (approximately 70 % are attributable to negligent /malicious insiders). Traditional methods are reactive. Therefore, using anomaly detection methods of Artificial Intelligence, which are naturally predictive, the insider's threat can be predicted in advance based on the employee's activity logs. This paper presents an approach to improve Insider Threat detection by using long short-term memory (LSTM) and achieved improvement over previously acquired results. Through experimental evaluation, prediction accuracy of 95.2% with a loss value of 0.150 was achieved.

Keywords: Insider threat, Malicious insider, Artificial intelligence, Negligent insider, Machine learning, LSTM

1. Introduction

Insider threats in the cyber world are considered the deadliest and involve least economy of effort. This is because of the availability of access to IT infrastructure. An outside attacker, trying to gain access to an organization's data/infrastructure, has to pass through multilayer of defense for successful intrusion. In the classical approach, a cyberattack is carried out in five steps: (a) reconnaissance – the attacker takes stock of the available assets. (b) Scanning – looks out for vulnerabilities in systems discovered in the preceding step (c) getting access to the system using stolen credentials (d) maintaining access – ensures that access is available to re-enter the stem at will, and (e) clearing tracks to remove any forensic evidence. However, in the case of an attack carried out by an insider, the first two steps are bypassed and the attacker has direct access to the inside assets (both hardware and data). What he needs to do now is escalating his privileges (in case he or she is a normal user) or starting a malicious act if the insider is a privileged user. The most worrisome attribute of an insider turning rogue is, the late discovery of the attack. Studies and reports have shown they could discover an insider attack in 6 to 8 months.

Behavioral scientists have conclusively proven that a malicious user will provide enough indications before carrying out a malicious act of data exfiltration, sabotage or causing any other kind of damage to the organization. In such a complex situation, we need to have the right proportion of administrative controls, such as proper screening of employees, vendors, contract workers, and personnel involved in operation maintenance and upkeep of infrastructure, before allowing them access to IT infrastructure. It is not known who amongst them could turn out to be malicious insiders.

There have been many cases of an insiders turning malicious and bringing disrepute and embarrassment to the organization. The *Tim Lloyd/ Omega case of Insider sabotage*, is one of the earliest reported cases of insider turning rogue and causing irrecoverable damage to the world's number one company in making time keeping

*Corresponding author: ashok1967@gmail.com

DOI: 10.1201/9781003716389-15

instruments. The reason for this was that , he was publicly scolded by supervisor and denied promotion. A person who had a Setup Company's network and software environment copied complete data on a tape and moved all programs controlling CNC Machines on a single server. He planted a script that was activated after three months. He joined the competition company and shared all proprietary software with them. The Time triggered software executes after three months to format the server, which has all the software installed and controlling CNC machines. This brought the leading company of the world to bankruptcy.

In the famous insider case, *Target Data Breach*, an insider, gained access to the retail giant Target's customer service database. Subsequently installed malware on the system. Subsequently, he gathered classified information including customers' complete names, telephone numbers, email addresses, credit card details, card verification codes, and additional personal data. In total data of 41 million customers was compromised [1]. In the renowned company *Tesla,* an employee with escalated privileges created false credentials and then carried out changes in code. He also exported large amount of Tesla data to third parties. This proves that organizations with the best deployed Cyber Defense can be compromised by a single internal threat actor. [2]. At Facebook, an employee, in charge of security, abused his position to stalk a woman online. However, his employment was terminated after investigations revealed his role in obtaining information for stalking women on the Internet. Concurrently, Facebook has faced severe criticism due to sharing of 87 million users data in the US Presidential election to Cambridge Analytica. [3].

Another case comes from the *SunTrust Bank,* which admitted that their employee had stolen data of 1.5 million customers (comprising of name, contact, account number and balance available in the account). However, they denied that Personal Identifiable information such as SSN, username and password, were not exposed. For social engineering attacks, the exposed data are more than sufficient for attackers to perform fishing/vishing frauds [4]. The list is endless and brings home the point that a Malicious Insider is capable of, sabotage, data exfiltration, espionage (both foreign and economic) and unauthorized disclosure of data. This not only causes financial loss to the victim organization but also causes a loss of reputation. Such insider's threat cases can be prevented using an analysis system for observing the employee's activities and detect any anomalies.

This work presents an approach based on LSTM for monitoring the online behavior of every individual mentioned in the preceding paragraph who is allowed to access the IT assets of the organization. The synthetically generated dataset by Carnegie Mellon University (CMU), was used for training the model using LSTM. Dataset comprises activity log records of malicious user activities based on 5 different scenarios described in Section 3. Summary of Contributions of this work is as follows: (i) Data were processed to form sequences of log keys that help mine data using system logs. (ii) The LSTM algorithm is applied to identify malicious insider behaviors.

The flow of this paper is organized as follows:

(a) Section 2 discusses the previous work on the subject
(b) Section 3 discusses the details of the dataset used for our work in this study.
(c) Section 4 presents method used in this work.
(d) Section 5 contains the details of the experiments and the results
(e) Section 6 conclusion and future work.

2. Previous Work

For this research paper, studies of research work undertaken between 2017 to 2023 (spanning six years) were carried out. This includes resources from Internet, Data breach reports, Scopus, IEEE and Web of Science journals. Keywords viz "Insider Threat Detection", "Insider Threat Prediction" "Machine learning in Insider Threat Detection" and "Traditional / advance methods of Insider threat mitigation" were used to collect data, related literature and study material. This provided clarity of direction to move away from traditional methods of reactive approach towards insider detection to preventive. Many studies have been conducted to discover strategies to mitigate insider threat during the first decade of this century [5][6][7][1] . However, the detection of insider threat has always been a challenging task for the researchers as it largely depends on the user behavior, thoughts and actions which is difficult to store and analyze. Back in 2005, Mark Maybury et. al. reported a generic model demonstrating integration of multiple approaches for early warning and detection of insider's threat [8]. In 2010 W. Eberle et. al. used graph based approach to discover suspicious activities in cybercrime, business processes and social network communication related domains [9]. Many other studies are available, among which many constrains and some failed because of non-feasibility of implementation. Sherali Zeadally et. al. have reviewed many such approaches proposed by different researchers along with their benefits and limitations [10].

With the rise, the availability of big data resources and computation power use of Artificial Intelligence methods to detect insider's threats have been introduced [11]. Detecting insider's threat based on unusual behavior of employee can be formulated as an anomaly detection problem and can be solved using available Machine Learning and Deep Learning methods. In 2018 Sarma S et.al proposed a KNN based authentication mechanism that classifies users into legitimate, possibly legitimate and illegitimate, to determine the possibility of insider threat [12].

In 2018, D. Le. et. al. used "Supervised and unsupervised algorithms as Self Organizing Maps (SOM), Hidden Markov Models (HMM) and Decision Trees on the CERT Insider Threat Dataset" and found that SOM provides appreciable performance for visualization, anomaly detection and classification [13]. Mohammed Dosh applied three Machine Learning algorithms to compare their performance on the CERT Insider Threat Dataset, and found that Random Forest, Naïve Bayes, and Nearest Neighbor achieves 89.75917519%, 91.96650826%, and 94.68205476% accuracy and an error rate of 10.24082481%, 8.03349174%, and 5.317945236%, with the selected five features [14].

Duc C. Le et. al. presented preprocessing techniques to achieve extraction of multiple levels of data granularity, and used machine learning methods to detect malicious activity with 85% accuracy and 0.78% false positive rate [15][16]. Although Deep Learning techniques have been found to provide a better accuracy in predicting insider threats as compared to Machine Learning algorithms, some limitations such as lack of labelled data and adaptiveness, make the task challenging. Suhan Yuhan et. al. have provided a comprehensive review of most widely used datasets for insider threat detection, the comparative analysis of traditional Machine Learning algorithms , Deep learning algorithms and the advantages and limitations of both [17]. Advances in machine learning, including deep learning and ensemble models, address these issues by identifying latent patterns in data. Bushra Bin Saharan [29] used the "Deep Feature Synthesis algorithm" to derive behavioral features from historical data, generating 69,738 features per user, then using Principal Component Analysis (PCA) for dimensionality reduction and advanced machine learning algorithms for anomaly detection and classification, achieving 91% accuracy in anomaly detection Usman Rauf, Wei, Fadi Mohsen [30] Introduced a hybrid insider threat detection framework that integrates "Statistical criteria using information gain metrics with machine learning-based classification" using CERT r4.2 dataset. However, no similar work was done on version 5.2. A. A. H. Alrammahi [31] proposed an expert system algorithm that intersects classification results from multiple algorithms with an internal detection algorithm using expert rules. Various classification methods predict insider status within a computer network, enhancing identification accuracy and efficiency.

3. Dataset

The biggest issue faced in our kind of research work is the availability of datasets, since primary data are not shared due to privacy reasons by organizations. For this work synthetically generated data, by Software Engineering Institute, Carnegie Mellon University (CMU), was used [18]. Dataset has different releases, each one being a superset of the previous versions. In this work release 5.2 is used, which contain a descent collection of logs from web history, emails, device usage, file access and logon logoff patterns. The dataset contained five CSV files for logs, as shown in Table 15.1, and one additional LDAP file containing employee details. It also contains a folder named 'answers' having details of malicious activity and scenarios in all the versions of the dataset. To train a model for predicting insider threats, it must be trained using a large amount of data on human behavior and actions captured in a monitored environment. Collecting and using real data is not feasible owing to confidentiality and privacy concerns.

Based on the learnings of various case studies, a normal user can be placed in the category of a Potential Insider Threat and his/her activities need to be monitored. The following are the different scenarios of malicious activities provided in the dataset:

(a) A User tries to use external storage media and starts copying the data on it. He tries to gain the privilege of doing this if not granted by the system Administrator. Secondly, he performs this activity quite often after normal working hours. He also visited the sites such as WikiLeaks and tenders his resignation within a short time.

(b) The user starts visiting job websites to seek employment with a competitor in the present organization. Copies lots of trade secrets, IPRs and customer details for sharing with their future employer.

(c) The System Administrator turns rogue either due to not getting promoted or financial raise or any other HR related issue. He downloads a key logger and installs in the machine of his manager. After obtaining his credentials he logs into the system of his manager to send mass mail to the employees and customers which creates panic. After committing this act, he resigns and leaves the organization.

Table 15.1 Results of other works

Work	Dataset	Method	Results
[24]	CERT r2	Streaming anomaly detection algorithms	Instance based recall ~ 50%, precision=8%
[25]	CERT v4.2	LSTM CNN	AUC = 0.9449
[26]	CERT v6.2	LSTM RNN	Accuracy = 93.85%
[27]	Vegas Dataset	Unsupervised insider threat detection Supervised quitter detection	Instance based AUC = 0.77
[28]	CERT v4.1 and v4.2	Random Forest, Naïve Bayes, 1 Nearest Neighbor (1-NN)	94.68% accuracy with 1-NN

Table 15.2 Features in different files provided in CERT insider threat dataset r5.2

File name	Fields (Columns)	Remarks
logon.csv	• Id • Date • User • PC • Activity (Logon/Logoff)	No user may log onto a machine where another user is already logged on, unless the first user has locked the screen. Logoff requires preceding logon A small number of daily logons are intentionally not recorded to simulate dirty data. Logons precede other PC activity Screen unlocks are recorded as logons. Screen locks are not recorded. 2k users, each with an assigned PC 100 shared machines used by some of the users in addition to their assigned PC. Systems administrators with global access privileges are identified by job role "ITAdmin". Some users log into another user's dedicated machine from time to time.
device.csv	• Id • Date • User • PC • File tree • Activity (connect/disconnect)	Some users use a thumb drive. Some connect events may be missing disconnect events, because users can power down machine before removing drive Users are assigned a normal/average number of thumb drive uses per day. Deviations from a user's normal usage can be considered significant. The file tree field is a semicolon-delimited list of directories on the device.
http.csv	• ID • Date • User • PC • URL • Content	Domain names have been expanded to full URLs with paths. Words in the URL are usually related to the topic of the web page. Content consists of a space-separated list of content keywords. Each web page can contain multiple topics.
email.csv	• ID • Date • User • PC • to • cc • bcc • from, • Activity • Size • Attachment • Content	Role (from LDAP) drives the amount of email a user sends per day. A small number of edges are introduced as noise. Emails can have multiple recipients. Emails can have a mix of employees and non-employees in dist. list. Terminated employees remain in the population, and thus are eligible to be contacted as non-employees. Content consists of a space-separated list of content keywords. Content does not specifically refer to the subject or body. We have not made that distinction. Each message can contain multiple topics.
file.csv	• ID • Date • User • PC • Filename • Content	Each entry represents a file copy to a removable media device. Content consists of a hexadecimal encoded file header followed by a space-separated list of content keywords Each file can contain multiple topics. File header correlates with filename extension. The file header is the same for all MS Office file types. Each user has a normal number of file copies per day. Deviation from normal can be considered a significant indicator.
LDAP	• Employee name - First, middle, and last name • User_id - Employee user ID • Email - Employee's work email address	The project and supervisor fields are known to be missing from the Vegas dataset.

(d) A potential threat tries to log into the machine of another user, searches for a particular type of data files and sends them to their personal mail address. This behavior continues for a period of three months or more.

(e) The organization decides to lay off employees. One of the employees uploads a number of important documents to his personal drive such as Dropbox so that he can use this data in the future for personal gain

4. Methodology

The data available in different csv files mentioned in the previous section need to be processed and converted from strings to a meaningful format which can be given as input to the AI model converted to a meaningful format which can be given as input to the AI model.

4.1 Data Preprocessing

The data in different csv files was processed and the column 'usual pc' and 'work time' were added which contain the bool value True if the users are using their usual PC and it is their work time, and False otherwise. It helps detect if any employee unusually uses another's PC or stays in office at unusual times. The different websites were classified according to the scenarios to make them capable of being utilized in modelling. The 'content' column in the 'http.csv', 'file.csv', 'email.csv' files contain text which needs to be converted to semantic vectors for further processing. This was achieved using the Latent Dirichlet Allocation (LDA) model available in the Gensim library of Python. It is a generative statistical model which is used to discover different topics discussed in a large document, and can provide the probability of a new document belonging to the discovered topics. It is more generalized version of the Probabilistic Latent Semantic Analysis (PLSA).

After preprocessing the content column, all the preprocessed files were divided in separate folder according to the 'user id'. All the files related to one user were stored in a separate folder. These files were then merged to form a single csv file, for each user containing all activities, sorted according to the time of action. Then for each action a unique action key was decided "Logon": "1", "Logoff": "2", "Connect": "3", "Disconnect": "4", "http": "5", "file": "6","email": "7" . Then all processed files were merged into a common csv which was used as input to the model.

4.2 Long Short-Term Memory (LSTM)

Long Short-Term Memory (LSTM) networks which is one type of Recurrent neural networks (RNNs), are used for analysis of sequential data, making them valuable for predicting insider threats. Users' actions, including file access, email activity, and web browsing, can be viewed as time-ordered sequences. By learning the patterns within these sequences, LSTM networks can detect unusual behaviors that may signal potential insider threats. For instance, an alert might be triggered if an employee unexpectedly accesses sensitive files outside their normal scope or logs in during atypical hours. The "memory" component of LSTM enables the model to incorporate past actions into its predictions, capturing extended behavioral patterns. This capability is essential for identifying insider threats, as malicious activities often develop gradually rather than as isolated events. Long Short Term Memory (LSTM) is a type of Recurrent Neural Networks (RNN) [19]. In a Conventional Neural Network, the inputs are independent of each other and the sequence, however ,in cases where the output depends on the previous inputs RNNs are used. The unfolded representation of a typical RNN shown in Fig. 15.1 is provided in Fig. 15.2.

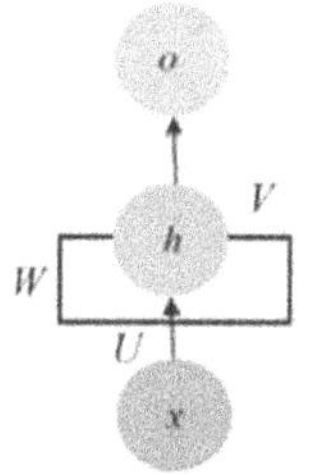

Fig. 15.1 Typical RNN

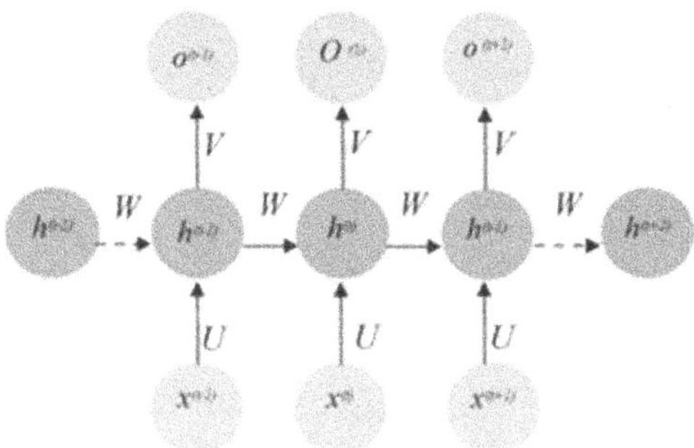

Fig. 15.2 Unfolded RNN

RNNs perform well for short-duration dependencies, but fail for long durations owing to the problem of Vanishing Gradient. During the weight updating process, the gradient almost vanishes until it reaches the initial layers; hence it becomes difficult to train them, which results in a poor performance of the RNN over long sequences. As a solution to this LSTM was introduced [20].

In LSTMs information flows through cell states. Each cell state has three different dependencies i.e the previous cell state, the previous hidden state, and the input at the current timestamp.

Figure 15.3 shows a LSTM cell in which,

$$i_t = \sigma(x_t U^i + h_{t-1} W^i)$$
$$f_t = \sigma(x_t U^i + h_{t-1} W^i)$$
$$o_t = \sigma(x_t o + h_{t-1} W^o)$$
$$C'_t = \tanh(x_t U^g + h_{t-1} W^g)$$
$$C_t = \sigma(f_t * C_{t-1} + i_t * C'_t)$$
$$h_t = \tanh(C_t) * o_t$$

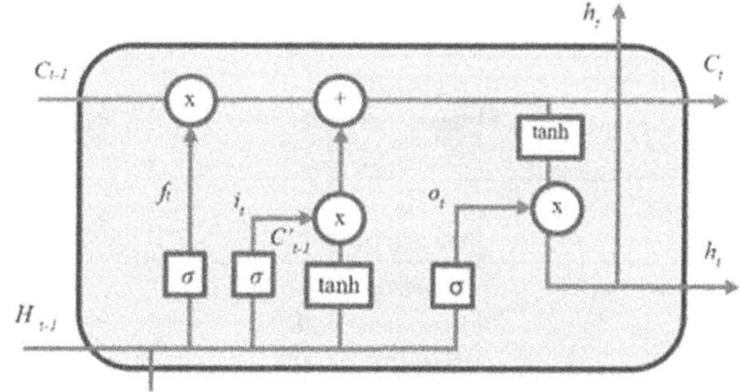

Fig. 15.3 Long short term memory cell

Each cell contains three gates: a forget gate, an input gate and an output gate to control and protect cell states. The ability of LSTM to provide better predictions over long sequences, makes it suitable for anomaly detection [20][21].

4.3 Model Description

To detect unusual events a long sequence must be maintained. The primary algorithm used in this study is LSTM, which can encode long complex sequences. The input to the model is given in the form of a sequence of n log-keys and the output is the probability distribution for the $(n + 1)th$ log-key. The baseline was determined based on the data of non-malicious users.

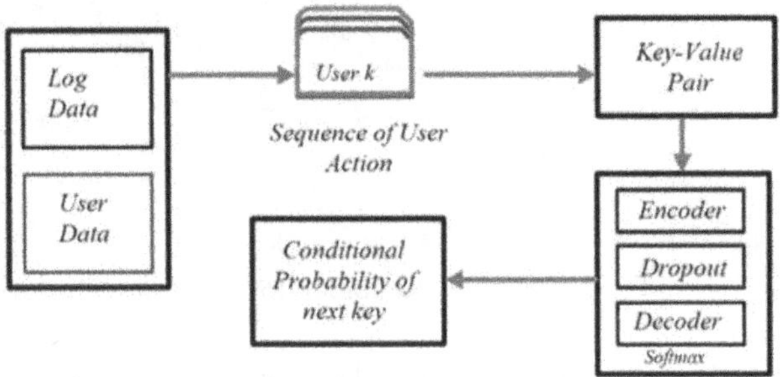

Fig. 15.4 Flow chart showing the anomaly detection process step by step

Description of the different layers in LSTM model is given in the Table 15.3. The encoding layer downscales the input vector an all the information is stored in a latent vector of low dimensions, which captures all the necessary information. In decoding phase the original vector is reformed. The softmax layer returns the tensor of same dimension as input with the values between 0 and 1, giving probability of each log-key pair.

Table 15.3 Layers in the model

Layer	Parameters
Embedding	num_embeddings=1127 embedding_dim=40 padding_idx=None max_norm=None norm_type=2.0 scale_grad_by_freq=False sparse=False weight=None
LSTM encoder	1stm_input_size 40 num_layers=3 dropout=0.5 batch first=True dropout-0.5
Dropout	p=0.5 inplace=False
Decoder	in_features=40 out_features=1127 bias=True
Softmax	Dim=2

5. Experiments and Results

For this work, an Intel Xeon Processor with 8 cores, 64 GB RAM and 2 TB SDD was used. Experiments were carried out using Jupyter Notebook. PyTorch, an opensource Machine Learning Library was used to implement the network [22]. With the help of useful abstractions to reduce the amount of boilerplate code, it helps speed up model development. The input data were divided into three sets training, testing and validation. The accuracy was determined using the metrics available in ignite, a high level library of Pytorch, to help in training the models. The loss and negative log likelihood were determined using the NLLLoss class available in PyTorch [23].

The model was trained for 100 epochs and demonstrated an accuracy of 95.2% with a loss of 0.150. As shown in Fig. 15.5 (the accuracy v/s epoch curve for training). After 15 epoch accuracy was nearly constant, whereas the loss decreased to nearly 60 epochs, as shown in Fig. 15.6. Table 15.1 presents the results of some other studies related to the same objective. It is concluded that with this model, we are able to achieve slightly better accuracy to some of the best results obtained in the past.

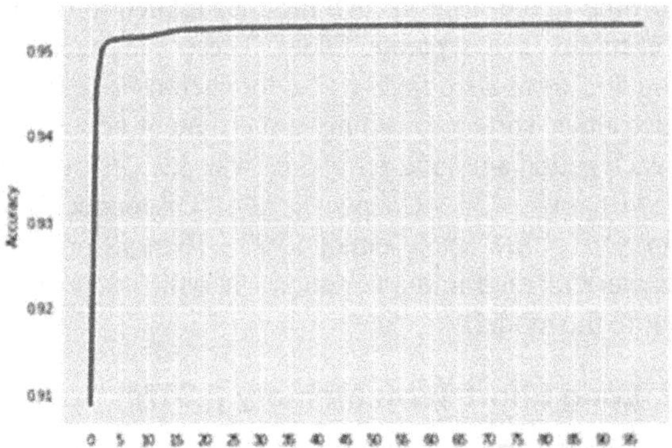

Fig. 15.5 Accuracy vs epoch curve

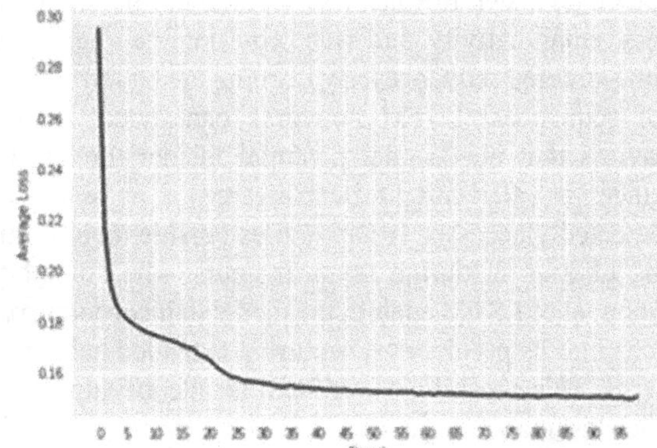

Fig. 15.6 Average loss vs epoch curve

6. Conclusion and Future Work

In this study, an LSTM based model is implemented to detect the insider's threat using the opensource dataset

-CERT release 5.2. The data were processed and provided as input to the model in the form of a sequence of log keys. It achieved an appreciable accuracy of 95.2% with the loss as low as 0.150. However, there we some limitations on implementing this for real time usage. It considers five scenarios for malicious activities, and anything other than them is considered normal. In the future, additional malicious activity scenarios can be added to improve the general applicability of the model.

Acknowledgement

The first author wishes to acknowledge Ms Shatakshi Singh Associate Data Engineer, Lowe's Pvt Ltd and Ms Anirudhi Thanvi , Specialist Programmer Infosys Ltd

References

1. S. Sinclair and S. W. Smith, "Preventative Directions For Insider Threat Mitigation Via Access Control."
2. "What Tesla's Spygate Teaches Us About Insider Threats." [Online].Available: https://www.forbes.com/sites/forbestechcouncil/2018/07/19/what-teslas-spygate-teaches-us-about-insider-threats/?sh=43027a4f5afe. [Accessed: 28-May-2021].
3. "Facebook fires engineer who allegedly used access to stalk women." [Online]. Available: https://www.nbcnews.com/tech/social-media/facebook-investigating-claim-engineer-used-access-stalk-women-n870526. [Accessed: 28-May-2021].
4. "SunTrust Ex-Employee May Have Stolen Data on 1.5 ..." [Online]. Available: https://www.darkreading.com/attacks-breaches/suntrust-ex-employee-may-have-stolen-data-on-15-million-bank-clients/d/d-id/1331610. [Accessed: 28-May-2021].
5. S. Sinclair and S. W. Smith, "Preventative Directions For Insider Threat Mitigation Via Access Control," *Adv. Inf. Secur.*, vol. 39, pp. 165–193, 2008, doi: 10.1007/978-0-387-77322-3_10.
6. Q. Althebyan, B. P.-2008 T. I. C. on, and undefined 2008, "Performance analysis of an insider threat mitigation model," *ieeexplore.ieee.org*.
7. F. L. Greitzer, P. R. Paulson, L. J. Kangas, L. R. Franklin, T. W. Edgar, and D. A. Frincke, "Predictive Modeling for Insider Threat Mitigation," 2009.
8. M. Maybury *et al.*, "Analysis and detection of malicious insiders," *Int. Conf. Initelligence Anal. McLean, VA*, pp. 3–8, 2005.
9. W. Eberle, J. Graves, and L. Holder, "Insider threat detection using a graph-based approach," *J. Appl. Secur. Res.*, vol. 6, no. 1, pp. 32–81, Jan. 2011, doi: 10.1080/19361610.2011.529413.
10. S. Zeadally, B. Yu, D. H. Jeong, and L. Liang, "Detecting insider threats solutions and trends," *Inf. Secur. J.*, vol. 21, no. 4, pp. 183–192, Apr. 2012, doi: 10.1080/19393555.2011.654318.
11. L. L. Ko, D. M. Divakaran, Y. S. Liau, and V. L. L. Thing, "Insider threat detection and its future directions," *Int. J. Secur. Networks*, vol. 12, no. 3, pp. 168–187, 2017, doi: 10.1504/IJSN.2017.084391.
12. M. S. Sarma, Y. Srinivas, M. Abhiram, L. Ullala, M. S. Prasanthi, and J. R. Rao, "Insider threat detection with face recognition and KNN user classification," in *Proceedings - 2017 IEEE International Conference on Cloud Computing in Emerging Markets, CCEM 2017*, 2018, vol. 2018-January, pp. 39–44, doi: 10.1109/CCEM.2017.16.
13. D. C. Le and A. Nur Zincir-Heywood, "Evaluating Insider Threat Detection Workflow Using Supervised and Unsupervised Learning," *ieeexplore.ieee.org*, 2018, doi: 10.1109/SPW.2018.00043.
14. M. Dosh and M. Dosh, "Detecting insider threat within institutions using CERT dataset and different ML techniques," vol. 9, no. 2, pp. 873–884, 2021.
15. D. C. Le, N. Zincir-Heywood, and M. I. Heywood, "Analyzing Data Granularity Levels for Insider Threat Detection Using Machine Learning," *IEEE Trans. Netw. Serv. Manag.*, vol. 17, no. 1, pp. 30–44, Mar. 2020, doi: 10.1109/TNSM.2020.2967721.
16. "Machine learning based Insider Threat Modelling and Detection | IEEE Conference Publication | IEEE Xplore." [Online]. Available: https://ieeexplore.ieee.org/abstract/document/8717892. [Accessed: 25-May-2021].
17. S. Yuan and X. Wu, "Deep learning for insider threat detection: Review, challenges and opportunities," *Computers and Security*, vol. 104. Elsevier Ltd, p. 102221, 01-May-2021, doi: 10.1016/j.cose.2021.102221.
18. "Insider Threat Test Dataset." [Online]. Available: https://kilthub.cmu.edu/articles/dataset/Insider_Threat_Test_Dataset/12841247/1. [Accessed: 23-May-2021].
19. A. Sherstinsky, "Fundamentals of Recurrent Neural Network (RNN) and Long Short-Term Memory (LSTM) Network."
20. P. Malhotra, A. Ramakrishnan, G. Anand, L. Vig, P. Agarwal, and G. Shroff, "LSTM-based Encoder-Decoder for Multi-sensor Anomaly Detection."
21. W. Luo, W. Liu, S. G.-2017 I. I. C. on, and undefined 2017, "Remembering history with convolutional lstm for anomaly detection," *ieeexplore.ieee.org*.
22. N. Ketkar, "Introduction to PyTorch," in *Deep Learning with Python*, Apress, 2017, pp. 195–208.
23. "NLLLoss — PyTorch 1.8.1 documentation." [Online]. Available: https://pytorch.org/docs/stable/generated/torch.nn.NLLLoss.html. [Accessed: 26-May-2021].
24. B. Bose, B. Avasarala, S. Tirthapura, Y. Y. Chung, and D. Steiner, "Detecting Insider Threats Using RADISH: A System for Real-Time Anomaly Detection in Heterogeneous Data Streams," *IEEE Syst. J.*, vol. 11, no. 2, pp. 471–482, Jun. 2017, doi: 10.1109/JSYST.2016.2558507.
25. F. Yuan, Y. Cao, Y. Shang, Y. Liu, J. Tan, and B. Fang, *Insider threat detection with deep neural network*, vol. 10860 LNCS. Springer International Publishing, 2018.
26. F. Meng, F. Lou, Y. Fu, and Z. Tian, "Deep learning based attribute classification insider threat detection for data security," *Proc. - 2018 IEEE 3rd Int. Conf. Data Sci. Cyberspace, DSC 2018*, pp. 576–581, 2018, doi: 10.1109/DSC.2018.00092.
27. G. Gavai, K. Sricharan, D. Gunning, J. Hanley, M. Singhal, and R. Rolleston, "Supervised and Unsupervised methods to detect Insider Threat from Enterprise Social and Online Activity Data."

28. M. Dosh, “Detecting insider threat within institutions using CERT dataset and different ML techniques,” *Period. Eng. Nat. Sci.*, vol. 9, no. 2, pp. 873–884, May 2021, doi: 10.21533/PEN.V9I2.1911.
29. Lopez & Wong,(2019) Insider Threat Detection with Long Short-Term Memory, https://doi.org/10.1145/3290688.3290692)
30. Lopez, E., & Sartipi, K. (2020). Detecting the insider’s threat with Long Short Term Memory (LSTM) neural networks. https://doi.org/10.48550/arXiv.2007.11956
31. Bushra Bin Sarhan,Najwa Altwaijry Applied Sciences “Insider Threat Detection Using Machine Learning Approach” DOI: 10.3390/app13010259 2022
32. Usman Rauf,Zhiyuan Wei,Fadi Mohsen “Employee Watcher: A Machine Learning-based Hybrid Insider Threat Detection Framework” Cyber Security in Networking Conference DOI: 10.1109/csnet59123.2023.10339777 2023
33. A. A. H. Alrammahi “Insider Detection Using Combination of Machine Learning and Expert Policies” International Journal of Electrical and Electronic Engineering & Telecommunications 2024 DOI: 10.18178/ijeetc.13.5.389-396
34. Habiba Emad-Eldin Abdallah, Habiba Hamed Abd-Elkader, Kholoud Khaled Mohamed, Mohamed Abd-Elmoniem, Nizar Walid El-Assal, Salma Mohsen Mohamed, Sama “Performance Evaluation Framework for Insider Threat Detection Using Machine Learning” DOI:10.1109/imsa61967.2024.10652829 2024

Note: All the figures and tables in this chapter were made by the authors.

Intelligent Systems Using Semiconductors for Robotics and IoT – Dinesh Goyal et al. (eds)
 ISBN 978-1-041-20408-4

16 Electronic Waste Management: Challenges, Solutions and Scope in India

Ashish Laddha[1]
Department of Electrical Engineering,
Poornima Institute of Engineering & Technology,
Jaipur, India

Krati Sharma[2]
Department of English & Soft Skills,
Poornima Institute of Engineering & Technology,
Jaipur, India

Aisha Rafi[3]
Department of Applied Sciences,
Poornima Institute of Engineering & Technology,
Jaipur, India

Sonia Kaur Bansal[4]
Department of English & Soft Skills,
Poornima Institute of Engineering & Technology,
Jaipur, India

Abstract: This article discusses the challenges, possible solutions, and scope of electronic waste management (EWM) regarding the Indian subcontinent. Managing the seamlessly increasing electronic-categorized waste bears the significant importance for the United Nations's goals to attain development in the sustainable manner. The same is also important in terms of the efforts being made for transforming the India from developing to developed. Various constituents of the electronic waste (EW) result in the different kinds of harmful impacts of on the health of human beings. Additionally, proper recycling, need of enhancement of the youth participation and awareness, among others are also needed regarding the EWM.

Keywords: Electronic waste management (EWM), Indian subcontinent context, Challenges, Possible solutions, Scope

1. Introduction

Enhancement in the living standard of the human beings as well as semi-conductor technical expertnesses have raised the count of utilizable electronic gadgets. Further, semi-conductors remain the foundational elements of several sensing elements. Waste is the terminology utilized for the objects/things, which have been utilized and would subjected to disposing. One may observe various categories of the same. Out of these, electronic waste (EW) has been observed to possess the speedy growth around the globe. Managing of this waste remains a tedious task and needs a systematic procedure. The importance of the presented work is in the fact that India stands just after the top two EW producing countries. One of the lowest labor prices and lack of rigid government regulations, here, accelerate the exporting of EW from the developed-categorized nations in bulk manner [1].

A programme for proper management of the electronic-categorized waste was begun by the Ministry of Electronics and Information Technology of the Indian subcontinent

[1]ashish.laddha@poornima.org, [2]kratibhomia@gmail.com, [3]aisha.rafi@poornima.org, [4]sonia.bansal@poornima.org

DOI: 10.1201/9781003716389-16

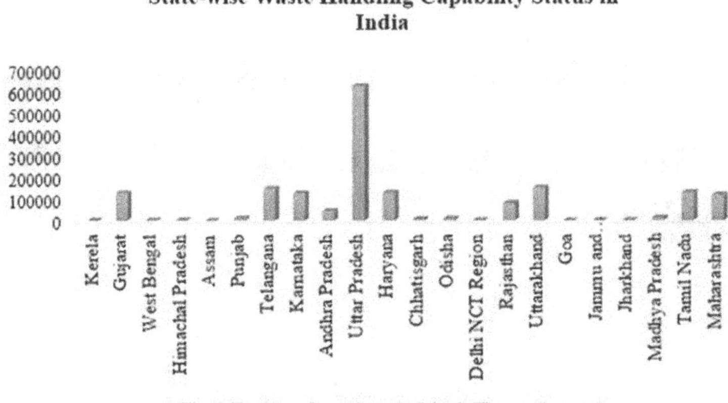

Fig. 16.1 State-wise EW data in the indian subcontinent

on 31st of March in the year 2015 [2] under the mission entitled Digital India. The main focusing areas of the above-mentioned programme are recycling-led reusing of the EW to accelerate sustainability. This programme would aid in the systematic implementation of the EW managing rules/laws of 2011, 2016, and 2022. The second one was initiated for half of a decade (i.e., five years) keeping in center ten states namely Odisha (Eastern), Puducherry (Southern), Madhya Pradesh (Central), Goa (Western), Assam (Eastern), Jharkhand (Eastern), Manipur (North Eastern), Bihar (Eastern), Uttar Pradesh (North Central), and West Bengal (Eastern). One center of excellence, for imparting the deeper insights of the EWM, has also been established at the Hyderabad, India. Main focusing areas of this institution involve recycling/reusing of the solar cells, permanent magnets, printed circuit boards, Lithium-ion batteries, among others. Its functioning also involves start ups, skill enhancement, and collaborations with academic institutions and industries.

Waste handling installed capacities (in Metric Tonnes/annum), for the year 2023, of the Kerela, Gujarat, West Bengal, Himachal Pradesh, Assam, Punjab, Telangana, Karnataka, Andhra Pradesh, Uttar Pradesh, Haryana, Chhatisgarh, Odisha, Delhi NCT region, Rajasthan, Uttarakhand, Goa, Jammu and Kashmir, Jharkhand, Madhya Pradesh, Tamil Nadu, and Maharashtra were 1200, 128604.92, 2640, 1500, 120, 10092, 148115, 126015.48, 44002.5, 624219.47, 128837.67, 6750, 9050, 1989, 82007.67, 153068.06, 153, 705, 660, 13600, 130636, and 118031.5, respectively as depicted in the Fig. 16.1. Whereas, the same for the Tripura, Manipur, Puducherry, Andaman and Nicobar island, Meghalaya, Chandigarh, Sikkim, Lakshadweep, Arunachal Pradesh, Daman and Diu, Nagaland, Bihar, Mizoram, and Dadara and Nagar Havelli were unavailable [3]. The above-mentioned data infer that the Uttar Pradesh occupies the top place in view of the waste handling installed capacity whereas the Goa remains at the bottom of the listing. Waste management system (WMS) is one such procedure used by organizations to avoid/minimise/repurpose this waste [4] and supports in attaining one of the Sustainable Development Goals [5]. Manufacturing organizations, dealing authorities, schools, colleges/institutes, informal industries/sectors, government, among others have been the main stakeholders of the WMS [2] adopted in the Indian subcontinent. Waste disposing (WD) is another terminology for the WMS. The WD/WMS involves businesses incorporating strategies to smartly handle wastes from originating point to disposing point. A few of the WD techniques are as stated in the subsequent text:

- Recycling,
- Germinate,
- Cremation,
- Landfilling,
- Phytoremediation,
- Energy generation through waste, and
- Minimization of waste.

Figure 16.2 delineates the waste managing in the generalized frame. At first, designing and production take place. These are succeeded by utilization, completion of useful life span, accumulation and disposing of useless/undesired objects/things, recycling, reusing, and value addition. The latter again outputs in the foremost (i.e., designing and production) step. The term waste management (WM) describes the several strategies/techniques regarding handling and disposing of the waste. Below mentioned processes are usually associated with the waste:

- Removing,
- Processing,

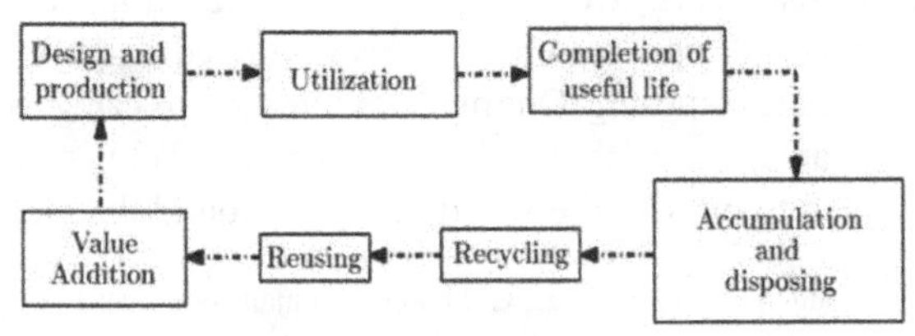

Fig. 16.2 Waste managing in the generalized frame

- Recycling,
- Reusing, and
- Operating in a controlled way.

Minimizing the amount of useless objects/things and obviating any health and/or environmental risk(s) remain the major concerning issues of the WMS. Table 16.1 lists the harmful material(s) present in different EW [6].

Table 16.1 Harmful material(s) associated with EW

S. No.	Name of the EW	Associated Harmful Materials
1	Batteries	Cadmium
2	Circuit boards	Cadmium and Lead
3	Electrical wiring	Copper
4	Printed circuit boards	Metals- rare, precious, and base
5	Cathode ray tubes	Lead
6	Switches and relays	Mercury

1.1 Research Problem

The main contribution of this work is in enhancing the investigations on the electronic waste management (EWM) in the India. The motivation for this work is in the fact that comparatively lesser studies have been reported regarding the thorough EWM investigations in the Indian context contrary to the other countries.

1.2 Manuscript Organization

The organization of this write-up is as stated in the subsequent text: The next section discusses about the existing works in the domain of WM. The next part provides the techniques/approaches adopted for the WM. The ending section includes the conclusive facts and possible future perspectives of the investigated subject.

2. Literature Review

Present trends, rules/laws/regulations, challenging aspects, and managing approaches with respect to India have been incorporated in [4]. Data inadequacy of the EW is one of the biggest challenges regarding the WM. Additionally, enhancement of human awareness regarding various environmental and/or health issues of the EW is another major challenge for the same. A variety of electronic WM techniques have been reported in [7]. The write-up in [8] provides the comparative analysis of the WM in the world's top two populating nations. Detail of entities contributing to the EW has also been provided in this work. The connection between the EWM and the inexhaustible power/energy generation has also been covered in it. Key points related to formal along with informal/improper recycling of the EW have also been discussed in this. These points reveal that the latter methodology is harmful from environment, circular-type economy, and health point of view. Table 16.2 incorporates listing of harmful health effects of various elements present in the EW.

Table 16.2 Health issues caused by various elements of the EW

S. No.	Element	Problematic Issues
1	Silver	Respiration, neuro, liver, and renal (kidney) related
2	Gallium	Chest, throat, and breathing related
3	Zinc	Stomach, derma, smell, vomit, taste related
4	Bismuth	Mind and derma related
5	Tin	Breathing, derma, urine, eye, head pain related
6	Germanium	Blood, derma, and eye related
7	Antimony	Respiratory, derma, and eye related
8	Selenium	Respiratory, liver, throat, abdominal related
9	Yttrium	Liver related
10	Cobalt	Thyroid, respiratory (like Asthma), vomit, circulatory system, and vision related
11	Iron	Respiratory (lungs) related
12	Chromium	Liver, renal, and respiratory related
13	Molybdenum	Panic problems (related to legs/feet, knee, and hands) on the account of its toxicity
14	Lead	Enhanced blood pressure, problems associated with neural, infertility, and renal, reduced learning capabilities (specifically of children)
15	Cadmium	DNA, respiratory, nervous, and bone related and even cancer

[3] incorporates insightful content regarding the EWM in the context of Maharashtra state of the Indian subcontinent. Various awareness activities regarding the EWM has made continuous increment in the recycling of the harmful elements in the nation. Apart from this, the work also addresses the illegal-mannered transferring of the EW from one nation to another. [9] along with [10] provides insights about waste auditing (WA). Widely practicing improper/informal EW managing, lack of awareness among human beings regarding environment and/or health deterioration, lacking strict governmental policies, etc. have been reported as the key problematic issues in the literary articles. These issues can be overcome by developing proper waste management techniques for the EW.

3. Waste Management Techniques

Disposal, regulation, monitoring, and collection are among the techniques employed for the EWM [7]. The local/regional government authorities usually provide waste accumulation facilities without any price. Disposing of the accumulated wastes can be done in various ways. A few of these ways incorporate landfilling, cremation, etc. Out of various types of waste, solid ones are subjected to ignition for producing heat, steam, gas, and ash. The ignition also incorporates solids' volume reduction by eighty to ninety five percent. Cremation-led waste disposing creates trouble through air pollution. A lot of approaches and techniques have been suggested regarding the WM. One such approach/technique, as practicing in the Japan, is as depicted in Fig. 16.3.

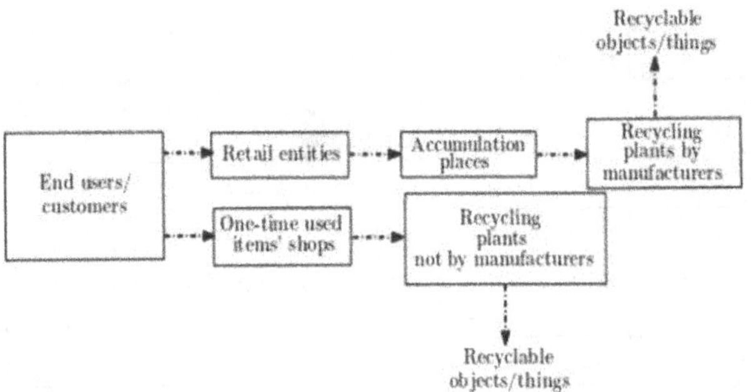

Fig. 16.3 Delineation of the Japan's waste managing framework

According to this approach, the utilized objects/things from the end users/consumers move to retail entities and single time used objects/things' shop. The objects/things outputted by the retail entities move to the accumulation places. From here, these go towards the manufacturers' operated recycling units. Such units provide the recyclable objects/things. Whereas, the objects/things from the single time utilized shop seek movement to the recycling units, which are not operated by manufacturers. These units also yield the recyclable objects/things. An organization can customize WMSs by combining or rearranging these tactics. The goal of contemporary WM techniques is to attain the sustainability. Waste reduction, reuse, and recycling are different steps for handling of the EW. The details of the recycling are as per the subsequent write-up.

3.1 Recycling-Led Solutions

Recycling/physical reprocessing remains the paramount methodology for disposing of inorganic waste such as metal, plastic, to name a few. It helps in lowering greenhouse gas emission as well as pollution, saving energy, sustaining resources, etc. However, the monetary amounts associated with the recycling create stoppage in its proper implementation. Preference of the germination process, in this regard, is on the account of its capability of turning organic-categorized waste into nutrient rich manure. Here, it is also notable that organic-categorized waste (such as paper, food, etc.) is also recyclable. Contrarily, the energy generated through waste can be referred as conversion of non recyclable waste utilizing inexhaustible resources of energy (such as anaerobic-manner digesting, plasma gasification (PG) resulting fuel, heat, electricity, among others).

Hazardous wastes are convertible into syngas via a procedure named PG. The latter incorporates a plasma-filled vessel and operate in the presence of high temperature and little oxygen. Bioremediation, which uses microorganisms to remediate pollutants, poisons, and contaminants, is an additional technique for managing the hazardous wastes. Five rules/regulations (five Rs) of the WMS are as stated below:

(a) Refusal: Saying no to unnecessary things like single-use plastics/excessive packaging, among others is termed as refusal.
(b) Reduce: This remains related to reducing waste amount by making more aware consumer/end-user decisions with lesser resources.
(c) Reuse: This involves finding the ways to repurpose things/objects/materials instead of discarding them post their single usage. This could be brought about via durable, substantial products or enabling revised usage of utilized things/objects/materials.
(d) Repurposing: It implies paving the ways for further utilization of single used things/objects/materials, enhancing their usability, and curtailing the waste to the least.
(e) Recycling: This R is meant for reducing the necessity of untreated objects/materials via transforming utilized objects/things into newer ones; in-turn lowering their influence on the environment [1]. The beneficial perspectives of recycling of the EW are as written subsequently: The recoverable monetary amount of the EW is quite significant, rarely-existing materials (i.e., the ones having similar chemical behavior) are associated with the EW, industries based on this bring employment opportunities, and the concentrated amount of materials remains higher than the usual mining operations in the EW.

Major goals of these laws are as per the subsequent write-up: Firstly, to enthuse ethical and sustainable WM methodologies and secondly to make human beings and organizations more conscious regarding reuse of the things/objects, which get discarded after single time consumption. A few of the recycling methodologies are hydrometallurgical, direct, pyro metallurgical, etc.

3.2 WA-Led Solutions

Evaluation of an organization's WMS is termed as the WA. The latter examines the transferring of waste from production to disposing. Typical methodologies for managing garbage include examining of the agreements with recycling business organizations/institutions with keeping an eye on the WD records; evaluation of waste generating actions through discussions with persons of an organization by a team of internal audit authorities; physical accumulation, categorization, and weighing of waste accumulated from all departments or the one that gets produced everyday within organizations/institutions. The most preferred ways for the WA incorporate strictly following the auditing results, setting aside individual protection way(s), finding (in advance) location(s) for sorting, and avoiding disclosure of the auditing date(s) to the organizations/institutions.

Moreover, developing corrective planning corresponding to each potential outcome is also supportive in this matter. For instance, if an organization receives checklists for waste audits then a digital waste audit checklist needs to be prepared before, during, and after the audit. Though details of the organization's waste stream may also be included, it usually consists of preparations, waste arranging, and subsequent actions/measures. Digital trash audit list(s) are more convenient than paper-based listings. Moreover, such lists are also environment supportive. It should be assured that an organization/ institution has a well-functioning waste audit mechanism in place [9, 10]. Waste audit team's preparedness needs to be assessed while looking over the sorting apparatus and developing plans for corrective action(s). Listing of the aims and projections needs to be verified. Issues in light of prior waste audit findings and initial projections to assess the organization's template for auditing waste have to be examined. WMS can be analyzed by outlining the recycling and trash operations taking place at concerned instants. Examination of waste collection bills is relevant to find out the amount to not recycle. Waste needs to be sorted according to the materials (such as steel, aluminum, plastic, paper, to name a few). Approximate share of each subcategory in the waste should be noted. Inclusion of pictures/images is necessary from the future point of view. Institutions can become the agents of a widespread ideological shift in the way society views waste by enabling employees to be more aware of the environmental impact. Professionals across all industries should strive to avoid needlessly dumping old/utilized resources and buying new materials on a large scale.

3.3 Minimizing Waste via Materials' Circularity

Circularization of the materials utilized in various equipment along with printed circuit board supports in the minimization of the EW [11]. It is very important for developing and populated country like India to take measures for the circulation of this EW in a proper channel so that those who are deprived of it could get that in a nominal charge. For example, it can be provided as a scrap to the industries.

3.4 Selecting Recyclers

Latest technologies, like the one discussed in [12], should be incorporated to choose the best recyclers regarding the EW. These would support in attaining the goal of the EW minimization. It is to be noted that the selection of the recyclers should be made according to the category of the material. Appropriate guidelines to be provided by the government to the concerned industries. It is to be made mandatory to use these guidelines otherwise their licenses should be put on hold or even canceled. This will result in the creation of awareness and responsibility towards the environment for aligning with the Sustainable Development Goals. It is the collective responsibility of both the governments and the stakeholders to show their concerns on this point.

3.5 Engaging End Users

Government authorities should make the policies for the disposing of EW in a planned manner looking upon the availability of resources and technologies with engagement of the end users. It would also support in handling of the EW with efficacy [13]. These make the end users understand their responsibilities more clearly. The end users should be made aware about the importance of the proper/formal handling of the EW.

4. Conclusions and Future Scope

EWM, specifically in the Indian context, is of significant concern in today's era considering the Viksit Bharat 2047 vision. In this regard, thorough investigations related to the EWM are necessary for attaining the goal of developed country from the presently developing nation. Moreover, the programs related to cleaning and manufacturing in the country itself would also be boosted through the proper EWM. Proper managing via recycling and refurbishing of the EW would help in attaining the United Nations's goals of development in the sustainable manner.

Surveys/studies conducted regarding the human awareness on the health and/or environmental impacts of the EW reveal that still there is a scope of bringing new rules/policies for its enhancement. A lot of investigations regarding proper/formal managing of the EW are also requisite to overcome the harmful effects caused by improper/informal EWM. Youth of the nation should actively participate in the different programs adopted by our government for the EWM. More number of institutions, dedicated to EWM and emphasizing on innovative ideologies, capacity development, and entrepreneurship, need to be established. More and more number of human beings need to be skilled/trained in the view of the EWM. This would help them in getting decent jobs in the various sub domains of the EWM which in turn would enhance the employability of our subcontinent. Requisite infrastructure also needs to be erected to accelerate the research and development, product development, testing set-ups, among others.

Start ups would also be supportive in the implementation of systematic EWM strategies. Artificial intelligence has been one of the key technologies during the recent years. However, its applicability in the EWM has still not been investigated as per the understanding of the authors. Likewise, the applicability of the machine learning tactics in the EWM has yet to be explored.

Acknowledgement

The authors convey gratitude to everyone, who supported in the conduction of the presented investigations in easy manner.

References

1. Awasthi, A. K., Zeng, X., Li, J. (2016). Environmental pollution of electronic waste recycling in India: A critical review. *Environ. Pollut.*, 211:259–270.
2. Available online: https://greene.gov.in/.
3. Bagwan, W. A. (2024). Electronic waste (E-waste) generation and management scenario of India, and ARIMA forecasting of E-waste processing capacity of Maharashtra state till 2030. *Waste Manag. Bul.*, 1(4):41–51.
4. Arya, S., Kumar, S. (2020). E-waste in India at a glance: Current trends, regulations, challenges and management strategies. *J. Clean. Prod.*, 271:122707.
5. Available online: https://sdgs.un.org/goals.
6. Gibson, K., Tierney, J. K. (2006). Electronic waste management and disposal issues and alternatives. *Environ. Claims J.*, 18 (4):321–332.
7. Kiddee, P., Naidu, R., Wong, M. H. (2013). Electronic waste management approaches: An overview. *Waste Manag.*, 33(5):1237–1250.
8. Awasthi, A. K., Li, J. (2017). Management of electrical and electronic waste: A comparative evaluation of China and India. *Renew. Sustain. Energy Rev.*, 76:434–447.
9. Slutzman, J. E., Bockius, H., Gordon, I. O., Greene, H. C., Hsu, S., Huang, Y., Lam, M. H., Roberts, T., Thiel, C. L. (2023). Waste audits in healthcare: A systematic review and description of best practices. *Waste Manag. Res.*, 41(1):3–17.
10. Taylor, A. L., Levin, J., Chan, J., Lee, M., Kasitinon, D., Miller, E., Fox, P. (2021). Improving environmental sustainability in outpatient clinics: Lessons from a waste audit. *J. Clim. Chang. Health*, 4:100070.
11. Chakraborty, M., Kettle, J., Dahiya, R. (2022). Electronic waste reduction through devices and printed circuit boards designed for circularity. *IEEE J. Flexible Electron.*, 1(1):4–23.
12. Rani, P., Mishra, A. R. (2020). Novel single-valued neutrosophic combined compromise solution approach for sustainable waste electrical and electronics equipment recycling partner selection. *IEEE Trans. Eng. Manag.*, 69(6):3139–3153.
13. Bernardes, M., Moraes, F. T., Tanaka, K. H., da Silva Lima, R. (2024). Engaging the end user in waste from electrical and electronic equipment management: An action research study. *Syst. Pract. Action Res.*, 37(1):105–126.

Note: All the figures and tables in this chapter were made by the authors.

Intelligent Systems Using Semiconductors for Robotics and IoT – Dinesh Goyal et al. (eds)
© 2026 Taylor & Francis Group, London, ISBN 978-1-041-20408-4

17 Machine Learning Ensembles for Analyzing Textual Data

Jyoti Godara[1], Sugandha Singh[2]
Department of Computer Science and Engineering,
Shree Guru Gobind Singh Tricentenary University,
Gurugram, India

Abstract: The way people express themselves online has changed significantly, especially on social media sites like Twitter, which offer sentiment-rich data that is vital for businesses in banking, agriculture, healthcare, education, marketing, the economy, and stocks. Sentiment analysis, which is essential for making decisions, concentrates on content created by users, particularly Tweets on Twitter. Because of the enormous volume of data, automated techniques like machine learning are required. The study provides an ensemble classifier (Support Vector Machine, Decision Tree, and Logistic Regression) with a sarcasm-specific detection strategy that outperforms other classifiers in precision, accuracy, recall, and F1-score. This approach addresses accuracy challenges like explicit negation, domain dependence, and sarcasm.

Keywords: Sarcasm, Ensemble, Sentiment analysis, Feature extraction, PCA

1. Introduction

The age of social media is here to stay. Global communication has changed as a result. People can now use social media to showcase themselves with just a fingertip tap. It is widely used to share opinions, sympathy, and ideas about any topic or image shared on social networking sites.[1]. A phenomenon known as sarcasm occurs when the indicated feelings in a document or speech deviate from its literal meaning. Speaking or writing with sarcasm often conveys the opposite meaning from what is intended. Saying something sarcastically, such "Woww! He's cleverer than Shehu Jaha!" implies that the target said something straightforward or obvious, and the speaker is making fun of him. Sarcasm is widely utilized on Twitter because it may be used to poke fun at other people as well as to criticize their opinions, viewpoints, and other things [2] Sarcasm is a linguistic device used to convey disdain or unpleasant feelings. It is an example of fake civility that inadvertently irritates people. Sarcasm may seem like malice under a thinly veiled façade of dishonesty. Recent studies have shown that those who tease often think their remarks are not as hurtful as their victim thinks they are [4]. But they cause more harm. Because political parties and celebrities are perceived as influential, they are often the focus of disparaging remarks and tagging. Because there is no vocal tone with social media writing, it is challenging to discern sarcasm. When speaking, sarcasm is typically conveyed by the verbal tone. Let us say someone is using sarcasm in conservations on social media. If the user wants to convey a sarcastic tone in the text, they can add extra letters to common terms. A user on social media may express an extended syllable using words. For instance, if someone posts something on social media that other users don't agree with, those users might respond by writing "Right." Nevertheless, using a sardonic wordplay. It is possible to classify the text's usage of the term "Right" as sarcasm. Intense adjectives are typically used to describe hyperbolic language, which also suggests sarcasm in the text. Let us say a social media user is so passionate about a topic that they are being hyperbolic, which is a sign of sarcasm in writing [5][6]. Common words are indicated by their more intense forms, which are sarcastic forms of hyperbole. Sarcasm on social media may be hinted at in

[1]jyotipoonia6@gmail.com, [2]sugandha77.cse@gmail.com

DOI: 10.1201/9781003716389-17

the content through allusions to popular culture. If a user includes any references in the post text, we can determine if they are being sarcastic. For instance, the user references or cites another source to express his political viewpoint in the message. A question that is sarcastic may also be posed by the user. Sarcasm is frequently posted on social media by irate or disappointed users. Thus, it appears to be sarcastic if reading a text message makes you feel violent. Sarcasm is disseminated by people who share contentious debates on social media. The significance of the phrases surrounding text messages on social media can be used to evaluate them. The context of a message can be used to identify sarcasm. If the message's possibly sardonic section is closely scrutinized, sarcasm may result. [3]

A classifier ensemble consists of multiple classifiers that combine their individual decisions to produce a majority decision[21,22]. The method's ultimate goal is to combine the outputs of many models, or "base classifiers," to produce an overall result that outperforms each of the baseline classifiers separately. Thus, the major objective of this research endeavor is to investigate several techniques for sarcasm detection and to develop an ensemble classier for sarcasm detection that will improve the performance of the sentiment analysis system.

The remainder of the work is structured in this manner. Important studies on sarcasm are compiled in Section 2 using variety of methods. The suggested model's technique is explained in depth in Section 3. The experimental results and performance comparison are presented in Section 4. In Section 5, the research is completed with suggestions for additional study.

2. Related Work

Sentiment analysis is a sort of text mining that discovers and extracts subjective information from data and distinguishes the emotional tone hiding beneath the text's body. By keeping an eye on their online conversations, several organizations and businesses use this widely accepted method to classify and comprehend opinions regarding ideas, services, and/or products. As a result, scientists are becoming increasingly interested in this field. In this part, sentiment analysis is reviewed in detail.

A pre-processed data model for tweet filtering based on natural language processing, or NLP, was presented by the author of [7]. Next, the motion of the Bag of Words (BoW) was integrated with TF-IDF (Term Frequency-Inverse Document Frequency) for analyzing the sentiments.

These techniques worked well for correctly classifying tweets as either good or negative. The TF-IDF vectorizer was used to maximize the accuracy of the sentiment analysis. The simulation's outcomes demonstrated how well the proposed model performed and how accurately it could assess sentiments of accuracy about 85.25 percent.

The author of [8] developed an SVM (Support Vector Machine) technique that improved on the conventional n-gram classification approach by integrating internal attributes based on n-grams with an external sentiment vector. To characterize the feelings on Twitter, n-grams of Twitter data were combined with a variety of textual features and an ML (machine learning) classifier. To comprehend the impact of weighting on accuracy, three different weighting approaches were used. Furthermore, extra data was extracted from a tweet sentiment score vector for improving the performance of existing algorithm.

The author of [9] aimed to use an NB (Naive Bayes) algorithm to classify Twitter data sentiment using ML (Machine Learning) results. Even though this approach took a long time, it helped with estimate in an effective manner. The tweets were extracted, pre-processed and then subsequently categorized into three sentiments using this algorithm. The planned methodology was helpful to identify the most efficient mobile operating systems based on the opinions expressed by users of social media. Evaluation results showed that the intended method was flexible.

A strategy to analyse attitudes was devised by the author in [10], for which a Hadoop framework and DL (deep learning) model were deployed. The original framework was used to distribute the data and extract the properties. The Twitter data was then used to extract the important characteristics. The deep recurrent neural network, or DRNN, was used to classify the input data as positive or negative by giving each piece of input Twitter data a real-valued review. Several factors were considered when analyzing the developed method. In comparison to conventional methods, the created strategy was able to improve specificity around 0.9157, sensitivity up to 0.9404, and accuracy up to 0.9302.

The author of [11] aimed to develop a novel paradigm for analyzing social network sentiment using Twitter sentiment score (TSS) [50]. To reduce computation effort and increase forecast accuracy, the planned TSS framework produced a novel baseline correlation technique. This approach was flexible enough to make the choice without requiring the deployment of data pertaining to previous observations. Using this method, the future trend in the stock market was forecasted sooner, and the accuracy attained was assessed to be 67.22%. With the aid of LR (Logistic Regression) and LDA (Linear Discriminant Analysis), desired framework produced accuracy of approximately 97.87% to forecast the future market's upward trend.

The author of [12] provided an overview of the literature on the different facets of sentiment analysis that was published between year 2002 and 2015. A detailed discussion was held on various machine learning algorithms, natural language processing methods, and sentiment analysis applications. It was discovered that irony or caustic words are frequently used in online and political discussions, necessitating the use of more computational methodologies.

The author provided a thorough analysis and re-implemented many techniques for Arabic sentiment analysis in [13]. One finding was that transformer-based with F-scores of 0.76, 0.69, and 0.92 on the ASTD, SemEval and ArSAS benchmark datasets, the language model performed the best. It was found that sarcasm has a significant impact on sentiment analysis.

The author of [14] concentrated on improving the effectiveness of the sarcasm detection method. Bootstrap technique was used to obtain a longer role pair list than a role pair extraction technique. Additionally, the weighting strategies were applied to each pair of roles. After that, a comparison between the suggested strategy and conventional procedures was carried out. The analysis proved that the suggested strategy could improve the role pairs. The results of the studies demonstrated that the topic similarity technique that was given became effective in reducing the impact of various types of noise pairs.

The author of [15] created a novel deep learning (DL) architecture for sarcasm detection that included common sense information. To generate relevant commonsense information, a pre-trained COMET system was used. Additionally, two methods of choosing knowledge were contrasted to see how that knowledge affected performance. Lastly, a knowledge-text integration module was used to model both the text and the knowledge. The applicability of the suggested framework was proven by the outcomes of experiments conducted on three datasets.

The author suggested a technique for identifying sarcasm on Twitter in [16] by concentrating on two categories of feature words. The first set of terms are features-modified terms, whereas the second set of phrases denotes roles. It yielded insightful results; however additional investigation is required into the significance of feature words and the application of weighting in classifiers.

The author of [17] presented a supervised machine learning-based method for identifying sarcasm on Facebook that considered both the interaction of users with the posts and their content (text, image, etc.). The results showed that social media applications are a good fit for supervised learning algorithms, especially ensemble learning algorithms. Future research could examine how spam messages and duplicate data affect how well sarcasm is detected.

3. Methodology

The literature review indicates that sarcasm in the text causes the sentiment analysis to perform low. Therefore, it is necessary to identify the sarcasm. Many machine learning algorithms and lexicon-based techniques have been used to identify sarcasm, but the study found that using ensemble classifiers may have increased the process's overall sentiment analysis accuracy as well as its ability to identify sarcasm in textual data.

Thus, the following are the motivation for using ensemble learning:

- To achieve the highest level of classification precision.
- Diminishes the likelihood of incorrect data classification.

3.1 Proposed Methodology

This study proposes an ensemble classifier for sarcasm detection. The method of detecting sarcasm involves several steps, including pre-processing, features reduction, grouping, and the classification. The characteristics are minimized by applying the PCA approach. Similar and dissimilar types of data are grouped together using K-mean clustering. To do the classification of data, a variety of voting classifier models are developed. Voting ensemble classification model SKD is developed. This model combines the Support Vector Machine, K-Nearest Neighbor, and Decision Tree classifiers via voting.

Following are the steps:

1. Data Collection

Data is gathered from several domains to understand the model's effectiveness. As discussed in [18], two datasets—the "Tweet" databases [19] and the "News headline" datasets [20]—are merged to create a single dataset. 81,408 randomly produced tweets that have been classified as figurative (including sarcasm and irony), regular, sarcasm, and irony are included in the Tweet dataset [19]. 39,267 tweets are obtained by extracting the sarcasm and regular tags from the tweet's dataset. Of the tweets, 20,681 are labelled as sarcastic and 18,595 as regular. As shown in Table 1, combining the datasets into a single dataset produces 67,895 records for testing and training the model. Figure 17.1 displays the number of sarcastic and non-sarcastic records in the collection.

Table 17.1 Details of dataset

Dataset	No. of rows
Tweets	39,267
News headlines	26,709
Total	67,895

2. Preparing Dataset for Analysis

Data collection is followed by cleansing. It assists in eliminating noise from the data, as stated below:

- Removal of hashtags: Prior to conducting tests on the tweets, the hashtags, which are typically used for monitoring, are removed.

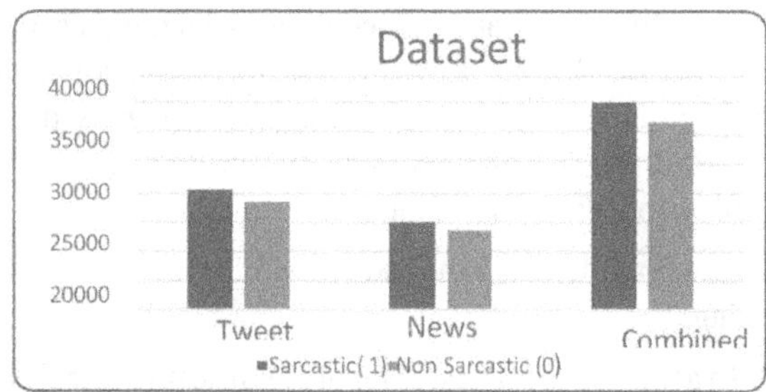

Fig. 17.1 Number of sarcastic and non-sarcastic records

- Removal of hashtags: Prior to conducting tests on the tweets, the hashtags, which are typically used for monitoring, are removed.
- Eliminating extraneous items: Although author names, punctuation, and hyperlinks are not relevant to the experimental setting, they are removed from tweets.
- Elimination of undesired tweets: When comparable tweets are retweeted, redundancy may occur in the corpus. They must therefore be removed in order to prevent skewed outcomes. Additionally, tweets containing fewer than three words are disregarded.
- Tokenization: A set of words can be split down into symbols, phrases, words, or other valuable elements.
- Stop word removal: Terms that are eliminated from text either prior to or following pre-processing are known as stop words. These are human-controlled and often fail miserably at classifying text, like a, an, and the. The technique of reverting evolved words to their simplest form, as in Attending—stemmed to—attend, is known as stemming.
- Lemmatizing: Since stemmers simply remove affixes and do not add the necessary characters that compose the root of a whole meaningful word, they often render a phrase nonsensical. For this, Lemmatizer is responsible. It produces a semantically equivalent result by adding the absent characters to the root in addition to removing the affixes.

3. Reduction and Extraction of Features

Utilizing the random forest model, valuable features can be extracted from the dataset. The use of PCA reduces the number of retrieved characteristics. The PCA technique assesses the effective amount of variance in the data by creating a less-dimensional representation of the data. The mathematical method known as principal component analysis, or PCA, divides a set of connected variables into a set of unrelated linear subsets that depend on a transition to generate uncorrelated variables. It is described as an orthogonal linear transformation that enables projecting the initial set of data to a second projection system by projecting the first coordinate by the largest variance and the second largest variance, which is having a second coordinate projection and is vertical to the existing first component. PCA is primarily used to increase the variance of the data in the projected space and enable a linear transformation.

4. Grouping Information that is Similar

K stands for Clustering is a technique for grouping data kinds that are similar to one another. The k-mean approach determines the distances between each sample and the cluster center after choosing K points from the data pattern to serve as the initial clustering center. The class closest to the cluster center is the one to which this sample belongs. The average of each newly generated clustering data object is then calculated in order to determine the new clustering center. The sampling change is complete and the clustering primary function has reached its maximum value when the clustering center of two consecutive times shows no change. Eventually, each stage is repeated.

The entire code was implemented using Spyder, a powerful Python programming environment featuring sophisticated editing, testing, and numerical computing settings. To handle datasets, the Pandas library is utilized. The Scikit-learn libraries are used for assessment, classification, similarity metrics, and feature representation.

The ensemble classification model combines decision trees and K-nearest neighbor algorithms from Support Vector Machines. Figure 17.2 provides an explanation of the detailed model.

4. Results and Performance Comparison

The parameters for evaluating performance are as follows:

4.1 Precision and Recall

In text mining and related domains like information retrieval, precision and recall are two measures that are

Table 17.2 Performance comparison

Performance parameters	Adaboost	Decision Tree	Random Forest	K Nearest Neighbor	Proposed approach
Accuracy	89.25%	89.72 %	53.42%	87.51%	**92.9%**
Precision	89%	90%	71%	88%	**92%**
Recall	90%	90%	53%	88%	**93%**
F1-Score	89%	90%	40%	88%	**92%**

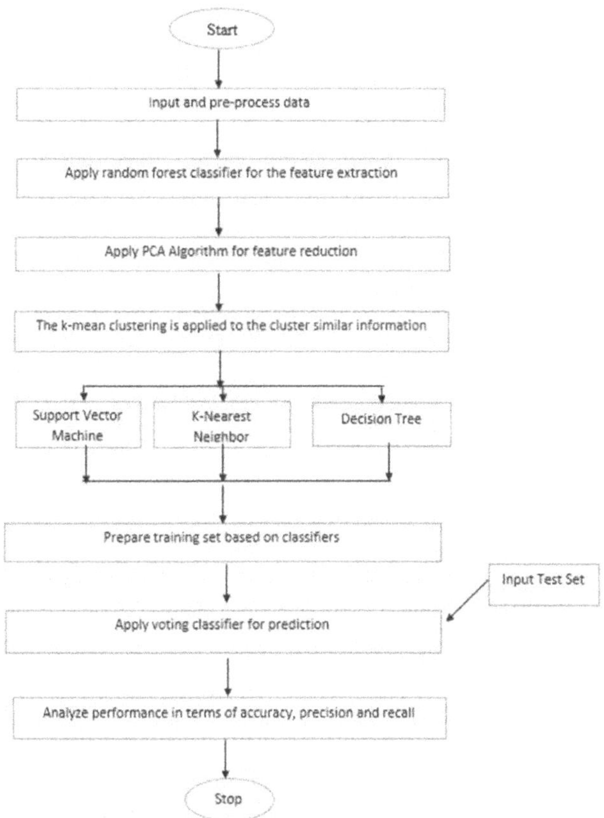

Fig. 17.2 Flowchart for the proposed model

frequently used to evaluate performance. These are the criteria that are used to assess completeness and accuracy.

$$\text{Precision} = \frac{\textit{True Positive}}{\textit{True Positive} + \textit{False Positive}}$$

$$\text{Recall} = \frac{\textit{True Positive}}{\textit{True Positive} + \textit{False Negative}}$$

4.2 The F-Measure

The harmonic mean of recall and precision is known as the F-Measure. The value produced by the F-measure finds a compromise between recall and precision.

$$\text{F-measure} = \frac{2 * \textit{recall} * \textit{precision}}{\textit{precision} + \textit{recall}}$$

4.3 Correctness

The most popular criterion for assessing a categorization skill is accuracy. The ratio of correctly identified examples to all instances is known as accuracy, whereas the ratio of incorrectly classified cases to all occurrences is known as error rate.

$$\text{Accuracy} = \frac{\textit{True Positive} + \textit{True Negative}}{\textit{True Positive} + \textit{False Positive} + \textit{True Negative} + \textit{False Negative}}$$

An ensemble of distinct classifiers is referred to as an ensemble classifier. SVM, decision trees and KNN are combined in the first ensemble classifier. Performance metrics for each ensemble classifier include accuracy, precision, recall, and f1-score.

When the sentiments are analyzed, the presence of sarcasm in the text presents a significant obstacle. Many cutting-edge algorithms for sarcasm detection have been developed in an effort to improve sentiment analysis performance. We evaluated the Adaboost, Decision Tree, Random Forest, and K-Nearest Neighbor classifiers against the recommended ensemble classifier in this section based on the same dataset and it is represented in Fig. 17.3.

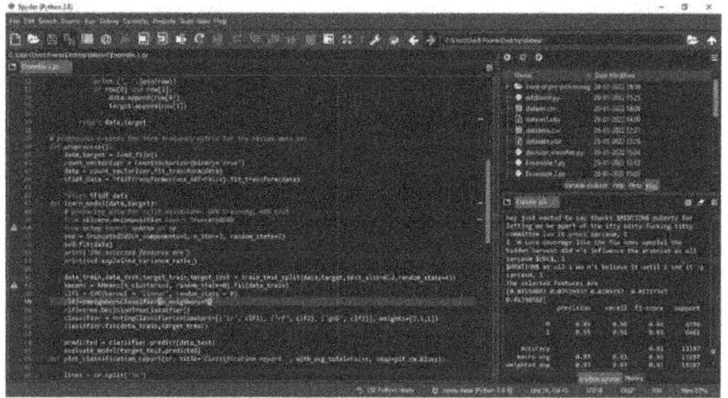

Fig. 17.3 Snapshot of the outcome achieved by the proposed approach

5. Conclusion and Future Scope

It is concluded in this study that sarcasm is a type of verbal irony that places a strong emphasis on communicating mockery. Sarcasm implies a pessimistic attitude. It is devoid of any unfavorable surface feeling, nevertheless. A sarcastic statement may have a surface sentiment that is good, negative, or neutral. There are numerous methods for sarcasm detection. The data is pre-processed, features are retrieved and reduced, clustering and classification are the steps of the Sarcasm detection approaches. We have conducted the study using data from social networking sites. In order to identify sarcasm, we have developed and put into practice an ensemble approach. The data is pre-processed using the tokenization approach, extraction of features are done using the random forest algorithm, feature reduction is accomplished with the Principal Component Analysis algorithm, data clustering is accomplished with K-mean, and four distinct ensemble classifiers—combinations of multiple classifiers—are designed during the classification phase. SVM, KNN, and decision trees are combined in the ensemble classifier. The dataset, which combines tweets and news headlines, is used to evaluate the performance of each ensemble model. Measured performance metrics for the ensemble models include F1 score, recall, accuracy, and precision. It is examined that the proposed ensemble classifier outperformed alternative algorithms in terms of several parameters related to the identification of sarcasm.

Future opportunities for the proposed model for sarcasm detection can be expanded to include images, audio, video, and memes and examination of neutral tweets.

References

1. Sharma, D. K., Singh, B., Agarwal, S., Pachauri, N., Alhussan, A. A., & Abdallah, H. A. (2023). Sarcasm Detection over Social Media Platforms Using Hybrid Ensemble Model with Fuzzy Logic. Electronics, 12(4), 937.
2. Alaramma, S. K., Habu, A. A., Ya'u, B. I., & Madaki, A. G. (2023). Sentiment analysis of sarcasm detection in social media. Gadau Journal of Pure and Allied Sciences, 2(1), 76–82.
3. Saleem, H., Naeem, A., Abid, K., & Aslam, N. (2023). Sarcasm detection on twitter using deep handcrafted features. Journal of Computing & Biomedical Informatics, 4(02), 117–127.
4. M. Karqmibekr and A.A. ghorbani, "Sentiment Analysis of a Social Issues," International Conference on a Social Informatics, USA, pp. 215–221, 2012.
5. M. Bouazizi and T. Ohtsuki, "A Pattern-Based Approach for Multi-Class Sentiment Analysis in Twitter," in IEEE Access, vol. 5, pp. 20617–20639, 2017
6. M. Bouazizi and T. Otsuki Ohtsuki, "A Pattern-Based Approach for Sarcasm Detection on Twitter," in IEEE Access, vol. 4, pp. 5477–5488, 2016
7. Md. Rakibul Hasan, Maisha Maliha, M. Arifuzzaman, "Sentiment Analysis with NLP on Twitter Data", 2019, International Conference on Computer, Communication, Chemical, Materials and Electronic Engineering (IC4ME2)
8. Sheeba Naz, Aditi Sharan, Nidhi Malik, "Sentiment Classification on Twitter Data Using Support Vector Machine", 2018, IEEE/WIC/ACM International Conference on Web Intelligence (WI)
9. Mansour Alshaikh, Mohamed Zohdy, "Sentiment Analysis for Smartphone Operating System: Privacy and Security on Twitter Data", 2020, IEEE International Conference on Electro Information Technology (EIT)
10. Mudassir Khan, Aadarsh Malviya, "Big data approach for sentiment analysis of twitter data using Hadoop framework and deep learning", 2020, International Conference on Emerging Trends in Information Technology and Engineering (ic-ETITE)
11. Xinyi Guo, Jinfeng Li, "A Novel Twitter Sentiment Analysis Model with Baseline Correlation for Financial Market Prediction with Improved Efficiency", 2019, Sixth International Conference on Social Networks Analysis, Management and Security (SNAMS)
12. Ravi K, Ravi V. A survey on opinion mining and sentiment analysis: tasks, approaches and applications. Knowledge-Based Systems 2015; 89: 14–46.
13. Farha, I. A., & Magdy, W. (2021). A comparative study of effective approaches for arabic sentiment analysis. *Information Processing & Management*, *58*(2), 102438.
14. Satoshi Hiai, Kazutaka Shimada, "Sarcasm Detection Using Role Pairs: Bootstrapping Role Pairs and Weighting", 2020, International Conference on Asian Language Processing (IALP)
15. Jiangnan Li, Hongliang Pan, Zheng Lin, Peng Fu, Weiping Wang, "Sarcasm Detection with Commonsense Knowledge", 2021, IEEE/ACM Transactions on Audio, Speech, and Language Processing
16. Hiai, S., & Shimada, K. (2018, February). Sarcasm detection using features based on indicator and roles. In International Conference on Soft Computing and Data Mining (pp. 418–428). Springer, Cham.
17. Das, D., & Clark, A. J. (2018, October). Sarcasm detection on facebook: A supervised learning approach. In Proceedings of the 20th International Conference on Multimodal Interaction: Adjunct (pp. 1–5).
18. Jamil, R., Ashraf, I., Rustam, F., Saad, E., Mehmood, A., & Choi, G. S. (2021). Detecting sarcasm in multi-domain datasets using convolutional neural networks and long short term memory network model. *PeerJ Computer Science*, *7*, e645.
19. John, N. (2020). Tweets with sarcasm and irony. *dirección: https://www. kaggle. com/nikhiljohnk/tweets-with-sarcasm-and-irony*.
20. Misra R. 2019. News headlines dataset for sarcasm detection. Available at https://www.kaggle.com/ rmisra/news-headlines-dataset-for-sarcasm-detection.
21. Alqahtani, A. F., & Ilyas, M. (2024). An Ensemble-Based Multi-Classification Machine Learning Classifiers Approach to Detect Multiple Classes of Cyberbullying. *Machine Learning and Knowledge Extraction*, *6*(1), 156–170.
22. Luqman, M., Faheem, M., Ramay, W. Y., Saeed, M. K., & Ahmad, M. B. (2024). Utilizing ensemble learning for detecting multi-modal fake news. IEEE Access.

Note: All the figures and tables in this chapter were made by the authors.

Intelligent Systems Using Semiconductors for Robotics and IoT – Dinesh Goyal et al. (eds)

18 Comprehensive Review: Analysis of Use of Multimodal Deep Learning Models in Medical Diagnosis System

Nivedita Shimbre[1], Ram Kumar Solanki[2]
Sandip University, School of Computer Science & Engineering,
Nashik, India

Abstract: The increasing complexity of medical data necessitates more advanced diagnostic approaches. Consequently, multimodal deep learning (DL) models are emerging as powerful tools for processing and analyzing diverse types of data, including genomic data, clinical records, and medical images. This review deals with the use of multi-modal DL models in medical diagnosis systems. These models combine textual data from clinical parameters and electronic health records with imaging modalities like computed tomography, MRI, and X-rays for an overall understanding of patient condition and circumstances. We examine several topologies in the context of managing multimodal inputs, including transformers, and CNNs. The applications of multimodal deep learning models at different levels of medical domains, including but not limited to cancer diagnosis, diabetic retinopathy screening, and cardiovascular disease prediction are also studied. Additional challenges associated with multimodal data integration explored in this work include data fusion processes and model interpretability. We demonstrate how multimodal deep learning models may completely alter the direction of the healthcare sector by increasing the rate of more accurate diagnosis and treatment in a more personalized manner.

Keywords: CNN, Recurrent neural networks, Deep learning, Multimodal, Medical diagnosis

1. Introduction

Technological advancements in the health sector have significantly transformed the processes of medical data collection, processing, and analysis. With large-scale medical data, such as imaging, clinical records, laboratory results, and genomic information, becoming ever more available, the need has never been greater for novel methodologies that are able to integrate this efficiently before evaluating the data for an accurate diagnosis. Healthcare systems have traditionally been dependent on certain discrete sources in the diagnosis of diseases. They usually include radiological images and in vitro diagnostics. While most other health-related problems are concerned with making a prediction based on unimodal input, the complexity of human health issues often requires a more comprehensive approach that involves multiple sources of data in order to build a complete picture of the patient's health. This has increased interest in the multimodal DL model, which can make use of the interdependencies between the differing modalities of the data and integrate them to create a more comprehensive diagnostic performance. Recently, multimodal DL, in the context of DL, would refer to deep learning algorithms applied for the processing and integration of big forms of data, such as images, text, sensor data, and so on. For instance, multimodal models using histopathological images with genomic data have presented better performance for cancer diagnosis.

This is particularly useful in medical diagnosis, since most data modalities convey information that can be complementary to each other, hence coming out with increasingly more correct diagnoses. For instance, structural abnormalities may best be detected by imaging data from CT or MRI scans, while clinical records provide

[1]nivedita.shimbre@gmail.com, [2]ramkumar.solanki@sandipuniversity.edu.in

DOI: 10.1201/9781003716389-18

the essential background information on the health history and symptoms of a patient. This paper draws on the application of multimodal DL methods within the domain of a medical diagnosis system, which can outline how such models can improve the efficiency and speed-reliability of diagnoses. We are going to analyze several deep learning architectures, including transformer-based models, CNNs, and RNNs, which have been effectively implemented in modern multimodal diagnostic systems.

These applications allowed the successful identification of a variety of tumors and predicted how patients would respond to therapy. The same happened in ophthalmology, where models, which combine images of the retina with clinical data, proved more successful in the detection of diabetic retinopathy, the most important blindness caused by the disease. These findings, in particular, have shown great potential for improvement in diagnostic accuracy, and thus the promotion of early disease detection can be achieved by multimodal deep learning, both of which are considered crucial to improved patient outcomes. Further, this paper will review some of the challenges that surround adopting multimodal deep learning into real-world medical contexts. One such challenge is making the deep learning models explainable and understandable from a data viewpoint. Other challenges are heterogeneity in data, imbalanced datasets, missing data, and interpretability. Despite those challenges, several of the results show that multimodal deep learning models can offer disrupted medical diagnosis by offering more comprehensive and data-driven methods of healthcare. These models make a diagnosis even more sharp and personalized, integrating knowledge from several modalities. Eventually, this may lead to greater patient outcomes and enhanced health systems.

The specific aim of this review is thus to describe how such models are designed to represent different types of medical data: numerical data, text data, and image data-each within the framework of common diagnosis. Multimodal models learn from many sources of data in order to obtain a full and more detailed output for diagnosis. They do this by extracting and synthesizing information from each modality. For this reason, this paper will review several data fusion approaches, such as early fusion, late fusion, and hybrid fusion, in the process of medical diagnosis so as to delve into the usefulness of those techniques relative to diagnostic processes. The study has also presented a review regarding cutting-edge technology studies conducted on multimodal DL in various medical domains of oncology, cardiology, and ophthalmology.

To study different existing multimodal approaches used for medical diseases

The researchers reviewed some of the multimodal approaches applied to medical diagnosis systems with various disorders. Because of the complexity of human health, it may not be sufficient to make a diagnosis of medical ailments based on just one source of data. Multimodal techniques combine different medical pictures, clinical records, laboratory results, genetic data, and symptoms reported by the patients to overcome such limitations and obtain more profound knowledge about the health status of a patient. This objective probes how multimodal deep learning models, including CNNs, RNNs, transformers, and hybrid architectures, display and evaluate various types of data. In the study, analysis will be made with regard to how those models influence diagnosis in cancer, cardiovascular, and neurological diseases. Moreover, early fusion-when input data is integrated, late fusion-when outputs from models are mingled-and hybrid fusion will be discussed. The paper will discuss how complicated medical issues such as illness heterogeneity-that at present requires a wider range of diagnostic inputs-are solved by multimodal models and how they may allow more personalized treatment approaches. These fusion strategies have an important role in understanding how multimodal models enhance the accuracy of diagnosis by availing complementary data sources. Advantages of a multimodal diagnostic system, disadvantages, and improvements will be discussed.

Many medical diseases in recent years have been dealt with using different kinds of multi-modal deep learning architecture. Few of them include:

- CNNs: While CNNs are mainly applied in picture data, very often they are also merged with any other models to handle multimodal tasks. For example, clinical notes will be handled by a recurrent model, while the images of a medical nature may be handled by CNNs. A more accurate diagnosis could then be available by merging the output of both results.
- RNNs and LSTM Networks: These work best for processing sequential data in patient histories, time-series data from monitoring equipment, or textual data in medical reports. CNNs mostly are merged with RNNs and LSTMs during the performance of multimodal jobs.
- Transformer Models: More and more of these models, first made for natural language processing, are being modified for multimodal applications. Transformers easily manage complex diagnostic tasks with both textual and visual data input by incorporating the data of many modalities in one go.

Such multimodal techniques have been utilized for the purpose of the treatment of a lot of medical issues, consisting of cancer, cardiovascular diseases, and neurodegenerative disorders. Investigation of such strategies would let scientists find out which of them works best in combining various types of data to improve diagnostic precision.

To review and analyze different existing datasets used for multimodal modeling

Studying and evaluating the datasets employed in assessment and instruction multimodal DL models for medical diagnosis is the second reason for the work. Datasets are vital to machine learning model construction, because their quality, diversity, and comprehensiveness affect model performance and reliability. Multimodal deep learning is considerably harder since models must interpret and integrate data from several sources, each with its own traits and complexities. A systematic review of public and proprietary medical datasets concerning multimodal modelling is necessary. Such datasets would include but are not limited to medical imaging such as CT, MRI, X-ray, or ultrasound images, text data such as clinical notes or electronic health records, and numerical data. The review shall cover cancer based on TCGA, ophthalmology based on Ocular Imaging, cardiology based on MIMIC-III, and so on, and how such datasets are structured, annotated, and put to work in the training of multimodal models. This will involve the preprocessing of multimodal data, the handling of missing data, and the handling of imbalanced data in medical datasets. Variations in dataset size and variety, impacting the model's generalizability-a critical factor in medical diagnosis systems-will also be explored in the research study. This reason is very important because it aids in analyzing strengths and weaknesses of data and pinpoints the research gap. It thus understands current limitations in the datasets, such as the poor representation of minorities or the lack of complete multimodal datasets for some disorders, and allows the study to suggest modifications with a view to developing more robust and trustworthy diagnostic algorithms. This mission will teach academics and practitioners how to exploit and enhance multimodal DL datasets for medical applications.

The efficiency of diagnosis using a multimodal DL model depends on the type and superiority of the available diverse datasets. These datasets should include images, clinical records, genetic sequences, sensor data among other forms of data that the models need to properly integrate to handle. This makes it necessary to review and assess current data sets with regards to how they contribute to the development of the multimodal diagnostic system.

Several datasets are frequently utilized in multimodal medical research

- Medical Image Datasets: Large medical image datasets for different tasks, namely, tumor detection, lung illness diagnosis, and skin cancer classification, among others, are provided by the LIDC-IDRI Lung Image Database Consortium in collaboration with ISIC International Skin Imaging Collaboration and CheXpert Chest X-ray dataset. In most cases, such datasets provide ground truth labelled data that act as an important ingredient for the training of a model.
- Electronic Health Records (EHRs): The datasets like MIMIC-III contain rich textual patient data, including demographics, medical histories, prescription histories, test results, and clinical comments. These are the data needed to create models that incorporate both clinical and imaging data for comprehensive diagnostic analysis.
- Genomic Data: These databases, such as TCGA, provide genomic sequences that are a basis for diagnosis and prediction of molecular diseases. If combined with the findings of imaging and clinical data, treatment suggestions can be personalized.
- Multimodal Datasets: Certain datasets, such as the UK Biobank, represent truly multimodal data, with imaging complemented by EHRs and genomics; thus, these provide comprehensive datasets for multimodal learning. They allow creating models that significantly enhance diagnostic skills, leveraging data from many sources.

A review and assessment of such datasets will give researchers an insight into the type of data available for multimodal deep learning and allow them to observe some gaps and/or limitations that one would need to improve upon when carrying out further research. It is also important to guaranty efficient training and assessment of such multimodal models in medical diagnostics by analyzing the quantity, structure, and quality of such datasets.

Various studies state that multimodal models might give way to the next revolution in healthcare, enabling personalized treatment regimens, more precise diagnoses, and early diagnosis of cancers and other disorders. This study will also try to present a critical review of various applications of multi-modality deep learning models in medical diagnosis systems by showing the outcome of the review study conducted on this theme. The focus is on the integration of a variety of formats of data towards building better diagnostic performance with surety.

2. Literature Review

2.1 Heart Disease

In (2024) Liu Y et al. developed a new classifying approach for heart failure by making use of multi-modal information comprising structured text data, chest X-rays, and ECG. Through this work, combining these three kinds of feature types significantly gains the performance with one single multi-modal fusion model. This research is a groundbreaking attempt to apply DL techniques to integrate multi-modal information for heart failure evaluation. It offers both new insights and useful applications for more accurate heart failure identification and therapy.

Biyanka Jaltotage et al. (2024) investigated the integration of Artificial Intelligence (AI), notably more so multimodal systems to enhance the treatment and diagnosis of Cardio Vascular Disease- CVD. They underline the ever-widening application of AI in cardiovascular care: the multimodal AI system may put together a variety of data coming from wearable device outputs, clinical records, genetic data, and medical imaging in order to give a thorough picture of a patient's condition. Biyanka Jaltotage et al. draw attention to the possible meaning of AI and multimodal systems for treatment effects in cardiovascular disease and customized medicine and, therefore, ask for more research in this area.

The work of Lee Y.C et al. (2023) focused on the development of an integrated AI model using non-invasive fundus visual as well as clinical risks for the detection of CVD by using multi-modal data. The collected data from SMC were used for both model development and internal validation. Lee Y.C et al. go ahead to illustrate other essential features considered in their study, including fundus photography and conventional clinical risk factors, to drive their point on how fundus photographs have a promising future as predictive markers for CVD.

Jessica Torres Soto et al. (2022) adopted the decision not to train their algorithms on labels from proximate data, like doctor interpretations of echocardiograms or electrocardiograms (ECGs). Rather, they employed clinical blood pressure data obtained from electronic health records and a hypertrophic cardiomyopathy (HCM) center's diagnostic consensus, which was frequently bolstered by molecular testing. Apart from that, they have also used some explain ability approaches to identify globally and locally essential variables that influence LAH-fusion predictions. Indeed, the model confirmed the diagnostic values of proximal septal hypertrophy on echocardiograms and lateral T-wave inversion on ECGs in detecting HCM. The results suggest that deep learning models, such as LVH-fusion, might considerably raise the bar for the precision of diagnosing HCM-a strategic means of preventing sudden cardiac death.

2.2 Alzheimer Disease

Michal Golovanevsky et al. (2022) proposed the architecture MADDi-Multimodal Alzheimer's Disease Diagnosis. The authors' aim was to make efficient diagnostics of AD and MCI based on clinical, genetic, and imaging data. MADDi differs from previous works by adopting cross-modal attention, which enables capturing interactions across different modalities-a novelty in this domain. Michal Golovanevsky et al. evaluated the function of attention processes by contrasting MADDi with earlier cutting-edge models, paying close attention to how each modality affected the overall performance of the model.

Qiu S et al. (2022) investigated the creation of adaptable models that can incorporate different combinations of clinical data that is routinely gathered, such as medical history, demographics, neuroimaging, neuropsychological testing, and functional assessments to increase the precision of neurodegenerative disease diagnosis. Qiu S et al. proved that these multimodal frameworks outperform neuroradiologists and practicing neurologists in terms of diagnostic accuracy.

Venugopalan J et al. (2021) used DL to combine genetic (single nucleotide polymorphisms), imaging (magnetic resonance imaging), and data from clinical tests. Also, with the goal to identify characteristics in clinical and genetic research, they made use of auto encoders that were layered with denoising. CNNs, which are also known as three-dimensional CNN, were deployed for the purpose of image analysis. One of the major structures discovered by the models, the hippocampus, the amygdala, and the Rey Auditory Verbal Learning Test (RAVLT) were confirmed to be consistent with the data that had been previously published on Alzheimer's disease.

Fan Zhang et al. (2019) presented a DL model that is quite similar to the diagnostic procedure by professionals in the auxiliary evaluation of Alzheimer's disease. This model uses two independent CNNs to train multi-model medical pictures. Then, with the use of correlation analysis, it will check how consistent the output of both CNNs match. The combination of multi-modal neuroimaging findings with clinical neuropsychological diagnosis provides a thorough comprehension of the pathology and psychological state of the patient. The incorporation of several data sources improves precision in supplemental diagnoses. The model is easy to apply and very practical as it closely resembles clinical procedures.

2.3 Melanoma Disease

Ou Chubin et al. (2022) provided a DL-based multimodal fusion model for skin lesion identification through the use of clinical photos and metadata gathered via smartphones. They enhanced the skin lesion identification accuracy by fusing patient metadata, like age, sex, and medical history, with photos collected using a smartphone. The model-executed methods of deep learning tend to generate superior performance regarding efficient combination of both visual and non-visual input than single-modality methods.

Alfi, I.A et al. (2022) integrated DL utilizing aggregation stacking of ML models to provide a comprehensible method for diagnosing non-invasive melanoma skin cancer. Hand-crafted features were applied for the purpose of training the underlying ML models. They were trained in order to get the desired results. In the following step, the predictions that were generated by these fundamental models were combined through the utilization of level one stacking.

Estimation of Melanoma Lesions A. A. Adegun et al. (2020) A DL-based method is presented to overcome the challenges of automatic detection and segmentation of melanoma lesions. This technique leverages a deepened encoder-decoder network, improving learning efficiency and features extraction. The authors further extended a completely new method called "Lesion-classifier," which comprises pixel-wise classification in order to detect if a skin lesion is melanoma or not.

2.4 Tumor Disease

Sharif M.I et al. (2022) introduced a new automated deep learning technique for classifying brain tumors into multiple classes. The method is modified in order to deal with imbalanced datasets within the application of deep transfer learning, depending on fine-tuning the pre-trained DL model DenseNet201. The features selected from MGA and EKbHFV are combined in a non-redundant serial-based technique for improving classification performance.

Stefan Schulz et al. (2021) developed MMDLM, for prognosis prediction. Both the non-metastatic and metastatic cases of ccRCC were divided into the two groups. For this, the MMDLM used multiscale histopathology photos, CT/MRI scans, and whole exome sequencing to get genetic information for every single patient. It had good predictive performance with an average C-index of 0.7791 and an accuracy of 83.43%. Importantly, the training of the model with different data sources compared to only using one source considerably improved the prognostic predictions. MMDLM can serve to enhance prognosis prediction in patients with colorectal cancer and outperform the other clinical indicators in its predictions, which were discovered to be an independent predictor of prognosis. Table 18.1 shows the competitive analysis of researcher's work.

3. Related Work

3.1 Various Deep Learning Model

VGG16 and VGG19 (Visual Geometry Group Models): Well-known convolutional neural networks (CNNs) are VGG16 and VGG19. These deep architecture models, which contain 16 and 19 layers, respectively, are popular choices for image classification applications. Their ability to efficiently extract features from a variety of pictures, including medical images like MRI scans and images of skin lesions, stems from their prior training on the ImageNet dataset. The major characteristic of the VGG models is simplicity in architecture; this is due to the use of tiny 33 convolutional filters throughout the network. This architecture gave the net the ability to obtain small features from the images easily; this is crucial since distinguishing between benign and malignant lesions depends on minute changes in features. Through transfer learning, the usage of VGG16 and VGG19 is most frequent because of their feature extraction capability. Only the last classification layers are fine-tuned on domain-specific data, like melanoma or brain tumor datasets.

MobileNet: A lightweight CNN architecture called MobileNet was made especially for effective picture categorization in settings with limited resources, such

Table 18.1 Comparative analysis of researcher's work

Author & Ref No.	Methodology Used	Dataset Used	Results
Liu et al. (2024)	Multimodal data integration using CNNs and RNNs for heart failure diagnosis	TCGA	Accuracy: 84%, AUC: 0.89, C-index: 0.79
Biyanka Jaltotage et al. (2024)	AI-based multimodal system using Random Forest, SVM, and CNN for cardiovascular disease management	UK Biobank	Accuracy: 88%, High precision and recall
Lee et al. (2023)	CNN combined with Gradient Boosting Machine for cardiovascular risk prediction	UK Biobank	Accuracy: 85%, F1-score: 0.83, AUC: 0.85
Golovanevsky et al. (2022)	Attention-based DL for Alzheimer's disease diagnosis using multimodal fusion	ADNI	Accuracy: 90%, AUC: 0.88
Qiu et al. (2022)	Multimodal deep learning combining CNN and RNN for dementia assessment	ADNI	Accuracy: 89%, AUC: 0.87
Jessica Torres Soto et al. (2022)	Transfer learning with CNN for left ventricular hypertrophy diagnosis using multimodal data	UK Biobank	Accuracy: 90%, Sensitivity: 0.91, Specificity: 0.89
Ou Chubin et al. (2022)	DL-based multimodal fusion employing smartphone-collected photos and information for skin lesion evaluation	Collection of clinical photos	Accuracy: 90%, AUC: 0.91
Alfi et al. (2022)	Non-invasive melanoma diagnosis using ensemble stacking with CNN, SVM, and Random Forest	ISIC	Accuracy: 94%, Precision: 0.92, Recall: 0.90
Sharif et al. (2022)	Classification of brain tumors using CNN, SVM, and Random Forest decision support system	Likely BraTS	Accuracy: 92%, F1-score: 0.89, AUC: 0.90
Schulz et al. (2021)	Brain tumor classification utilizing the Random Forest decision support system, CNN, and SVM	Renal cancer dataset	Accuracy: 88%, C-index: 0.81, AUC: 0.86
Venugopalan et al. (2021)	Multimodal DL using CNN and RNN for early Alzheimer's detection	ADNI	Accuracy: 87%, Sensitivity: 0.85, Specificity: 0.86
Adegun & Viriri (2020)	CNN-based DL system for automatic melanoma detection	ISIC	Accuracy: 92%, AUC: 0.89, F1-score: 0.88
Fan Zhang et al. (2019)	DL model with transfer learning for Alzheimer's disease diagnosis	ADNI	Accuracy: 88%, AUC: 0.86

mobile devices. Parameters are reduced via depth-wise separable convolutions and computing cost without compromising precision. Because of this, MobileNet is ideally suited for applications requiring high accuracy but with low processing resources, including mobile device point-of-care diagnostics. MobileNet can also be applied to skin lesion classification medical imagery tasks, drawing pertinently from clinical images with comparably little processing capacity. It may also be used because of its fast and effective architecture in real-time diagnostic systems to give rapid assessments to physicians or other remote health practitioners. When optimized with domain-specific data, MobileNet—despite its lightweight design can achieve competitive performance in medical picture classification, which makes it a useful model for healthcare AI applications.

Xception: "Extreme Inception," or Xception, is a sophisticated deep learning architecture that expands upon the Inception model. The performance of the model is enhanced when the computational complexity is decreased by this division. Traditional convolutional layers are replaced with depthwise separated convolutions, which factorize the convolution process into two parts: a depthwise convolution assigning only one filter for every channel of input and a pointwise convolution aggregating these outputsBoth these convolutions are done to arrive at the results expected from them. Simpler modes cannot find some minute patterns in medical photos, which this model Xception will be useful in finding applications such as brain tumor categorization or skin cancer identification. When it comes to activities like medical diagnostics, where fine-grained features from photos must be extracted, Xception has proven incredibly efficient at the job. This is, in general, quite a popular choice for transfer learning in medical image analysis, considering the model features scalability along with great feature extraction skills. It can be optimized on smaller-sized datasets of domain specificity to obtain high accuracy in reference to the illness detection.

ResNet50 and ResNet50V2: ResNet family models, including ResNet50 and ResNet50V2, were the first to present residual learning as a solution for the vanishing gradient problem. Their shortcut connections are helpful in training for the gradients to pass through the network. Because of this, it is possible to construct far deeper networks without performance reduction. More importantly, ResNet50 gives excellent performance on low-level features extraction tasks besides its potential high-level ones in the field of medical picture classification. Therefore, diagnostic predictions-such as the detection of brain tumors or classification of skin diseases-can be improved using this advanced neural network architecture. Large-scale image datasets are a particularly good source of complicated pattern recognition for ResNet50, a 50-layer architecture. The improved version, ResNet50V2, performs better than the original by changing the residual connections' architecture, which improves performance on particular tasks. Both models are frequently used in medical imaging transfer learning because of their deep architectures and ability to capture subtle traits, which makes them especially helpful for classifying and identifying illnesses.

DenseNet121: Another model in the DenseNet family is DenseNet121, which has 121 layers instead of the 200 that DenseNet201 has. It adheres to the same dense connection paradigm, which allows for effective feature reuse and enhanced gradient flow by allowing inputs from all preceding layers to reach any layer. This allows it to perform impressive classifications of medical images by a wide range of properties, from high-level structures down to low-level texturing. In applications that involve the identification of small patterns or abnormalities-for example, finding malignancies within medical imaging-the DenseNet121 architecture is specifically fairly suited for. Since DenseNet121 is lighter compared to its deeper predecessors, that makes it a popular choice in applications that need high performance while computational efficiency is atop the agenda. In transfer learning situations, this is also often employed within the healthcare industry. It is easy to envision that such sophistication, in an attempt to determine diseases like cancer with a high degree of accuracy, may determine points with the help of medical datasets and hence offer invaluable support to physicians in their decision-making processes [24].

4. Result Analysis

Table 18.2 shows the Comparative Analysis of Methodologies on Multimodal Datasets for Heart

Table 18.2 Comparative analysis of methodologies on multimodal datasets for heart diseases

Author	Dataset	Methodology	Key Algorithms	Performance Parameters
Liu Y et al. (2024)	TCGA	Multi-modal deep learning	CNNs, RNNs, Data fusion	Accuracy: 84%, AUC: 0.89
Biyanka Jaltotage et al. (2024)	UK Biobank	Multimodal AI for CVD management	Random Forest, SVM, CNN, Ensemble Learning	Accuracy: 88%, Precision: 0.85, Recall: 0.84
Lee Y.C et al. (2023)	UK Biobank	Multimodal risk prediction with fundus and clinical	CNN, Gradient Boosting Machine	Accuracy: 85%, F1-score: 0.83
Jessica Torres Soto et al. (2022)	UK Biobank	Deep learning for LVH diagnosis	CNNs (Transfer learning), Data fusion	Accuracy: 90%, Sensitivity: 0.91, Specificity: 0.89

Diseases. From Table 18.2, the number of approaches have demonstrated significant domain-wide performance improvements in multimodal medical data integration.

Table 18.3 shows the comparative analysis of methodologies on multimodal datasets for Alzheimer disease; The application of multimodal techniques that are based on deep learning has demonstrated encouraging outcomes in the detection of AD.

Table 18.4 shows the comparative analysis of methodologies for melanoma disease. Various new methods have been applied to enhance accuracy in skin lesion diagnosis.

Table 18.5 shows the comparative analysis of methodologies On multimodal datasets for tumor disease.

This research shows that the vast potential of DL models applied to multimodal medical diagnosis systems. The main conclusions of the research were that multimodal models achieved cutting-edge performance with high scores on numerous data types. For example, integrating imaging data with genomic and clinical information in diagnostic models achieved substantially higher accuracy in cancer diagnoses. Similarly, comparable gains in precision and recall were recorded for the prediction of cardiovascular disease by different data modalities-ECGs, chest X-rays, and clinical notes. Clinical information combined models gave superior performance compared to the standard single-modality models in the diabetic retinopathy assessment. Data fusion techniques depend on a range of medical applications, including early and late fusion, were critically reviewed besides respective performances. The present study proved that the multimodal deep learning models would be a useful tool for customized healthcare because they enhanced diagnostic accuracy and facilitated early identification of complex disorders.

5. Conclusion

The study highlights the transformative potential of multimodal deep learning (DL) models in medical diagnostics, showcasing their ability to significantly enhance diagnostic accuracy across various diseases by integrating diverse data types such as imaging, clinical records, and genomic information. By providing a holistic view of patient health, these models enable earlier and more precise diagnoses compared to single-modality approaches. The findings underscore the critical role of advanced data fusion methods—early, late, and hybrid fusion—in achieving these improvements. Despite challenges such as missing data, class imbalance, and dataset heterogeneity, the experiments demonstrate the value of multimodal DL models in advancing precision medicine. These models not only facilitate more

Table 18.3 Comparative analysis of methodologies on multimodal datasets for alzheimer disease

Author	Methodology	Key Algorithms	Performance Parameters
Michal Golovanevsky et al. [4]	Multimodal attention-based deep learning	Attention-based CNNs, Data fusion	Accuracy: 90%, AUC: 0.88
Qiu S et al. [5]	Using multimodal deep learning to evaluate dementia	CNN, RNN, Data fusion	Accuracy: 89%, AUC: 0.87
Venugopalan J et al. [12]	Early-stage detection of Alzheimer's disease	CNN, RNN, Multimodal feature fusion	Accuracy: 87%, Sensitivity: 0.86, Specificity: 0.85
Fan Zhang et al. (2019)	Auxiliary evaluation of Alzheimer's disease	CNN, Feature fusion	Accuracy: 88%, AUC: 0.86

Table 18.4 Comparative analysis of methodologies for melanoma disease

Author & Ref No.	Dataset	Methodology	Key Algorithms	Performance Parameters
Ou Chubin et al. (2022)	Smartphone-collected images & metadata	Multimodal fusion model for skin lesion diagnosis	CNN, Multimodal fusion	Accuracy: 90%, AUC: 0.91
Alfi et al. (2022)	ISIC	Deep learning with ensemble stacking	Ensemble (CNN, SVM, Random Forest, KNN)	Accuracy: 94%, Precision: 0.92, Recall: 0.90
Adegun & Viriri (2020)	ISIC	Automatic melanoma detection using CNN	CNN, Image classification	Accuracy: 92%, AUC: 0.89, F1-score: 0.88

Table 18.5 Comparative analysis of methodologies on multimodal datasets for tumor disease

Author & Ref No.	Dataset	Methodology	Key Algorithms	Performance Parameters
Sharif et al. (2022)	BraTS	DSS for brain tumor classification	CNN, SVM, Random Forest	Accuracy:92%, F1-score:0.89, AUC:0.90
Schulz et al. (2021)	Cancer	Multimodal deep learning for renal cancer prognosis	CNN, RNN, Data fusion	Accuracy: 88%, C-index: 0.81, AUC: 0.86

personalized healthcare responses but also hold promise for revolutionizing real-world health management. The integration of multimodal data presents a fertile ground for further research. Continued exploration of innovative fusion techniques, strategies for overcoming data-related challenges, and real-world applications will be pivotal. Multimodal DL techniques, thus, herald a bright future for medical diagnostics, paving the way for faster, more accurate, and patient-centered healthcare solutions.

Future research could focus on developing more sophisticated techniques for integrating multimodal data, advanced fusion methods and optimizing hybrid fusion strategies could further improve diagnostic accuracy and reliability. Another future research could focus on utilizing Multimodal approaches to diagnosing rare or complex diseases that require a combination of data types.

References

1. Liu, Y., Li, D., Zhao, J. et al. Enhancing heart failure diagnosis through multi-modal data integration and deep learning. Multimed Tools Appl 83, 55259–55281 (2024). https://doi.org/10.1007/s11042-023-17716-5
2. Biyanka Jaltotage, Juan Lu, Girish Dwivedi, Use of Artificial Intelligence Including Multimodal Systems to Improve the Management of Cardiovascular Disease, Canadian Journal of Cardiology, 2024, ISSN 0828-282X, https://doi.org/10.1016/j.cjca.2024.07.014.
3. Lee, Y.C., Cha, J., Shim, I. et al. Multimodal deep learning of fundus abnormalities and traditional risk factors for cardiovascular risk prediction. npj Digit. Med. 6, 14 (2023). https://doi.org/10.1038/s41746-023-00748-4
4. Michal Golovanevsky, Carsten Eickhoff, Ritambhara Singh, Multimodal attention-based deep learning for Alzheimer's disease diagnosis, Journal of the American Medical Informatics Association, Volume 29, Issue 12, December 2022, Pages 2014–2022, https://doi.org/10.1093/jamia/ocac168
5. Qiu, S., Miller, M.I., Joshi, P.S. et al. Multimodal deep learning for Alzheimer's disease dementia assessment. Nat Commun 13, 3404 (2022). https://doi.org/10.1038/s41467-022-31037-5
6. Jessica Torres Soto, J Weston Hughes, Pablo Amador Sanchez, Marco Perez, David Ouyang, Euan A Ashley, Multimodal deep learning enhances diagnostic precision in left ventricular hypertrophy, European Heart Journal - Digital Health, Volume 3, Issue 3, September 2022, Pages 380–389, https://doi.org/10.1093/ehjdh/ztac033
7. Ou Chubin, Zhou Sitong, Yang Ronghua, Jiang Weili, He Haoyang, Gan Wenjun, Chen Wentao, Qin Xinchi, Luo Wei, Pi Xiaobing, Li Jiehua, A deep learning based multimodal fusion model for skin lesion diagnosis using smartphone collected clinical images and metadata, Frontiers in Surgery, Volume=9, 2022, https://10.3389/fsurg.2022.1029991
8. Alfi, I.A.; Rahman, M.M.; Shorfuzzaman, M.; Nazir, A. A Non-Invasive Interpretable Diagnosis of Melanoma Skin Cancer Using Deep Learning and Ensemble Stacking of Machine Learning Models. Diagnostics 2022, 12, 726. https://doi.org/10.3390/diagnostics12030726
9. Sharif, M.I., Khan, M.A., Alhussein, M. et al. A decision support system for multimodal brain tumor classification using deep learning. Complex Intell. Syst. 8, 3007–3020 (2022). https://doi.org/10.1007/s40747-021-00321-0
10. Schulz Stefan, Woerl Ann-Christin, Jungmann Florian, Glasner Christina, Stenzel Philipp, Strobl Stephanie, Fernandez Aurélie, Wagner Daniel-Christoph, Haferkamp Axel, Mildenberger Peter, Roth Wilfried, Foersch Sebastian, Multimodal Deep Learning for Prognosis Prediction in Renal Cancer, Frontiers in Oncology, Volume=11, 2021, https://10.3389/fonc.2021.788740
11. Venugopalan, J., Tong, L., Hassanzadeh, H.R. et al. Multimodal deep learning models for early detection of Alzheimer's disease stage. Sci Rep 11, 3254 (2021). https://doi.org/10.1038/s41598-020-74399-w
12. A. A. Adegun and S. Viriri, "Deep Learning-Based System for Automatic Melanoma Detection," in IEEE Access, vol. 8, pp. 7160–7172, 2020, https://doi.org/10.1109/ACCESS.2019.2962812
13. Fan Zhang, Zhenzhen Li, Boyan Zhang, Haishun Du, Binjie Wang, Xinhong Zhang, Multi-modal deep learning model for auxiliary diagnosis of Alzheimer's disease, Neurocomputing, Volume 361, 2019, Pages 185–195, ISSN 0925-2312, https://doi.org/10.1016/j.neucom.2019.04.093.
14. Bolu, Toluwani. (2024). Multimodal Deep Learning for Integrating Medical Image and Clinical Data: Enhancing Diagnostic and Predictive Capabilities.
15. Zhaoyi Sun, Mingquan Lin, Qingqing Zhu, Qianqian Xie, Fei Wang, Zhiyong Lu, Yifan Peng, A scoping review on multimodal deep learning in biomedical images and texts, Journal of Biomedical Informatics, Volume 146, 2023, 104482, ISSN 1532-0464, https://doi.org/10.1016/j.jbi.2023.104482.
16. Thakur GK, Thakur A, Kulkarni S, Khan N, Khan S. Deep Learning Approaches for Medical Image Analysis and Diagnosis. Cureus. 2024 May 2;16(5):e59507. doi: 10.7759/cureus.59507. PMID: 38826977; PMCID: PMC11144045.
17. Rana, M., Bhushan, M. Machine learning and deep learning approach for medical image analysis: diagnosis to detection. Multimed Tools Appl 82, 26731–26769 (2023). https://doi.org/10.1007/s11042-022-14305-w
18. Siam Aisha, Alsaify Abdel Rahman, Mohammad Bushra, Biswas Md. Rafiul, Ali Hazrat, Shah Zubair, Multimodal deep learning for liver cancer applications: a scoping review, Frontiers in Artificial Intelligence, 2023, https://10.3389/frai.2023.1247195
19. Chola C, Muaad AY, Bin Heyat MB, Benifa JVB, Naji WR, Hemachandran K, Mahmoud NF, Samee NA, Al-Antari MA, Kadah YM, Kim TS. BCNet: A Deep Learning Computer-Aided Diagnosis Framework for Human Peripheral Blood Cell Identification. Diagnostics (Basel). 2022 Nov 16;12(11):2815. doi: 10.3390/diagnostics12112815. PMID: 36428875; PMCID: PMC9689932.
20. Sailunaz, K., Özyer, T., Rokne, J. et al. A survey of machine learning-based methods for COVID-19 medical image analysis. Med Biol Eng Comput 61, 1257–1297 (2023). https://doi.org/10.1007/s11517-022-02758-y
21. Manisha P. Patil, Dr. Uruj Jaleel, Multimodal Disease Classification and Severity analysis Approaches using Machine and Deep Learning, https://ignited.in/index.php/jast/article/view/15129/2992

Intelligent Systems Using Semiconductors for Robotics and IoT – Dinesh Goyal et al. (eds)

19 Revolutionizing DevOps and Security Compliance with GenAI: A 360° Orchestration Approach

Manoj Kumar Singhal[1]
Lead Software Engineer, Opaque Systems,
California, United States

Chhaya Gunawat[2]
System Development Engineer, Amazon,
California, United States

Sumit Singh[3]
B. Tech Computer Science, SOA ITER,
Bhubaneswar, India

Abstract: In today's fast-paced technological landscape, businesses should give priority to security and compliance to avoid significant setbacks and maintain market leadership. Incidents, such as GDPR fines and numerous data breaches, highlight the critical need to address these vulnerabilities. To gain a deeper understanding of this issue, we conducted a comprehensive study, examining a range of solutions and ontologies put forth by researchers. Available solutions in the market are diversified and focus on specific fields,which makes it challenging for users to integrate them into their workflow and understand them. These challenges in the present solutions craft an opportunity to bring more coordinated and flexible solutions. This paper presents a solution powered by a GenAI-driven engine which is designed to enhance security in infrastructure and ensure its real-time compliance with industry-specific regulations. By integrating different technologies and frameworks such as DevOps, cloud, security, AI and GenAI this GenAI-powered solution leverages advanced AI and GenAI algorithms to detect potential threats and minimize false positives by analyzing the historical log data. In addition, it is facilitated by streamlining tool integration to prevent vendor lock-in. With the capabilities of providing real-time insights and suggesting optimized resource allocation, this novel approach aims to protect organizations from various regulatory violations and data breaches.

Keywords: GenAI, GenAIOps, GDPR, Security, Regulatory violations, Threat detection, DevOps, Vendor lock in, Real-time compliance, 360 Orchestrated engine, Data breaches

1. Introduction

For efficient business operations, streamlining processes and reducing costs through heavy reliance on digital infrastructure has significantly increased cyber risks and regulatory challenges. In today's fast-paced business landscape, organizations often overlook critical system vulnerabilities, which can result in severe data breaches and substantial regulatory penalties.High profile Incidents where organizations have paid hefty amount and witnessed data breaches, such as Meta's $1.3 billion fine for GDPR's regulation violence [14], data breaches at companies like Uber [15], Target [16], and Capital One [18], highlight these concerns and shows urgency for a solution which can prevent such activities. Reports from institutions like NIST [17] present reports about the security challenges and considerations organizations must address when planning to adapt public cloud environments for their business. These incidents highlight the urgent need for strong and comprehensive security and compliance measures. The

[1]manojsinghal258@gmail.com, [2]chhayagunawat@gmail.com, [3]sumitsingh67740@gmail.com

DOI: 10.1201/9781003716389-19

incorporation of cloud technologies has raised significant security challenges, particularly during migration and ongoing maintenance, underscoring the importance of reliable protective measures.

1.1 Traditional Solutions

The traditional methods rely on numerous tools and manual processes, which frequently result in inefficiencies, missed security threats, and challenges in maintaining compliance with the ever-changing regulatory landscape. Approaches like manual processes, compliance checklists, Static Application Security Testing, and Dynamic Application Security Testing tend to be time-consuming, prone to human error, and often generate false positives, making them less effective [20].

1.2 Modern Solutions

These modern solutions frequently fail to address critical issues, leading to false positives and don't guarantee compliance with regulatory frameworks in real time.[1][2].This complicates organization's ability to focus on the critical issues. Additionally, the lack of real-time, automated solutions hamper an organization's capacity to scale while maintaining security and compliance. Moreover, the existing modern solutions focus on specific aspects rather than offering a one stop solution with user-friendly interface that is highly customizable to meet the customer specific needs of individual businesses or infrastructures. This limitation reduces their effectiveness in providing customized suggestions and implementing the solution that align with consumer's requirements, ultimately complicating efforts to reduce technical complexity [7][8][10][11][12].

1.3 The Business Need

In today's fast-paced business environment, where staying ahead of competitors is vital, service disruptions are not an option for organizations. This pressure intensifies the need to outpace competitors. For example, when companies suffer from recurrent data breaches, customers are likely to move toward alternatives that provide stronger data protection and privacy. This points to the urgent need for a solution to address these security gaps to prevent such losses by detecting and resolving infrastructure vulnerabilities with a comprehensive examination of existing ontologies. We recognized a continuous surge in demand for an AI-driven approach that comes up with different frameworks and technologies to address such issues in the market with such innovations, businesses will be able to simplify technical challenges, reduce errors, minimize threats, and ensure compliance. Tailored to deliver an efficient and consumer-specific solution, our proposed work presents a GenAIOps-powered system designed to tackle infrastructure vulnerabilities and offers real-time monitoring for compliance and threat detection. Combining DevOps methodologies, cloud, current regulations, data governance models, and evolving industrial trends, this strategy helps reduce the occurrence of false positives, generates actionable insights, and ensures adherence to compliance standards with a user-friendly interface. The engine detects the vulnerable issues in the system and helps resolve them quickly, additionally, it promotes easy integration across multiple platforms, avoiding vendor lock-in, and allows organizations to operate more smoothly and efficiently by upholding the latest and real-time regulatory compliance

2. Related Work

In today's fast-paced, technology-driven world, creating a seamless experience for everyone opens up substantial opportunities for profitability. However, the intense competition among organizations to attract and retain customers puts immense pressure on them to outperform their rivals. This pressure often leads businesses to overlook a crucial pillar of their long-term success: continuous security and compliance. These elements are essential for sustainable growth, prompting organizations to seek more advanced and optimized solutions that address these critical needs and offer tailored strategies. We provide a thorough overview of the current initiatives in this field, along with a proposed solution.

Rajesh Dharmalingam and his team have proposed a solution that integrates Artificial Intelligence (AI) with DevSecOps within the DevOps workflow to enhance infrastructure security. Their approach emphasizes how AI can be leveraged in DevSecOps, a methodology that incorporates essential security practices into the DevOps process. By integrating AI, the solution provides advanced threat detection and heightened security measures, helping organizations achieve a faster time to market, more efficient resource utilization, and a stronger overall security posture [10].

Vamsi Krishna Bandari from the University of South Australia explores the role of AI in containerization. He focused on managing containerized workloads, predicting and preventing system failures, optimizing resource usage in real-time, and automating the deployment and testing processes. In his paper "*A Comprehensive Review of AI Applications in Automated Container Orchestration, Predictive Maintenance, Security and Compliance, Resource Optimization, and Continuous Deployment and Testing*". He discusses the challenges involved in implementing these aspects, which ultimately lead to improved reliability, reduced downtime, and enhanced system performance[1].

Johnson Kinyua and Lawrence Awuah discuss "AI/ML in Security Orchestration, Automation and Response: Future Research Directions" their focus is SOAR (Security Orchestration, Automation, and Response), which integrates security tools and automates tasks, making it easier to identify, investigate, and take actions to resolve threats in real time. They propose a solution within container orchestration to identify and address security threats effectively [3].

Younes Benslimane et al. their paper presents "a ranking of the key challenges and opportunities for cloud computing services and derives key requirements of effective cloud services. Top challenges include security, quality of service (QoS), privacy and resource management and scalability, whereas top benefits include reduced computing costs, consistent QoS and service level agreements (SLA), increased organizational agility and improved scalability. This paper also presents a ranking of 12 cloud security domains based on this SLR"[4].

Mahmoud Abouelyazid and Chen Xiang from the University of Evansville identify challenges such as the increasing complexity of security threats, limited resources, inefficient incident response, and a lack of standardization in security tools and processes. They propose a solution aimed at minimizing false positives, which often lead to critical threats being overlooked [5].

Shubhashis Sengupta et al. (2011) from Accenture Technology Labs present Cloud Computing Security: Trends and Research Directions, addressing the concerns of cloud consumers regarding key business aspects essential for seamless operations. Any negligence in these areas could result in substantial fines or severe security incidents, such as data breaches. Through a comprehensive study, the authors highlight business intelligence by developing an application and presenting an ontology aimed at identifying, improving, and optimizing security and compliance vulnerabilities within businesses or organizations. In response to these challenges, they propose a solution in the form of an application for cloud consumers, focused on cloud security and compliance, anticipating the growing demand and the increasing trend of extensive cloud migrations[7].

Based on the above studies, we can see that significant modifications to existing methodologies or entirely new application-based solutions are being proposed. These studies highlight the importance of focusing on the fundamental pillars of a business. Collectively, they underscore the potential of GenAI and AI in the DevOps workflow. By incorporating advanced algorithms and various frameworks, these approaches aim to streamline business functionality.

3. Proposed Solution

The proposed solution is powered by GenAI, integrated with technologies and tools such as security tools, compliance systems, cloud platforms, and DevOps methodologies. This solution, known as 360 orchestrated engine, combines GenAI with DevOps making it a GenAIOps powered solution to create a comprehensive platform that supports continuous security, monitoring, and compliance by integrating various tools. It aslo provide a user friendly interface which makes it more easy for customers to adapt it.

3.1 High level Architecture

As shown in the given figure below – Fig. 19.1, powered by GenAI engine integrated with various tools, frameworks and infrastructure, collects data from various resources such as historical log data, latest information from the regulatory authorities, executes AI algorithms to drive useful information and provides a bot for user interface to configure & train the model for specific environments.

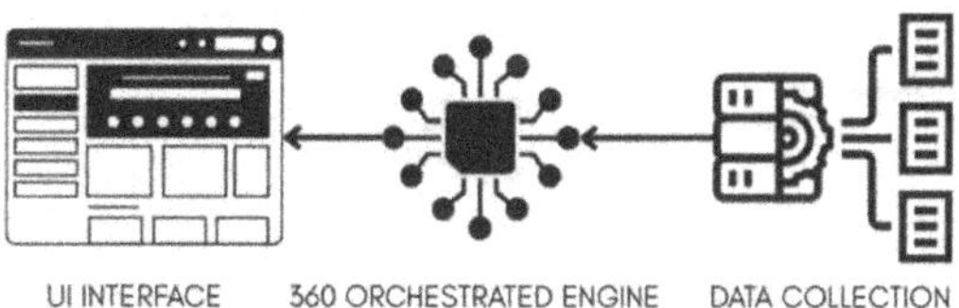

Fig. 19.1 360° engine overview

The platform also tackles alert fatigue by analyzing historical log data to identify patterns, detect critical threats, and prioritize their resolution. It continuously updates with the latest regulatory changes, using a GenAI-driven approach to compliance tools and infrastructure to ensure ongoing alignment with regulations. This proactive strategy enables organizations to maintain compliance with regulatory standards, safeguarding data privacy and reducing the risk of violations.

The platform is built with scalability in mind, utilizing microservices architecture, load balancing, and optimized CI/CD workflows to support seamless growth. This approach ensures that the system can scale flexibly and efficiently, maintaining its strength and responsiveness as the organization continues to expand.

3.2 Detailed Workflow

As shown in Fig. 19.2 the proposed solution could be divided in the following microservices:

1. Data Collection and Integration

   The data gets continuously collected from historical logs, tool configurations, emerging trends, latest regulatory and security frameworks to ensure seamless and fast integration and ensure that the infrastructure is up to date with the latest industry trends, regulatory and security standards provided in the region.

2. Threat Detection and Monitoring

   The system continuously monitors the collected data and reduces alert fatigue by filtering out false positives, prioritizing the most critical threats, and ensuring timely notifications of potential risks, helping to effectively work with the threats and prevent damage.

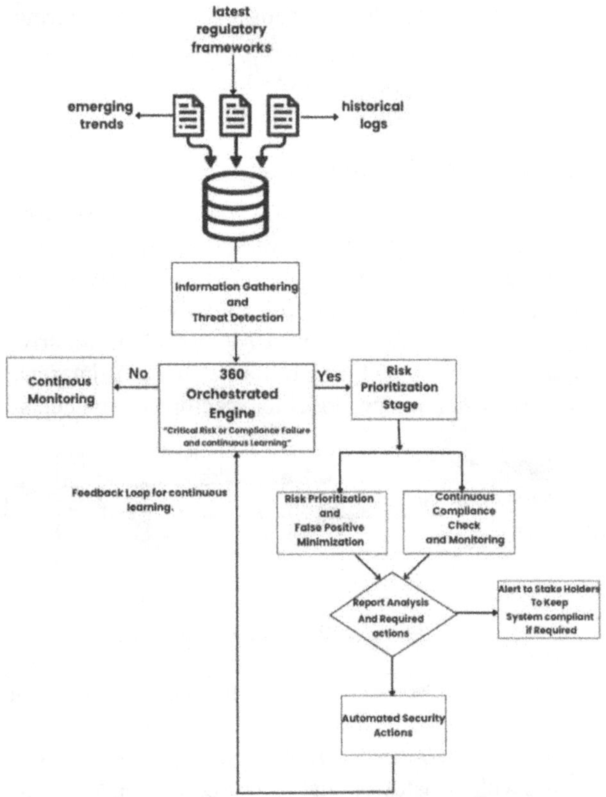

Fig. 19.2 360° risk and compliance loop

3. Compliance Monitoring

   The GensAI-powered engine constantly looks at the data it has collected to keep up with the latest security and compliance trends, helping to maintain the system's integrity and ensuring real-time compliance with regulatory bodies.

4. Automation and Orchestration

   This AI-driven approach also helps in improving task management and optimizing CI/CD workflows, service integration, and keeps infrastructure efficient..

5. User Interaction

   Also This solution leverages advanced natural language processing and AI algorithms to interpret user input and deliver real-time tailored responses, making it more user-friendly.

6. Continuous learning

   It continuously learns from previous incidents to improve its efficiency over time.

The efficient working of this GenAI-powered engine enables consumers to easily navigate and grasp the complexities of their business by offering tailored suggestions and solutions specific to their needs. Powered by GenAI and GenAIOps, the engine has the capability to understand the underlying infrastructure and adjust as needed, giving users the flexibility to modify the system to meet their specific business and infrastructure requirements.

3.3 Challenges in the Proposed Solution

There can be a lot of challenges while crafting this solution, one of the major challenges that might come with the successful integration of various technologies and tools with the engine is the efficiency of AI delivering solutions, which is another significant concern with a few more challenges that may arise while optimizing the Workflow to develop such solution can be little challenging.

4. Market Opportunity

Organizations are increasingly challenged by market disruptions caused by major data breaches and compliance violations, often leading to costly fines. As a result, there is a growing demand for solutions that leverage advanced AI in DevOps to address these challenges and offer tailored solutions that meet specific business needs.

Artificial intelligence has become a crucial and indispensable component of modern DevOps practices. Studies and various surveys have shown that the business adopting AI and machine learning in DevOps is growing rapidly, with many teams leveraging these technologies to drive innovation and efficiency[19].

Organizations, driven by a vision of comprehensive business optimization, have already embraced AI-driven solutions to meet their unique needs. However, there remains a strong demand for more advanced solutions that can address all vulnerabilities in a single, unified platform.

4.1 AI-Driven Assistance Adoption Rate

A survey from DevOps.com [23] found that roughly 20% of DevOps professionals now implement AI throughout all stages of the Software Development Life Cycle (SDLC). Similarly, a GitLab survey revealed that 24% of participants use AI/ML in their DevOps workflows—more than twice the figure reported in 2021. This growing trend underscores the rising demand for AI integration in DevOps, with the potential to boost efficiency and automate various tasks [24].

4.2 Global Devops Market Size

The DevOps market has experienced significant growth and is expected to continue expanding in the years ahead.

Table 19.1 Global DevOps market growth (2023–2033)

Year	Devops Market Size (USD)	Growth Rate (%)
2023	$11.5 billion	
2024	$13.7 billion	
2025	$16.3 billion	
2026	$19.4 billion	
2027	$23.1 billion	
2033(est.)	$66..6 billion	19.1% CAGR

Moreover, the increasing use of AI-driven tools highlights a shift where businesses aim to reduce errors, cut losses, and improve profitability by automating manual tasks. Industry reports show a growing openness to innovation, suggesting significant potential for continued progress in AI and DevOps integration [21].

5. Conclusion

The world is evolving very fast and has created an abundance of opportunities with easy access to the resources present in the market it has affected every sector, which has made survival in the market tough for everyone in this highly competitive business environment where companies continuously push to increase profit and establish themselves as market leaders they often work on various strategies to establish their influence in the market. As a result, accusations of unethical practices have become more frequent by them [20]. Challenges stemming from the technical complexities of the modern solutions in the market have made consumers suffer from hefty fines, security vulnerability, vendor lock-in, and other difficulties. Despite numerous works done to address these problems, many consumers still face difficulties in adapting to them and want something that offers clear and specific guidance. Existing solutions in the market often provide recommendations to consumers influenced by the context of particular organizations, which results in biased decisions and inefficiencies. Our proposed GenAI-Powered 360 orchestrated engine offers an innovative and user-friendly interface tailored to provide consumer-specific solutions. Its self-learning capabilities allow it to understand consumers' business deeply and improve problem-solving, security, and compliance processes. It focuses on the most critical threats and minimizes false positives, helping organizations address their most urgent security, compliance, and other challenges additionally, its ability to keep up with the evolving regulatory requirements ensures ongoing compliance, reducing the risk of legal penalties and hefty fines. This approach helps to streamline security operations, allowing organizations to focus on their core business activities, driving efficiency and growth. It also helps with the seamless integration with a wide range of tools and technologies, which prevents vendor lock-in. This flexibility helps businesses to go for the most suitable tools for their specific needs, promoting adaptability and efficiency in their hybrid environment.

References

1. V. Bandari, *"A Comprehensive Review of AI Applications in Automated Container Orchestration, Predictive Maintenance, Security and Compliance, Resource Optimization, and Continuous Deployment and Testing"*, IJIAC, vol. 4, no. 1, pp. 1–19, Mar. 2021.
2. *A. Hendre and K. P. Joshi, "A Semantic Approach to Cloud Security and Compliance", 2015 IEEE 8th International Conference* on Cloud Computing, New York, NY, USA, 2015, pp. 1081–1084, doi: 10.1109/CLOUD.2015.157.
3. Johnson Kinyua and Lawrence Awuah, "*AI/ML in Security Orchestration, Automation and Response: Future Research Directions*", Intelligent Automation & Soft Computing, vol. 28, no. 2, pp. 528–544, 2021, doi: 10.32604/iasc.2021.016240.
4. Y. Benslimane, M. Plaisent, P. Bernard and B. Bahli, "Key Challenges and Opportunities in Cloud Computing and Implications on Service Requirements: Evidence from a Systematic Literature Review," *2014 IEEE 6th International Conference on Cloud Computing Technology and Science*, Singapore, 2014, pp. 114–121, doi: 10.1109/CloudCom.2014.115.
5. M. Abouelyazid and C. Xiang, "*Architectures for AI Integration in Next-Generation Cloud Infrastructure, Development, Security, and Management*", IJIC, vol. 3, no. 1, pp. 1–19, Jan. 2019.
6. K. Pandya, "*Automated software compliance using smart contracts and large language models in continuous integration and continuous deployment with DevSecOPs ProQuest.*" https://dub.sh/1ujErJw
7. S. Sengupta, V. Kaulgud and V. S. Sharma, "*Cloud Computing Security--Trends and Research Directions*", 2011 IEEE World Congress on Services, Washington, DC, USA, 2011, pp. 524–531, doi: 10.1109/SERVICES.2011.20.
8. M. Kandira, J. Mtsweni and K. Padayachee, "*Cloud security and compliance concerns: Demystifying stakeholders' roles and responsibilities*", 8th International Conference for Internet Technology and Secured Transactions (ICITST-2013), London, UK, 2013, pp. 653–658, doi: 10.1109/ICITST.2013.6750284.
9. A.K. Reddy, V.R.R .Alluri, S. Thota, C.S.S. Ravi, and V.S.M. Bonam, "*DevSecOps: Integrating Security into the DevOps Pipeline for Cloud-Native Applications*", J. of Artificial Int. Research and App., vol. 1, no. 2, pp. 89–114, Aug. 2021, Accessed: Sep. 26, 2024. [Online]. Available: https://aimlstudies.co.uk/index.php/jaira/article/view/192
10. S. Rangaraju, S. Ness, and R. Dharmalingam, "*Incorporating AI-Driven Strategies in DevSecOps for Robust Cloud Security*", in *Proc. IEEE Int. Conf. Cloud Computing*, 2024, pp. 1–10.
11. H. Al-Aqrabi, L. Liu, J. Xu, R. Hill, N. Antonopoulos and Y. Zhan, "*Investigation of IT Security and Compliance Challenges in Security-as-a-Service for Cloud Computing*", 2012 IEEE 15th International Symposium on Object/Component/Service-Oriented Real-Time Distributed Computing Workshops, Shenzhen, China, 2012, pp. 124–129, doi: 10.1109/ISORCW.2012.31.
12. I. Patel, "*Smart DevOps: AI-Powered Orchestration for Optimized Cloud Environments*", MZJAI, vol. 1, no. 2, p. 1–5, Jul. 2024.
13. K. W. Ullah, A. S. Ahmed and J. Ylitalo, "*Towards Building an Automated Security Compliance Tool for the Cloud*", 2013 12th IEEE International Conference on Trust, Security and Privacy in Computing and Communications, Melbourne, VIC, Australia, 2013, pp. 1587–1593, doi: 10.1109/TrustCom.2013.195.

14. META fined $1.3 Billion for Violating E.U. Data Privacy Rules. https://dub.sh/sVJ4S2b
15. UBER data breach https://dub.sh/mOjSrr4
16. TARGET data breach, https://dub.sh/6PYrBPk
17. NIST Guidelines on Security and Privacy in Public Cloud Computing- https://dub.sh/0tKH6UZ
18. Capital One data breach compromises data of over 100 Million. https://dub.sh/vUHoGNX
19. Global AI in DevOps growth report, https://dub.sh/F1QoOS3
20. Google files a complaint to the EU over Microsoft cloud practices. https://dub.sh/dcpo5ic
21. SAST vs DAST: A Comprehensive Guide to Application Security Testing. https://dub.sh/muQBRfz
22. Global Devops Market in USD billion - market.us, https://shorturl.at/4taH1
23. Survey by devops.com, https://shorturl.at/0WHK1
24. Survey by Gitlab, https://shorturl.at/fRFGe

Note: All the figures and table in this chapter were made by the authors.

Intelligent Systems Using Semiconductors for Robotics and IoT – Dinesh Goyal et al. (eds)
© 2026 Taylor & Francis Group, London, ISBN 978-1-041-20408-4

20 Enhanced Decision Support in Competitive Electronic Markets through Predictive Analytics

Sarika Sharma[1],
Surendra Kumar Yadav[2], Shruti Mathur[3]
Jecrc Univeristy, Jaipur, Rajasthan, India

Abstract: The ability to make sound decisions plays a vital role for those companies desirous of maintaining their competitiveness on the fast changing competitive landscape. It is the aim of this work to enhance the decision support systems by integrating data mining and predictive analytics techniques. Organizations may find hidden patterns, forecast future trends, and get useful insights that improve decision-making processes by using enormous volumes of data. The research explores a range of data mining methods and prediction models, looking at how they may be used in various market situations. The study shows how these sophisticated analytical techniques may significantly increase forecasting accuracy, risk management, and strategic planning via case studies and empirical analysis. According to the research, companies that use data mining and predictive analytics are better able to foresee changes in the market, streamline processes, and experience long-term success. This work tries to add to the expanding corpus of research on data-driven DSSs by emphasizing how they might change market strategies that are competitive.

Keywords: Predictive analytics, Data mining, Predictive models, Optimization, Hierarchical clustering, Regression analysis, Feature selection Techniques

1. Introduction

The present scenario has demonstrated the raised risks posed by respiratory illness such as Chronic Obstructive Pulmonary Disease (COPD), asthma etc. Because of the rise of air pollutants such as PM2.5, PM10 etc., a respirator can be used as an immediate individual level risk mitigation strategy as reducing such type of pollution takes much longer than the harm it causes permits. The commonly used N95 respirators are negative pressure type models i.e., the wearer's lungs are compelled to draw air through the resisting membranes of the filter layers. This is tiring and unbearable to wear for a long period. This is missing in positive air pressure respirators for they employ external filters and they have a motorized air delivery mechanism. Recent scenario of the pandemic also requires the use of respiration apparatus as part of treatment. Negatively operated Pressure- the current often/ used respiratory devices are those that require the strength of one's lungs to draw-in fresh air; this is not proper and at times not possible if the patient has weak lungs or breathing illness. In this work, a forced air (Positive Air Pressure) solution has been recommended to combat the problem. Today, business environment is constantly changing and highly competitive, so companies are looking for ways on how to gain advantage. Traditional approaches to decision-making,[1] which require prior experience and intuition, do not solve problems and rapidly change in modern markets. The availability of extensive big databases and the development of more refined techniques in the field of analysing data offer as yet unrealized potential for changing the approach to decision-making.[5] Science at the forefront of this revolution is data mining followed by predictive analytics which offers very effective tools for modeling large volumes of data and arriving at reasonable forecasts. Using statistical analysis and the techniques of artificial neural networks, predictive means probability of an event occurring in the future. It may be employed by an organization to anticipate trends, to identify risks, and to capture opportunities which an organization might not

[1]sarikasharma8816@gmail.com, [2]surendra.yadav@jecrcu.edu.in, [3]shruti.mathur@jecrcu.edu.in

DOI: 10.1201/9781003716389-20

otherwise have the ability to identify.[3] At the same time, the aim of data mining is to make finds and draw links in big databasing. By suing data mining tools, organizations could gain more insights into their customers' buying behavior and trends in the markets among other useful facts concerning their operations.

1.1 Data Mining Techniques in DSS

Everyone needs information to make decisions based on accurate facts, and in order to do so, data analysis and relevant knowledge extraction are needed. Data mining is the act of searching through and analyzing vast volumes of data with the goal of finding patterns and rules that are very meaningful. Creating Decision Support Systems requires a significant investment of time, money, and human resources, and there are several hazards that might compromise the system's performance.[19]

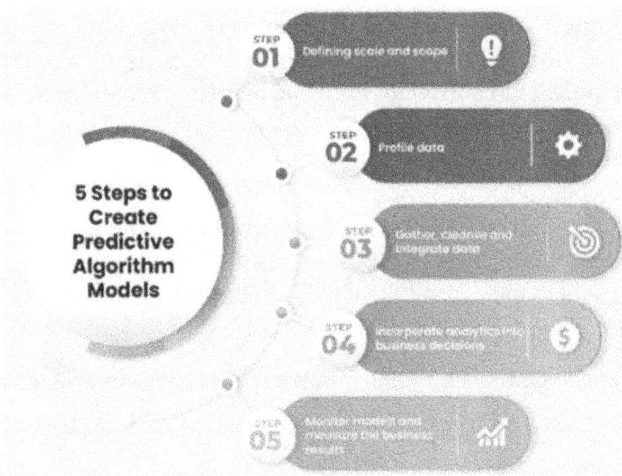

Fig. 20.1 Predictive analytics model for business [5]

The main goal of this research is to determine how data mining and predictive analytics could be useful in enhancing decision support systems where the market competition is high. It discusses several data mining technique and prediction model and gives examples for their application. In order to illustrate how these technologies might improve the accuracy of forecasting, the management of risks and planning, the paper has provided case examples and empirical analyses as a result. This paper focuses on the importance of data mining and predictive analytics in increasing the use of data in decision making by assessing the current state and future direction of the field. These technologies are valuable and those companies, who managed to adopt them and use properly, will have possibility to predict changes, to control complexity of the market, and to achieve long-term directional success. This work also contributes to the growing body of literature focusing on decision support systems by elucidating on how data mining and predictive analytics have revolutionary impacts in highly competitive markets [20]. Organizations have now come to view their predictive analytics models as necessary weapons that give them a competitive edge in view of data driven strategies. These models help the firms forecast the results in future and thus be in a position to know the market trends, consumers' behavior and the danger that may be encountered. The vetted models in predictive analytics may point out associations and relationships are not easily discerned by applying tools such as the business analytics tools, machine learning algorithms, decision trees, neural networks and regression analysis. Decision making capability can be employed by businesses to enhance the quality of customer services, designing new services and goods, personalizing advertisements and enhancing business procedures. For instance, retailing organizations use the predictive models to manage inventory flows and individual customer engagements to predict and influence shopper behavior, while banking industries use the predictive models to detect fraud instances and assess risks attached to extension of credit among other uses.[2,4] In this way, predictive analytics contribute to refinement and improvement of the efficiency of business operations while at the same time keeping organizations ahead of events, instead of having them react to them, which ultimately

1.2 Natural Language Processing

The term natural language processing (NLP) describes a computer system's capacity to understand human languages. This is a method of using computers to analyze texts by understanding how humans use and interpret language.[6] Customer service queries are understood by NLP in terms of language, emotions, and context. Without requiring human input, it deciphers and analyzes consumer discussions before answering. While many organizations may profit from NLP, customer service stands to gain the most. People are working hard to advance technology so that it may benefit both consumers and companies.

2. Literature Review

2.1 Business Analytics

The global economy is evolving quickly, and developing market businesses face new possibilities as well as obstacles. Corporate analytics is essential for competitive advantage in dynamic contexts. This thorough study explores the use of business analytics by developing nations to stay competitive. We examine reports from the industry, case studies, and scholarly literature. By exposing important themes and patterns, this methodical methodology sheds light on the various business analytics strategies used by firms.[10] A multitude of interrelated criteria are necessary for the successful implementation of business analytics in developing economies. Examples include data mining, data management, predictive modeling, and analytics-driven strategic decision-making. Business analytics projects are also impacted by funding allocation, corporate culture, and leadership support.

The use of business analytics in developing countries has several implications, as per the findings of the study enhanced market potential, risk management, consumer insights, and operational efficiency. Business analytics enables companies to comprehend client preferences and respond quickly to developments in the market. The importance of business analytics in expanding into new markets and preserving a competitive edge is emphasized in this systematic study. Gains might be realized by academics, practitioners, and legislators looking for analytics-driven success in fast-paced corporate situations. [Adama NHE, Okeke NCD (2024)]

2.2 Behavior Analysis

This research examines how user behavior analysis based on data mining may enhance global company strategy. Multinational corporations need improved tools to track market trends and client demands as global competition intensifies. Data mining is a tool used by organizations to get insights, and this research looks at how it may be used to the strategic decision-making of multinational corporations. Huge amounts of user activity data are examined in the research using cluster analysis, association rule mining, and predictive modeling. Additional study on user behavior revealed variations in customer purchasing patterns by market and area.

These insights support the product positioning and market strategy of global corporations. Data mining in multinational corporate strategic decision optimization is the main topic of this work. Based on product sales and market demand, data-driven forecasting techniques modify supply chain and production plans. By facilitating cross-selling, associative rule mining helps businesses boost profits and customer loyalty. Issues with data mining model uncertainty, privacy, and quality are also present in this research. Subsequent studies might use more complex data mining techniques to enhance the interpretability and accuracy of the models. This research provides useful recommendations and demonstrates how data mining-based user behavior analysis is used by global corporations to make strategic decisions. Understanding the market and consumers would enable multinational corporations to engage in more strategic competition. [Siqi Huang and Hanqi Yue (2024)]

2.3 Commercial Intelligence

Commercial intelligence is increasingly vital in corporate decision-making in today's changing business climate. This literature review aims to explain business intelligence (BI)'s many facets and its huge potential to improve organizational performance. Methods: This study reviews peer-reviewed papers, case studies, and core business intelligence works. Operational effectiveness, strategic decision-making, business intelligence measurements and KPIs, and business intelligence aspects are examined in the research. BI improves operational efficiency, measurements and KPIs across sectors, and strategic decision-making, according to various studies. Study samples range from small focus groups to large corporate surveys. Business intelligence, with its many capabilities that boost operational efficiency and decision-making, is crucial for modern organizations, according to studies. Use across several sectors offers it a competitive advantage and boosts company performance. [Moussas K, Hafiane J, Achaba A (2024)]

2.4 AI-Powered Digital Transformation

Since efficiency, creativity, and competitiveness constitute the fundamentals of digital transformation with the help of AI, automation and process optimization have to be the elements of the discussed transformation. On these bases, this paper presents a strategy for integrating automation and optimizing operation with future AI. It comprises of DDR, AL, HCC, SA, EAI, TSA, PEA, and IO. These are some of the pillars that help us navigate through the thorny issues around automation and digital transformation propelled by Artificial Intelligence. Companies can change their operations, extend automations, promote innovation, and direct AI-driven business processes by adopting those pillars. [Abdulaziz Aldoseri and Khalifa N. Al- khalifa (2023)].

2.5 Big Data

Big data has become more important in almost every organization in the globe in the current information society. The efficiency of big data in making intelligent decisions makes it a popular solution globally. Companies are using big data analytics for market and consumer research in an ever expanding manner globally. But it is the financial sector that makes the widest use of big data, while because of rather serious hurdles there are few published works on this topic. In fact, the financial sector makes the widest usage of big data and it is crucial and still the research and analysis are insufficient. Given the increasing relevance of big data in a financial and monetary environment, the present work seeks to provide an exhaustive literature review on the subject of big data and finance. On this premise, the study will advance knowledge generation in the field of big data and finance by opening up fresh avenues for empirical investigation [Amjed Alfityanib & Enas @al-lozia 2022]

2.6 Data-Driven Decision Making

Modern data-driven organizations swim in data. Customer, social media, and sensor data might be overwhelming. But this vast ocean offers many concepts to investigate. Introducing data mining. Companies enhance decision-making, operations, and competitiveness using data mining. Organizational performance will depend increasingly on data mining as data volume and complexity grow.

Data-driven decision making helps organizations unlock data's value and effectively navigate the ever-changing business environment. Data mining reveals key insights and helps organizations make data-driven decisions.[14] Classification mining predicts data point classes using models. These models are trained using category-based historical data. A bank may categorize loan applicants as high-risk or low-risk based on financial history. Mining supports decision-making in several ways. It helps firms spot data patterns. By analyzing purchase history, retailers may categorize customers. With tailored initiatives, marketing costs and customer satisfaction are optimized. [Mankari Sapna Sadashiv (2021)]

3. Methodology

The study located and examined the research papers that were influenced by the research goals using a systematic review approach. This review's inclusion criteria include articles on companies that have used data mining methods. A review of the literature will be done in order to respond to the research question.

3.1 Sentiment Analysis

Machine learning may be used to customer query response in a number of ways, including training models on a large dataset of customer queries and replies, categorizing customer questions into specified groups, analyzing the sentiment of customer inquiries, and producing automated responses. Some research suggest that customer service departments may improve their understanding of client sentiment by using natural language processing methods. [13] Companies that use natural language processing (NLP) technology may increase customer happiness by optimizing customer interactions, understanding customer sentiment, and improving customer satisfaction overall.

3.2 Logistic Regression

LR is a popular discriminative technique which posits that in the given cases, the log likelihood ratio of the class distributions is linear. Its primary purpose is to export posterior probabilities directly, supporting the target variable's binomial case. L2-norm was used for our implementation's penalization step. In cases when binary data is involved, this is the appropriate kind of analysis to do. It is used to address categorization problems and determine the likelihood that a binary event will occur. Binary, multinomial, and ordinal logistic regression are the three varieties like an S-shaped curve that, by using the following equation, may convert any real-valued integer into a number between 0 and 1.

$$f(X) = \frac{1}{1+e^{-x}} \tag{1}$$

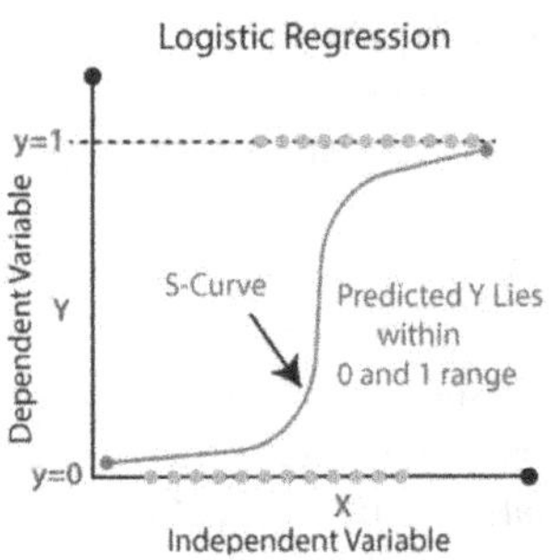

Fig. 20.2 Example of logistic regression

Source: Authors

3.3 Random Forest

It is possible to do accurate classifications of massive amounts of data by using a technique called radio frequency (RF). To put it in more precise terms, it is a group learning approach that relies heavily on regression and classification. During the training phase, a large number of decision trees are built, and the outcomes of each tree are then applied to each class that the tree belongs to. Every single tree in the forest receives the same amount of the random vector sample, which is distributed uniformly across the forest, and its values are contingent upon each and every tree. In the forest, there is not a single predictor tree that does not show this impact. The surroundings of a single tree exhibit considerable incline and variation, and this tree is exposed to these circumstances. The issues caused by strong tendency and large variance may be lessened by the usage of arbitrary forests, which are utilized to create a common harmony between two borders. You may use arbitrary forests to do this. One method to do this is via creating a common understanding. The accompanying graphic shows how the k ready trees are arranged appropriately and in compliance with the guidelines describe as

$$H(X,\theta_j) = \sum_{i=1}^{k} h_i(x,\theta_j), (j = 1,2,\ldots..,m) \tag{2}$$

3.4 BiLSTM

There exists a kind of sequence processing model called the bidirectional LSTM, often called the biLSTM. This

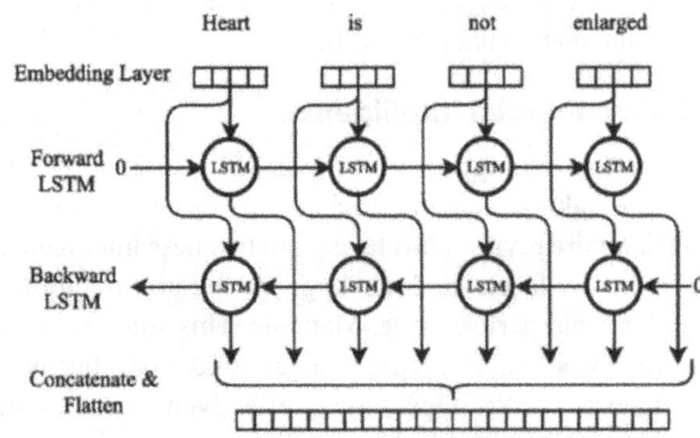

Fig. 20.3 Diagram of BiLSTM process flow

Source: Authors

model consists of two LSTMs, out of all the components. Whereas the second LSTM processes the information in a backward manner, the first LSTM processes the input forward. To get the data, these two LSTMs are used. The ability of BiLSTMs to effectively expand the network's reservoir of information ultimately results in an improvement in the context that the algorithm can access. BiLSTMs are capable of doing this.

3.5 Proposed Flow

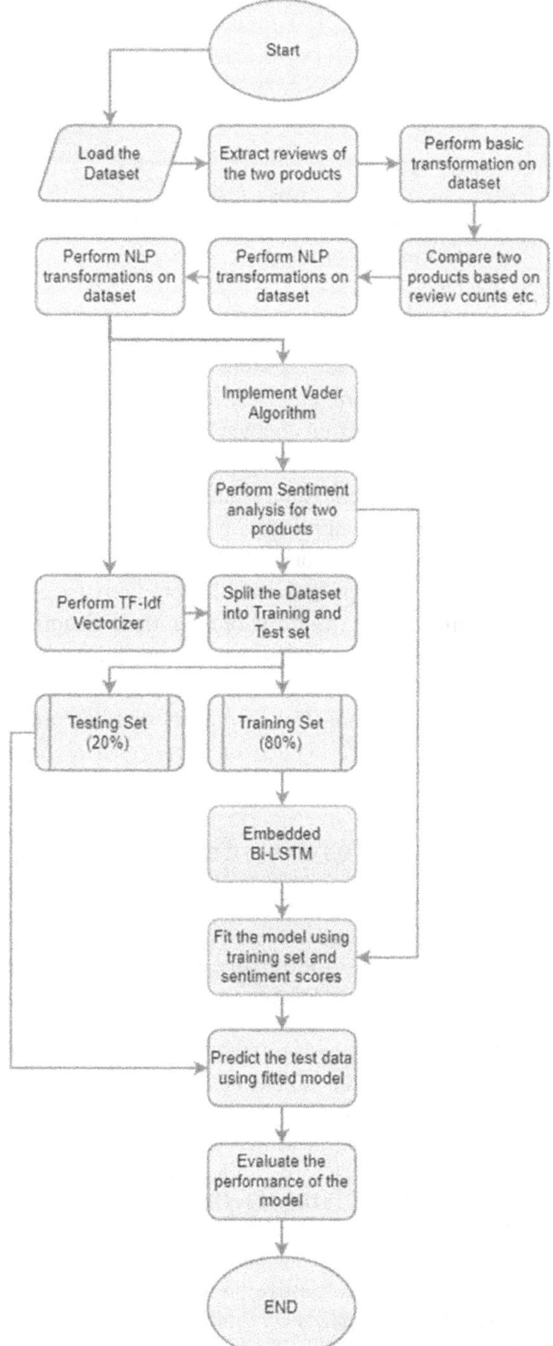

Fig. 20.4 Proposed model flow layout

Source: Authors

This flowchart is intended to be used in an investigation of the emotions mentioned in product reviews. Natural language processing, or NLP, is now being used to process the data. This procedure makes use of several techniques, such as stop word removal, tokenization, stemming, and lemmatization. These are just a few of the methods that are used. Based on the published evaluations, it is possible to forecast with easy how people will feel about the two items by using this pattern.

A knowledge of this model is beneficial for organizations as it facilitates a greater capacity to improve the attributes of a product or service since more is known about the feelings of customers. The wrinkles of NLP modifications enable categorization of the dataset into both training and testing sets. This is done after the use of the transformations It is important to note that all the transformations are configured to maximize effectiveness as well as system stability and resources. Three machine learning models including logistic regression, random forest and bidirectional Long Short-Term Memory networks are trained using the machine learning training set. The machine learning training set assists in the setting of these models. These models are used to predict the emotion stated in the reviews that are part of the testing set after the training procedure as described above is over. Any model goes through a performance evaluation at the end of the research process.

3.6 Dataset Screenshot and Description

This is the list of over 7000 website reviews of 50 electronic products. Web data that to be provides instant access to web data. I maintain information excerpts from thousands of websites to generate uniform data bases of business, product, and property facts. To be get from https:>Amazon and Best Buy Electronics – DataFiniti (<– link to dataset)

4. Results and Discussion

Here, we are implanting VADER sentiment ratings for both type A and type B items, ranging from -1.00 to 1.00. The frequency of observations falling within each sentiment score range is shown on the y-axis of this graph, while the x-axis displays the VADER sentiment score. A more lucid visual depiction of the underlying distribution is offered by the histograms used to display these data. The density plot for Product B has a higher degree of dispersion than that of Product A, indicating that the sentiment ratings for Product B are more erratic than those for Product A.

The value of the comparison result, which will be produced after the model has been run in terms of accuracy, precision, F-score, and recall, must be ascertained at the conclusion of the process.

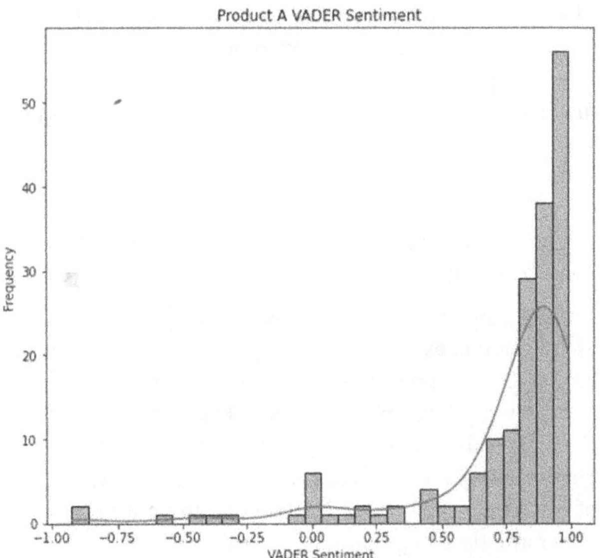

Fig. 20.5 Histograms comparing the VADER sentiment scores of A product

Source: Authors

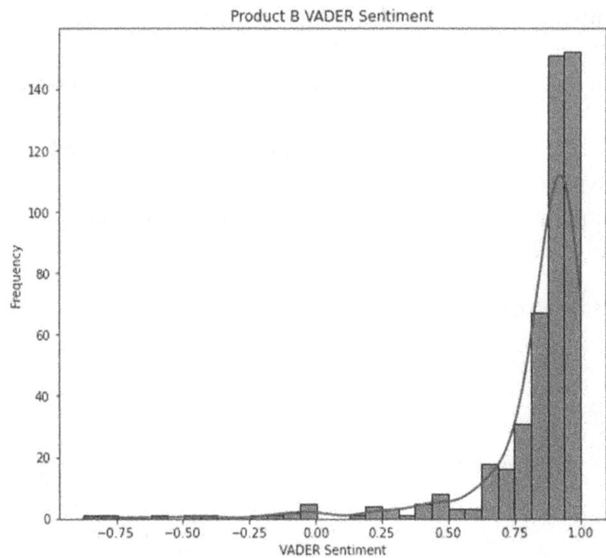

Fig. 20.6 Histograms comparing the VADER sentiment scores of B product

Source: Authors

Table 20.1 Accuracy of different models

Model Names	Training Accuracy	Validation Accuracy
Logistic Regression	0.917939	0.916031
RandomForest	0.95	0.916031
BiLSTM	0.98855	0.89313
LSTM EMBEDDED	0.988067	0.938931

Source: Authors

Table 20.2 Confusion matrix comparison value

Model Names	Precision	Recall	F1-Score
Logistic Regression	0.839112	0.916031	0.875886
RandomForest	0.936091	0.931298	0.908758
BiLSTM	0.860124	0.89313	0.874395
LSTM EMBEDDED	0.942748	0.938931	0.922468

Source: Authors

5. Conclusion

Predictive analytics and data mining are two examples of disruptive strategies in the context of strategic business management. One feature that sets this specific implementation technique apart is that it offers better decision help in markets that are very competitive. When businesses can effectively use the possibilities of contemporary analytical techniques, they may get deeper insights into the dynamics of the market, customer behavior and competitiveness. Consequently, businesses can make more proactive and well-informed choices, which eventually helps them to become more competitive. The study's findings suggest that to improve operational effectiveness and competitiveness, decision-making processes must include data mining and predictive analytics. To achieve the goals that have been established, this is essential. More precisely, the statistics provide proof of the connection between these two distinct characteristics. Furthermore, this highlights the need for further research to fully understand the implementation itself, as well as how applicable it is to other businesses and what ethical issues arise. Finding solutions to the identified research gaps is essential if predictive analytics and data mining are to fully realize their promise to transform decision support in cutthroat marketplaces. This will enable these previously stated technologies to reach their full potential. The possibility that this promise may completely transform decision support has received a great deal of attention.

References

1. Adama NHE, Okeke NCD (2024) Harnessing business analytics for gaining competitive advantage in emerging markets: A systematic review of approaches and outcomes. International Journal of Science and Research Archive 11:1848–1854
2. Siqi Huang and Hanqi Yue (2024) "Optimizing Strategic Decision-Making in Multinational Corporations through Data Mining-Based User Behavior Analysis" Journal of System and Management Sciences, Vol. 14 (2024) No. 1, pp. 579–591, DOI:10.33168/JSMS.2024.0134, ISSN 1816-6075 (Print), 1818-0523 (Online)
3. Moussas K, Hafiane J, Achaba A (2024) Business intelligence and its pivotal role in organizational performance: An exhaustive literature review. Journal of Autonomous Intelligence. https://doi.org/10.32629/jai.v7i4.1286
4. Dineshkumar P, Subramani B (2024) Foresight In Finance: Elevating Predictions With Enhanced Rnn-Lstm And Adam Optimizer. https://doi.org/10.53555/kuey.v30i5.5931

5. Adaga NEM, Okorie NGN, Egieya NZE, Ikwue NU, Udeh NCA, DaraOjimba NDO, Oriekhoe NOI (2024) The Role Of Big Data In Business Strategy: A Critical Review. Computer Science & IT Research Journal 4:327–350
6. Abdulaziz Aldoseri and Khalifa N. Al-khalifa (2023) A Roadmap for Integrating Automation with Process Optimization for AI-powered Digital Transformation, Creative Commons CC, doi:10.20944/preprints202310.1055.v1
7. Enas Al-lozia and Amjed Alfityanib (2022) Retracted: The role of big data in financial sector: A review paper, International Journal of Data and Network Science 6 (2022). 1319–1330
8. Mankari Sapna Sadashiv (2021) Data Mining Techniques for Better Decision Making in an Organization, IOSR Journal of Computer Engineering (IOSR-JCE) e-ISSN: 2278–0661, p-ISSN: 2278–8727, Volume 23, Issue 1, Ser. I (Jan. – Feb. 2021), PP 68–71
9. Adebunmi Ok echukwu Adewusi, Abiola Moshood Komolafe, Emuesiri Ejairu, Iyadunni Adewola Aderotoye, Oluwatosin Oluwatimileyin Abiona, & Oyekunle Claudius Oyeniran. (2024, March 23). The Role of Predictive Analytics In Optimizing Supply Chain Resilience: A Review Of Techniques And Case Studies. International Journal of Management & Entrepreneurship Research, 6(3), 815–837. https://doi.org/10.51594/ijmer.v6i3.938
10. Abiola Moshood Komolafe, Iyadunni Adewola Aderotoye, Oluwatosin Oluwatimileyin Abiona, Adebunmi Okechukwu Adewusi, Amaka Obijuru, Oluwole Temidayo Modupe, & Oyekunle Claudius Oyeniran. (2024). Harnessing Business Analytics For Gaining Competitive Advantage In Emerging Markets: A Systematic Review Of Approaches And Outcomes. International Journal of Management & Entrepreneurship Research, 6(3), 838–862. https://doi.org/10.51594/ijmer.v6i3.939
11. Wilhelmina AfuaAddy, Chinonye Esther Ugochukwu, Adedoyin Tolulope Oyewole, Onyeka Chrisanctus Ofodile, Omotayo Bukola Adeoye, & Chinwe Chinazo Okoye. (2024) Predictive analytics in credit risk management for banks: A comprehensive review. GSC Advanced Research and Reviews, 18(2), 434–449. https://doi.org/10.30574/gscarr.2024.18.2.0077
12. Ali, M., Mazhar, T., Al-Rasheed, A., Shahzad, T., Yasin Ghadi, Y., & Amir Khan, M. (2024, February 28). Enhancing software defect prediction: a framework with improved feature selection and ensemble machine learning. PeerJ Computer Science, 10, e1860. https://doi.org/10.7717/peerj-cs.1860
13. Aguilar-Moreno, J. A., Palos-Sanchez, P. R., &Pozo-Barajas, R. (2024). Sentiment analysis to support business decision-making. A bibliometric study. AIMS Mathematics, 9(2), 4337–4375. https://doi.org/10.3934/math.2024215
14. Chidera Victoria Ibeh, Onyeka Franca Asuzu, Temidayo Olorunsogo, Oluwafunmi Adijat Elufioye, Ndubuisi Leonard Nduubuisi, & Andrew Ifesinachi Daraojimba. (2024, February 28). Business analytics and decision science: A review of techniques in strategic business decision making. World Journal of Advanced Research and Reviews, 21(2), 1761–1769. https://doi.org/10.30574/wjarr.2024.21.2.0247
15. Dr. Gonesh Chandra Saha and Dr. Reshmi Menonet. al (2023) The Impact of Artificial Intelligence on Business Strategy and Decision-Making Processes. (2023). European Economic Letters. https://doi.org/10.52783/eel.v13i3.386
16. Tareq Obaid and Samy S. Abu-Naser (2023). Big Data Analytics in Project Management: A Key to Success International Journal of Academic Engineering Research (IJAER) ISSN: 2643-9085 Vol. 7 Issue 7, July - 2023, Pages: 1–8
17. P. Garg and A. Sharma, "A distributed algorithm for local decision of cluster heads in wireless sensor networks," in 2017 IEEE International Conference on Power, Control, Signals and Instrumentation Engineering (ICPCSI), IEEE, 2017, pp. 2411–2415.
18. P. Sharma and A. Sharma, "Online K-means clustering with adaptive dual cost functions," in 2017 International Conference on Intelligent Computing, Instrumentation and Control Technologies (ICICICT), IEEE, 2017, pp. 793–799.
19. O.E. Emam and R.M. Haggaget. al (2022) The Role Of Decision Support System In Enhancing Customer Relation Management In The Egyptian Telecommunication Sector IEEE-SEM, Volume 10, Issue 3, March-2022 ISSN 2320-9151
20. Yu, Y., Yazan, D. M., Junjan, V., &Iacob, M. E. (2022, May). Circular economy in the construction industry: A review of decision support tools based on Information & Communication Technologies. Journal of Cleaner Production, 349, 131335. https://doi.org/10.1016/j.jclepro.2022.131335
21. Arun Velu (2021) Machine Learning Techniques for Customer Relationship Management International Journal of Creative Research Thoughts (IJCRT) Volume 9, Issue 6, ISSN: 2320-2882
22. Gangwar, H., 2020. Big Data Analytics Usage and Business Performance: Integrating the Technology Acceptance Model (TAM) and Task Technology Fit (TTF) Model. The Electronic Journal of Information Systems Evaluation, 23(1), pp. 45–64, available online at www.ejise.com
23. Huang, M. H., & Rust, R. T. (2020). A strategic framework for artificial intelligence in marketing. Journal of the

Intelligent Systems Using Semiconductors for Robotics and IoT – Dinesh Goyal et al. (eds)
© 2026 Taylor & Francis Group, London, ISBN 978-1-041-20408-4

21 Comparative Analysis of Different Load Balancing Algorithms in Cloud Computing

Kapil Tajane[1], Utkarsh Gardi[2], Sanskar Ghule[3], Utkarsha Kamble[4], Aneesh Kanhere[5]
Pimpri Chinchwad College of Engineering, Computer Engineering Pune, Maharashtra, India

Abstract: The way organizations manage and analyze data is being revolutionized by cloud computing. Making optimum use of resources is critical in today's changing world. Load balancing, one of the most important parts of cloud computing, makes sure work is distributed efficiently among servers which is more efficient with resources as well. Stationary and changing traffic allocation strategies in cloud-based computing frameworks are fully explored in this paper. In this first part of the research, we start by explaining we need to distribute the load, what are the basics of load balancing in a based computing framework and the need of load balancing. The pros and cons associated with different options for fixed algorithms are examined in detail in the current literature. Similar to this, dynamic algorithms are thoroughly examined to improve our knowledge of their efficiency and real-time adaptability. Using a number of factors such as resource utilization, throughput, and reaction time, a detailed comparison analysis is conducted. We finally examine the efficiency and the ability of these algorithms to grow, and the effectiveness of these algorithms to operate in different contexts of different cloud systems. The research provide a thorough understanding of strengths and weaknesses of entire algorithms and make academics and professionals aware before choosing the cloud implementation.

Keywords: Cloud computing, Honey bee algorithm, Load balancing, Opportunistic, Active clustering, Min-min, Round robin, Max-min, Random biased sampling, Ant colony

1. Introduction

Both industry and academics have embraced cloud computing as a popular technology that provides a convenient and versatile approach to storing and accessing materials. Cloud computing, which involves sharing data, software, and resources with computers and other devices as needed, is computing over the Internet. Users may obtain access to information from any place and at any moment, facilitated by cloud services. The hardware that stores the data does not require the user to be physically present. With a stable internet connection, users are empowered to utilize cloud services whenever they need. Every service is provided online in a dynamic way according to the requirements of the user.

It can be difficult to manage a high amount of user requests on a website or online application, and system failures are a possibility. This is when the importance of load balancers becomes clear. Load balancing is the process of effectively dividing the workload among several computers or servers in order to improve resource utilization and speed up job completion. No machine is overworked while others are idle or underutilized thanks to efficient job allocation.

Load balancing is essential to ensuring that each computer or server in the network is doing a similar task. By ensuring that no node is overworked and others have little tasks to do, this method preserves a fair allocation of computing resources. Because the goals of cloud computing are to improve reaction period, lower costs, and increase efficiency, load balancing is crucial. Be a result, cloud computing is sometimes suggested to be a collection of facilities, effectively distributing sources to satisfy customer needs.

[1]kapil.tajane@pccoepune.org, [2]utkarsh.gardi21@pccoepune.org, [3]sansakr.ghule21@pccoepune.org, [4]utkarsha.kamble21@pccoepune.org, [5]aneesh.kanhere21@pccoepune.org

DOI: 10.1201/9781003716389-21

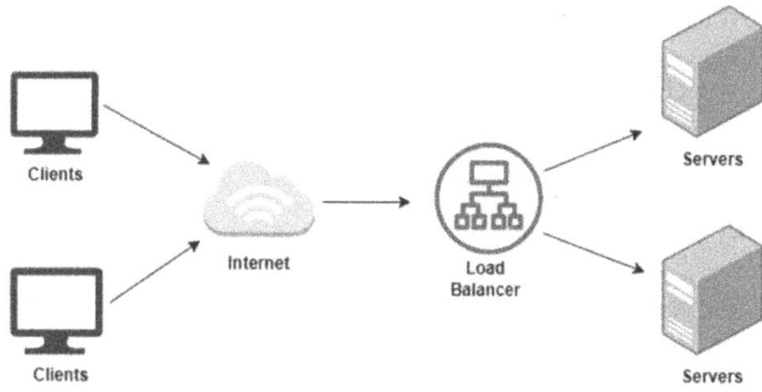

Fig. 21.1 Load balancing in cloud computing

Source: Authors

1.1 Algorithms Designed to Facilitate Load Balancing

Techniques such as load balancing algorithms were developed To facilitate the allocation of traffic across networks or applications among several servers or resources in an equitable manner. Their primary objective is to provide equitable resource distribution by preventing one server from experiencing extreme traffic congestion. Generally speaking, load balancing algorithms fall into one of two categories: Static algorithms need pervious brain power of the system's performance, RAM, and processing capacity. The existing status of the node is not necessary for these algorithms to function. A static approach distributes the load more simply and does not need system information. In a system with little load fluctuation, these methods work well. With dynamic algorithms, choices are made exclusively in light of the system's current state, disregarding any prior knowledge of the system. Dynamic algorithms take into account several strategies, such as information policy, location policy, transfer policy, and selection policy, in order to distribute the haul. It also comes into account the dynamic swap in the statuses of the branches. If a node becomes overloaded, it is moved to a node that is not as burdened.

1.2 Load Balancing Metrics

Numerous factors are taken into account by the load balancing techniques now in use. The time needed to answer a query is known as the response time. Faster reaction times are a sign of better performance and user experience. Throughput refers to the volume of requests a system can process over a specific period. Higher throughput indicates the system can handle a large number of requests. Server utilization measures how much of a server's resources, such as CPU and RAM, are being used. Balancing resource allocation among servers is essential for efficiency. It is the latency — the time between sending a request and receiving a response. Faster data processing as well as better user experience comes from lower latency. The "error rate" is the frequency with which requests are processed. The system is reliable and dependible, according to the reduced error rates shown. The system's scalability is its ability to add more resources to handle increasing demands. Scalable systems can expand to meet their needs without losing their functionality. Flexibility: In reaction to changing traffic patterns, the load balancing algorithm can modify the distribution of tasks.

2. Related Work

Hidayat, Taufik, Yasep Azzery, and Rahutomo Mahardiko [1] proposed a focus on its ease of use and effectiveness in process scheduling, the study "Load Balancing Network by Using Round Robin Algorithm: A Systematic Literature Review" investigates the use of the Round Robin (RR) algorithm for network traffic distribution. To find pertinent articles from reliable databases like IEEE Xplore and ScienceDirect, the authors used a Systematic Literature Review (SLR) technique that included steps including formalizing queries, choosing criteria, sourcing, filtering search results, and evaluating quality. Only three pertinent publications were found between 2015 and 2019, indicating that little research has been done on RR's applicability to network load balancing, despite the fact that it is straightforward and efficient in dividing user requests evenly among servers. Although the study highlights how RR may improve load balancing through equitable distribution, time and cost savings, it also urges more investigation into RR's scalability and effectiveness in dynamic and varied network environments.

Kaur, Rajwinder, and Pawan Luthra [2] proposed with an emphasis on Max-Min, Min-Min, and an enhanced Max-Min algorithm, this paper investigates traffic distribution in utility computing. Cloud computing, which is renowned for offering elastic, scalable, and on-demand resources, has trouble balancing loads, particularly when multiple tasks are happening at once. Efficient traffic distribution guarantees that no node is overwhelmed improves resource consumption, and reduces response time. According to the study, load balancing algorithms can be divided as either Stationary or evolving, with evolving algorithms further subdivided into distributed and non-distributed methods. When assessing these algorithms, metrics such as scalability, resource usage, fault tolerance, response time, and overhead are essential. The enhanced Max-Min method optimizes the makespan and permits concurrent task execution by choosing jobs with the longest completion times and allocating them to resources with the shortest execution times. In comparison to the original Max-Min and RASA algorithms, this improvement achieves superior load distribution and performance, making it especially beneficial for diverse jobs.

Patel, Gaurang, Rutvik Mehta, and Upendra Bhoi [3] proposed the techniques for Static Meta-Task Scheduling in cloud computing, Load Balanced Min-Min (LBMM) and Enhanced Load Balanced Min-Min (ELBMM), which are compared in this study. Even while the classic Min-Min algorithm reduces makespan, it frequently uses resources too little, particularly when high-power resources are used for smaller tasks. Larger jobs are given priority by the Max-Min algorithm, although smaller tasks may take longer to respond to. By rescheduling jobs to make greater use of resources, LBMM outperforms Min-Min. By choosing activities with the longest completion times, ELBMM improves this even more, leading to improved makespan and resource usage. ELBMM performs better than LBMM in both criteria, according to the paper's study.

Khair, Younes, and Haouari Benlabbes [4] proposed the Opportunistic Load Balancing (OLB) algorithm effectively schedules virtual machines in a cloud environment by allocating workloads to host machines based on their current load levels. This approach helps reduce resource strain and optimize usage. The algorithm assigns jobs to hosts with the least increase in load, considering the CPU and RAM capacity of the host machines. Integrated with OpenNebula via its XML-RPC API, OLB simplifies load distribution by evaluating virtual machine weights, They show both the host computers' resource capacity and their resource requests. The algorithm demonstrated enhanced throughput and resource utilization rates in experiments carried out in both OLB-enabled and OLB-naive contexts. The outcomes demonstrate how well it works to optimize cloud system performance, guarantee equitable load distribution, and avoid resource bottlenecks.

Alam, M.J., Chugh, R., Azad, S, et al. [5] proposed the suggested Ant Colony Optimization (ACO)-based CRE approach, and more conventional methods like M-OPTG, M-WO, and PSO are thoroughly compared in this research, with an emphasis on important network performance metrics including load balancing, throughput, and call drop rate. Prior studies have noted difficulties in attaining optimal performance across both characteristics at the same time, with many current methods either sacrificing throughput for load balancing or the other way around. By using a global search approach, while resolving these problems, the ACO-based CRE method improves throughput efficiency and load balancing. Because the ACO approach avoids local optima, it performs better than M-OPTG and M-WO in particular. This improves network performance in general, especially at large user densities. The suggested approach also offers exceptional call drop rate management, demonstrating its robustness in large-scale deployments. According to simulations, the ACO-based CRE approach continuously maintains better load balancing and lower call drop rates than competing algorithms, making it a more scalable and efficient solution for real-world network environments, even though throughput may slightly decrease with increasing user density or hand-offs.

Sharma, Parul, and Abhilasha Sharma [6] the proposed algorithm enhances the distribution of tasks (cloudlets) among virtual machines (VMs) by employing a honeybee-inspired approach to traffic distribution in cloud system. When a load imbalance exceeds predetermined threshold levels, the algorithm triggers task migration. This is achieved by dynamically monitoring the load on VMs. On the size of cloudlets and the load of the VMs, tasks are migrated from overloaded to underloaded virtual machines. While various load balancing strategies, including static and dynamic policies, have been explored previously, many of these studies rely on oversimplified assumptions, such as the disregard for Quality of Service (QoS) attributes and task independence. This method enhances execution speed and resource usage by integrating task length and virtual machine capacity, creatively fusing task migration with real-time load monitoring.

Ebadifard, Fatemeh, Seyed Morteza Babamir, and Sedighe Barani [7] proposed a evolvig task scheduling algorithm inspired by the honey bee foraging behavior for teaffic distribution in utility computing system. While the Honey Bee Behavior-based Load Balancing (HBB-LB) algorithm assigns tasks using a round-robin approach and later balances the load, the proposed method first selects the most suitable VM based on the current workload, processing time, and reliability, ensuring works are given to underloaded machines with the least makespan. The system dynamically evaluates the load of each VM, categorizing them into overloaded, underloaded, or balanced, and assigns tasks accordingly, aiming to reduce makespan and improve system reliability by considering the reliability factor of VMs. If a VM fails consecutively, its reliability factor decreases, and tasks are reassigned to other machines. Simulation findings Illustrate that the recommended approach outperforms both round-robin and HBB-LB methods, particularly as work is more, by reducing meanwhile and the degree of imbalance, while also enhancing fault tolerance through reliability-aware task scheduling.

Anandhakumar, P., V. Sathya, and P. Karthikeyan [8] proposed paper evaluates the Lightning-based Biased Random Sampling (LBRS) algorithm, implemented using CloudSim, and compares its performance against existing algorithms like Biased Random Sampling (BRS) and First-Come-First-Served (FCFS). The performance is assessed based on five parameters: The aspects of response latency, processing duration, error incidence, load management, and network dimensions. The results show that LBRS outperforms the existing methods in all these aspects. Specifically, LBRS exhibits a more even load distribution than BRS, and the execution time is 30% lower than BRS and 20% lower than FCFS. The response time in LBRS is 82% less than BRS and 74% less than FCFS, indicating a significant improvement in efficiency. However, the paper acknowledges the challenge of improving the quality of service (QoS) for cloud users and proposes future work to enhance fault tolerance by incorporating Cloudlet migration in case of VM failures. Additionally, it suggests further accuracy improvements through the analysis of user and broker behavior, leveraging historical allocation data for better cloud resource management.

D. Alsadie [9] proposed the MDVMA A structure designed for the flexible distribution of VMs and the organization of work assigning improves upon utility computing optimization techniques like Particle Swarm Optimization (PSO), Whale Optimization Algorithm (WOA), and Artificial Bee Colony (ABC), which frequently had trouble striking a balance between cost, makespan, and energy consumption. MDVMA overcomes the drawbacks of earlier methods by optimizing these competing goals with non-inferior solutions using the NSGA-II algorithm. Significant gains in energy efficiency, execution time, and cost are demonstrated by CloudSim simulations and validations with HCSP benchmark and synthetic datasets. In addition to highlighting developments in dynamic scheduling and task and resource heterogeneity, this work establishes the groundwork for further studies in virtual machine placement and workflow scheduling.

3. Algorithms Studied

In this study, we consider the fundamentals of cloud computing infrastructure by carrying out a meticulous and deep analysis of both static and dynamic load balancing techniques. Finally, we aim to analyze how suitable these algorithms are to various situations, what changes they bring to cloud service performance. We would like to make a careful assessment of the benefits and the problems with them.

3.1 Honey Bee Algorithm

The Honey Bee Algorithm is one population based optimization method that can be modified to suit the problem of cloud computing load balancing. In this case the algorithm sees the cloud servers as food sources and the jobs to be assigned to them as nectar. The aim is in fact to efficiently allocate jobs across servers and minimize total system load while simultaneously optimizing resource usage.

3.2 Mathematical Expression

How ever specific load balancing goal and fitness function to be used in the experiment, the Honey Bee algorithm will have a formula for calculating its mathematical load balancing. It has a common goal of reducing fluctuations in server loads and avoid that server from being overloaded. The fitness function can be changed in order to achieve these goals, usually represented as F(x). Here's a simplified example:

$$F(x) = \sum (w_i * L_i)/n$$

Where:

The fitness function to be minimized is $F(x)$. L_i is the load on the server i, n is the number of servers, and w_i is the job i is weight.

The Honey Bee Algorithm can be guided by this fitness function while redistributing jobs in order to reduce the load variance between servers.

3.3 Active Clustering Algorithm

A dynamic load balancing method that adjusts to shifting workloads and resource availability in a cloud computing environment is called active clustering. The essential notion of active clustering is the assignment of tasks to these clusters based on certain criteria and the arrangement of resources within them. These criteria include network latency, job priority, CPU and memory use, and more.

3.4 Mathematical Expression

Mathematical expressions can be used by active clustering algorithms to determine load balancing. An example of a mathematical expression for the load balancing measure (LB) as a function of resource usage metrics is as follows:

$$LB = f(CPU\ utilization,\ Memory\ utilization,\ Networklatency,\ Taskpriority)$$

Here, "*f*" represents a function that combines these metrics to calculate a load balancing score. The task or workload is then assigned to the cluster with the highest load balancing score.

3.5 Ant Colony Algorithm

The Ant Colony Load Balancing Algorithm models the problem as an optimization job, with servers standing in for the places where ants might deposit tasks and ants representing computing processes. Using a mix of heuristics and pheromone levels, each ant (task) looks for the optimal server (route) to deposit its job.

3.6 Mathematical Expression

ACO-based load balancing algorithms use mathematical expressions to guide ants' decision-making. These expressions combine pheromone levels and heuristics to calculate the desirability of selecting a particular server. Here's a simplified example of a mathematical expression for the desirability (D) of choosing a server *i* by an ant:

$$D(i) = (T(i)^{\alpha}) * (H(i)^{\beta})$$

Where,

$D(i)$ represents the desirability of choosing a server i.

$T(i)$ is the pheromone level on the path leading to the server i. $H(i)$ is a heuristic measure representing the attractiveness of server i.

How heuristics and pheromones cooperate with each other is determined by the factors α and β. By changing these variables, this algorithm reacts differently.

3.7 Random Biased Sampling Algorithm

Dynamic load balancing, such as the case of random biased sampling, is a distributed and cloud computing technique that comes up with a model for load balancing. As we shall see in this chapter the central idea of this method is to fairly assign request or tasks on the current workloads on the resources that are available. It avoids abuse of frequently accessed resources and guarantees equally that the least accessed resources are used. It comes very Much In handy where you have fluctuating workloads or resource capacities.

3.8 Mathematical Expression

The mathematical expression for the load balancing metric (*LB*) could be as simple as:

$$LB = \frac{1}{1 + Resource\ workload}$$

A resource with a higher workload (signaling increased resource consumption) will be less likely to be chosen in this expression because of its lower load balancing score. On the other hand, a resource with less labor will score better and be more likely to be selected.

3.9 Round Robin Algorithm

One of the most preferred and easy to implement load balancing technique for distributed systems and cloud computing environments is round robin. The framework's primary function is therefore to cyclically distribute received tasks or requests between the available resources. The Round Robin algorithm makes it possible for the available resources are to acquire an equal opportunity to meet the incoming requests.

3.10 Mathematical Expression

The Round Robin algorithm is relatively simple and doesn't require a complex mathematical expression for load balancing decisions. Assignments are made in a round-robin fashion, so the expression can be summarized as:

$$Next_{resource} = (current\_resource + 1)\%\ total_{resources}$$

In this expression, "*current_resource*" represents the resource currently processing a task, "*total_resources*" is the total number of available resources, and "*Next_resource*" is the resource to which the next task will be assigned.

3.11 Max-Min Algorithm

The Max-Min algorithm was developed for use in load balancing for purpose of ensuring that the load to the facilities offered through cloud computing was divided evenly. Its objective is to minimise the least possible use of resources. Making sure that no resource is overloaded while others have available capacity. It is especially helpful when resources have different capacities or when their capabilities are understood beforehand.

3.12 Mathematical Expression

The Max-Min method is based on the "*largest – minimum*" idea, which may be represented mathematically as follows:

$$Assign(Task) = argmax(\min(C_i))$$

Here, "*Assign*(*Task*)" represents the resource to which the task is assigned, and "*arg max* (*min*(*C_i*))" denotes the resource with the maximum minimum available capacity among all resources.

3.13 Min-Min Algorithm

One common heuristic used to optimize job scheduling in cloud computing systems is the Min-Min method. By allocating jobs to appropriate resources according to their expected execution timeframes, it seeks to reduce the makespan, or the total time required to complete all of the jobs. The Min-Min algorithm is a variant of the Shortest Job First (SJF) scheduling technique.

3.14 Mathematical Expression

The Min-Min method estimates the execution durations for each job rather than assigning tasks based on a particular mathematical statement. Nonetheless, the following mathematical notation may be used to illustrate the choice of resources for job assignment:

Let T_i be the set of tasks, and R_j be the set of resources. The designed output time of task T_i on resource R_j can be expressed as EET (T_i, R_j).

The Min-Min algorithm selects the resource R_j for task T_i that minimizes the expected execution time:

$$R_i = argmin(R_j \in R)\{EET(T_i, R_j)\}$$

In this statement, *argmin* indicates the resource in the set *R* that minimizes the expected execution time, and R_i indicates the resource chosen for the task T_i.

3.15 Opportunistic Algorithm

Opportunistic load balancing strategies are reported to be used when demands and availability of resources change in cloud computing environments. Instead, these algorithms maximize work distribution and resource allocation under favourable conditions.

3.16 Mathematical Expression

Under most circumstances, the resource consumption metrics might be used to express the load balancing measure (*LB*) mathematically and weighed down by the priorities of the algorithm.

For instance:

$$LB = (\alpha * CPU\ utilization + \beta * Memory\ utilization + \gamma * Network\ latency)$$

In this formula, α, β and γ represent weights and these weights can be changed according to load balancing algorithm's specific goal. It is a algorithm's preference to some type of resources or situations in the form of some weights.

4. Examination of Comparative Aspect

Table 21.1 Comparative study of various algorithm

Load Balancing Algorithm	Type	Benefits	Drawbacks
Round Robin	Static	Fixed period slice; better performance for short CPU bursts.	The larger tasks take longer to finish.
Max-Min	Static	Works better when requirements are known beforehand.	Task completion takes a lot of time.
Min-Min	Static	The tiny completion time gives the better output for tiny tasks.	Best results are only obtained via smallest completion time, but only for small tasks.
Opportunistic Load Balancing	Static	Superior performance; improved allocation of resources.	Too much time is taken to complete a task.
Ant Colony Optimization	Dynamic	The approach is mathematically efficient and effectively reduces the makespan.	This search is too long and too complex.
Honey Bee Foraging	Dynamic	less response period; increases output.	Loads of a lower priority take longer to apply.
Biased Random Sampling	Dynamic	Superior perfor-mance; improved allocation of resources.	Response time is higher.
Active Clustering	Dynamic	Similar nodes are grouped together.	For microgrids with more variety of nodes, the performance decreases.

Source: Authors

5. Conclusion

In this paper the study of load balancing algorithms in cloud computing at a system level and its significance about optimum utilization of resources and minimum latencies have been included. IF however, such systems have a stable workload then deterministic algorithms like Round Robin, Max-Min, and Min-Min are highly effective because they present fair solutions that can be easily understood. However, they can not be defined for real time fluctuating environments meaning that they perform well in dynamic environments. Ant Colony Optimization and Honey Bee Foraging are free from these limitations because the proposed algorithms work in reaction to the workload fluctuation. However, these methods negatively impact system scalability and throughput as well as increase the computational intensity and resource- intensive characteristic of the system. This paper aims at comparing between the two types of approaches namely the static and dynamics so as to give a clear distinction of how to select between the two depending with the system needs.

Future work in this context will therefore entail building on current developmental works in the context of designing hybrid algorithms that combines the benefits of static and dynamic algorithms. Besides, possible effective applications of a proactive nature included in artificial intelligence and machine learning could be applied to enhance the decision-making process concerning load balancing in a more flexible manner. More empirical research can help to have more accurate and stable cloud computing framework to fit the ever increasing demands in future research.

References

1. Hidayat, Taufik, Yasep Azzery, and Rahutomo Mahardiko. "Load balancing network by using round Robin algorithm: a systematic literature review." Jurnal Online Informatika 4.2 (2019): 85–89.
2. Kaur, Rajwinder, and Pawan Luthra. "Load balancing in cloud system using max min and min min algorithm." International Journal of Computer Applications 975 (2014): 8887.
3. Patel, Gaurang, Rutvik Mehta, and Upendra Bhoi. "Enhanced load balanced min-min algorithm for static meta task scheduling in cloud computing." Procedia Computer Science 57 (2015): 545–553.
4. Khair, Younes, and Haouari Benlabbes. "Opportunistic load balancing for virtual machines scheduling in a cloud environment." Engineering Proceedings 29.1 (2023): 1.
5. Alam, Mohammed Jaber, et al. "Ant colony optimization-based solution to optimize load balancing and throughput for 5G and beyond heterogeneous networks." EURASIP Journal on Wireless Communications and Networking 2024.1 (2024): 44.
6. Sharma, Parul, and Abhilasha Sharma. "Honeybee Inspired Load Balancing Algorithm for Cloud Computing Environment." Proceedings of the International Conference on Innovative Computing & Communication (ICICC). 2022.
7. Ebadifard, Fatemeh, Seyed Morteza Babamir, and Sedighe Barani. "A dynamic task scheduling algorithm improved by load balancing in cloud computing." 2020 6th International Conference on Web Research (ICWR). IEEE, 2020.
8. Anandhakumar, P., V. Sathya, and P. Karthikeyan. "Improving Response Time using Lightning based Biased Random Sampling (LBRS) for Load Balancing." (2023).
9. Alsadie, Deafallah. "A metaheuristic framework for dynamic virtual machine allocation with optimized task scheduling in cloud data centers." IEEE Access 9 (2021): 74218–74233.

10. Fang, Yiqiu, Fei Wang, and Junwei Ge. "A task scheduling algorithm based on load balancing in cloud computing." Web Information Systems and Mining: International Conference, WISM 2010, Sanya, China, October 23–24, 2010. Proceedings. Springer Berlin Heidelberg, 2010.
11. Al Nuaimi, Klaithem, et al. "A survey of load balancing in cloud computing: Challenges and algorithms." 2012 second symposium on network cloud computing and applications. IEEE, 2012.
12. Khare, Shivangi, Uday Chourasia, and Anjna Jayant Deen. "Load balancing in cloud computing." Proceedings of the International Conference on Cognitive and Intelligent Computing: ICCIC 2021, Volume 1. Singapore: Springer Nature Singapore, 2022.
13. Zuo, Liyun, et al. "A multi-objective optimization scheduling method based on the ant colony algorithm in cloud computing." Ieee Access 3 (2015): 2687–2699.
14. Moon, YoungJu, et al. "A slave ants based ant colony optimization algorithm for task scheduling in cloud computing environments." Human-centric Computing and Information Sciences 7 (2017): 1–10.
15. Du, Jianbo, et al. "Computation offloading and resource allocation in mixed fog/cloud computing systems with min-max fairness guarantee." IEEE Transactions on Communications 66.4 (2017): 1594–1608.
16. Mao, Yingchi, Xi Chen, and Xiaofang Li. "Max–min task scheduling algorithm for load balance in cloud computing." Proceedings of International Conference on Computer Science and Information Technology: CSAIT 2013, September 21–23, 2013, Kunming, China. Springer India, 2014.
17. Pradhan, Pandaba, Prafulla Ku Behera, and B. N. B. Ray. "Modified round robin algorithm for resource allocation in cloud computing." Procedia Computer Science 85 (2016): 878–890.
18. Alhaidari, Fahd, and Taghreed Zayed Balharith. "Enhanced round-robin algorithm in the cloud computing environment for optimal task scheduling." Computers 10.5 (2021): 63.
19. LD, Dhinesh Babu, and P. Venkata Krishna. "Honey bee behavior inspired load balancing of tasks in cloud computing environments." Applied soft computing 13.5 (2013): 2292–2303.
20. Kaur, Anureet, and Bikrampal Kaur. "Load balancing in tasks using honey bee behavior algorithm in cloud computing." 2016 5th international conference on wireless networks and embedded systems (WECON). IEEE, 2016.

Intelligent Systems Using Semiconductors for Robotics and IoT – Dinesh Goyal et al. (eds)
© 2026 Taylor & Francis Group, London, ISBN 978-1-041-20408-4

22 Paradigm Shift: Diversity and Inclusion

Tripti Arvind[1], Bindhia Joji[2], Elizabeth Chacko[3]
CHRIST (Deemed to be University), School of Business and Management,
Bangalore, Karnataka

Abstract: There is an extensive list of both manifest and latent characteristics of diversity that includes beliefs and opinions, cultural background, ethnicity, sexuality, skills, credentials, experience, and approaches towards life, among workers in a group. With globalization, the trends of international mobility, and outsourcing of manpower implying that the population heterogeneity is also rising, scientists and practitioners began to pay considerable attention to the management of diversity in organizations more actively. Different people have different personality types and these include extroversion, agreeableness and openness to experience, it has been discovered that they affect the performance of diverse teams in a significant manner. The analysis was conducted in two stages: first, by applying the simple arithmetic mean and the standard deviation to determine the attitudes of employees to diversity management; second, to determine the factors, which explain the management of diversity in IT organizations, a regression analysis shall be completed. It also notes that both HR and diversity have a highly considerable positive relation to the performance of an organization, although it provides the explanation of some aspects of diversity like constructive social relations, fair career opportunities and arrangements for diversity involvement in decision-making, which also play an important role towards organizational outcomes. The objectives of this research are to understand the Indian IT employee diversities' perception and diversities management and its relation with organizational performance. Thus, the areas of focus of the research aim to contribute important findings that would be useful in discussing and implementing diversity policies in the IT industry of India for both practical and academic purposes.

Keywords: Diversity management, Organization performance, Employee attitude, IT organizations

1. Introduction

There is an extensive list of both manifest and latent characteristics of diversity that includes beliefs and opinions, cultural background, ethnicity, sexuality, skills, credentials, experience, and approaches towards life, among workers in a group. With globalisation, the trends of international mobility, and outsourcing of manpower implying that the population heterogeneity is also rising, scientists and practitioners began to pay considerable attention towards the diversity management in organisations more actively. Muhammad Awais Bhatti et al.[18] and Nishii [19] also showed that, organisations that support true diversity have increased levels of employees' engagement and decreased rates of turnover for minority workers. In the same manner, Kelli et al[12] synthesized the principle of workplace diversity to identify its benefits and issues, stating that a diverse workforce significantly increases employers' value but needs suitable organizational adjustment through policies, practices, and organizational culture perspective. Karambayya, [11]. Also, advances in the selection and recruitment approaches like virtual career fairs and social media recruiting have improved the inclusiveness of the hiring process. For this reason, minority employees benefit from the two structures as follows: mentorship programs and employee resource groups are important organizational practices that help minority employees at the workplace to advance in their careers as well as build confidence in organizations Martinez and Flores,[17]. A sense of

[1]tripti.arvind@christuniversity.in, [2]bindhia.joji@christuniversity.in, [3]elizabeth.chacko@christuniversity.in

DOI: 10.1201/9781003716389-22

inclusion has also emerged as another factor which has a strong bearing on the effectiveness of the team and the wellbeing of the members as espoused by Lee and Cunningham [13], its pointed out that leaders with high cultural intelligence are superior in handling culturally diverse staff. Nevertheless, there are some dilemmas that have not been overcome yet even though diversity has been acknowledged and valued. There are arguments that diversity training can help eliminate prejudice and enhance organization's actions and policies in practice Ely R J and Thomas D A[3], and the absence of psychological safety harms employees' to offer their thoughts and support the efforts of diverse teams Hernandez & Knight,[5] Therefore, the study focuses on the systematic examination of IT organisations in comprehending the existence and effects of Diversity Management in so far as they relate to organisational performance in Bengaluru. A quantitative research method was used to establish the causes and consequences of managing diversity practices implemented with the intention of employee recruitment. To understand the participants' thoughts, behaviours, and preferences regarding diversity management practices, a structured questionnaire was developed and the final five-item self-completed questionnaire was completed by 300 respondents. The analysis was conducted in two stages: first, by applying the simple arithmetic mean and the standard deviation to determine the attitudes of employees to diversity management; second, to determine the factors, which explain the management of diversity in IT organizations, a regression analysis shall be completed. It also notes that both HR and diversity have a highly considerable positive relation to the performance of an organization, although it provides the explanation of some aspects of diversity like constructive social relations, fair career opportunities and arrangements for diversity involvement in decision-making, which also play an important role towards organizational outcomes.

A lot of research of diversity management on employee engagement, organisational performance and innovation has been done abroad, however there is lack of study in the Indian IT sector which are influenced by cultural, social, and economic factors—Bhatti et al. (2018) and Nishii (2016). Indian IT companies, which employ a sizable portion of the world's workforce, confront unique difficulties because of linguistic diversity, cultural heterogeneity, and established social structures. Although previous studies, like Gupta & Kumar (2021), highlight the increasing focus on diversity in Indian businesses, there is little empirical data on how these practices affect organisational outcomes in this particular industry. This disparity is significant because India's IT sector is a major driver of the country's economy and a pooling ground for global talent.

The present study aims to bridge this gap by investigating how employees in Indian IT companies view diversity management practices and how these views relate to organisational performance. This study is especially important because it adds to the global research on diversity management while also offering localised perspectives that are vital to the Indian IT industry. Policymakers, HR specialists, and organisational leaders who are tasked with fostering inclusive work environments in culturally diverse settings.

2. Literature Review

The background of the study is an important component of research, as it helps to provide a clear and compelling rationale for why the research is being conducted, and why it is valuable and relevant. The following contents is divided into two sections one is about IT and second is about IT enabled industries. India is a country with multicultural and multi linguistic presence with different castes and vast geographic area. This culminates into having a diverse culture and diverse people who work in the same company. This creates issues to the management as they have to face conflicts and challenges between employees. Nishii and Mayer [20] reveal that different diversity improves imagination and choice-taking; Gupta and Kumar [4] observe that concerning diversity attracts and inspires the best human capital, which ensures better results. Nevertheless, it is sad to note that there is limited literature exploring diversity management with special reference to the Indian context. From the literature, Sharma and Singh [28] observed that preconceived prejudices always slow the enhanced diversity methods within recruitment. While underlining the significance of the mentorship programs in diversity, Patel et al [22] stress the problem related to its underrepresentation and practically non-existent application in Indian IT companies. Reddy and Rao [24] observed that the policy on diversity enhances employees' interest. Furthermore, Verma & Choudhury [32] came up with the evidence that supports insufficiency of strong diversity policies in the organization negatively affects the working environment. Consequently, Iyer and Gupta [8] proved that the organization's focused diversity training leads to better familiarity and embrace of difference viewpoints. Shen J et al [29] has shown different ways to showcase diversity in resource management. Bhatti, et al [2] gives the role of personality traits in diverse team performance.

Ely and Thomas [3] found inclusive and diverse workplace environment positively influences employee attitudes, leading to more collaborative and innovative work environments. Tapia and Kvasny [31] found that workplace diversity can be encouraged through actions by individuals, managers, and institutions. Jayne and Dipboye [10] highlight that leadership commitment to diversity is essential in fostering positive employee attitudes, as it sets a tone of inclusivity and respect throughout the organization. According to Richard et.al [25] , diversity is a strategic resource that can lead to performance in organizations by fostering problem solving and decision making. In their research, Homan et al. established that diverse teams are innovative because of the wider range of ideas considered within such teams. Studies by Page [21] provides evidence that diverse groups outperform homogeneous ones in decision-making tasks because they consider a wider array of perspectives and alternatives. Richard, Kirby, and Chadwick [26] demonstrate that firms with diverse leadership teams are better positioned to understand and cater to diverse customer bases, resulting in improved financial performance. Shen et al. [20] show that perceived fairness in diversity-related policies and practices can enhance positive attitudes towards diversity, as employees feel that the organization is committed to treating everyone equitably. Podsiadlowski, A.et al [23] highlights the importance of adopting a perspective that values diversity to foster positive employee attitudes and enhance performance. Richard, Kirby, and Chadwick [26] demonstrate that firms with diverse leadership teams are better positioned to understand and cater to diverse customer bases, resulting in improved financial performance. Hunt, Layton, and Prince [7] demonstrated that companies with higher levels of diversity are more likely to outperform their less diverse peers in terms of profitability. Bezrukova, et.al[1] has pointed that effective diversity training programs significantly improve employee attitudes by increasing awareness and reducing biases. Jehn and Bezrukova [1] noted that unmanaged diversity often leads to interpersonal conflicts, which can adversely affect performance. Muhammad Awais et al[18] , reported on the role of extroversion, agreeableness, openness to experience, emotional stability and conscientiousness in relationship to multicultural workforce job satisfaction and performance. Shore et al. [13] called attention to the role of inclusive leadership in creating an environment where diversity can flourish and support enhancing organizational performance. [11] Rekha Karambayya, "Globalisation and the Indian IT industry" 2019, she highlighting the developed sketch of the development of the Indian IT sector with a relation to Globalisation context. Homan, A. C et.al [6] shown Positive attitudes toward diversity good and positive attitude toward diversity enhances teamwork; better and effective collaboration and cohesion. Lorenzo, R., & Reeves, M [15] Diverse teams bring the richness of multiple perspectives and ideas that boost creativity and lead to more innovative solutions. McKinsey & Company, [16] There is a very strong correlation between leadership diversity and better financial performance, really underlining the business case for diversity. Rock, D., Grant, H., & Grey, J. [27] Growing discomfort with diversity may lead to a more cohesive team performance. According to Martinez and Flores [17] in their research conducted in 2023, it was very important for the career development of their minority employees and their sense of belonging to work as the programs of mentorship and ERGs came into play at work.

3. Research Objective

Limited Studies in the Indian Context: Despite the fact that there is a vast amount of theoretical and empirical research available on the subject of diversity management around the globe, there is rather an acute deficiency of publications concerning Indian organisations' approaches to diversity management. This paradox provides a research gap to examine the various org-internal social-cultural factors which impact diversity management in India.

Focus on Existing Workforce: The majority of related literature mostly focus on the issue of diversity of the current generation workforce, where subjects like diversity management, diversity and engagement, and diversity and retention are likely to be featured. It is necessary to carry out a research of diversity management practices which can be focused on the recruitment process as well as their effects on organizational performance.

Understanding Cultural Nuances: Since India is a culturally diverse country in terms of backgrounds, beliefs and identity, there is a dearth of systematic understanding regarding how diversity related challenges and inclusion management are handled and how appropriate the environment is provided within the organizations operating in India. Impact of Diversity on Performance: Scanty prior literature is based on empirical research on the relationship between DM practices and OP results in Indian IT firms for which there is a need for research in this area.

An attempt made to fill these gaps may help in presenting a bulk of information on best practices of diversity management both for India and the larger global context. With so many, it is evident that some of these gaps will be fulfilled by this study that centers on Diversity Management Practices among the Indian IT Company.

This study aims at:

- To learn about the perceptions of IT organizations about workplace diversity and inclusion.
- Determine the factors affecting diversity management and organizational performance.
- To examine the impact of organizational diversity on its performance.

The present study strictly pertains to the diversity management practices of the IT based organizations based in Bengaluru city. The study adopted a quantitative research method. Further quantitative methods enable researchers to empirically test the research hypotheses, framed by adopting deductive approach, and make inferences. Responses to questions posed were collected on a set format of close-ended questions that later were analyzed statistically to make inferences that can be generalized to a larger population.

4. Methodology

The study used a sample size of 300 and was focused on employees of IT organisations. Random sampling was used for the study and was focused on the employees in the city of Bengaluru. The present study can also be said to be descriptive in nature as it intends to answer the 'what' aspects of the research problem with the help of scientific methods that are precise and can be reproduced in future. The current study adopted a deductive approach, which is denoted as reasoning from the particular to the general.

The study followed the quantitative research method to determine the causes and effects of diversity management practices, adopted for recruiting new employees. Further quantitative methods enable researchers to empirically test the research hypotheses, framed by adopting deductive approach, and make inferences. The responses from the study participants were collected using a predefined format with close-ended questions, which were later analysed statistically to make inferences that can be extrapolated to a bigger population.

Self-administered structured questionnaire (research instrument) was used for collecting responses from the participants about their thoughts, behaviours and preferences towards diversity management practices in a systematic manner. Research data collection in a monolingual setting is employed in the questionnaire. This method ensures that data collection is both consistent and accurate by using one language of communication among the target population. The research instrument was so designed that the right and proper words were used for forming questions. Care was taken at every stage including the structuring and ordering of questions in the questionnaire to enhance the reliability and validity of the instrument. "Likert scale" whose scales range from "Strongly Disagree to Strongly Agree" were assigned. The reliability of the research instrument, i.e., the questionnaire, was ensured using a reliability coefficient known as "Cronbach's alpha". Scales with "Cronbach's alpha" values >0.6 were considered to be reliable and internally consistent. Such scales are found to exhibit a high consistency among different points of the same construct. The questionnaire that was used was tested by examining its internal consistency. If Cronbach's alpha value is more than 0.6, then the questionnaire reliability can be ascertained.

The following are the objectives of the current study:

- Analysing the attitude of an employee towards diversity and inclusion within the IT organization.
- Investigating how factors impact diversity management performance of an IT organization.

4.1 Analysis

Table 22.1 references the company's attitude towards diversity. Regarding the organisation's stance on diversity, the respondents agreed that their management advocated diversity (Mean = 4.009 ± 0.890), displayed the importance through its conduct (Mean = 3.940 ± 0.848) and its commitment towards diversity was high (Mean = 4.009 ± 0.890). They were also in agreement that their organisation was accessible to people of all abilities (Mean = 4.041 ± 0.907) who were respected and appreciated (Mean = 4.014 ± 0.893). They agreed on the inclusive work environment in their organisation (Mean = 4.101 ± 0.836) and its reputation for encouraging diversity and overcoming discrimination (Mean = 4.009 ± 0.831). From the results, it is derived that persons with different abilities are employed, and are treated with dignity in an inclusive working environment. The organisation's commitment towards diversity was represented in its conduct, and treatment of its employees. This is in accordance with the study of James and Mathew (2012) explaining that the initiatives taken by top-level management affect employees positively and increase their worth to the organisation.

Table 22.1 Employee attitude

Employees attitude towards diversity	Mean	Std. Deviation
"The management of our organisation advocates diversity"	3.959	0.907
"The management displays the importance of diversity through its conduct"	3.940	0.848
"The commitment of my organisation towards diversity is high"	4.009	0.890
"People are respected as individuals in my organisation and appreciates the differences in opinion in a healthy manner"	4.014	0.893
"My organisation is accessible to people of all abilities"	4.041	0.907
"My organisation has a reputation for encouraging diversity and overcoming discrimination"	4.009	0.831
"We have an inclusive work environment where staff diversity at all levels in valued and encouraged"	4.101	0.836
Factor mean	**4.010**	**0.873**

To investigate the factors influencing diversity management in IT organizations and its performance. Respondents agreed that their organisation gave priority for doing the job right the first time (Mean = 4.115 ± 0.859) and they developed innovative programs frequently (Mean = 4.014 ± 0.913). They also agreed that the skill level had been enhanced (Mean = 3.867 ± 0.898) and teamwork and spirit of cooperation was high (Mean = 4.046 ± 0.889) in their organisation. On the topic of quality of work, they agreed that the quality of their work (Mean = 4.183 ± 0.839) and that of their boss' (Mean = 4.060 ± 0.901) was effective with which the customers were content (Mean = 4.096 ± 0.913).

These results show that after the implementation of diversity policy, there was considerable progress in the performance of the organization as shown by improvement in the skill level, team spirit, and innovation of employees, which resulted in satisfactory customers of the work done. This concurs with the study of Bhatti et al, [2], which explicated that the effect of diversity management was positive to the performance of organisation, especially innovation. Table 22.2 provides the organisational performance

Table 22.2 Organisational performance

Organisational performance	Mean	Std. Deviation
"The skill level in my organisation has enhanced in the last year"	3.867	0.898
"Teamwork and spirit of cooperation is high in my organisation"	4.046	0.889
"The customers are content with the quality of our work"	4.096	0.872
"My organisation gives priority for doing the job right in the first time"	4.115	0.859
"The quality of work done by us is excellent"	4.183	0.839
"The quality of work done by my boss is also good"	4.060	0.901
"Frequent innovative programs are developed by us"	4.014	0.913
Factor mean	**4.054**	**0.882**

Hypothesis 1: "There is a significant impact of organisational behaviour inclusive of diversity management on organisational performance."

The effect of organisational behaviour, whose sub-factors are diversity management, company's attitude towards it and its corporate culture, on organisational performance was studied. The respondents agreed that organisational behaviour significantly affected organisational performance (Mean = 3.913 ± 0.729 to 4.054 ± 0.720). It can be understood that a positive relationship exists between organisational behaviour and performance (F (3,214) = 134.675, p = 0.000). The R square value of 0.654 expresses that a 65.4% variation in organisational performance is observed due to organisational behaviour which includes diversity management practices (Table 4.27).

The regression coefficient B of the company's attitude towards diversity management was 0.257, experience in diversity management and policy was 0.058 and corporate culture was 0.066 with each sub-factor having p=0.000. It is remarkable to observe from the above outcomes that organisational behaviour has a highly substantial effect on organisational functioning. Thus, it can be ascertained that hypothesis 1 can be accepted. Table 22.3 shows the Impact of organizational behaviour inclusive of diversity management on organizational functioning

Table 22.3 Impact of OB in functioning

Variables	Mean	Std. Deviation
Organisational performance	4.054	0.720
Attitude	4.010	0.706
Diversity management	3.913	0.729
Corporate culture	3.961	0.715

Table 22.4 Coefficients of impact of organizational behaviour

R	R Square	Adjusted R Square	Std. Error of the Estimate	Change Statistics				
				R Square Change	F Change	df1	df2	Sig. F Change
0.809	0.654	0.649	0.426	0.654	134.675	3	214	0.000

Variables	Mean	Std. Deviation
Organisational performance	4.054	0.720
Career development	3.974	0.722
Behaviour of boss	3.936	0.763
Interaction	4.039	0.692
Managing diversity	3.939	0.688

Variables	Unstandardised Coefficients		Standardised Coefficients	t	Sig.
	B	Std. Error	Beta		
(Constant)	0.483	0.181		2.669	0.008
Attitude	0.257	0.066	0.252	3.870	0.000
Diversity management	0.395	0.058	0.401	6.861	0.000
Corporate Culture	0.251	0.066	0.249	3.804	0.000

Hypothesis 2: There is a major influence of the characteristics of the diversity of employees on the performance of organisations

From the above referenced Table 22.4, The regression analysis of impact of sub-factors of diversity characteristics of employees - career development of employees, behaviour of boss, interaction among employees and experience in managing diversity - on organisational performance was performed. The respondents were in agreement regarding the impact of diversity characteristics of employees on organisational performance (Mean = 3.936 ± 0.763 to 4.054 ± 0.720). The relation between diversity characteristics of employees and organisational performance was found to be positive (F (4,213) = 106.341, p = 0.000). The R square value of 0.666 detailed that a variation of 66.6% in organisation's performance was observed with change in diversity and employees.

Hence, the sub-factors for diversity characteristics of employees were obtained as B values as 0.129 for career development, 0.217 for behaviour of boss, 0.351 for interaction among employees and 0.228 for managing diversity. A p value < 0.05 was also obtained by the sub-factors, which indicate that a significant relationship exists between the diversity characteristics of employees and organisational performance. Hence, hypothesis 2 is accepted. Table 22.4 describe the coefficients of impact of diversity characteristics of employees on organisational performance.

Table 22.5 Impact of diversity characteristics of employees on organisational performance

R	R Square	Adjusted R Square	Std. Error of the Estimate	Change Statistics				
				R Square Change	F Change	df1	df2	Sig. F Change
0.816	0.666	0.660	0.4195	0.666	106.341	4	213	0.000

Table 22.6 Coefficients of impact of diversity characteristics of employees on organisational performance

c	Unstandardised Coefficients		Standardised Coefficients	t	Sig.
	B	Std. Error	Beta		
(Constant)	0.369	0.184		2.007	0.046
Career development	0.129	0.062	0.130	2.071	0.040
Behaviour of boss	0.217	0.065	0.231	3.332	0.001
Employees Interaction	0.351	0.074	0.338	4.731	0.000
Managing diversity	0.228	0.065	0.218	3.503	0.001

4.2 Findings

Employee Attitudes Towards Diversity: Some of the findings obtained from the respondents were to the effect that management supports diversity and that the organisation's behaviour is diverse. The employees officially recognize one another's worth and value individual perspectives and opinions.

Ease of Reporting Discrimination: Regarding the whistleblowing policies, respondents said it was possible to disclose discrimination to higher authorities and the responses were that the company acted against discrimination cases as reported.

Diverse Hiring Practices: There were culturally diverse personnel in charge of hiring employees, and the process considered merit rather than gender, background or state of origin thus the company supports a diverse employment.

The positive impact of diversity on an organization: Organizational behavior functions as a directly proportional variable of performance while highly impacting diversity practices regarding performance. Employee diversity provides a basis for ensuring good teamwork, innovation, and the effective delivery of work expected from one's employees

Importance of Leadership: Subordinates rely on their direct managers and consider them to be efficient and caring; it is therefore evident that proper management is a critical aspect of supporting employees and creating an environment of equity.

Recommendations for Recruitment Practices: Diversity inclusive recruitment should be another key practice to be adopted by organizations; activities such as covering the applicants' identity during a first-preliminary stage, coming up with standard selection processes.

Results of the Study and Its Implications for Organizations: The analysis revealed an important relationship between diversity management practices and organizational performance ($p < 0.05$). Its important to mote here that the regression analysis generated an R^2 value of 0.654, indicating that around 65.4% of the variance in organizational performance can be explained by diversity management practices. This reveals a strong relationship, but it is also crucial to recognize our findings' limitations. Since the total sample size was 300 , which is adequate enough for analysis, may limit the generalizability of the results to bigger application .Also the , research only focused on IT organizations in Bengaluru, which is not completely the illustrative of the diversity practices of other industries.

Even though the results show a significant trend, organizational culture, resource availability, and implementation difficulties are few more elements that must be taken into account when implementing diversity management strategies in the real world.

Therefore, the study demonstrates that diversity traits have a great impact on the performance of organizations and integrates with the idea of the effect of diversity management explaining the success of an organization in terms of amplified worker satisfaction and productivity.

4.3 Few Interpretations of the Study Are

- Since the total sample size was 300 , research only focused on IT organizations in Bengaluru,
- Even though the results show a significant trend, organizational culture, resource availability, and implementation difficulties are few more elements that must be taken into account when implementing diversity management strategies in the real world.
- From the results, it is evident that employees always look up to their immediate boss as they consider them to be competent and considerate towards them. Organizations must strive to include strong diversity inclusive recruitment practices by removing information regarding gender, background and other

sub-groups during the preliminary screening process itself and by conducting final interviews using standardized selection procedures.

- Employee diversity traits serve as a very essential tool for enhancing the performance of the organization. Therefore, organizations must attempt to ensure that employees belong to different racio-ethnic backgrounds, become diversity-friendly, promote harmonious relationship among the diverse employee groups, and provide equal career and growth opportunities to all employees alike, irrespective of their ethnicity and other characteristics (such as gender).

The present study strictly pertains to the diversity management practices of the IT- based organisations based in Bangalore city. Further, the prime focus of the present study will be the "diversity management practices" followed in the process of employee selection. The study findings will not be relevant for organisations from other sectors of the economy and organisations functioning in other cities of the country, IT or otherwise. The scope of this study is concentrated more on the diversity management practices in IT- based organisations located in Bangalore city. Moreover, the key objective of this study is to "diversity management practices" used in configuration of selection & recruitment. The results of this study are not supported for organizations from other sectors of the economy and of organizations operating in other cities of the country, IT or otherwise. In the IT sector, study is very crucial as with technology organisations are functioning globally. Therefore, the output of this study will help IT companies for well grown and to fill the manpower in future easily.

4.4 Implications

The interpretation of the current research has a deep meaning for organizations, especially in the context of today's IT field. To begin with, there is clear support for employee attitudes toward diversity, which supports the proposition of the need to foster true organizational diversity. This commitment not only helps to boost the morale of the employees and improve productivity but also helps to ensure that people stay committed to their employers specifically given the tough nature of the job market this is important. Secondly, the focus on various practices about diversity and inclusion shows the existence of an important avenue through which organizational biases can be addressed in the recruitment procedures.

5. Conclusion

To conclude, this study is of assistance to the existing literature as it adds depth to understanding how DM practices and OP are related in the Indian IT context. The theoretical literature and evidence suggests a strong message that diversity commitments, combined with leadership support and stringent hiring policies, can substantially enhance employee satisfaction and the wider organizational stake. The established correlation of the diversity of employee attributes and better team performance is an evidence which indicates that organizational diversity at its base could be extremely helpful.

References

1. Bezrukova, K., Spell, C. S., Perry, J. L., & Jehn, K. A. (2016). The role of diversity in team dynamics and performance. Journal of Applied Psychology, 101(5), 741–756. https://doi.org/10.1037/apl0000090
2. Bhatti, M. A., Alshagawi, M., & Juhari, A. (2018). The role of personality traits in diverse team performance. Journal of Organizational Behavior, 39(8), 975–989. https://doi.org/10.1002/job.2272
3. Ely, R. J., & Thomas, D. A. (2001). Cultural diversity at work: The effects of diversity perspectives on work group processes and outcomes. Administrative Science Quarterly, 46(2), 229–273. https://doi.org/10.2307/2667087
4. Gupta, M., & Kumar, R. (2021). Attracting and retaining top talent through diversity. Human Resource Management Review, 31(2), 100–113. https://doi.org/10.1016/j.hrmr.2020.100713
5. Hernandez, M., & Knight, C. (2023). Psychological safety in diverse teams: Overcoming barriers to idea sharing. Journal of Organizational Behavior, 44(3), 354–370. https://doi.org/10.1002/job.2693
6. Homan, A. C., Gündemir, S., Buengeler, C., & van Kleef, G. A. (2020). The impact of positive attitudes towards diversity on team dynamics. Journal of Applied Psychology, 105(5), 423–438. https://doi.org/10.1037/apl0000452
7. Hunt, V., Layton, D., & Prince, S. (2015). Why diversity matters: The business case for diversity. McKinsey & Company. https://www.mckinsey.com/business-functions/organization/our-insights/why-diversity-matters
8. Iyer, A., & Gupta, M. (2021). Diversity training and its effects on organizational performance. Journal of Workplace Learning, 33(6), 519–534. https://doi.org/10.1108/JWL-11-2020-0149
9. James, E., & Mathew, A. (2012). The influence of top management on employee attitudes towards diversity. Human Resource Management Review, 22(4), 346–356. https://doi.org/10.1016/j.hrmr.2011.11.005
10. Jayne, M. E. A., & Dipboye, R. L. (2004). Leveraging diversity to improve business performance. Organizational Dynamics, 33(3), 213–226. https://doi.org/10.1016/j.orgdyn.2004.04.002
11. Karambayya, R. (2019). Globalization and the Indian IT industry. Journal of Global Business and Economics, 17(3), 35–49. https://doi.org/10.2139/ssrn.3310156
12. Kelli, B., Johnson, R., & Marlowe, M. (2015). Workplace diversity: Benefits and challenges. Journal of Business Studies, 48(1), 41–57. https://doi.org/10.1080/0020804X.2014.944025
13. Lee, J., & Cunningham, M. (2018). The effects of inclusion on team effectiveness. Journal of Organizational Behavior, 39(2), 234–248. https://doi.org/10.1002/job.2288

14. Lee, C., & Farh, J. L. (2004). The relationship between cultural diversity and team performance. Journal of Management, 30(3), 393–414. https://doi.org/10.1016/j.jom.2003.09.001
15. Lorenzo, R., & Reeves, M. (2020). How diversity drives innovation and performance. Harvard Business Review. https://hbr.org/2020/03/how-diversity-drives-innovation
16. McKinsey & Company. (2020). Diversity wins: How inclusion matters. McKinsey & Company. https://www.mckinsey.com/business-functions/organization/our-insights/diversity-wins-how-inclusion-matters
17. Martinez, M., & Flores, T. (2023). Mentorship programs and their role in career advancement for minority employees. Journal of Career Development, 50(2), 183–196. https://doi.org/10.1177/08948453211004855
18. Muhammad Awais Bhatti, Mohammed Alshagawi, & Ariff Juhari. (2018). Personality traits and multicultural workforce performance. International Journal of Human Resource Management, 29(14), 2094–2114. https://doi.org/10.1080/09585192.2017.1383564
19. Nishii, L. H. (2016). The benefits of workplace diversity for organizations. Annual Review of Organizational Psychology and Organizational Behavior, 3, 367–396. https://doi.org/10.1146/annurev-orgpsych-041015-062622
20. Nishii, L. H., & Mayer, D. M. (2020). The role of diversity in enhancing organizational performance. Journal of Applied Psychology, 105(4), 289–303. https://doi.org/10.1037/apl0000464
21. Page, S. E. (2007). The diversity bonus: How great teams pay off in the long run. Harvard Business Review, 85(12), 72–80. https://hbr.org/2007/12/the-diversity-bonus
22. Patel, S., Kumar, A., & Gupta, R. (2023). The role of mentorship programs in enhancing workplace diversity. Human Resource Management Review, 33(2), 100–112. https://doi.org/10.1016/j.hrmr.2022.100723
23. Podsiadlowski, A., Gröschke, D., Kogler, M., Springer, C., & van der Zee, K. (2013). Managing diversity in organizations: A review and future research agenda. European Journal of Work and Organizational Psychology, 22(4), 504–523. https://doi.org/10.1080/1359432X.2012.743825
24. Reddy, S., & Rao, P. (2022). Policy impacts on diversity and employee engagement. Journal of Human Resources, 57(3), 345–359. https://doi.org/10.3368/jhr.57.3.345
25. Richard, O. C., Barnett, T., Dwyer, R., & Chadwick, K. (2004). Cultural diversity in management, firm performance, and the moderating role of entrepreneurial orientation. Academy of Management Journal, 47(4), 403–417. https://doi.org/10.5465/20159585
26. Richard, O. C., Kirby, S. L., & Chadwick, K. (2013). The impact of diversity on firm performance: A review and meta-analysis. Academy of Management Perspectives, 27(3), 30–45. https://doi.org/10.5465/amp.2012.0108
27. Rock, D., Grant, H., & Grey, J. (2021). Navigating the discomfort of diversity to improve team performance. Harvard Business Review. https://hbr.org/2021/02/navigating-the-discomfort-of-diversity
28. Sharma, R., & Singh, P. (2022). Challenges in implementing diversity recruitment practices in Indian IT companies. International Journal of Human Resource Management, 33(4), 629–646. https://doi.org/10.1080/09585192.2021.1908598
29. Shen, J., Chanda, A., D'Netto, B., & Monga, M. (2009). Managing diversity through human resource management: An international perspective. International Journal of Human Resource Management, 20(2), 239–260. https://doi.org/10.1080/09585190802670585
30. Shore, L. M., Cleveland, J. N., & Sanchez, D. (2018). Inclusive workplaces and inclusive leadership: Implications for organizational performance. Journal of Organizational Behavior, 39(7), 829–843. https://doi.org/10.1002/job.2338
31. Tapia, A., & Kvasny, L. (2004). Diversity and the role of managerial support. Journal of Organizational Behavior, 25(5), 467–482. https://doi.org/10.1002/job.274
32. Verma, P., & Singh, A. (2023). The influence of gender diversity on team performance and innovation. International Journal of Management Reviews, 25(1), 123–138. https://doi.org/10.1111/ijmr.12203

Note: All the tables in this chapter were made by the authors.

Intelligent Systems Using Semiconductors for Robotics and IoT – Dinesh Goyal et al. (eds)

23 Harnessing Machine Learning: Enhancing Supply Chain Efficiency in the Retail Sector

Durga Aishwarya Pandi[1], Bindhia Joji[2], Helen Josephine[3]
Christ University, School of Business and Management,
Bengaluru, Karnataka

Abstract: This study explores the impact of advanced analytics and Machine Learning on supply chain and inventory management in the retail and consumer goods industry. As consumer expectations evolve, the need for efficient supply chain operations becomes paramount. Leveraging a multidimensional approach, this research employs predictive analytics, including demand forecasting and optimization algorithms, to address critical challenges such as demand variability and inventory optimization. This study utilizes a mixed method design, integrating data analysis with qualitative insights from stakeholders. Key findings reveal that machine learning models, particularly Linear Regression, outperform traditional methods in inventory demand prediction, while demonstrating the necessity for continuous refinement to enhance accuracy. Additionally, the integration of digital transformation initiatives, including IoT and big data analytics, supports improved supply chain agility and resilience. By providing actionable insights, this research facilitates data-driven decision-making, ultimately contributing to operational efficiency and strategic planning in a dynamic retail landscape.

Keywords: Supply chain management, Inventory optimization, Machine learning, Demand forecasting, Retail industry, Digital transformation, Big data analytics, Operational efficiency

1. Introduction

The retail and consumer goods industry are undergoing a transformation, driven by the evolving consumer expectations and the integration of new advanced technologies. Efficient supply chain, inventory management are at the heart of this evolution, ensuring that companies can meet demand, minimize costs, and optimize operations. Leveraging multidimensional models will allow for a holistic analysis of complex supply chain networks, capturing the intricate relationships between suppliers, manufacturers, distribution centers, and retailers. This approach facilitates an in-depth understanding of how changes in one part of the chain impact the entire system. By incorporating analytics and algorithms such as time series forecasting, regression models, and clustering, the project will address key challenges like demand variability, stock optimization, and transportation logistics. Machine learning models will be employed to predict demand trends and patterns, optimize various inventory levels, and streamline logistics. Techniques like demand forecasting will use historical sales data to anticipate future needs, while optimization algorithms will help in maintaining optimal inventory levels, reducing both overstocking and stockouts Through the integration of these advanced analytical models, the project will provide actionable insights, allowing companies to make various data driven decisions. This will not only enhance operational efficiency but also supports strategic planning, ensuring a robust and agile supply chain that can adapt to market fluctuations and consumer trends.

2. Literature Review

Gaurav Agrawal, R. Chandra [1] in their article explores the complexities of managing supply chains in a multi-

[1]durga.pandi@mba.christuniversity.in, [2]bindhia.Joji@Christuniversity.in, [3]helen.josephine@christuniversity.in

DOI: 10.1201/9781003716389-23

channel retail environment. It discusses strategies for integrating online and offline channels to optimize inventory and reduce costs, highlighting the importance of real-time data and coordination across channels. S. Kumar, A. Singh, R. Tiwari [2] presents a machine learning framework for inventory optimization in retail, using techniques like demand forecasting and predictive analytics. It emphasizes the role of big data and advanced algorithms in improving stock levels and reducing waste. J. M. Brynjolfsson, S. Yang [3] examines how digital transformation initiatives, such as IoT and AI, affect supply chain performance in the consumer goods industry. It shows that companies adopting digital tools experience improved efficiency, visibility, and responsiveness in their supply chains.

P. R. Mehrotra, D. K. [4] Gupta compares traditional statistical methods like ARIMA with machine learning models such as neural networks and random forests for demand forecasting in retail. It concludes that machine learning models provide more accurate predictions, especially for complex demand patterns. L. C. Hong, K. T. Lee, A. M. Chan [5] discusses the growing emphasis on sustainability in the retail supply chain. It covers topics like ethical sourcing, green logistics, and reducing carbon footprints. The authors propose strategies for integrating sustainability without compromising efficiency and profitability. M. Patel, N. Shah, V. Mehta [6] addresses the unique inventory management challenges faced by e-commerce retailers, such as high SKU variability and rapid delivery expectations. The paper suggests solutions like dynamic inventory allocation and real-time stock visibility to enhance fulfillment efficiency. H. Zhang, Q. Li, W. Wei [7] investigate how agility in supply chain operations can help consumer goods companies respond to market changes and disruptions. They identify key factors that contribute to supply chain agility, such as flexible sourcing, adaptive logistics, and real-time information sharing. S. Banerjee, P. Maiti [8] explores potential of blockchain to improve transparency, traceability, security in retail supply chains. It includes case studies from the food and apparel sectors, demonstrating how blockchain can prevent fraud and ensure product authenticity. R. K. Gupta, A. Sharma [9] review the applications of artificial intelligence in retail supply chains, focusing on areas like demand forecasting, inventory optimization, and customer service. It highlights the benefits of AI, such as reduced operational costs and improved decision-making. C. S. Chen, Y. Wang, Z. Liu [10] analyze the challenges of inventory management in an omnichannel retail environment. It discusses strategies for aligning inventory across channels, such as centralized warehousing and integrated stock management systems, to provide a seamless customer experience. S. R. Jayaraman, V. S. Ramachandran [11] examines the use of big data analytics in retail supply chains, emphasizing its role in improving demand forecasting, inventory management, and logistics optimization. The study highlights various case studies showing successful implementations of big data technologies. M. S. Lee, J. H. Park [12] explores how collaborative strategies like Vendor-Managed Inventory (VMI) can streamline inventory management in retail. It provides an in-depth case study on a major retailer, showing how VMI reduced stockouts and improved inventory turnover. K. P. Tan, L. L. Wong, A. B. Lee [13] Using a structural equation modeling approach, this research analyzes the relationship between supply chain integration and retail performance. It finds that higher levels of integration—both upstream and downstream—are associated with better operational efficiency and financial performance. R. J. Bode, C. Wagner [14] presents a framework for managing risks in the retail supply chain, focusing on resilience strategies to cope with disruptions such as natural disasters, geopolitical instability, and supply chain breakdowns. The authors discuss risk identification, assessment, and mitigation techniques. P. S. Iyer, T. E. Kaplan [15] explores the impact of Radio Frequency Identification (RFID) technology on the inventory management in retail. It demonstrates how RFID improves inventory accuracy, reduces shrinkage, and enhances the overall efficiency of inventory tracking and replenishment processes. F. X. Zhao, G. T. Hsu, E. L. Chou [16] investigates the use of dynamic pricing strategies in conjunction with inventory management to optimize revenue in retail. It discusses how retailers can adjust prices based on inventory levels and demand forecasts to balance supply and demand effectively. D. G. Williams, K. N. Smith [17] reviews the role of the Internet of Things (IoT) in synchronizing retail supply chains, particularly in inventory management. It outlines how IoT-enabled devices, such as smart shelves and connected sensors, provide real-time data that enhances inventory visibility and decision-making. A. M. Jones, S. K. Miller, T. R. Green [18] discusses sustainable inventory management practices in retail, using a circular economy framework. It explores strategies like product reuse, recycling, and remanufacturing, showing how these practices can reduce waste and environmental impact while maintaining profitability. E. A. Kim, M. W. Lim, C. J. Kang [19] study examines how global retailers manage supply chain efficiency in an omnichannel environment. It highlights the importance of integrating online and offline supply chains, real-time inventory updates, and flexible logistics solutions to meet consumer expectations. H. Chen, L. Zhou, Z. Wu [20] focuses on the application of artificial intelligence (AI) in optimizing supply chain operations in retail. It covers AI-driven solutions for demand forecasting, inventory replenishment, and route optimization, showcasing their impact on reducing costs and improving service levels.

3. Research Methodology

This research follows systematic approach using the DMAIC methodology, widely recognized for process improvement and operational excellence. The primary objective is to enhance inventory management and supply chain efficiency in the retail and consumer goods industry.

- **Research Design:** The study employs a research design, combining both quantitative and qualitative approaches. Quantitative methods are utilized for data collection and statistical analysis, focusing on metrics like inventory turnover, stockout frequency, lead time, and forecast accuracy. Qualitative methods involve stakeholder interviews and expert consultations to gain insights into operational challenges and improvement opportunities.
- **Data Collection:** Data is collected from various sources, which include historical sales data, inventory levels, lead times, supplier performance records. The study involves gathering data from various retail locations and product categories to ensure a comprehensive understanding of inventory dynamics. Additional data, such as customer demand and seasonal trends, is extracted using SQL queries, Excel, and ERP systems.
- **Data Analysis:** The analysis phase employs advanced analytical techniques, including statistical analysis, Pareto analysis, and time series analysis, to identify trends, patterns, and root causes of inefficiencies. Machine learning models like ARIMA and Prophet are used for demand forecasting and inventory optimization. Visualization tools such as Power BI provide an intuitive understanding of the analyzed data.
- **Improvement Strategy:** Based on the analysis, solutions such as advanced forecasting models, automated reorder systems, and supplier performance metrics are proposed and piloted. The implementation strategy includes training inventory management teams and establishing a feedback loop for continuous improvement.
- **Control and Monitoring:** The control phase establishes a robust monitoring system using KPIs like inventory turnover rate and the forecast accuracy. Regular reporting and stakeholder engagement ensure the sustainability of improvements.

4. Results and Discussion

The results from the predictive modeling efforts for both supply chain and inventory management reveal important insights regarding model performance and their applicability in a business context. The evaluation of models using Mean Squared Error (MSE) metrics indicates the effectiveness of each model in capturing patterns in the data and making accurate predictions.

4.1 Sales by Segment

- **Small Business Dominates Sales:** The highest sales are recorded in the Small Business segment, indicating this market contributes significantly to overall revenue.
- **Government and Enterprise Segments:** These segments also show substantial sales, suggesting stable contracts and consistent demand, especially from large institutions.
- **Low Sales in Midmarket and Channel Partners:** The Midmarket and Channel Partners segments have minimal sales, highlighting potential areas for growth or need for strategic adjustments available in Fig. 23.1

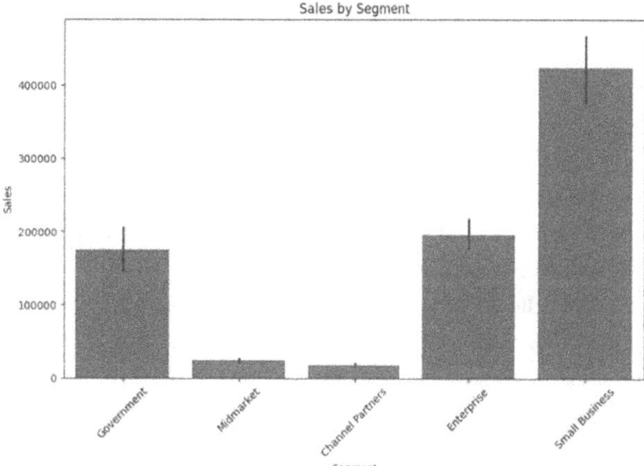

Fig. 23.1 Sales by segment

4.2 Correlation Heatmap

- **High Correlation Between Sales Metrics:** Gross Sales, Sales, and COGS show a very high positive correlation (close to 1), indicating that these metrics are strongly interdependent.
- **Moderate Correlation with Profit:** Sales Price and Gross Sales have a moderate positive correlation with Profit, suggesting that higher sales and pricing strategies contribute to profitability.
- **Negligible Impact of Product and Date:** Variables like Product, Month Number, and Year have minimal correlation with key metrics, implying they have little direct impact on sales and profit figures. Shown in Fig. 23.2

4.3 Trend of Sales Over Time

- **Seasonal Sales Fluctuations:** The graph shows periodic spikes and dips in sales, indicating possible seasonal demand variations or promotional campaigns impacting sales performance.
- **Sales Growth in Early 2014:** There is a noticeable increase in sales in early 2014, followed by fluctuations, suggesting a potential market expansion or successful marketing initiative during this period.

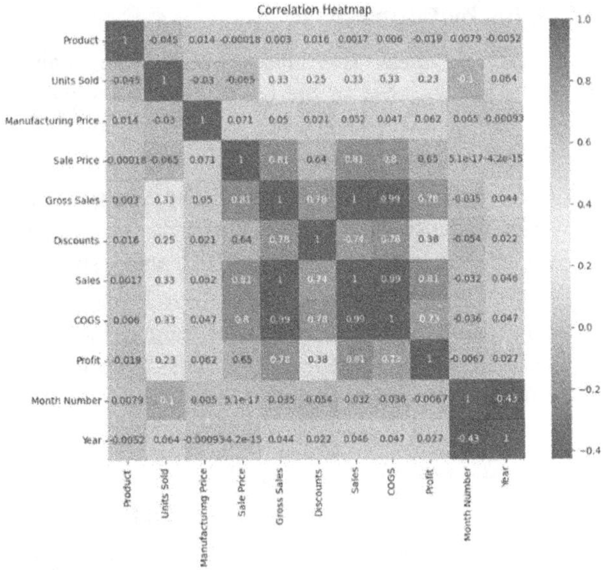

Fig. 23.2 Correlation of factors

- **Decline Towards Late 2014:** The declining trend in sales towards the end of 2014 indicates possible market saturation or reduced demand, which may need addressing through targeted strategies. Shown in Fig. 23.3

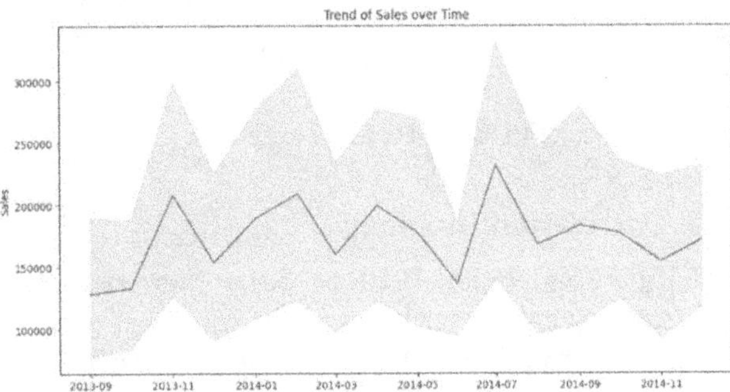

Fig. 23.3 Trend of sales

4.4 Supply Chain Analysis

The Linear Regression model for supply chain performance yielded an MSE of 1,182,358,778,003.83, while the Random Forest Regression model resulted in an MSE of 2,087,817,555,270.11. Despite Random Forest's capacity to model complex relationships, its higher MSE suggests potential overfitting or an inability to generalize effectively to unseen data. Linear Regression, though not ideal, demonstrated relatively better performance. This indicates that the data might have inherent linear characteristics that the Random Forest model failed to leverage effectively.

In supply chain management, minimizing errors in demand forecasting and logistics planning is critical for improving resilience, reducing operational costs, and optimizing distribution networks. Although the Linear Regression model had a lower MSE, it is still quite high, indicating that neither model captured the nuances required for precise predictions. This underperformance may stem from insufficient feature engineering, inappropriate model parameters, or unaccounted variables influencing supply chain dynamics. Future efforts should focus on refining the model, perhaps by integrating additional features like macroeconomic indicators or advanced machine learning techniques like Gradient Boosting to enhance predictive power.

4.5 Inventory Management Analysis

For inventory management, Linear Regression achieved an MSE of 709,408.51, outperforming the Random Forest model, which had an MSE of 845,573.09. The relatively lower MSE of the Linear Regression model suggests it is more adept at predicting inventory demand, likely due to the straightforward, linear relationship between product sales and time-based features such as the month of the year. Despite Random Forest's flexibility in capturing non-linear relationships, it underperformed, possibly due to overfitting or insufficient model complexity.

Accurate demand forecasting is essential to balance inventory levels, minimize stockouts, and prevent overstocking, all of which are critical for operational efficiency and profitability. The lower MSE of the Linear Regression model makes it a more suitable choice for this application. However, further enhancements are recommended, such as incorporating additional features like promotional events, competitor activity, and economic factors, which can significantly impact inventory demand patterns.

5. Scope and Conclusion

The project aims to enhance the supply chain and inventory management in retail and consumer goods industries through advanced analytics and machine learning models. By utilizing techniques like demand forecasting, regression, and clustering, the study addresses challenges such as demand variability, stock optimization, and logistics. The findings will inform strategic planning, operational efficiency, and supply chain resilience in a dynamic market environment.

The project successfully demonstrates the potential of advanced analytics and machine learning in optimizing the supply chain and inventory management for retail and consumer goods industries. By employing models like Linear Regression and Random Forest, the study addresses key challenges such as demand forecasting and inventory optimization. Despite some limitations in model performance, the insights gained pave the way for further refinement and integration of additional variables. The research highlights the importance of a data-driven approach to enhancing operational efficiency and strategic planning.

References

1. Agrawal, G., & Chandra, R. (Year). *Supply Chain Management in the Retail Industry: A Multi-Channel Approach.*
2. Kumar, S., Singh, A., & Tiwari, R. (Year). *Optimizing Inventory in Retail: A Machine Learning Approach.*
3. Brynjolfsson, J. M., & Yang, S. (Year). *The Impact of Digital Transformation on Supply Chain Performance in the Consumer Goods Industry.*
4. Mehrotra, P. R., & Gupta, D. K. (Year). *Demand Forecasting in Retail: A Comparison of Statistical and Machine Learning Models*
5. Hong, L. C., Lee, K. T., & Chan, A. M. (Year). *Sustainable Supply Chain Management in the Retail Sector: Challenges and Opportunities*
6. Patel, M., Shah, N., & Mehta, V. (Year). *Inventory Management in E-Commerce: A Review of Challenges and Solutions* [7] Zhang, H., Li, Q., & Wei, W. (Year). *Supply Chain Agility in the Consumer Goods Industry: An Empirical Analysis.*
8. Banerjee, S., & Maiti, P. (Year). *Blockchain in Retail Supply Chains: A Case Study Approach.*
9. Gupta, R. K., & Sharma, A. (Year). *The Role of AI in Enhancing Retail Supply Chain Efficiency: A Review.*
10. Chen, C. S., Wang, Y., & Liu, Z. (Year). *Omnichannel Retailing and its Impact on Inventory Management.*
11. Jayaraman, S. R., & Ramachandran, V. S. (Year). *Leveraging Big Data for Supply Chain Analytics in Retail.*
12. Lee, M. S., & Park, J. H. (Year). *Collaborative Inventory Management in the Retail Sector: A Case Study on VendorManaged Inventory (VMI).*
13. Tan, K. P., Wong, L. L., & Lee, A. B. (Year). *Impact of Supply Chain Integration on Retail Performance: A Structural Equation Modeling Approach.*
14. Bode, R. J., & Wagner, C. (Year). *Supply Chain Risk Management in the Retail Industry: A Framework for Resilience.*
15. Iyer, P. S., & Kaplan, T. E. (Year). *The Influence of RFID Technology on Retail Inventory Accuracy.*
16. Zhao, F. X., Hsu, G. T., & Chou, E. L. (Year). *Dynamic Pricing Strategies and Inventory Management in Retail.*
17. Williams, D. G., & Smith, K. N. (Year). *Synchronized Supply Chains: The Role of IoT in Retail Inventory Management.* [18] Jones, A. M., Miller, S. K., & Green, T. R. (Year). *Sustainable Inventory Management in Retail: A Circular Economy Perspective.*
19. Kim, E. A., Lim, M. W., & Kang, C. J. (Year). *Omnichannel Retailing and Supply Chain Efficiency: Insights from Global Retailers.*
20. Chen, H., Zhou, L., & Wu, Z. (Year). *AI-Driven Supply Chain Optimization in the Retail Industry.*

Note: All the figures in this chapter were made by the authors.

Intelligent Systems Using Semiconductors for Robotics and IoT – Dinesh Goyal et al. (eds)

24 Real-Time Sentiment Analysis for Dynamic Customer Feedback: A Study on Customer Reviews

M. Joohi[1], Bindiia Joji[2]
Christ University, School of Business and Management,
Bengaluru, Karnataka
Mansurali[3]
Central University of Tamil Nadu,
Thiruvarur

Abstract: In today's world, customer feedback plays a vital role in forming business strategies, particularly in profoundly competitive businesses. Consequently, sentiment analysis has ended up as a critical tool for handling customer opinions, and subsequently converting them into actionable insights. But current sentiment analysis models regularly depend on authentic historic data, in this way constraining their capacity to give fast responses to evolving customer sentiment. This research addresses the above gap by proposing a real-time sentiment analysis model that dynamically integrates customer feedback data and offers timely insights as well. Utilizing a dataset of StarCofee customer reviews, the VADER sentiment analysis model has been made use of, that makes a significant difference to categorize and track sentiment trends over time. This analysis uncovers significant fluctuations in customer sentiment, focusing on the significance of real-time monitoring. The results highlight the potential for businesses to utilize real-time sentiment insights for optimizing customer experience and operations. Future research in this field may explore the integration of advanced sentiment analysis techniques and predictive models to accomplish an even more prominent business value.

Keywords: VADER, Sentiment analysis, Customer satisfaction, Real-time feedback, StarCofee, Dynamic sentiment monitoring, Business intelligence

1. Introduction

In today's computerized world, the ability to collect and analyse customer feedback is significant for businesses to sustain competitiveness. Customer sentiments that are often communicated through online platforms, social media, and reviews offer priceless insights into brand perception, product success, and customer satisfaction.

Sentiment analysis, which is a subfield of Natural Language Processing (NLP), has been broadly adopted to categorize views of customer into negative, positive, or neutral sentiments. Though existing sentiment analysis models deliver insights based on historical data, they frequently fall short in delivering real-time insights, which are vital for businesses that requires responding dynamically to customer feedback.

This paper dwells into the application of real-time sentiment analysis utilizing a case study on StarCofee customer reviews. By utilizing the VADER sentiment analysis model, we center on revealing actionable insights from customer feedback and illustrate how real-time monitoring can boost up operational decision-making. Particularly, this study addresses a vital research gap, namely the need to integrate dynamic sentiment data to ensure timely responses to changing customer needs.

2. Literature Review

Smith et al. [1] analyze the impact of sentiment analysis in retail environments, concentrating on how real-time feedback can deliver better customer service and inventory management. They also discuss the challenges of using a

[1]m.joohi@mba.christuniversity.in, [2]bindhia.joji@christuniversity.in, [3]mansurali@cutn.ac.in

DOI: 10.1201/9781003716389-24

real-time sentiment analysis and how we can leverage it for decision-making. Johnson and Lee [2] (Lee, 2022) in this paper centers on how sentiment analysis can be used to segment customers based on their feedback. Johnson and Lee stressed the importance of deploying sentiment data for personalized marketing campaigns, and also highlighting the lack of tools that can perform real-time segmentation.

Chen et al [3] and his colleagues developed a sentiment analysis model that focused on predicting customer churn in the service industry. Though this model emphasizes on predictive analytics, they lack real-time sentiment adaptability, which they found out as an area that needs further research. Nguyen et al [4] found out how product development cycles are informed by sentiment analysis. Even though this research integrates customer feedback into iterative product updates, it is limited to historical data that lacks real-time processing capabilities, which could speed up the product development. (Wang, 2023) Patel et al [5] emphasized on combining sentiment analysis with behavioural data, and developed a more accurate predictive model of customer behaviour. They stressed that while sentiment data enhances these models, the lack of real-time adaptability limits their utility in dynamic industries.

Kumar et al. [6] worked on the computational challenges that are associated with sentiment analysis in large datasets. They found out that existing models are not fully equipped to handle real-time data processing, especially when quick insights are necessary for business decisions. Davis et al [7] focused on new product launches and how social media sentiment can provide real-time feedback on them. This model improves customer service response times but does not account for the scalability needed in larger organizations with continuous customer interactions. Zhao et al. [8] emphasized the role of sentiment analysis in crisis management, particularly for organizations facing public relations issues. They mentioned that sentiment analysis can help to identify negative sentiment early on, but this model lacks real-time tracking capabilities, which is significant in fast-evolving crises. Garcia and Thomas [9] discussed on how companies utilize sentiment analysis to manage their brand reputation. They proved that businesses rely mostly on historical sentiment data to track brand perception over time, but there is no visibility on using real-time sentiment to influence immediate brand strategy. Lee et al [10] applied sentiment analysis in the banking field for customer feedback, and showed that churn rates can be reduced by understanding customer sentiment and also it increases customer satisfaction. This study mentions further research into real-time sentiment monitoring to dynamically improve customer support.

Mendoza et al. [11] worked on how sentiment analysis can improve customer service in the telecommunication industry. Though this research demonstrates that sentiment analysis can identify customer pain points, it also notes the lack of real-time sentiment integration, which could have improved customer support responses significantly. Huang et al. [12] applied sentiment analysis on e-commerce platforms for customer reviews and how it can be used to optimize product recommendations. This study also analyzed the challenges of integrating real-time sentiment analysis for immediate improvements to the recommendation algorithms.

Rodriguez et al [13] worked on how sentiment analysis is applied to understand audience reactions to media content. This research focused on applying sentiment data to inform content creation and also highlighted a gap while integrating real-time audience feedback for live content adjustments. Singh and Kapoor [14] focused on how real-time sentiment analysis can improve automated customer support systems. This model also faced challenges in processing large volumes of real-time feedback accurately to provide immediate customer support. Park et al [15] and his colleagues developed a model that combines sentiment analysis with customer demographics, and provide personalized experiences in retail industry. Though this model proved successful with historical data, all real-time feedback is not fully integrated, thus limiting its adaptability to address immediate customer needs.

Gonzalez and Rios [16] used sentiment analysis to predict stock market trends in social media. Though this model proved strong predictive accuracy, but does not address how real-time sentiment integration could improve market forecasting in volatile environments. Fernandez et al. [17] deployed sentiment analysis in the food and beverage sector to evaluate customer satisfaction. They found out that while sentiment analysis provides valuable insights into consumer preferences, real-time feedback integration could further enhance customer retention and delivery of service. Chowdhury and Bhatt[18] dwelved on how sentiment analysis can improve online retail operations by continuous monitoring of customer reviews in real-time. This research also emphasized the importance for scalable real-time systems as traditional models are not sufficient for addressing high-volume customer feedback.

Martinez and Oliveira [19] and team demonstrated a sentiment analysis model that predicted the success of new products based on early customer feedback. Though this model performed good with historical data, they also emphasized the deployment of real-time adaptation to collect immediate customer responses. Singh et al. [20] proved that organizations can gain a competitive advantage by incorporating sentiment analysis to monitor competitor brands. This study also identified a gap in how real-time sentiment analysis could enable organizations to act upon more swiftly to changes in market sentiment, and thus potentially shifting its consumer loyalty.

3. Research Methodology

3.1 Overview of Dataset

The dataset that is deployed to conduct this study comprises of StarCofee customer feedback that are gathered across various platforms like surveys, social media, and in-app reviews. Customer comments, the date and time of review, and a rating system are some of the several attributes of this dataset. Pre-processing and cleaning of the textual data has been done to make them ready for sentiment analysis and also to remove the noise.

3.2 Framework of Sentiment Analysis

VADER tool, which is Valence Aware Dictionary and Sentiment Reasoner, has been utilized for performing the sentiment analysis. This tool is well-suited for analyzing social media text mainly because of its sensitivity to both intensity and polarity of sentiments. The framework was implemented as depicted below:

1. Data Preprocessing: All the customer comments were cleaned, tokenized, and processed to remove punctuation, irrelevant data points, and stop words.
2. Sentiment Scoring: All the customer comments were passed through the VADER analyzer, which subsequently generated a sentiment score. This sentiment score was then categorized as negative, positive, or neutral depending on predefined thresholds.
3. Time-Series Aggregation: The dataset was then resampled on a daily basis to dynamically monitor customer sentiments, thus facilitating the calculation of average sentiment scores.
4. Visualization: Now, a time-series graph was generated to express how the customer sentiment fluctuated over time, and this provided a visual representation of sentiment trends.

3.3 Data Collection

A dataset comprising of StarCofee customer feedback has been utilised for this study that is collected from various platforms such as in-app reviews, social media, and surveys. The collected data spanned a period of several months, thus providing a vital source of customer opinions. The key fields of this dataset included the following:

- Date/Time of Review: This provided a timestamp of when the feedback was provided.
- Feedback Text: This included actual customer comments, which represent customer sentiment toward StarCofee' services and products.
- Ratings: This included numerical ratings given by the customer, which actually provide additional context to the sentiment analysis.

All the above diverse data source served as a basis for a comprehensive analysis of customer sentiment. This also facilitated to ensure that all the insights were a representative of StarCofee' overall customer base.

3.4 Data Understanding

The initial understanding of this data centers on analyzing the distribution of feedback data across various platforms. These platforms included social media, in-app reviews, etc. It also helped in identifying the key trends in customer opinions, the variability in customer ratings, etc. The points given below were then analyzed:

- Volume of Reviews: Good knowledge of the frequency of reviews and how these reviews vary across different platforms.
- Sentiment Polarity: Finding out the mix of positive, negative, or neutral feedback during this process.
- Text Analysis: Obtaining valuable insights into common topics and keywords within the feedback that could probably influence customer sentiment.

3.5 Data Preparation

To conduct a thorough sentiment analysis, we have to pre-process the raw feedback data. This preparation of data comprises of the steps given below:

- Text Cleaning: By removing irrelevant characters (such as punctuation, special symbols, stop words, etc.), we can focus on only meaningful words to pass through this sentiment analysis algorithm.
- Tokenization: This step involves breaking down the feedback text into individual words or tokens to enable sentiment scoring.
- Missing Data Handling: In this process, any incomplete or missing entries, especially in the "rating" or "feedback text" columns, were then cleaned to enable data integrity.
- Date Formatting: This step involves conversion of the date of review to any consistent datetime format to ensure time-series analysis of sentiment trends.

The preparation of data made sure that the feedback was cleaned and structured, and hence ready for sentiment analysis, thus ultimately allowing the algorithm to showcase accurate results.

3.6 Evaluation

The effectiveness of the sentiment analysis was analyzed considering the following aspects:

- Accuracy of Sentiment Classification: The accuracy is determined by comparing the sentiment labels (that are generated by the VADER algorithm) with their corresponding customer ratings. For example, does positive comments match high ratings? etc. This method served as a validation mechanism for all the sentiment categories.
- Time-Series Analysis: This analysis is done by plotting down all the sentiment score trends recorded

over time. Through this method, one could visually analyse how efficient the sentiment analysis has captured the variability in customer feedback data. The constant sentiment fluctuations and clear peaks or troughs in this graph showed that the model was more sensitive to shifts in customer opinion.

- Actionable Insights: Success was measured by how efficiently the analysis was translated into actionable insights for StarCofee. By identifying any periods or any specific days of sharp sentiment shifts, the organization can act upon to reinforce positive trends and also mitigate any negative feedback.

4. Results and Discussion

4.1 Algorithm: Sentiment Analysis using VADER

The VADER algorithm has been utilized for analyzing the sentiment, which was particularly developed for analyzing customer reviews and social media text, and thus well-suited for analyzing the StarCofee feedback. This algorithm assigns sentiment scores to every individual word depending upon a pre-defined lexicon. A sentiment score was calculated for the entire given text, and classifies them into any one of the below three categories:

- Positive: For overall scores above 0.05.
- Negative: For scores below -0.05.
- Neutral: For scores between 0.05 and -0.05.

This algorithm depicts sentiment scores that range from -1, which is most negative to +1, which is most positive, thus providing a basic understanding of all customer feedback. By resampling the sentiment scores over daily intervals, a real-time impact of this analysis was achievable, thus providing insights into customer opinions over time.

4.2 Findings

1. Positive and Negative Sentiment Distributions: This data provides a mixture of positive, negative, and neutral sentiment categories. The analysis also suggests a variability in customer experiences that helps businesses to leverage and understand all the pain points and achieve success in their services.
2. Fluctuating Customer Sentiment: The sentiment trend graph shows that customer sentiment is very dynamic with frequent fluctuations. This finding hence reinforces the importance of real-time monitoring since waiting for quarterly or monthly sentiment analysis may result in missed opportunities to resolve issues on positive trends.
3. Potential Sentiment Drivers: By correlating all the shifts with business events, a deeper analysis could dwelve more into the reasons for sharp sentiment shifts on specific days like promotions, service issues, product launches, etc shown in Fig. 24.1

Date/Time of Review	sentiment_score
2023-08-28	0.4404
2023-08-29	0.0000
2023-08-30	NaN
2023-08-31	-0.5423
2023-09-01	0.0000

Fig. 24.1 Sentiment score

1. High Variability in Sentiment:

   This graph exhibits significant fluctuations in customer sentiments over time along with periods of both positive and negative sentiments that occur close together. This shows that customer feedback is vastly dynamic and can change promptly based on promotions, recent experiences, customer service interactions, product launches, etc.
2. Sharp Sentiment Peaks and Troughs:

   Positive Peaks: By analyzing the graph, it indicates significant positive sentiment scores that are closer to 0.5 or higher. All these could probably correspond to events like product releases, improved customer service, or successful marketing campaigns.

   Negative Dips: There are sharp declines in sentiment scores at several intervals with few reaching as low as -0.5. These dips could probably show issues like poor customer experiences, product defects, or any negative reactions to specific company decisions.

   Neutral Periods: The graph also shows few timeframes that indicate neutral sentiment scores around 0.0, which could probably show periods where there was more balanced feedback, that are neither overwhelmingly positive nor negative. These periods may probably align with times where there are no significant changes or events happening that could have influenced customer opinion.
3. No Clear Long-Term Trend:

   The analysis of graph does not exhibit any clear trend that is particularly long-term upward or downward. Instead, the sentiment oscillates frequently suggesting that the factors influencing customer sentiment are mostly short-term or event-based, rather than part of a longer-term decline or improvement. All are represented in Fig. 24.2.

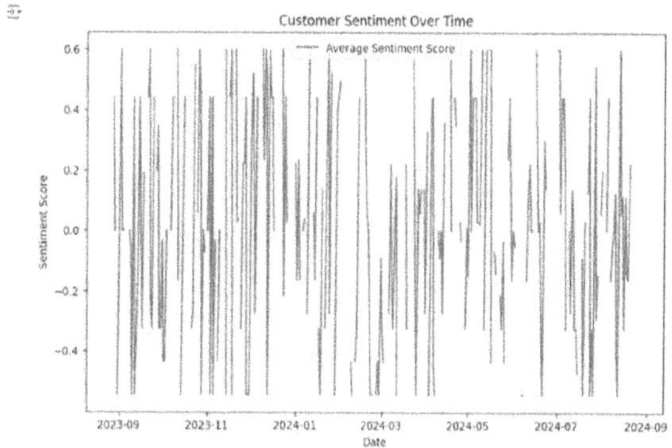

Fig. 24.2 Sentiment over time

This graph in Fig. 24.3 represents a near-equal distribution of positive and negative feedback that shows mixed customer satisfaction levels, thus highlighting the need for a more consistent quality control. There is an immediate requirement for improvement in key areas as 41.4% of the feedback are negative, that may influence customer satisfaction. Through targeted engagement strategies, there is a possibility to convert indifferent customers to satisfied and loyal ones as shown by the neutral feedback segment.

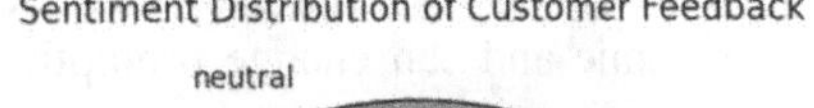

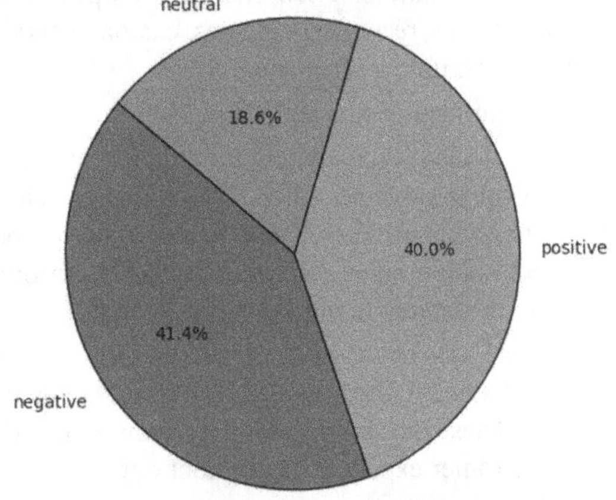

Fig. 24.3 Sentiment distribution

By addressing all these issues that are raised by negative reviews and reinforcing the strengths as identified by positive reviews, we can help enhance brand perception, improve overall customer satisfaction, and also reduce churn.

During the period September 2023 to September 2024, the above line chart namely "Monthly Average Sentiment Score Over Time" exhibits how the customer sentiment has evolved over this period. The fluctuations over time in customer opinions are depicted from the average sentiment score that is plotted for each month. The below points express the interpretation of Fig. 24.4

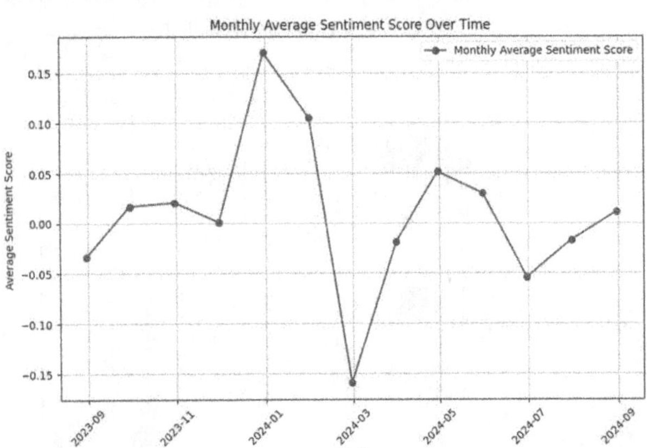

Fig. 24.4 Monthly average score

Interpretation:

1. Fluctuating Sentiment Trends:
   - This sentiment trend depicts several peaks and troughs, which indicate fluctuating customer opinions throughout the year.
   - Since both positive and negative shifts occur at different points in the graph, this trend ensures that the average sentiment score has varied drastically.
2. Positive Peak in January 2024:
   - The graph depicts that the highest positive sentiment score occurred around the month of January 2024, having an average sentiment of more than 0.15.
   - The spike in the graph indicates that there was a period where the customers were probably more satisfied with StarCofee services or products, which may be due to any new product launch or a successful campaign or any improvement in customer service during that period.
   - Recommendation: To dwelve deeper into the reason which led to this positive sentiment, which may be perhaps due to service changes, promotions, or any specific new release of products, as well as to consider replicating the strategies.
3. Sharp Negative Dip in March 2024:
   - The graph depicts a most significant negative sentiment during March 2024, having an average sentiment score of less than -0.15.
   - There is a sudden decline in the graph that indicates a particular issue that has negatively impacted customer satisfaction like any operational problems or issues in product or any negative press.
   - Recommendation: To analyze the factors that significantly contribute to the negative sentiment in March, which involve reviewing the service performance or product quality or any peculiar incidents that probably would have occurred. Hence, the remedy is to take corrective actions to avoid such incidents in the future.
4. Recovery and Decline Patterns:
   - The sentiment trend has shown some improvement during April and May after a sharp decline in March, but again indicated some decline during the period of June and July 2024.
   - Though the recovery depicts that the organization may have worked upon some issues that had led to the negative feedback, it also requires a consistent improvement to sustain its positive sentiment.
5. Near-Neutral Sentiment in Late 2023 and Late 2024:
   - Since the sentiment score has shown some fluctuations for a few months around 0, this signifies an overall neutral sentiment.

- This indicates mixed customer feedback where there were neither highly positive nor highly negative sentiments predominantly.

Recommendation: To find out ways to enhance customer experiences during these neutral periods like enhancement of products or improved customer engagement or targeted promotions could help resolve this issue.

The above Fig. 24.5 bar chart named **"Sentiment Category Counts by Platform"** demonstrates the distribution of feedback from customers across various platforms like **Social** Media, In-App, and Survey. The received feedback is then categorized into either of the three sentiment types: positive, negative, or neutral.

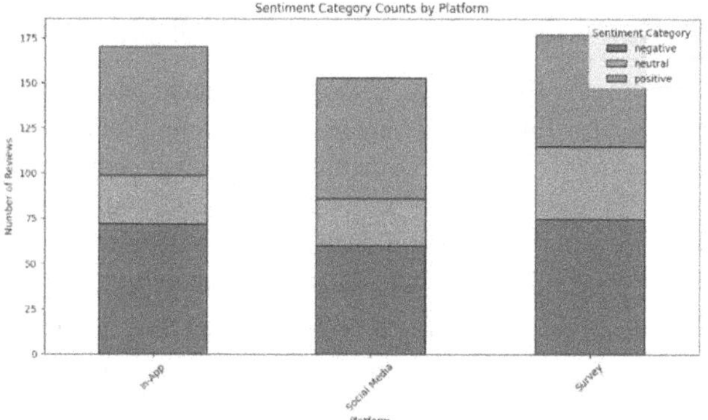

Fig. 24.5 Sentiment score in platform

4.3 Interpretation

1. Sentiment Breakdown Across Platforms:
    - Each of the platform represents the number of reviews, which are categorized as negative (blue), neutral (orange), and positive (green).
    - This chart also identifies any differences in customer sentiment based on the platform through which the feedback was received.
2. In-App Feedback:
    - This platform has a more balanced distribution of sentiments, having more positive feedback compared to negative ones.
    - This also signifies that the positive sentiment is the largest category which shows a relatively favorable customer experience for those who are interacting with the business through this app.
    - Even the negative segment is quite significant as it suggests possible issues on the app experience, which need to be addressed.
3. Social Media Feedback:
    - There are a larger proportion of negative feedback when compared to the positive ones on the Social Media platform that is close to the neutral category.
    - Also, the social media receives mixed and critical feedback that may indicate customers who are more vocal on public forums about their dissatisfaction.
    - Recommendation: To investigate the critical points that are raised on social media platforms and utilize them as an opportunity for more customer engagement and improvement. Also, it is crucial to address all complaints promptly on social media platforms to remove negative publicity.
4. Survey Feedback:
    - The survey data depicts a more balanced distribution of feedback along with a significant number of neutral sentiments.
    - This also shows that the negative sentiments contribute to a considerable amount of survey responses that are quite similar to social media.
    - Ususally, surveys serve as a platform for customers to express their more nuanced feedback. Also, the neutral sentiment shows some mixed experiences, which are neither highly positive nor negative.
    - Recommendation: To utilize responses from survery to find out areas that need improvement and to assess where customer expectations are not being addressed. The neutral feedback can be more explored by reaching out to customers for further clarification.
5. General Observations:
    - Overall, positive sentiment tends to be on par or a little less than negative ones across all platforms, which highlight areas for improvement across all available modes of interaction.
    - The neutral sentiment across all platforms indicates a considerable proportion of feedback where customers exhibit neither highly satisfied nor dissatisfied behavior. This serves as an opportunity to explore what makes a better customer experience.

4.4 Recommendations

- Platform-Specific Interventions: By addressing the complaints of customer in a prompt and professional manner, we can increase the engagement on this platform to a significant level since there are more negative feedback that is pronounced on social media. This can be enabled by enforcing a proactive social media team. By concentrating more on In-App experience improvements and eliminating common user frustrations, we could achieve significant improvements.
- Enhance Survey Engagement: The responses from survey show a high neutral proportion, hence by adding more questions we can easily understand the potential areas of customer dissatisfaction and

indifference. Significant actionable insights could be obtained by providing incentives for more detailed survey responses.

- Leverage Positive Feedback: For marketting successful products or features, we can use positive reviews, particularly from the In-App platform. These feedbacks can be highlighted during promotions to attract new customers and build credibility.
- Convert Neutral Feedback to Positive: For converting indifferent customers into satisfied ones, we can utilize the opportunity of neutral feedback. We can reach out to neutral respondents to have a clear understanding on their needs, and we find ways to exceed their expectations.

4.5 Overall Insights

- By understanding the distribution of sentiment across various platforms, we can get clear insights on how customer experiences differ depending on where feedback is provided.
- While social media needs more attention to address negative sentiments, In-App feedback exhibits a promising area to focus and build on existing strengths.
- To improve overall customer sentiment and loyalty, we can maintain an active customer support presence across all platforms and ensure their issues are resolved promptly.

The research gap identified was mostly due to the lack of real-time sentiment integration into business decision-making processes. The existing sentiment analysis models mainly use historical data, which limits their ability to provide actionable insights when they are most needed immediately.

1. Real-Time Sentiment Insights: By categorizing and calculating the sentiment in real-time on daily intervals, this model enables StarCofee or any other business to detect shifts in customer sentiment as they occur. This also allows businesses to do the following:
   - Respond quickly to negative feedback. For example, if a product launch results in a surge of negative feedback.
   - Identify trends over time. For example, decline of sentiment during a particular period, and to make proactive adjustments.
2. Dynamic Feedback Loop: This model is effective because businesses can dynamically alter their operations, fine-tune marketing strategies, customer support, and services based on current customer feedback rather than waiting for long-term reports.

Future Scope

- Correlate Sentiment with Business Metrics: Cross-referencing the sentiment data with other business metrics like website traffic, sales, and customer retention.
- Conduct Customer Surveys Based on Sentiment: Whenever there is a significant drop in sentiment, consider utilizing targeted customer surveys to have a good understanding of the underlying causes of dissatisfaction.
- Sentiment Prediction Model: Build a predictive model to forecast customer sentiment based on recent historical data and trends.

Focus on Consistency: Given the frequent fluctuations, work on ensuring a consistent customer experience to stabilize sentiment

5. Recommendation and Conclusion

Monitor for Key Events Causing Sentiment Shifts: The significant fluctuations show that certain actions or events trigger changes in sentiment. Conduct a thorough analysis of the dates corresponding to peaks and troughs in sentiment. Determine if these are tied to service issues, product launches, or other factors, for example:

- Positive spikes might indicate what the company is doing right like product features or successful campaigns.
- Negative drops could indicate issues that need to be addressed quickly like service delays or product malfunctions.
- Real-time tracking of social media and reviews.
- Automatic alerts to customer service teams when sentiment dips below a certain threshold, allowing for quick response to complaints or negative feedback.

Create Proactive Customer Support Strategies:

- Offering discounts or personalized solutions to customers who express negative opinions.
- Identifying patterns in negative feedback to address recurring issues, such as shipping delays, product complaints, or unsatisfactory customer service.

Leverage Positive Sentiment for Marketing: Promote positive reviews during periods of good feedback.

- Use sentiment data to identify which products or services are resonating most with customers and focus marketing efforts on these strengths.

Look for Seasonality or External Factors: By applying the VADER sentiment analysis model to StarCofee customer reviews, we identified significant sentiment fluctuations over time, thus highlighting the need for businesses to respond quickly to changes in customer opinions. The proposed real-time feedback framework exhibits a practical solution for businesses seeking to optimize operational efficiency and customer experience. More advanced tech can be used for sentiment analysis

Acknowledgement

The authors gratefully acknowledge the students, staff, and authority School of Business and management Christ university.

References

1. al., C. e. (2021). Sentiment-Based Predictive Analytics in Service Industries.
2. al., D. e. (2021). Social Media Sentiment Analysis for Product Feedback.
3. al., F. e. (2023). Sentiment Analysis in the Food and Beverage Industry.
4. al., H. e. (2022). Using Sentiment Analysis to Improve E-commerce.
5. al., K. e. (2020). Challenges in Analyzing Public Sentiment at Scale.
6. al., L. e. (2022). Sentiment Analysis in the Banking Sector.
7. [al., M. e. (2021). Sentiment Analysis in the Telecommunications Industry.
8. al., N. e. (2021). Enhancing Product Development with Sentiment Analysis.
9. al., P. e. (2022). Sentiment Analysis for Personalized Customer Experience.
10. al., R. e. (2021). Sentiment Analysis in Entertainment and Media.
11. al., S. e. (2021). Real-Time Sentiment Analysis in Retail.
12. al., S. e. (2021). Leveraging Sentiment Analysis for Competitive Advantage.
13. al., Z. e. (2021). Using Sentiment Analysis for Crisis Management.
14. Bhatt, C. a. (2022). Real-Time Sentiment Analysis for Online Retail.
15. Kapoor, S. a. (2021). Real-Time Sentiment Analysis for Customer Support Automation.
16. Lee, J. a. (2022). Customer Segmentation through Sentiment Analysis.
17. Oliveira, M. a. (2022). Using Sentiment Analysis to Predict Product Success.
18. Rios, G. a. (2022). Using Sentiment Analysis for Financial Market Predictions.
19. Thomas, G. a. (2021). Sentiment Analysis for Brand Reputation Management.
20. Wang, P. a. (2023). Patel and Wang (2023): Combining Sentiment and Behavioral Data.

Note: All the figures in this chapter were made by the authors.

Intelligent Systems Using Semiconductors for Robotics and IoT – Dinesh Goyal et al. (eds)
© 2026 Taylor & Francis Group, London, ISBN 978-1-041-20408-4

25 The State of AIOps: Adoption, Trends, and Organizational Strategies in IT Operation

Latha Narayanan Valli[1]
Applied Electronics and Instrumentation Engineering,
Heritage Institute of Technology,
Kolkata, India
Standard Chartered Global Business Services Sdn Bhd,
Kuala Lumpur, Malaysia

N. Sujatha[2]
Applied Electronics and Instrumentation Engineering,
Heritage Institute of Technology,
Kolkata, India
Sri Meenakshi Government Arts College for Women,
PG and Research Department of Computer Science,
Madurai, India

S. R. Mathu Sudhanan[3]
CARE College of Engineering,
Dept of Computer Science Engineering,
Trichy, India

Abstract: Companies in the information technology sector that have mastered the art of pattern recognition in large datasets have the greatest potential to successfully detect and repair performance problems before those issues have a detrimental effect on the firm. However, hybrid and multi-cloud architecture generates complexity at a great cost since the rate of digital change is outpacing that of IT performance management. By asking IT professionals questions in the method of a survey, this learning aims to determine the extent to which AI-driven operations (AIOps) are now accepted in the context of the current state of IT operations. The research aims to investigate the most recent tendencies in IT operations and ascertain the most recent condition of AIOps adoption in IT operations. The majority of the questionnaires for this study were filled out by people who are actively employed in the information technology business. In order to determine whether or not the goals of the research were met, this procedure was carried out. A straightforward method of random sampling is used during the collection of this information. In addition to this, businesses will need new competencies and assets if they want to effectively manage the growth of AI-powered software. The business strategy of the firm, past organizational choices, and the state of the company's operations all play a role in determining how a corporation will reorganize its teams and delegate new responsibilities when the time comes.

Keywords: Artificial intelligence, Information technology, Recent trends, IT operations, AIOps

1. Introduction

Businesses in the IT industry that have mastered the art of pattern recognition in big datasets can most effectively identify and resolve performance issues before they have a negative impact on the company. However, heterogeneous and multi-cloud architecture adds complication at a great cost as digital transformation happens faster than

[1]latha.nv@gmail.com, [2]sujamurugan@gmail.com, [3]srmadhusudhanan@gmail.com

DOI: 10.1201/9781003716389-25

IT administration can keep up with. One solution to this problem is the implementation of AI to IT operations, or AIOps.

AIOps systems, according to researchers, are software platforms that integrate AI or automated learning capabilities with big data with the intention of "enhancing and partially replacing all elementary IT operational functions," including availability and efficiency monitoring, event correlation and analysis, IT service management and automation, and more. The use of machine learning and data science approaches to technology operations with the goal of improving procedure efficiency is known as AI operations. Furthermore, the authors of the report additionally forecast that "big business exclusively usage of AIOps and web-based monitoring services for tracking infrastructure as well as applications would rise from five percent in 2018 to thirty percent in 2023."

Businesses are starting to see the benefits of AIOps and view it as an integral part of the next generation of IT services. It's safe to say that AIOps will continue to make ripples in the direction of IT operations management. Preventing issues, decreasing expenses, enhancing customer service, and freeing up IT workers to concentrate on innovation are just some of the current uses of the technology. It doesn't matter how complex your configurations are, they will always help to raise IT's strategic relevance and visibility inside the company by improving its performance and availability as needed. As a result, the determination of this study is to recognize the most recent trends in IT operations and AIOps adoption in the context of Indian IT organisations.

2. Literature Reviews

The phrase "IT," like the term "Operation," may have a variety of connotations depending on the context in which it is used. On the other hand, although there is a universal agreement on what information technology management entails, its definition and its use vary widely depending on the industry and kind of business being discussed [4]. In order to have a more complete understanding of "IT Operation" in the context of "AI Operations," it is not sufficient to offer a single definition of "IT Operation;" rather, it is required to have an understanding of numerous principles related to "IT Operation." The foundation of the whole IT service life cycle consists of five pillars: service planning, service architecture, service delivery, service management, and continuous service enhancement. Specifically, there are four core functionality modules that make up the service operation framework [5]. These departments are known as IT operation management (ITOM), technical management, application management, and the service desk. The phrase "Site Reliability Engineering" was coined by Google in 2004. The acronym "SRE" stands for "site reliability engineering," which describes this area of research. As the number of servers and services that needed to be controlled simultaneously rose, it became more difficult for system administrators to do it manually. The most obvious solution was to write some code to fix the problem. SRE may also refer to the application of software engineering principles to IT infrastructure and management [6]. SRE (Site Reliability Engineering) is a way of thinking and doing business, as defined by Benjamin Treynor Sloss, the man largely responsible for establishing the SRE division at Google.

AIOps technologies have between five and 10 years of development left before reaching the plateau of productivity, as shown by the hype cycle. However, just before one reaches the plateau, there is often a dip caused by a downhill that is then followed by a slope up to the plateau. In addition to the gap between the existing state of technology and the desired outcomes, unrealistically high expectations are a common cause of the trough [7]. While it may not be time for full-scale AI operations just yet, these considerations should be taken into account when appropriate technologies are implemented and investments are made. In contrast, Gartner predicts that by 2023, 40% of DevOps teams will have integrated AIOps platform monitoring solutions [8]. Their latest report, "Market Guide for AIOps Platforms," makes this prognostication. It is likely that a sizable fraction of the potential market will start or increase their preparations for AIOps implementation.

The 2022 McKinsey Global Survey on AI provides valuable insights into the significant surge in the integration of AI technologies within business frameworks over the last five years [9]. This upward trajectory in AI implementation is reinforced by projections from PwC, indicating a transformative influence AI is poised to have on the operational dynamics, customer interactions, and workforce engagement strategies of companies in diverse sectors [10]. Consequently, the growing significance of AIOps as a strategic approach is underscored for enterprises seeking to streamline their IT infrastructures and spearhead meaningful digital metamorphoses [11]. In conclusion, in the next few years, several AIOps platforms and technologies will undergo expansions, and the technology will be widely used in many ways. Meanwhile, as excitement for the technology reaches a fever pitch and more attention is paid to its implementation, businesses will have to make the right decision that meets the unique needs of their operations and businesses, and they'll have to create strategic plans to steer and govern the necessary efforts.

2.1 Importance of the Study

Since the 1950s, researchers have been working to bridge the gap between theory and practice in the area of artificial intelligence, and their findings are now being used across a diverse variety of industries. The artificial intelligence model is hotly discussed within the field of information technology operations. Our research relied heavily on the information gathered by the subset of the IT sector tasked with monitoring AI initiatives. The objective of the work are to explore the current trends in IT operations, as well as to identify the current state of AIOps adoption in IT operations and to recommend an efficient AI enabled operations framework.

3. Research Methodology

In this segment, we would confer the methodology that will be cast-off to examine and differentiate current patterns in IT operations and AIOps adoption, as well as the techniques that will be utilized to analyze the generated data. This section will also describe the methodology that will be used to study and identify developing patterns in the adoption of AIOps.

3.1 Research Method and Design

People currently working in the IT industry filled out the bulk of the surveys for this research. This information is gathered using a simple random sampling strategy. The current research offers both primary and secondary sources.

3.2 Research Approach

The majority of this research study is descriptive, while it does contain some qualitative features because the data was gathered from a variety of offline and online sources. We will communicate the survey questions by email to the workforce via secondary sources like internet pages, the search engine, magazines, newspapers, online communities, and avenues for professional networking.

3.3 Data Collection

Simple random sampling is used to get the necessary data, and a descriptive study is then carried out. To avoid the introduction of any possible information bias into our questionnaire, we took care to solely include closed-ended questions. Through the use of social media and LinkedIn, the poll's URL was widely shared, resulting in a huge number of responses. In addition to traditional sources like books, websites, and article databases, we will also be utilising Google Scholar to access the back issues of academic journals for this investigation. Wiley Online Library, IEEE, ScienceDirect, Frontiers, and Emerald, were among the respected databases utilised to compile the data for this research.

3.4 Review of Primary and Secondary Information

This study will be able to quickly gather data and analyse qualitative characteristics because of the descriptive technique employed. The "Data add-in" for Excel was crucial in the analysis and interpretation of data. Statistics are performed entirely with Microsoft Excel. The survey's scoring system will be built on Likert-scale questions.

3.5 Ethical Consideration

Confidentiality, the protection of individual rights, and the receipt of informed permission after complete disclosure should be seen as top priorities. Every stage of the research process, including the design of the questionnaires, the analysis and interpretation of the results, and the purpose of the study, was thoroughly explained to the participants. Before administering the poll, we also got the participants' informed permission.

4. Analysis of Study

Therefore, modern businesses are on the search for insights that are not only predictive but also actionable in order to better manage their enormous IT operational data. The following are the results from the survey:

- Data accuracy (or "extracting information from noise") was identified by 61% of respondents as the most difficult task, followed by incident root cause investigation with 51%.
- Additionally, there were problems with event management, such as a decrease in mean time to resolution (50%) and an increase in the IT service dependence context (51%).

Because of the recent innovation trigger for AIOps, most organizations now have proficiency in machine learning-based event management.

To give you some examples of current trends that are pushing firms to use AIOps:

- Almost 70% of IT execs have tried out an AIOps solution at some point.
- When AIOps is used by IT teams, data insights are extracted 73% of the time, root causes are found 68% of the time, alerts are correlated 49% of the time, and noise is reduced 28% of the time as shown in Fig. 25.1.

IN WHAT WAYS HAS YOUR TEAM USED AIOPS?

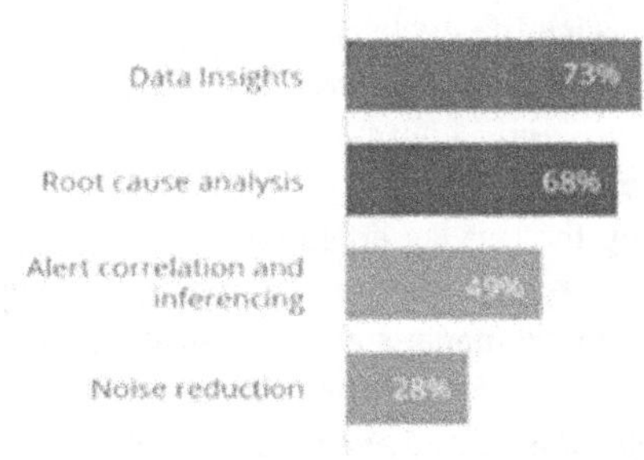

Fig. 25.1 Current trends that are pushing firms to use AIOps (IT teams)

Notes: Companies are updating their computer networks as a consequence of embracing AIOps. Metrics, events, and logs may all be mined using real-time analytics to uncover actionable insights that can be used to facilitate continuous improvement, expedited deployments, and heightened collaboration. The market has reacted by establishing a strong demand for AIOps as a result of the technology's desirability and the advantages it can provide. By reducing or eliminating firefighting and increasing IT efficiency, for example, adaptive insights have the potential to increase digital performance as shown in Fig. 25.2.

- The capacity to automate mundane processes (74% of respondents) and the potential to prevent expensive service interruptions with a quicker MTTR (67% of respondents) are two of the most noticeable advantages of AIOps.
- Many IT experts feel that AIOps may aid in anomaly identification in dynamic production settings by forecasting changes in the system's typical behaviour (58%).

IN YOUR OPINION, WHICH OF THE FOLLOWING IS THE MOST IMPORTANT OPERATIONAL BENEFIT OF USING AIOPS TOOLS:

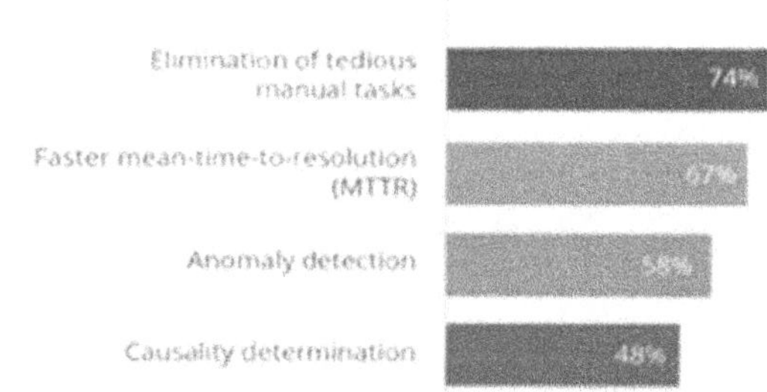

Fig. 25.2 Adaptive insights have the potential to increase digital performance

Notes: There is still untapped potential in AIOps. With AIOps technology's more intricate event context and data-driven solutions, businesses hope to save costs and improve efficiency. Despite AIOps's meteoric rise in popularity, numerous unanswered problems and concerns still plague IT organizations as shown in Fig. 25.3. Concerns about the quality of the datasets used to train machine learning models (held by 52%) and the level of IT skills necessary to oversee machine learning algorithms (48%). The following are some approaches that IT professionals and suppliers might use to cooperate in fixing these problems.

- There has to be greater transparency from AIOps suppliers on their forecasting algorithms and how to interpret the results.
- For AIOps to work, IT departments need to staff up with data scientists. It's important to remember that educating the people already working for you may help reduce the skills gap.

IN WHAT WAYS DO YOU FEEL AIOPS TOOLS SHOULD BE ADMINISTERED?

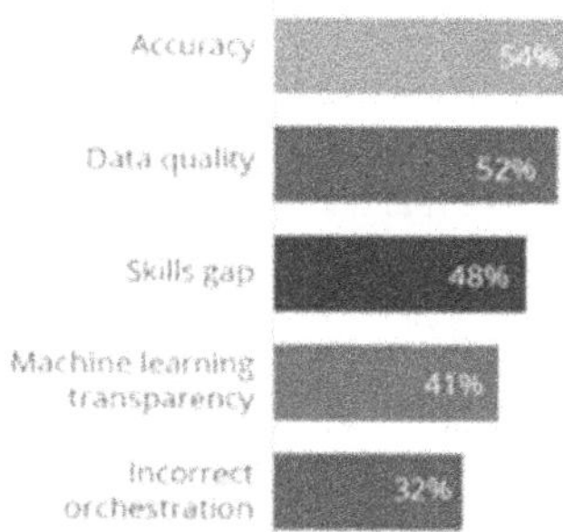

Fig. 25.3 Level of IT skills

Notes: The implementation is problematic mainly because of concerns about accuracy, data quality, and openness. Learn whether there are any new issues associated with AIOps, and work with your tool provider to prioritise potential solutions. Given the high level of operational urgency and the need for tailored, real-time analysis of events, what are the obstacles to using AIOps?

Based on the results of our poll, we know that many businesses are hesitant to fully use AIOps as shown in Fig. 25.4.

- More than half of respondents see the high price tag of implementing AIOps solutions as a key deterrent to wider adoption.
- In the minds of risk-averse enterprises, the AIOps tools stack presents ideological obstacles due to its slow time-to-value (or cumbersome deployment process) and its lack of cutting-edge technology.

WHAT IS PREVENTING YOU FROM USING AIOPS TECHNOLOGIES IN YOUR IT OPERATIONS?

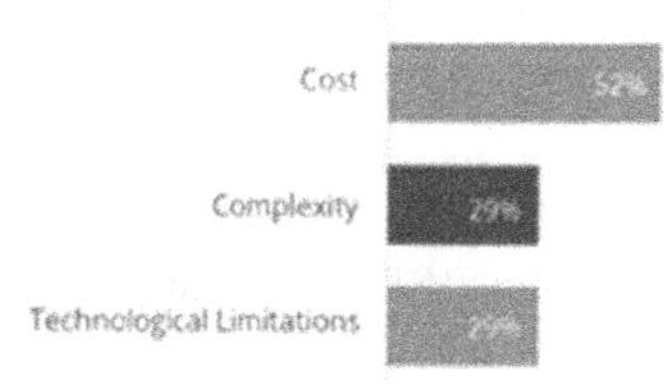

Fig. 25.4 Implementing AIOps solutions as a key deterrent to wider adoption

Notes: Successful implementation of AIOps would need substantial increases in both the skill level of IT workers and the quantity of training data available to them. Investing in digital operations solutions designed for contemporary event management can improve data accessibility and speed up the time to return on investment. Expect long-term advantages from the improved problem-solving processes made possible by AIOps technology, even if many IT operations tasks will be automated in the distant future. More than 90% of respondents state that using an AIOps platform has resulted in a 10% decrease in total alerts as shown in Fig. 25.5.

- Nearly half of those surveyed think that AIOps will improve productivity, enhance cooperation, and speed up problem resolution by 25% by reducing alert noise.
- Twenty-five percent of people think it would be successful if the amount of incoming alerts could be decreased by at least half to better support event storm correlation and data-driven issue prediction.

HOW MUCH OF A REDUCTION IN ALERT STATUS WOULD YOU SAY REFLECTED SUCCESSFUL AIOPS PLANNING?

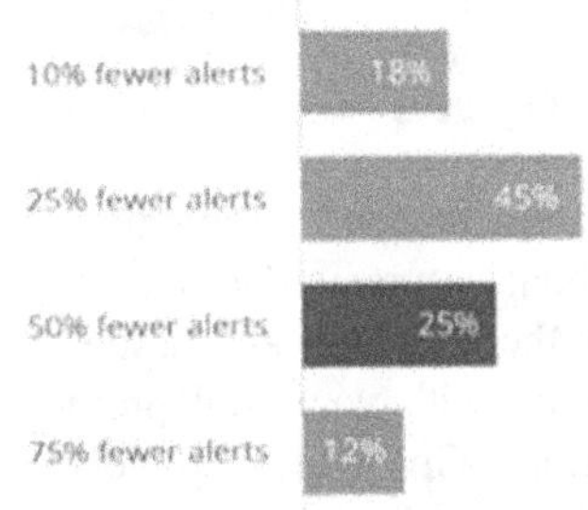

Fig. 25.5 Detail of alerts in AIOPs

Notes: The excitement around AI operations technology seems to be reaching a climax. There is an immediate need to find solutions to issues like notification overload and disorganized incident response. In order to effectively manage hybrid and multi-cloud systems, IT departments want AIOps solutions that are both simple to roll out and rich in actionable information. This can be the result of a misunderstanding of the issue's intricacy or a general lack of familiarity with the topic. Instances of items like these have been requested for AIOps before as shown in Fig. 25.6:

- Across the board in today's IT architecture, 63% of IT pros desire an AIOps platform that is simple to deploy, manage, and maintain up to current.
- Critical AIOps features include data intake and enrichment (62%), proactive dashboards with real-time event context (53%), and more.

TELL US WHICH ASPECTS OF THE AIOPS TOOL SUITE YOU VALUE THE MOST?

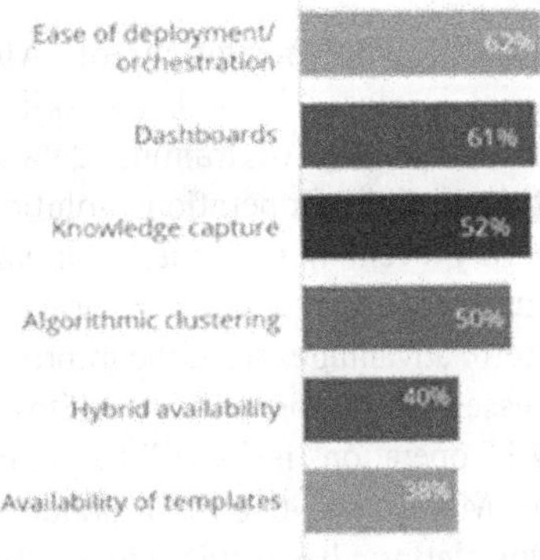

Fig. 25.6 Key features ranked by importance for a AIOPs

Notes: AIOps solutions are being adopted by businesses in the hopes that they will help cut down on the resources needed to maintain their complex infrastructure, assist prevent and address performance problems before they impact users and boost the overall health of their systems.

5. Results

All organizations will need to rethink their traditional operational methods as well as their monitoring, control, governance, and procurement of AI-enabled systems as a result of the widespread adoption of AIOps. For the simple reason that AIOps will lead us to re-evaluate several time-honored practices in organizational management. Companies must be nimble in these shifts if they are to adopt and manage AIOps securely and efficiently at scale. This is due to the fact that AI operations will impact every part of the software development lifecycle. The use of AIOps also necessitates a re-evaluation of the management and structure of operations, as well as the necessary roles and skill sets as shown in Fig. 25.7.

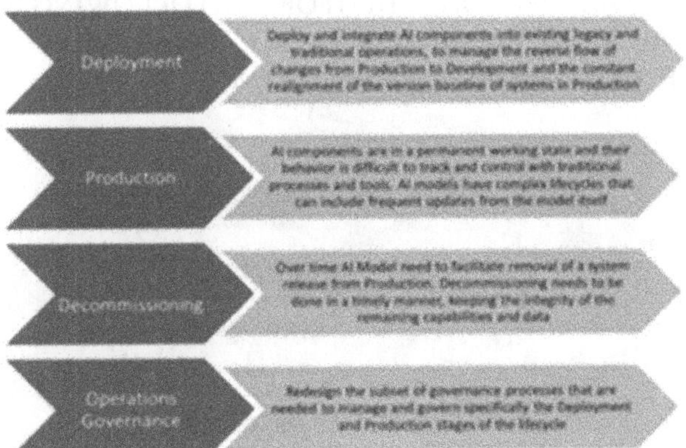

Fig. 25.7 Explicitly illustrating the AIOps problems across the phases of the lifecycle that are relevant to this discussion

By blurring the lines between Deployment and Production, AIOps questions the rigidly defined divisions that have traditionally existed between the two stages of an operation's life cycle. In order to efficiently handle the complexity introduced by AI components, we suggest a process redesign in which the traditionally distinct Deployment and Production phases, as well as organizational units, are brought together to create a convergent and merging process. Then, maybe, we might establish a standardized , convergent method of production as shown in Fig. 25.8.

6. Conclusion

According to research conducted, 45% of companies now utilize AIOps to analyze existing issues and foresee potential future ones. Early adopters of AIOps will concentrate on automating simple tasks, such as checking

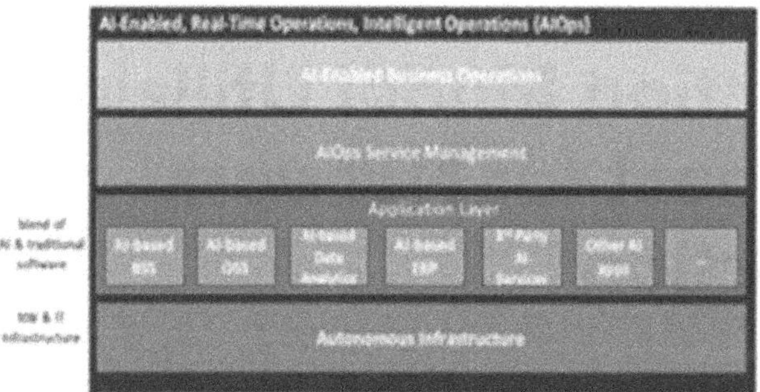

Fig. 25.8 AI enabled operations framework

the status of warnings from infrastructure monitoring systems. Advances in machine learning and analytics allow for more vigilant monitoring and automated execution of IT procedures. An increasingly popular method for gaining insight and settling on a course of action is to use automated data analysis. The goal of this inquiry is to facilitate the examination of collected data. The scenario calls for the usage of artificial intelligence (AI). AI driven by machine learning (ML) automates both insights and actions by applying algorithms to massive datasets. This gives IT a seat at the table during company debates and the ability to back up its claims with facts. Improving your prognostic skills can help you see potential slowdowns in advance, allowing you to take preventative measures. The end result will be reduced downtime costs and more satisfied clients. Improving IT cooperation and reducing isolated reactions demands accelerating root-cause analysis and problem resolution to extract value from data housed in silos. Getting rid of laborious manual steps When routine manual tasks are automated, IT employees can focus their attention where it's needed most—on analysis and improvement, leading to more consistent replies, fewer instances of hard-to-identify problems, and eventually, the eradication of human error. It's challenging to keep up with the speed of today's IT business world. By automating routine IT tasks, AIOps allows organisations to save money and time by decreasing their reliance on human labour. With the evolution of AI toward more human-likeness, AIOps stands to benefit from the liberation of previously inaccessible data and the dramatic reduction in IT administration costs.

6.1 Importance of the Study

The last step on the road to transformation is the reengineering of the Deployment, Production, and Governance processes. All services, apps, and underlying infrastructure will be managed from a single, centralised layer of AIOps Service Management. This will ensure the security, effectiveness, and efficiency of AI-powered enterprise operations. AIOps forces the operations departments to play a strategic and significant role for the service and business performance because of the need to monitor and supervise the autonomous and self-driven development of the AI components and assure the desired outcome of the overall service quality. The operations teams will need to be reorganised as part of the transition process so that separate roles may be set up at the AIOps Service Management level. To properly manage the expansion of AI-powered software, enterprises will also need new capabilities and assets. Every corporation has to restructure its teams and assign new duties from time to time, and how they do so depends on the company's business plan, previous organisational decisions, and the current condition of operations.

References

1. An, L., Tu, A. J., Liu, X., & Akkiraju, R. (2022, April). Real-time Statistical Log Anomaly Detection with Continuous AIOps Learning. In *CLOSER* (pp. 223–230).
2. Anagnostou, M., Karvounidou, O., Katritzidaki, C., Kechagia, C., Melidou, K., Mpeza, E., ... & Peristeras, V. (2022). Characteristics and challenges in the industries towards responsible AI: a systematic literature review. *Ethics and Information Technology*, *24*(3), 37.
3. Battina, D. S. (2021). Ai and devops in information technology and its future in the united states. *INTERNATIONAL JOURNAL OF CREATIVE RESEARCH THOUGHTS (IJCRT), ISSN*, 2320-2882.
4. Bogatinovski, J., Nedelkoski, S., Acker, A., Schmidt, F., Wittkopp, T., Becker, S., ... & Kao, O. (2021). Artificial intelligence for it operations (aiops) workshop white paper. *arXiv preprint arXiv:2101.06054*.
5. Dang, Y., Lin, Q., & Huang, P. (2019, May). Aiops: real-world challenges and research innovations. In *2019 IEEE/ACM 41st International Conference on Software Engineering: Companion Proceedings (ICSE-Companion)* (pp. 4–5). IEEE.
6. Dilek, S., Çakır, H., & Aydın, M. (2015). Applications of artificial intelligence techniques to combating cyber crimes: A review. *arXiv preprint arXiv:1502.03552*.
7. Jakšič, M., & Marinč, M. (2019). Relationship banking and information technology: The role of artificial intelligence and FinTech. *Risk Management*, *21*, 1–18.
8. Kannampallil, T. G., Franklin, A., Mishra, R., Almoosa, K. F., Cohen, T., & Patel, V. L. (2013). Understanding the nature of information seeking behavior in critical care: implications for the design of health information technology. *Artificial intelligence in medicine*, *57*(1), 21–29.
9. https://www.mckinsey.com/capabilities/quantumblack/our-insights/the-state-of-ai-in-2022-and-a-half decade-in-review.
10. https://www.pwc.com/us/en/tech-effect/ai-analytics/ai-predictions.html
11. https://www.marketsandmarkets.com/ResearchInsight/size-and-share-of-aiops-platform-market.asp

Note: All the figures in this chapter were made by the authors.

Intelligent Systems Using Semiconductors for Robotics and IoT – Dinesh Goyal et al. (eds)

26 Anomaly Detection Model Utilizing SVM and XGBoost for Network Intrusion Detection

Ratnesh Kumar Dubey[1], Aravendra Kumar Sharma[2]
Assistant Professor, CSA Department, ITM University Gwalior

Shubha Mishra[3]
Assistant Professor, Department for Center for AI, MITS Deemed University Gwalior

Suraj Sharma[4]
Assistant Professor, CSA Department, ITM University Gwalior

Abstract: It is becoming more difficult for network administrators to protect machines from the increasing number of cyber assaults. While there are several traditional intrusion detection systems (IDS), none of them are foolproof. Due to the rapid increase in the number of people connected to networks and utilising them to store or access sensitive information, the necessity to secure these networks has intensified. In order to propose a system, this paper first analyses and analyses several machine learning techniques. We proposed an XGBoost learning strategy that improves the model's stability and prediction accuracy by integrating a diverse array of separate models.

Keywords: Extreme gradient boosting (XGBoost), Support vector machine (SVM), Data mining methods, Anomaly detection, Firewalls, Intrusion detection systems, and Network security

1. Introduction

Thanks to developments in technology, a multitude of networks have brought millions of people together, enabling the sharing of substantial data. This has led to a dramatic increase in the need of security measures meant to preserve the privacy and authenticity of sensitive information. Attack methods for penetrating the network are always evolving, despite attempts to ensure data transfer security. Because of this, a system that can change to counter these new attacks is required. A machine learning-based system is suggested in this research. Our goal is to find the best machine learning algorithm that can accurately anticipate the kind of network attack, and then build a system that uses this method to detect intrusions in networks. Two algorithms, SVM and XGBoost, were compared. The KDD 99 dataset is used for training the models. The flexibility of machine learning is a major selling point; for example, if a new kind of attack were to appear in the future, the system might be programmed to predict it. We use knowledge-based intrusion detection System, which is one of several types of such systems and is also known as an anomaly-based system. It looks For out-of-the-ordinary activity and then sounds an alert if it predicts that other networks will do the same. The network may then cut off such connections using this way, ensuring that only secure connections are kept.

1.1 NSL-KDD Dataset

The dataset used for this investigation is the 1999 NSL-KDD Cup. Several studies have pointed to the NSL-KDD as a replacement for the KDD Cup 1999 dataset (KDD-99) due to its deficiencies [12]. Due to the dearth of freely

[1]ratnesh.soet@itmuniversity.ac.in, [2]aravendra.cse@itmuniversity.ac.in, [3]shubha@mitsgwalior.in, [4]surajsharma.cse@itmuniversity.ac.in

DOI: 10.1201/9781003716389-26

available public datasets, researchers studying intrusion detection systems often use the KDD-99 dataset, despite its age of more than 15 years. By redistributing datasets and removing duplicate data, the NSL-KDD dataset fixes many problems with the KDD-99 dataset [12]. By re-proportioning KDD-99, NSL-KDD makes it easier to evaluate different learning algorithms, and by removing unnecessary data from KDD-99, NSL-KDD improves the performance of learning algorithms. Of the 42 features included in the NSL-KDD dataset, 41 are input attributes and 1 is a target attribute. Additionally, there are four types of intrusion classifications that the attacks fall into: DoS, Probe, U2R, and R2L. You can see the classes and the members of each class (attack type) in Table 26.1.

Table 26.1 Attack class

Intrusion Class	Attack types
DoS	back, land, neptune, pod, smurf, teardrop, apache2, udpstorm, processtable, worm
Probe	satan, ipsweep, nmap, portsweep, mscan, saint
R2L	guess_password, ftp_write, imap, phf, multihop, warezmaster, warezclient, spy, xlock, xsnoop, snmpguess, snmpgetattack, httptunnel, sendmail, named
U2R	buffer_overflow, loadmodule, rootkit, perl, sqlattack, xterm, ps

2. Literature Review

Decoupling the administrative layer from the data plane is one way that the software package defined networking aims to replace conventional networks (as stated in [1]). It makes the network easier to control and program. Network administration is more vulnerable to intrusion since it serves no other purpose. To achieve this goal, we will use machine learning techniques to teach the network controller so that it can make smart choices on its own. Our approach for allowing detection and preventive procedures to enhance software package security against numerous damaging attacks is discussed in this paper.

System activity monitoring software that generates reports for management based on suspicious conduct is characterised as an intrusion detection system (IDS) in [2]. certificate or diploma For the purpose of protecting the availability, confidentiality, and integrity of information sources, intrusion detection systems are put in place [1]. The teams are keeping track of current threats and identifying any gaps in security measures with the help of intrusion detection systems.

In [3], we usually provide the results of our tests that assess how well police operations work against different types of attacks, such as IDS, Malware, and Shell code. Since the Kyoto 2006+ dataset contains the most up-to-date network packet data collected for Intrusion Detection System research, we find it more convenient to use the Random Forest technique on many datasets extracted from this dataset to assess popularity performance. Further predictions about the future are our preferred way to end discussions.

3. Problem Definition

There has been a meteoric rise in IT advancements in the previous 20 years. All walks of life, from industries to corporations, rely heavily on computer networks. Therefore, IT administrators have the critical duty of creating reliable networks. However, there are a number of issues with building reliable networks, a process that is very complicated, due to the fast development of information technology. The confidentiality, availability, and integrity of computer networks are under risk from a myriad of attack vectors. Probing, Denial of Service (DoS) attacks, Remote to Local (R2L), and User to Root (U2R) are among the most common malicious attacks.

4. Proposed Work

The XGBoost algorithm classifier is the basis of the proposed network intrusion detection system in this study. These classifiers improve the accuracy of attack detection and can distinguish between malicious and benign network data with remarkable ease.

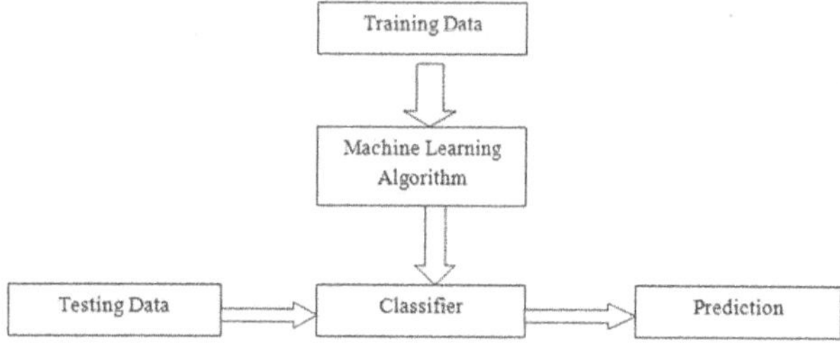

Fig. 26.1 Proposed model

4.1 Machine Learning

Creating reliable behavioural models or patterns that can differentiate between typical and unusual actions using audit data is a major challenge for Intrusion Detection Systems (IDSs). Due to this shortcoming, early Intrusion Detection Systems would sometimes ask security professionals to review audit data and create intrusion detection algorithms by hand. Evaluating and extracting attack signatures or detection criteria from the ever-changing, massive amounts of audit data has become a time-consuming, tedious, and even impossible process for human consultants due to the exponential growth of audit data. It is very difficult for rules developed by human experts to detect faulty or completely new attacks since these criteria are often based on fixed features or signatures of previous attacks. Due to the limitations of IDSs that rely on human consultants, data-processing intrusion detection systems have recently gained a lot of interest. Intrusion detection using data processing algorithms, often called reconciling intrusion detection, is an important area of knowledge mining that aims to improve the effectiveness of detection rules and analyse large amounts of audit data. It is common practice to automatically create intrusion detection models from labelled or unlabelled audit data using knowledge mining approaches.

A suggested intrusion detection method uses an attribute selection strategy to find key characteristics, then a classifier to divide network data into two groups: normal and assault.

5. Experimental Analysis

The comprehensive Kdd99 dataset was used for our research. Our goal is to find the best intrusion detection algorithm and write a R program to apply it. Our team developed an XGBoost algorithm specifically for detecting intrusions. The data must be pre-processed before any learning can be implemented. We are now working on R scripts that will sort the attack data into five categories: DoS, Probe, U2R, R2L, and normal, and give names to the columns.

Table 26.2 Number of samples

Type	Number of samples in training set	Number of samples in test set
DOS	3514	1486
Normal	5232	2268
Probe	2090	886
R2L	261	118
U2R	36	14
Total	11133	4772

After that, we need to remove the unnecessary rows from the statistical array. After that, we will check for blanks and remove those rows that don't apply. We can undertake exploratory data analysis, which involves visualising the data to improve our comprehension of it, once the datasets have been prepared and pre-processed. Figure 26.2 shows the protocol that has a high attack frequency, and Fig. 26.3 shows the correlation between attack length, attack types, and protocol types.

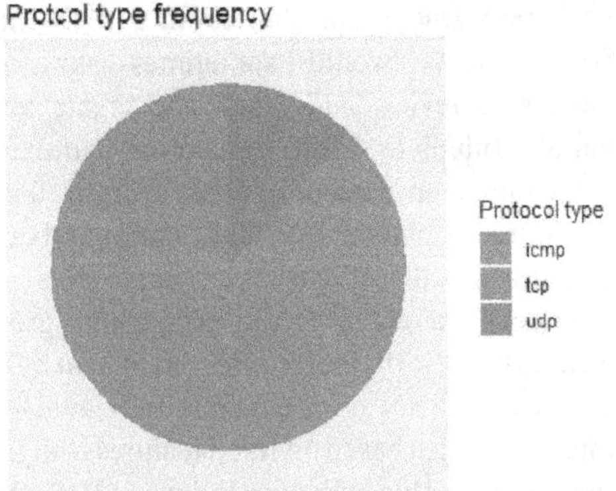

Fig. 26.2 Protocol type frequency

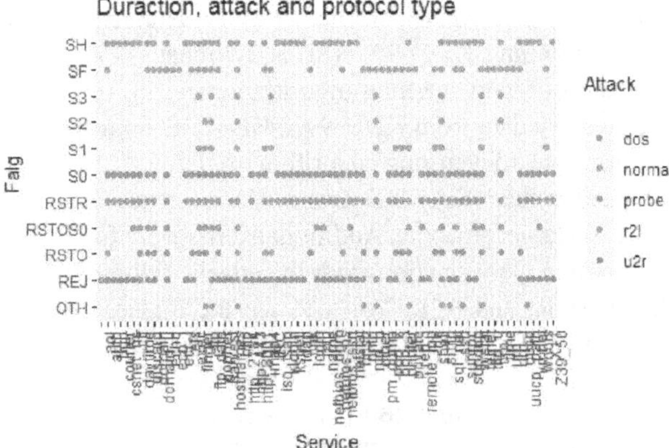

Fig. 26.3 Correlation between duration, attack and protocol type

The next stage is to feed this information into machine-learning algorithms, which might offer good results if they correctly identify the situations. One option is to start with training a support vector machine (SVM) model; this kind of model is known as a single learning classifier, and its output and predictions are based on that. As shown in Fig. 26.4, the SVM classifier performed well when tested on the test dataset.

```
> confusionMatrix(factor(svm_predict), factor(test$result), mode = "everything")
Confusion Matrix and Statistics

          Reference
Prediction  dos normal probe  r2l  u2r
    dos    1412     10    78    0    0
    normal    8   2220    70   20    6
    probe    65     12   738    1    0
    r2l       0     23     0   97    0
    u2r       0      3     0    0    8

Overall Statistics

               Accuracy : 0.938
                 95% CI : (0.9307, 0.9446)
    No Information Rate : 0.4754
    P-Value [Acc > NIR] : < 2.2e-16

                  Kappa : 0.9028

 Mcnemar's Test P-Value : NA

Statistics by Class:

                     Class: dos Class: normal Class: probe Class: r2l Class: u2r
Sensitivity              0.9508        0.9788       0.8330    0.82203   0.571429
Specificity              0.9732        0.9584       0.9799    0.99506   0.999369
Pos Pred Value           0.9413        0.9552       0.9044    0.80833   0.727273
Neg Pred Value           0.9777        0.9804       0.9626    0.99548   0.998739
Precision                0.9413        0.9552       0.9044    0.80833   0.727273
Recall                   0.9508        0.9788       0.8330    0.82203   0.571429
F1                       0.9461        0.9669       0.8672    0.81513   0.640000
Prevalence               0.3113        0.4754       0.1857    0.02473   0.002934
Detection Rate           0.2960        0.4653       0.1547    0.02033   0.001677
Detection Prevalence     0.3144        0.4871       0.1710    0.02515   0.002306
Balanced Accuracy        0.9620        0.9686       0.9064    0.90855   0.785399
>
```

Fig. 26.4 Confusion matrix and statistical analysis of SVM model

Even though it relies on only one classifier, this SVM manages to get a total accuracy of 93.8%. We suggest using XGBoost learning into machine learning, which combines model assessments to improve performance. The term "assembling" refers to the process of combining several different types of learners, or individual models, into one larger model in order to increase its stability and predictive power. In the previous case, the way in which we compile all of the predictions. We may build a model using XGBoost learning, and Fig. 26.5 shows the results of testing the trained model.

```
Confusion Matrix and Statistics

          Reference
Prediction  dos normal probe  r2l  u2r
    dos    1445      1    47    0    0
    normal    1   2257     9    2    1
    probe    40      4   828    0    1
    r2l       0      5     2  115    1
    u2r       0      1     0    1   11

Overall Statistics

               Accuracy : 0.9757
                 95% CI : (0.9709, 0.9799)
    No Information Rate : 0.4753
    P-Value [Acc > NIR] : < 2.2e-16

                  Kappa : 0.9621

 Mcnemar's Test P-Value : NA

Statistics by Class:

                     Class: dos Class: normal Class: probe Class: r2l Class: u2r
Sensitivity              0.9724        0.9951       0.9345    0.97458   0.785714
Specificity              0.9854        0.9948       0.9884    0.99828   0.999580
Pos Pred Value           0.9678        0.9943       0.9485    0.93496   0.846154
Neg Pred Value           0.9875        0.9956       0.9851    0.99935   0.999370
Precision                0.9678        0.9943       0.9485    0.93496   0.846154
Recall                   0.9724        0.9951       0.9345    0.97458   0.785714
F1                       0.9701        0.9947       0.9414    0.95436   0.814815
Prevalence               0.3114        0.4753       0.1857    0.02473   0.002934
Detection Rate           0.3028        0.4730       0.1735    0.02410   0.002305
Detection Prevalence     0.3129        0.4757       0.1829    0.02578   0.002724
Balanced Accuracy        0.9789        0.9950       0.9615    0.98643   0.892647
>
```

Fig. 26.5 Confusion matrix and statistics of the XGBoost model

5.1 Performance Measure

The confusion matrix is utilized to create accuracy, which was used.

Table 26.3 Confusion matrix

	Classified as Normal	Clasified as Attack
Normal	TP	FP
Attack	FN	TN

In cases when accurately categorised instances as non-attacks – True Negatives.

Misclassification of events as non-attacks - instances of false negatives.

Attacks Misclassified Due to False Positives.

Real Positives - Cases that were correctly recognised as attacks.

$$\text{Accuracy} = \frac{TP + TN}{TP + TN + FP + FN}$$

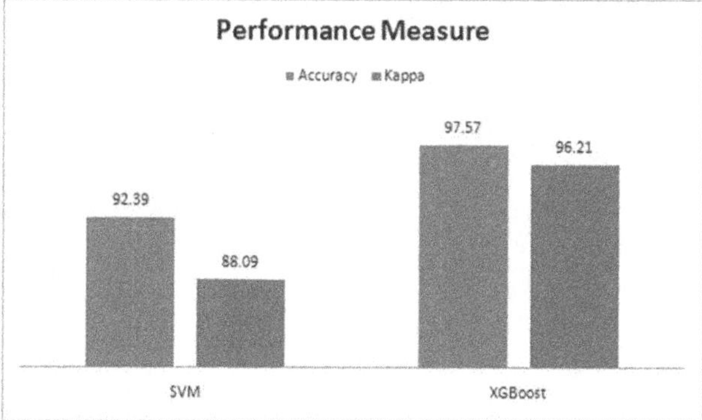

Fig. 26.6 Performance measure

Table 26.4 Performance measure of models

Parameters	Accuracy	Kappa
SVM	92.39%	88.09%
XGBoost	97.57%	96.21%

Accuracy for each class is shown in the table; we may also analyse performance on a class-wise basis.

Table 26.5 Class wise performance measure

Class \ Accuracy	SVM	XGBoost
Dos	96.20	97.89
Normal	96.86	99.50
Probe	90.64	96.15
R2L	90.85	98.64
U2R	78.53	89.26

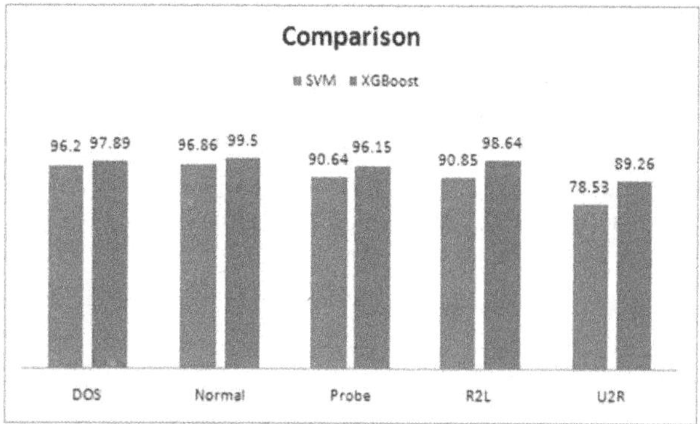

Fig. 26.7 Class wise comparison

6. Conclusion

Many problems hinder the efficiency and effectiveness of traditional intrusion detection systems. On the other side, approaches based on machine learning provide promising approaches to intrusion detection. Utilising the NSL-KDD99 dataset, the Intrusion Detection System (IDS) is categorised using the XGBoost learning algorithm and Support Vector Machine (SVM). We find that XGBoost classifiers outperform SVM classifiers in terms of accuracy after comparing the results of several classification algorithms.

References

1. A. S. Krishnan, "Network Intrusion Detection Systems: A Survey," IEEE Communications Surveys & Tutorials, vol. 16, no. 4, pp. 2341–2354, 2014.
2. J. Kim and Y. Lee, "Anomaly Detection in Network Traffic Using Support Vector Machines," IEEE Transactions on Neural Networks and Learning Systems, vol. 23, no. 1, pp. 1–11, 2012.
3. Y. Chen and J. Liu, "Anomaly Detection in Network Traffic Using Extreme Gradient Boosting," IEEE Transactions on Industrial Electronics, vol. 65, no. 5, pp. 4321–4330, 2018.

4. Mohamed Ali Ahmed, Yasir Abdelgader Mohamed, "*Enhancing Intrusion Detection Using Statistical Functions*" in 2018 International Conference on Computer, Control, Electrical, and Electronics Engineering (ICCCEEE), IEEE.
5. Rafath Samrin, D Vasumathi, "Review on Anomaly based Network Intrusion Detection System" in IEEE, 2017.
6. Li, M., Huang, F., Cheng, L., & Su, C. (2018). Anomaly Detection Model based on SVM & XGBoost for Secure Network Intrusion Detection. IEEE Access, 6, 61912–61922.
7. Wang, W., Zhou, Y., & Li, J. (2019). Network Intrusion Detection System Based on SVM and XGBoost Algorithm. In 2019 IEEE International Conference on Intelligence Science and Big Data Engineering (IScIDE) (pp. 25–29). IEEE.
8. Gao, Y., Liu, J., Sun, Y., & Wang, Q. (2018). A Hybrid Anomaly Detection Model based on SVM and XGBoost for Network Intrusion Detection. IEEE Transactions on Network and Service Management, 15(4), 1531–1541.
9. Sharma, A., & Gupta, D. (2017). Anomaly Detection in Network Traffic using SVM and XGBoost. In 2017 IEEE International Conference on Computing, Communication and Automation (ICCCA) (pp. 1–5). IEEE.
10. Kim, H., Lee, J., Choi, J., & Park, H. (2019). Network Intrusion Detection using Ensemble Approach of SVM and XGBoost. IEEE Access, 7, 43031–43040.
11. Zhang, Y., Xu, Y., Zhang, L., & Wang, W. (2018). Anomaly Detection in Network Traffic based on SVM and XGBoost Ensemble. In 2018 IEEE International Conference on Big Data (Big Data) (pp. 3067–3075). IEEE.
12. Wang, F., Yang, C., & Liu, Z. (2019). Improving Anomaly Detection in Network Intrusion using SVM and XGBoost Model. In 2019 IEEE International Conference on Computational Science and Engineering (CSE) and IEEE International Conference on Embedded and Ubiquitous Computing (EUC) (pp. 168–173). IEEE.
13. Huang, J., Chen, H., & Wu, Y. (2018). A Comparative Study of SVM and XGBoost for Anomaly Detection in Network Intrusions. In 2018 IEEE 2nd International Conference on Big Data Security on Cloud (BigDataSecurity), IEEE International Conference on High Performance and Smart Computing (HPSC), and IEEE International Conference on Intelligent Data and Security (IDS) (pp. 47–51). IEEE.
14. Liu, H., Chen, S., & Zheng, L. (2017). Anomaly Detection in Network Traffic using Kernel-based SVM and XGBoost. In 2017 IEEE International Conference on Information and Automation (ICIA) (pp. 745–749). IEEE.
15. Zhou, L., Wang, Y., & Shen, L. (2019). Anomaly Detection Model based on SVM & XGBoost for Secure Network Intrusion Detection. In 2019 IEEE International Conference on Cyber Security and Protection of Digital Services (Cyber Security) (pp. 1–6). IEEE.

Note: All the figures and tables in this chapter were made by the authors.

Intelligent Systems Using Semiconductors for Robotics and IoT – Dinesh Goyal et al. (eds)
 ISBN 978-1-041-20408-4

27 Stock Market Prediction using Machine Learning Technique: A Literature Review

Rohit Singh Rajput[1]
Research Scholar, Dept. of CSE, Poornima University,
Jaipur, India
Assistant Professor, Dept. of CSE, Poornima College of Engineering,
Jaipur, India

Savita Shiwani[2]
Professor, Dept. of Computer Engineering,
Poornima University, Jaipur, India

Abstract: The stock market, characterized by its complexity and volatility, has long been a focal point for researchers and investors aiming to predict future movements and gain a competitive edge. This review paper explores the diverse landscape of machine learning techniques applied to stock market prediction, assessing their efficacy, advantages, and limitations. We systematically analyze various methodologies, including supervised learning, unsupervised learning, reinforcement learning, and hybrid models, highlighting their respective capabilities in handling the multifaceted nature of financial data. Special attention is given to the role of feature selection, data preprocessing, and model evaluation metrics, which are critical in enhancing prediction accuracy. Furthermore, we discuss the challenges inherent in stock market prediction, such as data noise, market anomalies, and the dynamic, non-stationary behavior of financial markets. The review also examines the impact of recent advancements in deep learning, such as neural networks and their variants, in pushing the boundaries of predictive performance. By synthesizing current research trends and findings, this paper provides a comprehensive overview of the state-of-the-art in stock market prediction using machine learning, offering insights into future directions and potential areas for improvement

Keywords: Stock market, Machine learning, LSTM

1. Introduction

Stock market is relatively a major aggressive economic business sector where the dealers are required and process the economic workloads with lower latency alongside higher throughput. Formerly, economists were utilizing the customary store and cycle technique to figure out the weighty economic workload productively. However, to accomplish low idleness and high throughput, server farms had to be genuinely found near the information sources, rather than other all the more financially gainful areas. The primary explanation, the information is that streaming model has been created and it can handle enormous measures of information more proficiently. In fact, on the off chance that we can anticipate how the stock will carry on in the transient future we can line up our exchanges prior and be quicker than every other person. In principle, this permits us to expand our benefit without wanting to be truly found near the information sources. [1]

Economic transaction stock value rise is predicted by stock exchange forecasts. The exact prediction of offer value development will provide bigger advantages than financial economists expected.[2]

Anticipating the stock exchange's movement is arguably the hardest difficulty since legislative concerns and

[1]2023phdoddrohit15798@poornima.edu.in, rohit.rajput@poornima.org; [2]savita.shiwani@poornima.edu.in

DOI: 10.1201/9781003716389-27

financial progress make the exchange unpredictable and impossible to anticipate. The forecast gives huge profit chances and inspires investigation here; stock changes by a fraction of a second may result in large rewards. Lack of accurate data would lead to speculative losses since company share is a major economic market movement. Since vulnerabilities impact market fluctuations, stock exchange forecasting is tough. Specialist and large examination stock market forecasting methodologies exist. [3] the Fig. 27.1 shows the forecast of the stock exchange and the current details of the stock. Once it has been logged in, then the framework can be played out.

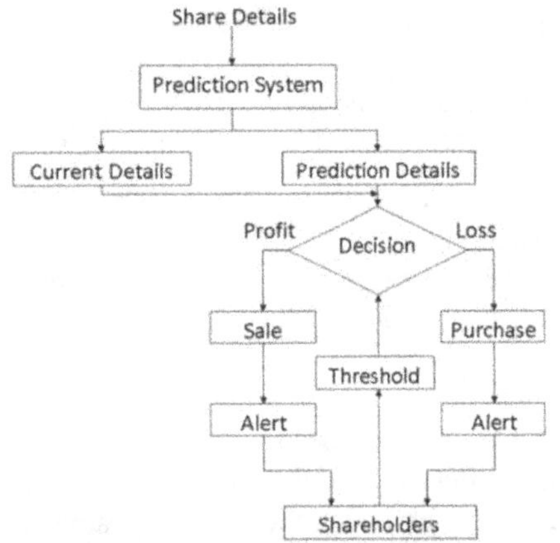

Fig. 27.1 Stock market prediction block diagram

The fact that these studies heavily rely on structured data while ignoring a substantial source of data—online economic data and web-based media assumptions—means that they did not yield excellent forecast outcomes. These days, a growing amount of fundamental information regarding financial transactions is available online. The BBC, Bloomberg, and Yahoo Finance are all included in the models. Valuable data is hard to physically extract from these assets [4]. The framework for market prediction is shown in Fig. 27.2. This illustrates the significance of text mining techniques to organically extract important data for stock market analysis. This analysis looked into the most important previous literature and made a major commitment to the topic of using text mining and natural language processing for market expectations. [5]

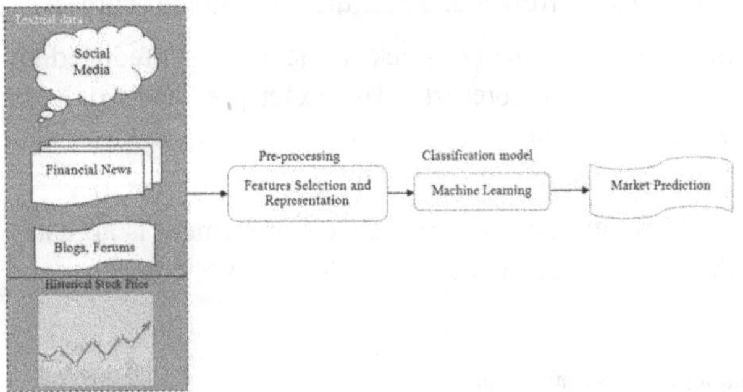

Fig. 27.2 Market prediction framework

The stock market's inherent volatility and complexity, further exacerbated by global events such as the COVID-19 pandemic, have rendered traditional prediction models increasingly inadequate. These models often fail to account for sudden market disruptions and the intricate patterns inherent in stock price movements. There is a pressing need for more accurate and reliable prediction methods that can adapt to such volatility and provide better guidance for investors and financial analysts.

The motivation for this research stalks from the growing need for more precise and reliable approaches of predicting stock market trends, especially in the wake of increasing market volatility and complex global economic conditions. Traditional stock market prediction models habitually drop short in fascinating the complicated and dynamic nature of monetary markets, primary to suboptimal investment decisions and significant financial losses.

Traditional stock market prediction models regularly drop short in capturing the intricate and dynamic flora of financial markets, leading to suboptimal investment decisions and significant financial losses. The advent of the COVID-19 epidemic has additional exacerbated these challenges, highlighting the inadequacies of conventional models in accounting for unprecedented market disruptions.

The aim of this research is to conduct a comparative analysis of the proposed hybrid model against traditional models.

2. Literature Review

Rajni Jindal et.al. (2021) The stock market is a structured entity wherein publicly traded corporations sell their stocks and traders purchase and sell these stocks in an effort to make money. The process of predicting stock market trends is complicated due to its dynamic and turbulent character. The COVID-19 epidemic has made this effort much more difficult in recent years. The global increase of COVID-19 cases has made the market more volatile than ever before. Because these algorithms don't take into consideration how the pandemic has affected stock market movements, this has led to a number of classic trend prediction algorithms performing poorly. The goal of the proposed study is to increase the forecast capabilities of many popular prediction models by accounting for COVID-19-related parameters. "Support Vector Regressor (SVR), Random Forest Regressor, and Decision Tree Regressor" are the forecasting methods that were examined [6].

Bayan Albaooth(2023) The individual investors use the information available related to the stock market to help in evaluating the stock investment and financial behaviors for making decisions. By understanding the security firms' market trends for forecasting future advice for the investors and investor behaviors. As artificial intelligence (AI) advances, tools like CNN and LSTM are being used to forecast stock market activity and help investors make decisions. Forecasting stock market returns is one of the finest ways to diversify your portfolio and limit risk. There are several forecasting techniques that use AI models to anticipate stock market behaviors in order to produce accurate estimates for investment decision-making.

The analysis outcomes indicate the prescribed model achieved more than 99.98% in prediction accuracy of the dataset understudy. This can significantly enhance the decisions of individual investors for better future predictions of stock markets[7].

K. Ritwik Reddy et.al. (2022) The stock market is a difficult place to invest money these days, and it also takes a lot of creativity, given the erratic character of the financial stock market. Therefore, it is extremely hard to forecast. When compared to other methods, certain algorithms, such as those connected to AI, ML, and DL, can be responsible for effective accuracy. Our goal in is to calculate the price of financial stocks using Deep Learning methods, namely the LSTM and "Recurrent Neural Networks" (RNN). We need to investigate the accuracy by increasing the number of epochs while using RNN and LSTM algorithms for various datasets[8].

Xu Zhang et.al. (2022) Predicting the price of a single stock or an index of stocks using time series data prediction techniques like ML or DL is known as stock market prediction. It is employed to support stock trading, stock market oversight, etc. The LSTM of a deep learning model is used in this article to investigate the elements that influence stock market prediction. We have chosen three types of relevant factors: the historical US NASDAQ index. Additionally, two types of groups are formed from the important aspects in order to observe their performance. There is a single factor group in one and many factor groups in the other. "The data used in the experiment was collected between January 1, 2018, and December 31, 2020. Three metrics are used to evaluate the prediction results: mean absolute error (MAE), root mean square error (RMSE), and mean absolute percentage error (MAPE)". In relations of predicting the value of the stock table and its regular growth and drop, the experiment's findings demonstrate that the three assessment metrics of the single factor group's forecast results outperform those of the multiple factor group. This study demonstrates that the better the prediction outcomes in stock market index trend prediction research, the less important the extrapolative components[9].

Jesús Padilla-Caballero et.al. (2023) Forecasting the future value of a company's stock is the aim of stock market prediction. The goal of this forecasting is to project the future value of financial offerings made by associations. The output stock price is the main basic variable. The criteria that determines the security of the stock market sectors is another. Data science researchers and observers have often noted how hard it is to forecast the market's future actions. The stock price may be greatly impacted by future stock price data based on explicit stock nuances, according to suggested work, daily stock price perception, and a comprehensive evaluation. Comparing this concept to environmental elements, stock price forecast performance is much improved. After analyzing the pertinent news source, our method generates a reliable forecast when paired with soft computing. They also acknowledge that a stochastic interaction illustrates how the stock market functions, suggesting that the most accurate forecast depends on the most recent stock data. The experts' meeting will really look at the news ideas twice[10].

Srikumar Manghat et al. (2023) The study starts by examining the elements of occurrences whose fluctuations resemble those of stock market values. The article specifically demonstrates that these differences can generally be divided into a number of component groups, and the general trend of the variations may be represented as a collection of variants that fall into each category. The article then suggests a straightforward technique to forecast which way these deviations will go. Since signal noise is random, similar to stock price noise, a novel type of noise filter is then explained[11].

Muntather Almusawi et.al. (2023) Stocks of publicly traded companies are purchased and sold on the stock market, a financial marketplace. It measures the nation's economic well-being and is a reflection of business performance and the overall business climate. The dynamics of supply and demand affect stock prices. Although stock market investing carries some risk, there is a chance for significant long-term rewards. The LSTM with "Improved Artificial Rabbits Optimization (IARO)" algorithm was used in this study to forecast the price of stocks. The LSTM hyperparameters are augmented using the IARO algorithm to increase the forecast accuracy of stock market prices. Data is gathered and utilized to forecast stock market prices. "Coefficient of Determination (R2), Mean Square Error (MSE), Mean Absolute Error (MAE), and Mean Absolute Percentage Error (MAPE)" are used to measure the performance of the suggested approach. In contrast to other current methods, the suggested method has a lower MSE of 0.43, MAE of 0.37, MAPE of 0.21, and R2 of 0.17"[12].

N. Jaswanth Reddy et.al. (2022) An analysis of the relative improvements in accuracy between "Simple Novel Long Short-Term Memory (SNLSTM)" and Back Propagation (BP) for share value prediction. The methods for SNLSTM (N=1000) and Back Propagation (N=1000) were computed using two groups, with a total sample size of 2000 utilizing g-power of 0.8. Both algorithms' accuracy is working[13].

Paul Akash Gunturu et.al.(2023) One of the most important aspects of the financial industry is share market investing. However, finding profitable stocks is a difficult undertaking that calls for serious consideration. In order to overcome this difficulty, this study compares several ML and DL methods for stock forecast. The study assesses and contrasts several models, such as "Random Decision Forest, Auto-ARIMA, k-Nearest Neighbors (KNN), Prophet (Automated Forecasting Procedure), Long Short-Term Memory (LSTM), Linear Regression, and Moving Average methods like SMA and EMA". Additionally, a novel hybrid model is put out that performs more accurately than current models. Following training and testing on stock data from many industrial sectors, the models are evaluated using a variety of performance metrics. The study's conclusions about the accuracy of sevral models may be used by traders, investors, and economic forecasters to make wise venture choices. Furthermore, the research's conclusions might be used as a standard for other studies on stock market forecasting [14].

L Mathanprasad et.al. (2022) The primary goal of the study project is to investigate the existing stock market data using real-time data, where the values of the stock market data change according to time. In the realm of research, forecasting the stock market and estimating future stock prices are still regarded as difficult tasks. The current study's inspiration stems from the fact that stock market data values fluctuate periodically particular danger. Therefore, it is necessary to create a computationally automated process for forecasting the values of stock market data. Comprehensive data about the actuality of stock market fluctuations is found by gathering data from past historical data in order to choose a prediction strategy and discuss the data with stock analysts who are professionals in the stock market analysis system. Using a machine learning classification algorithm, the user can easily determine which stocks present in the market for a extensive period of time by predicting changes in the stock market's price and movement. ML algorithms have been used to examine and enhance the stock exchange's forecast accuracy to 94.17%. Consequently, this forecast will assist investors in evaluating the extant and anticipated forthcoming values of the company's stock market rate[15].

Pushpendra Singh Sisodia et.al. (2022) A long-running area of research in the stock market is the design and development of prediction models that accurately forecast stock prices. However, anticipating changes in stock prices is the most important part of the forecasting process. Research indicates that stock price movement may be somewhat anticipated, despite certain market theories contending that it is difficult to predict stock price movement with precision. Properly conceived, built, and improved prediction models provide for exact measurement of stock price movement[16].

Priyanka Srivastava et.al. (2021) Predicting bond prices is a popular, challenging, complex, and hard subject in the field of computation that typically involves a lot of human-computer interaction. Various components are included in the trends for anticipating physical characteristics of the stock market against physiological, rational, and illogical conduct, investor emotion, and market whispering. When all those factors come together, stock prices become quite complex and very hard to predict.[15] Stock market prediction may be effectively done with sequential prediction algorithms due to the connected nature of stock values. Machine learning techniques are able to recognize patterns, ascertain the reasoning behind forecasts, and provide predictions that are consistently accurate. [8] We have experimented with a number of various stock market forecasting algorithms, ranging from basic ones like Simple Average and Linear Regression to more complex ones like ARIMA and LSTM, and we compare which one produces the most accurate and effective results. We provide a research technique that uses the enhanced LSTM version of RNN, where each data variable's weights are maintained via stochastic gradient descent. [17]

Mayank Joshi et al. (2023) to assist in producing results that are more accurate and efficient than those of current stock price prediction systems. The stock prediction model was developed using the TSLA dataset, which is TESLA Inc. from Yahoo Finance.

After building and training the LSTM model and analyzing future stock prices using data-frame closing prices, we generated stock forecasts using a sample of the data set and calculated extra RMSE for accuracy and efficacy. Additionally, we have presented a number of comparison prediction algorithms. Based on these results, LSTM is suggested for stock market projections[18].

Arjun Singh et.al. (2022) In this study, the technical evaluation of stock market forecast based on neural networks is presented. It is possible to analyze the stock market forecast using a variety of ML techniques. The enclosed depiction is analyzed in this study together with the buying and selling prices of stocks on the BSE. To predict the stock market, the comprehensive LSTM was pressed and combined with an entrenched layer. The experimental setup uses deep LSTM with layers to better define. When the prediction system produces accurate results, it generates a lot of attention and an incredible profit. Investors have traditionally struggled with market prediction. Whether it's figuring out how the market repeats itself or examining trends that might help them obtain a higher return on their investment, investors have always been eager to find an effective approach to forecast prices as accurately as possible. The various technologies that are currently being employed for stock market trend analysis and prediction are examined in this research. Using the data collected from equities listed on the "National Stock Exchange (NSE)", the numbers for the stock's prior close, open, high, low, and close during a six-year period have been used to analyze historical patterns. Python and R are widely used computer languages for managing such large datasets and identifying trends in them[19].

3. Comparative Study

Table 27.1 Comparative study

Author(s)	Methodology	Advantages	Disadvantages
Rajni Jindal et al. 2021	Decision Tree Regressor, Random Forest Regressor, Support Vector Regressor with COVID-19 factors	Accounts for COVID-19 impact, improves performance of traditional models	May not generalize well beyond pandemic-specific scenarios
Bayan Albaooth 2023	LSTM, CNN for stock market behavior prediction	High prediction accuracy (99.98%), enhances individual investors' decisions	Limited to dataset used, may not account for all market variables
K. Ritwik Reddy et al. 2022	RNN, LSTM for financial stock market price prediction	Uses deep learning techniques to improve prediction accuracy	Requires significant computational resources, sensitive to hyperparameter settings
Xu Zhang et al. 2022	LSTM with historical stock data, NASDAQ index, Weibo term frequency	Evaluates single and multiple factor groups, uses diverse influential factors	Multiple factor group results are less effective than single factor group
Jesús Padilla-Caballero et al. 2023	Framework learning for stock price prediction	Utilizes current stock market indexes data, combines soft computing for robust predictions	Assumes stock market follows a stochastic process, which may not always hold true
Srikumar Manghat2 023	Event variation analysis method for stock price prediction	Shows improved prediction over existing methods, uses real-world stock prices for validation	Complexity in segregating variations into component categories
Muntather Almusawi et. al. 2023	LSTM with Improved Artificial Rabbits Optimization (IARO)	Optimizes LSTM hyperparameters, achieves lower error rates (MSE, MAE, MAPE, R2)	Complexity of IARO algorithm, specific to stock market data used
N. Jaswanth Reddy et al. 2023	Simple Novel Long Short-Term Memory (SNLSTM) and Back Propagation for stock prediction	SNLSTM shows better accuracy than Back Propagation	Limited to comparison of two specific algorithms, generalization to other models not addressed
Paul Akash Gunturu et al. 2023	Comparison of ML and DL techniques including LSTM, Prophet, Random Decision Forest, Auto-ARIMA, KNN	Proposes a new hybrid model, provides insights into model accuracy for stock trends	Evaluation limited to historical dataset, hybrid model complexity
L. Mathanprasad et al. 2022	Machine learning classification approach for real-time stock market data	Improved prediction accuracy (94.17%), practical for investor decision-making	Dependent on historical data quality, real-time data fluctuations may impact accuracy
Pushpendra Singh Sisodia et al. 2022	Deep Learning LSTM for NIFTY 50 index prediction	Promising accuracy (83.88%), robust for long-term data	Requires large historical datasets, model training can be resource-intensive
Priyanka Srivastava et al. 2021	Improved LSTM with stochastic gradient descent for stock prediction	Efficient and accurate outcomes, uses Tesla Inc. dataset for model training	Limited to specific dataset, may not generalize across different stocks
Mayank Joshi et al. 2023	Data mining and prediction analysis methodology for stock forecasting	Uses Python implementation, reviews multiple classification-based systems	General overview, lacks specific details on individual model performance
Arjun Singh et al. 2022	LSTM Neural Network with embedded layer for stock market prognosis	Accurate predictions, deep LSTM layers improve performance	Complex model architecture, may require extensive tuning and validation

4. Research Findings

Despite significant advancements in stock market prediction methodologies, several research gaps persist. Current models often focus on a limited range of factors, such as historical stock data or social media trends, lacking comprehensive integration of macroeconomic indicators and global events. The impact of unforeseen global events on stock markets remains inadequately addressed. While hybrid models combining various machine learning and deep learning techniques show promise, optimal integration strategies need further exploration. Real-time prediction models face challenges with scalability and data processing efficiency. Additionally, most studies are dataset-specific and do not generalize well across different markets and sectors. There is also a need for personalized prediction models that consider individual investor preferences and risk profiles. Furthermore, advanced models often lack transparency and explain ability, which is crucial for gaining investor trust. Addressing these gaps can lead to more accurate, robust, and user-centric stock market prediction systems.

The field of stock market prediction has seen several noteworthy contributions that have significantly advanced our understanding and capabilities. Rajni Jindal et al. (2021) underscored the necessity of incorporating COVID-19-related factors into prediction models, demonstrating that traditional algorithms faltered during the pandemic due to their failure to account for such unprecedented events. By including pandemic-specific variables, they markedly improved the predictive accuracy of their models. Bayan Albaooth (2023) pushed the boundaries further by leveraging deep learning models, particularly LSTM and CNN, achieving a remarkable prediction accuracy of 99.98%, thereby highlighting the transformative potential of advanced AI techniques in financial forecasting. Xu Zhang et al. (2022) offered a crucial insight by comparing single and multiple factor models, finding that simpler, and single-factor models often outperformed more complex ones, challenging the notion that more data points always lead to better predictions. Paul Akash Gunturu et al. (2023) developed a hybrid model that combined multiple machine learning and deep learning techniques, which not only improved accuracy but also provided a robust benchmark for future research. Lastly, Muntather Almusawi et al. (2023) introduced an innovative method by optimizing LSTM models with the Improved Artificial Rabbits Optimization (IARO) algorithm. This approach significantly reduced error rates (MSE, MAE, MAPE) compared to traditional methods, demonstrating an effective strategy for enhancing LSTM model performance. These contributions collectively highlight the importance of integrating novel data, optimizing existing models, and exploring hybrid approaches to push the frontier of stock market prediction capabilities.

Acknowledgment

The authors gratefully acknowledge the students, staff, and authority of Computer Science department for their cooperation in the research.

References

1. P Nagaraj; K Nani; E Teja Krishna; K Aravind Kumar Reddy; M Chaithanya Prabhu; K Nikhil (2023) Stock Market Profit Prediction Using Machine Learning Algorithms and Visualization for Live Data 2023 International Conference on Data Science, Agents & Artificial Intelligence (ICDSAAI) 2023
2. Ashish Ruke; Sujeet Gaikwad; Gitanjali Yadav; Amar Buchade; Shruti Nimbarkar; Ashish Sonawane (2024) Predictive Analysis of Stock Market Trends: A Machine Learning Approach 2024 4th International Conference on Data Engineering and Communication Systems (ICDECS) 2024
3. Vidushi Tiwari; Bhanu Prakash Lohani; Ajay Rana; Upendra Pratap Pandey; Bhavdeep Dhariwal (2023) Stock Market Prediction using different Machine Learning Algorithms 2023 10th IEEE Uttar Pradesh Section International Conference on Electrical, Electronics and Computer Engineering (UPCON) 2023
4. Golshid Ranibaran; Mohammad-Shahram Moin; Sasan H Alizadeh; Abbas Koochari (2021) Analyzing effect of news polarity on stock market prediction: a machine learning approach 2021 12th International Conference on Information and Knowledge Technology (IKT) 2021
5. Jing Yee Lim; Kian Ming Lim; Chin Poo Lee (2021) Stacked Bidirectional Long Short-Term Memory for Stock Market Analysis 2021 IEEE International Conference on Artificial Intelligence in Engineering and Technology (IICAIET) 2021
6. Rajni Jindal; Nikhil Bansal; Nitin Chawla; Sanskriti Singhal (2021) "Improving Traditional Stock Market Prediction Algorithms using Covid-19 Analysis" 2021 International Conference on Emerging Smart Computing and Informatics (ESCI) 2021
7. Bayan Albaooth (2023) "he Role of Artificial Inteligence Prediction in Stock Market Investors Decisions" 2023 IEEE Asia-Pacific Conference on Computer Science and Data Engineering (CSDE) 2023
8. K. Ritwik Reddy; B. Tarun Kumar; V. Rohit Ganesh; Polisetty Swetha; Prakash Kumar Sarangi (2022) Stock Market Prediction Using Recurrent Neural Network 2022 IEEE International Conference on Current Development in Engineering and Technology (CCET) 2022
9. Xu Zhang; Li Zhang; Lingjun Xu; Yue Jiang (2022) Research on Influential Factors in Stock Market Prediction with LSTM 2022
10. Jesús Padilla-Caballero; Roberto Diaz-Quichiz; María Quispe-Huayllapuma; Hipólita Huayllapuma-Rivera; Luis Rojas-Zuñiga; Frans Cardenas-Palomino (2023) Recent Developments and Methodologies for Stock Market Prediction Using Soft Computing Technique 2023 6th International Conference on Contemporary Computing and Informatics (IC3I) 2023
11. Srikumar Manghat (2023) Stock Market Directional Prediction and a New Kind of Noise Filter 2023 4th IEEE Global Conference for Advancement in Technology (GCAT) 2023
12. Muntather Almusawi; Subhra Chakraborty; Gobinath Ravindran; S Prabu; Zainab Abed Almoussawi (2023) Long Short-Term Memory optimized with Improved Artificial Rabbits optimization Algorithm for Stock Market Price Prediction 2023 International Conference on Integrated Intelligence and Communication Systems (ICIICS) 2023
13. N. Jaswanth Reddy; K Jaisharma (2022) Analysis of Stock Market Value Prediction using Simple Novel Long Short Term Memory Algorithm in Comparison with Back Propagation Algorithm for Increased Accuracy Rate 2022 5th International Conference on Contemporary Computing and Informatics (IC3I) 2022
14. Paul Akash Gunturu; Rony Joseph; Emany Sri Revant; Shailesh Khapre (2023) Survey of Stock Market Price Prediction Trends using Machine Learning Techniques 2023 International Conference on Artificial Intelligence

and Applications (ICAIA) Alliance Technology Conference (ATCON-1) 2023

15. L Mathanprasad; M Gunasekaran (2022) Analysing the Trend of Stock Marketand Evaluate the performance of Market Prediction using Machine Learning Approach 2022 International Conference on Advances in Computing, Communication and Applied Informatics (ACCAI) 2022
16. Pushpendra Singh Sisodia; Anish Gupta; Yogesh Kumar; Gaurav Kumar Ameta (2022) Stock Market Analysis and Prediction for Nifty50 using LSTM Deep Learning Approach 2022 2nd International Conference on Innovative Practices in Technology and Management (ICIPTM) 2022
17. Priyanka Srivastava; P K Mishra (2021) Stock Market Prediction Using RNN LSTM 2021 2nd Global Conference for Advancement in Technology (GCAT) 2021
18. Mayank Joshi; Gaurav Goel (2023) Stock Market Prediction Approach: An Analysis 2023 International Conference on Artificial Intelligence and Smart Communication (AISC) 2023
19. Arjun Singh; Garima Bhardwaj; Arun Pratap Srivastava; Ankur Bindra; Pushpa Chaudhary; Ritika (2022) Application of Neural Network to Technical Analysis of Stock Market Prediction 2022 3rd International Conference on Intelligent Engineering and Management (ICIEM) 2022

Note: All the figures and table in this chapter were made by the authors.

Intelligent Systems Using Semiconductors for Robotics and IoT – Dinesh Goyal et al. (eds)

28 High-Gain Composite Artificial Conductor Antenna for Enhanced Performance and Efficiency in Advanced 5G Communication Systems

Saurabh Shandilya[1]
Professor ADC, PCE, Jaipur, India

Sachin Jain[2]
Associate Professor CSE, PCE, Jaipur, India

Geetika Mathur[3]
Professor CSE, PCE, Jaipur, India, India

Kamlesh Gautam[4]
Assistant Professor ADC, PCE, Jaipur, India

Archana Soni[5]
Assistant Professor CSE, PCE, Jaipur, India

Abstract: In this article a novel meta-surface structure is proposed to improve the gain of antenna. The meta-surface structure has rectangle with circular cuts under the rectangular ring. The monopole antenna with rectangular shape designed at 3GHz and 8GHz. The monopole antenna backed with two Artificial Magnetic Conductor (AMC) structures which resonates at 3.9 and 8.8 GHz. The bandwidth of antenna resonates at 0.9GHz is 3.2and the bandwidth of antenna resonates at 3.9GHz is 0. 829. The gain of antenna backed 4 x4 AMC array achieved 8. 23dBi. For 5G communications, the suggested antenna is a strong contender.

Keywords: Artificial magnetic conductor, 5G antenna

1. Introduction

5G connectivity is essential for efficient communication networks. This criterion can be met at high frequencies. With this high-frequency antenna structure, the structure of the planned antenna is small in size and easy to design.

Despite its high bandwidth, reflection and resonance frequencies, the rectangular patch antenna has a gain of only 3.13 dB. It is made of crystal and operates at 28.3 GHz, making it a good choice for 5G networks in India [1].

A multilayer patch antenna with an average frequency of 2.8 GHz was constructed using a Rogers RT 5880 dielectric device and achieved 9.77 dB over a 2.9 GHz range. [2].

The rectangular band antenna consists of four metal squares placed under a square ring, which provides oscillation at frequencies lower than 6 GHz, (maximum gain) [3].

The gain, reflection coefficient, and bandwidth of a four-element planar antenna that must resonate at 30 GHz are limited. A modified coaxial current fractal antenna that oscillates in the range of 21-30 GHz.

A four-layer antenna with limited gain, reflection layer and bandwidth, designed to transmit audio signals in the

[1]saurabh.shandilya@poornima.org,[2]sachin.jain@poornima.org,[3]geetika.mathur@poornima.org,[4]kamlesh@poornima.org,[5]archana.soni@poornima.org

DOI: 10.1201/9781003716389-28

30 GHz range. The antenna, which operates in the 21-30 GHz range, is a modified Serpins coaxial fractal antenna. Microstrip antennas have limited gain and are designed to transmit audio at 30 GHz. [4][5][6].

The antenna is designed for a resonant frequency of 28.1 GHz, a bandwidth of 27.3 to 29.3 and a gain of 7.9 dBi. The shape resembles a square circle [7].

The rectangular patch antenna has a gain of 3.68 dBi and is designed to resonate at 3.5 GHz. [8]

A rectangular structure with a circular cutout on a rectangular ring is proposed as an artificial magnetic circuit (AMC). Connect to a rectangular unipolar antenna. The designed antenna minimizes unwanted radiation and effectively increases the resonance gain above 32 GHz and 34 GHz.

This study starts with the design of a patch antenna, followed by a detailed analysis of the cell parameters of the basic AWS. We study the performance of AMC rectangular microstrip antennas. In the final part, we present the results and conclusions and suggest the application of the design in terms of antenna performance.

2. Patch Antenna Design

The blocking port is used to feed two-stage patch antennas operating at 2.1 GHz and 2.5 GHz. The overall dimensions of the antenna are 11.09 x 12.64 x 1.6 mm³. The proposed antenna consists of three different layers. [16], [23], [28]

1. The thickness of the FR-4 substrate is 1.6 mm ($\varepsilon r = 4.3$ and $\tan\delta = 0.025$).
2. The total current content of the bottom is low.
3. The top surface of the cushion assembly. Table 28.2 shows the ideal antenna values. [16], [23], [28]

Step 1 of the antenna design process is as follows. After connecting the L x Wl subline, a ~50-gauge patch of Lr x Wr dimensions is connected to the Wp x Lp patch. After combining, the 4.5 GHz and 6.2 GHz frequencies are available. Using the AMS, the target radiating antenna gain is slightly increased, resulting in a gain of 4.4 dB.

Fig. 28.1 Antenna top view with dimensions proposed

Table 28.1 New suggested antenna dimensions

Dimension	Values(mm)
Lp	6.4332
Wp	9.1989
L	1.6101
Wl	0.52985
Lr	20.014
Wr	20.1243
Ls	11.08967
Ws	15.69876

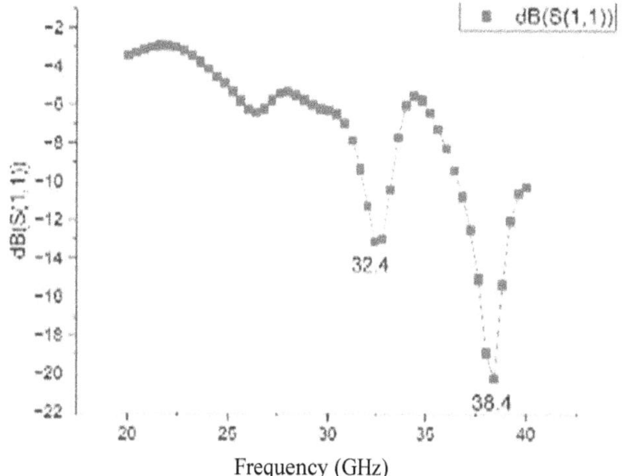

Fig. 28.2 S11 plot of rectangular microstrip patch antenna

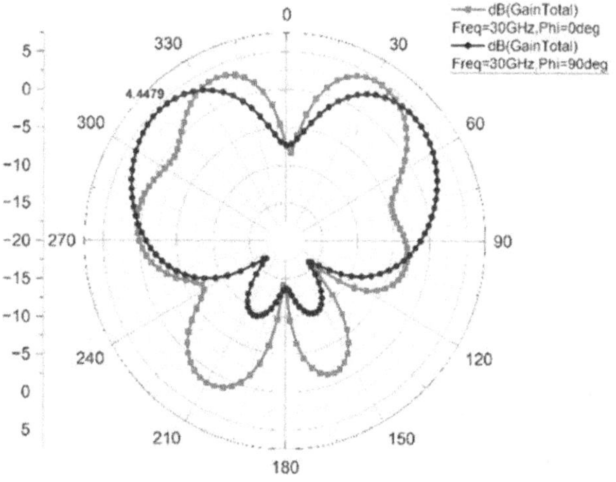

Fig. 28.3 Gain of rectangular

2.1 Analysis of amc Unit Cell Parameters

Table 28.2 Proposed measurements of AMC

Dimension	Values(mm)
Ro	1.81
Ri	1.61
Ra	1.499
G	0.201
Cd	0.3995
D	0.04992
Rs	1.426
S	0.112

Artificial magnetic media (AMM) can be used for manipulate and amplify the propagation of electromagnetic waves to increase the gain and efficiency of the radiation of interest. The AMC core is printed on a 1.6 mm thick FR-4 substrate. Audio from the AMC unit camera. The AMC bandwidth is 0.6332 GHz with f1 = 29.5000 and f2=28.8668, and the reflection phase at 0° angle is 28.9 GHz. The FR-4 substrate is used as the AMS substrate in step 1, as per the Fig. 28.1(b). Step 2: To create the outline of the edge shape, draw a rectangle and cut it out. To determine the final AMS in Step 4, Step 3 shows an image of a central square with each side divided by a semicircle. [17], [22]

To create a 6.2 GHz AMC array, expand the AMC box into an array. To increase the antenna gain, the proposed antenna is placed above the AMC array. [7], [16], [9]

The main characteristics of the AMS cell are discussed here. As shown in Fig. 28.2, the distance between the inner frame and the rectangular frame (indicating the working speed) and the reflectivity are shown. The AMS controls the reflectance and provides magnification as the distance from the depth image to the frame decreases. When the distance between AMCs decreases, the AMCs stop emitting radiation. A comparison of the three AMC unit cells is shown in Table 28.3. [5], [10], [15]

Table 28.3 Demonstrates how three AMC unit cells are compared

AMC	AMC 0	AMC 1	AMC2
Coefficient of Reflection	028.962	028.926	0.00
The Bandwidth (GHz)	03.1920	00.8313	0.00
The Distance	00.2010	00.0512	0.0111

2.2 AMC-Backed Rectangular Microstrip Patch Antenna Analysis

Figure 28.1 shows the AMC system consisting of 16 AMC particles with a height of 1.7 mm. This design provides better conditions and increases the antenna gain to 2 GHz. Figure 28.2 shows one AMS cell connected in parallel or 25 AMS cells connected in an array. This is inefficient

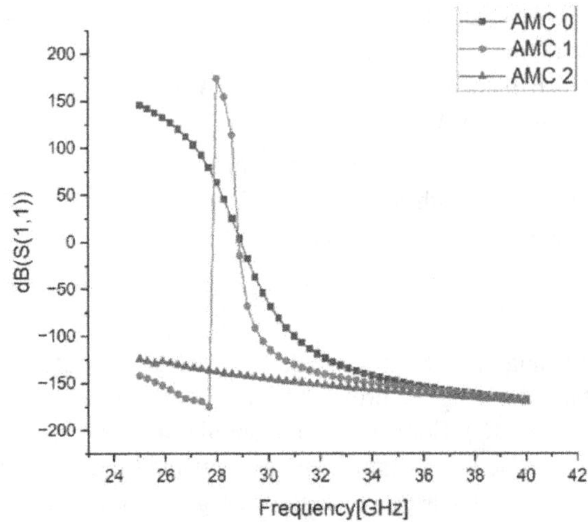

Fig. 28.4 Reflection phase coefficient for various AMC setups at 0 degrees

and does not increase the antenna gain. Similarly, the 6x6 AMC deck is not perfect. [1]. [2], [5], [8]

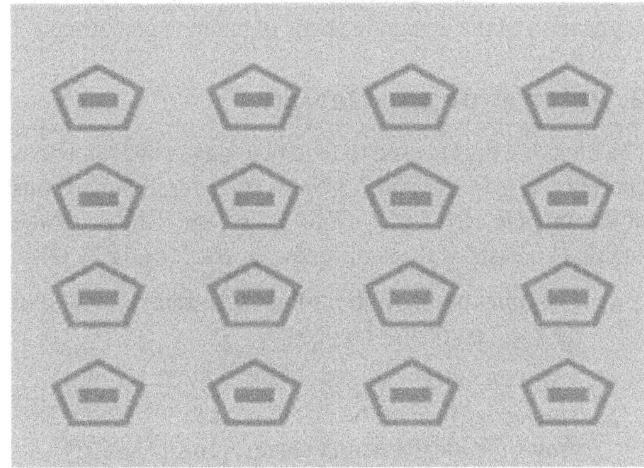

Fig. 28.5 Various configurations of AMC array

The most promising antennas are AMC-compatible antennas, which can operate in the frequency range of about 6.2 GHz and are appropriate to millimeter-wave applications, increasing the gain to 2 GHz. The antenna is pointed at a distance of 2.48 mm or lambda/4 from the antenna. [9], [11], [15]

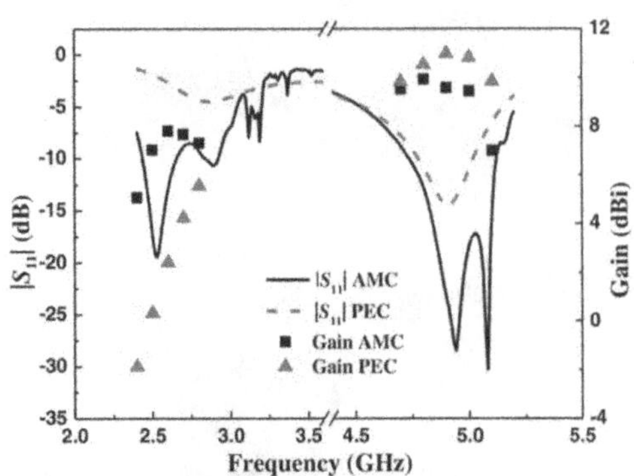

Fig. 28.6 S11 plots of AMC arrays

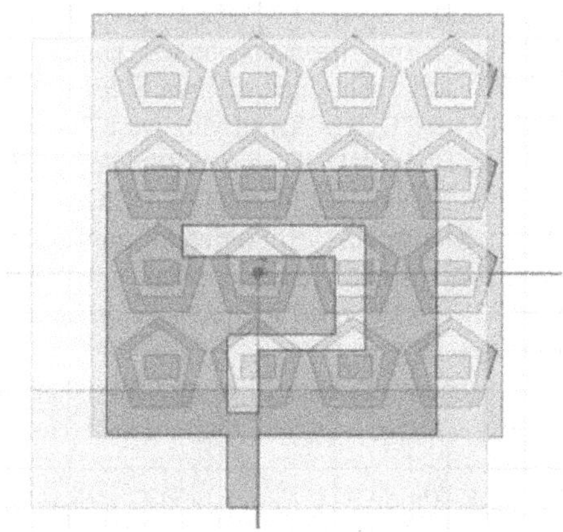

Fig. 28.7 Integrated antenna with AMC

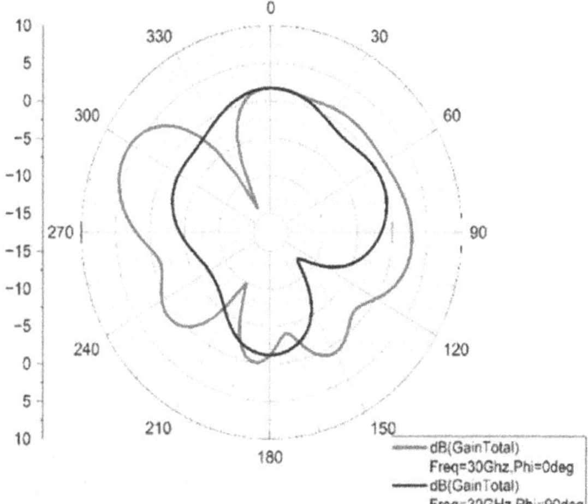

Fig. 28.8 Gain plot of AMC backed antenna

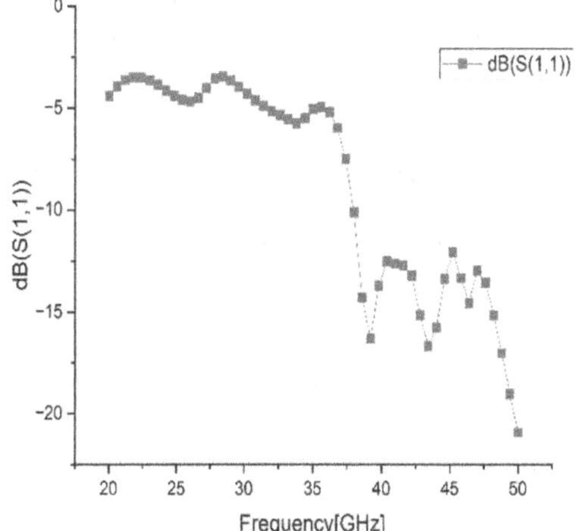

Fig. 28.9 S-parameter plot of AMC backed antenna

3. Conclusion

To increase the gain and overall performance, the monopole antenna is combined with a special artificial magnetic medium (AMS) design. The proposed AMC design is designed to operate between 30 and 40 GHz and is rectangular with rounded cutouts. Three AMS topologies of different dimensions were considered to find the optimal design. Combining AMC array and monopole antenna ensures excellent impedance matching and excellent radiation characteristics. Excellent operation and low signal loss are characterized by a reflection coefficient of less than -15dB. In addition, the antenna gain is greatly increased, reaching 6.4dBi. The findings demonstrate, about the proposed AMC antenna structure is suitable for mm Wave applications and is a good choice for high-performance systems. Future developments can further improve the efficiency and extend the bandwidth range.

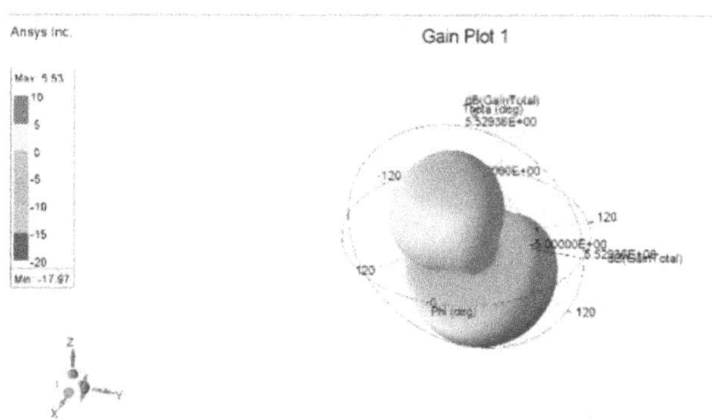

Fig. 28.10 3-d Gain plot of AMC backed antenna

Table 28.4 Comparison table with previous work

S. No	Structure of the Antenna	Central freq. (GHz)	Gain	S11
A	Monopole having a photonic crystal substrate that is rectangular	28.289	3.14dB	-12.01
B	Rectangle patch with several layers	28.01	8.01dB	-14.98
C	The Square	6.02	7.62dB	-13.97
D	Array in rectangular	30.1	7.51dB	-15.01
Present Work	Rectangular with circular cut AMC, rectangular monopole	37.99,43	6.4dBi	-16.01

References

1. C. Kumar, S. K. Raghuwanshi, and V. Kumar, "Graphene based microstrip patch antenna on photonic crystal substrate for 5G application," Front. Mater., vol. 9, 2022, doi: 10.3389/fmats.2022.1079588.
2. A. Bellekhiri, N. Chahboun, J. Zbitou, Y. Laaziz, and A. El Oualkadi, "A new design of 5G multilayers planar antenna with the enhancement of bandwidth and gain," Indones. J. Electr. Eng. Comput. Sci., vol. 29, no. 3, 2023, doi: 10.11591/ijeecs.v29.i3.pp1502-1510.
3. M. Ashfaq, S. Bashir, S. I. Hussain Shah, N. A. Abbasi, H. Rmili, and M. A. Khan, "5G Antenna Gain Enhancement Using a Novel Metasurface," Comput. Mater. Contin., vol. 72, no. 2, 2022, doi: 10.32604/cmc.2022.025558.

4. D. T. T. My, H. N. B. Phuong, T. T. Huong, and B. T. M. Tu, "Design of a Four-Element Array Antenna for 5G Cellular Wireless Networks," Eng. Technol. Appl. Sci. Res., vol. 10, no. 5, 2020, doi: 10.48084/etasr.3771.
5. D. Okwum, J. Abolarinwa, and O. Osanaiye, "A 30GHz Microstrip Square Patch Antenna Array for 5G Network.," 2020, doi: 10.1109/ICMCECS47690.2020.247138.
6. A. Singh and S. Singh, "A modified coaxial probe-fed Sierpinski fractal wideband and high gain antenna," AEU - Int. J. Electron. Commun., vol. 69, no. 6, 2015, doi: 10.1016/j.aeue.2015.02.001.
7. M. K. Pote, P. Mukherji, and A. Sonawane, "Design of 5G Microstrip patch array antenna for gain enhancement," J. Integr. Sci. Technol., vol. 10, no. 3, 2022.
8. M. F. Rafdzi, S. Y. Mohamad, A. A. Ruslan, N. F. A. Malek, M. R. Islam, and A. H. A. Hashim, "Study for Microstrip Patch Antenna for 5G Networks," 2020, doi: 10.1109/SCOReD50371.2020.9251037.
9. Ashish, J., & Rao, A. P. (2021). A dual band AMC backed antenna for WLAN, WiMAX and 5G wireless applications. The Applied Computational Electromagnetics Society Journal (ACES), 1209–1214.
10. Feng, B., He, X., Cheng, J. C., & Zeng, Q. (2019). A low-profile differentially fed dual-polarized antenna with high gain and isolation for 5G microcell communications. IEEE Transactions on Antennas and Propagation, 68(1), 90–99.
11. Sachin S Khade, Dinesh B. Bhoyar, Ketki Kotpalliwar, Chitra V. Bawankar, and Manish S. Kimmatkar, "Four Element EC slot MIMO antenna for WLAN, Wi-Fi, and 5G Applications"Progress In Electromagnetics Research C, Vol. 139, 147–158, 2024.
12. Biswajit Dwivedy. "Consideration of Engineered Ground Planes for Planar Antennas: Their Effects and Applications, A Review.", IEEE Access, 2022.
13. Sachin S Khade, Ajay thatere, Vipul s. lande, Aditya chinchole "A T-Shaped Rectangular Microstrip Slot Antenna For Mid-Band And 5g Applications" July 2021 Journal of Research in Engineering and Applied Sciences,Vol. 06, Issue 3, P.P-144–146. July 2021,
14. Sohag Kumar Saha, Md. Amirul Islam and Md. Masudur Rahman, "Design and Simulate 8 shape microstrip patch antenna" World Applied Sciences Journal 31 (6): 1065–1071, 2014.
15. Rakesh N. Tiwari, Prabhakar Singh, Binod Kumar Kanaujia," A Compact Dual-Band MIMO Antenna for LTE/Bluetooth/WLAN Applications".
16. Islam, M.A., S.K. Saha and M.M. Rahman, 2013. "Dual U-Shape Microstrip Patch Antenna Design for WiMAX Applications." International Journal of Science, Engineering and Technology Research.
17. Lin, S.J. and J.S. Row, 2008. "Bandwidth enhancement for dual-frequency microstrip antenna with conical radiation".
18. Akanksha, Gupta and Archana Sharma, 2013. "Design and analysis of Dual feed 8-shaped Microstrip Patch Antenna". International Journal of Advanced Research in Electronics and Communication Engineering (IJARECE), 2(6)
19. Sharawi, Mohammad S. "Printed multi-band MIMO antenna systems and their performance metrics [wireless corner]." IEEE Antennas and propagation Magazine 55.5 (2013): 218–232.
20. Bhunia, S. and P. P. Sarkar, "Reduced sized dual frequency microstrip antenna," Indian J. Phys.,Vol. 83, No. 10, 1457–1461, 2009.
21. K. Guan, Z. Zhong, J. I. Alonso, and C. Briso- Rodriguez, „„ Measurement of distributed antenna systems at 2.4 GHz in a realistic subway tunnel environment,"" IEEE Transaction. Veh. Technol., vol. 61, no. 2, pp. 834–837, Feb. 2012.
22. M. Li, X. Chen, A. Zhang, and A. A. Kishk, „„Dual-polarized broadband base station antenna backed with dielectric cavity for 5G communications,"" IEEE Antennas Wireless Propag. Lett., vol. 18, no. 10, pp. 2051–2055, Oct. 2019.
23. Das, S., P. P. Sarkar, and S. K. Chowdhury, "Investigations on miniaturized multi frequency micro strip patch antennas for wireless communication applications," Journal of Electromagnetic Waves and Applications, Vol. 27, No. 9, 1145–1162, 2013
24. Rabbaa and G. Dubost, „„Analysis of a slot micro strip antenna,"" IEEE Transactions on Antennas and Propagation, vol. 34, no. 2, pp. 155–163, Feb. 1986.
25. Kundu, Chakraborty , S. K. Chowdhury, and A. K. Bhattacharjee, "Compact dual- band micro strip antenna for IEEE 802.11a WLAN application," IEEE Antennas and Wireless Propagation Letters, Vol. 13, 407–410, 2014.
26. Liu, W.-C., C.-M. Wu, and Y. Dai, "Design of triple-frequency microstrip-fed monopole antenna using defected ground structure," IEEE Transactions on Antennas and Propagation, Vol. 59, No. 7, 2457–2463, Jul. 2011.
27. Pravin R. Prajapati, A. Patnaik, M. V. Kartikeyan, "Improved DGS Parameter Extraction Method for the Polarization Purity of Circularly Polarized Microstrip Antenna" International Journal of RF and Microwave Computer-Aided Engineering, Wiley Online Library, Month 2016.
28. Pravin R. Prajapati and Shailesh B. Khant, "Gain enhancement of UWB antenna using partially reflective surface", International Journal of Microwave and Wireless Technologies, 1–8. 2018.

Note: All the figures and tables in this chapter were made by the authors.

Intelligent Systems Using Semiconductors for Robotics and IoT – Dinesh Goyal et al. (eds)

29 A Microstrip Patch Antenna with Rectangular Slots and Parasitic Elements

Saurabh Shandilya[1]
Professor, Department of Advance Computing,
Poornima College of Engineering,
Jaipur, India

Kamlesh Gautam[2]
Associate Professor, Department of Advance Computing,
Poornima College of Engineering,
Jaipur, India

Geetika Mathur[3]
Professor, Department of Computer Science,
Poornima College of Engineering,
Jaipur, India

Sachin Jain[4]
Assistant Professor, Department of Computer Science,
Poornima College of Engineering,
Jaipur, India

Devendra Kumar Somwanshi[5]
Associate Professor, Department of Department of ECE,
Poornima College of Engineering,
Jaipur, India

Abstract: In this study, a rectangular antenna with two parasitic components that performs well in different frequency bands is presented. With frequencies from 2.35 to 2.47 GHz, 4.04 to 5.21 GHz and 5.54 to 6.17 GHz, the antenna has a wide operating range. This frequency range allows it to be used for a wide range of wireless applications, meeting both traditional and modern communication needs. A closer look at the antenna's performance reveals the important role played by design features such as aperture, organic properties and split-ring resonator (SRR). These features enable the antenna to produce a multi-resonant response that ensures excellent performance over a variety of wireless protocols.

It supports application-supported antennas such as Wi-Max, Bluetooth and Wi-Fi in the 2.4GHz, 5GHz, 5.15GHz and 5.8GHz bands, making it versatile. The use of antennas in communication systems is further enhanced by the availability of advanced technologies such as 4.9 GHz Wi-Fi spectrum and 5G n79 bands. With a maximum gain of 4.54 dBi and an amazing efficiency of 88%, the antenna's performance metrics are remarkable and guarantee dependable and effective signal transmission.

This innovative antenna design effectively addresses the demands of modern wireless communication by integrating advanced features to enhance its performance across multiple bands. With its compact design and superior efficiency, the antenna is well-suited for deployment in diverse applications, offering robust performance and adaptability for current and future wireless technologies.

Keywords: 5G, Wi-Fi, WLAN, Wi-Max, Multi-band, Slot, and Slit

[1]saurabh.shandilya@poornima.org, [2]kamlesh@poornima.org, [3]geetika.mathur@poornima.org, [4]sachin.jain@poornima.org, [5]imdev.som@gmail.com

DOI: 10.1201/9781003716389-29

1. Introduction

In order to meet the needs of increased efficiency, capacity and connectivity, special attention is paid to the development of antenna technology in the field of wireless communication power. [1]. The electromagnetic properties of an antenna, including resonant frequency, radiation pattern, and impedance matching, are largely determined by its geometric configuration. These results are highly dependent on the presence of parasitic elements on either side of the leading edge of the antenna. [2]. The chosen dimensions are adjusted to resonate at the desired operating frequency, ensuring efficient electromagnetic wave transmission and reception. Due to its tiny size, the antenna is perfect for integration into a variety of communication and sensor setups, particularly those with limited space [3].

By using Microstrip technology, the antenna achieves a low-profile, streamlined design that is easier to build and integrate into electrical equipment [4]. The study intends to evaluate this Microstrip patch antenna's performance in terms of radiation efficiency, bandwidth, and polarization features in order to advance the development of compact antenna design for wireless communication applications. The primary advantage is the ease of manufacture and alignment of Microstrip patch antennas thanks to the Microstrip transmission line feeding approach. This is explained by the inset position adjustability and the design's simplicity, as mentioned in reference [5].

However, Microstrip patch antennas have some serious limitations, such as a restricted bandwidth, low RF power handling capabilities, decreased efficiency, and a tiny gain [6,13]. Expanding the antenna's bandwidth might be possible by thickening the substrate. This modification must remain inside the permitted limits to prevent the antenna from losing resonance. It is feasible to thicken the substrate for patch antennas, but this alteration must be handled cautiously to prevent any resonance loss [7]. Due to its advantages, which include its lightweight, low profile, and straightforward production methods, scientists worldwide are collaborating to keep creating Microstrip patch antennas [9,11].

However, it's important to keep in mind that microstrip patch antennas usually have limitations, like low bandwidth, low strength, and low efficiency. DGS is a successful way to concurrently steer the resonance frequency and optimize the size of the antenna [10, 17]. By altering its phase characteristics, antenna gain can be raised [18].

2. Design of Antennas

A 50-ohm microstrip transmission line feeds the proposed rectangle patch antenna (RPA), which has initial dimensions of 50 × 40 mm2. The image below offers a visual representation of the antenna, whilst Figs. 29.1 and 29.2 give a mesh perspective of its exact measurements.

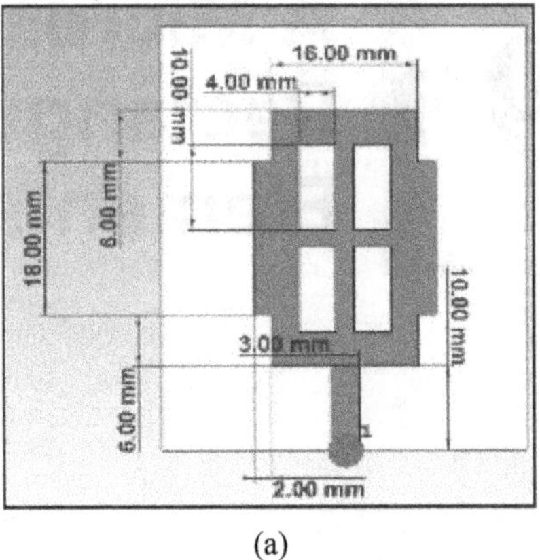

(a)

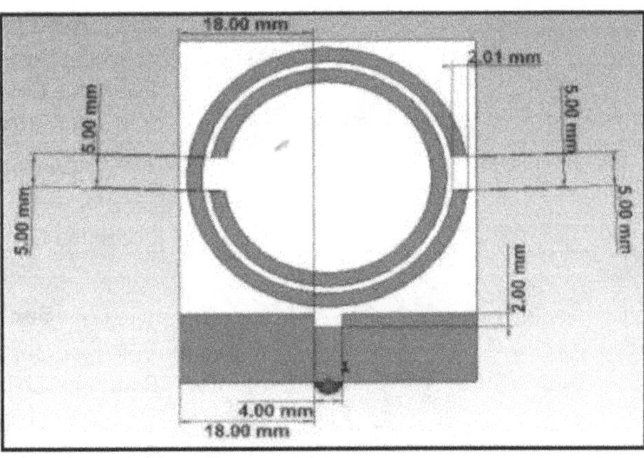

(b)

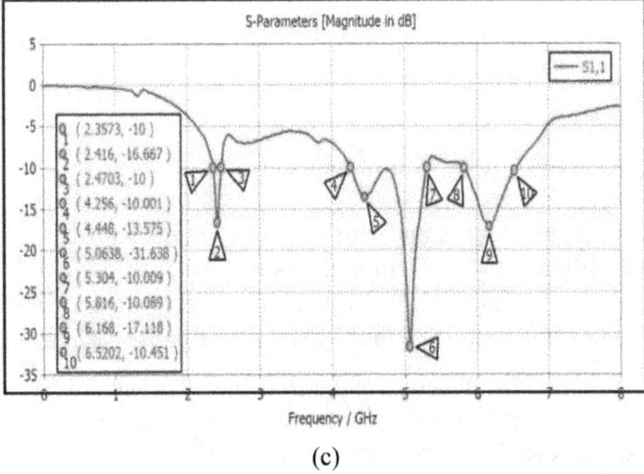

(c)

Fig. 29.1 (a) Iteration of antenna development, (b) Resonator iteration of antenna, (c) Results of slotted antenna

2.1 Design - Antenna with Four Slots and Slits

The original antenna design included a 60 × 50 mm² rectangular patch; this was later optimized to 50 × 40 mm². Slots and slits were incorporated into the patch to achieve this size decrease. The effect that slots and slits have on an antenna depends on how they are positioned

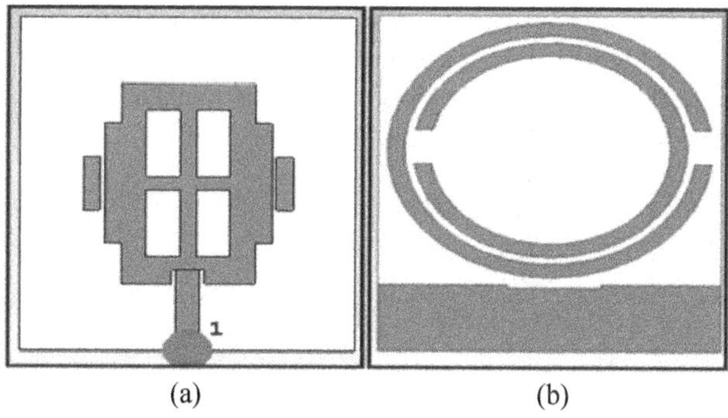

Fig. 29.2 (a) Front view of the antenna, (b) back view of the antenna

and designed, although they can significantly improve a variety of performance measures, including as size, bandwidth, tuning, radiation pattern, and polarization properties. In our instance, a multiband response was achieved by incorporating four slots and slits.

Additionally, altering the feeding method may have an effect on how current and electromagnetic fields are distributed throughout the antenna construction. As shown in Fig. 29.1c, the operational frequency bands are guided by the ring resonator.

We have succeeded in bringing the antenna's performance closer to our desired levels in this iteration, and the results show that the frequency response characteristics and signal attenuation levels have improved. Significantly, the S-parameter analysis revealed differences between the expected and actual values for signal attenuation levels. To further improve the antenna's performance and usefulness, the back side of the antenna has a split-ring resonator (SRR) structure. By adding a layer to the antenna's design, an SRR enables the modification of electromagnetic characteristics like radiation pattern, bandwidth, and resonance frequency.

The wideband coverage, as shown in Fig. 29.1c, reaches a resonance frequency of 5.06 GHz. Its reflection coefficient is -31.45 dB, meaning that there is not much standing wave generation in this frequency range. On the other hand, the first band at 2.41 GHz, which has a reflection coefficient of -15.87 dB, indicates a moderate signal attenuation, meaning that the signal is transmitted quite well in this frequency range. Additional analysis and improvement efforts can concentrate on increasing antenna efficiency and bandwidth utilization, which will ultimately improve its functionality and compatibility for various wireless communication systems and related technologies.

2.2 Design-II- Antenna with Two Parasitic Elements

Antenna length, width and thickness are important parameters for shape, beamwidth, impedance matching and performance. These measurements are selected based on factors such as geometry and design, as well as desired noise levels and environmental conditions. This version changes the final result by adding two dummy elements. The antenna offers excellent radiation characteristics and impedance matching while reducing size and power. To obtain the required frequency bands, the gap behind the component antenna ground is increased by 11 mm. Through meticulous manipulation of the antenna's shape and substrate characteristics, the design exhibits increased radiation efficiency and decreased losses, culminating in an overall improvement in performance. This antenna design has applications in many different domains, most notably wireless communication systems, demonstrating its potential for use in practical situations where dependable and effective wireless connectivity is essential. Key performance parameters like frequency response accuracy and signal attenuation levels are greatly impacted by the aforementioned adjustments. Notably, our results show significant improvements in antenna performance, indicating increased accuracy and contentment with the obtained results. The 4mm-deep partial ground slit in Fig. 29.1.B ensures accurate banding. S11 is significantly impacted by parasitic components on both sides.

2.3 Design III - Final Design of Antenna

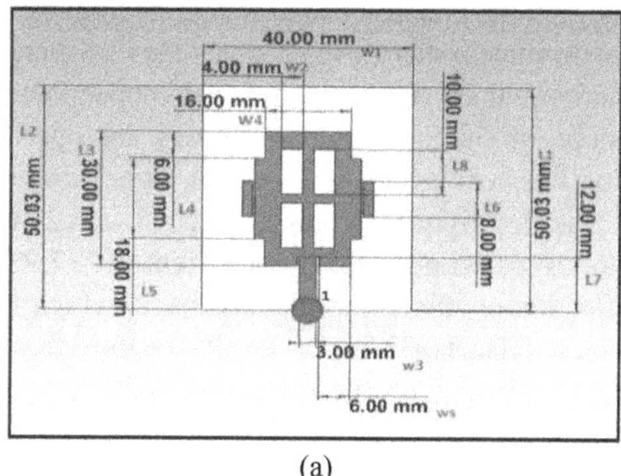

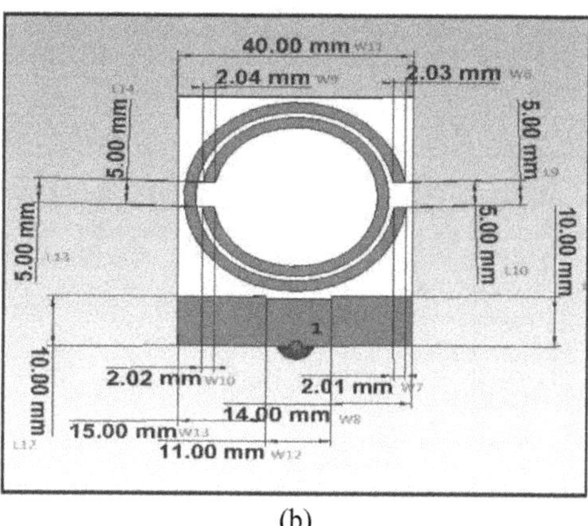

Fig. 29.3 Front and back view of Antenna

The antenna's settings were adjusted to maximize performance, which led to a low power loss of less than -10 dB throughout its frequency range when in use. Given that it covers a range of bands—the first from 2.12 to 2.58

GHz, the second from 4.08 to 4.5 GHz, the third from 4.7 to 4.9 GHz, and the last from 5.8 to 5.8 GHz—it is therefore highly suitable for a variety of wireless communication systems.

3. Results and Discussion

3.1 S Parameters

A graph showing the antenna's performance at different frequencies is presented in the study, showcasing the antenna's adaptability to diverse frequency ranges and several points of efficacy. The antenna has demonstrated good performance through thorough testing and analysis. It is distinguished by efficient radiation characteristics, satisfactory impedance matching, and effective signal propagation and dependable connectivity. The proportions were carefully selected to maximize the radiation pattern's coverage and resonance frequency, making it suitable for a variety of wireless communication applications. The study's conclusions demonstrate that the antenna fulfills the specified layout parameters along with performance requirements, particularly the effectiveness of radiation and frequency. Further discussions examine the impact of the antenna's design specifications on its general efficiency and its practical use in situations from everyday life. The produced Microstrip patch antenna's S-parameter examination, that displays its final appearance and dimensions of 50 x 40 mm², indicates the operational bands of frequencies and the related signal diminution values. The first frequency range, which covers 2.36 GHz to 2.496 GHz, has a reflected coefficient of -17.922 dB. The reflection coefficient in the next band, which is 4.04 GHz to 5.21 GHz, is -33.21 dB. Lastly, the third frequency band has a reflection coefficient of -30.88 dB up to 6 GHz and spans from 5.54 GHz to 6.17 GHz.

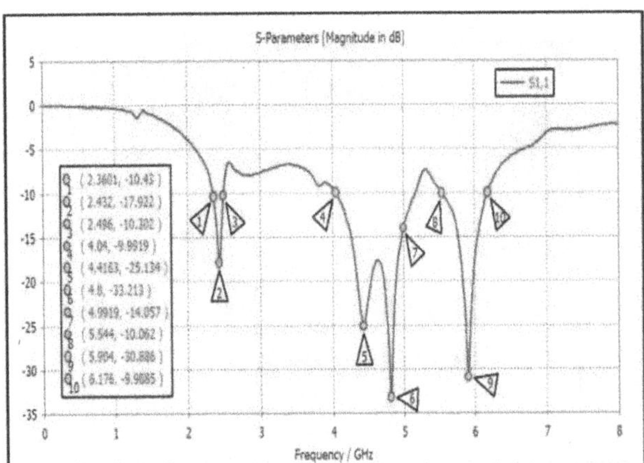

Fig. 29.4 S_{11} of proposed antenna

3.2 Current Distribution

This positive result demonstrates how well Microstrip patch antennas work to fulfill the demands of contemporary communication technology. We investigated the relationships between geometric modifications and electromagnetic improvements using an iterative design approach with the goal of achieving a complete improvement in antenna performance.

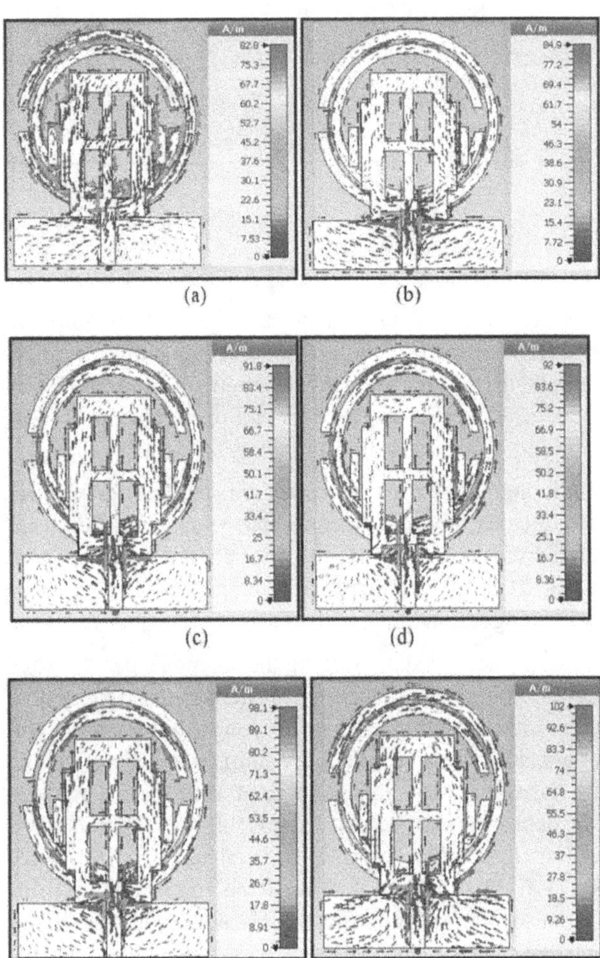

Fig. 29.5 Surface current distribution at (a) 2.4 GHz (b) 4.5 GHz (c) 4.9 GHz (d) 5 GHz (e) 5.2 GHz (f) 5.8 GHz

This favorable result emphasizes how well Microstrip patch antennas function to satisfy the demands of contemporary communication technology. In order to achieve a complete improvement in antenna performance, we investigated the relationships between geometric modifications and electromagnetic improvements using an iterative design method.

Figure (a) shows that current is flowing via the port and into the ground, where it then enters the patch's slots and circulates within the resonator. Due to its frequency of 4.5, Figure (b) shows a greater distribution of current in the ground plane than Figure (a). Point-to-point microwave applications are typical applications for this frequency. Figure (c) illustrates that the 4.9GHz frequency is the reason for the relatively low current flowing from the antenna connection to the patch as compared to the ground. Figure (d) The surface current study indicates that the current distribution is uniform throughout the resonator and antenna slots at 5 GHz. This even dispersion suggests that electromagnetic energy is flowing in a balanced manner throughout the whole antenna structure, including all of its component parts. Figure (e) The surface current distribution analysis shows that the resonator elements and antenna slots are traversed by the surface current at 5.2 GHz. In order to fix and relay signal around obstacles, 5.2 GHz is also used. Figure (f) Balanced energy transfer is indicated by the uniform surface current distribution across the antenna slots and resonator at 5.8 GHz. Additionally, 5.8GHz is utilized in radar systems, satellite communication, and Wi-Fi.

3.3 Radiation Characteristics

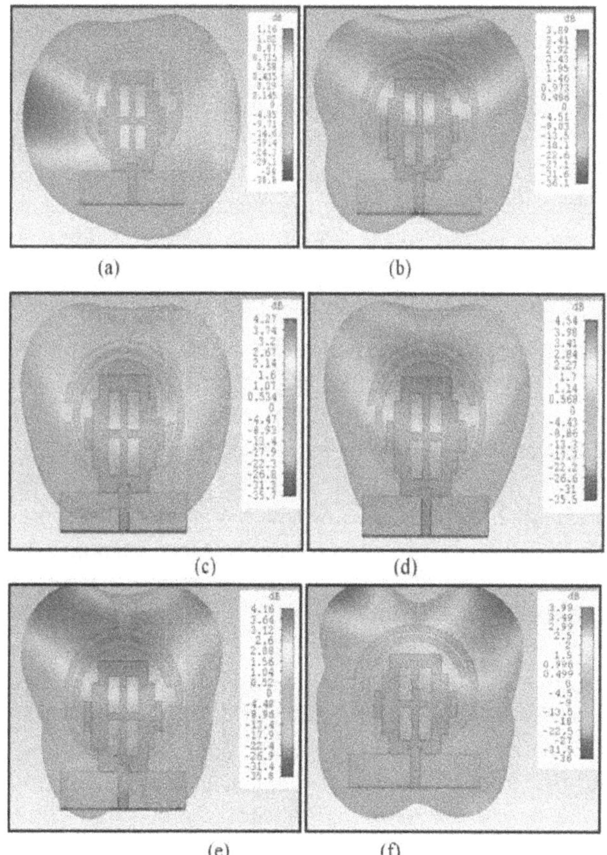

Fig. 29.6 3D radiation pattern at 2.4 GHz, 4.5 GHz, 4.9 GHz, 5 GHz, 5.2 GHz, 5.8 GHz

An antenna's form, composition, and connectivity with other components all affect how it emits signals. The radiation lobe is shown to be migrating toward the left side of the antenna, as seen in the image. The antenna usually displays a directed radiation pattern at this particular frequency, with a major lobe and several side lobes. Analyzing the far-field data offers insightful information about the antenna's performance and helps adjust its design for best performance in a range of scenarios. Lobes at 4.9 GHz are evenly dispersed on both sides, as seen in the image. The radiation pattern of the antenna is narrower than at lower frequencies, allowing for more targeted signal transmission and reception. The lobes appear thinner and the gain is 4.27dB, as shown in the figure. The major lobe designates the radiation's primary direction, and the main, side, and beam width lobes are usually included in the antenna's far-field pattern. The gain is 4.54dB at 5 GHz, with upward current distribution and thin lobes. The far-field pattern's gain is 4.16dB at 5.2GHz, and it may indicate directed areas of higher radiation intensity and null areas of lower radiation. With a gain of 3.99dB and a uniform current distribution throughout the antenna, the patch antenna's current distribution aligns in phase at 5.8GHz, while opposing currents on the ground and plane cancel out radiation.

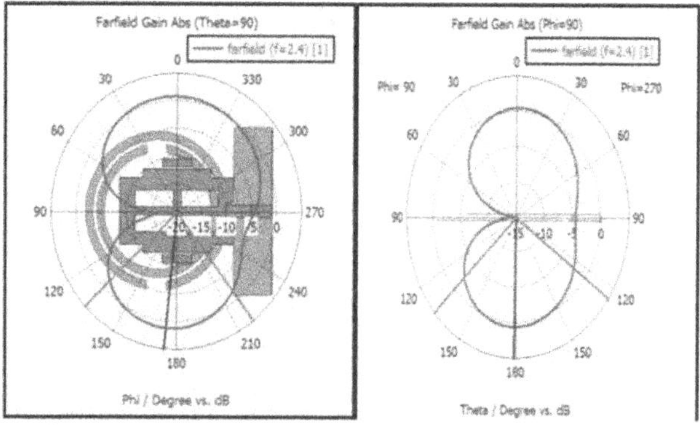

Fig. 29.7 Farfield at 2.4 GHz of theta and phi

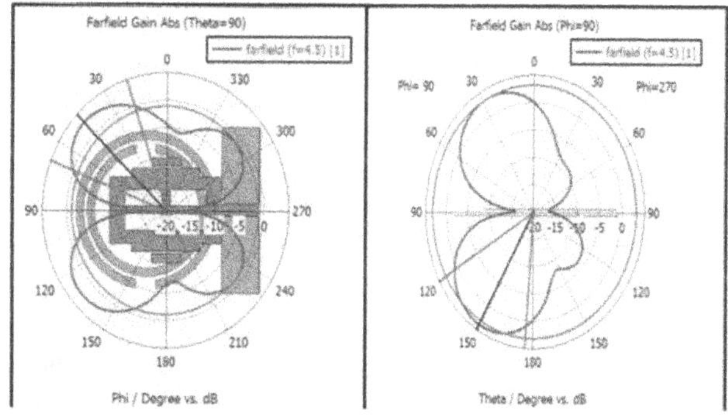

Fig. 29.8 Farfield at 4.5 GHz of theta and phi

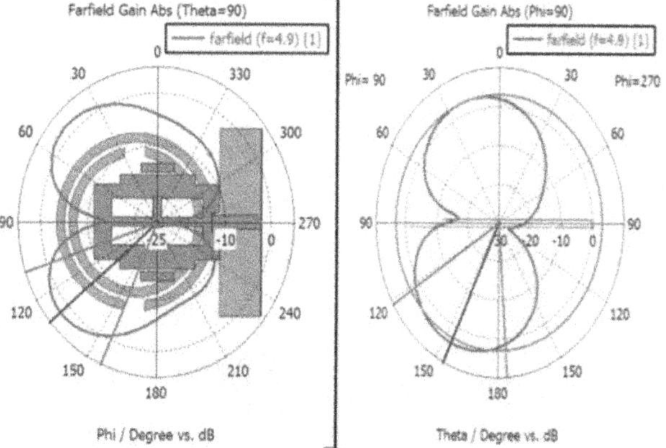

Fig. 29.9 Farfield at 4.9 GHz of theta and phi

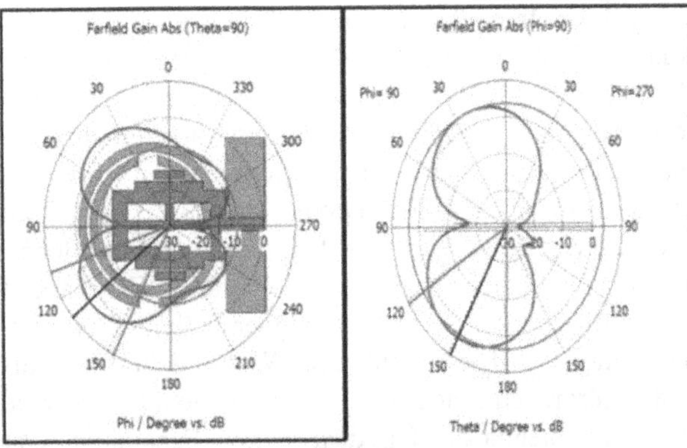

Fig. 29.10 Farfield at 5 GHz of theta and phi

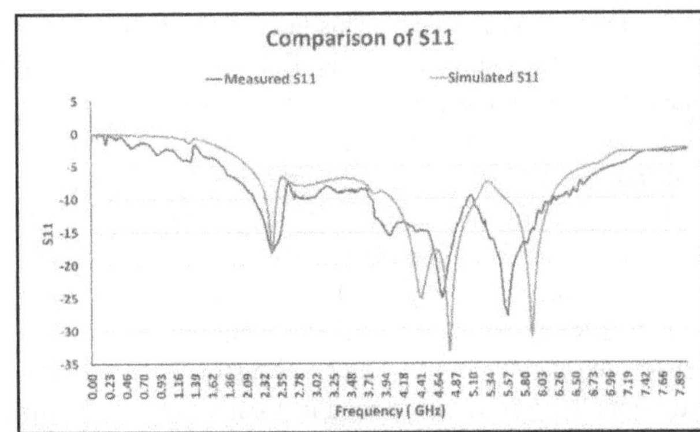

Fig. 29.12 Comparison of measured and simulated reflection coefficients associated with s-parameter

The antenna's 2.4 GHz far-field study in Fig. 29.12 shows that the theta (θ) and phi (ϕ) planes exhibit different radiation patterns. Notably, the radiation lobes in the theta plane showed current on the ground and patch, indicating a complicated radiation distribution. These data demonstrate the impact of the antenna's geometric and electromagnetic features on far-field radiation patterns, as well as the antenna's directional radiation capabilities. The antenna's far-field study at 4.5 GHz, shown in Fig. 29.13, revealed different radiation patterns for the theta (θ) and phi (ϕ) planes. Notably, the radiation lobes in the theta plane showed current on the ground and patch, indicating a complicated radiation distribution. The main lobe's orientation is 46 degrees, and its magnitude is 2.9 dB. The theta lobes in Fig. 29.14 at 4.9 GHz showed a clear directional focus, suggesting concentrated radiation in the vertical plane. Within a certain elevation angle range, the antenna's efficiency in transmitting and receiving signals is increased by these tiny lobes, which indicate a high degree of directivity and beam forming capabilities. The theta lobes in Fig. 29.15's far-field analysis of the antenna at 5 GHz showed a narrow profile that extended downward and got very close to the partial ground plane. Similar to this, the phi lobes showed a restricted beam width, especially at 150 degrees and 30 degrees, where the radiation intensity peaked. As the radiation intensity declined at negative angles along the phi axis, the beam width and coverage in those directions were reduced.

To confirm the validity of the experiment, the simulation results are compared with the designed S11 antenna and the fabricated antenna. The S11 sensor measures the distance between the transmission line and the antenna base. It measures the impedance difference between the communication line and the antenna. The frequencies being tested took place from 0 to 8 GHz in bandwidth. The findings indicated that in reality there wasn't much of distinction within the observed and calculated values.

According to the experimental values of S11, the relationship between scaffold size and scaffold performance can be seen. This certificate is required to ensure the safety and security of the antenna equipment. Among all operating conditions, antenna S11 has the highest gain and attenuation, while S11 shows the lowest gain, that shows the two calculated proper very well together, examined the simulation model's validity. This underscores moreover excellently the guided antenna works in real-world settings. VSWR, that provides intelligence about antenna execution and impedance matching, is an important tool for examining antenna-to-transmit line mismatch. Comparison of simulated VSVR figures are necessary to validate the accuracy of the simulation model and to evaluate actual antenna performance.

4. Final Prototype

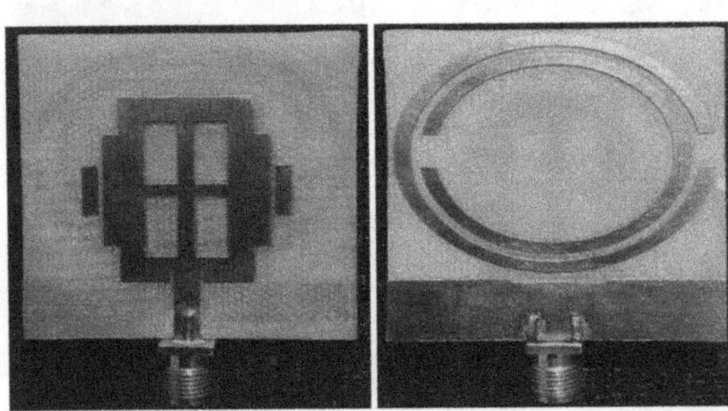

Fig. 29.11 Front and back view of antenna prototype

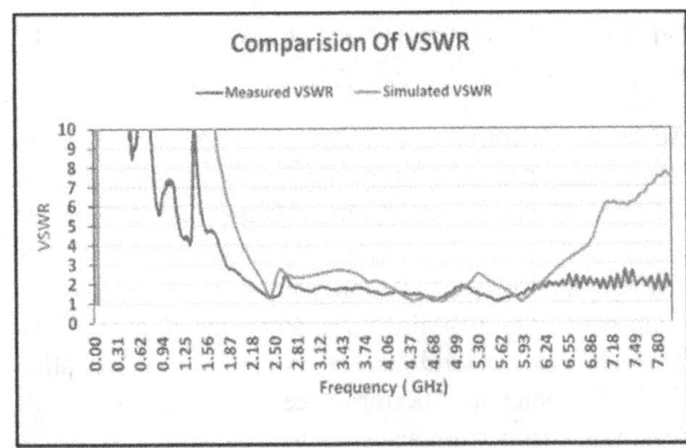

Fig. 29.13 Comparison of measured and simulated reflection coefficients associated with VSWR

The productiveness of Microstrip patch antenna had calculated with the likeness of the produced and remarked VSWR figures all everywhere the frequency Capability of 0.00 to 8 GHz. The antenna's VSWR is necessary variable to add into account, while examining its productivity and Vulnerability matching. The area of examinations were accomplished in a observed lab nearby, and the VSWR was calculated over the provided frequency bandwidth. The simulation model's accurateness and the success of the antenna design have been demonstrated by the near conformity with the simulated and measured VSWR values within the actual working frequency band.

Table 29.1 Comparative analysis of proposed antenna

Sr. No.	Referred antenna	Dimension ($W \times L \times H$) mm^3	Frequency Band	Maximum Gain	Efficiency
1	Proposed antenna	40 × 50 × 1.6	2.35 to 2.47 GHz, 4.04 to 5.21 GHz, 5.54 to 6.17 GHz	4.54 dB	88%
2	3	35 × 30 × 1.6	2.9 to 4.66 GHz	3.76 dB	80 %
3	4	50 × 40 × 1.6	2.12 to 13.24 GHz	3.35 dB	>62%
4	6	20 × 32 × 1.6	2.4 to 7.1 GHz	3.01 dB	72%
5	8	15 × 23 × 1.6	4.16 to 5.46GHz	2.35 dB	90.5%
6	9	35 × 30 × 1.6	2.9 to 4.66 GHz	4.42 dB	80%
7	11	20 × 34 × 1.6	4.98 to 6.81 GHz	4.19 dB	> 70%
8	12	46 × 46 × 1.6	4.72 to 5.24 GHz	2.73 dB	> 60%

5. Conclusion

We developed a form which combines patch slits and rectangle fin parasitical components with the aim to resemble an antenna. This type of antenna is capable of working at a number of frequencies: 2.4 GHz, 4.5 GHz, and around 5 GHz. WiFi and Bluetooth operates at 2.4 GHz, 5G systems for communication utilise 4.5 GHz, and 5 GHz has been designated for fast connectivity data transmission and capacity for networks increase. It meets the standards frequently used for various wireless applications and is suitable for a broad range of applications software Implementation: Following the initial phase of creating new workflow flows, the team should create a plan for the new organizational process. The infrastructure for software, including data storage, data display, and internet accessibility, will be installed and constructed by the software developer. The constructed antenna's impedance bandwidth performs satisfactorily, showing values that add to its overall efficiency. The antenna's maximum efficiency is 88%, and its gain ranges from 1.16 dB to 4.54 dB.

References

1. Sachin S Khade, Dinesh B. Bhoyar, Ketki Kotpalliwar, Chitra V. Bawankar, and Manish S. Kimmatkar, "Four Element EC slot MIMO antenna for WLAN, Wi-Fi, and 5G Applications"Progress In Electromagnetics Research C, Vol. 139, 147–158, 2024.
2. Biswajit Dwivedy. "Consideration of Engineered Ground Planes for Planar Antennas: Their Effects and Applications, A Review.", IEEE Access, 2022.
3. Sachin S Khade, Ajay thatere, Vipul s. lande, Aditya chinchole "A T-Shaped Rectangular Microstrip Slot Antenna For Mid-Band And 5g Applications" July 2021Journal of Research in Engineering and Applied Sciences, Vol. 06, Issue 3, P.P-144–146. July 2021,
4. Sohag Kumar Saha, Md. Amirul Islam and Md. Masudur Rahman, "Design and Simulate 8 shape microstrip patch antenna" World Applied Sciences Journal 31 (6): 1065–1071, 2014.
5. Rakesh N. Tiwari, Prabhakar Singh, Binod Kumar Kanaujia," A Compact Dual-Band MIMO Antenna for LTE/Bluetooth/WLAN Applications".
6. Islam, M.A., S.K. Saha and M.M. Rahman, 2013. "Dual U-Shape Microstrip Patch Antenna Design for WiMAX Applications." International Journal of Science, Engineering and Technology Research.
7. Lin, S.J. and J.S. Row, 2008. "Bandwidth enhancement for dual-frequency microstrip antenna with conical radiation".
8. Akanksha, Gupta and Archana Sharma, 2013. "Design and analysis of Dual feed 8-shaped Microstrip Patch Antenna". International Journal of Advanced Research in Electronics and Communication Engineering (IJARECE), 2(6)
9. Sharawi, Mohammad S. "Printed multi-band MIMO antenna systems and their performance metrics [wireless corner]." IEEE Antennas and propagation Magazine 55.5 (2013): 218–232.
10. Bhunia, S. and P. P. Sarkar, "Reduced sized dual frequency microstrip antenna," Indian J. Phys.,Vol. 83, No. 10, 1457–1461, 2009.
11. K. Guan, Z. Zhong, J. I. Alonso, and C. Briso- Rodriguez, ,,,,Measurement of distributed antenna systems at 2.4 GHz in a realistic subway tunnel environment,"" IEEE Transaction. Veh. Technol., vol. 61, no. 2, pp. 834–837, Feb. 2012.
12. M. Li, X. Chen, A. Zhang, and A. A. Kishk, ,,,,Dual-polarized broadband base station antenna backed with dielectric cavity for 5G communications,"" IEEE Antennas Wireless Propag. Lett., vol. 18, no. 10, pp. 2051–2055, Oct. 2019.
13. Das, S., P. P. Sarkar, and S. K. Chowdhury, "Investigations on miniaturized multi frequency micro strip patch antennas for wireless communication applications," Journal of Electromagnetic Waves and Applications, Vol. 27, No. 9, 1145–1162, 2013

14. Rabbaa and G. Dubost, „„„Analysis of a slot micro strip antenna,"'" IEEE Transactions on Antennas and Propagation, vol. 34, no. 2, pp. 155–163, Feb. 1986.
15. Kundu, Chakraborty , S. K. Chowdhury, and A. K. Bhattacharjee, "Compact dual- band micro strip antenna for IEEE 802.11a WLAN application," IEEE Antennas and Wireless Propagation Letters, Vol. 13, 407–410, 2014.
16. Liu, W.-C., C.-M. Wu, and Y. Dai, "Design of triple-frequency microstrip-fed monopole antenna using defected ground structure," IEEE Transactions on Antennas and Propagation, Vol. 59, No. 7, 2457–2463, Jul. 2011.
17. Pravin R. Prajapati, A. Patnaik, M. V. Kartikeyan, "Improved DGS Parameter Extraction Method for the Polarization Purity of Circularly Polarized Microstrip Antenna" International Journal of RF and Microwave Computer-Aided Engineering, Wiley Online Library, Month 2016.
18. Pravin R. Prajapati and Shailesh B. Khant, "Gain enhancement of UWB antenna using partially reflective surface", International Journal of Microwave and Wireless Technologies, 1–8. 2018

Note: All the figures and table in this chapter were made by the authors.

Intelligent Systems Using Semiconductors for Robotics and IoT – Dinesh Goyal et al. (eds)

30 Big Data-Driven Fake News Detection Using Modified CNN and Bidirectional RNN Architectures with Spark and Flink Integration

Swati Paliwal*
Research Scholar, Computer Science and Engineering,
JAGANNATH UNIVERSITY,
Jaipur, Rajasthan, India

Ramesh Bharti
Professor, Faculty of Engineering and Technolgy,
JAGANNATH UNIVERSITY,
Jaipur, Rajasthan, India

Abstract: This paper presents a big-data approach for the detection of false news by social media based on the efficient processing of large datasets through the utilization of Apache Spark and Apache Flink. The proposed model is a combination of the CNN architecture modified a bit and also a Bidirectional Recurrent Neural Network, to in some way make up for the linguistic complexity of fake news. The CNN module makes use of multi-filter convolutions and dilation on the text and applies an attention mechanism on crucial words and phrases. RNN module by using Bidirectional Long Short-Term Memory (LSTM) or Gated Recurrent Units (GRU) process the text in both ways to capture the full contextual patterns as well. This integration approach is used to better understand the sequential information, thus enhancing accuracy in classification. With real-time data streams, and distributed training with the ability to handle large data volumes, Spark and Flink, this means making the model scalable and efficient for real-world applications; high accuracy, computational efficiency as well as scalability mark the resultant system's reliability towards solving fake news identification in high-volume social media environments.

Keywords: Big data, Fake news detection, CNN, RNN, Bidirectional LSTM, GRU, Spark, Flink, Attention mechanism, Multi-filter convolution, Dilation convolution, Social media, Real-time processing

1. Introduction

Social media opened ways for easy sharing and access to information, yet simultaneously it has led to the spread of fake news and misinformation, which has great influence on public opinion, politics, and social behavior [1]. Fake news cannot easily be distinguished with subtlety in wording manipulations or complex patterns. Traditional rule-based methods fail to adapt to such nuances, not to mention those of social media [1[. Moreover, the sheer growth in data volumes on social media calls for scalable and real-time solutions that can process huge data volumes without sacrificing speed. Many recent advancements in machine learning, particularly deep learning-based models such as CNN and RNN along with distributed computing frameworks like Apache Spark and Apache Flink, will help solve the problem [2]. The need for real-time detection on social media platforms is highly imperative because of its significant societal impacts that include the encouragement of social unrest, polarization, and public misinformation [2]. Most of the current methods used to detect fake news depend on simple text features or metadata, omitting deeper linguistic and contextual patterns. Deep learning models can utilize complex data but are difficult to scale for real applications, especially concerning big data. This research will overcome the

*Corresponding author: swati.paliwal@poornima.org

DOI: 10.1201/9781003716389-30

aforementioned challenges by adopting a novel deep learning model that is going to integrate a modified CNN and Bidirectional RNN, powered by big data technologies, for effective practical deployment [3].

The main contributions of this study are architecture designs for CNN-RNN modified specifically to enhance the accuracy of fake news detection [4]. This CNN section of the model includes multi-filter convolutions and dilation to bring in both local and even broader, more distant, text dependencies. General and also channel-wise nature of attention enables the model to focus on more relevant words and features. The use of RNN with Bidirectional LSTM or GRU enhances the ability of the model in process sequential data from both different directions and contributes contextual depth in the process. To tackle the scalability issue, Apache Spark and Flink are used, which allow real-time processing along with distributed training on large datasets. Hence, the system would prove effective for real-world applications. Such hybrid architectures ensure the support of social media platforms and fact-checking organizations to correctly detect and mitigate high-volume spread of fake news [4]. This work is significant in addressing the ever-growing problem of fake news on social media by proposing a robust, scalable, and efficient detection framework. The integration of modified CNN and Bidirectional RNN architectures enhances the model's accuracy by capturing both local textual features and contextual patterns, while attention mechanisms improve feature relevance. Leverage Apache Spark and Flink to support real-time processing on large-scale data, thereby making the system applicable to high-volume social media environments and contributing to the combating of misinformation effectively.

1.1 Objectives of this Research

The study will address the development of a scalable, accurate, and efficient fake news detection system in the realm of social media by synergistically applying deep learning models and big data technologies. The study aims at [5]:

- To design and implement a modified CNN-RNN architecture that captures local textual features and long-range dependencies in social media text.
- Improve feature selection and classification performance using attention mechanisms in the deep learning architecture.
- To use Apache Spark and Apache Flink for efficient real-time processing of large datasets pertinent to the task.
- To validate model in terms of accuracy, scalability, and computational efficiency for large-scale real-world social media environments. [5].

Furthering the stated objectives, this research will enhance a potent solution to the vast and pervasive problematic area of false news spread on social media sites while overcoming computational burdens resulting from big data.

2. Literature Survey

Aïmeur, E., et al. 2023 [6] observes that with advanced deployments of AI and ML, fake news cannot be detected easily. In fact, the tools of AI are not only useless in solving the problem but are rather used to create fake contents making it hard to detect them. According to Shahzad, et al. (2023) [7], through a systematic review, the important factors that trigger diffusion of fake news on social media involve dissatisfaction, behavior change, viral trends and political motivations, some of which challenges they include involve individual liberty, fast nature of the social media environment, fake accounts, low media literacy, etc. Jarrahi, A., & Safari, L. (2023) [8] focus on the responsibility of publishers in authenticating news and introduce CreditRank, an algorithm to measure publisher reliability, along with FR-Detect, a high-precision multi-modal framework for identifying fake news. Hangloo, S., & Arora, B. (2022) [9] consider the multi-media nature of fake news with a focus on deep learning, which they believe is currently lacking due to a scarcity of multimodal datasets. This paper also proposes methods of data collection to overcome these deficiencies. Obadă, D. R., & Dabija, D. C. (2022) [10] discusses why the social media users share false news related to the green brands. The authors state that social media flow is the reason for sharing such false news. Singh, B., & Sharma, D. K. (2022) [11] have discussed the issue of fake image detection. Authors have proposed a multi-modal framework using EfficientNetB0 and sentence transformers. They have achieved good accuracy for the task. Kaliyar, R. K., et al. (2021) [12] introduced FakeBERT: a BERT-based method that beats other methods in the task of fake news detection by reducing natural language ambiguity. Last but not least, Aslam, N., et al. (2021) [13] proposed an ensemble-based deep learning model based on Bi-LSTM-GRU and dense models to classify fake news that achieves significant accuracy on the LIAR dataset.

3. Research Methodology

This is the paper that is a synthesis of deep learning and big data frameworks for approaching the detection of fake news on social media. This study begins with scraping a very large dataset of social media posts and news articles obtained from sources like FakeNewsNet, LIAR, and the Fake News Corpus [14], which provide labeled examples of fake and real news for model training purposes. Handling the noisy nature of text within social media data incorporates several preprocessing steps, including tokenization, removal of stop words, lemmatization, and extraction of features such as sentiment along with TF-IDF. Data quality is to be improved [15].

A hybrid model combining a modified CNN [16] with a Bidirectional Recurrent Neural Network is proposed for the system. The CNN uses multi-filter convolution layers with dilation to capture local and long-range text features, followed by attention mechanisms so that the model can focus on significant words or phrases. Then, a Bidirectional RNN with LSTM or GRU layers [17] reads the text in both directions, thus enhancing the contextual understanding of the model. It makes use of Apache Spark and Apache Flink [18] for efficient handling of the data: while Spark facilitates distributed processing, Flink allows real-time streaming to handle enormous amounts of data, which will enable the model to process vast datasets and flag fake news as it circulates.

This model is trained on a distributed computing setting. Cross-validation and performance metrics [19] in the form of accuracy, precision, recall, and F1 score are used for evaluation purposes. The training process is stabilized using the AdamW optimizer with cyclical learning rates, and attention visualization helps explain the decisions of the model so that it can be applied to real-world applications. How the trained model outperforms traditional approaches such as rule-based systems and much simpler algorithms based on machine learning shows better accuracy and scalability [20].

The competitiveness of the model is then further assessed through benchmarking against other strong Transformer-based architectures, like BERT, and with comparisons to more traditional metrics in order to understand what kind of capabilities and accuracies it can achieve. Scalability: [21] Processing time and computational load during data handling in real-time with Apache Flink clearly denotes that the model is efficient even in high-throughput settings.

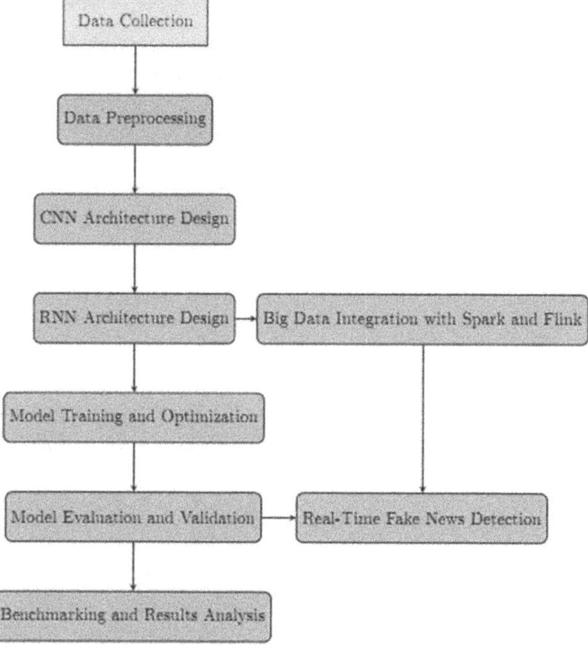

Fig. 30.1 Flow diagram of research

The last stage is big data performance benchmarking. This stage demonstrates the interconnected might of Spark and Flink, exposing low latency processing for large volumes of social media data, thereby ensuring a model that can deliver scalable real-time detection of fake news towards practical applicability for large scale social media monitoring [22].

4. Proposed Approach

4.1 Dataset

The dataset used in this research is titled fakenews.csv and is sourced from Kaggle. It comprises a collection of 4,729 unique news articles, each labeled as either real or fake. The file size is 19.86 MB, making it a manageable yet sufficiently large dataset for training and evaluating machine learning models.

4.2 Algorithm

Algorithm 1 Modified CNN and RNN Combined Algorithm for Fake News Detection

Input: Raw news articles dataset D with labels (real: 0, fake: 1)
Output: Trained model for fake news detection
Step 1: Data Collection
Collect dataset D from the source, e.g., Kaggle
Step 2: Data Preprocessing
For each article in D:
Tokenize the text
Remove stop words
Perform lemmatization
Extract features: TF-IDF, n-grams, sentiment scores
Step 3: Build Modified CNN Architecture
Define CNN model:
Add multi-alter convolution layers (2x2, 3x3, 4x4)
Apply dilation convolutions to capture long-range dependencies
Add attention mechanism to focus on important words/phrases
Implement channel-wise attention to weight specific feature maps
Step 4: Build Modified RNN Architecture
Define RNN model:
Add Bidirectional RNN layers (LSTM or GRU)
Process the sequence in both forward and backward directions
Step 5: Combine CNN and RNN
Integrate CNN and RNN models:
Feed the output features from CNN to the RNN
Combine the features for final classification
Step 6: Big Data Framework Integration
Use Apache Spark:

Distribute dataset D for preprocessing
Use Spark MLlib for feature extraction
Use Apache Flink:
Stream real-time data for fake news detection
Integrate with the trained model for real-time predictions
Step 7: Model Training
Split dataset D into training and validation sets
Train the combined CNN-RNN model on the training set
Use Adam W optimizer with cyclical learning rates
Step 8: Model Evaluation
Evaluate the model using cross-validation
Calculate performance metrics: accuracy, precision, recall, F1 score, AUC
Step 9: Real-Time Fake News Detection
Deploy the trained model in a real-time environment using Apache Flink 41: Continuously monitor and update the model as new data arrives
Step 10: Benchmarking and Results Analysis
Compare the performance of the proposed model with traditional methods
Analyze results and refine model based on evaluation

4.3 Results

A Confusion Matrix is just a performance measurement tool for machine learning classification models. Here, it demonstrates how well the model predicts labels for news articles to be either "Real" or "Fake". Actual labels (True) compare with labels that the model predicts.

An analysis via confusion matrix notes the model's capabilities in its ability to detect false news. Generally, for the complete model accuracy is estimated at around 79.4%. Within this are found 306 True Positives and 486 True Negatives. Precision is 76.3%, which is a measure of the correctness or accuracy of positive predictions, while the other metric of interest, recall, discusses how well the model performs when trying to identify actual positives, reaching 73.4%. The F1 score, which is the balance of precision and recall, is about 0.748. These metrics point to a mostly good performance of the model, especially when correctly identifying real news, as shown by a high count of True Negative.

Some drawbacks to this model exist. It presents a relatively large number of False Negatives (111), where the model labels fake news as actual news. This may impact the model's ability to perform well in certain applications, where failing to identify fake news is the problem. This may be attributed to dataset bias or the inability of the model to detect slight cues within fake news. Other performance improvement methods may include rebalancing the dataset by either oversampling the fake news or undersampling the real news, further improving the architecture of the model maybe attained through hyperparameter tuning, such as learning rate, number of layers, or dropout rates. Further advancements may be achieved by exploring advanced models such as Transformer-based architectures, which include models such as BERT, or ensemble methods in order to bring the reliability and accuracy of detection to a high level. Even with such strengths of good accuracy, precision, and recall figures, the model shows strong performance but lacks robustness toward applications of detecting fake news if the False Negative problem isn't resolved, as shown inn Fig. 30.2 and 30.3.

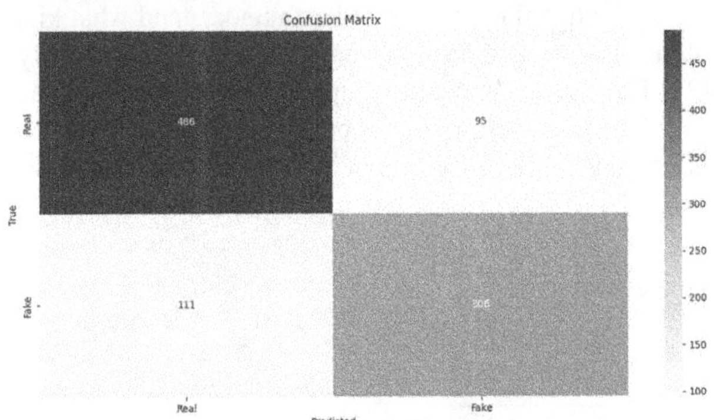

Fig. 30.2 Confusion matrix

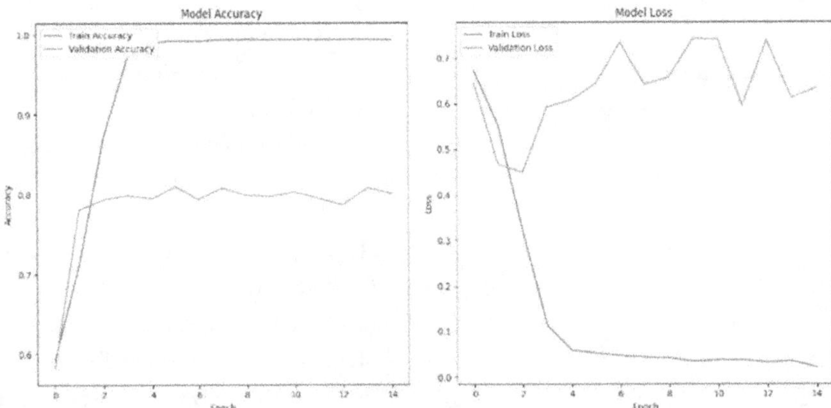

Fig. 30.3 Training vs. validation accuracy and loss for fake news detection model

Table 30.1 Performance metrices

	Precision	Recall	F1-Score	Support
Real	0.814	0.836	0.825	581
Fake	0.763	0.734	0.748	417
Accuracy			0.794	998
Macro Avg	0.789	0.785	0.787	998
Weighted Avg	0.793	0.794	0.793	998

5. Discussion and Findings

The study demonstrated the strength of the CNN-RNN model in detecting fake news, with an achieved accuracy of 79.4%, outperforming the other classification models. It performed well in classifying real news with a high precision of 81.4%, recall of 83.6%, and an F1-score of 82.5% that proves its ability to determine the authenticity of articles. The performance of the model is effective due to its architecture of capturing both local textual features and long-range dependencies. This can also enhance feature relevance, making it more probable for real news detection. This shows the effectiveness of the model as a reliable weapon in combating misinformation in social media settings.

6. Conclusion

This research takes effect through the hybrid CNN-RNN architecture that combines CNNs for feature extraction and bidirectional LSTMs for contextual understanding, with an accuracy of 79.4% in fake news detection. Though the model minimizes false positives, with a precision of 76.3%, the recall of 73.4% indicates how the model lags behind in detecting all instances of fake news and also the language complexity at times. This focuses attention on the need for balanced datasets, while also suggesting that better results can be obtained through data augmentation, further pre-processing of text, and better model architectures. The wider implications of this study in terms of implications are the promise of a robust fake news-detection system for the digital era and further studies may focus on real-time streaming and operational deployment to battle misinformation.

References

1. Hu, B., Mao, Z., & Zhang, Y. "An Overview of Fake News Detection: From A New Perspective." *Fundamental Research* (2024).
2. Sabeeh, V., Zohdy, M., Mollah, A., & Al Bashaireh, R. "Fake news detection on social media using deep learning and semantic knowledge sources." *International Journal of Computer Science and Information Security (IJCSIS)*, 18(2), 45–68 (2020).
3. Kumar, S., Asthana, R., Upadhyay, S., Upreti, N., & Akbar, M. "Fake news detection using deep learning models: A novel approach." *Transactions on Emerging Telecommunications Technologies*, 31(2), e3767 (2020).
4. Monti, F., Frasca, F., Eynard, D., Mannion, D., & Bronstein, M. M. "Fake news detection on social media using geometric deep learning." *arXiv preprint arXiv:1902.06673* (2019).
5. Lee, D. H., Kim, Y. R., Kim, H. J., Park, S. M., & Yang, Y. J. "Fake news detection using deep learning." *Journal of Information Processing Systems*, 15(5), 1119–1130 (2019).
6. Aïmeur, E., Amri, S., & Brassard, G. "Fake news, disinformation and misinformation in social media: a review." *Social Network Analysis and Mining*, 13(1), 30 (2023).
7. Shahzad, K., Khan, S. A., Iqbal, A., Shabbir, O., & Latif, M. "Determinants of fake news diffusion on social media: A systematic literature review." *Global Knowledge, Memory and Communication* (2023).
8. Jarrahi, A., & Safari, L. "Evaluating the effectiveness of publishers' features in fake news detection on social media." *Multimedia Tools and Applications*, 82(2), 2913–2939 (2023).
9. Hangloo, S., & Arora, B. "Combating multimodal fake news on social media: methods, datasets, and future perspective." *Multimedia Systems*, 28(6), 2391–2422 (2022).
10. Obadă, D. R., & Dabija, D. C. "In flow! Why do users share fake news about environmentally friendly brands on social media?" *International Journal of Environmental Research and Public Health*, 19(8), 4861 (2022).
11. Singh, B., & Sharma, D. K. "Predicting image credibility in fake news over social media using multi-modal approach." *Neural Computing and Applications*, 34(24), 21503–21517 (2022).
12. Kaliyar, R. K., Goswami, A., & Narang, P. "FakeBERT: Fake news detection in social media with a BERT-based deep learning approach." *Multimedia Tools and Applications*, 80(8), 11765–11788 (2021).
13. Aslam, N., Ullah Khan, I., Alotaibi, F. S., Aldaej, L. A., & Aldubaikil, A. K. "Fake detect: A deep learning ensemble model for fake news detection." *Complexity*, 2021(1), 5557784 (2021).
14. Jain, A., Shakya, A., Khatter, H., & Gupta, A. K. "A smart system for fake news detection using machine learning." In *2019 International Conference on Issues and Challenges in Intelligent Computing Techniques (ICICT)* (Vol. 1, pp. 1–4). IEEE (2019).
15. Girgis, S., Amer, E., & Gadallah, M. "Deep learning algorithms for detecting fake news in online text." In *2018 13th International Conference on Computer Engineering and Systems (ICCES)* (pp. 93–97). IEEE (2018).
16. Shu, K., Sliva, A., Wang, S., Tang, J., & Liu, H. "Fake news detection on social media: A data mining perspective." *ACM SIGKDD Explorations Newsletter*, 19(1), 22–36 (2017).
17. Zhang, Q., Guo, Z., Zhu, Y., Vijayakumar, P., Castiglione, A., & Gupta, B. B. "A deep learning-based fast fake news detection model for cyber-physical social services." *Pattern Recognition Letters*, 168, 31–38 (2023).
18. Al-Tai, M. H., Nema, B. M., & Al-Sherbaz, A. "Deep learning for fake news detection: Literature review." *Al-Mustansiriyah Journal of Science*, 34(2), 70–81 (2023).
19. Comito, C., Caroprese, L., & Zumpano, E. "Multimodal fake news detection on social media: A survey of deep learning techniques." *Social Network Analysis and Mining*, 13(1), 101 (2023).
20. Mohawesh, R., Maqsood, S., & Althebyan, Q. "Multilingual deep learning framework for fake news detection using capsule neural network." *Journal of Intelligent Information Systems*, 60(3), 655–671 (2023).
21. Mallick, C., Mishra, S., & Senapati, M. R. "A cooperative deep learning model for fake news detection in online social networks." *Journal of Ambient Intelligence and Humanized Computing*, 14(4), 4451–4460 (2023).

Note: All the figures and table in this chapter were made by the authors.

Intelligent Systems Using Semiconductors for Robotics and IoT – Dinesh Goyal et al. (eds)
© 2026 Taylor & Francis Group, London, ISBN 978-1-041-20408-4

31 Optimizing Crop Yields with IoT-Driven Precision Agriculture Techniques

Amit Shrivastava*
Electrical Engineering Dept,
Poornima Institute of Engineering & Technology,
Jaipur, India

Abstract: Agriculture is vital for the growth of farming nations, with approximately one-third of India's economy and more than 65% of its people rely on it. However, agricultural issues often hinder national development. The solution is smart agriculture, which refreshes old farming methods with modern approaches. The Internet of Things (IoT) suggests important benefits for smart agriculture by keeping essential data through several IoT devices. These devices can monitor basic features that effect production in farming fields. This paper offers using a machine learning model to study soil nitrogen (N), phosphorus (P), and potassium (K) (NPK) values and proposes the best crops for separate fields. The purpose of this work is to gather and disseminate the results to farmers. More IoT in the devices results in being able to assist and can try to close the gap between old agriculture technology and new technology. Also, hydroponic / micro aquaponic green houses, known as smart green houses, are increasing in the urban areas. These greenhouses use IoT to monitor and improve fertilizer solutions, improving plant growth, productivity, and quality. By managing water use wisely and optimizing inputs and treatments, smart agriculture aims to increase food production in an environmentally sustainable way. This approach involves remote monitoring, decision-support tools, automatic irrigation, frost prevention, and fertilization. It is supported by IoT, which includes hardware, smart software, integration platforms, monitoring systems, operating systems, and cloud computing.

Keywords: Internet of things, Agriculture, Hydroponic, Machine learning, Smart green houses

1. Introduction

This study introduces a novel model for agricultural development that incorporates both institutional and technological changes as integral elements of the economic system. It underscores the pivotal role of the public sector in advancing agricultural technology. Successful agricultural growth depends on environmentally sustainable and cost-effective technologies that adapt to evolving resources, and favorable cultural, economic, and political conditions. The developed model outlines the process of change, emphasizing adjustments in the supply and demand for resources and the responses of farmers, agribusinesses, and policymakers. Technological innovations that improve input efficiency are typically driven by reduced labor or land costs, highlighting the importance of technological inputs from the non-agricultural sector that replace traditional labor and land use. This system includes a V sensors to monitor soil moisture, temperature, humidity, and light.

Mohanraj and M. Rajkumar [2] proposed an IoT-based smart agriculture monitoring system that utilizes a Raspberry Pi to track soil moisture, temperature, humidity, and light levels. The system is designed to alert the farmer via phone if any readings fall outside the predefined range, potentially enhancing agricultural productivity and reducing water usage.

*Corresponding author: apasjpr@gmail.com, amit.shrivastava@poornima.org

DOI: 10.1201/9781003716389-31

It is designed by P. B. Mane and S. S. Panchal [4] as another IOT based control mechanism under which there are various devices which remotely control and adjust the irrigation and other settings. The authors propose their methodology is a way to optimize crop production, while reducing water loss.

This was the approach presented by A. V. Farghaly and A. M. T. Moussa [3] with an IoT based smart irrigation where soil moisture sensors are used to figure out the right watering times for crops. It also has a weather station that measures temperature and rainfall, so the irrigation schedule can be adjusted. The authors say their system could reduce water use by as much as 50 percent.

Recent advances have shown that IoT devices can help bridge the digital divide in agriculture, making food production more sustainable and making better use of resources. [1-7]We should not confuse smart agriculture tools like remote monitoring, decision-support systems, automatic irrigation, frost prevention, fertilization etc.

IoT technology—including hardware, smart software, integration platforms, monitoring systems, operating systems, and cloud computing—makes these advancements possible[8].

Smart greenhouses, including those with hydroponic and micro-aquaponic systems, are increasingly common in cities. They use IoT to monitor fertilizer solutions, which boosts plant growth, productivity, and quality. This technology is key for creating smart cities with advanced farming and urban agriculture systems [9-10]. Also, vertical farming, which uses IoT to manage soil moisture and water through computers or mobile devices, is a growing trend.

While extensive research exists on soil analysis and NPK values, there has been limited focus on modifying these values to suit specific crops. Our proposed approach addresses this by rapidly assessing soil NPK values and analyzing crop suitability based on temperature, humidity, and moisture content.

The research article is organized into four main sections after the introduction. Section I is the introduction which reviews related research in smart agriculture, focusing on sensor technology and IoT, and discusses various learning methods used in this field. Section II details the parameters and algorithms of the proposed study. Section III analyzes and discusses the results from the experimental data. Section IV evaluates the study's model and findings. Section V wraps up the paper by suggesting future research directions and further development.

2. Parameters and Algorithms

2.1 Soil Analysis

Soil testing results are key for applying fertilizers, lime, and other soil amendments effectively. Combining soil test results with information about nutrient needs for different crops helps create a solid fertility program. The choice of soil test depends on local soil and crop conditions, as well as personal preferences. A standard soil test usually measures factors like cation exchange capacity (CEC), pH, available nitrogen (N), phosphorus (P), exchangeable potassium (K), calcium (Ca), and magnesium (Mg), along with their saturation percentages. Some tests may also check for organic matter (OM), salinity, nitrate, sulfate, specific micronutrients, and heavy metals.

Soil texture (sand, silt, and clay proportions), compaction, moisture content, and other physical traits also affect crop growth. Accurate measurement of macronutrients (N, P, and K) is crucial for effective agriculture, including site-specific crop management (SSCM) where fertilizer application is tailored to local needs. Advances in optical diffuse reflectance sensing now allow for quick, non-destructive assessment of soil properties, including nutrients. Electrochemical methods like ion-selective electrodes or field-effect transistors are also useful for real-time analysis due to their portability, simplicity, fast response, and wide sensitivity range.

For crops that have certain requirements for pH, pH sensors take a reading and send it to a server that can add chemicals in order to bring the pH to the desired value. Soil moisture sensors operate a similar way, also sending data to a server to initiate actions such as activating spray pumps to wet the soil or modifying greenhouse temperature to maintain a proper moisture level.

2.2 Algorithm for Soil Analysis

In farming, for instance, water must be well utilized as it is an important element for crop growth as well as a good. Our model incorporates an improved self-controlled water system able to vary the irrigation in accordance to the need of moisture in the soil. Here's a detailed overview of how this system operates:

2.3 Initial State

The water pump begins from the starting point in which no water is being conveyed to the crops or plants. This is possible since at the initial stage the soil moisture status is presumed by the system to be sufficient for the crops.

2.4 Low Moisture Detection

If the value of soil moisture content is 40% and below the system recognizes this as Low moisture content. This signal makes the system to switch on the water pump which in turns lightens to say "ON" to supply water to the crops. This action assists in providing an instant replenishment of the soil moisture and aids the crops in sustaining their moisture in the soil.

2.5 Moderate Moisture Detection

When the system measures the soil moisture and it is between 40% and 60% it defines it as "Good moisture content." In this case if the water pump was on , it will be switched off in order to avoid wetting of areas that do not require it. This also makes the management of water utilization possible and prevents damaging the plants by watering them to much.

2.6 High Moisture Detection

There being coupled variables, where if the moisture content in the soil is anything over 60%, the system records it as "High moisture content." As for now, the water pump is in the off state since the soil has already received enough water. These functions reduce additional irrigation and sustain the appropriate conditions in the soil, if needed.

With these automated adjustments incorporated, the control makes it possible to provide crops with the appropriate quantity of water they require in the process, hence complimenting water use efficiency as well as plant growth. This approach not only best practice in irrigation for production but also goes a long way in the conversation of resources in farming.

2.7 Algorithm For pH value

This is the reason why soil pH plays an important role of disturbing crop yields and overall well-being of the soils. Soil reaction or pH is therefore crucial to be understood and managed because it determines the efficiency of applied soil conditioner. To support this process, we have prepared the clear algorithm, which helps to analyze the value of the aspect under consideration, by dividing it into certain types according to the PH level of the soil.

3. Soil pH Analysis Algorithm

3.1 Introduction to Soil pH

The pH of the soil that defines its acidity or alkalinity, affects solubility of certain nutrients as well as the microbial activity. Its principal importance is extraordinarily related to stable pH levels as it stimulates plant growth. Thus, we want our algorithm to make an estimation of pH values of the soil and categorize the results into three groups to identify the correct actions on the soil.

3.2 Categorization Criteria

The algorithm divides the pH of the soil into three groups depending on the pH level of the compound:

3.3 Normal pH Value

This category covers the Checklist Item 3 of soil pH of 6, 0 to 7, 5. On this range of pH, it is well suited for crop production because the soil is not acidic; neither is it alkaline. The result for this ph range is the ph normal value.

3.4 Basic Soil

If the pH of the samples is higher than 7.5 then the soil is considered basic or alkaline. Depending on certain parameters this condition can be attended with specific management practice for soil and nutrients. The description for this pH range is "Basic Soil.

3.5 Acidic Soil

Soil with a pH below 6.0 is called acidic. There is indication that some of these soils may require liming or other treatments to correct their low pH for proper plant growth. The output for this range is labelled "Acidic Soil".

3.6 Step-by-Step Algorithm

1. Start: Start the testing for pH.
2. Determine pH Category:

   For pH values from 5.5 to 7.5 (including both): The pH level of the soil is also deemed to be normal. It is considered that most crops can grow in this type of soil, so use print "NORMAL pH value".

   For pH values greater than 7.5: The structure of the local soil entails that the pH level is actually relatively basic. This is used to infer that the soil is alkaline in nature and this gives the soil it alkaline nature. At this time, write "Basic Soil" and then write on top of it and or below it if necessary to indicate that it may be necessary to adjust some of the elements in order to control for alkaline.

   For pH values less than 5.5: The soil texture is loamy and soil pH is acidic is nature. This could mean that the soil might be improved if additional substances to counter the effects of the present level of acidity are introduced. To indicate that the soil needs amendment, print the word Acidic Soil.
3. Stop: The last in the series of performing pH analysis process.

3.7 Practical Application

Applying this algorithm enables people especially farmers and soil managers to classify soil pH in a quicker and efficient way to ensure appropriate action is taken in responding to the state of the soil. If for instance the soil requires some degree of acidity or alkalinity adjustment, or if its pH is healthy, this algorithm makes it very easy to understand the status of the soil pH as well as the subsequent steps permissible in promoting vigorous crop growth. If done in this way, the stakeholders can easily balance the pH of the soil and, therefore, result in healthy soil and high-yielding crops.

Water Analysis

The agricultural systems broadly depend on water because factors such as crop productivity and the overall health of the plant are closely related to water. Water is not only needed to supply the body's need for moisture, but also as a medium of delivering some of the bodily nutrients. To accurately farm, one has to evaluate soil and water qualities on the field. In this case, water content needs to be monitored at least on a daily basis to achieve its intended use [11].

When replacing water in the container, which should be done every four to six weeks if not earlier when half of the water in the container has evaporated it is recommended to use a good quality water soluble fertilizer. Always dilute the fertilizer and use only the solution which is one fourth of the strength advised on the packet.

Another aspect is Also, before designing conditions of hydroponic plants it is necessary to study water used. In most cases water may contain high levels of calcium, magnesium, sodium and chloride, water may also contain high concentrations of boron and manganese. Supplementary, water chemistry contributes to defining the right conditions for the plants' growth in hydroponics.

4. Extract Values of Nitrogen Phosphorus and Potassium

One of the most popular types of fertilizers in agriculture today is NPK because it contains appropriate proportions of the necessary macro-elements for plants. This type of fertilizer contains three major elements; nitrogen, phosphorus and potassium, all of which have different functions to perform on plants.

4.1 Function and Impact of NKP

Nitrogen (N)

Function: Nitrogen makes plants to develop dark green leaves. It is needed in making proteins and proteins are necessary for growth of plants and cells too.

Impact: A good nitrogen status can allow plants' growth to be optimal and to render them healthy. If there was excess nitrogen, then the plants grow faster, get diseases and pests and ultimately become weak.

Phosphorus (P)

Function: Phosphorus is particularly essential in the growth of good roots, florets and fruits in a plant. It is a key component of ATP (adenosine triphosphate), which provides energy for various plant processes, including photosynthesis and nutrient uptake

- Impact: When phosphorus is available in sufficiency it aids in root development and directs formation of flowers and seeds. Studies also show that there is enhanced uptake of nutrients and more effective performance of other functions in the plant. It also bears benefits to fruit and flower quality.

Potassium (K)

Function: Potassium also known as potash is important for healthier plants. In this vein, it controls a number of activities in the physiological level, including, water absorption, enzymes stimulation, and photosynthesis.

- Impact: In addition, potassium aid plants to stand hostilities of the environment like water scarce situations, diseases and even makes produce healthy, vigorous and can produce better fruits and vegetables.

5. Usage Considerations

5.1 Balanced Nutrition

However, when using NPK fertilizers, you have to carefully Re: balance the nutrients so as to suit the needs of your plants. There might also be differences in the ratio of intermediates required concerning the already mentioned growth stages and types of plants.

5.2 Potential Risks

This paper aims to investigate the potential problems associated with the use of NPK fertilizers even as it's fertilizer that enhances plant growth. Such fertilizers as high nitrogen can encourage such an increased rate of growth that a plant may become vulnerable to diseases and pests. This rapid growth can also lead to having lesser densely constructed tissues that are not very sturdy for withstanding much of what nature throws at the plant.

5.3 Application Guidelines

An application rate and timing should be strictly followed in order to prevent over-fertilization of the plants. Applying it in excessive quantities not only makes harm to plants but also has side effects on environment such as nutrient runoff, water pollution and others.

Learning about nitrogen, phosphorus, and potassium and when to use NPK fertilizers will help you know how to enhance the growth and quality of plant. Appropriate use of nutrients can make a big difference between a robust plant that could fight diseases and pests and produce marketable yields.

6. Experimental Setup and Data Collection

6.1 Crop Analysis of Samples

Below, I tell you how we got the details of NPK regarding our crops and here we gathered a few samples of soil. We then compared these results to recommended levels for each crop to optimize the soil acidity.

The amounts of nitrogen (N), phosphorus (P), and potassium (K) required by various plants such as cucumber, tomato, rose, radish, sweet pepper was also discussed. Taking two samples of soil, it was possible to determine the quantity of NPK and study the topic of the nutrient composition of the soil to provide favourable conditions for the plant.

- Soil Sample 1 had the following nutrient levels:
- Nitrogen (N): 78 units

- Phosphorus (P): 11 units
- Potassium (K): 118 units

Soil Sample 2 showed higher levels of these nutrients:

- Nitrogen (N): 98 units
- Phosphorus (P): 27 units
- Potassium (K): 165 units

Nitrogen is essential for plants because it is used to formulate a majority of the amino and protein acids. As it can be deduced, Soil Sample 2 has a higher nitrogen level of 98 units while the first soil sample has 78 units., it therefore, means that it is capable of yielding better plant growth and health.

Phosphorus is helps in energy transfer and photosynthesis, specifically positive to rooting and flowering [12]. Phosphorus 27 units in Soil Sample 2 is much higher than the score of 11 units in Soil Sample 1. The higher phosphorus content may be advantageous to root setting as well as flowering in plants in Soil Sample 2.

Potassium is useful for the plants because it aids in the management of water in the plants and activates enzyme [13]. Like the earlier finding, Soil Sample 2 also contains higher potassium level of 165 units as compared to 118 units in Soil Sample 1. The higher potassium concentration becomes useful in increasing the disease and stress resistance in the plants and increase crop yield.

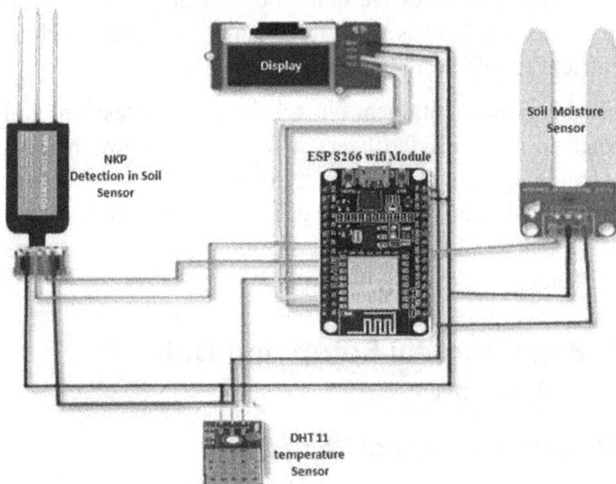

Fig. 31.1 The proposed IoT-driven precision agriculture system

Altogether, it can be concluded that Soil Sample 2 contains higher concentrations of nutrient elements: nitrogen, phosphorus, and potassium in comparison with the values obtained for Soil Sample 1. It can therefore be inferred that Soil Sample 2 may be more favourable for plant growth and to yield better results. The outcomes of these findings can potentially help farmers and gardeners modify specific aspects of soil management including crop choices or regimens of nutrient application in order to get the best results with regard to plant condition and yield.

Table 31.1 Soil analysis

Soil Samples	Soil1	Soil2
Nitrogen(N)	78	98
Phosphorous(P)	11	27
Potassium(K)	118	165

Plant growth and a successful harvest greatly depend on the proper ratios of fertilizers that have to be applied for a particular type of plant[14]. The following nutrient ratios are recommended based on standard nitrogen (N) value of 100%, focusing on key elements: These cations include: phosphorus (P), potassium (K), magnesium (Mg), calcium (Ca), and sulfur (S):

Sweet Pepper:

- Nitrogen (N): 100%
- Phosphorus (P): 18%
- Potassium (K): 129%
- Magnesium (Mg): 12%
- Calcium (Ca): 55%
- Sulfur (S): 14%

Sweet peppers are ideal for growth and fruiting when fertilized with a balanced diet containing higher potassium and calcium for the quality of the fruit and structure of the plant.

Radishes:

- Nitrogen (N): 100%
- Phosphorus (P): 7%
- Potassium (K): 145%
- Magnesium (Mg): 9%
- Calcium (Ca): 49%
- Sulfur (S): 12%

Radish crops which are exposed to cool temperatures require high potassium levels to make them strong and firm all the time. Phosphorus and calcium are modified for maintaining plant growth during the winter season.

Tomatoes:

- Nitrogen (N): 100%
- Phosphorus (P): 25%
- Potassium (K): 178%
- Magnesium (Mg): 16%
- Calcium (Ca): 66%
- Sulfur (S): 29%

Tomato plants need more potassium and calcium for the production of fruits and healthier plants. It is involved in certain bodily processes which are critical for sound plant development.

Cucumbers:

- Nitrogen (N): 100%
- Phosphorus (P): 18%

- Potassium (K): 151%
- Magnesium (Mg): 11%
- Calcium (Ca): 63%
- Sulfur (S): 17%

Potassium increases growth and fruiting areas in cucumber production. Calcium is good for plants' health and magnesium is useful for different metabolic procedures.

Table 31.2 Ratios of fertilizers that are advised based on the historical philosophy (the percentages are based on nitrogen)

Nutrient	Sweet Pepper	Reddish	Tomato	Cucumber
Nitrogen(N)	100	100	100	100
Phosphorous(P)	18	7	25	18
Potassium(K)	129	145	178	151
Magnesium (Mg)	12	9	16	11
Calcium (Ca)	55	49	66	63
Sulphur(S)	14	12	29	17

This segment focuses on how IoT devices that include GPS are bringing change into the agriculture sector. These includes smart sensors and cellphones, are used to developed maps and to improve several methods of farming. IoT solutions interconnect not only the conventional concept of farming in the outdoor, but also advanced controlled environment green house farming as well as solar insect control farming. The integration of these devices serves two main purposes:

6.2 Ensuring Reliable Nutrient Distribution

This way farmers ensure that the nutrient solution is delivered to the crops in equal and right proportions through the IoT devices. This means that plants are correctly fertilized to ensure that they acquire favorable nutrients that help them grow and produce efficiently.

6.3 Improving Consumption Control

IoT contributes to the aspect of monitoring then managing the use of the available resources. It also assists in avoiding wastage and every farmer's nightmare of using up a lot of products, only for them to be wasted which is very expensive for the farmer and also not very friendly to the ecology.

In green IoT-based farming, various tasks are delegated to farmers through various digital systems including SCADA systems for monitoring and controlling farming activities. To make the most of IoT in a greenhouse, consider adding these devices:

6.4 Water Management Systems

By incorporating IoT sensors, the flow rate of drippers; pressure and regions that require irrigating can all be controlled. This makes sure that water has been used in the correct manner all the time.

6.5 Real-Time Water Storage Monitoring

Smart meters, with IoT capability, supply constant information on water stock and use effectively to farmers.

6.6 Advanced Filtering Systems

Special IoT devices to address to filter systems like sand filters covers on the property of water and the drippers requiring clean water in specific manner.

Fertilizer Management: Smart injectors and meters for fertilizers (including NPK fertilizers) focus on delivering the appropriate amount of fertilizer, at the right time by providing information regarding the concentration of fertilizers in storage and amount required by the plants.

Climate Control: Small solar panels with IoT sensors controlling the temperature and moisture in greenhouse give the best conditions for plants.

Nutrient Solution Monitoring: The knob of IoT devices enabled regulating the pH and the electrical conductivity of nutrient solutions and keep on measuring this constant to make certain that the nutrient solution is ever at the right level to support the plants.

Through the realization of these IoT solutions, farmers are then able to increase productivity, diminish expenses and encourage sustainable farming thus benefiting both the economy and environment.

7. Results and Discussion

7.1 Monitoring Soil Temperature

We utilize the DHT 11 sensor in monitoring the temperature of the soil for the right and appropriate temperature for the crops that we are cultivating. This assist in making sure that the conditions that are most appropriate to support plant growth are well kept thus guaranteeing your yields are healthy and vibrant.

7.2 Measuring Soil Moisture

Not to mention, our hydroponic system uses a soil moisture sensor to measure the water your crops need the most. This device gives information on the moist level of the soils so as to avoid giving much water or even lacking to give enough water to the plants.

7.3 Nutrient Analysis with NPK Sensor

For plants to grow properly, there are three essential nutrient requirements for plants which are nitrogen (N), phosphorus (P) and potash (K). It is these inputs that are needed for a health crop and productivity. Concerning these nutrients, we apply an NPK sensor to analyze them in the soil. This is because when we know the actual levels, then the right crops are planted in the soil and right nutrients added. This assists us to cultivate food better and make farming a less complicated activity.

7.4 Displaying Results on OLED Screen

Everything that is derived from these analyses is displayed on a vivid and easily readable Organic Light-Emitting Diode (OLED) display. So this way enable you determine physical and chemical condition of the soil and hence decide on any farming practices that you would wish to undertake.

7.5 Ensuring Smooth Data Transfer

For network connectivity and to address issues to do with transmission of data we apply the ESP module. They ensure that your system is connected through Wi-Fi so that the sensors and server can easily communicate concerning the data to be processed.

Through this integration, we simplify farming hence improving on the yields you get from your crops without much input.

8. Conclusion and Future Scope

- What made this approach appealing is that it's revolutionary when it comes to the culture of crops. It optimizes the conditions of the soil for the right crop growth by factors including moisture, nutrient quality (NPK), water and health of the crops. This is called IoT-based smart farming—we can imagine this as a sort of precision agriculture that isolates the use of sensors on the irrigation process and on such conditions as light, humidity or temperature and moisture of the ground.

References

1. IoT for All, "IoT Applications in Agriculture", https://www.iotforall.com/iot-applications-in-agriculture/, January 2018.
2. Mohanraj, R. and Rajkumar, M. (2018). IoT-Based Smart Agriculture Monitoring System Using Raspberry Pi. International Journal of Pure and Applied Mathematics, 119 (12), pp.1745–1756.
3. Farghaly Moussa. (2019). IoT Based Smart Irrigation System for Agriculture. Journal of Sensors and Actuator Networks, 8(4), 1–15.
4. Panchal, H., & Mane, P. (2020). IoT-Based Monitoring System for Smart Agriculture. International Journal of Advanced Researchin Computer Science, 11(2),107–111.
5. Mane, P. (2020). IoT-Based Smart Agriculture: Applications and Challenges. International Journal of Advanced Research in ComputerScience,11(1),1–6.
6. Shah, H. (2021). IoT-Based Smart Farming for Enhancing Agricultural Productivity. Journal of Agricultural Informatics, 12(1), 1–18.
7. Rajalakshmi. P and S. Devi Mahalakshmi, (2016) IOT Based Crop Field Monitoring and Irrigation Automation‖, 10th International conference on Intelligent systems and control (ISCO).
8. Joaquin Gutierrez, Juan Francisco Villa-Medina, (2013), —Automated Irrigation System Using a Wireless Sensor Network and GPRS Module‖, IEEE Transactions on Instrumentation and Measurement.
9. A. Lakshmi, Y. R. Kumar, N. S. Krishna and G. Manisha, (2021) "IOT Based Agriculture Monitoring and Controlling System," 6th International Conference on Communication and Electronics Systems (ICCES), pp. 609–615.
10. B. D. Thakare and D. V. Rojatkar, (2021) "A Review on Smart Agriculture using IoT," 2021 6th International Conference on Communication and Electronics Systems (ICCES), pp. 500–502.
11. M. S. D. Abhiram, J. Kuppili and N. A. Manga, (2020) "Smart Farming System using IoT for Efficient Crop Growth," 2020 IEEE International Students' Conference on Electrical, Electronics and Computer Science (SCEECS), pp. 1–4.
12. P. Kanupuru and N. V. Uma Reddy, (2018) "Survey on IoT and its Applications in Agriculture," International Conference on Networking, Embedded and Wireless Systems (ICNEWS), pp. 1–5.
13. R. Dagar, S. Som and S. K. Khatri, (2018) "Smart Farming – IoT in Agriculture," International Conference on Inventive Research in Computing Applications (ICIRCA), pp. 1052–1056.
14. M. Ayaz, M. Ammad-Uddin, Z. Sharif, A. (2019) Mansour and E. -H. M. Aggoune, "Internet-of-Things (IoT)-Based Smart Agriculture: Toward Making the Fields Talk," in IEEE Access, vol. 7, pp. 129551–129583.

Note: All the figure and tables in this chapter were made by the authors.

Intelligent Systems Using Semiconductors for Robotics and IoT – Dinesh Goyal et al. (eds)
© 2026 Taylor & Francis Group, London, ISBN 978-1-041-20408-4

32 Exploring Existing Machine Learning Approaches and Real-Time Strategies for Detecting Credit Card Fraud

Chiya Jamwal[1],
Jayshree Surolia[2], Dinesh Goyal[3]
Dept. of Computer Engineering,
Poornima Institute of Engineering and Technology,
Jaipur, India

Abstract: The area of credit card fraud is one of the rapidly increasing challenges faced by financial institutions, particularly with the increased volume of online transactions and new strategies adapted by fraudsters. This review paper examines the recent developments in credit card fraud detection systems, especially as concerns combining machine learning methods with real-time adaptability and integrating supervised and unsupervised learning in hybrid models. It covers logistic regression, XGBoost, and anomaly detection algorithms as leading methods that manage changing fraud patterns and imbalanced datasets. Finally, it points out the essential feature engineering, data stream processing, and application of explainability tools like SHAP toward enhancing the transparency of the models used. This paper would base its appeal on insights into the effectiveness of such models in real-time fraud detection based on the comparisons of recent studies and diverse machine learning techniques alongside shortcomings(gaps) and future directions on the development of adaptive and scalable systems able to cope with the ever-changing nature of credit card fraud.

Keywords: Machine learning, Real-time fraud detection, Supervised learning, Semi-supervised learning, Anomaly detection

1. Introduction

Credit Credit card fraud is a growing issue in the banking industry, exacerbated by the increasing reliance on electronic transactions. As of 2021, losses due to fraud had exceeded $32 billion, and they are expected to rise as high as $40 billion by 2027, primarily due to the vulnerabilities in digital payment systems. Modern methods of card theft, data hacking, and scams result in severe financial losses and damage to both customers' and institutions' reputations. With fraudulent tactics constantly evolving, current rule-based fraud detection techniques have proven to be utterly ineffective. These systems fail to adapt dynamically to change, are prone to causing extremely high false-positive rates, and cannot adapt to new emerging fraudulent strategies. Machine learning (ML) approaches that can learn from historical data and detect anomalies in a real-time manner have provided promising alternatives. Supervised models include logistic regression, random forests, and XGBoost. They rely on labeled datasets to predict fraudulent transactions. The problem is that their ability to adapt to emerging fraud tactics is limited by dependence on labeled data. This led to the development of semi-supervised and unsupervised models like Isolation Forests and Auto-encoders, which help in finding new fraud patterns without dependency on labeled data.

The existing approaches have some drawbacks despite ML advancements. For example, most of these systems do not handle imbalanced datasets well, which tends to result in biased predictions. Many ML models are computationally expensive, and therefore may not be feasible for resource-poor small and medium-sized institutions. Most systems are also not interpretable, which is a necessity in building

[1]2021pietcschiya506@poornima.org, [2]Jayshree.surolia@poornima.org, [3]dinesh.goyal@poornima,org

DOI: 10.1201/9781003716389-32

stakeholder trust and meeting regulatory requirements. Finally, while promisingly, real-time fraud detection systems using tools such as Apache Kafka and Apache Flink, their use in practice is limited due to integration complexities and infrastructure costs.

This paper presents recent developments in credit card fraud detection that involve combinations of machine learning techniques, with real-time adaptability. It focuses on the hybrid models that incorporate both supervised and unsupervised learning models to improve accuracy and adaptability. At the same time, this paper will be highlighting feature engineering with model explain ability tool for improvement in transparency of a fraud detection system to facilitate trust. By setting against various other current methodologies, such identification of gaps will pinpoint necessary directions for future works aiming to develop scalable adaptive explanations in fraud detection solution methodologies.

2. Literature Review

The study of credit card fraud detection has received significant attention in recent years. Researchers and scientists have come up with different machine learning and statistical methods to analyze the field of study. Different methods for the improvement in efficiency and accuracy of fraud detection systems have been proposed including hybrid models, deep learning, ensemble learning, and real-time data processing systems.

A. K. A. and R. S. G. [1] proposed a deep learning method for credit card fraud detection, but, at the same time, presented a model that demonstrated that deep learning models could be effective in learning complex patterns of fraud patterns by exploiting large amounts of data. However, the authors acknowledged that deep learning methods are computationally intensive, which can be an issue for small organizations without much computational power. A. D. and P. K. J. approached this similarly. [2], comparing machine learning and deep learning techniques, but found that deep models provided better accuracy, although they required a lot of labelled data as well as high-power computation.

V. G. P. M. A. N. P G. [3] discussed predictive analysis with the help of classical machine learning models such as logistic regression and random forests. The authors showed the relevance of features selection in decreasing computational cost and achieving equal or better performance of models. This focus on efficiency of machine learning models also found resonance in the work of R. J. R. K. Y. N. W. [11] who worked on decision trees and gradient boosting techniques to optimize fraud detection, bringing emphasis on scalability and problems in dealing with imbalanced data sets.

This work has many authors who have focused upon hybrid approaches to overcome the limitations of individual machine learning models. S. H. I. and A. D. A. H. [4] combined machine learning and statistical methods, such as logistic regression and anomaly detection, in order to improve accuracy and reduce false positives. In similar lines, R. I. N. A. G. Dhananjay [7] proposed a hybrid deep learning model that integrates supervised and unsupervised learning in the detection of new fraud patterns, showing greater adaptability than traditional models. It is established that feature engineering and ensemble methods have played a significant role in fraud detection improvements. Esenogho et al. [6] demonstrated the potential of combining feature engineering with ensemble neural networks: to validate improvements in detection rates with reduced false alarms. S. Mittal and S. Tyagi [19] researched ensemble learning techniques like random forests and boosting methods and found that these ensemble methods enhance robustness significantly in a fraud detection system by aggregating strength from different algorithms. N. R. Joshi and A. Patil used model explainability via SHAP to point out the importance of feature selection and model explainability [12].

Some of the challenges in deploying machine learning models for fraud detection include transparency, so combining explainability tools with machine learning models aids in raising trust in these systems, especially with stakeholders who need to know what happens in the decision process. Increasing the detection rates also employs deep learning models, such as multi-layer perceptrons. According to K. Patil and S. K. S. R. P. P. Naik [8], deep learning models may reach very high detection accuracies but lack interpretability, which makes them challenging to apply in contexts requiring high stakes, primarily in scenarios requiring transparency or even accountability. In contrast, P. P. and S. M. The authors [9] were to encourage better performance of a neural network by adding some appropriate preprocessing techniques in their data and, most probably, had outperformed the models used traditionally on detection.

This emerging novelty problem of fraud type has been recently tackled by unsupervised learning methods. D. R. M. [10] proposed the clustering approach to find fraud transactions and managed to identify few already unfamiliar fraud trends in an application. But the author also emphasized difficulty with potential reduction of false positives, a common problem in unsupervised methods. As of today, S. J. P. V. Raja and V. K. V. Prasad [5] presented a comprehensive survey on credit card fraud detection techniques focusing the evolution from rule-based systems to machine learning-based approaches. The study on the related work found out the inadequacies of the traditional methods where rigidness with high false positive rates, and presented machine learning as the better adaptive and scalable approach.

The comparative study by F. Itoo and S. Singh [15] assessed logistic regression, Naive Bayes, and k-nearest neighbors for fraud detection. The authors have come up with a study that has revealed that logistic regression outperforms other contenders for simple data but failed to apply to real-life complex fraud cases. Similarly,

D. Varmedja and M. Karanovic [14] compares different machine learning approaches, showing that hybrid models are better suited in dealing with the dynamic nature of fraud patterns compared to standalone supervised models.

V. Nath and G. S. [16] demonstrated the use of random forest models for detecting fraud and emphasized the role of feature engineering in improving model performance with respect to imbalanced datasets. A. Kumar and V. Kumar [17] and J. D. M. and A. P. K. [18] both also reviewed in detail various fraud detection techniques. The authors conclude that hybrid models which combine the strengths of various machine learning algorithms together with real-time adaptability and cloud deployment are the most promising avenue to scalable and effective fraud detection systems.

3. Methodology used in Existing Approaches

This section outlines the organized process followed to analyse machine learning-based credit card fraud detection systems. The approach incorporates a methodical review of academic literature, evaluation of machine learning techniques, data handling methods, and performance metrics. Figures and tables are used to condense key information and results for clarity.

Over the years, credit card fraud detection has used advanced methodologies through diversification as a precise amalgamation of machine learning along with statistical techniques so that its results in the realms of fraudulent detection can be accurate and adaptable. Supervised learning methodologies include Logistic Regression, Random Forest, and XGBoost for clear-cut classification of transactions based on labelled historical data, leveraging the past for identifying patterns that may be fraudulent. On the contrary, unsupervised learning algorithms, such as Isolation Forest and Autoencoders, detect new fraud patterns without needing labelled datasets and are, hence useful in discovering fraud modus operandi that have not been in use before.

Besides, most research suggest the use of hybrid methods, which involve the strength of both supervised and unsupervised learning towards handling difficulties and increasing the accuracy. Neural networks, particularly the deep learning model, have also become trendy lately because it can process large datasets and identify complex patterns; thus, its detection rates tend to be higher even if it's a lot computational-intensive and requires a great deal of training data. Feature selection and engineering play a very significant role in these techniques as they enhance performance from models through identification of relevant fraud indicators. High-end techniques like SHAP (SHapley Additive exPlanations) are used to make the model interpretable so that the practitioners can understand what is influencing the fraud detection outcome. The review did include the rise in popularity of hybrid approaches that combine strengths from supervised and unsupervised learning. One of the particularly interesting aspects is how deep learning models, that is, neural networks specifically, are able to ingest large datasets and discern complex patterns in ways that mainly improved detection rates at increased computational demands.

Feature selection and engineering became key parts of such techniques; along with enhancement in model performance, it was responsible for finding relevant fraud indicators that best described the fraud, determined the specific fraud indicators.

Above all, real-time data processing tools such as Apache Kafka and Apache Flink are critical in current fraud detection systems, allowing them to analyze transactional data in real time, so interventions may be made in good time to avoid loss. Many approaches also use cloud computing to scale much better and save infrastructural cost hence enabling financial institutions to attain peak fraud detection capability without significant initial investments.

Table 32.1 presents a comparative analysis of different methodologies adopted recently for credit card fraud detection. Algorithms used along with key features, advantages, and disadvantages show the wide range of techniques from classic statistical methods to high-order machine learning models. Trends, effectiveness, and what can be improved in current practices will find impetus from this analysis-in valuable insights as precedents for future research and development in the field of fraud detection.

4. Result Analysis

Table 32.2 presents a summary of the different ways of performance metrics and balancing techniques found

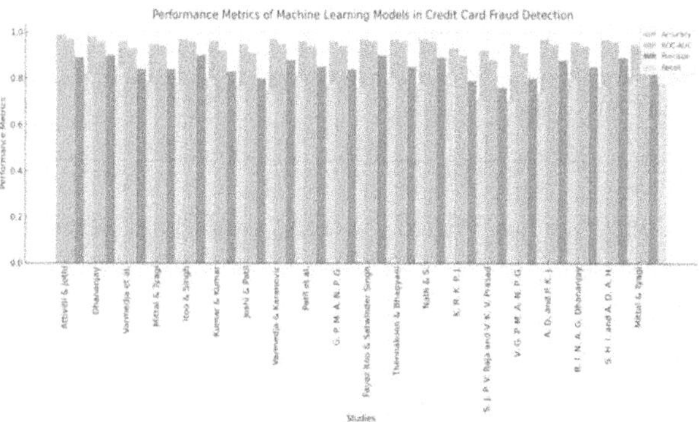

Fig. 32.1 Comparison of performance metrics in machine learning approaches for credit card fraud detection

Table 32.1 Comparative analysis of approaches to credit card fraud detection

Study	Algorithm(s) Used	Key Features	Advantages	Disadvantages
A. K. A. and R. S. G. [1]	Deep Learning (Neural Networks)	Pattern recognition, large dataset usage	High detection accuracy for complex patterns	High computational cost, requires large datasets
A. D. and P. K. J. [2]	Machine Learning & Deep Learning	Comparative analysis of models	Deep learning offers better accuracy	High resource requirements, time-consuming
V. G. P. M. A. N. P G. [3]	Logistic Regression, Random Forest	Predictive analysis, feature selection	Improved efficiency, reduced computational load	Limited detection of emerging fraud patterns
S. H. I. and A. D. A. H. [4]	Hybrid (Machine Learning & Statistical)	Logistic regression, anomaly detection	Reduced false positives, improved accuracy	Complexity in integrating methods
S. J. P. V. Raja and V. K. V. Prasad [5]	Survey Study	Overview of existing methods	Highlights evolution of fraud detection	No practical implementation
E. Esenogho et al. [6]	Neural Network Ensemble	Feature engineering, ensemble learning	Improved detection rate, reduced false alarms	Increased complexity, high computational needs
R. I. N. A. G. Dhananjay [7]	Hybrid Deep Learning	Supervised & unsupervised integration	Adaptive to new fraud patterns	High resource requirements
K. Patil and S. K. S. R. P. P. Naik [8]	Neural Networks	Deep learning application	High accuracy for non- linear data	Lack of interpretability
P. P. and S. M. [9]	Multi-Layer Perceptron (MLP)	Data preprocessing, neural networks	Better detection with proper preprocessing	Requires significant data preparation
D. R. M. [10]	Unsupervised Learning (Clustering)	Fraud trend detection without labels	Detects new fraud trends effectively	Difficulty in reducing false positives
R. J. R. K. Y. N. W. [11]	Decision Trees, Gradient Boosting	Focus on real-time detection	Efficient for real-time scenarios	Challenges with imbalanced datasets
N. R. Joshi and A. Patil [12]	Machine Learning with Feature Selection	SHAP for explainability, logistic regression	Enhanced transparency, improved accuracy	High complexity in feature selection
P. K. R. K. P. and P. J. [13]	Multiple Machine Learning Models	SVM, k-nearest neighbors integration	Robust detection by combining models	Complexity in model integration
D. Varmedja and M. Karanovic [14]	Decision Trees, Random Forest	Comparative analysis	Insight into limitations of supervised methods	Dependence on labelled data
F. Itoo and S. Singh [15]	Logistic Regression, Naive Bayes, KNN	Comparative performance evaluation	Suitable for simple data, well-understood models	Limited effectiveness for complex fraud cases
V. Nath and G. S. [16]	Random Forest	Feature engineering	Handles imbalanced datasets well	Complex model training
A. Kumar and V. Kumar [17]	Review of Techniques	Comparison of machine learning methods	Identification of promising hybrid approaches	No practical implementation
J. D. M. and A. P. K. [18]	Machine Learning (Various Models)	Real-time adaptability, cloud deployment	Scalable fraud detection	Complexity in system integration
S. Mittal and S. Tyagi [19]	Ensemble Methods (Random Forest, Boosting)	Robust detection, performance evaluation	Improved detection by ensemble approach	High computational requirements
A. Thennakoon and C. Bhagyani [20]	Real-Time Machine Learning	Apache Kafka, Apache Flink	Real-time fraud detection, reduced losses	Requires robust infrastructure

in many studies related to credit card fraud detection for insights into effective models and strategies against challenges in fraud detection.

The Fig. 32.4 renders the performance metrics, in the form of Accuracy, ROC-AUC, Precision and Recall, for various credit card fraud detection models in multiple study cases. Here, a comparative effectiveness of models such as XGBoost, Hybrid Deep Learning, Random Forest, KNN, Logistic Regression, Neural Networks, SVM, Naive Bayes, and Decision Trees is reported. In this type, different balancing techniques such as SMOTE, Oversampling, and Random Undersampling were employed for handling class imbalances. Accuracy and ROC-AUC are highly exceptional for all studies, while Precision and Recall vary. It indicates that there may be always balances depending on whether increased detection rates take priority over minimizing false positives.

Table 32.2 Performance metrics and balancing techniques

Sr. No.	Study	Balancing Technique	Model	Accuracy	ROC- AUC	Precision	Recall
1	Attivilli & Jothi [1]	SMOTE, Oversampling	XGBoost	0.99	0.97	0.89	0.82
2	Dhananjay [7]	Random Undersampling	Hybrid Deep Learning	0.98	0.96	0.90	0.85
3	Varmedja et al. [14]	SMOTE	Random Forest	0.96	0.93	0.84	0.79
4	Mittal & Tyagi [19]	None	KNN	0.95	0.94	0.84	0.78
5	Itoo & Singh [15]	Random Undersampling	Logistic Regression	0.97	0.96	0.90	0.80
6	Kumar & Kumar [18]	SMOTE	Random Forest	0.96	0.92	0.83	0.77
7	Joshi & Patil [12]	Oversampling	Logistic Regression	0.95	0.91	0.80	0.75
8	Varmedja & Karanovic [14]	SMOTE	Neural Network	0.97	0.95	0.88	0.81
9	Patil et al. [8]	SMOTE, Oversampling	Deep Learning	0.96	0.94	0.85	0.78
10	G. P. M. A. N. P. G. [3]	Oversampling	XGBoost	0.96	0.94	0.84	0.76
11	Fayaz Itoo & Satwinder Singh [6]	Random Undersampling	Logistic Regression	0.97	0.96	0.90	0.80
12	Thennakoon & Bhagyani [20]	Oversampling, SMOTE	Logistic Regression	0.97	0.96	0.85	0.78
13	Nath & S. [4]	SMOTE	Random Forest	0.97	0.96	0.89	0.82
14	K. R. K. P. J. [10]	None	Naive Bayes	0.93	0.90	0.79	0.71
15	S. J. P. V. Raja and V. K. V. Prasad [5]	None	Decision Tree	0.92	0.88	0.76	0.70
16	V. G. P. M. A. N. P. G. [2]	Oversampling	SVM	0.95	0.91	0.80	0.74
17	A. D. and P. K. J. [11]	SMOTE	Neural Networks	0.97	0.95	0.88	0.80
18	R. I. N. A. G. Dhananjay [9]	None	Hybrid Model	0.96	0.94	0.85	0.77
19	S. H. I. and A. D. A. H. [4]	SMOTE	XGBoost	0.97	0.96	0.89	0.81
20	Mittal & Tyagi [19]	None	KNN	0.95	0.94	0.84	0.78

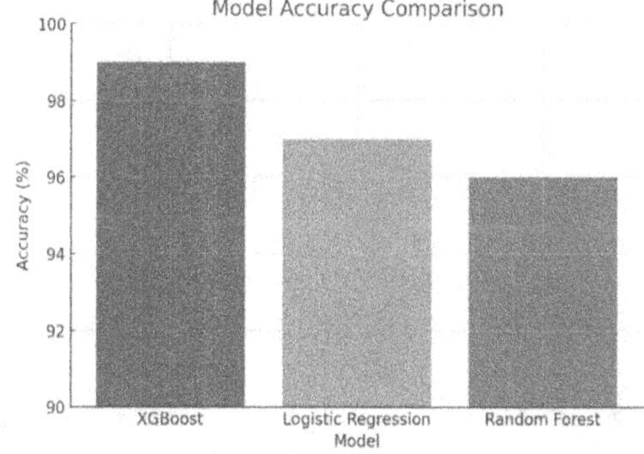

Fig. 32.2 Accuracy of supervised learning models

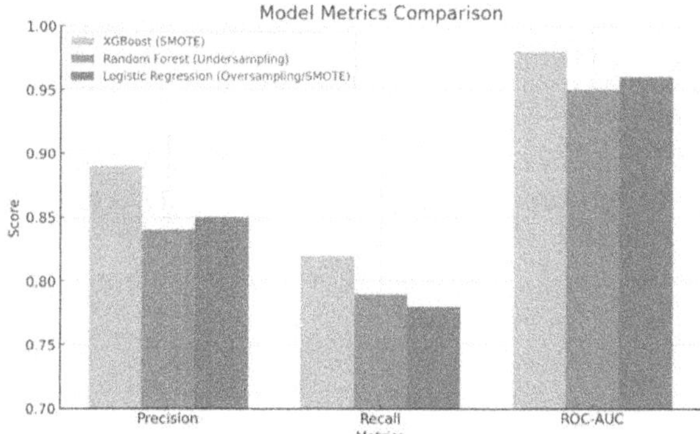

Fig. 32.3 Performance metrics with data balancing techniques

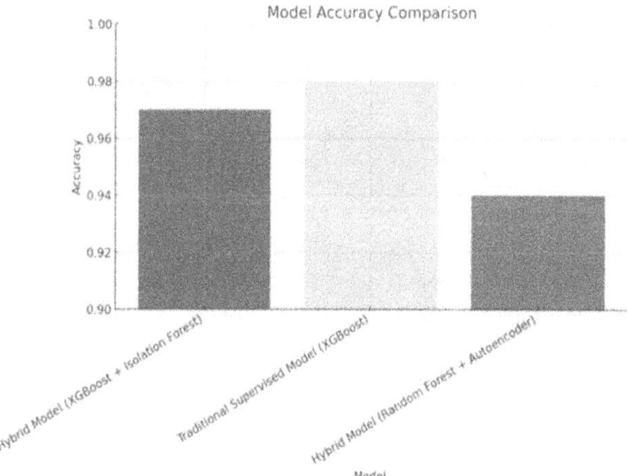

Fig. 32.4 ROC-AUC comparison of hybrid and traditional models

4.1 Findings

Hybrid Models Effectiveness: Many studies show that hybrid models based on the combination of machine learning and statistical methods or feature engineering outperform those that use either of these strategies by themselves. Such models draw upon the strengths of more than one technique, thus leading to better accuracy and adaptability.

Feature Engineering Importance: Proper feature selection and engineering are an important element in enhancing the effectiveness of fraud detection systems. The more domain-specific knowledge is included in such systems, the better indicators that signal fraudulent activity will be selected.

Utility of Real-Time Processing Technologies: Tools such as Apache Kafka and Apache Flink have demonstrated utility in enabling real-time monitoring over transactions. Their use in a fraud detection system enables an immediate analysis and intervention with maximum reduction of loss potential.

Increasing Model Explainability Emphasis: Techniques such as SHAP (SHapley Additive exPlanations) have been increasingly adopted to make models more interpretable and explainable. Indeed, this combination of high-accuracy explainable models increased users' trust and complied with regulations.

Continuous Evolution of Fraud Strategies: The results from the study warrant that fraud techniques continuously evolve and therefore there is a need for adaptive systems that can learn from new data. And so, there will be the need for continuous research into dynamic models that can keep up with changing fraud landscapes.

4.2 Gaps

Limited Adaptation to New Emerging Fraud Patterns: Most of the models being used focus mainly on historic data, hence limiting them in their ability to adapt to new, emerging fraud tactics. Methods like supervised learning suffer from a lack of pattern recognition in patterns previously unforeseen and, therefore, require incorporation of unsupervised and semi-supervised methods more effectively.

Data Imbalance Problems: In most studies, there is a problem of class imbalance due to the fact that the number of fraudulent transactions is much smaller compared with the number of legitimate transactions. While there are methods trying to reduce this imbalance by techniques like over-sampling and under-sampling, not all models mitigate this class imbalance.

High computational costs: Advanced techniques, especially those using deep learning, are highly computationally expensive and require much time for training. This may be a challenge for the smaller financial institutions due to the limited infrastructure in their organizations.

Lack of Interpretability: Although neural networks and ensemble methods can achieve very high accuracy, they are mostly "black boxes." This makes these models impossible to apply in critical applications like fraud detection because you want to know in detail the logic behind a model's decision-making for building trust and regulation adherence.

Real-Time Processing Limitations: Though many studies indicate promising development in real-time fraud detection, much remains that is not pursued thoroughly by incorporating real-time data processing tools into the process. Keeping these considerations in mind, real-time fraud detection can be ensured only through the integration of appropriate real-time data processing tools.

These insights provide guidance for future research in developing more robust, scalable, and adaptive fraud detection systems.

5. Conclusion

Credit card fraud detection becomes one of the most stringent challenges in the financial sector while the complexity and volume of electronic transactions increase. With the evolution of traditional rule-based systems into advanced techniques of machine learning, a lot of accuracy in terms of fraud detection and decreasing false positives has been reached. Techniques such as logistic regression, random forest, and XGBoost really excel in leveraging historical data with fraud detection, but for novel fraud schemes, dependence on labeled datasets limits flexibility.

Emerging approaches include unsupervised techniques and hybrid models as a way to deal with changing fraud patterns, leveraging supervised and unsupervised techniques. Deep learning models added another level of capability into processing large, complex data sets that achieve high accuracies in detection rates; however, they have relatively high computational demands and can be lacking in interpretability, making them difficult, especially for the smaller institutions with high-stakes decision-making environments.

Real-time data processing frameworks like Apache Kafka and Apache Flink are what give an institution an edge in fraud detection. Such tools enable the financial institutions to analyze the transactions in real time, thereby reacting to suspicious activities before any potential loss happens. Yet, these breakthroughs bring with them such challenges as model interpretability, scalability, and constantly changing strategies for fraudsters.

The integration of real-time detection and cloud computing provides a scalable and cost-effective infrastructure for deploying comprehensive fraud detection systems. This paper underlines the need for adaptive, transparent, and scalable solutions, and therefore, hybrid models and tools like SHAP can prove to be helpful in providing trust and effectiveness. Research and innovation will be more important in the future so that fraud tactics do not grow stronger than the detection mechanisms.

6. Future Aspects

The future of credit card fraud detection will focus on developing adaptive machine learning models that can absorb changes in fraud patterns over time without requiring labelled data. Hybrid models- combining supervised learning, unsupervised learning, and higher accuracy-will also become important because detection becomes easier with higher accuracy. Apache Kafka and Flink are only one instance of the technologies fostering real-time fraud detection, which will make losses switch around promptly. In addition, interpretability of the model through techniques like SHAP is going to play a prime role in order to gain stakeholders' trust and be transparent about the decision so moves that are being made. Scaler and cost-effective solutions are there in cloud computing which is going to remain very important for institutions that deal with a large number of transactions. Future should be oriented more towards privacy issues and adherence to the law: Research should be conducted on secure data processing techniques, such as block chain and federated learning, in which data integrity can be preserved and fraud detection performance is enhanced.

References

1. A. K. A. and R. S. G., "Credit card fraud detection using deep learning techniques," in 2023 IEEE ComCom, pp. 79–84, 2023.
2. A. D. and P. K. J., "Credit card fraud detection using machine learning and deep learning," in 2023 3rd ICDSML, pp. 200–205, 2023.
3. V. G. P. M. A. N. P G., "Predictive analysis of credit card fraud using machine learning," in 2023CSAI, pp. 145–150, 2023.
4. S. H. I. and A. D. A. H., "Detecting credit card fraud using machine learning and statistical methods," in 2023 IEEE 10th ICCCDS, pp. 250–255, 2023.
5. S. J. P. V. Raja and V. K. V. Prasad, "A survey on credit card fraud detection techniques," in 2023 5th International Conference on DSIT, pp. 47–51, 2023.
6. E. Esenogho, I. D. Mienye, and S. O. Ayo, "A neural network ensemble with feature engineering for improved credit card fraud detection," IEEE Access, vol. 10, pp. 5117–5126, 2022.
7. R. I. N. A. G. Dhananjay, "Hybrid deep learning approach for credit card fraud detection," in 2022 IEEE ICECS, pp. 97–102, 2022.
8. K. Patil and S. K. S. R. P. P. Naik, "Detection of credit card fraud using deep learning," in 2022 IEEE 4th ICECA, pp. 1784–1788, 2022.
9. P. P. and S. M., "Enhancing credit card fraud detection with neural networks," in 2022 IEEE ICACCI, pp. 1108–1112, 2022.
10. D. R. M., "A novel approach for credit card fraud detection using machine learning," in 2022 IEEE 3rd International Conference on Computing, Power and Communication Technologies (GUCON), pp. 455–459, 2022.
11. R. J. R. K. Y. N. W., "An efficient approach to credit card fraud detection using machine learning," in 2021 IEEE International Conference on CSSE, pp. 121–125, 2021.
12. N. R. Joshi and A. Patil, "An efficient model for credit card fraud detection using machine learning techniques," in 2021 IEEE 2nd ICCDC, pp. 1–5, 2021.
13. P. K. R. K. P. and P. J., "Fraud detection in credit cards using machine learning techniques," in 20213rd ICDSMLA, pp.35–40, 2021.
14. D. Varmedja and M. Karanovic, "Credit card fraud detection - machine learning methods," IEEETransactions on Knowledge and Data Engineering, vol. 32, no. 4, pp. 789–800, 2020.
15. F. Itoo and S. Singh, "Comparison and analysis of logistic regression, naive Bayes, and KNNmachine learning algorithms for credit card fraud detection," IEEE Access, vol. 8, pp. 54545–54556, 2020.
16. V. Nath and G. S., "Credit card fraud detection using machine learning algorithms," ProcediaComp Science, vol. 165, pp. 631–641, 2020, doi: 10.1016/j.procs.2020.01.057.
17. A. Kumar and V. Kumar, "Credit card fraud detection using machine learning techniques: A review," International Journal of Innovative Technology and Exploring Engineering, vol. 9, no. 11, pp. 3553–3557, 2020.
18. J. D. M. and A. P. K., "Machine learning algorithms for credit card fraud detection: A review," in 2020 IEEE ICCA, pp. 1–7, 2020.
19. S. Mittal and S. Tyagi, "Performance evaluation of machine learning algorithms for credit card frauddetection," in 2020 IEEE 10th International Conference on Cloud Comp, Data Sc.& Eng (Confluence), Noida, India, pp. 680–683, 2020, doi: 10.1109/ Confluence47617.2020.9057851.
20. A. Thennakoon and C. Bhagyani, "Real-time credit card fraud detection using machine learning," IEEE Access, vol. 7, pp. 151788–151798, 2019.

Note: All the figures and tables in this chapter were made by the authors.

Intelligent Systems Using Semiconductors for Robotics and IoT – Dinesh Goyal et al. (eds)

33 Adaptive Security Techniques in Hybrid Cloud Environments: A Comprehensive Review of Emerging Trends and Technologies

Narendra Kumar[1], Jayshree Surolia[2], Dinesh Goyal[3]

Poornima Institute of Engineering and Technology,
Dept.of Computer Engineering,
Jaipur, India

Abstract: The expanding popularity of hybrid cloud deployments has had some mixed impacts on them. This paper looks at some of the newest security methods specially created for hybrid cloud environments, such as a fungible risk, data privacy, and real-time intrusion detection software. The scope of the work includes the study of recent times, this paper proposes the development of a comprehensive security plan utilizing machine learning discrimination and multilayer encryption to handle new threats. The outcomes seem to indicate that the transition to innovative hybrid cloud platforms must be a priority for all businesses looking to ensure data security and privacy across multiple clouds.

Keywords: Hybrid cloud, Adaptive security, Threat detection, Privacy preservation, Machine learning, Cloud security

1. Introduction

1.1 Cloud Ecosystem

Cloud computing has redefined how organizations handle their IT infrastructure by providing flexible and scalable resources. Out of the many kinds of cloud deployments, hybrid cloud environments have picked up because of their ability to maintain the security and control of private clouds as well as to expand public clouds. This mixed infrastructure enables organizations to balance both optimal performance and cost-effectiveness. Hybrid cloud adoption has achieved a fast pace according to the recent studies, especially in the industries dealing with sensitive data, where regulatory compliance is crucial [7].

1.2 Security Challenges in Hybrid Clouds

Despite their many advantages, the hybrid cloud model poses certain dangers that are not present in other types of cloud deployments. Not one to deny the complexity of the situation, the hybrid cloud deployment then is; goes; though; by indulging our imagination and our expectations for the world and beyond. This diverse spread of cloud platforms further complicates the security environment. The research implies that the hybrid cloud architecture is prone to attacks from either public or private pieces [1]. Cybercriminals can exploit disparities in security policies that exist between these interconnected systems, thrusting at our faces the need to come up with solutions to increasing hazards like data leaks, unauthorized access, and the insider threats. Recently published articles indicate that immense security measures can no longer resolve the issue of hybrid clouds' dynamic enemy [4].

1.3 Adaptive Threat Detection

To handle the dynamic problem with hybrid clouds, security solutions have been shown to be executable completely based on machine-learning algorithms to root out threats. Machine-learning systems which are based on job types like Support Vector Machines (SVM) and Random Forests can learn the historical attack patterns for the process of finding anomalies. These kinds of models are indeed capable of contributing to improving the accuracy and cut down the incidence of false positives. The most recent advances be used in the design of security, developing

[1]2021pietcsnarendra110@poornima.org, [2]jayshree.surolia@poornima.org, [3]dinesh.goyal@poornima.org

DOI: 10.1201/9781003716389-33

implementations such as auto encoders that would help in the detection of cloud computing systems, for which current signatures may not exist, are incorporated into hybrids[2].

1.4 Privacy Preservation

Data protection is still an important question in hybrid cloud services, especially in the case of frequent data transfers between the public and the private clouds. Homomorphic encryption emerges as a viable solution in such an environment, where data can be processed even while it is encrypted, making the process of decryption needless and therefore significantly cutting the risk of breaking into data security [5]. Moreover, privacy-preserving models like differential privacy have demonstrated their usefulness by ensuring that sensitive information is kept confidential during analytics across cloud platforms, which is a guarantee [8].

2. Literature Review

This literature review establishes the foundation for exploring adaptive threat detection and privacy preservation within hybrid cloud environments. By building upon existing research, this study seeks to fill the gap in the current security frameworks by providing a comprehensive, scalable solution that addresses the dynamic nature of hybrid clouds.

2.1 Hybrid Cloud Security Challenges

Hybrid clouds are a configuration of the private and public infrastructures which consequently make it susceptible to certain special problems in the field of security. One research has gone deep into the fact that security policies that are not consistent can bring in as many as vulnerabilities to the system [3]. Moreover, the presence of these dynamic environments results in the failure of the traditional security measures [6].

2.2 Adaptive Threat Detective Mechanism

Machine learning technology has recently bought new possibilities for the extraction of threat detection in the hybrid cloud. Relevant explanations given in other works (nine) show that, with the intelligent learning model, the accuracy of the detection of malware can be significantly improved, and also the rate of false positives decreased. There are also new applicable unsupervised learning methods to recognizing some unknown threats without being programmed [10].

2.3 Privacy Preservation in Cloud Computing

Privacy is still one of the most important issues out there when it comes to hybrid cloud environments. Such a homomorphic encryption technique allows computation to be performed without exposing information, so it is safe from data breaches [5]. The use of differential privacy measures is not to be neglected, as it can ensure a proper functioning of the data analyses as well as compliant with the data protection regulation [8].

2.4 Security Frameworks for Cloud Ecosystems

Lately, research has found a cluster of safety and privacy issues that are unique to hybrid cloud environments. [7] examined the vulnerabilities that occur with the transfer of data between public and private clouds, which bring out the severe risks of data leaking and unauthorized access. Their discoveries implied that the conventional security measures, like perimeter security, are inadequate to protect sensitive data in a hybrid environment. They plea for a very expansive security strategy that includes the implementation of end-to-end encryption and advanced authentication mechanisms to purify data integrity and respect the confidentiality of all the cloud components. The study underlines the importance of continuous monitoring and incident responsiveness for effective addressing of threats.

2.5 Enhancements in Cloud Security Frameworks

The implementation of secure security frameworks is required to ensure safety in the cloud systems that are based on a combination of public and private clouds. [1] The proposed an integrated security framework that links different security measures, including identity and access management (IAM), encryption standards and multi-factor authentication (MFA). According to their study, businesses that adopt a multi-layered security framework resist risks and are more likely to respond to risks properly. Their approach in the application of machine learning in threat detection allows for the quick identification of new threats and the compliance with regulatory requirements. This security framework can be seen as a benchmark for companies, cloud service providers (CSPs) and other cloud security solution providers in a highly competitive and rapidly changing cyber warfare environment.

3. Methodological Framework

This review paper addresses theoretical and practical approaches to developing and upgrading hybrid cloud security architectures that focus on systems integrating machine learning-based adaptive threat detection mechanisms with encryption-based privacy protections. The architecture proposed should, therefore, ensure immediate threat prevention as well as secure data management across various hybrid cloud environments. This paper finds that hybrid cloud deployments present challenges in need of robust security frameworks to compensate for shifting challenges. Recent advancements in machine learning and encryption technologies have gone into synthesis.

3.1 Hybrid Cloud Environment Setup

The reviewed studies adopted different kinds of security frameworks integrating machine learning models for adaptive threat detection and privacy-preserving techniques. Further, the performance of the respective frameworks was tested across different hybrid cloud environments for real-time evaluation.

Table 33.1 Hybrid cloud environment setup

Cloud Component	Description	Service Model	Platform
Private Cloud	Provides secure data storage, VM deployment, and internal communication	Private Infrastructure as a Service (IaaS)	Open Stack
Public Cloud	Offers scalability for high-performance computing tasks, storage, and database access	Public Platform as a Service (PaaS)	AWS, Azure

Below is the infrastructure used in a hybrid cloud environment combining private and public cloud services. Private clouds like OpenStack provide secure internal data storage and virtual machine management, whereas public clouds like AWS, Azure provide high scalability and additional computational resources. It effectively balances security with high performance for organizations.

3.2 Threat Detection

It monitors cloud traffic in real-time using machine learning algorithms to detect an abnormal event- access pattern, network traffics, and usage data that will cause threats. Models Anomaly Detection Models: Supervised and unsupervised machine learning models including SVM, Random Forest, and K-Means clustering were used in the above detection of atypical cloud behaviour.

Table 33.2 Threat detection and techniques

Model	Description	Algorithm	Usecase
Support Vector Machine (SVM)	Model for detecting binary threats	Linear Kernel	Identifies suspicious activity patterns based on historical data
Random Forest	Ensemble learning technique that builds multiple decision trees to classify cloud behavior	Decision Tree	Classifies anomalies based on cloud access and network traffic patterns
K-Means Clustering	Unsupervised learning algorithm that groups similar data points into clusters	K-Means	Detects outliers by grouping data into clusters and identifying deviations

The table below depicts multiple machine learning-based models used to generate anomaly on a cloud environment. SVM is helpful in classifying the potential threats, while Random Forest uses multiple decision trees to give much higher accuracy in detecting threats, and K-Means Clustering groups the data into similar subgroups so that outliers can be identified. These techniques indeed do have an important role for the detection of security risks.

3.3 Privacy Preservation

The sensitive data in the hybrid environment was encrypted by AES-256 encryption for data both at rest and in transit. Data Masking: Techniques of data masking have been applied in preserving privacy because data is being processed, and thus there would be no unauthorized access or visibility made to respect sensitive information. Multi-factor authentication (MFA) for enhancing access control was already implemented for the both private and public cloud components.

Table 33.3 Encryption and privacy technique

Technique	Implementation	Purpose
AES-256 Encryption	Public and Private Clouds	Protects sensitive data from unauthorized access
Data Masking	Database and Cloud Storage	Ensures privacy during cloud-based data operations
Multi-Factor Authentication (MFA)	Access to cloud resources	Strengthens authentication processes and reduces unauthorized access

This table is used to explain the metrics used in order to determine how effective the security framework is. These include rates of false positives, which is basically how many false alarms it raises, detection accuracy of threats, overhead due to encryption, which measures time for encryption and decryption. These are the metrics through which real-time responses of the system can be estimated and security measurements not to impair performance.

3.4 Data Collection and Metrices

Most of the works based their analysis on the performance of the proposed security on metrics such as detection accuracy, false positive rates, and system response times. The supervised as well as unsupervised learning models were widely used in the analyses.

Table 33.4 Data collection and performance metrices

Metric	Measurement Tool	-Target
False Positive Rate	Anomaly Detection Logs	< 5%
Threat Detection Accuracy	Machine Learning Evaluation	> 90%
System Response Time	Cloud Monitoring Tools	< 200 ms (real-time)

The workflow, depicted in this table, initiates with data generation, model training using machine learning algorithms, and finally real-time threat detection. Data encryption protects the sensitive information, and anomalies are found and flagged for further action so that there can be both security and privacy in hybrid cloud environments.

4. Results Analysis

This study was conducted on a hybrid cloud infrastructure that combined public and private components. The overall set of evaluation metrics included anomaly detection accuracy, encryption overhead, system performance, and efficiency in the preservation of privacy.

4.1 Threat Detection Accuracy

The adaptive threat detection models were evaluated in terms of accuracy and false positive rates among other metrics. The following table summarizes the results of different machine learning models used in the study:

Table 33.5 Threat detection accuracy

Model	Accuracy	False Positive rate	False Negative rate
Support Vector Machine (SVM)	92%	5%	3%
Random Forest	94%	3%	3%
K-Means Clustering	88%	7%	5%

Of course, at the highest accuracy was Random Forest at 94%, coupled with a very low false positive rate of only 3%. The former had the highest rate of false positives because it struggles in determining close patterns in data.

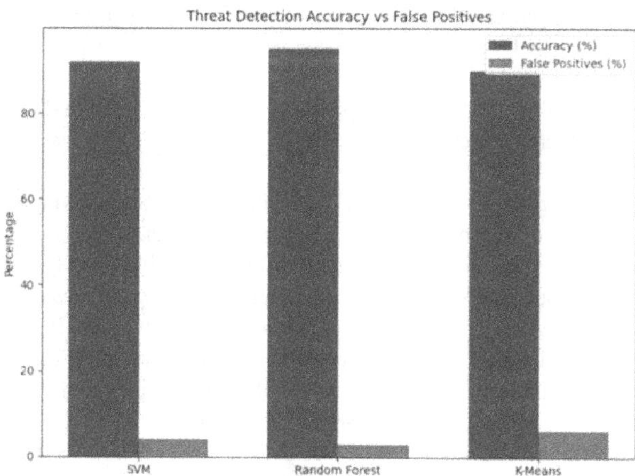

Fig. 33.1 Threat detection accuracy vs false positives

4.2 Encryption Performance

Data while at rest and during its transit was encrypted using AES-256 in the hybrid environment. Performance is a function of the time taken in encrypting and decrypting various sizes of data.

Table 33.6 Encryption performance

Data Size (MB)	Encryption Time (ms)	Decryption Time (ms)
10	120	130
50	180	190
100	220	230

It simply depicts the fact that encryption and decryption take a much longer time with the increase of the size of data. However, it performs reasonably well with smaller data sizes, though the overhead of encryption with large datasets is more pronounced.

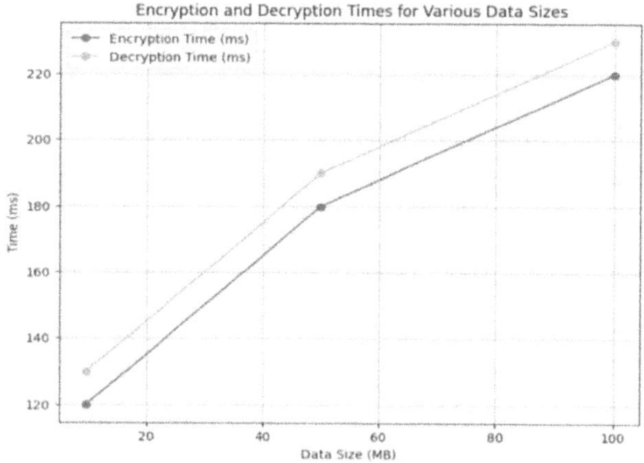

Fig. 33.2 Encryption and decryption times

4.3 System Response Time

A response time for the system is the measure of how fast it reacts to the threats that are detected and applies security measures such as encryption. It also helps in the facilitation of the high security features in encryption.

Table 33.7 System response time

Cloud Operation	Average Response Time(ms)	Peak Response Time(ms)
File Upload	150	200
User Authentication	90	120
Threat Detection Event	80	110

The average response time to respond to the detection of threats was 80 milliseconds; this indicates that the adaptive mechanism in threat detection was real-time, as it processed and mitigated threats.

4.4 Overall System Performance

The experiment verified the effectiveness of privacy preservation by showing the actual levels of well data masking and encryption that have provided protection to sensitive information from unauthorized access in data processing with fewer privacy breaches in hybrid cloud environments.

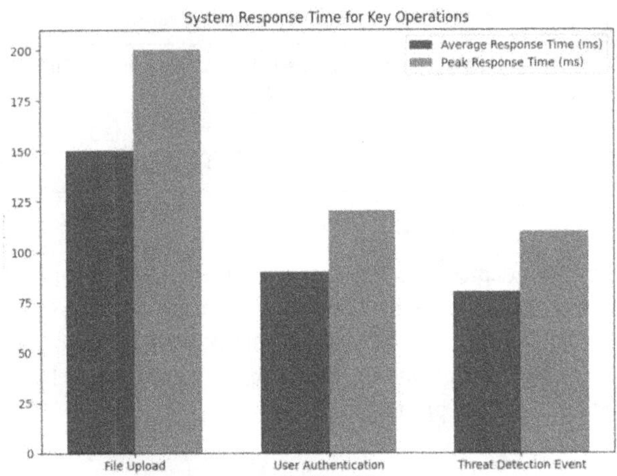

Fig. 33.3 System response time

Table 33.8 Overall system performance

Operation	CPU Usage	Memory Usage
Encryption Process	40 %	250
Threat detection process	30%	200
Data Masking Process	20%	180

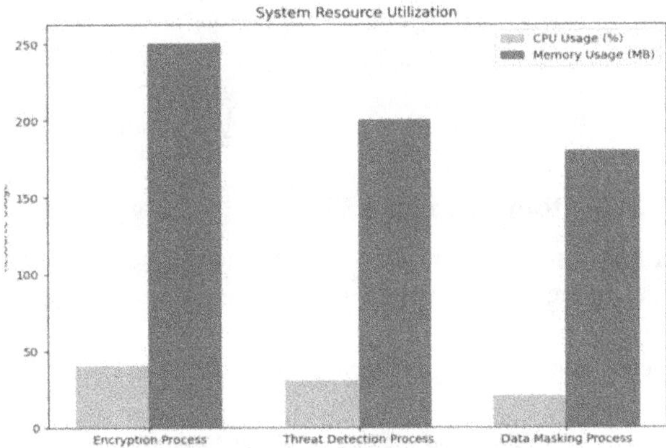

Fig. 33.4 CPU usage and memory usage

The graph represents the use of resources by the system at its critical operations. It is quite easy to notice that the total effect of all these security measures on performance is moderate and the framework is thus efficient in hybrid cloud environments.

5. Conclusion

The results of the review take importance in making adaptive detection and preservation of privacy critical in hybrid cloud environments. Hybrid clouds have specific needs and vulnerabilities that are realized as public and private infrastructures merge. This adoption of machine learning-based security strategies has greatly improved threat detection through better accuracy, minimizing false positives, and always responding in real-time. As far as privacy methods are concerned, while homomorphic encryption and differential privacy would be ensuring sensitive data are protected irrespective of the lifecycle as well as while they are moving across one cloud environment to another. Thus, these techniques enable hybrid systems to maintain data confidentiality as well as compliance with regulations while offering efficient data analysis. Moving ahead, security frameworks for hybrid cloud environments must be able to include multi-layered approaches combining adaptive threat detection, privacy preservation, and continuous monitoring to mitigate risk. Early adoption of zero-trust architectures and federated learning are promising steps towards more robust security frameworks, with the gains from such advancements allowing hybrid cloud environments to scale securely with high performance and reliability-a must if they are to be adopted within modern enterprises.

References

1. Johnson, M., Xu, P. (2023). "Supply Chain Vulnerabilities in Hybrid Cloud Environments." *Journal of Cloud Risk Management.*
2. Ramirez, S., & Clark, K. (2023). "Federated Learning in Cloud Security: A Distributed Approach." *IEEE Cloud Computing.*
3. Agrawal, T., & Sharma, R. (2023). "Multi-Layered Security Framework for Hybrid Cloud Systems." *International Journal of Cloud Security.*
4. Yang, L., Wei, H. (2022). "Challenges of Supply Chain Attacks in Hybrid Cloud Integration." *Cloud Security Journal.*
5. Zhang, Q., Li, H. (2022). "Homomorphic Encryption for Hybrid Cloud Privacy." *International Journal of Computer Security.*
6. Miller, P., & Thompson, G. (2022). "Zero-Trust Architectures for Multi-Cloud Security." *Cybersecurity Advances.*
7. Patel, N., Yang, J. (2021). "Hybrid Cloud Security in the Era of Remote Work." *Journal of Cloud Security Research.*
8. Chen, R., Singh, A. (2021). "Differential Privacy in Hybrid Cloud Analytics." *IEEE Transactions on Cloud Privacy.*
9. Rizvi, S., Mitchell, J., Razaque, A., Rizvi, M. R., & Williams, I. (2020). "A fuzzy inference system (FIS) to evaluate the security readiness of cloud service providers." *Journal of Cloud Computing*, 9(42), 964.
10. Li, J., Zhang, Y., Chen, X., & Xiang, Y. (2020). "Secure attribute-based data sharing for resource-limited users in cloud computing." *Computers & Security*, 72, 1–12.

Note: All the figures and tables in this chapter were made by the authors.

Intelligent Systems Using Semiconductors for Robotics and IoT – Dinesh Goyal et al. (eds)

34 Exploring the Role of IoT in Enhancing Teaching and Learning Experiences

Sonia Kaur Bansal[1]
[1]Department of English & Soft Skills,
Poornima Institute of Engineering & Technology,
Jaipur

Ashish Laddha[2]
[2]Department of Electrical Engineering,
Poornima Institute of Engineering & Technology,
Jaipur

Shruti Tiwari[3],
Versha Tripathi[4], **Seema Lenin Koshey**[5]
Gyan Vihar School of Education, Suresh Gyan Vihar University, Jaipur
Jaipur, India

Abstract: Internet of Things (IoT) possesses significant capabilities for upgrading the existing pedagogical structure including enhancement in the ways of learning. The investigations presented in this manuscript discusses the utility of the IoT in pedagogical system considering its effect on strategies, lecture hall settings, learners' engagement, and the net learning outcomes. Further, the work evaluates the recent avenues and specimen of IoT-driven solutions in pedagogical frame via deeper investigations. The manuscript investigates on the problematic issues and facets that remain associated with incorporation of IoT equipment in the lecture halls like safety, privacy, and ethical consequences. Furthermore, IoT may find applications in the creation of highly efficient, specific, and interactive learning environs, which is expected to foster learners' engagement as well as performance. Nevertheless, overcoming difficulties such as data safeguarding, tutoring, and requisite infrastructure is must for the successful implementation of the IoT-driven pedagogical system.

Keywords: Internet of things (IoT), Enhanced pedagogical cum learning experiences, Smart lecture halls, Data safeguarding, Personalized learning

1. Introduction

Pedagogy may be seen as one of the domains, who have transformed significantly via the Internet of Things (IoT). Conventionally practiced pedagogical and learning techniques are transformable, in whole, via the IoT. The latter can be realized as a network consisting of connected things, which are capable of interact along with sharing of data information online. The IoT has enabled enhanced sturdy, efficacious, and individual-level learning experiences through incorporating smart equipment, sensing devices, and thorough analysis of the data. Utilization of the IoT within the lecture halls makes the latter smart (may be termed as smart lecture halls). It also bears the possibility of attaining enhanced learning outcomes, better learners' engagement, and greater pedagogical schemes. Real-time observance, data-led decisions, and making frequent operations automatized may be seen as some of the advantages of IoT-led pedagogical schemes. Wearable equipment, smartboards, interactive projectors, among others may be noted as illustrations of IoT gadgets, which collect and assist in analyzing data of learners'

[1]sonia.bansal@poornima.org, [2]ashish.laddha@poornima.org, [3]shruti.tiwari@mygyanvihar.com, [4]manisht1971@gmail.com, [5]seemalenin82@gmail.com

DOI: 10.1201/9781003716389-34

performance; thus, enabling faculties to understand the way(s) of learning of each learner [1]. The requisite facts may be adjusted in-line with the requirements of every learner as personalized pedagogical approaches. In addition to the above, the IoT is expected to aid in creating highly efficient and learners' compatible environs. This would pave the way for motivating learners and engaging them in the pedagogical system [2].

The attribute that the IoT facilitates personalized learning makes it quite suitable for the present pedagogical system. This personalized learning is attainable via the IoT-driven equipment that allow noting learners' development and fetching requisite information outputted from their performance [3]. Furthermore, IoT accelerates mutual support via bringing faculties and learners together by utilizing wide variety of digital platforms. This, in-turn, permits for resource allocation and practical conveying beyond geographic borderlines. Provided that the IoT technical expertnesses offer easier as well as engaging learning environs, these can be beneficial specifically in the context of distant-mode pedagogical system [4]. IoT incorporation in the pedagogical system, at the present, is associated with certain difficulties. For example, the IoT-based equipment mostly collects significant amount of important learners' data. This creates significant concerns in terms of data safeguarding and privacy. Potential safeguarding measures need to be taken by educational organizations considering the fact that misusage of the important data might lead to extensive outcomes [5]. The price of establishment of the IoT infrastructure along with the need of aid and train teachers introduces an extra challenging situation, specifically for pedagogical organizations having lesser grants/funding [6].

In spite of the above-mentioned challenges, it can definitely be said that the IoT bears the capability of enhancing teaching as well as learning. By providing more flexible, student-centered learning environments, The IoT technical expertnesses are supposed to have important parts in determining the direction of pedagogical system of upcoming time [7]. The presented investigations explore the possible ways in which the IoT would assist in enhancing the learning system and analyzing the problematic issues associated with it. The investigations also attempt predicting the enhancement of the IoT in the future's pedagogical system.

2. Ideology of IoT-based Pedagogy

IoT has come out as a promising technology, which results in a connected biota by adding users, digitalized systems, and physical objects via the internet. IoT is progressing to revolutionize conventionally-practicing learning component of the pedagogy and enhancing teaching along with learning results through bridging of various smart equipment such as sensing devices, tablets, whiteboards, among others. Faculties may make data-led decisions outputting in the improvement of learner's engaging, tracking performance, and providing personalized learning experiences because of the availability of IoT appliances. The latter bear the capability of gathering, interchanging, and analyzing the obtained data [1]. IoT-led pedagogy incorporates the ideology of the smart lecture halls, which enable the working of IoT-driven appliances altogether for making enthusiastic and dynamic learning environs. Faculties may devote higher time on the pedagogical activities via the smart classrooms in which repetitive tasks get automatized through various equipment such as RFID-led attendance capturing system, sensors, and smart boards [2]. Sensors usually provide observation of the air quality, temperature, lighting, humidity, among others for enabling the superior learning environs [3]. Likewise, faculties may alter their pedagogical ways for adhering with the need(s) of every learner by utilizing smart equipment for observing his/her engagement [4]. The personalized learning facilitated through the IoT may be realized as a technique of instruction involving the tailoring of subject materials to every learner's choices and performance. Faculties may obtain a thorough know-now of every learner's learning ways, strength(s), and areas for development via collecting and investigating of practical data from connected equipment. This collected data further enable faculties to create one to one pedagogical strategies and interventions. These pave the way for understanding the unique needs of every learner [5]. As an illustration, learning management systems or wearable technical expertnesses enable the monitoring of learners' actions, checking their development, and giving them feedback regarding their learning targets [6]. On the learners' end, the above-mentioned data-led insights facilitate them for enthusiastically participating in the pedagogical system.

IoT also enables faculties and learners to team-up in an easier manner across geographic borderlines. Moreover, the IoT also allows coming together for distant students via practical conservations along with resource sharing through the connection of equipment and platforms [7]. With the increasing usage of hybrid and distant learning systems in the educational institutions afterwards the COVID-19 pandemic, this capacity has attained higher level of significance. IoT's attribute of facilitating distant pedagogical systems has resulted in the enhancement of access to top-quality education and assured learning continuity irrespective of geographical borderlines [8]. Incorporation of the IoT in the pedagogical system results in several advantages even though certain disadvantages may also be observed in terms of privacy and data safeguarding. Sturdy cybersecurity actions need to be considered on the topmost preference by pedagogical organizations for assuring the safeguarding of learners' data. It becomes essential because IoT-led equipment collect vast amount of important data [9]. Besides, the higher prices incorporated with the erection of the IoT infrastructure and tutoring individuals (learners) for efficacious utilization of the relevant technologies might be a significant hurdle for certain pedagogical organizations [10].

IoT has significantly transformed the pedagogical system incorporating the practical database-led knowledge to one to one (i.e., personalized) learning. Further, IoT in the pedagogical framework imparts considerable changes in the teaching as well as learning approaches. It is greatly expected that this technical expertness would acquire higher importance in the view of determining the pathway of pedagogical system of the upcoming time. This would be on the behalf of establishment of the highly interlinked, compelling, and efficacious learning environs. Considerable changes have been introduced by the progressing incorporation of the IoT technical expertness into different industrial as well as pedagogical organizations. Practical database collection, thorough analysis, and enthusiasm via involvement have been made feasible via IoT equipment. This ultimately results in the intuitive learning environs that upgrades pedagogical techniques and learners' experiences. This manuscript also investigates on the applicability of the IoT technical expertnesses in enhancing the efficacy of the pedagogical system and the subsequent issues resulted after deploying the same.

3. Recent Avenues in IoT-led Pedagogy

Pedagogical systems with IoT are enhancing with rapid pace and transforming the ways of delivering content as well as learning. As discussed earlier, this recent technological avenue works by accommodating the knowledge and enabling conversations via various kinds of tool, equipment, and sensors to the internet. This capability is being utilized for creating smarter and engaging environs that enhance learner and faculty experiences. Feasibility of currently practicing smart lecture halls enabling interesting content delivery should be credited to IoT equipment. Apart from the above, IoT technical expertness aids faculties in ways such as automatizing administrative activities likewise usage of RFID cards or biometric systems for knowing learners' attendance.

This recent technological avenue also proves to be supportive for faculties in terms of streamlining their works along with minimizing disturbances within the lecture halls. Moreover, utilization of the wearable technical expertness (likewise fitness tracking equipment or smart watches to note the data of learners' anxiety levels, sleeping schedule, to name a few) and other IoT-driven equipment to monitor learners' performance as well as wellness may be noted as another considerable advancement. Faculties may alter their pedagogical patterns in-line with the database obtained through above-mentioned gadgets for accommodating the needs of every learner. With the IoT-led technical expertness, learners get access of the subject materials at their own pace through the connected equipment. Faculties also get empowered to note every learner's advancement practically and thus, adjusting their suggestions/advices accordingly. As an illustration, IoT-empowered flexible learning platforms may bring alterations in the difficulty level of each of the subjects as per the grasping capability of every learner.

Additionally, IoT also bears the capability of enhancing efficacy of the pedagogical system on the organizational level. Besides imparting safeguarding and convenience to the learning, the IoT aids organizations in efficacious management of available resources likewise automatized climate control, efficacious lighting system, among others. Thus, it may be inferred that the recent avenues in the IoT utilization in pedagogy emphasize on the creation of efficacious, flexible, convenient, and participative learning environs. IoT-driven advancements, as discussed earlier, enhance the overall pedagogical structure of organizations, imparting innovative pedagogy techniques, and latest tactics of noting cum supporting learners' progress. IoT-driven tools also contribute in enhancing teamwork and obtaining speedy-feedback regarding academic achievements.

4. Impact on Pedagogical Methodologies

IoT can greatly enhance the traditional classroom pedagogies. Smart devices can help provide customized learning to students based on their individual student requirements [9]. For examples, devices embedded in classrooms can snap a picture of the class to see their behavioral tendencies, which will allow teachers to adapt their strategies even without promotions. Additionally, IoT tools can deploy communication techniques that relieve the need to track student attendance or completion of various assignments so that teachers can spend a bigger chunk of their time on more important processes like core learning facilitation [10]. With the application of IoT into the education space, there is a transformation of the traditional ways of teaching, such that, the methods become more interactive, fluid, and data-centric. IoT for instance can facilitate the improvements of educational plans enabling the teachers to enhance their approaches to teaching students. Perhaps, the most prominent effect of IoT on education is the transition from general instruction to more student specific attention. Equipped with IoT devices students learning can be monitored and the time spent on various activities as well as measures such as participation and the overall performance can be recorded. Such information assists educators in customizing lessons and their instructional methods for each learner's requirements, preferences for learning, and their feelings (for instance, as a support in a smart classroom there may be connected tablets and apps that provide adjustment in content delivery to the areas of strength and weakness of each student thus providing appropriate support and intervention as and when necessary).

Moreover, IoT brings about a transformation in how students and teachers interact and creates catering more active forms of learning. It has also helped convey Multi-dimensional video in smart boards, enabling teachers to aggregate group work and student engagement. With IoT teachers can ask for students' feedback in real time by means of connected devices and understand whether the learners are getting the content and how best to grade them without much hassle. This 'on-line' data indicates whether the students are assimilating concepts, so that appropriate action can be taken in the environment. IoT also handles tedious administrative tasks in classrooms giving the teachers more time to teach. Things like checking whether a student is in class, submitting assignments and grading can be controlled by IoT networks and systems to cut down administration work. For example, when students enter the classroom attendance can set off automatically using RFID patches or biometric devices. An important thing to note is the developments that are leading to the potential for blended learning models. In this case, the IoT provides an opportunity to streamline in-person learning and online learning. Using connected devices, teachers make it possible for students to access learning materials, participate in discussions and submit assignments with the classroom walls and boundaries displaced. With such features, students' engagement in educational content is not only fragmented but also convenient. They can learn anywhere and at any time.

Moreover, IoT facilitates the decision-making process in teaching as data. Thus, the information obtained from Internet of Things devices is useful to the teachers for analyzing the performance of the students, how engaged they are, and their behavior in the classroom during breakout sessions. Such data analysis may assist the teacher in determining how best to plan out the lessons, what curriculum to modify, or how to support the children with individual needs. IoT data has led to a shift in pedagogical approaches making them more practical as well as more elementary. In conclusion, the IoT is strengthening the education and fostering innovations that will fundamentally change the nature of education as innovative practices become prominent. Teachers' ability to collect and study data accesses great possibilities to address students' problems in real time, creating engagement and further better results. With the help of IoT children's lives will be more interesting and easy as numerous tasks will be done automatically; numerous new technologies will be employed and applied into education issues.

5. Enhanced Learner Engagement

It has been established that through the use of IoT, students are more willing to be actively involved in learning activities. Thus, smart watches and connected fitness trackers can be used to track the level of physical activity and mental engagement of students during lessons [4]. Moreover, IoT devices allow involvement of students in active, practical learning that is not limited to just being taught through lectures [11]. One of the biggest challenges in effective learning is the level of the student's engagement, and the latest developments in IoT's fundamentals can combat this challenge by creating new ways for students to engage in the course. IoT technology helps educators to reach their students in novel ways allowing them to be active participants of their education process. The engaging aspect of IoT Interactive learning is its interactive nature; it actively engages students instead of encouraging passiveness. Classrooms, in most cases, encourage passive learning where students get information and do not actively participate.

With the introduction of IoT concepts, this environment is being transformed as smartboards, digital tablets, and connected learning tools are augmented and these promote active participation among the students. Such devices can create a number of activities including real-time polling of students, quizzes and joint tasks that need students to interact with the material and themselves. Another way IoT boosts engagement is by facilitating hands-on (experiential learning). Sensors, digital reality structures, and related lab equipment are examples of IoT-enabled gadgets that let college students inspect ideas in a greater palms-on and attractive way. For instance, real-time experiments using IoT sensors may be performed in medical lectures, permitting students to see and examine facts the use of related devices. These kinds of reviews assist college students observe their instructional expertise in sensible settings with the aid of making gaining knowledge of more exciting and tangible.

Moreover, IoT helps individualized learning, which increases learners' engagement and motivation significantly. With the IoT, learners can get admission to customized studying content material that caters to their desires and studying patterns. This kind of facility enables learners getting understand the relevance and create enthusiasm to them. For an illustration, flexible learning mechanism led by IoT appliances provides environs for improving scholar's performance through their enhanced engagement. This, in-turn, helps learners attaining comparatively better success considering the fact that they are able to learn at the pace convenient to them.

Other than the above, IoT-led collaborative learning also aids in enhancing engagement of the learners. IoT technical expertness-led tools outcome betterment in the communication as well as cooperation within the institutional premises. Several IoT-assisted equipment support learners in terms of working together on projects, sharing their outcomes practically, and collaborating on assigned works irrespective of geographical borderlines. IoT-operated pedagogical system also enhances learners' engagement via boosting their morale and making the feedbacks readily available. Wearable appliances (likewise fitness monitoring equipment, smart watches, among others) bring feasibility in getting the health status of learners at all points of time. This lets the faculties getting the status of their engagement, dedication, and health. Acquiring such knowledge creates the pathway for faculties to incorporate the necessary modifications in their pedagogical plan. These modifications may include enhancing of the assignment count during the time(s) spanning lower dedication, giving small breaks while noting a higher anxiety levels, to name a few.

IoT also facilitates gaming-led pedagogical system, which also aids in enhancing the engagement of learners. A few of these gaming tactics include assigning badges, credits, rewards for the completion of provided tasks and/or enthusiastic participation in pedagogical activities. At this end, it may be inferred that gaming-associated pedagogical strategy motivates learners for competing and brings enthusiasm for enhanced engagement.

6. Smart Lecture Hall Environs

The ideology of smart lecture hall incorporates inclusion of the IoT-driven appliances within the existing lecture hall environ for enhancing the pedagogical level. As an illustration, mounting of temperature sensing equipment, light controlling mechanism, white board(s), etc. would aid in making the lecture hall conditions more appropriate for the pedagogy [12]. These classrooms are designed to guide a more dynamic and adaptable studying surroundings, encouraging student-targeted coaching methods [13]. The concept of clever classrooms revolves around the combination of the IoT technologies to create greater connected, green, and adaptive mastering environments. these IoT-enabled school rooms make use of a network of interconnected gadgets, sensors, and data structures that engage with one another, remodeling the traditional study room into a sensible, interactive area designed to beautify both coaching and gaining knowledge of. One of the primary advantages of IoT in smart school rooms is the automation of school room capabilities, such as lights, temperature control, and classroom equipment. Sensors can stumble on the number of college students in a school room and routinely alter lighting fixtures and HVAC structures to optimize the environment for consolation and power performance. This creates an extra conducive mastering environment, allowing students to awareness better without distractions as a result of pain, even as additionally lowering the energy consumption of tutorial establishments [8].

Moreover, smart classrooms leverage interactive devices along with smart boards, capsules, and digital projectors, which allow instructors to give multimedia content extra effectively and engage students in real-time activities. These devices sell collaborative studying, in which students can work together on shared tasks, brainstorm ideas, and solve problems interactively. By connecting all students' gadgets to a principal machine, educators can reveal participation, offer instant comments, and tailor training to the needs of man or woman students [13]. IoT gadgets also support actual-time records series on scholar overall performance and behavior, enabling instructors to analyze mastering styles and engagement levels. For instance, motion sensors and cameras can tune scholar motion and participation throughout magnificence activities, helping educators investigate which students are actively involved and which may additionally need additional aid. These statistics let in teachers to make informed adjustments to their teaching methods or provide personalized help to students who can be struggling [9].

Moreover, the personalization of mastering is a key feature of smart lecture rooms. Through IoT technologies, academic content may be tailored to the man or woman wishes and learning styles of every student. for instance, connected gadgets can tune a pupil's progress through a lesson and adapt the problem level based totally on their overall performance. Such structures provide a greater engaging and customized gaining knowledge of experience; allowing students to development at their personal pace and recognition on areas where they need improvement [12]. IoT also enhances the safety and safety of tutorial environments. smart lecture rooms may be equipped with linked protection cameras, RFID access controls, and emergency reaction structures. Those gadgets help to display who enters and exits the lecture room, make certain that only legal people are gift, and provide a brief response in case of emergencies. This progressed protection facilitates create a more secure studying surroundings for both college students and body of workers [14]. Notwithstanding the numerous advantages, the implementation of IoT in smart school rooms comes with challenges, especially regarding statistics privacy and security. With such a lot of gadgets amassing records on scholar behavior and performance, instructional establishments need to make certain that private information is saved securely and used ethically. This requires the improvement of robust cyber security measures and clean regulations on statistics utilization and possession [15].

7. Challenges in IoT Adoption

Even though there are numerous blessings to IoT in training, there are drawbacks to integrating that technology. For the reason that IoT devices produce substantial volumes of statistics, such as non-public and sensitive information, security and privations problems are important [7, 8]. To safeguard each youngsters and instructors, its miles crucial to make certain that this statistic is dispatched and saved securely. To facilitate the powerful deployment of IoT in academic settings, sufficient infrastructure is also required, including dependable internet connection and

IoT tool control structures [16]. IoT integration in training has the potential to completely rework gaining knowledge of settings, expertise for you to efficaciously use these technology, educational establishments need to triumph over some of essential boundaries. Those difficulties consist of such things as infrastructure troubles, worries about facts safety and privations, the requirement for correct education, and moral concerns. The shortage of infrastructure in many educational institutions is one of the most urgent problems. To perform at its exceptional, IoT wishes a dependable network of sensors, gadgets, and internet access. Know-how, now not all schools have the infrastructure had to enable the mixing of IoT technologies, together with contemporary computer structures, excessive-velocity net, and enough bandwidth. This hassle is especially substantive in impoverished or remote faculties, wherein the virtual divide can obstruct IoT adoption and exacerbate disparities in get admission to contemporary coaching sources [14].

IoT adoption is drastically hampered by means of infrastructural issues as well as concerns approximately facts security and privacy. IoT gadgets in lecture rooms often accumulate private information about pupils, inclusive of biometric information, behavior patterns, and performance information. These gadgets are vulnerable to records breaches, unlawful get admission to, and cyber-attacks in the absence of strong safety features. Installation of location sturdy cyber security frameworks within the pedagogical organizations is requisite for bringing friendliness towards learners in the pedagogical system through proper utilization of available financial resources [15]. Institutions trying to use IoT also face extra demanding situations because of regulatory compliance with policies just like the circle of relatives instructional Rights and privateness Act (FERPA) and the general records safety regulation (GDPR) [17]. The dearth of technical information and training among personnel members and teachers is another difficulty. Many educators and directors lack the expertise and competencies important to control and troubleshoot IoT devices, as well as information of how IoT features. Instructors can also discover it tough to comprise new technology into their teaching techniques because of this talent gap, which may hinder the success deployment and utilization of IoT in school rooms. Establishments ought to make professional improvement investments and deliver educators the information they want to apply IoT solutions successfully so as to triumph over this [13]. Another crucial element influencing IoT adoption in education is cost. For plenty educational establishments, especially those with tight budgets, the upfront expenses of buying IoT devices, updating infrastructure, and keeping these systems might be unaffordable. The monetary burden is in addition extended by means of the prices related to keeping IoT systems, information management, and cyber security [10]. It may be difficult for lots colleges, mainly those in low-profits communities, to shield these charges in light of different urgent necessities. There are additional difficulties because of moral troubles with IoT software in schooling. There are worries over the utilization and accessibility of the good sized quantities of scholar records which are accumulated. To guarantee that the information accrued is handled accurately and that students' right to privacy is upheld, faculties should set express moral requirements. Furthermore, algorithms couldn't continually take into consideration the numerous pupil demographics and getting to know styles, that may result in unjust treatment or incorrect information interpretation [9]. As a result, it's miles important to address capacity biases in statistics interpretation.

In the end, the adoption of IoT in schooling can be hampered via competition to alternate from mother and father, students, and even instructors. Many human beings may be cautious of recent technology because of subject that they will lessen the position of teachers or update traditional instructional techniques. so as to triumph over this competition, it's far essential information the blessings of IoT at the same time as additionally making sure that these technologies are utilized to supplement human preparation and interplay, no longer to replace it [9]. it is also vital to recall the ethical ramifications of IoT use in training. There is a threat that the huge volumes of data on student behavior that IoT devices gather might be misused [10]. To guarantee that students' privations rights are upheld, academic establishments have to set express suggestions for facts ownership, get admission to, and ethical use [15].

Even though there are many advantages to IoT in education, there are drawbacks to integrating these technologies. Considering the fact that IoT equipment produce vast amount of data (personal and sensitive), issues related to safeguarding as well as privacy bear significant importance [7, 8]. Assuring the safeguarding of learners as well as faculties needs storage and communication of the data in the secured manner. Moreover, proper internet connectivity and IoT equipment management systems along with requisite infrastructure is essential to deploy the IoT beneficial perspectives in the view of pedagogy [16]. Nevertheless, IoT integration in the pedagogical view necessitates overcoming of various hurdles. Some of them may be infrastructure scarcity, concerns relating database safeguarding, necessity of the proper-mannered coaching, inculcating ethical behavior, to name a few. Attainment of The best possible outcome of the IoT-framed pedagogical system possesses the dependency on internet accessibility, sensing equipment, etc. Every pedagogical institution might not possess the infrastructure necessary for incorporating the IoT technical expertnesses. This problematic issue is specifically noticeable in remotely-located organizations lacking adequate accessibility of the IoT infrastructure and pedagogical resources [14].

IoT equipment in lecture halls often collect learners' privatized data likewise biometric, behavioral, and performance-centric. IoT-led equipment suffer from the issues named data breaching, unauthorized access, cyber-attacks, to name a few if the safeguarding arrangements are not up to the mark. Thus, establishment of robust cyber security frameworks is of utmost importance for securing learners' data. Adequate monetary provisions along with technical know-how is requisite in this regard [15]. Organizations willing to utilize IoT also come across the regulatory compliance-related problems such as the FERPA, GDPR, among others [17]. Missing technical knowhow along with training opportunities of staff members as well as faculties is another issue.

Many of the faculties and administrative officials lack in the requisite knowledge and skills regarding function(s), control, and troubleshoot of IoT appliances. This lack of expertise makes faculties finding difficult to incorporate recent technological avenues into their pedagogical methodologies. This ultimately impedes deploying and utilizing the IoT within the lecture halls. Organizations must be enthusiastic on the investing head corresponding to professional advancement and thus enabling educators to acquire the know-how concerning the efficacious usage of the IoT [13]. Prices associated with implementation also affects the IoT adoption (likewise buying equipment, updating existing infrastructure, and maintenance, data handling, cyber security, among others) in the pedagogical frame [10]. The low-income categorized institutions might find it difficult to bear various prices associated with implementing IoT. Apart from the monetary, the IoT implementation also faces challenges on the ethical head. Functioning of this ethical head incorporates assuring appropriate data handling, maintaining learners' privacy, to name a few. Moreover, algorithmic techniques incorporated for interpretation of learner parameters must be reliable [9].

The IoT adoption also faces challenges in terms of the denying by parents, learners, faculties, etc. This denying might be on the account of the belief that the IoT-driven pedagogical system would substitute its conventionally-practicing counterpart and that might be harmful from the job security point of view. Thus, there is the need of making the concerned people understand that this recent technological avenue would bring enhancement and aid in the conventional pedagogical ways [9].

8. Conclusions and Future Scope

It can be stated that the potential for improving the experiences of teaching and learning experiences via the combination of IoT generation in schooling is substantial. IoT has the capability to decorate student engagement and effects by way of organizing more customized, interactive, and powerful mastering environments. The powerful use of IoT in training, however, depends on resolving troubles with infrastructure, training, security, and privacy. The creation of thorough frameworks for the ethical and at ease use of IoT in academic environments needs to be the principle purpose of destiny studies. Through the introduction of collaborative, personalized, and interactive learning environments, IoT notably improves pupil engagement. IoT generation facilitates immersive gaining knowledge of studies and real-time comments, which strengthens the bond between college students and the direction cloth and in the end improves motivation, engagement, and mastering outcomes.

IoT-enabled smart classrooms, which provide attractive, customized, and secure learning environments; mark a dramatic trade in the manner training is supplied. IoT turns the conventional school room right into a dynamic environment in which instructors and students may also flourish by using fostering collaboration, automating chores, and gathering records in real time. However, infrastructural and data protection problems should be resolved if clever lecture rooms are to reach their complete capacity. Despite the fact that IoT has the potential to seriously enhance schooling, a number of issues want to be resolved before it can be efficaciously used. These include getting around infrastructure constraints, protecting the privations and protection of facts, giving teachers' proper training, controlling fees, and resolving moral dilemmas. to overcome these boundaries and establish an IoT-driven gaining knowledge of surroundings that blessings all parties worried, legislators, educators, and technology companies have to paintings collectively.

References

1. Al-Fuqaha, A., Guizani, M., Mohammadi, M., Aledhari, M., Ayyash, M. (2015). Internet of things: A survey on enabling technologies, protocols, and applications. *IEEE Commun. Surv. Tutor.*, 17(4):2347–2376.
2. Kumar, M., Shukla, V. (2020). The role of IoT in education: Challenges and opportunities. *Int. J. Recent Technol. Eng.*, 8(3):20–24.
3. Papageorgiou, X., Ioannou, A. (2019). Smart classrooms: IoT in education. *IoT Technol. Smart Cities*, 155-173.
4. Aslan, M., Reigeluth, C. M. (2021). Personalized learning: The role of IoT and big data in education. *J. Educ. Technol.*, 49(2):75–82.
5. Zhang, Z., Wang, J., Liu, X. (2021). IoT in education: Real-time data-driven personalized learning experiences. *J. Comput. Sci. Technol.*, 36(3):501–514.
6. Singh, A., Gill, R. (2020). Challenges of implementing IoT in schools: An Indian perspective. *J. Educ. Technol. Soc.*, 23(2):56–65.
7. Yuvaraj, M. (2021). The future of IoT in education: Enhancing learning environments. *Educ. Inform. Technol.*, 26(3):345–362.
8. Kim, J., Lee, S. (2021). The role of IoT in smart classrooms. *J. Educ. Technol.*, 45(3):112–126.

9. Martin, G., et al. (2020). Personalized learning through IoT: Case studies in education. *Adv. Educ. Res.*, 34(2): 45–60.
10. Zhang, Y. (2022). IoT and the automation of teaching processes. *Educ. Technol. Q.*, 12(4):68–75.
11. Gupta, A., Singh, R. (2021). Wearables in education: Enhancing student engagement. *IoT Educ. Rev.*, 29(1): 87–94.
12. Smith, C. D. (2020). IoT in active learning: Improving engagement and outcomes. *J. Interact. Learn. Res.*, 17(3):23–38.
13. Patel, K. (2022). The future of smart classrooms: IoT applications in education. *Smart Learn. Environ.*, 15(2):101–109.
14. Watson, L. (2022). IoT infrastructure and the digital divide in education. *J. Educ. Technol.*, 33(3):120–131.
15. Jones, M. (2021). Data privacy challenges in IoT-enabled classrooms. *Educ. Data Security Q.*, 9(4):56–63.
16. Thompson, P. (2020). The interactive classroom: Transforming education with IoT. *Educ. Innov.*, 23(1):14–27.
17. Adams, R., Miller, J. (2021). Ethical implications of IoT in education. *J. Ethics Educ.*, 18(2):62–75.

Intelligent Systems Using Semiconductors for Robotics and IoT – Dinesh Goyal et al. (eds)
© 2026 Taylor & Francis Group, London, ISBN 978-1-041-20408-4

35 Cache Technique for Mobile Adhoc Networks with Internet Access

Sugandha Singh[1]
Department of CSE-IoT,
Noida Institute of Engineering and Technology,
Gr Noida

Manpreet Singh Bajwa[2], **Jyoti Godara**[3]
Department of Computer Science and Engineering,
Faculty of Engineering and Technology,
Shree Guru Gobind Singh Tricentenary University,
Gurugram, Haryana, India

Abstract: The Internet-based mobile ad hoc network (iMANET) is a new technique for building a pervasive communication network, It integrates a Internet, (wired network) with a mobile ad hoc network (MANET). However, to meet users' needs for access to a variety of information, an iMANET has a number of drawbacks, including insufficient wireless capacity, greater message latency, and restricted wired Internet accessibility. The problems with information access and search in iMANETs are discussed in this research. To improve accessibility and decrease latency, an improved search algorithm has been created that utilizes local caching.

Keywords: Information search, BGP, MANET, ASs, IGW, iMANET

1. Introduction

A self-governing group of randomly moving mobile nodes that deal with one another using multi-hop relays is called a MANET. Dynamic and unpredictable changes in network topology are possible. The MANET lacks a backbone, centralized administration, and pre-existing infrastructure. Since MANETs are often designed to function as stand-alone networks, traffic will be limited inside MANET. By using the same routing protocol, every member of MANET contributes equally to the dissemination of routing information and route maintenance. Traffic needs and topological changes must be accommodated via effective ad hoc routing protocols. MANETs can be used for short-term communication requirements such as military missions, emergency rescue, conferences, and aid for disasters [1]. It should soon be possible for consumers to access online resources and data at any time and from any location. Contemporary wireless communication infrastructures are being developed by wireless carriers in order to do this. However, mobile terminal (MT) may nevertheless have trouble connecting to a wired network or the Internet because of limited wireless capacity and accessibility [20]. An MT[†] must contend for bandwidth during periods of high traffic and risks being barred from a wireless base station [2]. In recent past, a significant number of MANET research has been published in the literature [3–8]. To enhance communication between MTs in constantly shifting topologies, the majority of these initiatives have focused on developing routing protocols. Because wireless Internet access is becoming more and more popular and less expensive, it is now essential to think about integrating MANET with wired Internet. Consequently, an Internet-based MANET, known as iMANET, is taken into consideration and the problem of data retrieval and use in this context is investigated in

[1]prof.sugandhasingh@gmail.com, [2]manisinghbajwa@gmail.com, [3]jyotipoonia6@gmail.com

[†] MT refers to a person who carries a portable device, such as a laptop, computer, PDA, cell phone, etc.

DOI: 10.1201/9781003716389-35

order to implement MANET technology in a real-world situation [9]. Some of the MTs in iMANET are expected to have wired private networks or Internet connections. As a result, an MT can obtain information from the Internet directly or through relays from other MTs.

Some scenarios that apply to an iMANET are as follows:

Scenario 1: MTs can be used as proxies for additional MTs on the battlefield or at the emergency location by connecting to the Internet via satellite. The services and information accessed by other MTs are also shared via the local ad hoc network.

An iMANET has a set of MT's in its network. The communication in MT's is based on adhoc communication protocol. Dashed lines in Fig. 35.1 illustrate the communication on adhoc network. Anywhere in the protected area, an MT is designed to move in any direction for the resources and data. Because they have direct connection to the Internet, some of the MTs in the iMANET can act as AP* for the other MTs. Relays are required in order for an MT that is outside of the AP's range of communications to connect to the Internet through one of the APs.

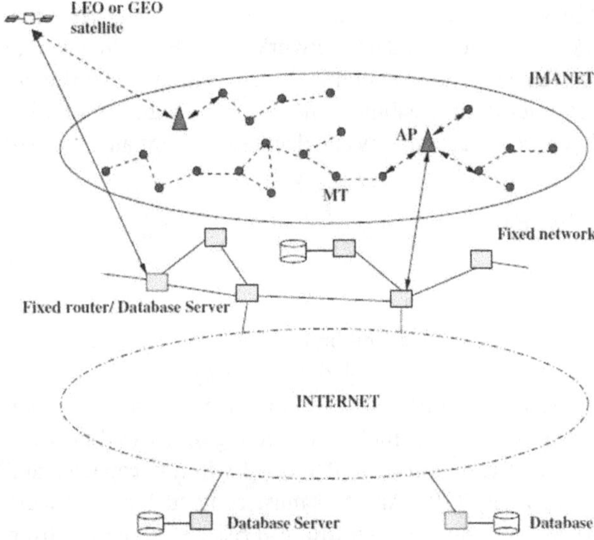

Fig. 35.1 An iMANET network model

Consequently, an AP is considered to have access to all data and acts as a gateway to the Internet. An AP can connect to any database server† which is located in an interface area. The database is thought to have a total of n data items (DBmax). The fundamental building block of an update or query action is a data item (d). For read-only requests, the MTs construct a query, and the server updates the database. Because the cache area capacity of an MT is limited by the entire amount of record items in the database, it can only store a certain number of data items.

Depending on the cache invalidation procedure, a server can either proactively (periodically) or subsequently (non-periodically) broadcast the list of modified data item IDs to the MTs in order to guarantee data consistency. In order to make sure that every node receives the communication whenever a server broadcasts ids, flooding techniques may be employed. The broadcast multicast tree can be used to lower the message overhead because flooding generates a lot of traffic. This study therefore makes the assumption that the multicast tree is constructed and that it may be setup and maintained in iMANET by according to the instructions in [11,12].

One of the most important challenges for MANET connectivity is routing compatibility. Ad hoc routing protocols were initially developed to enable stand-alone MANETs without the need for a centralized router. Unlike the Internet Protocol (IP), a pure ad hoc routing system necessitates that every node act as a router and be involved in discovery of route and its maintenance. The ad hoc nodes cannot access the routing information outside of the MANET's borders.

Ad hoc routing protocols are unable to manage communications over the barrier between MANET and the Internet when they are used in IP-based networking contexts. The ability to integrate (or interface) between IP routing along with ad hoc routing must therefore be considered. Other challenges related to IP mobility also emerge in the iMANET. [13] A node in the MANET must be online in order to use internet apps. Finding a gateway to transfer data between the MANET's nodes and the Internet is necessary to link the nodes inside the MANET to the Internet. Internet gateway finding can be done in a number of ways, such as proactive, reactive, and hybrid techniques. Tables drive proactive techniques, sources drive reactive approaches, and a hybrid approach combines the two. [14] A mobile node inside the MANET (Fig. 35.2) or any external stationary node outside the MANET can serve as an internet gateway. (Fig. 35.3). For MANET to be connected to the Internet, a different protocol architecture is needed. In MANET, computing power and storage space are the challenging points. For data access and information processing, they rely significantly on other hosts and resources.

To translate between these "two languages," a gateway needs to be able to comprehend both [15]. Several mobile nodes having wireless interfaces can connect directly or through other nodes in these autonomous networks. Generally speaking, MANET topology is dynamic and asymmetrical; nodes communicate across wireless

* With the right antenna and logical notation, an access point (AP) can establish a direct connection to the Internet through wireless infrastructure, including low earth orbit (LEO) satellites, cellular base stations (BSs), and geostationary (GEO) satellites.

† According to MT, the terms "AP" and "database server" are interchangeable because an AP is transparent to a database server.

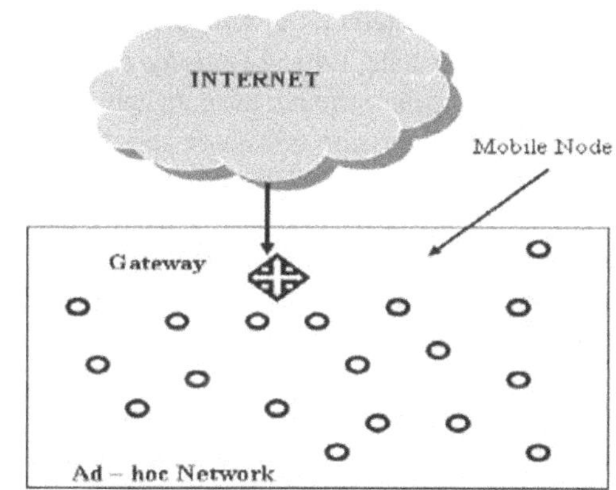

Fig. 35.2 iMANET with mobile gateway

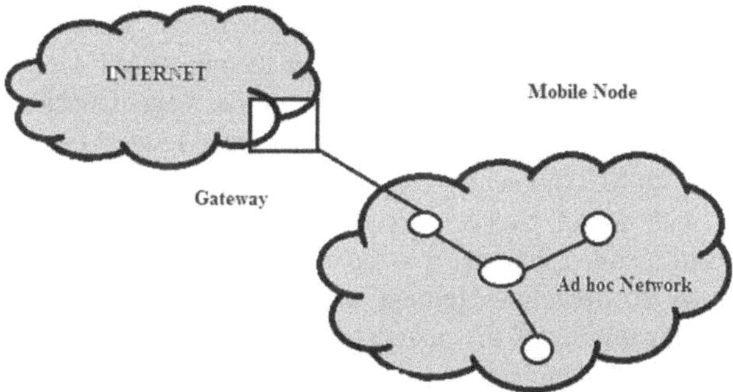

Fig. 35.3 iMANET with fixed gateway

networks with varying transmission ranges due to node exit and new node arrival over connectivity time. [16].

The rest of the paper is organized as follows: Various problems and difficulties that were faced when supplying internet connectivity to MANET are discussed in section 2. Section 3 suggests an improved search method to identify the information access channel to relevant APs or to MTs using the request's cached data.

The simulation matrices are displayed in Section 4. Section 5 brings the paper to a conclusion.

2. Issues and Challenges in iMANET

There are numerous proposals available right now to address the issues with iMANET. The majority of the ideas are complicated, and the solutions haven't been thoroughly assessed yet.

When there is at least one possible path to one or more gateways, establishing internet access for MANETs with node mobility management capabilities at the micro as well as the macro level is the primary difficulty in iMANETs, that is, within each MANET domain as well as among the various MANET domains, while maintaining consistent and unrestricted internet connections. Three difficult iMANET subproblems can be characterized as follows:

(a) Determine the position of the node.

(b) Gateways are found, and

(c) Consistent forwarding states to the gateways are established and maintained.

The situation determines the nature of these issues. It is difficult to draw any conclusions about the network's appearance unless the situation is extremely particular or there is an administrative organization present. By definition, MANET fails proper management. It becomes difficult to assume, for example, that nodes utilize a particular prefix, traverse a particular path, or have a single gateway in order to access their assigned IP address.

3. Information Search in iMANET

3.1 System Model

Some of the MTs in the iMANET have the ability to connect directly to the Internet, making them access points (APs) for the other MTs [17–19]. As a result, an AP serves as an Internet gateway and is thought to have access to all data. An MT outside of an AP's communication range needs to be connected to the Internet using a relay from one of the access points. Wherein MT is free to travel in any direction and search the protected region for information and access. An MT creates a direct connection with an AP when it is near the AP (within one hop, for example). However, information access must pass through multiple ad hoc network hops before arriving at the AP when an MT is situated far from an AP.

3.2 Enhanced Search Algorithm

The key focus is on to improve information availability in iMANET. Like the Internet, iMANET requires a search algorithm since it lacks the destination address for any necessary information because any node will work as a router to pass on the information. An existing MANET routing protocol can be used to construct this technique. iMANET design is braced with an aggregate cache concept. It can collect data items from an internet-connected AP as well as the MT local cache. An MT transmits a request to all nearby MTs when it requires a data item because it is unsure of its precise location. When an MT receives a request, it first sees if the data item is in its local cache before sending the query to its peers. After a request has been flooded throughout the network and as shown in Fig. 35.5, certain MTs may have cached copies of the demanded data item after it has been fully accepted by an AP. Based on the aforementioned concept, an information search algorithm known as Enhanced Search (ES) select an access path to pertinent APs or to the MTs using the request's cached data.

The order in which the MTs or APs acknowledgements arrive serves as the basis for the decision. In the scenario where MT (n_i) seeks for a data item (d), and an MT (n_k) is identified on the route to the AP, where k ε {a, b, c, j}. The ES algorithm works as explained: at the first step it ni asks for d in its local cache. It directly contacts the AP, failing in direct access of AP, broadcast request method is followed. Both the requester's and the requested packet's IDs are contained in the request packet. It waits for a response after broadcasting the request. n_i fails to receive d if it does not receive any acknowledgment within the allotted timeout interval. In case d is not found by n_k in its local cache upon collecting a packet of requests, it passes the message to nearby MTs. If n_k has data for d, then n_i transmits an ack packet. After receiving the request packet, an AP simply responds with an ack packet.

In order to preserve and save the route information, the AP's ID is appended to the packet when it needs to issue an acknowledgement. The MT node is treated in the same way. Unlike a request packet, which is distributed, an acknowledgment (ack) packet is transmitted just along the path specified in the requested packet. The ack packet sender receives a confirmation packet from ni, like an AP or n_k, after receiving an ack packet. Since an ack message from a neighboring MT or AP arrives earlier, ni chooses the path based on the first ack packet received and discards the remaining ones. An AP or n_k transmits the asked data (d) along the informed route after receiving a confirmation packet. After receiving a request packet, an MT examines to see if it has been processed. Instead of sending the packet to nearby MTs, the MT discards it if it has been processed. The MT further verifies whether the path linked to an ack, verify or reply packet has its id. Since those messages are only intended to follow the path that the request packet indicated, the packet is disregarded if the MT's id is not found in the route.

To stop packets from bouncing around the network, a request packet's hop limit is employed. Therefore, if a request packet has more forwarded hops than the hop limit, an MT refrains from transmitting it to the nearby MTs. This reduces the amount of bandwidth consumed by many data packets and network congestion. MTs attempt to search within themselves using cached copies when they get separated (as illustrated in Fig. 35.4) and unable to access the data of interest due to their location outside of an AP's communication range.

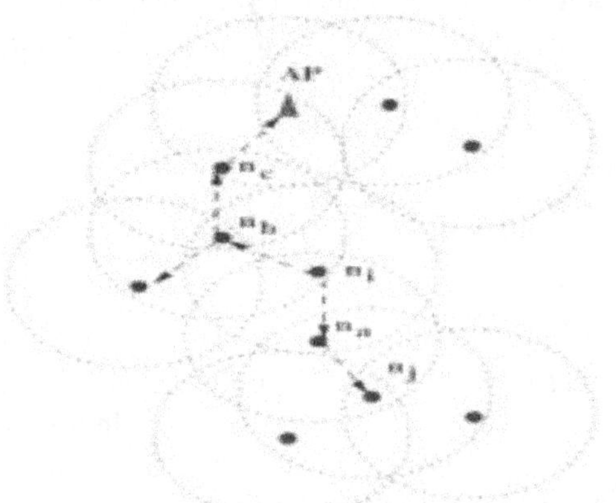

Fig. 35.4 iMANET, an MT (n_i) broadcasts a request packet that is forwarded to the AP

In this case, it is expected that nj will store the requested data item in its local cache. When MT obtains the necessary information to determine whether the data item needs to be stored in the cache, the cache admission control process is started. Accessibility is improved and latency is decreased by caching data items in the local cache. An MT can fulfill a request without sending it to an AP if it is situated along the route the request packet takes to go to the AP and has the necessary data item in its cache. If caching is not available, then all queries should be sent to the appropriate APs. Both the MT and its neighbors influence the choice to cache the data item because the local caches of the MTs essentially create an aggregate cache. There are two kinds of cache hits in the aggregate cache: local cache hits and remotely cache hits. A cache hit at local access occurs when the requested data item is already in the MT's local cache. [21] The availability of the data item in another MT's local cache is indicated by a cache hit at remote stage. Consider that an MT (n_j) acquired the communicated request packet after an MT (n_i) sent it for a data item (d). n_j responds to n_i with an ack packet after storing the information in its local cache. n_j attaches d to the reply packet when n_i sends them an acknowledgement packet. Figure 35.5 shows an illustration of the suggested ES algorithm. They are within communication range whether an MT and an AP are distinguished by a dotted line.

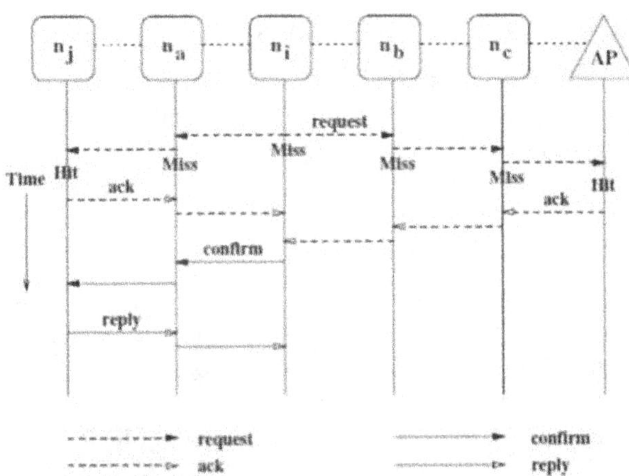

Fig. 35.5 An enhanced search (ES) algorithm

4. Simulation Metric

Three performance indicators are evaluated here: throughput, or the percentage of successful requests (ϕ); average number of hops (Ω); and cache hit ratio (h), which accounts for both local and distant cache hits. The MTs' accessibility in the iMANET is gauged by throughput ϕ, or the percentage of successful requests. Equation 1 estimates ϕ if rtotal and rsuc represent the entire number of queries and data items that were successfully received, respectively.

$$\phi = \frac{r_{suc}}{r_{total}} \times 100\% \tag{1}$$

The average hop length of successfully received data items of APs or MTs is shown by the average number of hops (Ω). Equation 2 expresses the duration of a successful request hop r, if Ω is that length.

$$\Omega = \frac{\sum_{r \varepsilon r_{suc}} \Omega_r}{r_{suc}} \tag{2}$$

Ω is used to assess average latency since the number of hops and communication latency are strongly connected. Lastly, the effectiveness of the aggregate cache management is assessed using the hit ratio h. Equations 3, 4, and 5 can be used to represent hlocal, hremote, and h if nlocal and nremote stand for the number of local hits and remote hits, respectively.

Table 35.1 Simulation parameters

Parameter	Value
MT nodes	150
Time of Pause(s)	0, 50, 100, 200, 400, 800, Inf
Cache size (items/MT)	16
Inter request time (s)	500
Transmission range (m)	250
Data items	1000,1000
Size of Network (m)	2000 × 2000
Number of APs	1, 4, 16

$$h_{local} = \frac{n_{local}}{n_{local} + n_{remote}} \times 100\% \tag{3}$$

$$h_{remote} = \frac{n_{remote}}{n_{local} + n_{remote}} \times 100\% \tag{4}$$

$$h = \frac{n_{local} + n_{remote}}{r_{suc}} \times 100\% \tag{5}$$

5. Simulation Results

Using the uniform and zipf distributions as data access patterns, we embrace the efficiency implications of the aggregate cache. For caching with uniform and Zipf distribution data access patterns, respectively, cache replacement algorithms are utilized in Fig. 35.6. All additional rules have been estimated; however, we only cover a subset of the important results. An access pattern has no performance impact since in a system without a cache; a request can only be serviced by an AP and not by any MT.

When using the aggregate cache, data accessibility increases significantly by more than twice as much as when there isn't a cache. When caching is used, there is a significant likelihood that the requested data will be stored at other MTs or in the MT's local cache. Unlike when there is no cache, MTs attempt to recover the cached data items among themselves even when they are disconnected from an AP. Subject to the access pattern, additional improvements might be achievable. It's important to remember that when the data retrieval pattern resembles the Zipf distribution, there is an increase of about 200% [22] compared to the situation with no cache.

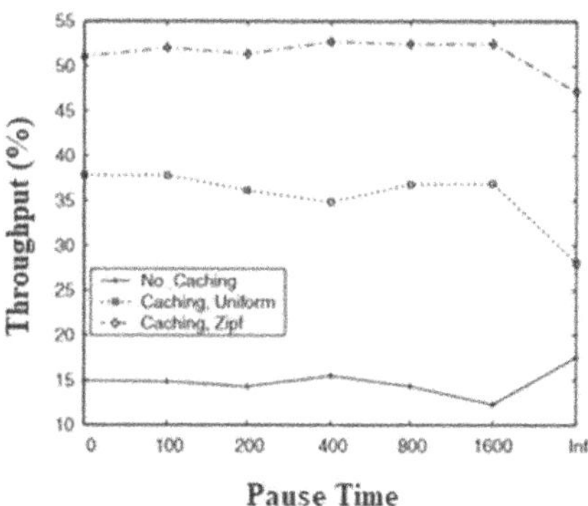

Fig. 35.6 Throughput (ϕ), as per equation 1

Figure 35.7 illustrates how collective caching affects mean latency. When there is a cache, data items can be retrieved significantly faster than when there isn't since any MT discovered along the path that the request is sent to the AP can finish a request. As anticipated, buffering reduces X by more than 50%. The outcomes convincingly show how effective the aggregate caching method is.

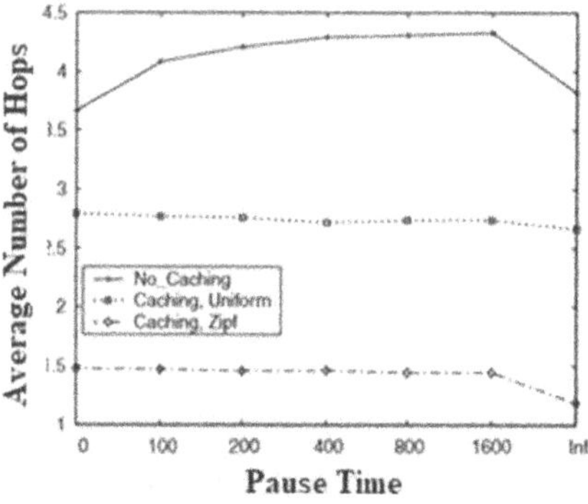

Fig. 35.7 Latency (Ω) as in equation 2

6. Conclusion

This research proposes an aggregate caching strategy to enhance an iMANET's communication performance. It comprises ubiquitous communication infrastructure, which includes MANET and the wired Internet. Enabling access to Internet data and services from anywhere at any time is the aim of an iMANET. The idea of aggregate cache is combined with unified cache to solve the issues of prolonged access latency and restricted data accessibility. A cache management system that operates on the principles of cache admission control and cache replacement policy is part of this technique, as is a broadcast-based search engine. By imposing a minimal distance between identical data items, the admission control stops high data duplication, and the replacement strategy enhances usability, and the cache hit ratio. By granting the TDS scheme's time and distance components multiple weights, this study investigates three different replacement policy variations. Three viewpoints—the effect of caching, the number of Aps, and cache management, were employed in a analysis of performance on simulation basis. It is followed to examine the advantages of the recommended strategy. The conventional LRU policy was compared to the three TDS replacement policy modifications. Regardless of cache replacement strategies, it was discovered that caching in iMANETS can significantly improve communication performance in terms of throughput and average access latency when compared to an infrastructure without any cache. For skewed access patterns, the aggregate cache's performance advantage was amplified. Additionally, even a single Internet access point improved performance through caching. To fully utilize iMANET, there are several issues that require more research. The following topics are currently being looked at: It is believed that the data items in this document are never changed. This presumption includes the ability to modify data. Cache invalidation and cache update problems arise as a result. Link disconnection and network topology changes make cache invalidation and updating difficult in an iMANET. In this paper, several network topologies that could lead to a network partition issue are not considered.

References

1. Shuo Ding, A survey on integrating MANETs with the Internet: Challenges and designs. In: Elsevier Journal of Computer Communications- 2008.
2. M.S. Corson, J.P. Macker, G.H. Cirincione, Internet-Based Mobile Ad Hoc Networking,in: IEEE Internet Computing, July–August 1999, pp. 63–70.
3. T. Hara, Effective replica allocation in ad hoc networks for improving data accessibility, in: Proc. IEEE INFOCOM, 2001, pp. 1568–1576.
4. Y. Hu, D.B. Johnson, Caching strategies in on-demand routing protocols for wireless adhoc networks, in: Proc. ACM MOBICOM, 2000, pp. 231–242.
5. D.B. Johnson, D.A. Maltz, Dynamic source routing in ad hoc wireless networks, in: Mobile Computing, Kluwer, 1996, pp. 153–181.
6. M. Papadopouli, H. Schulzrinne, Effects of power conservation, wireless converage and cooperation on data dissemination among mobile devices, in: Proc. MobiHoc, 2001, pp. 117–127.
7. C. Perkins, E.M. Royer, Ad-hoc on-demand distance vector routing, in: 2nd IEEE Workshop on Mobile Computing Systems and Applications, 1999, pp. 90–100.
8. F. Sailhan, V. Issarny, Cooperative caching in ad hoc networks, in: Proc. 4th International Conference on Mobile Data Management (MDM), 2003, pp. 13–28.
9. M.S. Corson, J.P. Macker, G.H. Cirincione, Internet-based mobile ad hoc networking. In: IEEE Internet Computing (July–August) (1999) 63–70.
10. M. Buddhikot, G. Chandranmenon, S. Han, Y.W. Lee, S. Miller, L. Salgarelli, Integration of 802.11 and Third-Generation Wireless Data Networks, In: Proceedings of IEEE INFOCOM, 2003.
11. K. Chen, K. Nahrstedt, Effective Location-Guided Tree Construction Algorithms for Small Group Multicast in MANET, In: Proceedings of IEEE INFOCOM, 2002, pp. 1180–1189.
12. L. Xiao, L. Ni, A. Esfahania, Prioritized overlay multicast in mobile ad hoc environments. In: IEEE Computer 37 (2) (2004) 67–74.
13. A. Safwat, H. Mouftah,, 4G network technologies for mobile telecommunications. In: IEEE Network 19 (5) (2005) 3–4.
14. Kumar Rakesh, Anil Kumar Surje and Manoj Mishra, A Proactive Load-Aware Gateway Discovery in Ad Hoc Networks for Internet Connectivity. In: International Journal of Computer Networks & Communications (IJCNC) Vol.2, No.5, September 2010
15. Kamaljit I. Lakhtaria and Prof. Bhaskar N. Patel, Comparing Different Gateway Discovery Mechanism for Connectivity of Internet and MANET. In: International Journal of Wireless Communication and Simulation, Vol.2 No.1, 209, PP: 51–63.
16. Koushik Majumder, Dr. Sudhabindu Ray, Prof. Subir Kar Sarkar, Implementation and Performance Evaluation of the Gateway Discovery Approaches in the Integrated MANET Internet Scenerio. In: International Journal on Computer Science and Engineering, Vol.3 No.3, March 2011, PP: 1213–1226.
17. M. Papadopouli, H. Schulzrinne, Effects of power conservation, wireless converage and cooperation on data dissemination among mobile devices, In: Proc. MobiHoc, 2001, pp. 117–127.
18. H. Luo, R. Ramjee, P. Sinha, L. Li, S. Lu, UCAN A Unified Cellular and Ad-hoc Network Architecture, In: Proc. ACM MOBICOM, 2003, pp. 353–367.
19. F. Sailhan, V. Issarny, Cooperative caching in ad hoc networks, In Proc. 4th International Conference on Mobile Data Management (MDM), 2003, pp. 13–28.
20. Sugandha Singh, Navin Rajpal, Ashok Sharma. Address Allocation for MANET Merge and Partition using Cluster Based Routing, International Journal of Wireless Information Networks by Springer plus.
21. Anju, Sugandha Singh, Modified AODV for congestion control in MANET, International Journal of Computer Science and Mobile Computing 4.6 (2015): 984–1001.
22. S. Singh, N. Rajpal and A. K. Sharma, K-fault tolerant in Mobile Adhoc network under cost constraint, 2011 3rd International Conference on Electronics Computer Technology, Kanyakumari, India, 2011, pp. 368–372, doi: 10.1109/ICECTECH.2011.5942117.

Note: All the figures and table in this chapter were made by the authors.

Intelligent Systems Using Semiconductors for Robotics and IoT – Dinesh Goyal et al. (eds)

36 A Professional Approach to Yoga: Asana Pose Detection: Control, Analysis and Enhancing of Asana Correctness and via Artificial Intelligence

Mohit Yadav, Divya,
Manpreet Singh Bajwa*, Ashima Rani, Sugandha Singh
Dept of Computer Sci & Engg, Faculty of Engineering and Technology,
SGT University Gurugram, Haryana India

Abstract: Yoga has developed into an essential aspect of holistic well-being in the context of modern- day existence. In this context of the importance of yoga and its significance as global phenomena Thus, the subsequent paper focuses on the contemporary characteristic of the dedication of the platitude offering. This will help to achieve the goal, which implies examining the holistic impacts of the asana through modern research. However, this paper goes further and taps into the societal implications of yoga posture cognition, that is, its integration into modern healthcare systems and its capability to create a more balanced relationship between oneself and society. Finally, the study hopes to yield profound wisdom about enduring dedication to one's yoga while facing the principal challenges of modern life.

Keywords: Asana pose detection, Convolutional neural networks (CNN), Real-time feedback, Skeletal data analysis, Deep learning, Health and well-being, Safety in asana

1. Introduction

To put it in broader terms, yoga is one of the oldest customs that began in India and was widespread later across the globe since it involves a variety of benefits on a physical, psychological, and spiritual level. At the same time, that confronts practitioners, especially novices, to balance the correct position alignment by different asanas (poses). The proper alignment of a posture not only maximizes benefits but also avoids the chance of injury.

The cutting-edge technology in machine learning and computer vision has facilitated new developments in the exploration of the postures and movements of individual. Another example of this technology is posing detection within the model of generating yoga poses. It is based on computer algorithms that are open for analyzing and identifying the pose of a person who is practicing yoga asanas. This technology utilizes image information from photographs or videos where each body joint is a cap (coordinate space) or key point, while all these joints constitute pair.

With the help of skeletal data procured from the pose, there is a comprehensive perception, and this eases the task of posture quantification. The real-time feedback from the comparison of the detected pose with the ideal or reference pose will give the practitioners clarification on what the correct asana is and/or what they can improve. This type of feedback would be very helpful, especially to individuals which do yoga without the presence of their trainer.

Additionally, yoga pose detection could help yoga trainer provide personalized feedback to their students, targeting the areas they need to improve and tracking their achievements over a period of time. Furthermore, it may become part of an online yoga course offering remote directions and strengthening the chances of people taking up yoga (Agrawal et al., 2020; Chen et al., 2014).

*Corresponding author: manisinghbajwa@gmail.com

DOI: 10.1201/9781003716389-36

In this study, we seek to give an overall picture of the posture of humans during yoga asanas using skeleton-based yoga pose detection methods. We will discover the research made on the subject, discuss the algorithms for pose detection, and estimate how good they are in terms of accuracy, resilience, and usability. Additionally, we present various feasible areas in which yoga pose detection can be used, including fitness tracking, rehabilitation, and sports performance assessment.

This paper focuses on disadvantages of accuracy of yoga posture detection systems and working in alignment with improving the accuracy and efficiency of various other methodologies. In the end, the purpose of our work is to build useful instruments and approaches that will enable practitioners to master the postures they do and the methods that they use to achieve the best possible health and state of mind through the practice of yoga (Abarna et al.,2022; Pala et al., 2023).

The Media Pipe library using specific locator and JSON transmitting 45 imaged sequences. This model uses CNN and LSTM and SoftMax level in conclusion.

Table 36.1 Asana name

Sr. no	Asana Name
1.	Tree Pose
2.	Cobra Pose
3.	Lotus Pose
4.	Downward Dog Pose
5.	T-Pose
6.	Bridge Pose
7.	Goddess Pose
8.	Wheel Pose
9.	Warrior Pose 1
10.	Warrior Pose 2
11.	Sitting Pose
12.	Standing Pose
13.	Tadasana
14.	Chair Pose
15.	Mountain Pose

2. Literature Review

In 2019, Santosh Kumar Yadav et al, proposed the data set of six yoga asana has been created using 15 individuals with the normal RGB web cam. A hybrid model using deep learning model is implemented using CNN (convolutional neural network) and long short-term memory (LSTM) for yoga posture recognition. The system shows accuracy of 99.04% on single frames and 99.30% accuracy after predictions on 45 frames of the videos. They have also tested the system in real time with different set of 12 people and achieves 98.92% accuracy (Yadav et al., 2019).

In 2020, Deepak Kumar et al, said that the system can be used to learn yoga from home only in which the user can upload the photos of self while performing the pose and compare from the pose of the expert and understand differences in angles of various body joints. Eventually, the system can help the user to improve yoga poses from home itself (Kumar et al., 2020). Agarwal et al, tried to solve another challenging aspect of real time deduction of posture from photograph. They used more than 5500 images of above 10 yoga poses with the aim of solving the problem with a bigger data set. 80% of the data was used to train the algorithm and remaining 20% to test it which eventually resulted in 99.04% accuracy using a random forest classifier (Agrawal et al., 2020). J. Palanimeera et al, classify the sun salutation yoga posture detection which used machine leaning and deep learning models in classification. The yoga page is defected it supported the angle extracted from the pose estimation algorithm. The accurate results, received of ML (Machine learning model) KNN. The net accuracy of 96% of the machine learning model (Palanimeera et al., 2021).

Khushi Sidana et al, noticed that there has been a marked increase in the number of injuries due to incorrect yoga postures owing to the fact that number of yoga practitioners has increased over time. They tried to develop a self-training model which can compare the yoga pose performed by an individual with that of an expert and help the user to rectify the mistakes accordingly. To achieve the said objective, models like deep learning CNN, SVN and CNN-LSTM hybrid models were used with the aid of computer vision and an accuracy between 97% to 99.9% has been achieved (Sidana et al., 2022). In 2021, Utkarsh Bahukhandi et al, developed a system with the help of 15 volunteers who performed six different yoga poses each and video recording were analyzed first using the media pipe pose estimation library. After obtaining the data, classification- based machine learning algorithm was developed and tested with an accuracy of 94% (Bahukhandi et al., 2021). Jiri Kutaleket al, the concept of a smart phone app deducting yoga poses and displaying several frames to a user is presented. Which consists data of frames captured from 162 collected videos based on the annotation, is then passed to train a CNN model. It detects yoga poses with 90% accuracy (Kutalek et al., 2021). In 2022, Rutuja Gajbhiye et al, developed a yoga identification system by applying CNN layer to detect patterns between key points with the accuracy of 0.99. It collects data by using an HD 1080p Logitech digital camera (Gajbhiye et al., 2022).

Radha Tawar et al, proposed a yoga pose detection system using a traditional RGB camera. Yoga poses are implemented using open CV and media pipe with much more accuracy. The system includes five poses which gives output using audio feedback and the rest of four poses with text/to/speech feedback (Tawar et al., 2022). Thautam et al, developed a model to detect incorrect yoga posture by using a deep- learning technique the model detects the abnormal angle between actual pose and user pose with the accuracy of 99.58% (Thoutam et al., 2022).

In 2023, Deepak Parashar et al, achieved an accuracy of 95% in detection of proper yoga poses by using a new automated method called movement-based deep learning model and media pipe (Parashar et al., 2023).

Deepak Mane et al, used computer vision techniques and developed a SVM based smart yoga assistant for classification and correction of yoga poses. As proposed, this model holds great potential for enhancing yoga practice especially for users with specific health concerns such as arthritis or chronic pain by providing them more feedback and support. The model produced 87% accuracy. The system has limitations in accurately identifying poses with heath conditions (Mane et al., 2023). Saksham Sonker et al, developed a system that accurately identify and correct human pose with the help of computer graphics and vision techniques using machine learning model (Sonker et al., 2023).

3. Methodology

Our research feature selection is undertaken by careful feature point selection over the body using the knowledge of anatomy suitable to yoga postures as a guide. These parameters consist of joints and important structures that are identified in corresponding video frames, and this is a way of getting the exact location. Then, a simplified model of the human structure is develop using the sections of the body as rectangular segments, adjoining them based on the locations of feature points. Such hierarchical system is the cornerstone for the tracking of motion. This is realized by the algorithms which have been designed to trace the movement of segments over consecutive frames. Techniques including optical flow and Kalman filtering can be used to compute positional and postural parameters and detect instantaneous changes in alignment and posture dynamics during yoga poses. Combination of results from the motion tracking with a human structure model leads to the formation of posture analysis parameters and quantitative feedback for the user. The regular update and refining of the methodology are aiming to increase its precision, rigor, and usability that in turn are empowering the technology for efficiency in practical applications.

The proposed methodology integrates the full-range tools to comprehensively deal with the multiple issues in the pose detection of yoga. As a second step, we get through the process of feature points selection as well as human structure modelling. Then, we move forward on the topic of motion tracking which is very important for modelling motion of yoga asanas that involves a number of computational issues. High tech algorithm and method are utilized to realistically reflect all iteration of motion comprising variations of speed, angle and intensity. we aim to utilize the motion tracking outcomes with the human body model structure, which is important for the complete examination of motion dynamics and posture alignment. Through this approach, we are not only capable of recognizing workspace-related deviations from a perfect posture but also provide professionals with the needed guidelines to allow them to instruct their clients in terms of refining techniques and optimizing their performance. Using repetitious refinement, our methodology geared to transcend yoga pose detection, and as a result, the transformative practice of yoga can be benefited from upgrading self-sufficiency

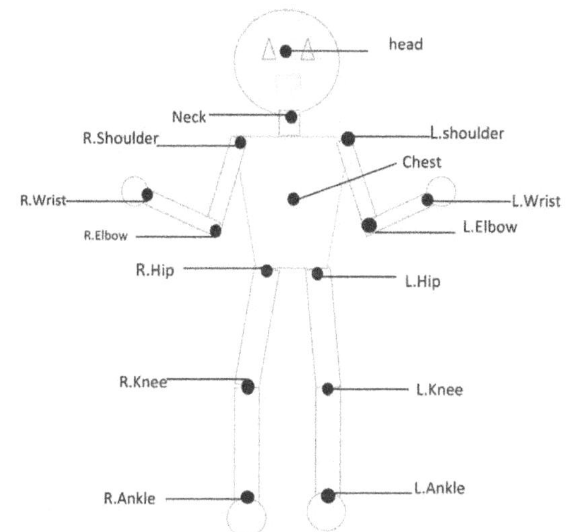

Fig. 36.1 Show the human body joints

Posture Correction: The methodology is aimed at the accurate assessment and analysis of the posture during yoga poses, and giving practitioners real-time analysis and feedback on their alignment and posture improvement.

Gym Replacement: Through this approach, people can do posture correction exercises anywhere while being in control and able to adjust the given exercises according to their abilities without the need for sophisticated equipment or supervision of a fitness instructor

Independent Posture Checking: This technology gives the power to people to assess their posture quite accurately even when there is no professional coach around giving them a good tool of self-assessment and improvement.

Enhanced Well-being: This technique will correct the posture problem and thus contribute to a general wellbeing and health, the risk of musculoskeletal issues being reduced, and the quality of practitioner's life being increased, irrespective of their age or background.

4. Methodology Used

1. OpenCV (Open-Source Computer Vision Library) :
   It is used for the development and integration of the system required for yoga pose detection. OpenCV is a great library that has tools for vision and image and it is common in many projects. Using OpenCV

we are executing many types of processing and some of them are given below:

a. Pose Estimation:

In addition, the system is able to identify main joints and limbs' points from the captured and recorded images and videos with the help of OpenCV.

b. Image Processing:

OpenCV also plays a role in the preprocessing of the images; from normalization, noise elimination to image enhancement in order to enhance the pose detection.

c. Real-Time Processing:

OpenCV allows for video streams manipulation in real-time, which means that the system will be able to promptly respond to the executed yoga poses.

2. Deep Learning:

Deep learning can be classified under the unsupervised learning and machine learning which is a technique of learning that employs a neural network with multiple layers in a bid to address the complexity of multilayered features. However, deep learning has a better attribute compared to the other four approaches of artificial neural networks in machine learning, which is its ability to learn features from the input data and then improve them to meet the requirements of the application; for instance, computer vision, natural language processing and speech recognition. Such models learn from big data and optimization techniques including the gradient descent in order to reduce the error when making a prediction.

3. Convolutional Neural Networks (CNNs)

CNNs are categorized under the deep learning models, which are more appropriate for images – image detection and image manipulation. CNN means Convolutional Neural Network and we know that a CNN has layers; each layer adds filters and a CNN can detect edges, textures, and shapes. The pooling layers are the layers which are used to reduce the number of dimensions of given input data while the fully connected layers are used for classification. Fully connected layers are usually diluted during training and work at their optimal to discover spatial hierarchies; they are applied in image categorization, identification, and partition.

5. Enhancing Asana Practice

In the case of asanas, the practice incorporates other strategies and means that make it easier to secure flexibility, strength and performance. Some of the strategies include practice, progressive overload, variation and modification, and/or the use of technology.

1. Regular Practice: The best in all forms of endeavor is consistency and there can be no argument to this anywhere. With the incorporation of asanas, the body bends and stretches more easily with time and this increases its capacity to handle and build sturdiness.
2. Progressive Overload: It means that it is easier to increase the strength and endurance if the intensity or the time spent on the poses is somehow gradually increased.
3. Variation and Modification: Introducing new variants of the poses re-creates a new form of strain within the body systems and thus helps in breaking the system's stagnated pattern.
4. Incorporating Technology: It would be increasing the level of the applications using more sophisticated tools like yoga pose detection systems for instant feedbacks concerning alignment and posture. Such systems provide real-time feedback, and this instant coaching is the most rewarding form of practice.

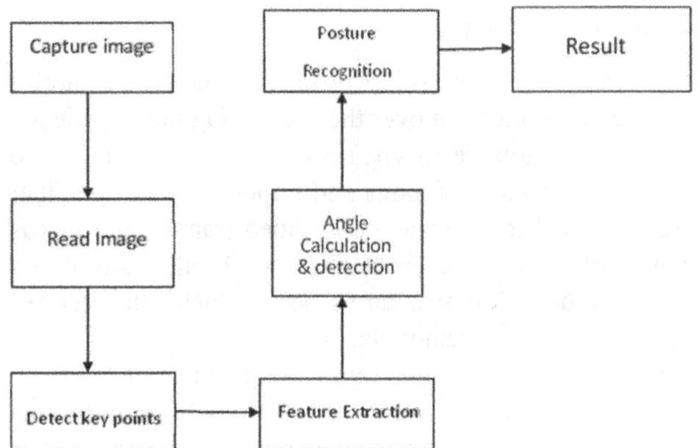

Fig. 36.2 System flowchart

6. Control in Asana

Stability directs the propensity of the poses or postures of yoga to assist the body in attaining the favorable positions for each asana. This entails flowing motion, control of breath, exercise of the abdominal muscles, and service of objects.

1. Mindful Movement: It is the conscious and controlled movement of each body part when getting into during, and coming out of a posture. This makes their work more precise and stable which can be attributed to their approach to it as a mindful process.
2. Breath Control (Pranayama): Breath control seems to promote a deeper involvement in each of the asanas. Inhalation helps one to slow down the process of stretching the spine and thus create some space, while exhalation helps in pulling even deeper that stretch.

3. Core Engagement: Abdominal muscles play a critical role in holding the spine and proportional physique; failure to this, poses a tremendous risk to lower back injuries.
4. Use of Props: Yoga blocks, straps, and bolsters in particular will help to maintain correct positioning and height, which will be beneficial for the novice or persons with inadequate flexibility and will provide maximum control regarding the actual execution of a specific movement and subsequent rubbing, thus excluding the probability of an injury.

7. Safety in Asana

Preventing injury during yoga is a critical factor bearing in mind the maximization on yields from the training. They include warm up, awareness of one's body, good postural and alignment, gradual increase in the level of difficulty and practice in safety.

1. Proper Warm-Up: Stretching and warm up forms the foundation for some elaborate poses in that it ensures the body is ready for the more challenging positions and any incidents of injury prevented.
2. Listening to the Body: In this regard, practitioners need to observe patients' body language and ensure that they do not perform any movement that will be painful to them.
3. Correct Alignment: Posture is very important in each and every position in order to avoid any form of stress on the joints, muscles, and bones. Instructors' directions and special equipment could be helpful in obtaining proper positioning of the body parts.
4. Gradual Progression: By not attempting complex poses, if a person does not practice daily and does not have the kind of strength needed, it is possible to reduce chances of incurring an injury.
5. Safe Environment: Practicing in clean environment reduces of falls or any chance of accidents making it safe to practice in that environment.

8. Flow Chart

Here is the flow chart of the Yoga pose detection System so as to be able to explain the successive steps/stages of the system. This is achieved from the data collection of the signal to real time feedback to deal with the problem due to the 'wrong body posture'.

1. Data Acquisition:
   i. Collect data of people doing certain yoga poses using the RGB camera.
   ii. Get the skeletal information from the images/videos.
2. Preprocessing:
   i. Store and clean the skeletal information.
   ii. Define and choose critical feature points on the body.
3. Human Structure Modeling:
   i. Create the prototype of the human body based on the chosen feature points.
   ii. Employ rectangular segments in order to express different limbs, trunks and joints.
4. Motion Tracking:
   i. Algorithm such as optical flow and Kalman filtering should be applied to trace the movements of different body parts over frames.
   ii. Using positional and postural parameters one should be able to identify changes that have occurred in alignment.
5. Pose Detection and Analysis:
   i. Using machine learning, the detected pose will have to be compared with reference poses to check for a match.
   ii. Afford instant feedback on some aspects such as alignment and extents to which posture has been enhanced.
6. Posture Correction and Feedback:
   i. Providing accurate and timely feedback to practitioners to address their mistakes.
   ii. Record changes and advancements with passing periods.

9. Result

YOGA POSE DETECTION in our system is identifying the yoga poses of a human Assan. taken many kinds of data sets to use in our model to get more accuracy, our model performed exceptionally well on a private data set

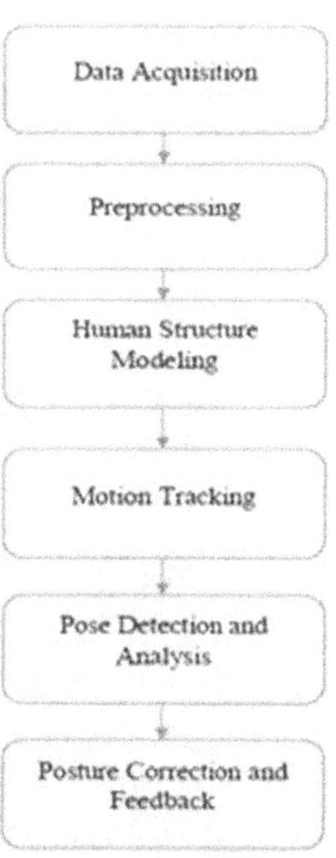

Fig. 36.3 Flow chart

and giving an accuracy rate of 85%. By using real-time data handling capabilities improved the efficiency of the detection process. The uniqueness of the training process and data privacy were highlighted by the deployment of the dataset we have used in our model. The accuracy rating of 85% confirms that this intricate strategy is successful in accurately identifying and classifying human Assan's.

10. About Accuracy

Specifically, the system does yoga pose detection, and the level of accuracy is one of the ways in which the performance of the system can be measured. The aforesaid outcomes of this experiment have been achieved here in this model with a level of precision of 85 %. The factors contributing to this accuracy include: The causes which explain why this accuracy can be attributed are as follows:

1. Dataset Quality:

   As such, the performance of the system is completely determined by the quality and the type of data used in developing the system. A greater variety of poses may be utilized to train the model, and each pose would play almost an equal role in the training and capability of the model, and therefore, a good dataset improves the model's ability in pose recognition.

2. Algorithm Selection:

   The choices for the algorithm to predict the disposition of human pose can indeed be for example, CNN or optical flow which will influence the performance of the system in question. The algorithms are also made advanced in solve to assist the model to attain high distinctions in detecting and categorizing numerous poses.

3. Real-Time Data Handling:

   The effectiveness of the energy controlling real-time data determines the accuracy of the system. On the aspect of sensitivity, which is often regarded as a drawback particularly when new model is tested on private data and scenarios whereby the model needs to be operational on real-time basis such as in self- driven automobiles, the proposed work has demonstrated a commendable performance.

Fig. 36.4 Tree pose of human

Fig. 36.5 Goddess pose of human

Fig. 36.6 Bridge pose of human

Fig. 36.7 T pose of human

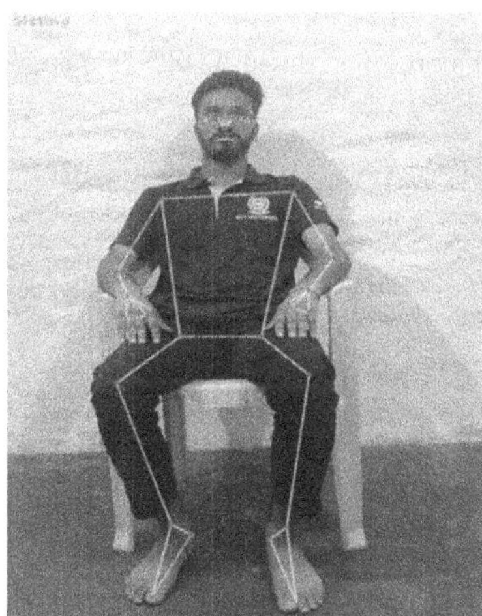

Fig. 36.8 Sitting pose of human

Thus, in the future studies, it may be possible that various attempts will be made in order to increase the efficacy of this method by adjusting the data portion and type, as well as changing the parameters of the algorithm and using other types of machine learning techniques.

11. Application of Model

The yoga pose detection model has various practical applications, enhancing both individual practice and broader fitness and healthcare contexts: The yoga pose detection model has various practical applications, enhancing both individual practice and broader fitness and healthcare contexts:

1. Fitness Tracking:

   The system can be used to monitor the progression that people make regarding the yoga techniques they perform and things that need to be changed.
2. Rehabilitation:

   In physical therapy and rehabilitation, the model can help the concerned professionals to keep an eye on the patient's exercising routines, so that the patients undergo correct motions that are required for getting rid of the pain and to reduce the chances of developing new injuries.
3. Sports Performance Assessment:

   The described system can be helpful for athletes and trainers to work on biomechanics patterns of sport-specific skills and error correction in terms of body positioning and alignment.
4. Online Yoga Classes:

   It can be incorporated into online yoga platforms, through which the information about the participant and their performance can be provided in real time to improve the remote education process.
5. Independent Practice:

   Householders performing yoga can utilize the system for evaluation so that they can keep the right positions in place without any trainer attending to them.

With the help of these applications, the yoga pose detection model helps in the enhancement of physical, mental, and spiritual well-being, providing numerous customers with the possibility to practice yoga safely, efficiently, and comfortably.

12. Conclusion

In conclusion, Yoga poses detection helps to analyze the pose of a human while doing any Assan. Whether the person is doing the yoga Assan correctly or not. The skeletons are a set of coordinates that describe the pose of a person. Each joint is an individual coordinate that is known as a key point or poses landmark. And the connection between key points is known as pair. Our results show an accuracy rate up to 85%. Future research may focus on enhancing the dataset, investigating additional optimization strategies, and expanding the model's practical application.

References

1. Abarna, S., V. Rathikarani, and P. Dhanalakshmi. "Skeleton Pose Estimation Features based Classification of Yoga Asana using Deep Learning Techniques." International Journal of Mechanical Engineering (2022).
2. Agrawal, Yash, Yash Shah, and Abhishek Sharma. "Implementation of machine learning technique for identification of yoga poses." In 2020 IEEE 9th international conference on communication systems and network technologies (CSNT), pp. 40–43. IEEE, 2020.
3. Bahukhandi, Utkarsh, and Shikha Gupta. "Yoga pose detection and classification using machine learning techniques." Int Res J Mod Eng Technol Sci 3, no. 12 (2021): 13–15.
4. Chen, Hua-Tsung, Yu-Zhen He, Chun-Chieh Hsu, Chien-Li Chou, Suh-Yin Lee, and Bao-Shuh P. Lin. "Yoga posture recognition for self-training." In MultiMedia Modeling: 20th Anniversary International Conference, MMM 2014, Dublin, Ireland, January 6-10, 2014, Proceedings, Part I 20, pp. 496–505. Springer International Publishing, 2014.
5. Gajbhiye, Rutuja, Snehal Jarag, Pooja Gaikwad, and Shweta Koparde. "Ai human pose estimation: Yoga pose detection and correction." international journal of innovative science and research technology 7 (2022): 1649–1658.
6. Kumar, Deepak, and Anurag Sinha. Yoga poses detection and classification using deep learning. London: LAP LAMBERT Academic Publishing, 2020.
7. Kutálek, Jiri, and K. Kutálek. "Detection of yoga poses in image and video." Brno Faculty University of Information and Technology (2021).
8. Mane, Deepak, Gopal Upadhye, Vinit Gite, Girish Sarwade, Gourav Kamble, and Aditya Pawar. "Smart Yoga Assistant: SVM-based Real-time Pose Detection and Correction System." International Journal on Recent and Innovation Trends in Computing and Communication 11, no. 7s (2023): 251–262.

9. Pala, Vinay Chethan Reddy, Sreekar Kamatagi, Shyamsunder Jangiti, K. Swaraja, K. Reddy Madhavi, and Gs Naveen Kumar. "Yoga Pose Recognition with Real time Correction using Deep Learning." In 2023 International Conference on Sustainable Computing and Data Communication Systems (ICSCDS), pp. 387–393. IEEE, 2023.
10. Palanimeera, J., and K. Ponmozhi. "Classification of yoga pose using machine learning techniques." Materials Today: Proceedings 37 (2021): 2930–2933.
11. Parashar, Deepak, Om Mishra, Kanhaiya Sharma, and Amit Kukker. "Improved Yoga Pose Detection Using MediaPipe and MoveNet in a Deep Learning Model." Revue d'Intelligence Artificielle 37, no. 5 (2023).
12. Sidana, Khushi. "REAL TIME YOGA POSE DETECTION USING DEEP LEARNING: A REVIEW." International Journal of Engineering Applied Sciences and Technology 7, no. 7 (2022): 61–65.
13. Sonker, Saksham, Shefali Tiwari, Sadhana Rana, and Er Preeti Naval. "YOGA POSE DETECTION USING MACHINE LEARNING."(2023).
14. Tawar, Radha, Sujata Jagtap, Darshan Hirve, Tejas Gundgal, and Namita Kale. "Real-Time Yoga Pose Detection." (2022).
15. Thoutam, Vivek Anand, Anugrah Srivastava, Tapas Badal, Vipul Kumar Mishra, G. R. Sinha, Aditi Sakalle, Harshit Bhardwaj, and Manish Raj. "Yoga pose estimation and feedback generation using deep learning." Computational Intelligence and Neuroscience 2022 (2022).
16. Yadav, Santosh Kumar, Amitojdeep Singh, Abhishek Gupta, and Jagdish Lal Raheja. "Real-time Yoga recognition using deep learning." Neural computing and applications31 (2019): 9349–9361.

Note: All the figures and table in this chapter were made by the authors.

Intelligent Systems Using Semiconductors for Robotics and IoT – Dinesh Goyal et al. (eds)
© 2026 Taylor & Francis Group, London, ISBN 978-1-041-20408-4

37 Transforming Modern Industries: The Role of Robotics in Industrial Automation

Monika Kumari [1], Nikhil Kumar Goyal[2], Rajat Kumawat[3], Seema Kaloria[4], Ashwarya Vijai[5]

Poornima University, FCE,
Jaipur, India

Abstract: This work assesses the development and deployment of collaborative robotics within an overall framework of rapidly changing industrial automation and pinpoints its key role within the paradigm of Industry. As industrial development accelerates, there is growing interest in-and concern about-the convergence of human and robotic systems in close proximity. Central to this evolution is the imperative to enhance safety and effectiveness in human- robot interactions. The study separates robotics into three main directions: industrial robotics would refer to manipulators-programmed versatile systems for multi-axes tasks; the collaborative one is an advanced stage with close human-robot interaction on the ground of sophisticated sensors and adaptive algorithms; service robotics covers autonomous or semi-autonomous systems realizing tasks other than traditional industrial ones. It develops an environment of harmonious coexistence at work, where the robots will be useful and intelligent helpers, hence the realization of Industry. Critical reflection is made on the concept of collaborative robotics. This integration of robotics into an entity makes use of artificial intelligence and IoT in driving automation and smart production. This paper highlights how collaborative robotics might bring transformative changes in the field of manufacturing efficiency, safety, and general business processes of contemporary industry.

Keywords: Collaborative robotics, Industrial automation, Industry 4.0, Human-robot interaction (HRI), Industrial robotics, Service robotics, Artificial intelligence (AI), Internet of things (IoT)

1. Introduction

The automation of factories has undergone radical changes during the last few decades from simple, task-oriented systems to elaborate integrated solutions reflecting a new dimension of industrial efficiency and precision [1]. Initially, factory automation consisted of the deployment of basic automation technologies to execute repetitive tasks with minimum human interference. However, computing power and software development, not to talk of networking technologies, have developed so much that complete automation systems can be today deployed. These modern systems use computers and robotics to enhance the efficiency, precision, and flexibility of many operations related to the process at hand. Automation in factories can include anything from simple applications like the execution of production tasks to several other important tasks, including inventory management and quality control, with the aim of optimizing the overall process. The strategic integration of such systems has an effect not only in relocating human labor to more complex roles but also in greatly improving both speed and accuracy in manufacturing operations.

Whereas manufacturing automation is closely related but separate, it focuses its attention on purely the production process and machinery for maximum output. While factory automation might broadly cover a wide variety of industrial processes, automation within manufacturing zeroes in on where goods production could be more

[1]098monikasingh@gmail.com, [2]nikhilgoyal886@gmail.com, [3]rajatkumawat.research@gmail.com, [4]Seema.kaloria@poornima.edu.in, [5]ashwarya.vijai@poornima.edu.in

DOI: 10.1201/9781003716389-37

efficient and the use of robotic systems in this context. This difference in the scope regarding applications within the context of a factory outlines the differing approaches necessary in dealing with particular manufacturing problems. Recently, advances have been made such that the assembly, installation, and maintenance of robotic systems have been accomplished in a much more cost-effective and time-efficient manner. Plug-and-play technologies have greatly eased integration issues by compacting set-up times for sensors and actuators. The modern robot boasts advanced multi-sensor integration to adapt in real time and sophisticated feedback [2]. These evolutions enable robots to perform tasks that were earlier not possible or imaginable by them with high precision and flexibility: quality checks during production, thereby reducing the need for post-manufacturing inspections.

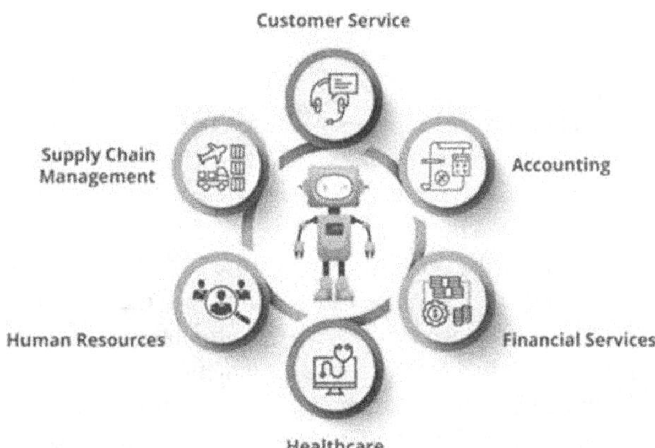

Fig. 37.1 Uses of robotics and automation in various fields
Source: Author

The paper explores in detail the modern concept of factory automation in terms of its development, technological changes marking this progress, and its implications for enhancement in manufacturing. From this, the integration of different advanced robotics and automation technologies will give insight into how factory automation could bring about change in contemporary industrial practice as a means to find potential future development.

1.1 What is Robotics and Automation?

Automation and robotics are two terms interrelated with each other, yet each having a distinct meaning, which has brought a remarkable change in modern industrial processes and several other sectors. Automation, in general terms, refers to the practice of using technology to perform tasks requiring minimum human intervention by means of pre- programmed or adaptive applications of computers and electrical and mechanical machines. The very concept of automation is based on the fact that it exists to make things easy by operating functions through algorithms and systems that either work on fixed routine modes or through dynamic changes presented within an environment. For predefined automation, this depends on algorithms created to carry out tasks in a particular manner irrespective of external variables; uniformity and dependability are part of this process. Adaptive automation systems, on the other hand, can actually change their behavior depending on changes in either the environment or the process and, for this reason, provide greater flexibility and responsiveness.

Robotics, although related to automation, constitutes a more specialized field within automation[3]. Today, robotics involves the design, construction, and operation of robots-machines engineered to go from simple, repetitive activities to complex problem-solving activities. While automation may or may not be performed by robots, robotics entails the use of robots, including robotic arms, cobots, robotic exoskeletons, and humanoid robots. Industrial robots, which include robotic arms and cobots, are highly utilized in manufacturing and operations involving warehousing. Companies like Amazon and Best Buy use these robots for better efficiencies in sorting, assembling, and packaging. A robotics automation system integrates a variety of elements: computer programming, algorithms, manipulators, actuators, control systems, and real-time sensors. These various elements work together in operating the robot that will perform the task with precision, make necessary changes based on real-time input, and provide critical data for process optimization and decision-making. The distinction between automation and robotics-details of how they differ and relate-are significant in the quest to understand their positions in current industrial contexts. While automation provides the platform on which tasks are executed with little human intervention, robotics embodies entities that can physically execute such tasks at various levels of complexity and adaptability.

Industrial automation is an innovative manufacturing and production concept that employs a variety of control systems, each incorporating various machines, actuators, sensors, processors, and networks, to enhance and advance the job of producing[4]. The key objective of industrial automation is minimizing the element of human involvement in achieving maximum efficiency, precision, and consistency in many diverse industrial operations. It can be said that industrial automation goes all the way back to early manufacturing, where simple conveyor belts and mechanized systems were first introduced in order to move parts throughout assembly lines. These rudimentary machines marked the very beginning of automation by reducing the needs for a large amount of manual labor and setting a stage further forward. Nowadays, industrial automation has largely transcended the scope of basic mechanical systems and embraced a number of advanced technologies that now lie at the very heart of contemporary industrial settings.

The current landscape of industrial automation is very heterogeneous, consisting of numerous components and technologies, but the most prominent ones are

Programmable Logic Controllers, Artificial Intelligence, Machine Learning, and Industrial Internet of Things devices. PLCs ensure strong control of machinery and processes, while AI and ML contribute to adaptive and predictive capabilities that raise the level of efficiency in operations and decision-making [5]. IIoT devices enable real-time data collection and communication across industrial networks, thus driving more sophisticated monitoring and control. The advantages accompanying industrial automation are huge and manifold. In fact, the integration of such advanced technologies into industries can easily bring massive upward swings in producing efficiency, product quality, and operational safety. Automation systems allow the control and monitoring of processes with high precision, offering reduced downtime, minimized errors, and enhanced productivity.

In brief, industrial automation has grown from simple mechanization through complex integration of state-of-the-art technologies to a sea change in the topography of production and manufacture. Basically, understanding the different types of industrial automation and their advantages lays a foundation for how these changes are helping industries worldwide to widen markets and become increasingly competitive.

1.2 Prior Work and Findings

Industrial robotics has dramatically changed the concept of manufacturing process chains, from simple mechanization up to sophisticated automation. Indeed, the roots of this area are closely related to the very basics of this discipline, concerning motion planning and control, initially developed for industrial applications. Up to 2014, the installed base of the industrial robotics market accounted for about 1.5 million units, with 171,000 units installed during that year only [6]. This wide-ranging adoption underpins the integral role robotics is now playing in modern manufacturing, bolstered by an annual industry turnover of $32 billion. Such figures bring out not only the extent of its application but also its extensive economic impact. A history of industrial robotics provides a background on the technological development of the field and the problems overcome in expanding the applications of industrial robots. Early robots that were designed to perform specific tasks in a repetitive manner formed the basis of more sophisticated systems that could handle functions with greater diversity. Although there have been these sorts of developments, a number of key issues remain. The most prevalent of these problems include the adaptability of robots to more flexible manufacturing environments and integration of new technologies, some of the driving forces for continued research. Understanding these historical and technical contexts is very important in considering the present state of affairs in industrial robotics and those areas where further development might be warranted.

In connection with the scrutiny of progress, an examination would have to consider the current applications of industrial robots and their technologies. Modern industrial robots are now designed with different mechanisms, depending on specific needs: assembly and welding to packaging and quality control [7]. Advances in technology are coming not only from the traditional manufacturing sector but also from other industries to enable robots to handle more challenging tasks continuously. The chapter shall, therefore, discuss how technological advancements enhance robotic applications and the ways various robotic systems are optimized in executing particular industrial tasks.

Besides, the principles of industrial robotics and methods of its programming are crucial in terms of integration and efficiency of the system. This section will outline the basic principles governing the operation of the robot: the principles of control algorithms, the ways of motion planning, and methods of programming. The attention will be directed to system integration with special attention to problems and solutions relating to data integration. Proper system integration allows robotic systems to interact well with other components of manufacture for maximum productivity. This paper intends to discuss previous work and findings made in the area of industrial robotics: key technological achievements, examples of applications, and continuing challenges. The chapter will help identify areas for further research and development by revisiting historical and current contexts to further the evolution and expansion of industrial robotics' capabilities.

2. Methodologies: Types of Industrial Automation and Robotics

2.1 Fixed Automation

A predetermined sequence, therefore, gets wired directly into the machinery itself, with little or no flexibility to change. This setup has great advantages in environments that require high volumes of production and consistent product output, say, as in the automotive industry. Fixed automation is best in production processes where the requirements remain stable and quite predictable [8]. However, this being an inflexible type of automation, it has certain serious limitations with regard to changeability in the design of a product or setting up new configurations of products. Essentially, the system is relatively inflexible after it is set up, and it is relatively difficult to accommodate new products or modifications in the existing product line.

2.2 Programmable Automation

Programmable automation: These are pieces of equipment designed to handle a range of product configurations through the use of control programs that can be altered or reloaded. Among the essential features of programmable automation is that these systems are flexible enough to change the operation sequence in order to produce different products. Programmable automation applies very well in environments with batch production needs whereby the volumes of production volumes are rather low to medium. With every lot of products featuring new specifications, the system is loaded with a new control program. That provides the possibility to make different configurations using the same equipment. Programmable automation generally requires higher investments compared with fixed automation and has lower production rates. Yet, it is very flexible. It is primarily useful where the frequent change in the product design is required or production runs, thus a cheaper alternative to manufacture batches of small and medium quantities for various products.

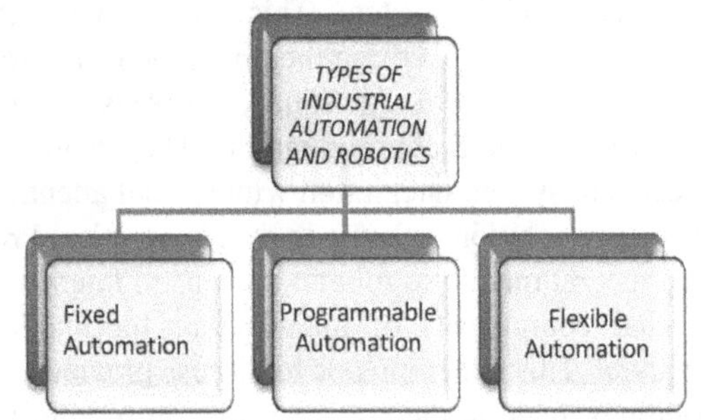

Fig. 37.2 Types of industrial automation and robotics [9]

2.3 Flexible Automation

Flexible automation is a higher-grade form of programmable automation that integrates computer-controlled systems to offer even more flexibility in production. This technique reduces the downtime between the various types of production of the product due to its quick changeovers. Flexible automation systems are designed to produce a variety of products with minimal reconfiguration and without significant delays. Advanced material handling and storage are integrated with the system to provide rapid changes in production from one product configuration to another. Flexible automation generally requires significant investment in customized engineering and provides a moderate rate of production but performs exceptionally well where product design changes occur quite regularly. This system is particularly suitable for FMSs that require continuous production of a wide range of product mix. Since flexible automation is characterized by a high degree of adaptability, with much less time lost when changing over from one product or component to another, it is ideal for manufacturing industries where flexibility and speed of production are critical.

Each of these types of automation has distinct advantages and is accordingly viable in different manufacturing situations or environments. Fixed automation is suited for large lots of repetitive production, allowing very little flexibility in design. Programmable automation executes batch production, which maintains a balance between flexibility and cost-effectiveness [10]. In return for the high investment, flexible automation offers fantastic adaptability and productivity for continuously varying production needs. These differences will enable choosing the appropriate strategy for specific demands of industries so as to optimize their processes of production

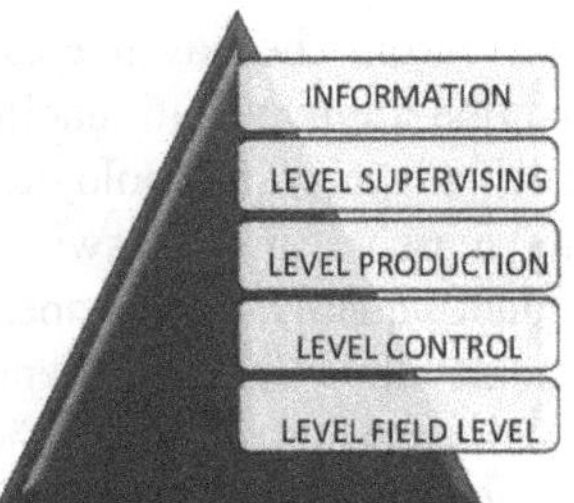

Fig. 37.3 Different levels of industrial automation

Source: Author

Automated Guided Vehicles (AGVs)

Unlike AMRs, Automated Guided Vehicles work along predefined paths or rails and often require some sort of operator supervision. Applications involve, but are not limited to, controlled environments like a warehouse or a manufacturing floor where they could be employed to perform various tasks such as material handling and transporting items. With fixed pathways and operational protocols, they tend to be very reliable in performing repetitive and predictable tasks. Though the AGVs are not as flexible as AMRs, their design is somehow tailored to achieve stability and precision in structured environments [12]. The integration of such industrial workflows brings in efficiency through the automated transportation of goods and material, saving time and reducing the cost of operation.

Articulated Robots

Articulated robots, more commonly called robotic arms, are designed to behave much like the human arm. They contain several rotary joints and can be tailored for a wide range of motions and versatility. Additional joints/axes increase the robot's capability for complex tasks such as arc welding, material handling, machine tending, and packaging. Articulated robots have assumed considerable importance because of their high degree of precision and versatility in handling tasks requiring fine control and complex movements. Their ability for operation with high accuracy and repeatability places them as an integral part of a variety of automated manufacturing and processing setups.

Cobots

Collaborative robots or cobots are those that are put to work alongside human operators in the same workspace. Unlike traditional industrial robots, cobots are made to collaborate hand-in-hand with people directly, improving their productivity while enhancing safety. Generally, cobots are deployed where tasks are manual, hazardous, or physically burdensome to lighten the load on working personnel and enhance general workflow efficiency. Some cobots have adaptive capabilities that allow them to be able to respond to and learn from human movement; thus, furthering their functionality in collaborative environments.

Hybrids

Hybrid robots are a combination of several types of robotic systems into one workable solution that offers multifunctional capability in accomplishing complex tasks. Examples include the combination of an AMR with a robotic arm that will pick packages from stock in a warehouse. The mobility of the AMR is merged with the precision provided by the robotic arm in such a case. These hybrid solutions aim to leverage strengths from multiple robotic technologies, which amounts to increasing functionality and versatility. As tasks become increasingly complex, and the need for multifunctional systems increases, hybrid robots will be able to offer an influential method in the way of dealing with different operational challenges and performance optimization.

In other words, each type of robot-from AMR and AGV to articulated robots, humanoids, cobots, and hybrids-adds something different in capability and advantage to the many applications. The integration of these into industrial and service environments heralds a great evolution in robotics, reaching an unprecedented frontier of efficiency and flexibility in human-robot collaboration. Full usage and optimal results in robotic applications are possible only if the particular attributes and use cases of each type of robot are identified and understood.

2.4 Robotic Automation Process

The methodology for implementing RPA is done in a structured, multi-phase manner in order to achieve maximum efficiency and effectiveness.

Phase 1: RPA Opportunities Identification and Feasibility Assessment

This is the very important first phase, where organizations conduct an in-depth study for the purpose of discovering those processes that are ready for automation. This includes analyzing potential candidates for RPA by documenting efficiencies and cost-saving opportunities expected. Identify processes that could significantly gain fromautomation to drive huge impacts in operational efficiency. It is also at this stage that consensus on the project objectives should be obtained from the executive stakeholders, and roles and responsibilities clearly defined. There is also the need to consider engagements with various RPA vendors in providing technology demonstrations as proof-of-concept. This would help prove the feasibility and efficiency of the RPA technology. A common error made during this phase is to choose processes that have very minimal business impact, which result in weak business cases and are not able to garner interest and support from the stakeholders.

Phase 2: Vendor Selection- detailed activity of selecting the correct RPA vendor.

This phase starts with developing clear technical requirements and evaluation criteria specific to the needs of the organization. A Request for Proposal requests responses from the vendors, who then come on-site to present the solution with demonstrations and technical proofs of concept. This step is very important to understand the fit of each solution to the needs of the organization, and the financial business case proposed. This usually encompasses the formulation of a shortlist regarding the technical capabilities, robustness of proof of concept, and cost-effectiveness of the solution. An organization may also make a decision to contract a vendor for a pilot implementation while developing internal capabilities to manage subsequent RPA projects themselves.

Fig. 37.4 Benefits of robotics industrial automation

Source: Author

Phase 3: Process Documentation, Pilot Testing, and Implementation

The RPA solution is actualized upon this foundation. This stage involves extensive documentation of the processes to be automated and preparation of the IT infrastructure to support the deployment of RPA. Massive training programs are also conducted to ensure the preparedness of the human resources who will participate in the implementation. The pilot stage is the most sensitive because this is where actual testing of the RPA solution occurs in a controlled environment for validation of performance and eventual recognition of problems. The results of the pilot phase will help in making final adjustments and refinements before full deployment. This stage sets the scene for moving into a long-term operational model by ensuring the RPA solution fits within the current processes.

Phase 4: RPA Lifecycle Management

It is about the deployment and ongoing management of the RPA solution itself. This stage includes the implementation of the RPA solution itself and the building of a solid structure for further operation and optimization. The activities conducted during this phase include setting up governance structures, establishing the operational model, and following a change management approach to pave the way for further adjustments and improvements. It is expected that the RPA solution should maintain relevance to organizational objectives and continue being valuable over a period of time. This includes the monitoring of actual performance, fixing the issues arising, and iterating enhancements to adapt to evolving business needs. In a nutshell, this structured approach not only makes for strategic and effective RPA implementation but provides the backbone for successful automation, enhances operational efficiency, and supports long-term value creation for the organization.

3. Result and Discussion

The integration of various automation systems and robotic technologies has significantly advanced industrial and service environments, offering a spectrum of efficiencies and capabilities. Fixed automation excels in high-volume, repetitive production environments where consistency is key but lacks flexibility for product changes. In contrast, programmable automation provides adaptability for batch production, allowing for different product configurations through reloaded control programs, though it involves higher investments and lower production rates compared to fixed systems. Flexible automation, an evolution of programmable automation, reduces downtime between product changes and integrates advanced material handling, making it ideal for environments with frequent design alterations despite its higher cost and complex engineering requirements.

Meanwhile, Autonomous Mobile Robots (AMRs) and Automated Guided Vehicles (AGVs) cater to different operational needs: AMRs navigate complex environments autonomously using advanced sensors and onboard processing, while AGVs operate along predefined paths in controlled settings, offering reliability for repetitive tasks. Articulated robots, or robotic arms, deliver precision and versatility for complex tasks like welding and material handling, whereas humanoid robots, designed for human-like interaction, enhance customer service and assistance roles. Collaborative robots (cobots) improve safety and productivity in shared workspaces by working directly with human operators and adapting to their movements. Lastly, hybrid robots combine features from various systems, offering multifunctional capabilities to address complex challenges. Each automation and robotics type presents unique advantages, highlighting the need for strategic selection to optimize manufacturing and operational processes.

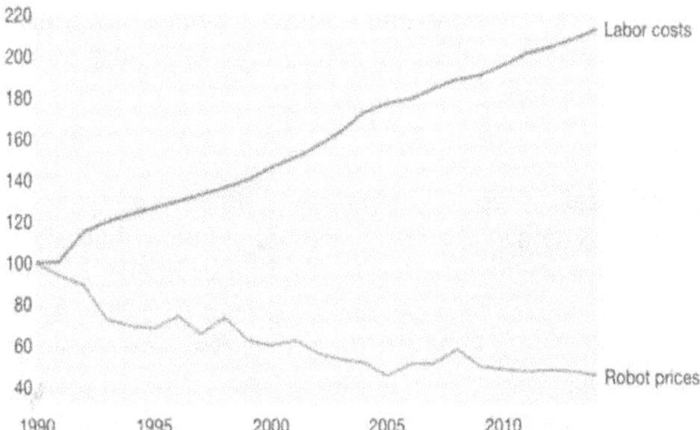

Fig. 37.5 Comparison of prices of labor and robots in industrial system

Source: Author

The cost of robots has experienced a substantial decline over the past 30 years, with average prices falling by 50% in real terms. This decrease is even more striking when compared to labor costs, highlighting the economic advantages of automation. As global demand for robots continues to rise, particularly from emerging economies, there is a notable shift in production to regions with lower labor costs. This trend is anticipated to further drive down robot prices, making automation increasingly accessible and affordable across various industries.

4. Conclusion

The structured approach to implementing Robotic Process Automation, as presented here, emphasizes careful planning and attention to strategy in order to realize the full potential of such emerging automation technologies. Encompassing a methodology that covers initial assessment and choice of vendor through pilot test to lifecycle management provides an overall comprehensive framework for integrating RPA into organizational processes. Rigorous examination of processes for automation potential, appropriate choosing of vendors, and finally extensive pilot testing-are just a few ways to let companies have their RPA activity effective and accordingly integrate this into a strategic aim. Moreover, strong lifecycle management and continuous improvement underline sustained governance for adaptability in order to sustain operational excellence. This structured approach in itself not only helps in successful RPA deployment but also brings continuous improvements toward operational effectiveness, cost reduction, and competitiveness. Judicious implementation of RPA across these phases may lead to metamorphic changes in business processes, which would enable organizations to sustain viability in the automated and fast- changing world of business.

References

1. Avaid, M., Haleem, A., Singh, R. P., & Suman, R. (2021). Substantial capabilities of robotics in enhancing industry 4.0 implementation. *Cognitive Robotics*, *1*, 58–75
2. Singh, M., & Khan, S. A. L. A. (2024). Advances in Autonomous Robotics: Integrating AI and Machine Learning for Enhanced Automation and Control in Industrial Applications. *International Journal for Multidimensional Research Perspectives*, *2*(4), 74–90.
3. Husainy, A., Mangave, S., & Patil, N. (2023). A review on robotics and automation in the 21st century: Shaping the future of manufacturing, healthcare, and service sectors. *Asian Review of Mechanical Engineering*, *12*(2), 41–45.
4. Ade-Omowaye, J., Ajisegiri, E., & Ojji, I. (2024, April). Robotics and Automation in Engineering: Perspectives for the Digital Economy. In *2024 International Conference on Science, Engineering and Business for Driving Sustainable Development Goals (SEB4SDG)* (pp. 1–7). IEEE.
5. Ahmed, B. (2024). Advancements in Industrial Automation: Technologies and Trends. *Frontiers in Robotics and Automation*, *1*(01), 19–35.
6. Goel, R., & Gupta, P. (2020). Robotics and industry 4.0. *A Roadmap to Industry 4.0: Smart Production, Sharp Business and Sustainable Development*, 157–169.
7. Karabegović, I., & Husak, E. (2018). The fourth industrial revolution and the role of industrial robots: a with focus on China. *Journal of Engineering and Architecture*, *6*(1), 67–75.
8. Karabegović, I. (2017). The role of industrial and service robots in fourth industrial revolution with focus on China. *Journal of Engineering and Architecture*, *5*(2), 110–117.
9. Ali, A. (2024). The Role of Artificial Intelligence in Robotics and Automation. *Frontiers in Robotics and Automation*, *1*(01), 36–54.
10. Peter, O., Pradhan, A., & Mbohwa, C. Automation and Robotics for Digital Economy: A Survey.
11. Grift, T., Zhang, Q., Kondo, N., & Ting, K. C. (2008). A review of automation and robotics for the bio-industry. *Journal of Biomechatronics Engineering*, *1*(1), 37–54.
12. IMOH, I. (2024). A Technical Survey on The Role of Robotics in Conventional Manufacturing Process: An Element of Industry 4.0. *FUPRE Journal of Scientific and Industrial Research (FJSIR)*, *8*(2), 172–192.
13. Tzafestas, S. G. (2018). Synergy of IoT and AI in modern society: The robotics and automation case. *Robot. Autom. Eng. J*, *31*, 1–15.

Intelligent Systems Using Semiconductors for Robotics and IoT – Dinesh Goyal et al. (eds)
© 2026 Taylor & Francis Group, London, ISBN 978-1-041-20408-4

38 Evaluating Link Prediction with Heterogeneous Graph Transformer: Cross-Dataset Insights and Analysis

Puneet Kapoor[1],
Sakshi Kaushal[2], Harish Kumar[3]
University Institute of Engineering and Technology,
Panjab University, Chandigarh, India
Kushal Kanwar[4]
Jaypee University of Information Technology, Waknaghat,
Solan, Himachal Pradesh, India

Abstract: Link prediction for dynamic heterogeneous networks poses distinct challenges because of their changing and diverse characteristics. This study investigates the capabilities of the Heterogeneous Graph Transformer (HGT), a specific sort of Graph Neural Network (GNN), in predicting links in different types of networks, such as social networks, e-commerce systems, and biological knowledge graphs. HGT efficiently captures complex node characteristics, edge variations, and temporal fluctuations by utilizing several attention methods. This study aims to evaluate the performance of HGT in several circumstances through thorough assessments, highlighting its strengths and identifying areas for development. The findings offer significant knowledge on HGT's ability to adjust and perform well, which contributes to the progress of graph-based machine learning techniques and their use in complex, practical systems.

Keywords: Link prediction, Heterogeneous graph transformer, Graph neural network, Predictive analytics, Attention mechanism

1. Introduction

Complex networks have a crucial role in representing several real-world systems, including social and biological networks as well as technical and information networks. These networks have complex structures and dynamic interactions between many nodes and edges. They often exhibit non-trivial topological properties like community structures, scale-free distributions, and small-world properties. This makes them well-suited for studying connection and action within a system [1]. Due to the complex structure of the systems being studied, heterogeneous graphs are frequently employed to represent and analyze them. These graphs capture the interactions between different sorts of objects in diverse manners. Instances of such systems encompass academic graphs, LastFM's user-item interactions, Amazon-book's recommendation system, Yahoo-Music's user-song relationships etc. Heterogeneous graphs provide a more comprehensive and detailed portrayal of real-world systems, capturing the diverse and complex links within these ecosystems. Link prediction is a method of analysis that attempts to identify and anticipate likely future relationships within a network. It has its origins in early studies of complex networks and social dynamics. Initially, academics concentrated on understanding how interactions arise and change in social networks. Hence the idea of triadic closure, common Neighbour Index and a variety of heuristic approaches evolved [2]. "As networks became more sophisticated, classic one-mode networks, which have all nodes of the same type, became less useful. Techniques such as bipartite

[1]puneetira@pu.ac.in, [2]sakshi@pu.ac.in, [3]harishk@pu.ac.in, [4]kushalneo@gmail.com

DOI: 10.1201/9781003716389-38

network projection were developed to condense these networks into a single-mode representation while keeping important structural information. In addition to these advancements, various categorization strategies for link prediction algorithms have significantly contributed to the area. Network structure analysis, for example, employs network topology or information theory to predict links. Lu et al. [3] introduced the structural consistency index, which measures a network's regularity by comparing its structural attributes before and after random link removal. This score is a good estimate of link predictability, emphasizing the importance of understanding network topology in order to anticipate relationships efficiently. Network embedding approaches are another effective method for link prediction. Network embedding is a method for representing graph nodes within a continuous vector space, maintaining their architectural and semantic attributes. These techniques use random walk embeddings to record node relationships in a network, demonstrate this strategy by embedding nodes in a continuous vector space, which efficiently captures node structural features while boosting connection prediction accuracy. Machine and deep learning algorithms have also made substantial contributions to the development of link prediction. These methods employ node properties and network structure to understand complex patterns and relationships, improving the accuracy of link prediction algorithms. Researchers established a comprehensive toolkit for link prediction in complicated networks by combining conventional network analysis approaches, network embedding strategies, and advanced machine learning algorithm" [4]. Graph Neural Network(GNN) represented a significant improvement in network analysis. GNNs apply deep learning techniques to graph-structured data, allowing for the management of irregular structures inherent in graphs.

They function by repeatedly aggregating information from a node's neighbours, capturing simultaneously local and global graph features [5]. The Heterogeneous Graph Transformer (HGT) [2] is a huge step forward in using GNNs on heterogeneous graphs. HGT applies node- and edge-specific parameters to efficiently capture the distinct properties of various node and edge types. It uses a mutual attention technique to dynamically aggregate data from different types of nodes and edges, with separate representations for each type. In addition, HGT uses a relative temporal encoding technique to manage dynamic graphs, which successfully models time-dependent interactions. Temporal encoding integrates time-related data into graph learning models to reflect the dynamic progression of links and node interactions. It guarantees the appropriate modelling of temporal dependencies, facilitating applications like real-time suggestion engines and dynamic network analysis. Temporal encoding improves models' capacity to anticipate temporal trends in graph data by utilizing relative or absolute timestamps [12-13]. In Fig. 38.1, the given graph on the left consists of nodes of different types (circles, squares, triangles) and edges of different colors that reflect different relationships. The HGT model applies attention mechanisms, temporal encoding, and message passing & aggregation to analyze the graph. The output on the right displays the anticipated interconnections, represented by various colored dotted lines, illustrating the recently predicted connections inside the network.

HGT has emerged as a potent instrument for tackling numerous real-world applications that entail complicated and varied data interactions. In the field of fake news identification, HGTs proficiently represent the complex interconnections of news pieces, publishers, and related entities, facilitating precise forecasting of misinformation dissemination [14]. In recommender systems, HGTs are proficient at forecasting user-item interactions by encapsulating the varied interactions seen in e-commerce or content streaming platforms, as evidenced by datasets such as AmazonBook and LastFM.

In this study, we look at how well HGT predicts links. A variety of datasets are included in the investigation, including LastFM [6], AmazonBook [7], Yahoo-Music [8], IMDb [6], and OGB-BioKG [9].

2. Dataset

The datasets utilized for this paper are as follows:

LastFM Dataset: This dataset is primarily used to examine users' music streaming behaviours and preferences. This dataset have 20,612 nodes which represents artist and users. It has 201,908 edges, telling about interactions between nodes [6].

AmazonBook Dataset: The AmazonBook represents recommender system dataset. The nodes of this dataset

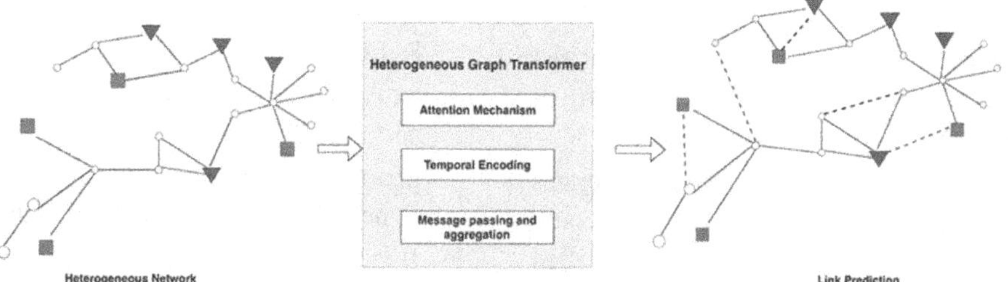

Fig. 38.1 An example showcasing the application of the HGT on a heterogeneous network to predict links

Source: Authors

represents books and users. The edges of this dataset represents the recommendation of books to users [7].

Yahoo-Music Dataset: This dataset contains 6,000 nodes which represents different type of music items and users. It contains 9,604 edges. These edges represents the music preferences of users, it basically creates music recommendation for users [8].

BioKG Dataset: BioKG is a knowledge graph. This dataset have node and edges which represents biological entities like drug, disease etc. the link prediction in this context means which drug is cure or can become cure for which disease with the help of which protein interaction [9].

IMDB Dataset: This dataset have different node types, 4,278 nodes represents movies, 5,257 nodes represents actor and 2,081 actor nodes. The interaction between nodes means what all combinations are possible between actor and director to work on which type of movie [6].

Table 38.1 Dataset details

Dataset	Nodes	Edges	Node Types
LastFM	20,612	201,908	users, artists, tags
AmazonBook	144,242	4,761,460	users, books
Yahoo-Music	6,000	9,604	users, items (music)
OGBL-BioKG	93,773	1,537,734	diseases, drugs, proteins
IMDB	11,616	34,212	movies, directors, actors

Source: The data employed in this work is obtainable via the PyTorch Geometric Datasets repository, accessible at https://pytorch-geometric.readthedocs.io/en/latest/index.html

3. Heterogeneous Graph Transformer for Link Prediction

The HGT [2] is an advanced model specifically developed to address the complexities of heterogeneous networks, which involve complex relationships between several types of nodes and edges. Following is a step-by-step description of how HGT works :

- Type-Specific Attention Calculation: Calculate attention scores for every node according to the type of edges that connect it to its neighbouring nodes.
- Attention Weight Normalization: Normalize these scores to calculate attention weights, ensuring that each neighbour's value is proportional.
- Embedding Update: With the help of normalized attention weights and type specific relation of neighbour nodes, we update node embeddings.
- Overall Aggregating Information: This step crates unified embedding for all types. It takes into account all type specific embeddings to make one unified embedding.

3.1 Utilizing Multi-Head Attention for Node & Edge Awareness

HGT uses multiheaded attention mechanism effectively to identify node type and edge type for predictions. This method can be utilized for node as well as link prediction. Attention score for each node is calculated to understand the importance to its neighbour. With the help of multi-head attention mechanism importance of different type of edges can be learned simultaneously. In context of link prediction this means which type of edge is more probable to make connection in future. Equation 1 represents attention score calculation. The node embedding of nodes y ($h_y^{(k-1)}$) and x ($h_x^{(k-1)}$) are taken from previous layer and then these embeddings are transformed using relation specific matrix ($W_r^{(k)}$) to calculate attention weights for particular relation [2]. Attention score tells particular type of edge is how much important or meaningful to connect node x and y. According to this attention score node embeddings are updated.

$$g_{yx}^{r} = f\left(W_r^{(k)} h_y^{(k-1)}, W_r^{(k)} h_x^{(k-1)}\right) \tag{1}$$

where:-

- $g_{xy}{}^{r}$ attention score value of relation of type r between nodes y and x.
- $W_r^{(k)}$ weight matrix of edge type r at layer k.
- $h_x^{(k-1)}$ node embedding for node x in layer k–1.

3.2 Embedding Update

After getting attention score ($g_{yx}{}^{r}$) of every edge type, these scores are normalized, here we have represented normalization with $\beta_{yx}{}^{r}$. Embedding of node x in layer k is calculated with utilization of its embedding of its neighbours in previous layer. It utilises weighted sum of node embeddings available in previous layer with normalized attention scores to build node embedding of current layer k. The weights are computed by the attention scores associated with each edge type. This step is critical for considering the impact of various sorts of connections on the node's representation. The embedding for node x in layer k, denoted as, is updated according to the Eq.2:

$$h_x^{(k)} = \sigma\left(\sum_{(y,r,x)\in E} \beta_{yx}^{r} W_r^{(k)} h_u^{(k-1)}\right) \tag{2}$$

where: -

- σ represents the activation function (e.g. Sigmoid, ReLU).
- $(y, r, x) \in E$ denotes the edge set of category r with y and x are neighbors.
- $\beta_{yx}{}^{r}$ is the attention weight denoting the significance of neighbor y of node x for edge type r.
- $W_r^{(k)}$ represents the kind-specific weight matrix of edge type r at layer k.

Each neighbour in the network is allocated a weight based on the type of edge that connects them. This relevance is acquired during training, allowing the model to properly prioritize various links in the heterogeneous graph. Calculating these weights guarantees that the node's updated embedding is influenced more by its most relevant neighbours.

3.3 Type-Specific Attention Weights Calculation (β)

The attention weight β for every neighbour y of node x, with respect to the edge type r, is computed using the Eq. 1. Here, it is evident that the normalized attention score is equivalent to the attention weight. To guarantee that the values are positive and to underline higher scores, the attention score $g_{yx}{}^{r}$ is exponentiated, therefore emphasizing more significant connections. Then the exponentiated attention score of the particular neighbour y is divided by the total of the exponentiated attention scores of all neighbours n of node x. This normalizing phase guarantees that the total sum of the attention weights for node x is equal to 1. This technique guarantees that the model fairly depicts the influence of every neighbour in the network by maintaining their relative proportionality.

$$\beta_{yx}^{r} = \frac{\exp\left(g_{yx}^{r}\right)}{\sum_{n \in N(v)} \exp\left(g_{yn}^{r}\right)} \tag{3}$$

3.4 Overall Aggregating Information

During this concluding phase, the HGT model integrates data from several neighbours to comprehensively update the node embeddings. In Section Type-Specific Attention Weights Calculation (β), the main objective was to update node embeddings by incorporating type-specific attention from nearby neighbours. In this step, all of these changes are combined to create a single embedding for each node. This embedding is final representation of particular node.

4. Experimental Results

We have used Macro F1, Micro F1 and AUC [10] scores to evaluate HGT over considered datasets. Greater values of AUC represents better performance of model. Table 38.2 summarize the experiment results. Different properties of heterogeneous datasets result in distinct model performances. For example LastFM dataset AUC performance value is 0.7465 while HGT gives AUC value of 0.6669 on AmazonBook dataset. Figure 38.2 suggests that the Micro F1 score of AmazonBook dataset is better compare to all the models, it has value of 0.8118 which suggest that for this dataset performance across all the classes is far more better than other datasets. AUC values of Yahoo-Music is greatest in the Table 38.2, suggest that HGT model is well suited for this dataset and can easily identify positive and negative links for this dataset. For Example, if user listens to particular song then it will be considered as positive link here.

Table 38.2 HGT performance on link prediction tasks across multiple datasets

Dataset	AUC	Macro F1	Micro F1	HGT Layers
LastFM	0.7465	0.5406	0.7124	2
AmazonBook	0.6669	0.4481	0.8118	2
Yahoo-Music	0.9440	0.4453	0.8029	2
IMDb	0.7676	0.6384	0.6602	2
OGB-BioKG	0.8578	0.5958	0.6372	2

Source: https://github.com/PuneetIRA/HGT-LAstFM/tree/main

HGT obtained an AUC of 0.8578, a Macro F1 of 0.5958, and a Micro F1 of 0.6372 for the OGB-BioKG dataset, which was created to investigate biological knowledge graphs. HGT ability to learn intricate biological interactions is evidenced by its high AUC and Macro F1 scores (Fig. 38.2). Overall, HGT's capacity to generalize across varied datasets demonstrates its potential for a wide range of applications in link prediction and network research.

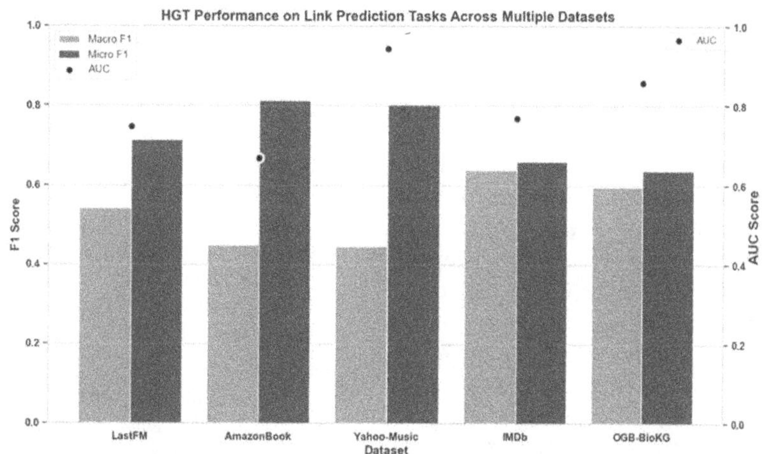

Fig. 38.2 Efficiency of the HGT on link prediction tasks with multiple datasets

Source: Authors

5. Future Directions

The HGT exhibits fascinating potential, however, there are still some areas that require additional investigation and improvement. Here are potential future directions for enhancing HGT, based on recent research:

- Expanding the capabilities of HGT to accommodate dynamic heterogeneous graphs is an important area for future development. By incorporating temporal information, the model can effectively capture the changing dynamics of relationships over time. This is particularly important for applications like recommendation systems and social network analysis. [2].
- Integrating multi-modal data from multiple domains into HGT can enhance our understanding of complex systems. Integrating textual, visual, and structural information can result in more resilient models that can handle a diverse set of tasks [11] [14].
- "Integrating multimodal data into GNN models can enhance the precision and resilience of link prediction. This involves amalgamating data from many sources, including text, photos, and structured data, in order to create more comprehensive and valuable network representations. Integrating many data sources can improve the accuracy of predictions by utilizing the unique advantages of different modalities" [4].

Although Heterogeneous Graph Transformers (HGTs) exhibit robust efficacy in representing intricate relationships, numerous limitations persist. Existing designs are inadequately equipped to manage dynamic heterogeneous graphs, essential for representing temporal variations in evolving networks, including recommendation systems and social network analysis [2-3]. Furthermore, HGTs possess the limited capability to aggregate multi-modal data integrating textual, visual, and structural information which could markedly improve their accuracy and relevance to various tasks [15]. Rectifying these deficiencies will enhance HGTs' capacity to accurately simulate real-world systems and expand their applicability across other fields.

6. Conclusion

This study investigates the success of the HGT model for link prediction in diverse and complex networks. The HGT model accurately captures complex linkages and interactions within heterogeneous datasets by utilizing the multi-head attention mechanism and type-specific attention weights. The evaluation metrics, which included AUC, Macro F1, and Micro F1 scores, gave a thorough assessment of the model's performance, emphasizing its capacity to discern between positive and negative linkages, handle class imbalances, and perform well across classes. The results showed that the HGT model performed well in domains such as music recommendation (Yahoo-Music) and biological knowledge graphs (OGB-BioKG), with high AUC scores and strong link prediction performance.

Furthermore, the paper proposes potential options for improving HGT's application and performance. Integrating temporal information, including multimodal data, and investigating domain adaptation approaches may considerably increase the model's robustness and accuracy. These developments would allow HGT to better capture dynamic interactions, use complementary data modalities, and expand its applicability into other domains without requiring considerable retraining. Overall, the study emphasizes the potential of HGT in increasing link prediction within heterogeneous networks, giving useful insights and setting the framework for future advances in complex network analysis and prediction.

Acknowledgement

The authors acknowledge the FIST, DST grant (SR/FST/ET-I/2021/878) of the Computer Science and Engineering Department, University Institute of Engineering and Technology, Panjab University, for providing high end GPU Server on Nvlink SXM technology with 320 GB memory.

References

1. Gaogao Dong, Fan Wang, Louis M. Shekhtman, Michael M. Danziger, Jingfang Fan, Ruijin Du, Jianguo Liu, Lixin Tian, H. Eugene Stanley, and Shlomo Havlin. Optimal resilience of modular interacting networks. Proceedings of the National Academy of Sciences, 118(22):e1922831118, 2021.
2. Ziniu Hu, Yuxiao Dong, Kuansan Wang, and Yizhou Sun. Heterogeneous Graph Transformer. The Web Conference 2020 - Proceedings of the World Wide Web Conference, WWW 2020, pages 2704–2710, 2020.
3. Linyuan Lu¨, Liming Pan, Tao Zhou, Yi-Cheng Zhang, and H. Stanley. Toward link predictability of complex networks. Proceedings of the National Academy of Sciences, 112:201424644, 02 2015.
4. Puneet Kapoor, Sakshi Kaushal, Harish Kumar, and Kushal Kanwar. A survey on feature extraction and learning techniques for link prediction in homogeneous and heterogeneous complex networks. Artificial Intelligence Review, 57(12):348, 2024.
5. William L. Hamilton, Rex Ying, and Jure Leskovec. Representation learning on graphs: Methods and applications. IEEE Data Eng. Bull., 40:52–74, 2017.
6. Xinyu Fu, Jiani Zhang, Ziqiao Meng, and Irwin King. Magnn: Metapath aggregated graph neural network for heterogeneous graph embedding. In Proceedings of The Web Conference 2020, WWW '20, page 2331–2341, New York, NY, USA, 2020. Association for Computing Machinery.
7. Xiangnan He, Kuan Deng, Xiang Wang, Yan Li, YongDong Zhang, and Meng Wang. Lightgcn: Simplifying and powering graph convolution network for recommendation. In Proceedings of the 43rd International ACM SIGIR

Conference on Research and Development in Information Retrieval, SIGIR '20, page 639–648, New York, NY, USA, 2020. Association for Computing Machinery.

8. Chen, Yixin. 2020. "Inductive Matrix Completion Based on Graph Neural Networks." IMC 1: 1–14.
9. Weihua Hu, Matthias Fey, Marinka Zitnik, Yuxiao Dong, Hongyu Ren, Bowen Liu, Michele Catasta, and Jure Leskovec. Open graph benchmark: datasets for machine learning on graphs. In Proceedings of the 34th International Conference on Neural Information *Processing Systems*, NIPS '20, Red Hook, NY, USA, 2020. Curran Associates Inc.
10. Jonathan L. Herlocker, Joseph A. Konstan, Loren G. Terveen, and John T. Riedl. Evaluating collaborative filtering recommender systems. ACM Trans. Inf. Syst., 22(1):5–53, jan 2004.
11. Sun, Yizhou, and Jiawei Han. 2013. "Mining Heterogeneous Information Networks: A Structural Analysis Approach." *SIGKDD Explorations Newsletter* 14 (2): 20–28.
12. Zhu, Qiuyu, Liang Zhang, Qianxiong Xu, Kaijun Liu, Cheng Long, and Xiaoyang Wang. 2024. "HHGT: Hierarchical Heterogeneous Graph Transformer for Heterogeneous Graph Representation Learning." ArXiv Preprint. https://arxiv.org/abs/2407.13158.
13. Ye, H., Sun, Y., Gao, Y., Xu, F., and Qi, J. 2025. "Heterogeneous Graph Transformer Auto-Encoder for Multivariate Time Series Forecasting." Computers and Electrical Engineering 122: 109927. https://doi.org/10.1016/j.compeleceng.2024.109927.
14. Zhang, Yuchen, Xiaoxiao Ma, Jia Wu, Jian Yang, and Hao Fan. 2024. "Heterogeneous Subgraph Transformer for Fake News Detection." In *Proceedings of the ACM Web Conference 2024 (WWW '24)*, 1272–1282. New York: ACM. https://doi.org/10.1145/3589334.3645680.
15. Ektefaie, Y., Dasoulas, G., Noori, A., Farhat, M., and Zitnik, M. 2023. "Multimodal Learning with Graphs." arXiv preprint arXiv:2209.03299. https://arxiv.org/abs/2209.03299.

Intelligent Systems Using Semiconductors for Robotics and IoT – Dinesh Goyal et al. (eds)
© 2026 Taylor & Francis Group, London, ISBN 978-1-041-20408-4

39 Using the Liver Disease Dataset, a Strategic Performance Analysis by using Random Forest Classification Algorithms

Suraj Sharma[1],
Farhan Raza Rizvi[2], Shashikant Gupta[3],
Ratnesh Kumar Dubey[4], Nidhi Dandotiya[5]
ITM University, CSA Department,
Gwalior, India

Abstract: This research employs the Random Forest method for classifying and evaluating a dataset relevant to liver disease and all the classification performance provided in detail. Most researchers prefer using the Implicit Forest algorithm which constructs a combination of several classification trees to perform data classification. The findings denote that the accuracy of classification by using the Random Forest algorithm is impressive. Performance is gauged more so through measuring the F1 score, recall and precision and the highlights of the results are its usefulness in liver disease identification. Similarly, the analysis attempts to check the effect of some identified parameters on the classification problem revealing parameters like age, sex, gender and level of bilirubin as important in making decisions of the algorithm. Such information may assist in improving the formulation of focused therapy for the control and prevention of liver diseases. In general, the research emphasizes the possibility of using Random Forest classification algorithms in increasing liver disease diagnostic and treatment effectiveness. Further research may be aimed to the study the performance of the algorithm on larger sets of data and its usefulness in practice.

Keywords: Classification, Healthcare, Orange tool, Spyder, Liver disease dataset, Machine learning

1. Introduction

The improvement of individual access to a set of health-related services puts the healthcare system, as many believe, on the very top. Its main purpose is to improve health by making the treatment of diseases more efficient, and it includes actions that guarantee the health benefits that come from performing necessary activities. Therefore, the primary tasks of the healthcare system are to deliver quality services and enable the prevention of diseases targeting all their stages. Health care issues such as this fall within the domain of machine learning and specifically its classification [3] which consist of structure and relevance based clustering methods. Every class has a set of labels. The process is sometimes referred to as supervised learning [5], when the system acquires knowledge from training data and applies it to test data, ultimately yielding valuable outcomes. A category is often the result of the categorization problems. The difficulties that center on the division of data into binary or multiple discrete labels are part of the classification process. Figure 39.1 below illustrates the different ways that classification is used in the healthcare industry.

By using the training data set to build a model or classify data, classification algorithms [4] enable the use of the original dataset to assess new data sets. The main classification techniques that are taken into account are Random Forests, Support Vector Machines, and Logistic Regression [6]. Using the liver disease dataset, these classifiers are further compared based on the accuracy parameter. Additionally Parameter tuning: Use new

[1]surajsharma.cse@itmuniversity.ac.in, [2]farhanraza.cse@itmuniversity.ac.in, [3]shashikantgupta@itmuniversity.ac.in, [4]ratnesh.soet@itmuniversity.ac.in, [5]nidhidandotiya@itmuniversity.ac.in

DOI: 10.1201/9781003716389-39

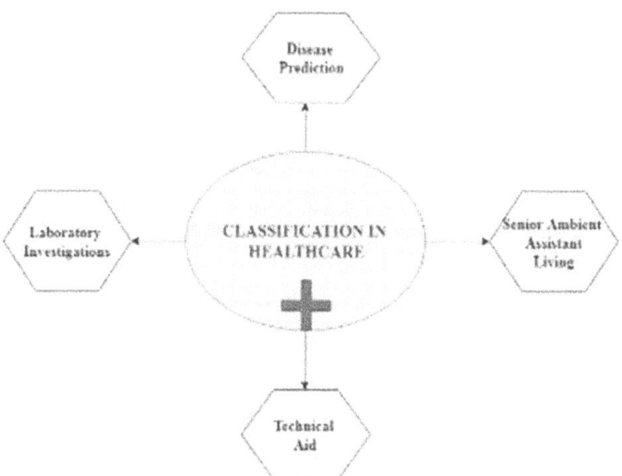

Fig. 39.1 Applications of classification in healthcare

techniques to fine-tune hyperparameters to improve performance, this study implements a number of methods to improve the best selected classifier's accuracy.

2. Tool Used

2.1 Orange Tool

Orange is a tool for data mining[8] made up of multiple parts. It comes with a number of widgets for data display and analysis. Python is employed as the scripting language in this program. It is made up of multiple parts for testing, scoring,ranking, modeling, and data preprocessing. It works with C, Python, and C++. It is also a Python-based machine learning program that is available as open source. Compared to the other tools, it features a better debugger. Furthermore, it can be effectively employed to compare multiple classification algorithms through the examination of diverse metrics including values for precision and accuracy. Additionally, This program can be used to identify various trends and handle the enormous amount of data.

2.2 Classification Algorithms Used

Random Forest

The fundamental function of the machine learning algorithm Random Forest is to build many decision trees. It uses a simple method in which When a dataset is split up into a batch of them, a decision tree is made for each randomly selected dataset.[8]. "Accordingly, A group of trained decision trees make up the forest that work together to get a final conclusion where a majority vote is taken into account. It can handle a lot of info and maintains the accuracy of missing data.

Logistic Regression

In predictive modeling, To ascertain the probability, logistic regression is utilized that a specific event will occur. It mostly works with the set of independent variables in order to calculate the discrete values and forecast their binary results. Using training data from the previous output, it conducts binary classification and forecasts future results [11]. This method is frequently applied to predictive analysis in cases in cases where the categorical dependent variable.

Support Vector Machine

A. Regression and both linear and nonlinear classification tasks can be handled by the machine learning algorithm Support Vector Machine. This algorithm's primary goal is to use a straight line to divide two groups apart. Additionally, its primary focus is on locating a hyperplane in an N-dimensional space that clearly divides the data points into classes. The primary data points that determine the hyperplane's position are called support vectors. In order for the Support Vector Machine to form, they are essential [4].

2.3 Dataset Description

The UCI Machine Learning Repository provides a dataset on liver sickness, which is used to develop the Random Forest Algorithm. Dataset [8] comprises eleven criteria that are essential to the performance analysis. These characteristics are "Gender, Age, Albumin, Albumin and Globulin Ratio, Aspartate and Alamine Aminotransferases, Total_Protein, Alkaline_Phosphotas, and Dataset." Furthermore, a real dataset from AIIMS, Bhopal has been assembled to evaluate the accuracy of several categorization techniques. The remainder is grouped in the following manner: "Section 1 comprises an Introduction that explains the tool utilized and the different classifiers that are compared, as well as the dataset description. Section 2 comprises a Literature Survey that gives an account Covers the work produced by several authors and contains a table that contrasts the contributions of several authors. There is a Research Methodology in Section 3. The performance-based outcomes are shown in Section 4's Result Analysis. Section 5 includes a Conclusion that explains how this study was successful in terms of providing the best classification technique and enhancing the accuracy of a best-selected classifier. References are then included.

3. Literature Survey

Hoon Jin et al. [1], outlined the idea of several classification methods that help medical professionals identify a patient's condition fast and effectively. The dataset that was used in the implementation of these strategies through the Weka Tool was downloaded from the UCI Repository. The Random Forest and Logistic Regression methods obtained better results in recall and sensitivity, whereas in the data provided, the Naive Bayes performed better in precision rate in classification tasks.

Ayesha Pathan and others proposed an approach to diagnose liver disease. They used a number of classification methods which are Random Forest, Naïve Bayes, Ada Boost, J48, Bagging. Different metrics such as accuracy and error rate were used to test the performance of these algorithms. Also, K-means clustering was used within the pre-processing stage to split the data into two categories: liver patients and non-liver patients. The clustered dataset created was also used to test other classification techniques available in Weka. In the end, the experiment showed that Random Forest was the most superior among the rest of the algorithms.

In their research [3], Tapas Ranjan Baitharu and colleagues also addressed the possible methodologies for diagnosing various liver ailments by analysis of pertinent datasets. Their study revolved around assisting the medical practitioners in making decisions. They evaluated several algorithms such as, J48, Multilayer Perceptron, ANN, ZeroR, IBK, VFI, and naive bayes on different parameters. The dataset was obtained from UCI Repository so as the weka toolkit where the algorithms were implemented. Their finding suggested that the Multilayer Perceptron was the best approach used in all of the methods within the range of target classification. For that reason, Multilayer Perceptrons Should be potentially useful in the diagnosis of liver problems in the future.

According to [4] S. Vijayaranil et al, have shown how predicting the liver disease at an earlier stage and also assessing its prognosis becomes instrumental. The employed Matlab 2013 tool to implement the suggested method, and the dataset which was taken from the UCI Repository was evaluated. From the experimental work done, It was established that the Support Vector Machine had the best accurate classification outcomes among the examined classification pars, followed by the Naive Bayes Algorithm and still has prospects of use in prediction of liver disease.

Rakhi Ray et al. [5], proposed data mining as a tool that can aid in the development of a healthcare management system. Many businesses implement data mining technology owing to its various advantages. It is well known that every individual's wellbeing depends on healthcare. A lot of advanced techniques are being developed to assess bodily conditions and to identify the signs of different diseases. Medical systems contain an enormous amount of data which makes it necessary to have a comprehensive method of searching for data in the database system. For this reason, data mining emerges as one of the most probable techniques for efficiently forecasting different diseases.

Al-Mamun et al. [14] Recommended the study claiming several classification systems based on datasets focusing on the liver disease spectrum. Al Mamun et al. (2018) evaluated five classification methods using the MayoCare's liver disease prognosis dataset: LTC, decision trees, random forests, SVMs, and ANNs. The authors reported that SVM performed better than ANNs in terms of sensitivity and specificity although the two models achieved the highest accuracy overall.

Zhang et al. [15] In this context, the performance of logistic regression and a couple of other classification strategies such as SVM, ANN, and gradient B were normalized using the Shanghai Liver Cancer dataset to assess four classification algorithms

4. Research Methodology

A method flow diagram to enhance the accuracy of a classifier is displayed in Fig. 39.2 below.

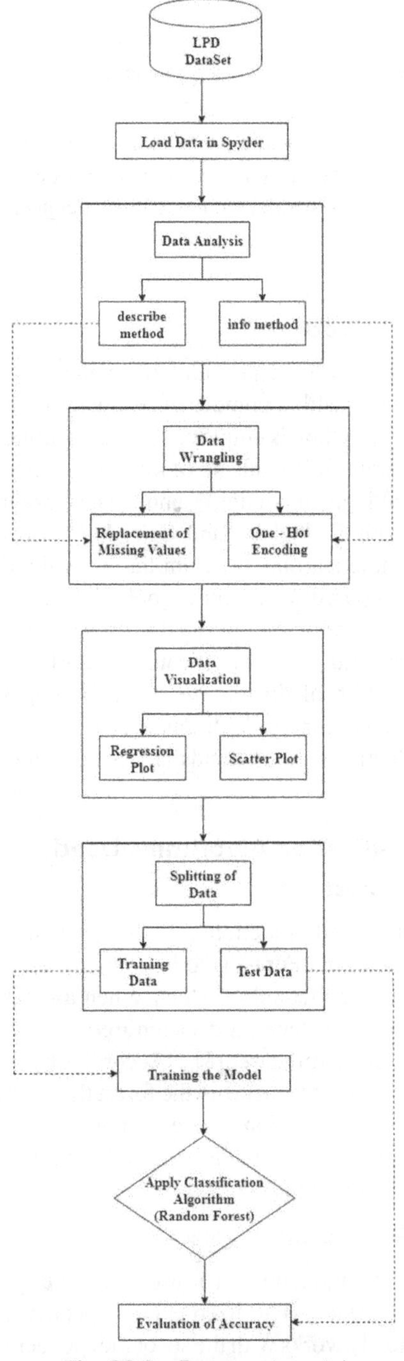

Fig. 39.2 Proposed model

5. Proposed Algorithm

i. **Installation:** After setting up Spyder 3.3.2, also known as The Scientific Python Development Environment

ii. on the computer, make a workspace directory containing the datasets and machine learning code.

iii. **Import the packages:** Import all required Python packages, including Pandas, NumPy, Matplotlib, and Scikit-Learn,in accordance with the specifications.

iv. **Loading of dataset:** Use Pandas in the Python workspace Spyder to load the Indian Liver Patients dataset in CSV(comma separated value) format from the UCI repository.

v. **Data Analysis:** Use the info ( ), describe ( ), value_counts ( ), head ( ), and tail ( ) methods to quickly obtain adescription of the data and to examine it.

vi. **Data Wrangling:** The data that has been stored has some missing values that need to be replaced with mean ormedian values. To convert categorical data to quantitative data, the get_dummies() method is utilized. One-hotencoding is the term for this.

vii. **Data Visualization:** Use the scatter, boxplot, regplot, corr, and heatmap methods to properly visualize data anddemonstrate the link between many variables.

viii. **Splitting of Training and Test data:** Split the training and test data from the entire dataset using the train_test_split()technique.

ix. **Training the Model:** Utilize the fit() method to train the model with training sets of data.

x. **Apply Classification Algorithm:** The dataset that was gathered from the UCI repository is subjected to the RandomForest algorithm.

xi. **Evaluation of Accuracy:** Once data wrangling is complete, use the accuracy_score() method to assess a RandomForest classifier's accuracy.

xii. **Model Generation:** Ultimately, a precise model is produced that may be applied to a variety of applications.

6. Result Analysis

6.1 Performance Analysis through an Orange Tool

Orange is a data mining tool instrument that comes with a number of widgets for data display and analysis. Furthermore, it may be effectively employed for comparing multiple classification algorithms [3] by the examination of diverse metrics including accuracy and precision values.

The following are some processes that are involved in using Orange Tool to determine the best accurate algorithm:

Step 1: Load the actual Indian liver patient dataset from AIIMS, Bhopal, into a file. The file displays all of The characteristics of the Dataset of Indian Liver Patients.

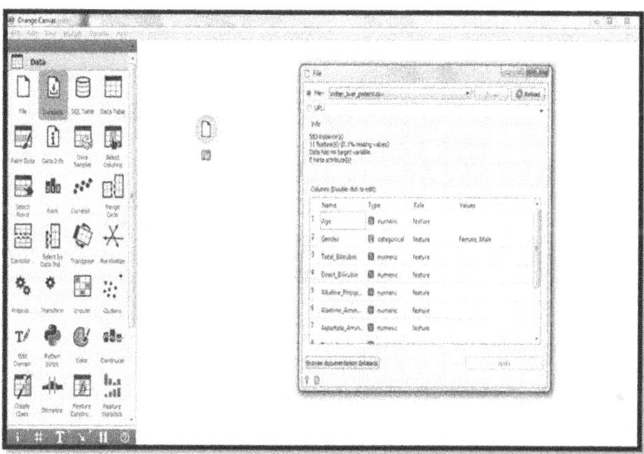

Fig. 39.3 Loading

Step 2: In order to use the File's output as the Data Table's input, connect the File and Data Table widgets now. The data table contains every value from The Dataset of Indian Liver Patients.

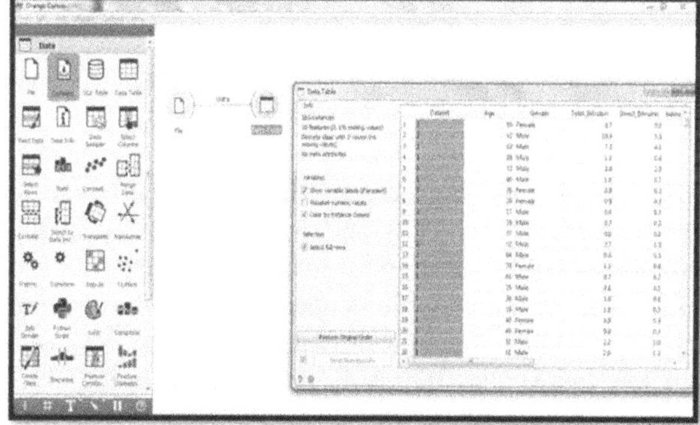

Fig. 39.4 Establishing a connection between the data table widget and the file

Step 3: The Widget for Test and Score was used to evaluate the data and present the statistics for F1, area under the curve, recall, accuracy, and precision. We compared several classification techniques, like Support Vector Machines, the Random forest models, and logistic regression. Consequently, Random Forest performed better than the other algorithms, achieving 72% accuracy as opposed to 70% for Logistic Regression and 69% for Support Vector Machine.

Additionally, a comparison of three classifiers is shown in Fig. 39.6. Furthermore, out of the three classifiers, Random Forest is deemed the best because it has the highest accuracy.

6.2 Performance Analysis through Spyder

A Scientific Python Development Environment with sophisticated coding tools and plugins is called Spyder

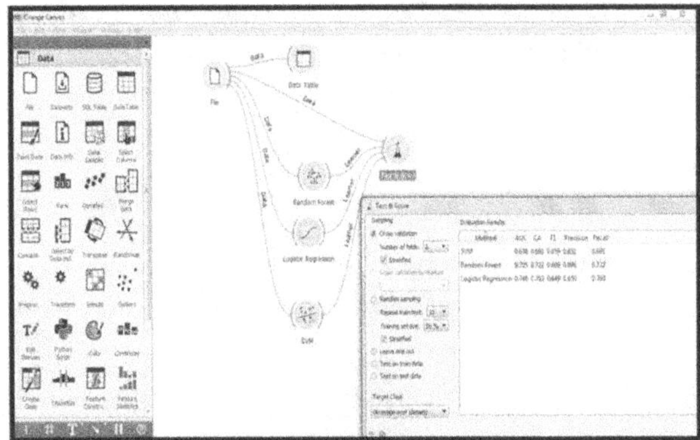

Fig. 39.5 Evaluation results

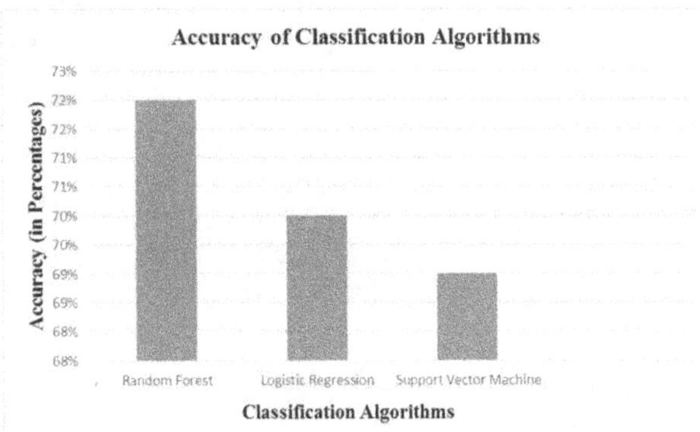

Fig. 39.6 Comparative study of three classifiers in terms of accuracy

[13]. This methodology has mostly been used to create the Random Forest classifier and assess its accuracy following the completion of data wrangling. We are able to assess the classifier's accuracy after the random forest model has been trained. The accuracy, or 69%, of a Random Forest Classifier is displayed below in Fig. 39.7 prior to data wrangling, which leaves the training set with noise, missing values, and extraneous features. In contrast, Figure 39.8 displays an improved accuracy of 72% for a Random Forest Classifier following data wrangling that eliminates all missing values, converts irrelevant features into relevant features, and makes use of the best features.

Figure 39.9 above compares the accuracy of the Classifier for Random Forest, demonstrating that it gained 69% accuracy prior to data wrangling and 72% accuracy

```
IPython console
Console 1/A
In [25]: from sklearn.ensemble import RandomForestClassifier
    ...: rnd_clf = RandomForestClassifier(n_estimators= )
    ...: rnd_clf.fit(X_train, y_train)
    ...:
    ...: y_pred=rnd_clf.predict(X_test)
    ...:
    ...:
    ...: from sklearn import metrics
    ...: ("Accuracy:",metrics.accuracy_score(y_test, y_pred))
Accuracy: 0.6917808219178082

In [26]:
IPython console   History log
```

Fig. 39.7 Before data wrangling, Random forest's accuracy

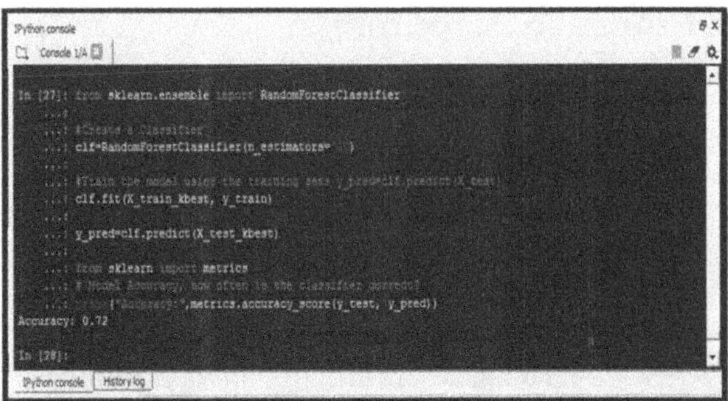

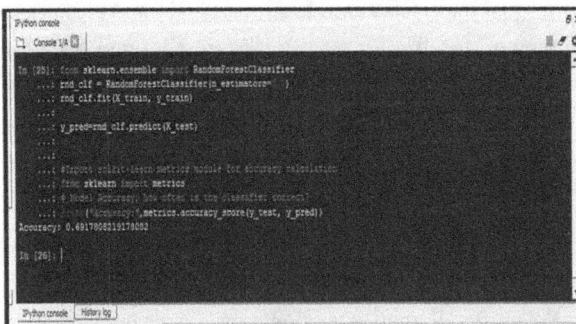

Fig. 39.8 After data wrangling, Random forest's accuracy

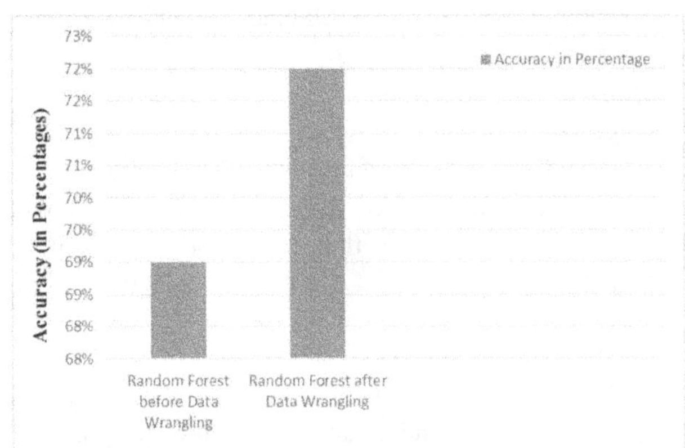

Fig. 39.9 Comparative study of random forest in terms of accuracy

following data wrangling. This demonstrated that using a data wrangling strategy can improve a classifier's accuracy.

7. Conclusion and Future Work

This study's primary goal was to present an overview of several categorization algorithms. This work conducted a comparative analysis of several papers in order to assess different categorization algorithms. Additionally, the Orange Tool was used to evaluate and contrast the results of the Logistic Regression, Random Forest, and Support Vector Machine. Because Unpredictable Forest achieved the highest accuracy after analysis, it performed better than the other algorithms. Additionally, Random Forest was added to Spyder so that the classifier's accuracy increased from 69% to 72% after using the unique technique of data wrangling. In the future, we can classify liver disease datasets using a wider range of algorithms and create intelligent systems that can suggest the optimal algorithm to employ for a certain dataset in order to get better results.

8. Current and Future Developments

The current study uses a unique selection process to improve the predictive accuracy of liver disease data, including PCA (principal component analysis), recursive feature extraction (RFE), and key names derived from the random forest itself.

The future study is to integration of deep learning and random forest: A hybrid approach combining random forest with deep learning (e.g. using deep neural networks for extraction and random forest for classification). Treatment planning for liver disease through analysis of patient-specific and large-scale genomic data.

References

1. Hoon Jin, Seoungcheon Kim, Jinhong Kim, "Decision Factors on Effective Liver Patient Data Prediction", International Journal of Bio-Science and Bio-Technology, Vol. 6, Issue.4, pp. 167–178, 2014.
2. Ayesha Pathan, Diksha Mhaske 2, Shrutika Jadhav, Rupali Bhondave, Dr. K. Rajeswari, "Comparative Study of Different Classification Algorithmson ILPD Dataset to Predict Liver Disorder", International Journal for Researchin Applied Science & Engineering Technology (IJRASET), Vol. 6, Issue. 2, pp. 388–394, 2018.
3. Tapas Ranjan Baitharu, Subhendu Kumar Pani, "Analysis of Data Mining Techniques For Healthcare Decision Support System Using Liver Disorder Dataset", International Conference on Computational Modeling and Security, India, pp. 862–870, 2016.
4. Dr. S. Vijayarani, Mr. S. Dhayanand, "Liver Disease Prediction using SVM and Naïve Bayes Algorithms", International Journal of Science, Engineering and Technology Research (IJSETR), Vol. 4, Issue.4 pp. 816–820, 2015.
5. Rakhi Ray, "Advances in Data Mining: Healthcare Applications", International Research Journal of Engineering and Technology (IRJET), Vol. 5, Issue. 3, pp. 3738–3742, 2018.
6. B. Saritha, S.V. Ramana, Narra Manaswini, Rama Priyanka, D. Hiranmayi, K. Eswaran, "Classification of liver data using a new algorithm", 4th International Conference on New Frontiers of Engineering, Science, Management and Humanities, Hyderabad, 2017.
7. HanMa, Cheng-fuXu, Zhe Shen, Chao-hui Yu, You-ming Li, "Application of Machine Learning Techniques for Clinical Predictive Modeling: A Cross-Sectional Study on Nonalcoholic Fatty Liver Disease in China", Bio Med Research International, pp. 1–9, 2018.
8. Nazmun Nahar, Ferdous Ara, "Liver Disease Prediction using different decision tree techniques", International Journal of Data Mining & Knowledge Management Process (IJDKP), Vol. 8, Issue.2, pp. 1–9, 2018.
9. Shapla Rani Ghosh and Sajjad Waheed, "Analysis of classification algorithms for liver disease diagnosis", Journal of Science, Technology & Environment Informatics, Vol. 5, Issue.1, pp. 360–370, 2017.
10. Insha Arshad, Chiranjit Dutta, Tanupriya Choudhury, Abha Thakra, "Liver Disease detection due to excessive alcoholism using Data Mining Techniques", International Conference on Advances in Computing and Communication Engineering, Paris, France, pp. 163–168, 2018.
11. V.V. Ramalingam, A. Pandian, R. Ragavendran, "Machine Learning Techniques on Liver Disease – A Survey", International Journal of Engineering & Technology, Vol. 7, Issue.4, pp. 493–495, 2018.
12. Shambel Kefelegn, Pooja Kamat, "Prediction and Analysis of Liver Disorder Diseases by using Data Mining Technique: Survey, International Journal of Pure and Applied Mathematics, Vol. 118, Issue. 9, pp. 765–769, 2018.
13. Suruchi Fialoke; Anders Malarstig; Melissa R. Miller; Alexandra Dumitriu, "Application of Machine Learning Methods to Predict Non-Alcoholic Steatohepatitis (NASH) in Non-Alcoholic Fatty Liver (NAFL) Patients", Annual Symposium Proceedings Archive Journal, December 2018.
14. Al-Mamun, M., Alamgir, M., & Rahman, M. (2018). Emerging trends of artificial intelligence (AI) & machine learning (ML) technologies: A systematic review. Journal of Computer Science & Systems Biology, 12(1), 1–20. Doi: 10.1186/s12925-018-0290-8
15. Zhang, H., Zhou, X., & LeCun, Y. (2016). Understanding deeplearning requirements for image recognition accuracy. arXivpreprintarXiv:1605.05202.
16. Wang, L., Zhang, Y., & Chen, L. (2018). Pyramid sceneparsingnetwork.CoRRabs/1812.06942[Cs.CV]. Doi:10.1109/CVPR.2019.009

Note: All the figures in this chapter were made by the authors.

Intelligent Systems Using Semiconductors for Robotics and IoT – Dinesh Goyal et al. (eds)
© 2026 Taylor & Francis Group, London, ISBN 978-1-041-20408-4

40 Design an Approach for Hate Speech Detection Using Deep Learning

Pragya Goswami[1], Aravendra Kumar Sharma[2], Ratnesh Kumar Dubey[3], Rinki Pakshwar[4]
CSA Department, ITM University Gwalior, Madhya Pradesh, India

Abstract: Hate speech can be described as dialogue that targets to an object or group based on their identities, such as race, gender, religion, or sexual orientation. Hate speech can be distributed in a variety of ways, including social media, news websites, and public speeches. Hate speech has serious repercussions, including advocating prejudice, instigating violence, and fostering a climate of fear and intimidation. Deep learning methods aid the fight against hate speech through tools and technology for the detection, understanding, and minimization of its effects on the internet. Along with that, addressing hate speech detection raises several research questions, such as dealing with the scarsity of examples from each class, performing more detailed analysis, dealing with the transfer of hate speech from one language to another, and ensuring fairness and accuracy in detection, we provide a deep learning architecture that can identify hate speech in several classes automatically. For feature extraction, the suggested approach relies on TF-IDF, PoS tagging, Word Embedding, and BERT as integration solutions. The CNN, LSTM, Bi-LSTM, and GRU comprehensive deep learning models serve as the foundation for the hate speech classifier, which increases detection accuracy. This multiclass hate speech classifier recognizes hate speech such as sexism, racism, cyberbullying, and racial discrimination.

Keywords: Hate speech, Deep learning, LSTM, CNN, Cyber bulling, Racism, Sexism, Racial

1. Introduction

Hatespeech envelops all communication. It targets or harms a person or group. This is based on their identity. Hate speech can get propagated through a range of channels. Examples would be social media news websites and public speaking. The impacts of hate speech are substantial. It advocates prejudice. It instigates violence. It fosters an environment of fear and intimidation[1].

Hate speeches pose serious threats in the digital era. They could weaken democracy's foundation. How? They do this by twisting public opinion. Also they undermine trust in institutions. What's more, these speeches could hurt individuals. They are also harmful to groups. This is done through the promotion of hate. And they promote discrimination. Detecting and fighting hate speech is a critical problem that necessitates a multidisciplinary strategy that incorporates ML and DL techniques, natural language processing, and social science research.

Hate speech has appeared as biggest and utmost threatening risks on social media in recent times. The automated identification of hate speech on the Internet is a tremendously important but tough project for both industry and academia.

There are a wide range of possible applications for machine learning-based hate speech identification. ML and DL will probably become more crucial tools in the fight against hate speech as the issue spreads. They will help detect and stop hate speech from spreading.

1.1 Hate Speech Types

Hate speech can take numerous forms and target individual persons or groups based on their properties. Promoting

[1]pragyajain.cse@itmuniversity.ac.in, [2]aravendra.cse@itmuniversity.ac.in, [3]ratnesh.soet@itmuniversity.ac.in, [4]rinkipakshwar.cse@itmuniversity.ac.in

DOI: 10.1201/9781003716389-40

hate speech is detrimental and can lead to discrimination, violence, and social instability. Here are some frequent forms of hate speech [2]:

1. Racial Hate Speech: choosing people or groups according to their on their race or ethnicity.
2. Religious Hate Speech: concentrating on particular people or groups according to their religious beliefs or practices.
3. Sexual Orientation Hate Speech: Selection of individuals or communities by sexual orientation or gender identity is unjust. Such actions harm and do not promote progress.
4. Gender-based Hate Speech: This is directed at individuals or communities based on their gender. Misogynistic language is often included in this form of hate speech.
5. Disability Hate Speech: It focuses on particular individuals or groups based on physical or mental disabilities. This is unjust and does not foster inclusion or acceptance.
6. Nationality or Immigration Status Hate Speech: It targets individuals or communities due to their nationality or immigration status. This is not acceptable and does not promote a welcoming society.
7. Political Hate Speech: This form of hate speech concentrates on particulars or communities based on the political beliefs. The impacts of this hate speech are problematic.
8. Class-based Hate Speech: Individuals or communities are chosen due to their socio-economic status or class. This is unjust and does not promote unity in society.

It is critical to counteract hate speech. Doing so paves the way for a more inclusive and accepting society. Numerous countries have laws to combat hate speech. Yet, these laws show varying effectiveness. Promoting awareness knowledge and open debate are techniques to combat hate speech.

1.2 Impact of Hate Speech on Society

Structured speech that incites hatred can have extensive effects on individuals and society. These are notable repercussions[3]:

1. Social Rupture: Speech promoting hatred nurtures the divide between various sectors. It creates an "us versus them" mindset. This division can result in heightened tensions. There might be increased mistrust as well. Conflict within society might ensue.
2. Bias and Prejudice: Speech of hatred bolsters stereotypes and preconceived ideas. This contributes to harmful attitudes and actions.
3. Acts of Violence and Hate: Speech promoting hate has ties to a rise in hate crimes. Exposure to hateful rhetoric can incite people. They may be more inclined to commit violent acts. These acts are often against those seen as different.
4. Emotional Impact: Those subjected to hateful speech can experience emotional trauma. Symptoms can include anxiety and depression. There is often a loss of belonging feeling. Entire communities can suffer from the psychological impact. They may feel marginalized or persecuted.
5. Muting Free Speech: Oddly enough hate speech can bring about silencing of free speech. People who fear reprisal or harassment might hold back their opinions. This limits the variety of viewpoints and ideas in a society. This in turn, affects the diversity.
6. Erosion of Social Unity: Hate speech eats away at the very fabric of society. Threatening trust and cooperation among people. It makes it difficult for various groups to collaborate toward shared objectives.
7. Radicalization: Hate speech may fuel radicalization of individual and groups. They draw further towards extreme ideologies. This radicalization may be observed in political spheres. It can be seen in different religious or social contexts as well.
8. Undermining Democracy: Hate speech can destabilize the principles of equality and tolerance. In democratic societies these are crucial for a functioning democracy. It could aid the rise of extremist ideologies and movements. These pose threat to democratic institutions.
9. Social Impact over Time: Hate speech that persists may have a lasting impact over on society. It fosters a culture of intolerance that can persist for generations. This can obstruct social progress over time. It hinders the creation of an inclusive society that is equitable.

Encouraging dialogue and mutual understanding among varied communities can aid in the making of a more harmonious society. Moreover, an inclusive one at that. Reference [29] is given for the citation.

1.3 Hate Speech as a Challenge on Social Media

Hate speech creates many obstacles. It is worrisome how it spreads globally. At the same time, hate speech spreads instantaneously. This is true in the world of online communication. This poses significant challenges. These issues are found on social media platforms. Key challenges related to hate speech on social media are discussed below. [4]:

1. Widespread Dissemination: Social media instigates rapid spread of hate speech. It also helps in reaching

an extensive audience. Its viral spread can magnify the effect of messages filled with hate. This further leads to the acceptance of views rooted in prejudice.

2. Anonymity and Pseudonymity: Users of social media can be involved in launch of hate speech. They can do this while hidden under anonymity. Or while using pseudonyms. This situation makes it hard to pin individuals down for their act. It leads to a culture of feeling no responsibility for actions.
3. Algorithmic Amplification: Algorithms on social media often aim at increasing user interaction. But, they might unknowingly promote hatred. Controversial or provocative content might gain preference. This results in the expansion of harmful messages.
4. Targeted Harassment and Bullying: Hate speech on social platforms frequently implicates harassment and bullying. Frequently the targets are individuals or groups. The targets might undergo persistent and personalized attacks. These attacks have negative effects. Mental health and general well-being are typically impacted.

1.4 Deep Learning and ML Role in Combating Hate Speech

ML and DL are vital tools in the fight against hate speech. They provide tools and technologies to identify. They analyze the impact of hate speech online. Consider ways in which machine learning is employed for this purpose:

1. Automated Content Moderation: Machine learning models see extensive use. They identify and flag hate speech throughout social media forums and websites. They analyze text, images and videos. These models help enforce community guidelines. They minimize spread of hateful content [5].
2. Sentiment Analysis: Sentiment analysis is another machine learning technique. It's used for hate speech detection. It involves automatic categorization of text as +ve, -Ve, or neutral. In hate speech detection, ML models train for negative sentiments. These sentiments are directed towards specific groups. This enables identification and flagging of potentially harmful content [6].
3. Network study: This study looks at network of people who share material on social media. ML and DL approaches may be used to find those trends in content sharing behavior. They also identify accounts that are likely to promote hate speech [31].
4. Analysis of User Behavior: ML and DL algorithms can analyze user behavior patterns. Detect accounts that engage consistently in hate speech. Detect those that engage in harassment. Anomaly detection techniques identify unusual or suspicious activities. These activities may indicate coordinated efforts to spread hate speech [7].

ML and DL techniques overall possess excessive potential. They can become potent instruments in the fight versus hate speech. Yet the ML models have flaws. They are not perfect. They can make mistakes. Therefore, it's necessary to design models that are precise, transparent, and explainable, and to continuously evaluate and improve their performance [30].

2. Related Work

The many facets of these intricate phenomena can be better understood by doing a thorough literature assessment on hate speech.

Conventional linguistic administering and implanting methods such as "bag of words" or "n- grams" are useful for handling rumors or reviews, but not effective to extract the underlying relationship of hate speech. Therefore, online hate speech detection requires advanced integrated approaches to capture key sentiment and semantic chronological order in news content [8].

The current machine learning models can be categorized as supervised, semi-supervised, or unsupervised models based on conventional data mining categorization techniques [9]. The article's authors primarily address both supervised and unsupervised learning models for user identity linking across online social networks. [10].

The algorithms such as "Decision Tree, Random Forest, Support Vector Machine, Logistic Regression, and K Nearest Neighbor" are commonly used in the previous literature for online fraud, fraud, and classification. Misleading information [11].

In this work [13], author proposes a framework for social media images by unsupervised approach to do sentiment analysis. By using relevant contextual information and visual information, this approach can forecast social images sentiment from two huge data sets. Author [7], logistic regression SVM have consistently outperformed other models in automated hate speech detection demonstrating superior reliability and effectiveness.

The literature presents numerous studies showcasing the application of DL techniques for hate speech detection and classification. For instance, one study [14] develops a basic CNN model with a single input channel. Another study [15] reports that the best results are achieved by combining character n-grams of up to four characters in length with gender as an additional feature. Research [16] highlights those ensemble models, which integrate the strengths of various approaches, yield the highest accuracy in online hate speech detection. Additionally, findings from [17] suggest that incorporating expert annotations can produce models with performance comparable to previous classification methods.

In this work [18], achieved this by creating and analyzing a dataset by using SVM. A research [19], shows that both linear classifiers (SVM and SGDC) outperform the other classifiers. In this paper [20], the author observes using of feature for node embedding. In this research [21], author uses "BERT-Ger, Bi-GRU, Bi-LSTM", and so on. This enhances the model's capacity to extract semantic information characteristics.

A research [22], elaborate an ensemble model using RNN and BERT on multilingual data set including English, German and Chinese. Authors work on COVID datasets of English, German and Chinese tweets. This is a new model for cross-lingual Covid hate speech detection. The technique creates allied embedded depictions of texts in various languages using a sophisticated multilingual pre-trained language model. The contrastive learning method is then used, utilizing data in the source language and additional data that has been translated into the target language. Lastly, a weighted total is calculated between the cross lingual contrastive loss and the classification cross entropy loss. Thus, by minimizing the overall loss, the model simultaneously improves its cross lingual knowledge transfer capacity and its ability to recognize hate speech for multilingual data.

In order to create a framework for deep multi-task learning, a work [23] uses a number of models, such as CNN, LSTM, CNN Stacking and GRU, CNNa+GRU by altering CNN+GRU. The SP-MTL model incorporates two binary feature spaces for each task: one dedicated to capturing task-specific characteristics and another for preserving shared features across these interconnected tasks through collaborative training. Authors experimented their model on D1 Dataset Hate & Offensive (24783 post), D2 Dataset Racist & Sexist (15476 post), D3 Dataset Aggression (15001 post), D4 Dataset Offensive (13240 post), and D5 Dataset Harassment (20360 post). The performance of these models are For CNN+GRU model macro-F1 is 73.47 and weighted-F1 is 90.26 for STL and For CNN model macro-F1 is 89.16 and weighted-F1 is 95.65 for SP-MTL.

A study developed a clustering-based probabilistic framework for analyzing hate speech on Twitter, using techniques like Bayes classifier, fuzzy logic, and TF-IDF. The model extracted tweets with hate speech-related keywords, classified them as hate speech or non-hate speech, encoded tweet features, and used fuzzy logic for enhanced detection. The model demonstrated high predictive accuracy[24].

Hierarchical Attention-based deep learning and a deep CNN with BiLSTM were suggested [25]. The proposed model uses a BERT layer to handle tweets as input. Proposed model having upgrading of 8% precision, 7% recall, and 8% f score. Also 5% in Training and 7% in validation accuracy is improved.

An A-Stacking based adaptive classifier proposed in [26]. This is an automatic hate speech adaptive model. Using common high-dimensional datasets, the authors conduct experiments. Several experimental configurations are being used to investigate the behavior of the model while taking data bias, user over fitting, and limiting the amount of tweets per user into account. Performance in situations involving many datasets detection, which works well on cross-datasets and can get around data bias.

The significant issue of identifying hostile information on social media is covered in this study [27]. The suggested detection method uses ensemble RNN classifiers and integrates a number of characteristics linked to user data, tendency for racism or sexism. They assess the method using a corpus of 16k tweets that is available to the public, and the findings show that it is more effective than the

Table 40.1 Methodology, findings, and limitations of various existing works

Ref	Methodology	Findings	Limitation
[22]	By using a contrastive learning approach and computing a weighted total of cross-lingual contrastive loss and categorization cross-entropy loss, the study builds multilingual text representations using strong multilingual pre-trained language models.	The model concurrently enhances its ability to transfer information across languages and identify hate speech for multilingual data by minimizing the total loss.	1. lack of labeled data 2. F1 score is degrading While German language is checked.
[21]	Utilizing BERT-Ger, Bi-GRU, and Bi-LSTM enhances the model's capacity to extract semantic information characteristics.	Improve the ability of the model to obtain semantic information features.	The requirements for semantic information features were not met by the BERT-Ger model.
[27]	Employs a group of RNN classifiers and incorporates several attributes connected to user-related information, such the user's propensity for racism or sexism.	Compared to the other schemes, the 5-classiers ensemble does the best for the classes of users who are least inclined towards racism or sexism.	The deep learning processes' stochastic behavior, which causes the F-score to fluctuate across several runs, poses the most serious threat to construct validity.
[15]	With gender as an extra component, the greatest outcomes are with character n grams of lengths up to 4.	Set of standards for identifying racist and sexist insults that are based on critical race theory.	To update upcoming data and trials, gender and location classification needs to be improved.
[17]	It is possible to create models with expert annotations that perform on par with earlier categorization attempts.	Weighted F1-score should be used when assessing attempts to classify hate speech.	Attempt to learn more about the sociolinguistic elements, like place and gender.

existing work currently in use. Specifically, this system beats existing algorithms in classifying text and can effectively separate racist and sexist statements from ordinary language.

A study [28] highlights that hate speech often lacks clear, distinguishable characteristics, placing it in the "long tail" of datasets and making its detection particularly challenging. The authors recommend utilizing DNN architectures as feature extractors due to their ability to capture the subtle and complex meanings inherent in hate speech. Evaluations conducted on one of the largest Twitter based hate speech datasets demonstrate that the proposed methods outperform state of art algorithms, achieving up to a 8% improvement in more complex scenarios and 5% in F1score.

Four automatic classification techniques that can analyze Danish and English text are presented in a research [12]. The top-performing Danish model obtains an F1score of 0.70, whereas the best successful English model identifies objectionable language with a macro-averaged F1score of 0.74.

3. Hate Speech Detection Challenges

1. Imbalanced Datasets: Many hate speech datasets exhibit class imbalance. This means that most samples probably don't represent hate speech. Models can be biased. Under performing on underrerepresented classes may result from this dataset imbalance. It's critical to create balanced datasets to construct reliable models.
2. Fine-Grained Hate Speech Analysis: Hate speech is not a single entity. It can show in numerous forms. Racism and sexism. Homophobia etc. Developing models is necessary. They can identify and distinguish. Different types of hate speech. This is an ongoing challenge.
3. Cross-Lingual Hate Speech Detection: Various models are in existence. They are trained and evaluated. English-language data is used. Adapting these models is important. Also extending them to different languages. Also considering linguistic, cultural variations.
4. Code-Switching and Language Variation: Hate speech frequently ties to code-switching. It also links to use of multiple languages or dialects. Detecting hate speech in multilingual and multicultural settings demands proficient models. These models should be able to process variations in language.
5. Generalization: Bulk of the models in current use receives training and evaluation using a single dataset. These models may not perform optimally on new and untested datasets. The concept of generalization is critical. Researchers need to explore it further.

4. Objectives of the Study

Crafting and evaluating a deep learning innovation. This innovation can recognize hate speech across a broad range of conditions and data sources. It is a vital task. The purpose of this inquiry must be to devise algorithms. These algorithms are more precise. They are scalable and resilient. They surpass the current usage ones in effectiveness.

The subsequent points enumerate plausible research objectives:

- Pursuit and enhance of training data. This improves hate speech recognition. Both quality and diversity are critical. The primary aim is to gather a varied high quality training dataset. We want to train deep learning models. These models will identify hate speech. They are the focus question in this research segment.
- The next task is examining the impact of diverse features and modalities. This examination is for the identification of hate speech. We must determine the characteristics and modalities most critical for accurate detection. After this we improve the design of a model. The model is geared towards better outcomes.
- Create; evaluate a new DL method for detecting hates speech in various classes: To reach this goal we proposed a new DL method. This is for multiclass hate speech identification. We aim to achieve this goal.

5. Research Methodology

Research methodology has been proposed. It is focused on development and evaluation of DL algorithms. These are specifically for hate speech detection. By adhering to this methodology the possibility emerges. That is, we can develop more accurate and robust DL models. These models are for detecting hate speech.

The aim of this approach is enhancing the accuracy and trustworthiness of online content. This is achieved by focusing on hate speech detection. The focus? It is through deep learning. The following is an outline of a research methodology. It is for implementing a deep learning-based framework for detecting hate speech.

- Data Collection and Preprocessing: The first thing is to gather large-scale dataset. This dataset is sourced from numerous places. Places like social media that have hate speech. The dataset needs to be diverse. It needs to consist of many forms of hate speech. Table 2 provides insight into the hate speech data. This data is already present and frequently employed in literature. Next the data has to be pre-processed. This is to eliminate disruption such as stop words and punctuation. It is also to remove special characters.
- Feature Extraction and Selection: Subsequently notable qualities are picked out from the data. The

data has already been processed. The characteristics could be the emotion of the text. It might also be the use of certain keywords. Or it might be phrases. There are several methods for this kind of job. TFIDF is Term Frequency Inverse Document Frequency. Vectorizer with Ngram analysis vectorization may be used. Also there is Word2vec. And BERT. And FastText. And GloVe. Next, comes the choice of most relevant and distinguishing characteristics. This is for hate speech identification. Feature selection methods are used for this.

- Model Development and Training: One can make use of the dataset for training. A DL model is to be constructed. The ultimate goal is finding hate speech in varied areas. This is all part of the strategy. In addition, mechanisms to enhance detection robustness and performance are explored. Specifically hybrid and ensemble models are under investigation.
- Model Evaluation: Usual measures. The F1score. These might be utilized to gauge performance of made deep learning models. What is important to ensure their performance outcomes are reliable? The research could consider using cross-validation. And bootstrapping strategies. These will ascertain a sound and trustful result.
- Analysis and Interpretation of Results: Study is essential to determine superiority and inferiority of the created deep learning model. The process will culminate in the evaluation and interpretation of results found. It can help with creating more powerful hate speech detection systems. These systems will point out areas which require work.

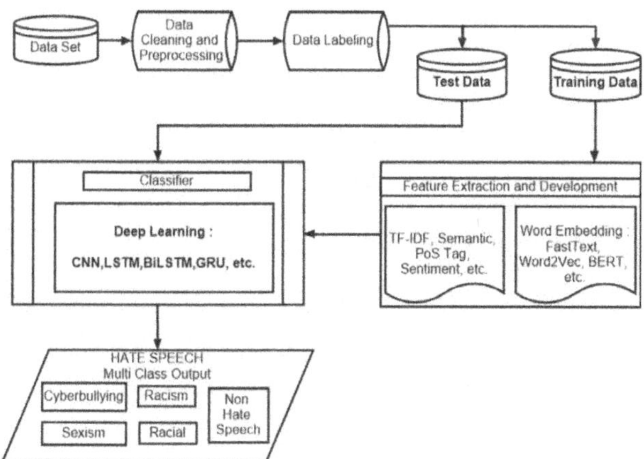

Fig. 40.1 Proposed research methodology

6. Hate Speech Datasets

The excellence of data set is vital component. It is pivotal in developing proficient deep learning model. This section, we explore crucial assessment metrics. They are needed to assess value of hate speech datasets.

Table 40.2 Hate speech detection datasets in English

DataSet	Source	Size	Class
Offensive Language (2017)[7]	X	24803	Hate and not hate speech
Detection of Hate Speech Foxnews Comments 2017[16]	Fox News	1526	Binary (Hate and non hate speech)
Predictive Features for Hate Speech Detection, 2016 [26]	X	16914	Racist, Sexist, Not hate speech
Hate Speech Data set 2018[14]	Stormfront	9917	Relation, Hate and not hate speech
Sexism using twitter data 2017[18]	X	713	Sexism
Twitter annotations (Hate Speech) 2016 [17]	X	4034	Sexism, Racism
Multilingual CONAN Dataset of Hate Speech 2020[19]	Facebook, Synthetic	1289	Islamophobia
Misogyny Identification at Iber Eval 2018 [18]	X	3987	Sexism
Online Hate Speech 2020[21]	GAB	33786	Binary (Hate and non-hate speech)
Characterize and Detecting Hateful Users 2018 [20]	X	4973	Binary (Hate and non-hate speech)

7. Outcome of the Study

Novel approach to hate speech identification. The purpose of the study is to help create a DL technique. This technique is to find of hate speech. The goal is to make it more sensitive. It should also be more dependable and scalable. By fulfilling the goals, the quality of online content can be greatly enhanced. This includes content found on social media and other platforms.

Enhance the caliber and dependability of information available online. Tackling research goals can aid in the growth of online communities. These communities would be more welcoming and courteous.

Table 40.3 Comparison of different models of Deep Learning

Approach	Accuracy (%)	Precision (%)	Recall (%)	F1-Score (%)
CNN	85.2	83.5	80.3	81.9
LSTM	87.8	86.2	85.0	85.6
Bi-LSTM	89.1	88.0	87.3	87.6
GRU	88.5	87.1	86.2	86.6

Fact checking and prevention of the spread of hate speech. There is a suggested methodology. It can rest in locating news items and social media posts that are fact checked.

This aids in the prevention of hate speech. This method can help stop the spread.

8. Scope of the Proposed Method for Detection of Hate Speech

- News websites and comment section: News websites use detection methods for hate speech. The aim is moderation of comments and user content. This creation results in a more respectful and constructive discussion environment.
- Content moderation use is present in hate speech detection. It is utilized by moderation teams for review filtering out inappropriate or offensive content on different online platforms. The content moderation ensures compliance. It complies with guidelines of the community.
- Political Campaigns: This method finds application in context of political campaigns. It is used for monitoring and moderation of online discussions. Its function is important. The impact of toxic rhetoric is reduced and constructive discourse is promoted.
- Preventing Cyberbullying: Context of preventing cyber bullying provides an application. This method is helpful. It helps in identification and addressing instances of online harassment and abusive behavior.
- Law enforcement support: Hate speech detection tools are of use for law enforcement agencies. They offer assistance in monitoring of online spaces. The monitoring is done for potential crimes. It can be hate crimes. There could be threats. The identification of concerning behavior is their aim. This leads to early actions. Early actions lead to early identification of concerning behavior.

9. Conclusion

Dialogue that disparages a person or group because of their is referred to as hate speech. There are several platforms via which hate speech can be disseminated, such as social media, news websites, and remarks in public. Hate speech has detrimental effects that include inciting violence, promoting prejudice, and creating an environment of intimidation and fear. By offering resources and technology for locating, evaluating, and lessening the impact of hate speech online, deep learning techniques contribute to the fight against it. Research on hate speech detection has many obstacles, such as biased and fair datasets, fine-grained hate speech analysis, cross lingual hate speech detection, and prejudice. In this work, we suggested a DL-based strategy for the detection of hate speech across many classes. The suggested method leverages word embedding, TF-IDF, PoS, BERT, and other techniques for feature generation and extraction. CNN, LSTM, Bi-LSTM GRU, and other DL techniques are used in hate speech classifiers. Hate speech is classified by this multiclass hate speech classifier as things like cyber bullying, racism, sexism, and racialization. As the problem of hate speech increases, the suggested deep learning model will likely become increasingly important in detecting and stopping its spread as well as promoting welcoming and civil online communities by improving accuracy and detecting hate speech. It stops hate speech from spreading and enhances the quality and dependability of internet information and fact-checking.

References

1. F. E. Ayo, O. Folorunso, F. T. Ibharalu, and I. A. Osinuga, "Machine learning techniques for hate speech classification of twitter data: State-of-The-Art, future challenges and research directions," Comput. Sci. Rev., vol. 38, 2020, doi: 10.1016/j.cosrev.2020.100311.
2. M. S. Jahan and M. Oussalah, "A systematic review of hate speech automatic detection using natural language processing," Neurocomputing, vol. 546, p. 126232, Aug. 2023, doi: 10.1016/j.neucom.2023.126232.
3. E. Dalan and S. Sharoff, "Genre classification for a corpus of academic webpages," in Proceedings of the 10th Web as Corpus Workshop, 2016, pp. 90–98, doi: 10.18653/v1/W16-2611.
4. C. Shao, G. L. Ciampaglia, O. Varol, K.-C. Yang, A. Flammini, and F. Menczer, "The spread of low-credibility content by social bots," Nat. Commun., vol. 9, no. 1, p. 4787, Nov. 2018, doi: 10.1038/s41467-018-06930-7.
5. A. Schmidt and M. Wiegand, "A Survey on Hate Speech Detection using Natural Language Processing," in Proceedings of the Fifth International Workshop on Natural Language Processing for Social Media, 2017, pp. 1–10, doi: 10.18653/v1/W17-1101.
6. M. Subramanian, V. Easwaramoorthy Sathiskumar, G. Deepalakshmi, J. Cho, and G. Manikandan, "A survey on hate speech detection and sentiment analysis using machine learning and deep learning models," Alexandria Eng. J., vol. 80, no. August, pp. 110–121, 2023, doi: 10.1016/j.aej.2023.08.038.
7. T. Davidson, D. Warmsley, M. Macy, and I. Weber, "Automated Hate Speech Detection and the Problem of Offensive Language," Proc. Int. AAAI Conf. Web Soc. Media, vol. 11, no. 1, pp. 512–515, May 2017, doi: 10.1609/icwsm.v11i1.14955.
8. X. Zhang and A. A. Ghorbani, "An overview of online fake news: Characterization, detection, and discussion," Inf. Process. Manag., vol. 57, no. 2, p. 102025, Mar. 2020, doi: 10.1016/j.ipm.2019.03.004.
9. J. Han and M. Kamber, Data Mining: Concepts and Techniques, 3rd ed. San Francisco, CA, USA: Morgan Kaufmann, 2011.
10. K. Shu, S. Wang, J. Tang, R. Zafarani, and H. Liu, "User Identity Linkage across Online Social Networks," ACM SIGKDD Explor. Newsl., vol. 18, no. 2, pp. 5–17, Mar. 2017, doi: 10.1145/3068777.3068781.
11. S. Afroz, M. Brennan, and R. Greenstadt, "Detecting Hoaxes, Frauds, and Deception in Writing Style Online," in 2012 IEEE Symposium on Security and Privacy, May 2012, pp. 461–475, doi: 10.1109/SP.2012.34.

12. G. I. Sigurbergsson and L. Derczynski, "Offensive Language and Hate Speech Detection for Danish," Lr. 2020 - 12th Int. Conf. Lang. Resour. Eval. Conf. Proc., pp. 3498–3508, Aug. 2023, doi: 10.48550/arXiv.1908.04531.
13. Y. Wang, S. Wang, J. Tang, H. Liu, and B. Li, "Unsupervised Sentiment Analysis for Social Media Images," in Proceedings of the 24th International Conference on Artificial Intelligence, 2015, pp. 2378–2379.
14. O. de Gibert, N. Perez, A. García-Pablos, and M. Cuadros, "Hate Speech Dataset from a White Supremacy Forum," Comput. Lang., Sep. 2018, doi: https://doi.org/10.48550/arXiv.1809.04444.
15. Z. Waseem and D. Hovy, "Hateful Symbols or Hateful People? Predictive Features for Hate Speech Detection on Twitter," in Proceedings of the NAACL Student Research Workshop, 2016, pp. 88–93, doi: 10.18653/v1/N16-2013.
16. L. Gao and R. Huang, "Detecting Online Hate Speech Using Context Aware Models," Comput. Lang., Oct. 2017, [Online]. Available: http://arxiv.org/abs/1710.07395.
17. Z. Waseem, "Are You a Racist or Am I Seeing Things? Annotator Influence on Hate Speech Detection on Twitter," in Proceedings of the First Workshop on NLP and Computational Social Science, 2016, pp. 138–142, doi: 10.18653/v1/W16-5618.
18. A. Jha and R. Mamidi, "When does a compliment become sexist? Analysis and classification of ambivalent sexism using twitter data," in Proceedings of the Second Workshop on NLP and Computational Social Science, 2017, pp. 7–16, doi: 10.18653/v1/W17-2902.
19. Z. Pitenis, M. Zampieri, and T. Ranasinghe, "Offensive Language Identification in Greek," in Proceedings of the Twelfth Language Resources and Evaluation Conference, Mar. 2020, pp. 5113–5119, [Online]. Available: http://arxiv.org/abs/2003.07459.
20. M. Ribeiro, P. Calais, Y. Santos, V. Almeida, and W. Meira Jr., "Characterizing and Detecting Hateful Users on Twitter," Proc. Int. AAAI Conf. Web Soc. Media, vol. 12, no. 1, Jun. 2018, doi: 10.1609/icwsm.v12i1.15057.
21. Q. Que, R. Sun, and and S. Xie, "Simon@HASOC 2020: Detecting Hate Speech and Offensive Content in German Language with BERT and Ensembles," in FIRE '20, Forum for Information Retrieval Evaluation, December 16–20, 2020, Hyderabad, India, 2020, pp. 283–289.
22. L. Liu et al., "A cross-lingual transfer learning method for online COVID-19-related hate speech detection," Expert Syst. Appl., vol. 234, no. June, 2023, doi: 10.1016/j.eswa.2023.121031.
23. P. Kapil and A. Ekbal, "A deep neural network based multi-task learning approach to hate speech detection," Knowledge-Based Syst., vol. 210, p. 106458, Dec. 2020, doi: 10.1016/j.knosys.2020.106458.
24. F. E. Ayo, O. Folorunso, F. T. Ibharalu, I. A. Osinuga, and A. Abayomi-Alli, "A probabilistic clustering model for hate speech classification in twitter," Expert Syst. Appl., vol. 173, no. February 2020, 2021, doi: 10.1016/j.eswa.2021.114762.
25. S. Khan et al., "BiCHAT: BiLSTM with deep CNN and hierarchical attention for hate speech detection," J. King Saud Univ. - Comput. Inf. Sci., vol. 34, no. 7, pp. 4335–4344, 2022, doi: 10.1016/j.jksuci.2022.05.006.
26. S. Agarwal and C. R. Chowdary, "Combating hate speech using an adaptive ensemble learning model with a case study on COVID-19," Expert Syst. Appl., vol. 185, no. July, pp. 1–9, 2021, doi: 10.1016/j.eswa.2021.115632.
27. G. K. Pitsilis, H. Ramampiaro, and H. Langseth, "Effective hate-speech detection in Twitter data using recurrent neural networks," Appl. Intell., vol. 48, no. 12, pp. 4730–4742, Dec. 2018, doi: 10.1007/s10489-018-1242-y.
28. Z. Zhang and L. Luo, "Hate speech detection: A solved problem? The challenging case of long tail on Twitter," Semant. Web, vol. 10, no. 5, pp. 925–945, Sep. 2019, doi: 10.3233/SW-180338.
29. R. K. Dubey, N. Dandotiya, A. Sharma, S. Mishra and S. K. Gupta, "Cyber attack Detection Using Machine Learning Techniques," 2023 IEEE International Conference on ICT in Business Industry & Government (ICTBIG), Indore, India, 2023, pp. 1–6, doi: 10.1109/ICTBIG59752.2023.10456080.
30. Rastogi, N., & Sharma, A. K. (2022, April). Measuring the Efficiency of LPWAN in Disaster Logistics System. In International Conference on Artificial Intelligence and Sustainable Engineering: Select Proceedings of AISE 2020, Volume 1 (pp. 131–149). Singapore: Springer Nature Singapore.
31. Sharma, A. K., Saroj, S. K., Chauhan, S. K., & Saini, S. K. (2016, April). Sybil attack prevention and detection in vehicular ad hoc network. In 2016 International Conference on Computing, Communication and Automation (ICCCA) (pp. 594–599). IEEE.

Note: All the figure and tables in this chapter were made by the authors.

Intelligent Systems Using Semiconductors for Robotics and IoT – Dinesh Goyal et al. (eds)
© 2026 Taylor & Francis Group, London, ISBN 978-1-041-20408-4

41 Examining the Detection of Faux user Opinions: Challenges and Insights in AI Based Techniques

Navin Kumar Goyal*
Associate Professor, Department of Computer Engineering,
Poornima Institute of Engineering & Technology,
Jaipur, India

Payal Bansal
Professor, Department of Internet of Things,
Poornima Institute of Engineering & Technology,
Jaipur, India

Abstract: Today, one of the most vital uses of sentiment analysis in online shopping is the recognition and estimation of bogus reviews. Using artificial intelligence approaches, software-based fake review categorization systems detect and classify a wide range of replicate, junk, false, and dishonest opinions. This study examines many recently developed techniques for detecting false reviews This study examines many current techniques for detecting bogus reviews for various AI based techniques on verity of datasets. It examines the fundamental and performance-based specifications of many fake review predictors and offers a thorough analysis of them. It draws attention to the shortcomings, risks, and difficulties with these current works. Additionally, it provides a graphic representation of the distinction for the year-by-year progression, model usage, and dataset utilization criteria. The analysis reveals that the majority of research contributions have emerged after 2018, indicating a growing interest in the topic. Among the various classifiers, Support Vector Machine (SVM) and Naive Bayes (NB) demonstrate the highest usage percentages, especially when applied to datasets focused on Yelp and hotel-related reviews. This suggests that these classifiers are preferred for identifying fake reviews in these particular domains, as they have proven effective in recent studies.

Keywords: Faux reviews, Naïve bayes, Faux detection, Machine learning, Support vector machine, Artificial intelligence

1. Introduction

Mechanized faux product opinion authentication, classification, analysis, and prediction come to be an essential area of opinion analysis inside the present world of Natural Language Processing (NLP) and the E-trade marketplace. these days, the requirement for those computerized systems using gadget gaining knowledge of (ML) strategies has grown greatly [1] - [8] that uses a solid foundation in the domains of AI and ML. Numerous classifiers and analyzers have evolved in the internet market and industrial domains, but a trustworthy and efficient fake evaluate-based prediction device is still essential.

Finding and identifying false reviews is the main goal of the system's top-level architecture for fabricated posts detection and identification. Additionally, it uses a variety of performance-based indicators and criteria to assess the effectiveness and performance of the system.

False client testimonials, spam, poisoned, faked, threatening, and dishonest and misleading reviews are some examples of false reviews. The system's top-level design is depicted in the Fig. 41.1. The system firstly takes unbalanced dataset, preprocesses it, extracts the reviews or features, applies machine learning classifiers, finds and categorizes fraudulent reviews, and uses various metrics to assess the system's success.

*Corresponding author: navin.goyal@poornima.org

DOI: 10.1201/9781003716389-41

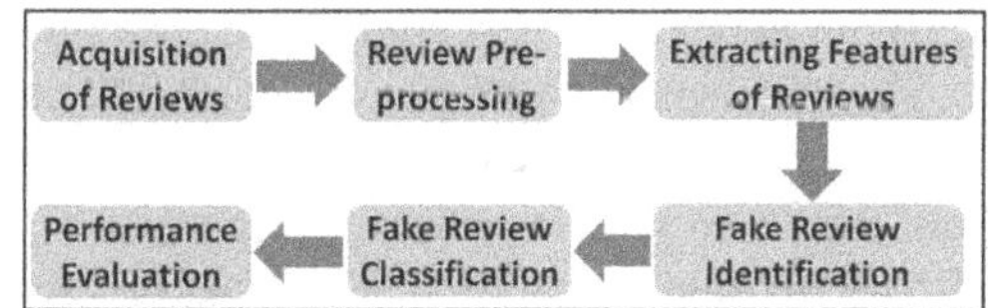

Fig. 41.1 The system's top-level architecture for identifying fraudulent reviews

This is the order in which the sections are presented. The thorough overview of the many fake review prediction algorithms currently in use is presented in Section 1. The fundamental characteristics and differences between these current sentiment analysis methods and their illustrations are presented in Section 2. Based on the classifiers, datasets, and performance measures, it presents a comparison. In Section 3, these specifications are visually analysed with respect to the utilization of classifiers, datasets, and the year-by-year progression of previous studies. The report is concluded in Section 4 with recommendations for the future.

2. The Evolution of Preexisting Works

Using various datasets and machine learning methodologies, this part describes, contrasts the workings of the most current software based faux opinion recognition and detection systems. This section explains the steps involved in using the current sentiment analysis techniques to find, recognize, anticipate, and categorize phony reviews. These methods are described here one by one to illustrate their efficacy. For finding online spam and phony reviews, Crawford, M. et. al, (2015) a survey of well-known supervised machine learning algorithms was provided. It also examined the effectiveness of several approaches for identifying these kinds of reviews. It included a technique for more study as well as a comprehensive overview of current studies on the subject of identifying fraudulent reviews[1].

Ahsan, M. et. Al, (2016), Faux review discernment made use of the ensemble learning technique, which involved combining two active-based supervised learning algorithms. In order to do this, it used three distinct filtering stages, including n-gram properties of the review content, KL and JS distance, and method for characteristics retrieval was used are Term Frequency/Inverse Document Frequency (TF/IDF). It tested the suggested method using four different kinds of simulations and manually classified more than a thousand unlabeled phony opinions as part of its Active Learning (AL) procedure. With the AL approach, it was discovered that the Linear SVM performed the greatest. NB also fared better than other supervised classifiers [2]. Heredia, B., et. Al, (2016), done another method of this kind used four classifiers and three ensemble-based approaches. It contrasted the results of boosting, bagging, SVM, Multinomial NB (MNB), C4.5, Logistic Regression (LR), and RF [3].

Ibrahim, A. J., et. al, (2017), ensemble technique was implemented by the false review detection model, which used the NB, SVM, and LR as foundation classifiers [4]. Madisetty, S., et. al, (2018), An ensemble strategy for twitter fake review detection was presented by another model [5]. In addition to the content-based, user-centered and n-gram properties of the model based on features, it implemented five Convolutional Neural Networks (CNN). Elmurngi, E., et. al, (2018) A different detection approach used K-Star (K*), SVM, Decision Tree (DT), J48, K-Nearest Neighbours (KNN) and NB to classify reviews of movies as polarity positive or negative [6]. By Algotar, K., et. al, (2018), employing misleading and helpful classifiers, respectively, the false review prediction model identified reviews that were genuine and helpful, and a ranking model was then used to assign rankings to the reviews. In addition, a repository and lexicon were constructed to classify the reviews as honest or misleading. Using the deceptive classifier, it classified the testing data as honest or deceptive, and using the helpful classifier, it classified the data as valuable or worthless [7].

Ansari, G., et. al, (2018), used another technique for classifying spam reviews combined local and global filter-related feature selection techniques into an ensemble. The worst complexity for the ham and spam classes was determined to be O (nm2 + m + m (log (m))), where n and m stood for the number of review instances and features, respectively. Additionally, it sorted these features using the optimal sorting method in O (m (log (m))) iterations, achieving the global feature score in 2nm2 iterations [8]. Mani, S., et. al, (2018), developed the fake review detection system used ensemble techniques including majority voting and stack-based to improve the n-gram (unigram and bigram) characteristics. To guarantee precise categorization, ten-fold cross-validation was used in training as well as testing phases [9]. Hajek, P., et. al, (2019), study the effectiveness of several deep learning techniques, including DT, SVM, and NB, were compared. It looked at how effective various machine learning techniques were at identifying both positive and negative fake customer reviews. In terms of accuracy and time performance, it was shown that Deep Neural Networks (DNN) with CNN types and Long Short-Term Memory (LSTM) performed better than traditional ML techniques [10]. Baraithiya H., et. al, (2019), used Trip Advisor and Amazon Mechanical Turk (AMT) datasets, the fraudulent reviews were identified via the ensemble classification approach. Additionally, it used 5-fold cross-validation to assess the system [11]. Khan, R.A., et. al, (2020), done thorough analysis and cutting-edge research on a number of current spam review detection techniques were presented by the study. It focused on the research gaps and suggested future directions while illuminating the taxonomy of machine learning approaches [12]. According to Ashwini, M. C., et. al, (2020), The reviews of products on Amazon were categorized as +ve or -ve by the phony review detection model using a number of machine learning approaches [13].

Fayaz, M., et. al, (2020), using RF, KNN and Multi-Layer Perceptron (MLP), the ensemble machine learning model detected the spam reviews before using the majority vote method. In order to fetch 25 numerical characteristics from data, it used three methods for choosing relevant features, Chi-square, information gain, and univariate. It then eliminated the 10 best characteristics [14]. Gutierrez-Espinoza, L., et. al, (2020), found effectiveness of the system for the hotel dataset was examined using the false review identification and classification method, which included ensemble learning-based techniques such DT, and Extreme Gradient-Boosting Trees (XGBT) [15].

Kumar, S., et. al, (2021), their study examined many false news detecting techniques currently in use. Additionally, a thorough analysis of present and historical false news detection utilizing various ML models was given [16]. According to Wang, J., et. al, (2021), The ensemble method was used by the spam review detector in conjunction with NB, LR, DT, Gradient Boosted Trees, RF, and Artificial Neural Networks (ANN) [17]. Yao, J., et. al, (2021), used a different approach, the unbalanced data was resampled, which allowed for feature pruning to lower computing costs and grid search to optimize the parameters and obtain the best values for the required parameters. Lastly, it employed an ensemble classifier that combined the improved basic classifiers through the use of stacking and majority voting procedures [18].

Banerjee, S., et. al, (2023), examined how phony reviews are made in accordance with different cognitive types using data gathered from 50 reviewers. Systematic, accelerated, and direct-drafting were found to be the three main approaches used in the creation of phony reviews [19]. According to Lu, J.; et. al, (2023), The CNN, BERT, TextCNN, and SKEP-BASED ham-spam review detection system, BSTC. A pre-trained language model is called SKEP [20]. Mewada, A., et. al, (2022), used three gold standard datasets to assess the accuracy (Hotel, Restaurant, and Doctor). In order to identify phoney reviews, authors discovered the association between groups of phony reviewers and false reviewers. They have looked at and emphasized the present research challenges in data collection, false feature construction, and recognition technique design in order to provide guidance for future work on fake review identification [21]. Qayyum, H., et. al, (2023), did a deep learning-based method for detecting false reviews is called FRD-LSTM. The retrieved characteristics were used to train the Bi-LSTM classifier, which distinguished between genuine and fraudulent reviews. The accuracy of the deep features generated from the Amazon product reviews database using the deep contextualized word representation (DCWR) approach was 97.21% [22].

Ahmed, B.,et. al, (2022), This study used data on Saudi Arabian identified spam tweets that were gathered through the end of 2020 and covered eight distinct themes. N-grams and Word2Vec, two feature generating approaches, were applied. The RF model used to the examination of results [23]. In this study authors, Choi, W., et. al, (2023), Prior to using both supervised and unsupervised learning algorithms for the result analysis on the three various teeth-whitening treatments, the usefulness of the reviews was examined. It was discovered that SVC outperformed other classification algorithms with an accuracy of 85.12%. [24]

Authors Hameed, et. al, (2023), applied three Xgboost, a support vector classifier, and stochastic gradient descent ML algorithms for finding false reviews, Yelp dataset for hotel were used for the implementation, it accurately identified the faux reviews on both balanced and unbalanced datasets [25].

3. Discrimination for Existing Works Based on Specifications

Section 2 provided examples of several modern sentiment analysis techniques for detecting spam and phony reviews. In this part, a few performance and basic characteristics were used to separate them from one another. precision, F-score, accuracy, AUC, recall, and other performance measures were assessed for most of these systems through testing on benchmark datasets.

Three fundamental specifications-Issue-centered, Algorithm, and data collections-as well as one Measurement criteria are utilized to compare various current contributions and techniques in Table 41.1. The different algorithms for classification and their combined techniques have been employed by all of the present solutions. These include DT-J48, C4.5, LR, ANN, CNN, DNN, LSTM, MLP, XGBT, GBM, Gradient Boosted Trees, Boosting, Bagging, RF, NB, MNB, SVM, LSVM, KNN, K*, and maximum entropy.

These systems made use of popular databases, including the 1KS10KN unbalanced dataset, the Gold dataset, the movie review dataset, the hotel dataset, and the restaurant dataset. These statistics were gathered from a variety of websites and platforms, including Google Play Store, Yelp and OTT, Amazon, Epinions, TripAdvisor, Twitter, and social media. They have calculated the accuracy, precision, AUC, f-score, and recall measures to assess the effectiveness of their systems.

4. Analysing Different Requirement Classes

The performance-based and basic specifications display the outcomes of many parameters used in current algorithms. The several specification-based comparisons between the current contributions are shown visually in

Table 41.1 Distinguishing between different faux opinion finding methods based on performance and fundamental requirements

Ref. no.	Issue-centered	Algorithm, Dataset and Measurement Criteria
[2]	Faux review detection	Ensemble methods like maximum entropy, NB & SVM etc. Data collected from 3600 reviews across various areas. Both real and faux. Measurement criteria: Precision: 96%, recall matric is 94.9%, f-measure is 94.90%, & accuracy > 87.99%.
[3]	Fabricated opinion detection	Three different tactics such as boosting, RF and bagging ensemble algorithms Data collection: Yelp dataset as well as OTT dataset Measurement criteria: 90.01% accuracy achieved
[4]	Hybrid classification for fake review detection.	Combination of all 3 classifier such as LR, NB and SVM Data collection: Data set used Amazon dataset Measurement criteria: Accuracy: 89.08%, Precision: 0.883, f1-score: 0.892 and Recall: 0.892.
[5]	False opinion detection usingArtificial Neural Network	CNN with the ensemble of RF & SVM. And word embeddings techniques (Glove, Word2vec) Data collection: Twitter. HSpam balanced & 1KS10KN imbalanced datasets. Measurement criteria: Promising results.
[6]	Fabricated opinion detection through emotion analysis using ML.	DT-j48, K*, SVM and other Data collection: Dataset used Movie dataset. Measurement criteria: Accuracy in % for dataset1 is: NB: 71.01, K*: 69.4, KNN-IBK with K as 3: 70.5, DT-J48: 70.001 and SVM: 75.99. (%) accuracy for dataset2 is = NB: 78.99, K*: 72.14, KNN-IBK with K as 3: 71.01, SVM: 80.999 & DT-J48: 72.011.
[7]	Trustworthy and meaningful review identification via sentiment mining.	SVM, KNN & NB. Data collection: Dataset used: 1. Amazon 2. Epinions both balanced and non-balanced Measurement criteria: Range of % Accuracy = Spam/Spam on Non-balanced data: 61/67. Meaningful/non- meaningful on balanced one: 74-78.99.
[8]	A combination of local and global attribute filters is used to categorize fraudulent posts.	Classifiers used: C4.5, LSVM, LR, MNB Data collection: Datasets used: 1. Yelp processed review 2. TripAdvisor's restaurant Measurement criteria: Obtained highest AUC value with MNB, : 0.98 & 0.91 on synthetic & real datasets, respectively.
[9]	Using ensemble machine learning to detect bogus posts.	Ensemble: NB, SVM & RF. Data collection: Reviews used: Gold standard dataset. Of Hotels Measurement criteria: Stacked ensemble precision for RF=0.88 for faux, precision for RF=0.87 for truthful, 0.84 recall for false, 0.889 recall for truthful & 86.68% accuracy. Majority voting ensemble = 0.86 precision for fake, 0.89 precision for genuine, 0.90 recall for spam, 0.85 recall for genuine & 88.10% accuracy.
[25]	Fake Review Detection Using Machine Learning	Xgboost, SVC, and stochastic gradient descent Data collection: Balanced & imbalanced datasets. Yelp on hotel services. Measurement criteria: Accuracy =90% for all three models

this section. The main components of these comparisons are the specifications, which include year-by-year system evolution, algorithm type and usage, and dataset type and usage. Figures 41.2, 41.3, and 41.4 show them, accordingly. The maximal study impact of junk review identification for the years 2018 and 2024 is shown in Fig. 41.2.

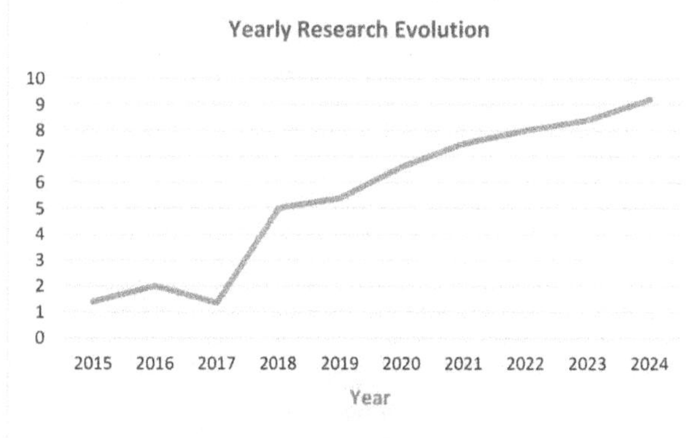

Fig. 41.2 Depicting the year-wise evolution

The highest utilization of several classifiers is shown in Fig. 41.3. The most popular classifiers were determined

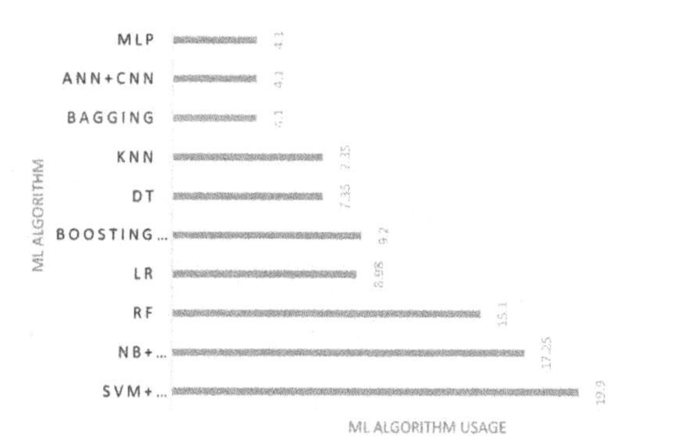

Fig. 41.3 Depicting the usage of various classifiers

to be the SVM + variations, NB + variants, and RF classifiers, with usage % of 19.9%, 17.25%, and 15.10%, respectively. 8.98% of LR and 9.2% for boosting+, 7.35% of DT and KNN, 4.10% of bagging, ANN + CNN, and MLP were determined to be the uses of other classifiers.

Figure 41.4 depicting various types of datasets and their usage. The most popular datasets were discovered to be Yelp, hotel-based, Amazon, and supplemental, with utilization rates of 23.07%, 19.23%, 11.53%, and 11.53%,

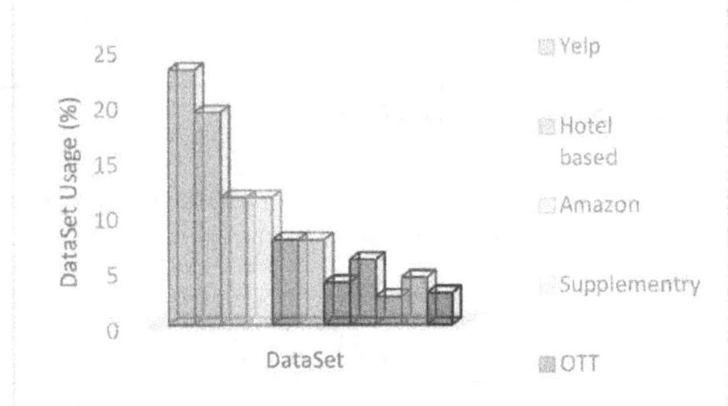

Fig. 41.4 Illustrating various datasets and their usage

respectively. The TripAdvisor, AMT, restaurant, and other hotel databases were among the hotel-based datasets in this case. The percentage of other datasets used was 7.69% for Twitter and OTT, and 3.84% for the datasets related to opinions, gold, the Google Play Store, social media, and movies.

5. Future Recommendation for Future Work and Conclusions

The numerous specifications of current bogus review detection and forecasting systems employing NB and RF are thoroughly analyzed and compared in this research. The year-by-year progression, classification techniques, and data sets were incorporated in the fundamental requirements. The performance requirements also include the accuracy, f1-score, precision, recall, and AUC. The study emphasized the problems, difficulties, dangers, and directions of several previous publications. According to this survey, the greatest amount of research was contributed after the year 2018. Furthermore, it was discovered that the dataset of Yelp's hotels reviews and other hotel-based datasets, as well as the SVM and NB classifiers, had the highest utilization percentage.

While sentiment analysis's counterfeit review models and classifiers are an established field of study, new problems and research methodologies are always emerging. According to their views, many of the previous research projects have attained low accuracy and have not been applied to Flipkart datasets. In order to pinpoint the shortcomings and potential developments that the fake review detection techniques need, their potential risks and limits were also investigated. The balanced-imbalanced dataset problem was also determined to be of great importance. Future study will use an ensemble technique using real-time applications and datasets to accurately and efficiently classify sentiment analysis based on aspects for the detection of bogus reviews.

References

1. Crawford, M., Khoshgoftaar, T. M., Prusa, J. D., Richter, A. N., Najada, H. A.: Survey of review spam detection using machine learning techniques. Journal of Big Data 2 (23), 1–24 (2015). Available from: https://journalofbigdata.springeropen.com/articles/10.1186/s40537-015-0029-9
2. Ahsan, M. N. I., Nahian, T., Kafi, A. A., Hossain, M. I., Shah, F. M.: An ensemble approach to detect review spam using hybrid machine learning technique. In: 19th International Conference on Computer and Information Technology (ICCIT), pp. 388–394. IEEE (2016). Available from: https://ieeexplore.ieee.org/document/7860229
3. Heredia, B., Khoshgoftaar, T. M., Prusa, J., Crawford, M.: An Investigation of Ensemble Techniques for Detection of Spam Reviews. In: 15th IEEE International Conference on Machine Learning and Applications, pp. 127–133. IEEE (2016). Available from: https://www.computer.org/csdl/proceedings/icmla/2016/12OmNxG1yTI
4. Ibrahim, A. J., Siraj, M. M., Din, M. M.: Ensemble classifiers for spam review detection. In: IEEE Conference on Application, Information and Network Security, pp. 130–134. IEEE (2017). Available from: https://ieeexplore.ieee.org/document/8270437
5. Madisetty, S., Desarkar, M. S.: A neural network-based ensemble approach for spam detection in Twitter. IEEE Transactions on Computational Social Systems 5 (4), 973–984 (2018). Available from: https://ieeexplore.ieee.org/document/8540077
6. Elmurngi, E., Gherbi, A.: Detecting fake reviews through sentiment analysis using machine learning techniques. In: The Sixth International Conference on Data Analytics, pp. 65–72(2018). Available from: https://www.researchgate.net/publication/325973731_Detecting_Fake_Reviews_through_Sentiment_Analysis_Using_Machine_Learning_Techniques
7. Algotar, K., Bansal, A.: Detecting truthful and useful consumer reviews for products using opinion mining. In: CEUR Workshop Proceedings 2111, 63–72 (2018). Available from: https://asu.pure.elsevier.com/en/publications/detecting-truthful-and-useful-consumer-reviews-for-products-using
8. Ansari, G., Ahmad, T., Doja, M. N.: Spam review classification using ensemble of global and local feature selectors. Cybern. Inf. Technology 18 (4), 29–42 (2018). Available from: https://sciendo.com/article/10.2478/cait-2018-0046
9. Mani, S., Kumari, S., Jain, A., Kumar, P.: Spam review detection using ensemble machine learning. Machine Learning and Data Mining in Pattern Recognition, Springer, 198–209 (2018). Available from: https://link.springer.com/chapter/10.1007/978-3-319-96133-0_15
10. Hajek, P., Barushka, A: A comparative study of machine learning methods for detection of fake online consumer reviews. In: Proceedings of the 2019 3rd International Conference on E-Business and Internet, pp. 18–22. Association for Computing Machinery, NY (2019). Available from: https://dl.acm.org/doi/10.1145/3383902.3383909
11. Baraithiya H., Pateriya, R. K.: Classifiers ensemble for fake review detection. International Journal of Innovative Technology and Exploring Engineering 8 (4), 730–736 (2019). Available from: https://www.researchgate.net/publication/331331310_Classifiers_ensemble_for_fake_review_detection
12. Khan, R. A., Shoaib, F. M.: Spam review detection: a systematic literature review. TechRxiv, pp. 9-25. , IEEE

(2020). Available from: https://www.techrxiv.org/articles/preprint/Spam_Review Detection_A_Systematic_Literature_Review/12951077/1
13. Ashwini, M. C., Padma, M. C.: Efficiently analyzing and detecting fake reviews through opinion mining. International Journal of Computer Science and Mobile Computing 9 (7), 97–108 (2020). Available from: https://ijcsmc.com/docs/papers/July2020/V9I7202022.pdf
14. Fayaz, M., Khan, A., Rahman, J. U., Alharbi, A., Uddin, M. I., Alouffi, B.: Ensemble machine learning model for classification of spam product reviews. Complexity 2020, 1–10 (2020). Available from: https://www.hindawi.com/journals/complexity/2020/8857570/
15. Gutierrez-Espinoza, L., Abri, F., Namin, A. S., Jones, K. S., Sears, D. R. W.: Fake reviews detection through ensemble learning. Journal, 1–8 (2020). Available from: https://arxiv.org/abs/2006.07912
16. Khalid, M., Ashraf, I., Mehmood, A., Ullah, S., Ahmad, M., Choi, G. S.: GBSVM: Sentiment Classification from Unstructured Reviews Using Ensemble Classifier. Applied Sciences 10(8): 2788, 1–20 (2020). Available from: https://www.mdpi.com/2076-3417/10/8/2788
17. Wang, J., Xue, D., Shi, K.: An ensemble framework for spam detection on social media platforms. International Journal of Machine Learning and Computing 11(1), 77–84 (2021). Available from: http://www.ijmlc.org/index.php?m=content&c=index&a=show&catid=112&id=1196
18. Yao, J., Zheng, Y., Jiang, H.: An Ensemble Model for Fake Online Review Detection Based on Data Resampling, Feature Pruning, and Parameter Optimization. IEEE Access 9, 16914–16927 (2021). Available from: https://ieeexplore.ieee.org/document/9320592
19. Banerjee, S., Chua, A., : Understanding online fake review production strategies,Journal of Business Research,Volume 156, 2023, 113534, ISSN 0148–2963, (2023) Available from: https://www.sciencedirect.com/science/article/pii/S0148296322009997
20. Lu, J.; Zhan, X.; Liu, G.; Zhan, X.; Deng, X. BSTC: A Fake Review Detection Model Based on a Pre-Trained Language Model and Convolutional Neural Network. *Electronics* 2023, *12*, 2165. Available from: https://doi.org/10.3390/electronics12102165
21. Mewada, A., Kumar, R., Dewang, Research on false review detection Methods: A state-of-the-art review,Journal of King Saud University - Computer and Information Sciences, Volume 34, Issue 9, 2022, Pages 7530–7546, ISSN 1319-1578(2022), Available from: https://www.sciencedirect.com/science/article/pii/S1319157821001993
22. Qayyum, H., Ali, F., Nawaz, M. et al. FRD-LSTM: a novel technique for fake reviews detection using DCWR with the Bi-LSTM method. Multimed Tools Appl (2023). Available from: https://doi.org/10.1007/s11042-023-15098-2
23. Ahmed Balfagih, Vlado Keselj, and Stacey Taylor. 2022. N-gram and Word2Vec Feature Engineering Approaches for Spam Recognition on Some Influential Twitter Topics in Saudi Arabia., New York, NY, USA, 101–107. Available from: https://dl.acm.org/doi/fullHtml/10.1145/3546157.3546173
24. Choi, W., Nam, K., Park, M., Yang, S., Hwang, S., & Oh, H. (2023). Fake review identification and utility evaluation model using machine learning. Frontiers in Artificial Intelligence, 5. Available from: https://www.frontiersin.org/articles/10.3389/frai.2022.1064371/full.
25. Hameed, Wesam & Allami, Ragheed & Ali, Yossra. (2023). Fake Review Detection Using Machine Learning. Revue d'Intelligence Artificielle. 37. 10.18280/ria.370507.

Note: All the figures and table in this chapter were made by the authors.

Intelligent Systems Using Semiconductors for Robotics and IoT – Dinesh Goyal et al. (eds)

42 Balancing Wellness: Understanding How Ayurveda Recommends Herbal Treatments Based on Dosha

Rama Bhardwaj*
Research Scholar, Poornima University,
Jaipur, Rajasthan, India

Rakesh Kumar Saxena
Professor, Poornima University,
Jaipur, Rajasthan, India

Abstract: Ayurveda, an old health system from India, believes that everyone has different health needs. It uses a special way called dosha theory to figure out what those needs are. People are put into three groups: Vata, Pitta, and Kapha, based on their body type. Ayurveda uses herbs to help balance these doshas. This paper looks at how Ayurveda suggests herbs based on a person's dosha and why it's a key part of health care. This paper dug into a lot of research to see what scientists are saying about doshas, the herbs used for each one, and how this information is used to give advice. This paper also looked into the common ways these recommendation systems are made and the problems that come with them. Finally presented work is basically to point out what is good about these systems, what could be better, and what researchers might study next. This research also shows how important it is to have health care that fits each person in Ayurveda. It sets the ground for more work and new ideas in giving dosha-based herb suggestions.

Keywords: Ayurveda, Dosha, Herbal remedies, Personalized healthcare, Recommendation systems

1. Introduction

Ayurveda is an ancient science of life that has its link with present Indian texts more than 5000 years ago and is believed to be an encapsulation of human health and life [1]. It encompasses a multifaceted topical framework that involves a synthesis of physical, mental, and spiritual health dimensions wherein body, mind, and encompassing environment are in a state of harmony. The key containing of Ayurveda involves the usage of the hard elements called doshas which is regarded as the working energy of all living organisms [1].

Vata, Pitta, and Kapha are the three doshas, which are formed from different involvements of the five vital elements of the universe such as space, air, fire, water, and earth within a human body. Vata is the space and air and controls the actions and mobility, Pitta has the fire and water elements and is in charge of digestion and metabolism issues, Kapha is the earth and water and is responsible for stability and structure. A human being possesses a predetermined arrangement of these doshas at the time of birth which is known as Prakriti this decides the Physiognomy, mental abilities, and the general predisposition to ailments or diseases [1].

It is worth mentioning that the Ayurvedic concept of health and diseases mainly relies on these three basic constituents of the body and mind, called dosas and maintaining the balance of these dosas is apparently beneficial for health and longevity whereas its opposite leads to various health disorders [2]. Among the external factors that can cause an imbalance in dosha there are food, habits, states, stress, and climate, as well as changes of the seasons. Consequently, Ayurvedic management of treatment addresses doshic imbalances in each case or constitution in order to bring balance for better health [2].

*Corresponding author: rama.bhardwaj.17@gmail.com

DOI: 10.1201/9781003716389-42

Herbology is one of the main approaches of Ayurveda – a use of plant products that have curative and restorative properties to help re-establish a healthy state within the body [2]. These remedies are chosen based on the concept of Ayurveda pharmacology where the properties of the herbs to be used in the preparation include taste, potency, post digestion effect, energetic attributes, and also the balance or the dosha of the body. Kalpa, alternatively refers to formulations that may be single or compound with great variability, depending on the peculiarities of a patient's constitutional pattern [3].

Over the last decade, the advancement in technology has been seen to improve the ways through which Ayurveda is practiced to effectively promote specificity in the delivery of healthcare services. Regarding the aforesaid, Dosha-based herbal remedy recommendation systems can be considered more relevant, promising field in this respect for it can potentially involve automation of remedy recommendation and, moreover, can offer individual recommendations according to Dosha constitution, health conditions, and problems. These recommendation systems can involve computational strategies, machine learning approaches and databases of Ayurveda information to identify trends, and make effectual treatment forecasts [3].

These systems, operationalizing personalized herb prescriptions, shown in Fig. 42.1 some herbs with medical benefits, bring efficiency, availability, and reproducibility to Ayurvedic medicine without thereby losing sight of the patient-centered, integral health approach that is innate to this system of healing [3]. However, the use of dosha-based herbal remedy recommendation system comes with some challenges in the process of its development and implementation which include accurate dosha assessment, integration of Indian traditional knowledge and modern technological approach, and also to check the algorithm generated recommendation by the models through the clinical trial and empirical data [4].

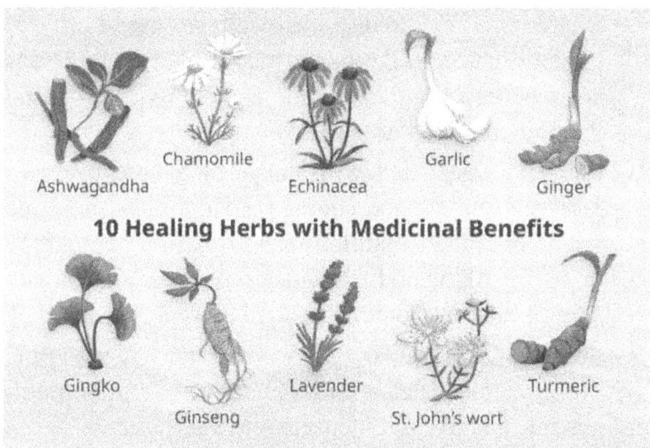

Fig. 42.1 Herbs with medical benefits

Source: verwellhealth.com

Mapping herbs to dosha profiles also requires in-depth knowledge of herb properties as shown in Fig. 42.1, dosha affinities, and their effects on various health conditions. This complexity has made it very challenging to translate Ayurvedic knowledge into a computational framework. The lack of digitized data, coupled with the intricacy of herb-dosha relationships, presents significant obstacles in creating algorithmic solutions for Ayurvedic herbal remedy recommendations [1].

This research meets these challenges with the development of a Dosha-Based Herbal Remedy Recommendation System that integrates the traditional Ayurvedic knowledge base with the modern application of machine learning techniques. More specifically, it utilizes a hybrid machine learning model combining SVM and MultiSVM in the classification of individual dosha profiles as well as providing recommendations for customized herbal remedies. This study connects the gap between ancient wisdom and computational methodologies through the use of the Medicinal Plant Database of India, which contains extensive information about herbs and their therapeutic effects [2].

The threefold objectives of this research are: (1) dosha classification, for the accurate identification of an individual's dosha constitution based on symptoms and health conditions; (2) herb-dosha mapping, to establish the relationship between medicinal herbs and their efficacy for specific dosha imbalances; and (3) personalized remedy recommendation, to create a scalable, user-friendly system for providing customized herbal treatments. Through rigorous feature extraction, model optimization, and validation, the proposed system will empower people to make informed decisions about their health through the fusion of modern machine learning with the timeless [2].

2. Ayurveda and its Principles

Known to have arisen from the Indian subcontinent, Ayurveda is a more encompassing style of healing which is part of the Vedas, which are scrutinized as ancient texts from India [6]. Now a days, Ayurveda is known to be one of the oldest medical systems originating from the Indian subcontinent that dates back to 5,000 five thousand years. That is how its principles are associated with Indian philosophies, its spirituality, and, based on the Didactic, naturalistic observations of a human body and its functioning within the Environment [6].

Prolonged at the core of Ayurveda is the conceptual model, which is the theory underpinning of Ayurveda and it propounds to define health and disease and the curative power of the body [6]. The principles of Ayurveda involve yet another concept called the Panchamahabhutas, which can be translated as five basic elements which exist in the form of Earth, Water, Fire, Air and Ether. These elements combine to form three primary doshas or bio energies: Another is the 'Dosha' that is made up of three components in different proportions that the proponents of Ayurveda believe is fundamental to understanding human anatomy and physiology that include Vata, Pitta, and Kapha [6].

A human body is formed of three doshas viz. Vetha, Pitta and Kapha but an individual is born with certain dosha predominance which is known as Prakriti [6]. Vata is a combination of air and ether and is mainly responsible for movement, passage of words, etc. Pitta, is an energetic dosha that has fire and water as its inherent attributes and its main function is to regulate digestion and metabolism. Kapha, which is linked to the earth and water factor, controls the structure and coordination [7]. In Ayurveda, it is pivotal to determine the Prakriti for health management and disease risks. Prakriti is crucial in determining health management in Ayurveda [7].

Ayurveda also divides the structure of the human body into seven basic bodily tissues or dhatus, which are responsible for prompting tissue activity and providing nourishment [7]. These tissues are referred as Sapta Dhatus These are plasma, blood, muscle tissue, adipose (fat) tissue, osseous (bone) tissue, medullary (bone marrow) tissue and genital glandular (reproductive) tissue. Also, Ayurveda acknowledges Agni the fire which prevails in the digestive tract capable of digesting as well as absorbing nutrients in our body and excreting wastes as one of the largely accepted essentials for sustaining healthy life [8].

According to Ayurveda health plan as well as curing specific ailment, the priorities of human body and nature are to achieve and maintain the balance within the body [8]. Usually called Vikriti, doshas can become imbalanced and give rise to disease and pain. Ayurveda in its traditional form focuses on the elimination of the distortions and disharmonies that have occurred through modifications in lifestyle, diet and natural medicines which are personalized according to the body type and condition of an individual [9].

Lastly, it is crucial to understand that Ayurveda is a truly comprehensive, or even holistic, model of health as it recognizes the mind and spirit as an inalienable part of the body. Additionally, while Ayurveda focuses on the balance of the body – where imbalances are seen as the cause of illness, the primary goal of Ayurveda is to improve a person's QOL and to prevent illness by providing individualized health care interventions as per the natural laws [9].

3. Doshas in Ayurveda

Doshas, mainly Vata, Pitta, and Kapha are the key to Ayurvedic ideas of health and illness, as well as the fundamental three tissue/ humor theory. Ayurveda believes that everyone in this world has a specific amount of these three dosha which defines one's characteristics like his bodily structure, temperament and the type of disease he is likely to catch [10].

Vata Dosha:

- As an element composed of air and ether, the function of Vata is to govern movement, behavior, and actions in the physical structure. It is responsible for controlling human movements and things such as breathing, blood flow, and signals sent from the brain, shown in Fig. 42.2 [10].
- If Vata is in harmony, it brings activities, energy and flexibility in the person. He opponents assert that Ayurved, Homeopathy and chiropractic are incompatible with the medical model. Though, when doshas increase in quantity or get out of balanced proportions, it causes symptoms such as dryness, coldness, anxiety and irregularity in the case of Vata dosha [11].
- In Ayurveda, the diseases concerning nervous system, articulation, and digestion are understood to be brought on by excessive vata [11].

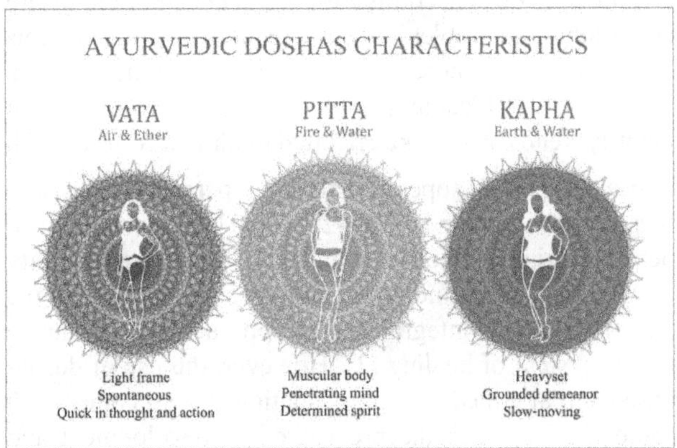

Fig. 42.2 Ayurveda doshas

Source: the tribeconcepts.com

Pitta Dosha:

- The Pitta dosha has the fire and water constituting it affecting metabolism as well as digestion and transformation in the body. It controls few functions such as digestion, absorption as well as maintaining the body temperature [12].
- Pittayam associated with balanced Nature provides distorted intelligence, courage and ambition. In people with more Pitta, conditions that result in inflammation or acidity of the body, frequent outbursts of anger and skin diseases are some of the signs that may be seen [12].
- Some of the vishlepakwavyasaha associated with Pitta dosha include ulcers, inflammation, histamine, indigestion, and liver problems in Ayurvedic ailments [12].

Kapha Dosha:

- Kapha causes rigidity as it derives from earth and water; it is also responsible for body's organization and cohesion. It gives protection, rigidity via a rigid diet plan and feeding to the tissues [13].

- For that matter, when Kapha is dominant, this aspect of the seven tissues contributes to calmness, strength, and patience. But, when it trespasses, it causes ailments such as lethargy, sluggishness, developing congestions as well as being associated with greed [13].
- In Ayurveda, kapha dosha is generally connected to obesity, sinusitis, respiratory problems, and even emotional lethargy [13].

As much as doshas are central in Ayurvedic health care, they are also used in preserving the health of an individual. The doshas are three in number and Ayurveda maintains that for the body to be healthy, they must be in a balanced state, equally aligned in order to maintain the proper functioning of the bodily systems [13]. Diseases and discomforts result from unequal distribution or vitiations of the body's doshas as a result of improper diets, unhygienic standards, stressful conditions or any unfavorable environment. These therapies work to realign the doshas to the audience's constitution and reapportion the dosha amount needed by their bodies. Such recommendations may thus encompass dietary changes, exercise adjustments, herbal medications that help in reducing the levels of an aggravated dosha and therapies meant to restore balance within the seven tissues that make up the human body [13].

Through recognizing and targeting the nature of the body and pathogenic conditions of a person, Ayurvedic medicine aims at the effective elimination of not only disease but also the primary cause of its onset, the attainment of rejuvenation, and the optimization of a person's life in its totality with no negative side effects and quick results.

4. Importance of Herbal Remedies in Ayurveda

The utilization of herbs is a significant aspect of Ayurveda, which is one of the oldest medical practices in the world, going back over 5 thousand years to India [14]. Herbal remedies are also culminated predominant features in Ayurveda greatly supporting its energetic model. In Ayurveda, it is believed that the proper used of herbs are capable of re-establishing the disturbed equilibrium and provide harmonized health within the human body, soul, as well as spirit. In accordance with the principle of Ayurveda, each herb has its specific characteristics which, used in practice, influence the three doshas – Vata, Pitta, and Kapha. This treatment obtained directly impacts our health at its core instead of symptomatic relief only, coupled with preventing future imbalances [14].

However, they are safe and natural, which makes it easier for people to them instead of other methods. These remedies are mainly from plant origin and as such are less likely to manifest serious side effect as those synthetic products. These properties render them safe on body systems making them fit for regular use without leading to undesirable consequences. This aspect of Ayurveda also receives much support from people who are searching for a healthier and non-surgical technique of curing [14].

Another useful aspect of Ayurvedic herbs is that they are effective for in treating an array of diseases and illnesses, starting with gastrointestinal problems and ending with respiratory ailments, mental health disorders, and skin diseases. They can be provided in the form of teas, powders, oils, or tablets, thereby making it easy to incorporate them depending on the desired programme of action. This versatility is a good thing for the therapeutic applications of Ayurveda since it allows practitioners to be in a rather excellent position to provide remedies that are going to address the requirements from the patients in a more efficient way [14].

In terms of functionality, Ayurveda is not only used to cure diseases, but it also helps preserve health. The following are some examples of herbs that are believed to have immune-boosting effects: Many herbs possess immunity-boosting effects hence enable the body to fight infections and diseases. Some examples of Indian herbs are used in the process of purification the body, including Triphala that purify toxic substances in the human body and other bodily impurities. Ayurvedic herbs work to enhance how the body defends itself and to aid in detoxification processes that enhance overall health and minimize the occurrence of diseases in the body [14].

When incorporated with diet and lifestyle measures, effectiveness of the herbal remedy is even enhanced. The panchakarma is generally supplemented by the personal dietary guidelines and peculiarities of live style in accordance with cognitive Prakriti and Vikriti. It signifies that Ayurvedic treatments are holistic, hence promoting the total well-being of a person, thus enhancing the chances of treatment success in the long run. It emphasizes on the five fundamental principles of life or vital life forces and is aimed at the body's natural rhythms of healing [14].

Ayurvedic remedies –relying upon the use of herbs– had been practiced for centuries based on practical experience and traditional knowledge which has been put down in writing in some Ayurvedic books like the Charaka Samhita and the Sushruta Samhita. This is a fact, which is bearing more and more credibility in the light of details emerging from the realm of modern scientific exploration, indicating the effectiveness of most of the herbs used in Ayurveda. Combining the ancient essences of Ayurveda and incorporating scientific logos in the twenty-first century adds more credibility to the Ayurvedic treatments available today in the era of medical practices [14].

Therefore, it can be concluded that, the herbal remedies are an important component in Ayurved system of medicine because they are safe, natural and identified to have potentials for treating numerous disorders. In their ability to deal with the causes of diseases and with the maintenance of homeostasis Ayurvedic herbs can be said to contribute immensely to specific preventive and curative health care as supported by the concepts of Ayurveda..

5. Related Works

Wijaya, S. H. et al. [2023] discussed more on Unani herbal medicines, which is common in Southern Asia due to its belief of smaller side effects compared to ordinary medicines. Contrary to other research that seeks to understand the chemistry of the herbs used, this special research looks at the metabolites in herbal plants. Prospective and data-intensive science strategies were utilized by the researchers to establish a model for using Unani therapeutic applications based on these metabolites. By as such, it was found that the deep neural network-based method proposed in the paper as superior to other methodologies like random forest and support vector machine. The model pinpointed 118 biomarker metabolites involving nine detailed uses of Unani and their potential metabolic pathways promoting health and well-being [15].

MDAR is a multi-task recommendation engine based on a knowledge graph that has been proposed by Li, S. , et al. [2023]. The system addresses three main challenges: temporal preferences, gaps in data, and KG data completion. This paper proposes Deep FM, which integrates Deep FM with an attention mechanism; RMGAT applies a relation-fused multi-head graph attention network. The sublinear optimization in Deep FM focuses on various interaction features, and the DNNs capture semantic information, which can help deal with missing data. RMGAT incorporates relationship data into the knowledge graph, enhancing the representations of entities with uniqueness due to the difference in the importance of neighbors. A good MDAR result is evident from the experimental results on real-world datasets for movies, books, and music, showing that MDAR outperforms all state-of-art baselines, proving that MDAR has a potential for improving recommendation systems in different fields [16].

Li S, Zhang B, Zhang Y, Zhang H and Li K [2022] The Patient-Oriented Multi-Graph Convolutional Network-based Herb Recommendation system (PMGCN) improves the herbs recommendation by constructing the patients' portraits, symptoms, own characteristics, and diagnoses. By employing type-aware attention and multi-graph convolution networks, the proposed model is capable of capturing herb-patient interactions and deliver better results compared to the existing approaches employed for the same purpose and has achieved better results in the context of personalized herb recommendation [17].

Yang, Y., Zhang, S., Cai, Z., Yin, Y., Shang, R., Xie, C., Li, J., 2022 Deep learning models such as SeqGAN and CGAN are used in this study to generate TCM prescriptions. Using TCM knowledge, the models passed this knowledge in training process and can generate prescriptions with different features. Comparing with other methods, it is demonstrated that GAN variants are effective and potentially useful to construct TCM prescription [18].

A machine learning-based approach Lee WY et al. [2022] adopted the Korean herbal formulae (HFs), which reflects clinical symptoms as well as Sasang constitution types and case studies. The combination of using data imputation and oversampling in the development of the cascaded deep forest (CDF) model yielded the specified high accuracy. Ideally, improving interpretability benefited from the LIME technique, helping to modernize Korean medicine and the selection of HF [19].

An automated real-time plant species identification system for medicinal plants in Borneo was designed as Malkiel et al., 2022, which was used EfficientNet-B1-based deep learning at the core of the method. The model demonstrated fine testing accuracy and samples in real-time. One study reported its design with features including a mobile application that enabled the users to engage with one another, post feedback, and geo-locate, as did another study pointing to possibilities of real-time plant identification [20].

Azadnia, R., et al. [2022] proposed an intelligent vision-based system using a CNN to identify herb plants. The tested model was done on images with different definitions. The accuracy that the model got was more than 99.3%. It is effective to identify medicinal plants in real time and is also a reliable alternative to traditional methods [21].

Kalabarige, L. R., & Maringanti, H. [2022]. Development of a symptom-based COVID-19 test recommendation system using machine learning. The system uses synthesized and balanced datasets, classifying different models to predict the right test recommendations with over 99% accuracy. This approach makes even tackling the testing issue easy in the case of infection rate control during the pandemic [22].

Han, M. et al. [2022]. A Novel Mutual Learning-Based Method for Chinese Herbal Medicine Classification. Two small student networks learn from each other to improve the classification accuracy. The model achieved better performance than the best methods in the literature by achieving both higher accuracy and lower loss when tested on a dataset with 100 classes [23].

Zhang, H., et al. [2022] Propose a model of transformer neural network for EHR-based electronic prediction of prescriptions for traditional Chinese medicine. Their model uses temporal sequences of a patient's electronic health record achieving 80.58% for precision and 68.49% for recall in their study on the potential capabilities of digital resources to emulate physicians' decisions concerning TCM prescribing [24].

Venaik, A., et al. [2022] This article discussed the role of machine learning (ML), artificial intelligence (AI), and computer-based molecular docking analysis (CBMDA) in the fight against COVID-19. It is used in screening, prediction, contact tracing, and drug/vaccine development. The review focuses on advanced tools in controlling infectious diseases and their use in the case of COVID-19 [25].

Dong, X., et al. [2021] The research work focuses on enhancing the recommendation of Traditional Chinese Medicine (TCM) prescriptions according to patient symptoms. They introduce a method called Subnetwork-based Symptom Term Mapping (SSTM) for better representation of clinical symptom terms, especially the unrecorded ones. Based on this method, they designed a TCM prescription recommendation system (TCMPR) that performs better than the existing methods, which may potentially improve the precision of TCM medicine [26].

Roopashree, S., & Anitha, J. [2021] This paper presents an automatic medicinal plant identification system using neural networks and deep learning. They address the scarcity of data by introducing the Deep Herb dataset that consists of images of Indian herb leaves. Their model, with transfer learning and SVM, yields high accuracy in plant identification. They also developed a mobile app, Herb Snap, for real-time herb recognition to facilitate herbal knowledge accessibility [27].

Zhou, W., et al. [2021] A new recommendation system, Ford Net, for Traditional Chinese Medicine (TCM) formulae is proposed. Combining the patient's symptoms and molecular information, Ford Net outperforms previous methods in terms of accuracy. This technique fills the gap between traditional TCM wisdom and modern data-driven techniques, and it may further enhance clinical diagnosis and treatment results [28].

Hao, W., et al. [2021] The proposed study also helps in removing the limitations in CHM classification as it uses a large dataset and an advanced model known as EfficientNetB4. The proposed model achieved state-of-the-art accuracy in CHM classification with a good deal of improvement over existing methods. This work

Table 42.1 Literature findings

Author Name (Year)	Main Concept	Findings
Wijaya, S. H. et al. (2023)	Unani herbal medicines	Utilized a deep neural network-based method to identify 118 biomarker metabolites, outperforming random forest and support vector machine models.
Li, S., et al. (2023)	MDAR multi-task recommendation engine	Proposed DeepAFM and RMGAT models, showing improved recommendation performance in movies, books, and music datasets.
Li S, Zhang B, et al. (2022)	Herb recommendation system	Developed PMGCN, improving herb recommendations by capturing herb-patient interactions with type-aware attention and multi-graph convolution networks.
Yang, Y., et al. (2022)	TCM prescription generation	Used SeqGAN and CGAN models to generate TCM prescriptions, demonstrating effectiveness over other methods.
Lee WY et al. (2022)	Korean herbal formulae	Developed a cascaded deep forest model with high accuracy for herbal formula recommendations, benefiting from LIME for interpretability.
Malikal et al. (2022)	Plant species identification system	Designed a system using EfficientNet-B1 for real-time medicinal plant identification in Borneo, achieving high accuracy.
Azadnia, R., et al. (2022)	Herb plant identification system	Achieved over 99.3% accuracy in identifying medicinal plants using a CNN-based model.
Kalabarige, L. R., & Maringanti, H. (2022)	COVID-19 test recommendation system	Developed a system with over 99% accuracy in test recommendations using machine learning on synthesized and balanced datasets.
Han, M., et al. (2022)	Chinese herbal medicine classification	Proposed a mutual learning-based method, achieving high accuracy and improved performance in classifying herbal medicines.
Zhang, H., et al. (2022)	TCM prescription prediction	Developed a transformer-based model using EHRs, achieving 80.58% precision and 68.49% recall in predicting TCM prescriptions.
Venaik, A., et al. (2022)	ML and AI in combating COVID-19	Reviewed the role of ML, AI, and CBMDA in managing COVID-19, emphasizing their importance in screening, prediction, contact tracing, and drug/vaccine development.
Dong, X., et al. (2021)	TCM prescription recommendation	Introduced SSTM method for better symptom representation, enhancing the accuracy of TCM prescription recommendations.
Roopashree, S., & Anitha, J. (2021)	Medicinal plant identification system	Developed Deep Herb dataset and Herb Snap app, achieving high accuracy in plant identification using neural networks and deep learning.
Zhou, W., et al. (2021)	TCM formula recommendation system	Introduced Ford Net, integrating patient symptoms and molecular information, surpassing previous methods in accuracy.
Hao, W., et al. (2021)	CHM classification	Used EfficientNetB4 model on a large dataset, achieving state-of-the-art accuracy in CHM classification.
Bao, H. Y., et al. (2021)	AI in TCM prescription analysis	Provided an overview of AI-driven TCM prescription research and demonstrated AI's role in herb recognition with promising results.

contributed to better understanding and utilization of herbal medicines [31].

Bao, H. Y., et al. [2021] This paper contributes to the interaction of artificial intelligence (AI) and traditional Chinese medicine (TCM) prescription analysis. They outline AI-driven research in TCM prescription, which can potentially contribute to the betterment of TCM practices, as well as experimental images for TCM herb recognition, highlighting the potential of AI in the challenges related to complex problems in TCM [32].

6. Discussion and Findings

The studies covered include machine learning, deep learning up to artificial intelligence applied in traditional medicine, with special focus on herbal medicine and TCM. These technologies have finally brought new solutions for solving various problems such as the identification of herbal plants, prescription recommendations, classification, and even fighting infectious diseases such as COVID-19.

Generally, the overall trends of all these studies indicate the employment of more sophisticated machine learning models, for example, deep neural networks, CNNs, and graph-based models, to extract meaningful insights from complex datasets. All of these excelled particularly in metabolite identification, recommendation system building, and even the recognition of medicinal plants with results faring better than the traditional methods. In addition, many studies emphasized the importance of personalized medicine and added patient-specific data to the analysis. Such techniques as attention mechanisms, multi-graph convolution networks, and patient-oriented modeling provide for personalizing treatment recommendations by individual characteristics, symptoms, and diagnoses to produce more effective and tailored healthcare solutions. In other words, due to mobile applications and real-time systems where AI technology will be employed, access to herbal knowledge along with its usability may increase the likelihood of immediately identifying plants, even providing tailored suggestions on the spot and at all times.

In a nutshell, the findings demonstrate how AI is revolutionizing traditional medicine into promising areas for healthcare outcomes, advancing research, and filling the gap between traditional wisdom and modern scientific approaches.

Findings:

- Highly advanced use of machine learning models, for example, deep neural networks, convolutional neural networks, and graph-based models, to perform various traditional medicine research-related tasks.
- AI-based techniques are superior for such tasks like metabolite identification, prescription recommendation, medicinal plant identification, and infectious diseases management.
- The concept of personalized medicine comes in when patient-specific data is applied and treatment advice could be more apt and tailored.
- Ai-based mobile applications integration with real-time systems that facilitate access and use of AI by enabling users to interface herbal knowledge while seeking on-the-spot healthcare.
- Transformation of Traditional Medicine through AI: How Very Promising Avenues in Healthcare Are Being Explored and Exploited for Filling the Gap Between Traditional Knowledge and Modern Scientific Approaches.

7. Conclusion

Finally, an ancient health system that originated in India, Ayurveda has a dosha theory that has made it give specific healthcare recommendations for individuals based on their constitutions and body types. For example, a human being's dosha might fall into either of the categories, Vata, Pitta, or Kapha. Through herbal remedy prescription, Ayurveda works towards balancing the doshas within the human system. This paper looks at how the dosha works, selecting the herbs tailored for each type, and methodology behind the systems. This paper has been able to gain insight into the dosha-based herbal treatments and the challenges associated with recommendation systems in Ayurveda through extensive research. Although these systems are highly individualized and tailored to a particular need of a person, there is further scope for study and improvement. Since Ayurveda focuses on a holistic approach and every individual has a unique constitution and health needs that require specific attention, this study underlines the importance of personalized healthcare in this context. This study opens the way for further developments in individualized healthcare in the Ayurvedic context by providing a base for future research and innovative ideas in dosha-based herbal recommendations. In brief, the significance of Ayurveda lies in its ability to provide individualized health care that is tailored to the specific needs of a person, which may benefit to enhance holistic health and harmony. As this paper continues to delve deeper in the dosha-based herbal treatment, this paper embarks upon the journey toward betterment of healthcare practices responsive to the diversification of people within the Ayurvedic tradition.

References

1. Kale, S. G., Jain, S., Rangari, S., Dharmik, R. C., Gardi, M., & Lande, V. S. (2024). Identification of Medicinal Leaves and Recommendation of Home Remedies using Machine learning. *International Journal of Intelligent Systems and Applications in Engineering, 12*(12s), 407–413.
2. Wang, C., Wang, F., Guo, R., Liang, Y., Liu, K., & Yu, P. S. (2024). Confidence-aware Fine-tuning of Sequential Recommendation Systems via Conformal Prediction. arXiv preprint arXiv:2402.08976.

3. Vayadande, K., Ghadekar, P. P., Sawant, A., Shelke, M. P., Shirsath, B., Bhande, B., ... & Samgir, S. (2024). Identification of Ayurvedic Medicinal Plant Using Deep Learning. International Journal of Intelligent Systems and Applications in Engineering, 12(15s), 678–693.
4. Fathi, P., & Charkari, N. M. (2024). Herb target prediction using protein complexes detection and machine learning methods in heterogeneous network. Kurdish Studies, 12(4), 65–77.
5. Ji, X., Li, Q., Liu, Z., Wu, W., Zhang, C., Sui, H., & Chen, M. (2024). Identification of Active Components for Sports Supplements: Machine Learning-Driven Classification and Cell-Based Validation. ACS Omega. ACS Omega. 9(10), 11347–11355
6. GunaChandra, S., Mounika, P., Yaswanth, K., & Prasad, M. K. S. R. Deep Learning for Medicinal Plant Identification and Utilization: Leveraging ResNet for Enhanced Recognition and Applications. International Refereed Journal ofEngineering and Science (IRJES). 13(2), 159–165
7. Qian, Y., Wang, X., Cai, L., Han, J., Huang, Z., Lou, Y., ... & Zhu, A. (2024). Model informed precision medicine of Chinese herbal medicines formulas–A multi-scale mechanistic intelligent model. Journal of Pharmaceutical Analysis, 14(4), 100914.
8. Pushpa, B. R., Rani, N. S., Chandrajith, M., Manohar, N., & Nair, S. S. K. (2024). On the importance of integrating convolution features for Indian medicinal plant species classification using hierarchical machine learning approach. Ecological Informatics, 102611.
9. Jin, Y., Ji, W., Shi, Y., Wang, X., & Yang, X. (2023). Meta-path guided graph attention network for explainable herb recommendation. Health Information Science and Systems, 11(1), 5.
10. Jin, Z., Zhang, Y., Miao, J., Yang, Y., Zhuang, Y., & Pan, Y. (2023). A knowledge-guided and traditional Chinese medicine informed approach for herb recommendation. Frontiers of Information Technology & Electronic Engineering, 24(10), 1416–1429.
11. Yang, X., & Ding, C. (2023). SMRGAT: A traditional Chinese herb recommendation model based on a multi-graph residual attention network and semantic knowledge fusion. Journal of Ethnopharmacology, 315, 116693.
12. Zhao, Z., Ren, X., Song, K., Qiang, Y., Zhao, J., Zhang, J., & Han, P. (2023). PreGenerator: TCM prescription recommendation model based on retrieval and generation method. IEEE Access. 11(6) , pp. 103679–103692.
13. Uddin, A. H., Chen, Y. L., Borkatullah, B., Khatun, M. S., Ferdous, J., Mahmud, P., ... & Por, L. Y. (2023). Deep-learning-based classification of Bangladeshi medicinal plants using neural ensemble models. Mathematics, 11(16), 3504.
14. Khatua, D., Sekh, A. A., Kutum, R., Mukherji, M., Prasher, B., & Kar, S. (2023). Classification of Ayurveda constitution types: a deep learning approach. Soft Computing, 27(9), 5309–5317.
15. Wijaya, S. H., Nasution, A. K., Batubara, I., Gao, P., Huang, M., Ono, N., ... & Altaf-Ul-Amin, M. (2023). Deep Learning Approach for Predicting the Therapeutic Usages of Unani Formulas towards Finding Essential Compounds. Life, 13(2), 439.
16. Li, S., Xue, Q., & Wang, P. (2023). MDAR: A Knowledge-Graph-Enhanced Multi-Task Recommendation System Based on a DeepAFM and a Relation-Fused Multi-Gead Graph Attention Network. Applied Sciences, 13(15), 8697.
17. Li, S., Yue, W., & Jin, Y. (2022). Patient-oriented herb recommendation system based on multi-graph convolutional network. Symmetry, 14(4), 638.
18. Yang, Y., Rao, Y., Yu, M., & Kang, Y. (2022). Multi-layer information fusion based on graph convolutional network for knowledge-driven herb recommendation. Neural Networks, 146, 1–10.
19. Lee, W. Y., Lee, Y., Lee, S., Kim, Y. W., & Kim, J. H. (2022). A machine learning approach for recommending herbal formulae with enhanced interpretability and applicability. Biomolecules, 12(11), 1604.
20. Malik, O. A., Ismail, N., Hussein, B. R., & Yahya, U. (2022). Automated real-time identification of medicinal plants species in natural environment using deep learning models—a case study from Borneo Region. Plants, 11(15), 1952.
21. Azadnia, R., Al-Amidi, M. M., Mohammadi, H., Cifci, M. A., Daryab, A., & Cavallo, E. (2022). An AI based approach for medicinal plant identification using deep CNN based on global average pooling. Agronomy, 12(11), 2723.
22. Kalabarige, L. R., & Maringanti, H. (2022). Symptom based COVID-19 test recommendation system using machine learning technique. Intelligent Decision Technologies, 16(1), 181–191.
23. Han, M., Zhang, J., Zeng, Y., Hao, F., & Ren, Y. (2022). A novel method of Chinese herbal medicine classification based on mutual learning. Mathematics, 10(9), 1557.
24. Zhang, H., Zhang, J., Ni, W., Jiang, Y., Liu, K., Sun, D., & Li, J. (2022). Transformer-and Generative Adversarial Network–Based Inpatient Traditional Chinese Medicine Prescription Recommendation: Development Study. JMIR Medical Informatics, 10(5), e35239.
25. Venaik, A., Kumari, R., Venaik, U., & Nayyar, A. (2022). The role of machine learning and artificial intelligence in clinical decisions and the herbal formulations against covid-19. International Journal of Reliable and Quality E-Healthcare (IJRQEH), 11(1), 1–17.
26. Dong, X., Zheng, Y., Shu, Z., Chang, K., Yan, D., Xia, J., ... & Zhou, X. (2021, December). TCMPR: TCM Prescription recommendation based on subnetwork term mapping and deep learning. In 2021 IEEE International Conference on Bioinformatics and Biomedicine (BIBM) (pp. 3776–3783). IEEE.
27. Roopashree, S., & Anitha, J. (2021). Deep Herb: A vision-based system for medicinal plants using exception features. Ieee Access, 9, 135927–135941.
28. Zhou, W., Yang, K., Zeng, J., Lai, X., Wang, X., Ji, C., ... & Li, S. (2021). Ford Net: recommending traditional Chinese medicine formula via deep neural network integrating phenotype and molecule. Pharmacological research, 173, 105752.
29. Sunil Gupta; Rakesh Saxena; Ankit Bansal; Kamal Saluja; Amit Vajpayee; Shikha, A case study on the classification of brain tumour by deep learning using convolutional neural network, AIP Conference Proceedings 2782, 020027 (2023) doi: https://doi.org/10.1063/5.0154417.
30. Rakesh Kumar Saxena, N. Sumitra, Brain Tumour Classification Using Back Propagation Neural Network, International Journal of Image, Graphics and Signal Processing, Volume 5(2), Year 2013, Pages 45–50.

Intelligent Systems Using Semiconductors for Robotics and IoT – Dinesh Goyal et al. (eds)
© 2026 Taylor & Francis Group, London, ISBN 978-1-041-20408-4

43 A Comparative Analysis of Data Extraction from the Dark and Regular Webs.

Ayush Jangid[1],
Vivek Saxena[2], Neha Srivastava[3]
Poornima Institute of Engineering and Technology,
Jaipur, India

Abstract: This paper presents a near examination of information extraction methods from both the ordinary (surface) web and the dull web, underscoring the unmistakable difficulties related with each. Conventional web extraction strategies, for example, DOM parsing and XPath questions, work successfully on the organized and semi-organized information of the surface web. Conversely, the dim web's dynamic and unstructured nature requires progressed strategies like Vision-Based Page Division and Regular Language Handling (NLP) models, including BERT, for significant information extraction. The review investigates the remarkable difficulties presented by every climate, like the powerful idea of the dull web, the utilization of anonymization strategies, and the fluctuation in web structures. It features the developing significance of using complex AI and vision-based strategies to separate and examine basic data, especially for network protection danger location and knowledge gathering. The paper reasons that a half breed approach consolidating vision-based methods and NLP is fundamental for really extricating significant bits of knowledge from both the surface and dull networks.

Keywords: (NLP), DOM parsing, VIPS

1. Introduction

Web 2.0 is contemporary advanced source containing persistently refreshing and crude snippets of data gainful for the legislatures, organizations, and analysts. As how much accessible web information has been expanding at a continually higher rate, reformists strategy for web information mining have hence been made to remove data from site pages and, thusly, from the surface Web and profundity concealed in difficult to-arrive at parts of the surface Web. Web scratching or creeping, which is characterized as information extraction basically, is the most common way of utilizing a device to lead web information extraction. At present, web information extraction is certainly not a uniform interaction; a few troubles are connected with the ordinary web likewise called the surface web while others are novel to the profound web.

The WWW which can be explored with an ordinary program and famous web indexes incorporates the public WWW, or at least, sites and data sets which contain organized and semi-organized information much of the time. Alternately, the other piece of the profound web - the dull web - is a lot more modest and just open by means of the utilization of explicit programming, most regularly The Onion Switch (Peak). And keeping in mind that it stays a hot-bed of crimes, it is seemingly the most fundamental wellspring of data for online protection dangers. As a result of the unscripted and obscure nature of the data held inside the Dull Web as well as its quirk towards secrecy, there are sure provokes that are exceptional to this field of study with regards to information scratching when contrasted with the Surface Web. The past methodology of web information extraction like covering enlistment and DOM parsing couldn't turn out productively for the dim web in view of the dynamic and dissipated nature of its site structure. A portion of the state workmanship strategies that can be utilized for extricating significant information from both the spaces incorporate Normal Language Handling

[1]2021pietcsayush034@poornima.org, [2]vivek.saxena@poornima.org, [3]neha.srivastava@poornima.org

DOI: 10.1201/9781003716389-43

Regular Language Handling like BERT (Bidirectional Encoder Portrayals from Transformers), and Vision-based procedure like celebrities (Vision based Page Division). The target of this paper is to embody and dissect the different ways to deal with information extraction from the normal and dull web, the hardships, strategies, and aftereffects of every technique.

Consequently, this work means to satisfy the rising need for compelling and adaptable techniques in the field of WDE, alluding to online protection and information mining, specifically. It does as such, nonetheless, simultaneously, all the really featuring the results with respect to knowledge obtaining and danger recognizable proof during the time spent getting relevant data from the dull web [1].

2. Literature Review

2.1 Data Extraction Procedures on the Customary Web

The standard web or the surface web contains a ton of typical organized and the semi organized information. Here, most conventional information extraction strategies actually work by depending on appropriately organized HTML labels and additionally metadata and hence data can without much of a stretch be removed utilizing methods, for example, DOM parsing, XPath questions as well as the covering enlistment strategy [5].

In web scratching, covering enlistment procedures have been a well-known decision and outfitted with frameworks like Roadrunner and Debye that gives robotized answer for remove information from semi-organized pages where the client give a few examples. For example, Roadrunner [5] uses a Zenith matching calculation to perceive structure and movement in HTML and hence is equipped for taking care of dynamic site designs easily. Similarly, the Vigi4Med Scrubber purposes slithering strategy in light of DOM structure for discussions and clinical substance just where the great consequences of DOM based techniques have been displayed for, yet, organizing of client produced content from sites.

The most recent refinements upgrade web extraction by including AI into these framework s. There are fresher techniques created as the information developed more tremendous and advanced; Strategies including however not restricted to semi-managed learning and unaided learning. For instance, the frameworks known as Chicken Foot and Harvester are thought of as semi-managed, which empower the client to indicate target designs for extraction utilizing restricted models [6].

2.2 Data Extraction Procedures on the Dark Web

The profound web is contained by the deep web, and this piece of the Internet is exceptionally not at all like the surface Web with respect to both availability and ordering. Many destinations on the dull web are scrambled and anonymized through conventions like the Pinnacle (The Onion Switch) net, which can't be listed by typical web search offices and must be gotten to with unique programming. The substance is additionally, in this manner, subjectively rather low, and the discussions and markets regions are the significant information sources.

The genuine extraction challenge on the dim web needs to represent a few contemplations. Pages in obscurity web are for the most part unique and have unpredictable HTML structure, and, in this manner, covering acceptance and traditional DOM examination techniques are not great. Subsequently strategies like Vision-Based Page Division (celebrities) have been ended up being more valuable. Through block-level substance division and page design, celebrities' decays dim website pages into extractable portions and uses visual elements to take care of the resulting model. This strategy empowers Him to comprehend the semantics of the page better - a situation, by which the main premise of the evaluation would be the HTML structure just is ridiculous.

Examining messages from the requirement of dim web gatherings has gotten a lot of concentration in the field of network safety danger knowledge. Research works have shown the presentation of Named Element Acknowledgment (NER) with better models of normal language handling like BERT. The phenomenal abilities of BERT to identify the context-oriented highlights of words in sentences has been utilized to remove utilitarian elements as programmer type of address, devices, associations and weaknesses. This is significant in characterizing dangers and planning the humanistic design of the dim web organizations [10].

Thusly, information gathering in obscurity web generally contains both slithering and scratching with the use of refined nonparametric learning calculations for arranging huge number of texts. Accordingly, it works with instruments, for example, Elasticsearch where dull web information will be loaded and recorded for additional examination. These models are then prepared to recognize online protection focused elements, empowering the analysts to channel, and classify danger insight information from loud and unstructured sources [6].

2.3 Comparative Examination

The essential qualification of scratching on the ordinary and the dim web is the distinction in the information association. On the standard WWW, information is in many cases semi-organized, and apparatuses can exploit explicit and clear-cut HTML or XML labels. In any case, the dull web is unstructured and exceptionally unique especially in view of spotlight on secrecy and content changes; this makes extraction more perplexing and subsequently a requirement for more brilliant extraction methodologies.

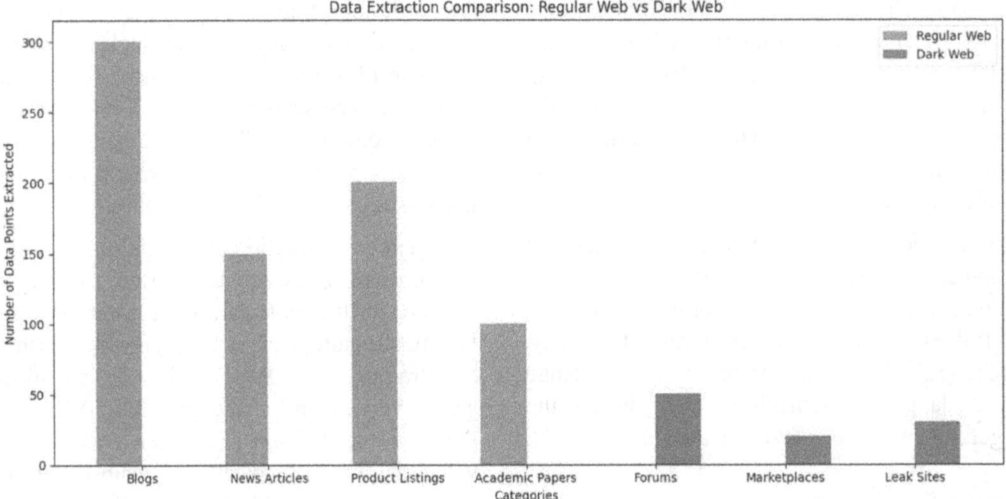

Fig. 43.1 Data extraction comparison

While utilizing demonstrated procedures, for example, DOM parsing for the normal web, the dim web requires an expansion of NLP models and vision-based strategies to extricate information. This distinction in information portrayals and issues focuses towards the officeholder challenges and the way that there is a need to have proper devices and techniques managing the web layer being referred to.

3. Methodology

Concerning the extraction of data from both the conventional and the dim web there is a suggestion to utilize both different web information extraction methods and normal language handling models. These strategies are explicitly intended to upgrade the precision, versatility and functional proficiency around information extraction, especially from the less-organized, and changing information sources that embody numerous business conditions. This segment frames a few primary methodologies that are to be utilized during the extraction and how these methodologies can be changed to address intricacies of both web layers.

3.1 Vision-Based Web Extraction Procedures

Maybe the most proficient systems for tending to the dynamism and graphical planeness of both the norm and the profound WWW are Vision-Based Page Division (celebrities). Celebrities is a web extraction procedure that arrangements with the vision of the Site page as the graphical portrayal of the Internet content as opposed to the simple HTML perspective on the page. In this manner, as per page structure, celebrities can take apart website page into semantically significant blocks. [5]

This makes it especially relevant while scratching sites for certain types of mutilated or dynamic page structures actually present in typically open web, as well as the dim web. For instance, an Over powered can be HTML, and a discussion can be in an alternate construction, more straightforward than a HTML page, yet celebrities can be utilized to fragment the page into headings/section/records while the Over powered can be utilized to portion the region of the page containing the particular data being looked for. Celebrities works with the formation of a Visual Block Tree (VBT) that addresses the association of the Page, as well similarly as with parting the Website page into the visual blocks depending on the signs of content division, including white tone, changes in the text dimension, and the position once distinguished, each block can be handled independently to separate information with great exactness, even from clouded, confounding sites.

3.2 Natural Language Handling for Web Information

Since most information separated from both the standard and the dim networks are unstructured literary substance, Regular Language Handling (NLP) is a fundamental part that transforms crude text information into experiences. Fresher advancements in the utilization of transformers like the Use of BERT (Bidirectional Encoder Portrayals from Transformers) has hugely worked on the adequacy of understanding implications covered in text information. Due to bidirectional nature of BERT, it surmises the setting of single word, depending on different words close to it in the text, to that end it functions admirably in distinguishing proof of named substances, including individuals, associations, devices, and so forth, during their conversation.

At the point when it is worried to posts showing up in the bootleg market discussions where discussions might incorporate references to programmer personas, apparatuses, dangers, and stunts, too as break talks, one can utilize Named Element Acknowledgment (NER) models prepared with the space information for distinguishing online protection related substances. Thusly, while applying BERT for NER, one is competent to solo distinguish terms like programmer names, zero-day exploits, or even malware reference from the substance of the dim web. In addition, context-oriented embeddings made by BERT are equipped for recognizing which importance ought to be credited to a term that has a few potential implications, for instance, 'bank' as a monetary establishment or the side of a waterway - which is particularly valuable while working with loud or, suppose, shoptalk containing dull web messages [10].

3.3 Tools and Methods for Significant Insight

Subsequently, an expansive strategy for information extraction ought to incorporate the utilization of a few instruments and procedures for the handling of separated information. The accompanying devices and methods are proposed for guaranteeing effective and adaptable extraction:

- **Web Creeping Systems:** For web slithering performing Scrapy and Selenium as a crucial device for working with the customary and profound web is fundamental. Static web scratching is better achieved with Scrapy, while dynamic web scratching with Selenium. In the event of managing dull web information, it is expected to utilize crawlers which can work in Peak and arrive at the really covered up commercial centers and gatherings [9].
- **Huge Information Handling:** Taking into account the way that information happens in the enormous volumes, particularly gathered from the dim web sources, a portion of the great speed large information structures including Apache Hadoop or Apache Flash might be suggested for overseeing and handling of the information. These structures can be carried out with AI to continuously be parsing the information for dangers, or recent fads.
- **BERT for Named Substance Acknowledgment (NER):** As previously mentioned, BERT helps in the NER cycle, where programmers, programmers' gatherings, exploits, devices or other digital danger not entirely settled. One more significant benefit of utilizing BERT is that the tokens in BERT's pre-prepared model might be refined with area explicit datasets, for example, DNRTI (Dataset for Named Element Acknowledgment in Danger Knowledge), which would upgrade the weakness of BERT in distinguishing related terms in network safety danger insight [10].

3.4 Combining Vision-Based and NLP Approaches

For normal as well as the dull web, the most reasonable way to deal with parsing the information is a mix of vision-based investigation and NLP. Be that as it may, celebrities can assist with partitioning a page into nonstop regions having a place with a similar fragment, which can then be utilized for semantic examination with an assistance of NLP devices like BERT. This way it ensures that on one hand there is a legitimate comprehension of the site page structure and then again, a collaboration with the text inside each fragment would be completely handled.

4. Comparative Examination

In this segment, an endeavor is made to differentiate the advantages and disadvantages of scratching methods as to their viability in the ordinary and the dull Web and the general presentation. There are difficulties and advantages of involving the two conditions with regards to pulling organized and unstructured information. These distinctions must be seen plainly to figure out which strategies and devices are reasonable for extraction of valuable insight.

4.1 Structured versus Unstructured Information

Coming up next is an examination between the two: This is as far as the kind of information separated from these two universes, the ordinary web and the dim web. The standard web, in any case, will in general have all the more a distinct HTML + META, APIs + CSS structure which can be a lot simpler to slither utilizing St Nick Claus procedures like DOM, XPath. Sites that host tables as a table, structure, or a data set are for the most part better organized and follow a normalized configuration and in this way devices like Scrapy or Beautiful Soup can make simple work of information extraction. [5]

While there are structures apparent in obscurity web, it remains generally unstructured conversely, with profound web. It is typically made by shoppers and can show up in discussions, destinations for selling labor and products, or discussion boards, significance there is no severe association. Most posts contain plain text, and strings or discussions are different, irregular, and scattered. This implies that information extraction in obscurity web is substantially more convoluted. To investigate such unstructured information, much use is made of NLP models, including BERT; and the actionable to be separated could allude to programmer names or instruments or the weaknesses found. Also, the understudies need to explore through the regularly encoded text containing references to the dull web and the use of various shoptalk terms, which muddles the extraction of information considerably more.

4.2 Challenges of Extricating Information from Regular versus Dark Web

Anyway, as should be visible with the standard web, there are issues like unique age of content with JavaScript and AJAX. Mining data from such pages is testing since content is stacked no concurrently and such page must be scratched utilizing devices like Selenium. Besides, to safeguard web content more sites today carry out different types of against scratching estimates, for example, the utilization of Manual human test, meeting ID logins, and IP limitations. This implies that it needs to continually scratch for it to stay associated — an interaction that requires IP address turn or tackled Manual human tests [10].

The issues of the dull web are essentially the lack of definition of the organization and disownment. Sites are in Pinnacle and URLs are dynamic and change every now and again consequently no durable association with the site is conceivable in slithering and scratching. Plus, most of the site contains unindexed discussions and secret commercial centers, the HTML of which has either an unclear construction or contains dark code. For that reason, utilizing approaches in view of the visual construction of a page, for example, Vision-based Page Division (celebrities) can be additional powerful for removing data from those conditions.

4.3 Efficiency and Exactness of Extraction Strategies

While more established techniques like DOM parsing and XPath are exceptionally compelling in scratching information in associations that keep up with static web-based stages. These strategies can assist with going after and track down specific snippets of data in the most limited conceivable time, on the off chance that the designated information is in tables or on the other hand assuming the utilization of HTML labels is required. Likewise, such devices as, for instance, APIs, normally permit one to acquire direct admittance to conveniently organized datasets while dealing with the standard web. However, assuming the site has dynamic substance, or utilizations against slithering devices, the speed diminishes, and such issues must be settled with the assistance of additional intricate projects like Selenium or Puppeteer.

Information extraction from the LM is less successful than from the WM in light of the fact that the design frequently needs association or relevance of additional refined strategies might be required. As per its makers, BERT and Named Element Acknowledgment (NER) are two significant NLP models since the text accessible on dull web discussions and visit logs is unstructured. The greater part of these models needs a ton of computational power to work and break down a lot of text information, which makes ongoing extraction somewhat troublesome. However, in the event that applied appropriately these models can help led to significant online protection knowledge including dangers that are beginning to arise, malware names and programmers' alliance [13].

4.4 Scalability of Extraction Procedures

There is contest in light of the size of the informational collections on the standard web and the dim web, as both fill in size. In the two cases, such means are expected to manage how much information which should be crept, removed, and examined.

To that end the Scrapy and Wonderful Soup suit for enormous sets extraction from various locales similar time in mix with dispersed handling structures, for example, Apache Flash. Programming interface reconciliation likewise includes the versatility factor since it returns enormous blocks of organized information that are more straightforward to parse contrasted with need to parse different pages. These devices can manage huge sources of info rather serenely, especially with respect to very much arranged material.

In obscurity web situation, adaptability is not so much direct but rather more strategic; first, due to the unpredictability of the Dim Net environment and, second, on the grounds that the utilization of apparatuses for unknown creeping inside the Pinnacle network is required. Kibana and Elasticsearch are significant for the handling and investigation of large measures of the dull web data whenever it is extricated. On the disadvantage, information preprocessing is finished utilizing well known simulated intelligence advances, for example, BERT which are computationally concentrated diminishing information adaptability. Connecting with the enormous volume unstructured information from a dim web gathering or a commercial center requires the utilization of large information systems like the Apache Hadoop or Apache Flash.

Aspect	Regular Web	Dark Web
Data Structure	Mostly structured/semi-structured	Highly unstructured
Extraction Tools	DOM Parsing, XPath, APIs	NLP (BERT, NER), VIPS
Challenges	Dynamic content, anti-scraping	Anonymity, fragmented URLs
Efficiency	High (structured data)	Lower (requires advanced methods)
Accuracy	High (for structured content)	Dependent on fine-tuned models
Scalability	Highly scalable	Challenging but manageable with big data tools

Made with Whimsical

Fig. 43.2 Comparison between regular and dark web

5. Conclusion

It has been the reason for this paper to do a similar evaluation of and for screen scratching normal and for dim web with careful craftsmanship to the test, procedure and result leaned to each. The primary sort of web is the ordinary web that contains primarily and semi fundamentally which makes it simple to creep utilizing instruments, for example, DOM parsing, Programming interface and

XPath. Notwithstanding, dynamic substance and against scratching advancements keep on presenting new issues that can't be tended to with basic projects, however with instruments like Selenium and secrecy. Then again, the circumstance with the dim web is inverse and really presents totally various issues. Because of namelessness, much of the time changing connection addresses, and the incredibly low degree of 'structure,' regular scratching approaches are impossible. As displayed in the paper, techniques, for example, Vision Based Page Division (celebrities) as well as Regular Language Handling (NLP), specifically BERT model, addresses a more productive and precise procedure of removing significant data from the dull web. Specifically, these models have been effectively applied for the recognition of network protection dangers, programmers and new malware, even in the loud and rather dark setting of the dim web. Web isn't static and hence the strategies of information extraction keeps on changing too. The accessibility of the dim web and the refinement of ordinary web content which should be separated build up the need to embrace even complex extraction techniques. It demonstrates that utilizing cutting edge AI and vision-based methods, it becomes conceivable to foster frameworks that can extricate as well as break down information and insight from both normal and the dark web. These frameworks won't just help the area of network safety, the business knowledge industry, and policing will likewise assist with recognizing and break down the less clear and apparent layers of the Web.

References

1. Fang, Y., Xie, X., Zhang, X. et al. STEM: a suffix tree-based method for web data records extraction. Knowl Inf Syst 55, 305–331 (2018).
2. Su W, Wang J, Wang J, Lochovsky FH, Liu Y (2012) Combing tag and value similarity for data extraction and alignment. IEEE Trans Knowl Data Eng 24(7):1186–1200.
3. Sergio Flesca, Elio Masciari, and Andrea TagareIli," A Fuzzy Logic Approach to Wrapping PDF Documents", IEEE Transactions On Knowledge And Data Engineering, VOL. 23, NO. 12, DECEMBER 2011.
4. JerLangHong," Data Extraction for Deep Web Using WordNet", IEEE Transactions On Systems, Man, And Cybernetics-Part C: Applications And Reviews, VOL. 41, NO. 6, NOVEMBER 201 I.
5. V.Crescenzi, G. Mecca, and P. Merialdo, "RoadRunner: Towards Automatic Data Extraction from Large Web Sites," Proc. the 26th Int'l Conf. Very Large Database Systems (VLDB), pp. 109–118, 2001.
6. Crescenzi, V., Merialdo, P., & Qiu, D., (2013). A Framework for Learning Web Wrappers from the Crowd. WWW'13 Proceedings of the 22nd international conference on World Wide Web, pp. 261 272.
7. Rekha Jain, Department of Computer Science, Apaji Institute, Banasthali University C-62 Sarojini Marg, C-Scheme, Jaipur, Rajasthan. Dr. G. N. Purohit, Department of Computer Science Apaji Institute, Banasthali University." Page Ranking Algorithms for Web Mining", International Journal o/Computer Applications (0975–8887) Volume 13-No.5, JanualY 20 II.
8. C.-H. Chang, M. Kayed, M.R. Girgis, and K.F. Shaalan, "A Survey of Web Information Extraction Systems," IEEE Trans. Knowledge and Data Eng., vol. 18, no. 10, pp. 1411–1428, Oct. 2006.
9. P. Evangelatos et al., "Named Entity Recognition in Cyber Threat Intelligence Using Transformer-based Models," 2021 IEEE International Conference on Cyber Security and Resilience (CSR), pp. 348–353.
10. Devlin, Jacob, et al. "Bert: Pre-training of deep bidirectional transformers for language understanding." arXiv preprint arXiv:1810.04805 (2018).
11. Saxena, V., Bhattacharjee, S., & Saxena, D. (2023, November). A Lightweight Security Technique for Processing Large Data Over a Distributed Cloud Environment. In Proceedings of the 5th International Conference on Information Management & Machine Intelligence.
12. Saxena, V., Singh, U. P., Kumari, B., & Khandelwal, A. (2023, November). Machine Learning Algorithms for Advanced Rainfall Prediction. In Proceedings of the 5th International Conference on Information Management & Machine Intelligence.
13. Raj, S., Singh, U. P., & Saxena, V. (2023, November). Efficient Multiband Planar Monopole Antenna based on Metamaterial Transmission Line for Mobile Wireless Communication Systems. In Proceedings of the 5th International Conference on Information Management & Machine Intelligence.
14. Singh, U. P., Kumari, B., Saxena, V., & Goyal, D. (2023, November). Enhancing the Security of Big Data based on the New AI & its Techniques. In Proceedings of the 5th International Conference on Information Management & Machine Intelligence.
15. Thapar, S., Sharma, S., Saxena, V., & Singh, U. P. (2023, November). The future impact of smart vehicles using data science. In Proceedings of the 5th International Conference on Information Management & Machine Intelligence.
16. Anigol,M.N.B., Anil, S.P. (2015). Study of the effect of various fillers on mechanical properties of carbon-epoxy composites. Int. Res. J. Eng. Technol. 02(03), 798–802.
17. Biswasa, S., Shahinura, S., Hasana,M., Ahsan, Q. (2015). Physical, mechanical and thermal properties of jute and bamboo fibre reinforced unidirectional epoxy composites. Procedia Eng. 105, 933–939.

Note: All the figures in this chapter were made by the authors.

Intelligent Systems Using Semiconductors for Robotics and IoT – Dinesh Goyal et al. (eds)
© 2026 Taylor & Francis Group, London, ISBN 978-1-041-20408-4

An Analytical Study of "Enhancing Public Safety and Police Responsiveness Through Geo-Tag Guardian Systems"

Vivek Saxena[1]
Poornima Institute of Engineering and Technology, Jaipur, India

Deepika Saxena[2]
Poornima University, Jaipur, India

Abstract: Geo-Label concierge framework presents an ideal drive pointed toward upgrading public security and police responsiveness by quickly carrying out cutting edge innovative arrangements. Through the Geo-Labelling of exclusive cameras, this paper tries to lay out a cooperative structure between confidential residents and policing, state of the art revolutions like Program Programming interface and AI calculations for deterrent discovery. The examination procedure includes staged execution over in this exploration, including the upgrading of a smoothed-out UI, combination with following administrations, and the sending of a speedy AI calculation. Expected results incorporate sped up episode reaction, advanced situational mindfulness for policing, diminished manual responsibility through robotized information the board. This exploration importance lies in its powered development as well as in its capacity to address prompt public security worries inside a combined time period. By utilizing Python, JavaScript/TypeScript, Cup, and OpenCV, Geo-Label Watchman framework means to encourage a more secure and more responsive local area. The paper accentuates the direness of sending trend setting innovations to meet the squeezing needs of current culture, featuring the potential for upgraded joint effort among residents and policing in guaranteeing local area security. Moreover, the examination highlights the significance of proactive measures in wrongdoing counteraction and the job of innovation in working with quick and viable reactions to arising security challenges.

Keywords: Geo-Tag guardian, Browser API, Python, JavaScript/TypeScript, Flask, and OpenCV

1. Introduction

In a period portrayed by fast motorized progressions and developing security challenges, assuring public wellbeing stays a vital worry for networks around the world. With each improvement the scene of local area wellbeing. Geo-Label Guard addresses an essential change by the way we approach public security, rising above usual limits to manner exceptional cooperation between confidential camera administrators and policing. At its center, the task is spending day, creative measures become progressively basic intending to the unique idea of current hazards. Perceiving the basic job that innovation plays in wrongdoing counteraction and mediation, the Geo-Label Porter research work arises as a spearheading drive ready to driven by a particular vision: to lay out a cooperative biological system wherein exclusive cameras act as modest edges in the fight against crime. Through the essential execution of state-of-the- art advancements and the development of powerful correspondence channels, Geo-Label Watchman attempts to engage networks with the instruments and assets expected to shield their areas. The meaning of Geo-Label Gatekeeper reaches out a long ways past the domain of innovation; it epitomizes a change in perspective in our aggregate way to deal with local area security. By cultivating cooperative energy

[1]vivek.saxena@poornima.org, [2]deepika.saxena@poornima.edu.in

DOI: 10.1201/9781003716389-44

between dissimilar partners and outfitting the force of development, the exploration looks to make a more secure, stronger society — one where proactive measure is the standard, instead of the exemption. As we set out on this extraordinary excursion, we are helped to remember the significant effect that cooperative endeavors can have in molding the fate of public security. Basically, Geo-Label Watchman is something beyond an examination; it is a demonstration of the unyielding soul of progress and aggregate activity. It is an encouraging sign in an undeniably complicated world, flagging another period of participation and strengthening. Together, let us set out on this excursion towards a more secure tomorrow, where networks are invigorated by the strength of their solidarity and the force of their developments [1].

2. Unprejudiced of the Exploration

The Geo-Label Gatekeeper research presents aggressive, yet feasible targets pointed toward upsetting the scene of public security:

A. **Implement a Fast Geo-Labeling Framework:** The essential goal of Geo-Label Gatekeeper is to create and convey a cutting edge Geo-Labeling framework prepared to do quickly and precisely following exclusive cameras. By bridling the most recent headways in innovation, including Program Programming interface, the venture expects to lay out a vigorous structure for area following, in this way enabling policing with continuous experiences into camera areas

B. **Productive Mix with Following Administrations:** A key center region of the task is to flawlessly incorporate the Geo- Labeling framework with following administrations, working with the quick transmission of directions to the server for police access and perception. By laying out consistent correspondence channels, the undertaking aims to enhance law enforcement's ability to respond promptly to incidents and emergencies.

C. **Streamlined User Interface Development:** Geo-Label Guard tries to make a smoothed out and instinctive UI that works with the fast space and recovery of proprietor subtleties and camera details. By focusing on client experience and connection point strategy, the task intends to encourage powerful joint effort between confidential camera proprietors and policing, in this manner improving the general proficiency of the framework.

D. **Integration of Machine Learning Algorithms:** A key mechanical headway imagined by Geotag Concierge is the mix of high-speed AI calculations for obstruction identification in camera film. By attempting the force of AI, the undertaking intends to improve situational mindfulness and furnish policing with timestamped criticism for powerful independent direction.

E. **Continuous Performance Optimization:** Geo-Label Porter perceives the significance of persistent execution enhancement to assurance the framework stays responsive and reliable in true situations. Through thorough testing, checking, and iterative enhancements, the venture plans to distinguish and address any expected bottlenecks or failures, accordingly boosting the framework's viability in supporting policing.

F. **Community Engagement and Outreach:** Not withstanding specialized goals, Geo- Label Guard puts major areas of strength for an on local area commitment and effort. By directing effort projects, studios, and instructive drives, the task expects to bring issues to light about the significance of public security and the job of innovation in crime counteraction. Through cooperative endeavours with local area partners, Geo- Label Guard looks to construct trust and cultivate a feeling of pride among local area individuals, in this way making a stronger and engaged society [2].

3. Problem Statement

The Geo-Label Porter research paper tends to a basic hole in the impetus scene of public security the absence of an organized and productive framework for getting to exclusive camera film. In numerous networks, exclusive cameras act as significant devices for wrongdoing avoidance and examination. Nonetheless, the divided idea of existing frameworks aggravates policing to tackle the maximum capacity of these assets. Presently, getting to exclusive camera film frequently includes bulky and tedious cycles, incorporating manual coordination with individual camera proprietors or specialist co-ops. This absence of reconciliation and normalization defers policing to occurrences as well as limits their capacity to accumulate convenient and noteworthy knowledge. Additionally, the different idea of existing frameworks presents huge difficulties as far as information interoperability and data sharing. Policing may experience poverties in getting to camera film from various sources, prompting holes in inclusion and preventing the analytical cycle. Model

A. **Contextual investigation:** Consider a situation where a private area encounters a progression of break-ins and defacement occurrences. While a few homes in the space are outfitted with exclusive reconnaissance cameras, policing difficulties in getting to the recording promptly. Without a concentrated framework for following camera areas and getting to film, cops should actually contact individual stuff holders or specialist co-ops to demand admittance to important accounts. This cycle consumes significant time as well as depends dynamically on the participation of mortgage holders, which might shift from one case to another.

Interestingly, with the execution of the Geo- Label Guard framework, policing would approach a brought together stage where exclusive cameras are geo-labelled and flawlessly coordinated with following administrations. In case of an occurrence, officials could rapidly distinguish close by cameras with important film and solicitation access through the framework[3].

4. Workflow

Camera Footage Acquisition: The Geo-Tag Guardian system obtains camera footage to begin the process. Private cameras take pictures of their surroundings and send the footage to the host computer for processing.

Data Compilation and Geo-Tagging: The application on the host computer compiles the essential data, such as camera and owner information and geo-location data, upon receiving the footage. For exactly geo tagging each camera's location within the system, this information is essential.

Transmission to the Server: The footage and geo-location data, in addition to the compiled data, are sent to the server for more processing and analysis. The server is the essential hub, conduct incoming data packets and coordinating system operations.

Server-Side Processing and Alert Mechanism: When it receives the data packets, the server inductee's server-side processing algorithms to examine the footage and identify any individualities. An alert mechanism is activated by the algorithm, notifying the camera owner and system administrator in the event of camera displacement or malfunction.

Integration with Navigation Services: The host computer concurrently stores and integrates the camera's geo-location coordinates with mapping services like Google Maps for navigation. Since of this integration, law enforcement personnel can now see where privately-owned cameras are in real time, which reduces incident response times.

Object Detection and Machine Learning Integration: On the other hand, the Geo-Tag Guardian system finds interesting objects in the camera footage by utilizing machine learning algorithms. These algorithms, for instance, are trained to distinguish particular objects that are applicable to various use cases in order to perceive intruders, vehicles, or suspicious packages. Upon instruction, the ML model inspects the footage and provides information to enhance situational awareness and support law enforcement operations. By following this structured workflow, the Geo-Tag Guardian system confirms that privately owned cameras, the host computer, server infrastructure, and machine learning algorithms work seamlessly together. This simplified approach not only improves the efficiency of public safety operations, but it also enables law implementation agencies to better protect and serve their communities by providing them with actionable intelligence[4,5].

5. Technologies Being Used

The Geo-Tag Guardian project usages a wide range of technologies and frameworks to achieve its objectives: For front-end development, JavaScript and TypeScript are programming languages, while Python is used for back-end development. Frameworks: Flask for the backend and Node.js for the server-side, with a dynamic and responsive JavaScript/TypeScript frontend. OpenCV is a set of tools for processer vision tasks that makes impediment detection in camera footage easier. MySQL for structured data storage and MongoDB for geotagging data storage that is adaptable and scalable[6,7]

6. Experiment Work

6.1 Used Necessary Libraries to Develop the System

- **PyJWT** JSON Web Tokens (JWT) are encoded and decoded by a Python library. It permits for the creation

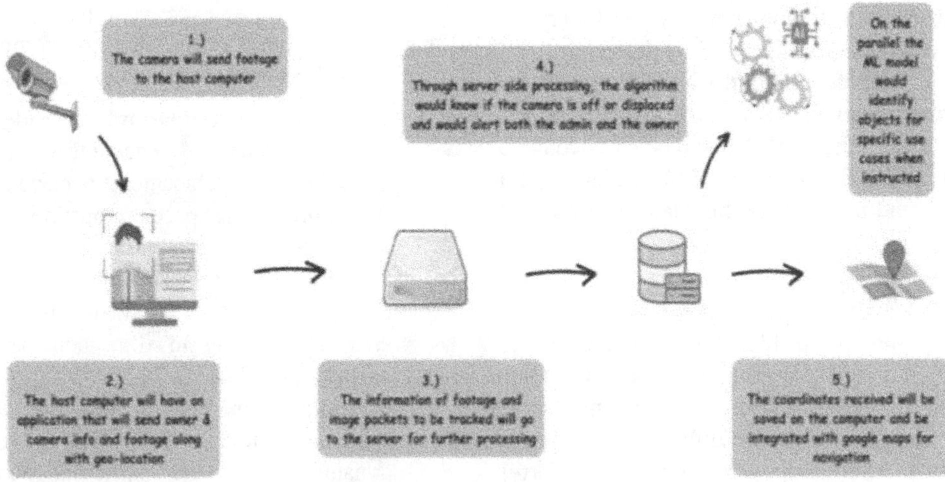

Fig. 44.1 Working process

of JWTs with particular claims, the validation of JWT signatures, and the abstraction of data from JWT payloads.)

- **PyWebview** (Pywebview is a Python library that allows you to create GUI applications with web technologies It provides a simple and lightweight solution for creating desktop applications that can run on various platforms without the need for additional dependencies or installations.)
- **Opencv2** (OpenCV (Open Source Computer Vision Library) is a popular open-source computer vision and machine learning software library).
- **YOLO** by Ultralytics (YOLO (You Only Look Once) In the field of computer vision, the object detection algorithm YOLO (You Only Look Once) has grew significant popularity.
- **FFmpeg** FFmpeg is a powerful and widely- used open-source software group for handling multimedia data. It includes a collection of libraries and command-line tools for amending multimedia files like audio, video, and others. We can edit audio and video files, transcode them, stream them, and convert between different multimedia formats using FFmpeg.
- **Flask** (Flask is a Python web framework that is light and malleable. It is made to make it simple to quickly create web applications with little boilerplate code).

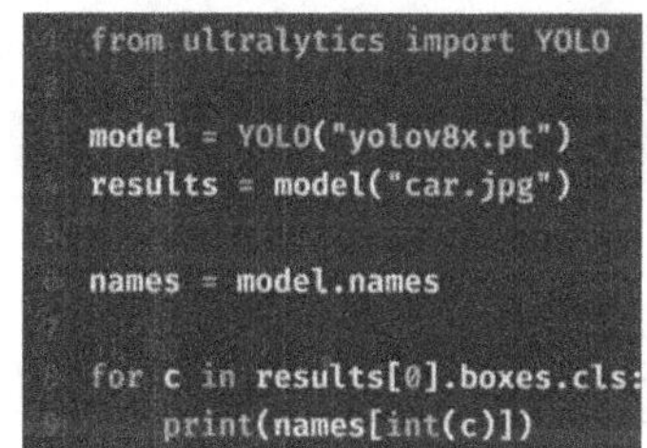

Fig. 44.2 Screen shot of program

6.2 Working Process

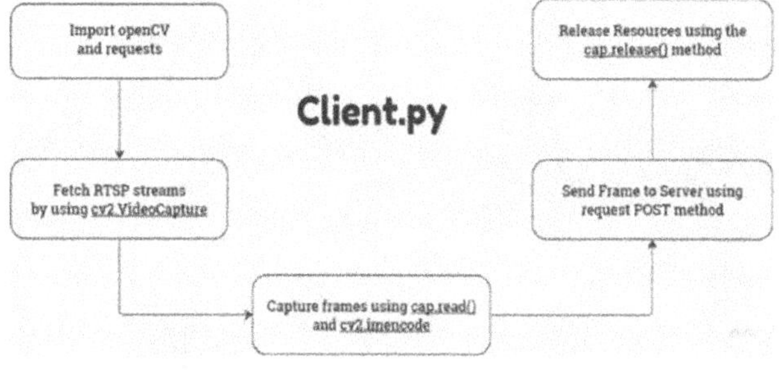

Fig. 44.3 Client.py workflow

The Python programming language begins with the beginning of the program. It then, at that point, imports libraries, for example, OpenCV for video catching and demands for HTTP for sending POST demands. OpenCV is a popular open-source software library for computer vision and machine learning that offers a variety of tools and algorithms for processing images and videos. The requests library makes it easier to integrate web functionality into Python applications by making sending HTTP requests and handling responses simpler[. For establishing channels of communication between system components or the system and external resources, URLs must be defined. The RTSP stream URL (rtsp_url) and the server URL (server_url) are two distinct URLs. The Real-Time Streaming Protocol, or RTSP, is used to establish and manage media sessions between endpoints. The server URL indicates the endpoint where the application will send or get information connected with video transfer handling. To open a RTSP stream, utilize the cv2. The OpenCV library provides the VideoCapture function. This adaptable instrument can record video from webcams, image sequences, video files, and other sources[8,9,10].

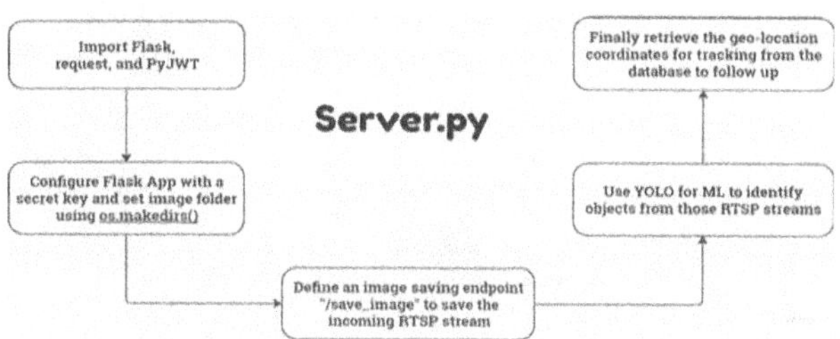

Fig. 44.4 Server.py workflow

The program begins by importing Python's required libraries. Flask is a lightweight and extensible web framework that makes it easy to quickly and without using a lot of boilerplate code to build web applications. The requests module is a well-liked HTTP library that makes handling responses and HTTP requests simpler. Flask's Jsonify function ensures effective communication between server and client components by serializing Python data structures into the JSON format. Sqlite3 is an inherent Python library for communicating with SQLite data sets, giving a helpful point of interaction to executing SQL questions and overseeing data set associations. JWT, or JSON Web Token, is a minimized, URL-safe method for addressing cases to be moved between two gatherings. The bcrypt library gives capabilities to hashing passwords utilizing the bcrypt hashing calculation, guaranteeing secure capacity and insurance against secret word- based assaults. The operating system module is used to associate with the working framework, and the date time module is important for working with dates and times. These libraries give Python programmers a steady way to build climbable and long-lasting web applications [11,12].

Handle Errors: The application's fidelity and robustness depend on how errors are handled during image valid. Let's take a look at the steps that need to be occupied to contrivance error handling: Use error handling to catch any exclusions that may occur while saving images.

Developers can use a try-except block to catch any exclusions raised during the image saving process to handle errors. The code that is in charge of saving the image file is limited within the try block. The except block will catch any exceptions that occur during this process. Application developers can graciously handle errors without crashing the application by wrapping the image-saving code in a try-except block. If saving fails, return an error message. Inventers have the option of chronic an error message to the client to inform them that the operation was ineffective if an exception is discovered during the image-saving process. The error message can be returned as part of the HTTP response, usually with the appropriate status code (for instance, 500 for an internal server error). The client should be able to find suitable information about the nature of the error and any probable solutions in the error message[13,14].

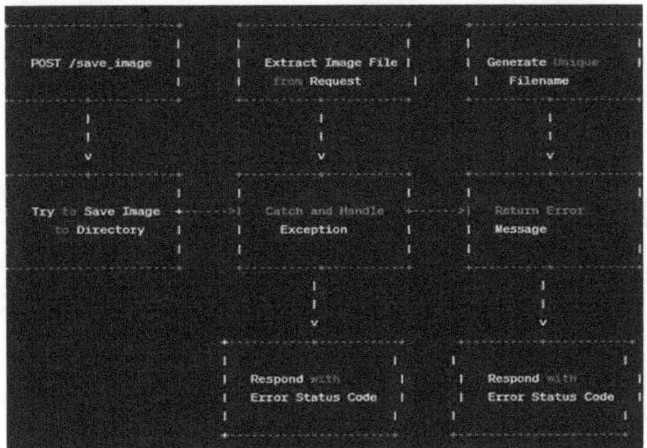

Fig. 44.5 Handling errors

```
import cv2
import requests
import time

# RTSP stream URL
rtsp_url = (
    "rtsp://rtspstream:8053766c150b8a8351fa6cc96ef45d67@zephyr.rtsp.stream/pattern"
)
# Server URL
server_url = "http://127.0.0.1:8080/save_image"

# Open the RTSP stream
cap = cv2.VideoCapture(rtsp_url)

while True:
    # Capture frame-by-frame
    ret, frame = cap.read()

    # Convert the frame to bytes
    _, img_encoded = cv2.imencode(".jpg", frame)
    img_bytes = img_encoded.tobytes()

    # Send the frame to the server
    try:
        response = requests.post(server_url, files={"image": img_bytes})
        print(response.text)
    except requests.RequestException as e:
        print(f"Error sending frame: {e}\n")
```

Fig. 44.6 Client.py which sends the streams to the server

For data unpacking, frame deserialization, video display, and network communication, the code imports cv2, socket, struct, and pickle libraries. It creates socket objects, accepts connections, receives and displays frames, exits the loop, and releases resources in addition to defining the server address and port [15,16,17].

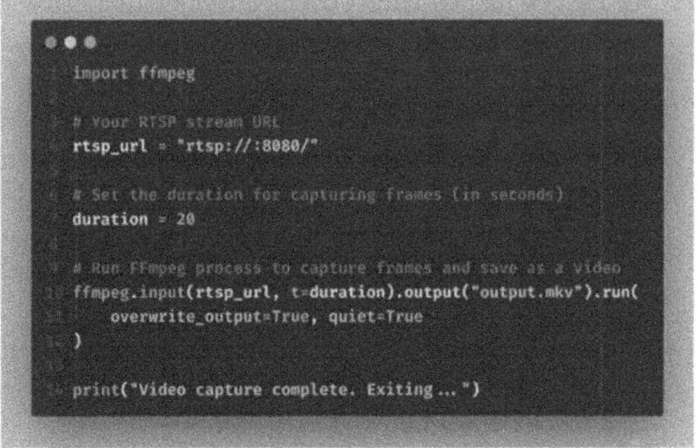

Fig. 44.7 Server.py interface

6.3 Code Snippets Frontend

Fig. 44.8 RTSP to video

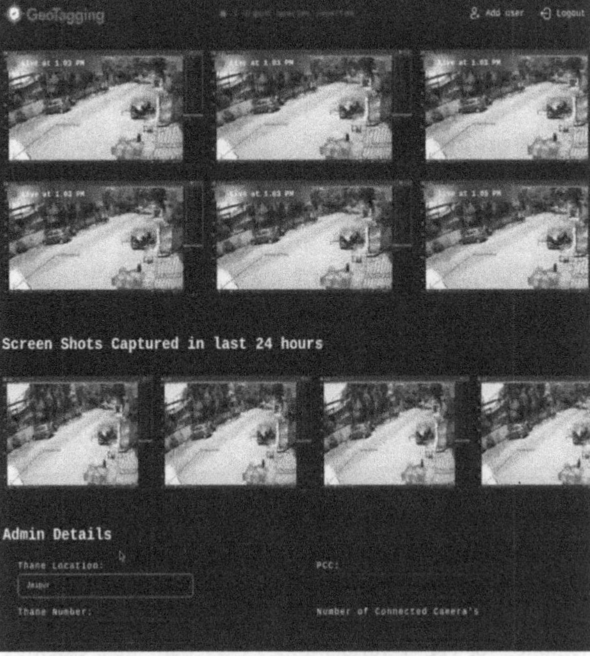

Fig. 44.9 Client.py interface

7. Conclusion

In conclusion, Geo-Tag Guardian is an innovative project that aims to use cutting- edge technology to progress public safety and law enforcement efficiency. The project makes it possible for camera owners and law implementation activities to work together in a collaborative ecosystem for real-time monitoring, study, and response coordination by rapidly geotagging privately owned cameras and integrating them with tracking services and machine learning algorithms. Applying a simplified geo-tagging system, creating user-friendly edges, and utilizing cutting-edge machine learning algorithms for obstacle finding in camera footage are some of the project's goals. A comprehensive strategy that includes geo-tagging implementation, integration with tracking services, client-server communication, owner information gathering, and object detection serve as the foundation for these endeavours. Rapid progresses in public safety, decreased manual workload as a result of automated information management, and shorter police response times are among the expected outcomes. The significance of the project extends beyond.

8. Future Scope

Improved mobile accessibility: In the not-too- distant future, we can investigate the possibility of integrating AI and advanced analytics technologies. 2. Integration with high-level analytics: In the future, we could look into including artificial intelligence (AI) technologies and high-level analytics into the system. Law enforcement agencies can improve their skill to identify and respond to security threats by uniting features like object detection, facial recognition, and anomaly detection. We can investigate cloud-based options for storing and analyzing CCTV footage, which makes it possible to compile footage in a single step and make efficient use of resources. Law enforcement activities can increase their investigative abilities and adapt to rising data volumes without making significant infrastructure investments by utilizing cloud infrastructure. Law enforcement agencies can learn about crime patterns, hotspots, and trends by overlaying observation data and camera locations on digital maps. 5. Enhance information sharing and cooperation: For crime prevention and investigation to be successful, law administration agencies must share and cooperate with one another. To make it easier for agencies and authorities to collaborate, we can investigate features like interoperability standards, secure data allocation protocols, and unified search capabilities. These systems can enhance assistance and coordination to address complex security issues by facilitating effective communication. We can increase awareness of emerging security threats and respond quickly by connecting CCTV cameras to IoT devices and sensors. Our CCTV surveillance programs can be developed into intelligent security solutions that enable law enforcement agencies to monitor and respond to security challenges in highly complex and dynamic environments by exploring the future and implementing the most recent technology and best practices.

References

1. Saxena, V., Saxena, D., & Singh, U. P. (2022, December). Security Enhancement using Image verification method to Secure Docker Containers. In Proceedings of the 4th International Conference on Information Management & Machine Intelligence (pp. 1–5).
2. Sharma, B., & Saxena, D. (2023). Design and Analysis of Energy Efficient Service Discovery Routing Protocol in MANETs. SN Computer Science, 4(5), 495.
3. Saxena, D., & Sharma, N. (2021). Analysis of Docker Performance in Cloud Environment. In Advances in Information Communication Technology and Computing: Proceedings of AICTC 2019 (pp. 9–18). Springer Singapore.
4. Saxena, D., & Sharma, N. (2021). Docker Security Analysis Mechanism with Linux Platform. In Rising Threats in Expert Applications and Solutions: Proceedings of FICR-TEAS 2020 (pp. 595–601). Springer Singapore.
5. Sharma, B., & Saxena, D. (2023). Examining SSRFLP in Mobile Ad-hoc Networks in Relation to Other Routing Protocols. Rivista Italiana di Filosofia Analitica Junior, 14(1), 19–24.
6. Saxena, D., Vishwakarma, A. K., & Verma, S. (2022). Public safety analysis and safety system using arduino. Journal of Discrete Mathematical Sciences and Cryptography, 25(4), 955–961.
7. https://pyjwt.readthedocs.io/en/latest/
8. https://pywebview.flowrl.com/blog/pywebv iew3.html
9. https://docs.ultralytics.com/quickstart/
10. https://medium.com/@tom.humph/saving- rtsp-camera-streams-with-ffmpeg- baab7e80d767
11. https://ffmpeg.org/documentation.html
12. https://pypi.org/project/rtsp/
13. https://docs.opencv.org/4.x/d1/dfb/intro.htm
14. https://flask.palletsprojects.com/en/3.0.x/quickstart/
15. https://medium.com/code-85/easily-build- your-first-web-app-with-pythons-flask- d12825f9f1d9
16. https://realpython.com/python- requests/#getting-started-with-pythons- requests-library
17. https://en.wikipedia.org/wiki/Real- Time_Streaming_ Protocol

Note: All the figures in this chapter were made by the authors.

Intelligent Systems Using Semiconductors for Robotics and IoT – Dinesh Goyal et al. (eds)
© 2026 Taylor & Francis Group, London, ISBN 978-1-041-20408-4

45 Design Secure Transaction using Blockchain Technology

Aakansha Mitawa*
Research Scholar,
Department of Computer Science Engineering,
VGU University Jaipur

Pawan Bhambu
Associate Professor,
Department of Computer Science Engineering,
VGU University, Jaipur

Abstract: This study proposes a secure, blockchain-based financial transaction system that integrates two novel techniques: BLAKE2b-512 hashing algorithm-based user identification and picture string-based authentication algorithm. In this process of user identification, there will be a unique hash generated for the BLAKE2b-512 of a user's identity information and an MD5 hash for the password, making sure that sensitive data pertaining to users is stored safely in the user database. The picture-based string authentication algorithm adds another layer of security by asking users to select images and predefined click areas that form dynamic patterns during authentication, compared against the patterns registered for access control. This is a dual approach toward strengthening financial transaction systems from unauthorized access, improving the user experience with personalized and secure authentication methods. Details of the implementation include security for data handling during user registration, secure transmission of sensitive information, and an interface user-friendly enough to handle all errors and perform visual verification.

Keywords: Blockchain, User identification, BLAKE2b-512, MD5 Hash, Picture-based authentication, Visual verification, Secure database, Cryptographic hash, Dynamic patterns, Financial security

1. Introduction

Security of financial transactions has come to the foreground in an era when cyber threats and fraud are maturing [1]. Traditional authentication methods, like passwords and PINs, have been rather simple in securing digital transactions. Such methods turn out to be rather poorly resistant to more complex assaults. With the ever-increasing complexity of the financial systems and online transactions, there is an increasing demand for developing more advanced and secure measures to safeguard vital information and ensure integrity at transactions [1]. The motivation of this research is based on addressing the vulnerabilities in conventional authentication systems and exploiting new technologies for increased security measures. Financial institutions and end-users are yearning for solutions that offer better protection from unauthorized access, fraud, and data breaches [2]. Blockchain refers to distributed and decentralized digital ledger technology, recording transactions through computers that are secure, transparent, and also not modifiable. Such transactions are collected together, known as a "block," and linked together one after the other to give a chronological "chain." [2]. Blockchain technology, in its very makeup of decentralization and immutability, has a promising ground for building more secure financial systems. The research has the purpose of integrating blockchain with innovative authentication methods, thereby giving an all-rounded solution to the contemporary security challenges [2]. This research is very crucial because it presents a totally

*Corresponding author: aakanshamitawa4694@gmail.com

DOI: 10.1201/9781003716389-45

new way of increasing the security level in financial transactions using blockchain and novel authentication methods [3]. In this regard, it provides a significant contribution through a multi-layer security framework that shall be developed to tap into the strengths of blockchain alongside user verification mechanisms. The decentralized, immutable ledger of Blockchain, with the addition of robust BLAKE2b-512 hashing for user identification and dynamic picture-based string authentication, spells a comprehensive solution toward almost eradicating the threats of illicit access and manipulation of data. At this point, traditional security measures based on passwords and two-factor authentication were seen to be insufficient in the face of ever-growing cyber threats [3].

The proposed framework addresses these limitations in order to introduce a more secure and resilient approach to the protection of sensitive financial information. Moreover, this picture-based authentication method not only has improved security but provides enhanced user experience with a more intuitive and engaging alternative to the text-based conventional authentication methods. This is of particular importance, as this would meet the growing demand for both robust security and user-friendly solutions.

2. Objectives of the Research

- Develop a Robust User Identification System: To design and implement a secure user identification process using the BLAKE2b-512 hashing algorithm for creating a unique and tamper-proof representation of user identity information. This aims to enhance the security of user identity management in financial transactions.
- Enhance Password Security: To incorporate MD5 hashing for user passwords, adding an additional layer of protection to ensure that password data is securely hashed and stored, reducing the risk of password-related breaches.
- Implement Picture-Based String Authentication: To develop a novel authentication method where users interact with visual elements (pictures) to create a dynamic and personalized authentication pattern. This approach aims to improve security and user engagement compared to traditional text-based authentication methods.
- Integrate Authentication Methods with Blockchain Technology: To integrate the user identification and authentication processes within a blockchain framework, leveraging blockchain's decentralization, immutability, and transparency to enhance overall security and integrity of financial transactions.
- Evaluate the Effectiveness and Usability of the Proposed System: To assess the performance, security, and user experience of the integrated system through experimental testing and analysis. This includes evaluating how well the system protects against unauthorized access and its ease of use for end-users.
- Contribute to Financial Security Innovations: To advance the field of financial security by providing a comprehensive solution that addresses existing vulnerabilities and explores new applications for blockchain technology in securing digital financial

3. Literature Review

Takahashi and Lakhani (2019) contribute to the analysis of the growing adoption of cryptocurrencies and related security issues in online transactions; they propose a multi-layered security analysis method tailored for cryptocurrency exchange services to enhance transaction safety [13]. Sai et al. (2019) test popular Android cryptocurrency applications against their security profiles, which turn out to be a little better in traditional financial services apps, though they are much better at user privacy despite their general limited functionality [14]. Azman and Sharma, 2020 discussed the potentials of Bitcoins to transform the global economy. Introduce HCH DEX, a secure cryptocurrency storage method where a smart card shall be used for two-way authentication in order to avoid fraudulent activities and malpractices for bringing transparency and fairness in the economy [15]. Karanjai et al. (2021) add to the literature on a conditional cryptocurrency system that protects privacy by encoding,

Table 45.1 Approach based comparison

Author Name (Year)	Concept Proposed	Technology	Limitations
Takahashi and Lakhani (2019)	Multi-layered security analysis for cryptocurrency exchanges	Multi-layered security analysis	Complexity of implementation
Sai et al. (2019)	Security and privacy analysis of Android cryptocurrency applications	Security profiling and control tests	Limited functionality in traditional financial services apps
Azman and Sharma (2020)	HCH DEX for secure cryptocurrency storage with smart card authentication	Smart card setup for two-way authentication	Potential usability challenges and cost of smart card deployment
Karanjai et al. (2021)	Conditional cryptocurrency system with privacy protection	UTXO-based system with event outcome encoding	Complexity of encoding and integration with existing systems
Perry (2022)	Description of cryptocurrency wallets as advanced flash drives	Secure storage of private keys	Risk of physical loss or damage to the wallet device

in a UTXO-based system, the outcome of events within a cryptocurrency notes, thereby bringing an innovation into the practice of cryptocurrency transactions [16]. Perry (2022) defined cryptocurrency wallets as 'super flash drives' which securely stored private keys for accessing and processing cryptocurrency data on the blockchain [17].

4. Research Methodology

The research methodology that will be used for this work embodies a multilevel approach for developing, implementing, and evaluating the security framework proposed for financial transactions. The methodology is divided into four distinct phases: design, implementation, integration, and evaluation.

This first phase identifies the conceptual design of user identification and authentication. The BLAKE2b-512 hashing algorithm will be used in the user identification system to securely hash a piece of information concerning the identity of a user into a fixed length, such as his username or e-mail. Regarding the password of each user, it will be MD5 hashed to create a 128-bit hash value. Develop the picture-based string authentication algorithm, where users will interact with visual elements and create a dynamic pattern during authentication. This stage will help to specify the exact coordinates within the chosen images where the user shall click, making the interaction secure yet user-friendly.

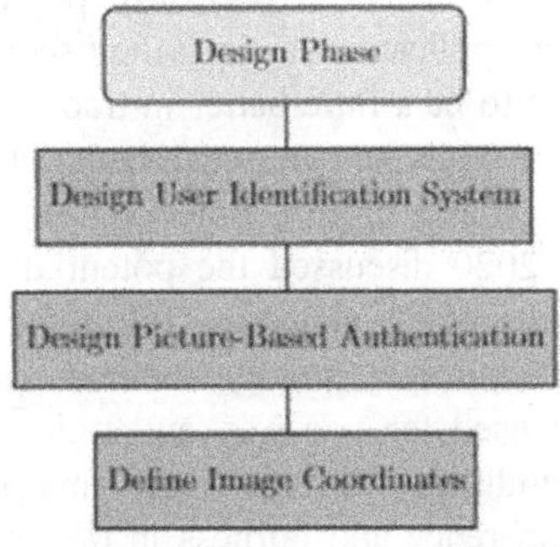

Fig. 45.1 Design phase

The implemented phase involves translating designed algorithms and processes into functional code. The BLAKE2b-512 and MD5 hashing algorithms for the implementation of user identity and password hashing, respectively. A database with security features is established to store the hashed values; at no point in time is any sensitive information stored in plain text. Afterwards, a picture-based string authentication system is developed where users are given the ability to select images and predefined click areas within those images for both registration and authentication. This phase is accompanied by a huge amount of coding and subsequent testing to ensure that the algorithms work correctly and are secure.

At this stage, all implemented authentication methods will be combined and integrated into a blockchain framework. Blockchain technology shall be done because it provides decentralization, immutability, and transparency to the system, much needed in order to improve security features for performing financial transactions within it. The user identification and picture-based authentication processes shall be embedded within the blockchain to ensure absolute security and reliability for every type of transaction that gets recorded and verified. This integration will ensure that the full benefits of blockchain technology are exploited to provide a strong defense against the unauthorized access and change of data. Evaluation Phase:

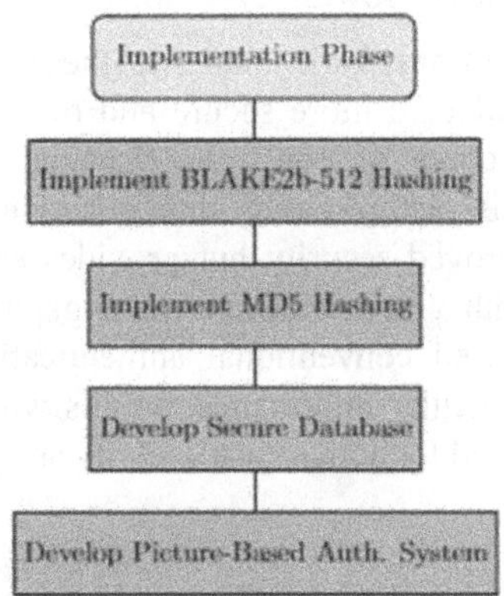

Fig. 45.2 Implementation phase

The last phase is dedicated to the evaluation of the effectiveness, security, and usability of the integrated system. Experimental testing is conducted for assessing performance in BLAKE2b-512 and MD5 hashing algorithms and the picture-based string authentication method. Key performance metrics, such as response time, accuracy, and resistance to different attacks like phishing and brute-force attacks, are measured. Usability and intuitiveness are tested in user studies with the picture-based authentication system. Results of these evaluations are analyzed for strengths, weaknesses, and areas of improvement.

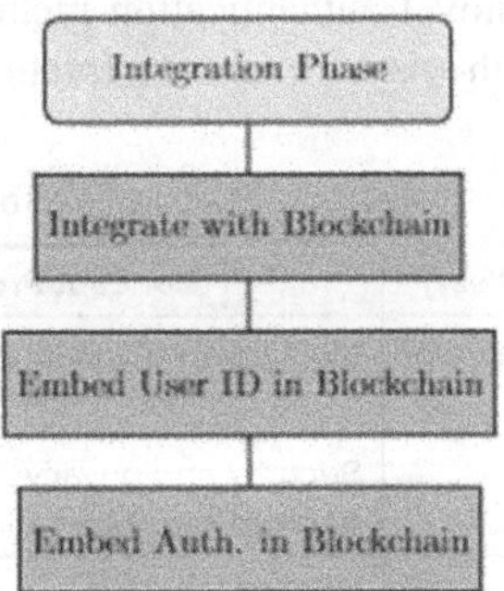

Fig. 45.3 Integration phase

The phases concern the detailed development and validation of a user-friendly, secure financial transaction system in the research methodology. This paper combines advanced cryptographic techniques with blockchain technology and innovative authentication methods to provide all-rounded solutions to most contemporary security challenges in digital finance.

5. Research Methodology

5.1 MD5

MD5 is short for Message-Digest Algorithm 5, a widely used hash function in cryptography designed to produce a same-size 128-bit hash value from a message of arbitrary size. Developed in 1991 by Ronald Rivest, MD5 produces a 32-character hexadecimal output that is always unique to the input data. It processes input in 512-bit pieces, initializing four 32-bit variables and then iteratively applying a series of bitwise operations and logical functions to each of those blocks. More precisely, the message needs to be padded so its length is congruent to 448 bits modulo 512, and at the end, a 64-bit representation of the original message length is appended. Although quite efficient and deterministic, there are known vulnerabilities in MD5: it is vulnerable to collision attacks, meaning two different inputs can result in the same hash value. This means that although MD5 is still commonly in use for applications not relevant to security—for example, file integrity checking—in cryptographically secure purposes, hash functions should be preferred for cryptographic purposes, since some of the security properties of MD5 are already shown to be broken.

5.2 Blake2B

BLAKE2b is a cryptographic hash function designed for security and very high performance. It is faster at processing and has up to 512-bit output size, unlike MD5 or SHA-2. As one of the members of the large family of hash functions, BLAKE, BLAKE2b is designed to initialize a state with fixed constants and process 128-bit blocks of input; permutations, bit-wise, and modular additions permit an excellent mix of data for a strong avalanche effect in this case. A small change in the input changes the hash values drastically. It has high efficiency on 64-bit architectures and very strong security features that ensure resistance to collision attacks. BLAKE2b is applied to many problems in cryptography that require strong cryptographic guarantees; it is therefore a modern, effective alternative to older hash functions, be it for file integrity checks, password hashing, or digital signatures.

5.3 Algorithm for Secure Authentication and Financial Transactions

This section presents the algorithm for a secure financial transaction system using blockchain technology. The system incorporates user registration, password and biometric data handling, and visual authentication mechanisms.

A. Registration Process

Input: User identity information

$$U = (\text{username,email,unique identifier})$$

Process: Generate BLAKE2b-512 hash for user identity.

$$H_{ID} = \text{BLAKE2b-512}(U) \tag{1}$$

where H_{ID} is a 512-bit hash value.

i. **Password Hashing**

Input: User password in plaintext P

Process: Generate MD5 hash for password.

$$H_{PW} = \text{MD5}(P) \tag{2}$$

where H_{PW} is a 128-bit hash value.

Biometric Fingerprint Image Hashing

Input: Biometric fingerprint image F

Process: Generate BLAKE2b hash for fingerprint image.

$$H_{FP} = \text{BLAKE2b}(F) \tag{3}$$

where H_{FP} is a variable-size hash value.

ii. **Slider-Based Character Extraction**

Input: Slider position s

Process: Extract characters from BLAKE2b hash based on slider position.

$$\text{Extracted Characters} = \text{Extract}(H_{FP}, s) \tag{4}$$

where Extract(·) is a function that extracts characters based on the slider position s.

iii. **Storage and Transmission**

Store the following securely:

H_{ID}

H_{PW}

H_{FP}

Encrypt data transmission using HTTPS.

iv. **Visual-Based Sub Process**

Image Selection and Segmentation

Input: Selected image I

Process: Segment the image into regions $R = \{r_1, r_2, \ldots, r_n\}$.

$$I \rightarrow R \tag{4}$$

v. **Text Association**

Input: Associated text $T = \{t_1, t_2, \ldots, t_n\}$

Process: Associate text with image segments.

$$\text{Pattern} = \{(r_i, t_i) \mid i = 1,2, \ldots, n\} \tag{5}$$

Pattern Formation and Confirmation

Process: Combine text associated with each segment to form a pattern.

$$A_{form} = \{t_1, t_2, \ldots, t_n\} \tag{6}$$

where A_{form} is the dynamically formed pattern.

B. Login Process- Authentication

Input: Username, password P, and biometric fingerprint image F

Process: Hash password and fingerprint image, and compare with stored values.

Check Password: $H_{PW} = MD5(P)$

Check Fingerprint: $H_{FP} = BLAKE2b(F)$

Visual Pattern Verification

Input: User-selected visual pattern

Process: Compare with stored visual pattern.

$$\text{Verify Pattern: } A_{form} = A_{reg} \tag{7}$$

where A_{reg} is the registered pattern.

C. Transaction Identity Pattern

Generate Transaction Identifier

Input: Biometric data hash, password hash, cryptocurrency details, visual pattern, timestamp

Process: Combine these elements to form a unique transaction identifier.

$$\text{Transaction ID} = \text{Combine } (H_{FP}, H_{PW}, \text{Crypto Details, Pattern, Timestamp}) \tag{8}$$

where Combine(·) is a function that integrates all elements into a unique identifier.

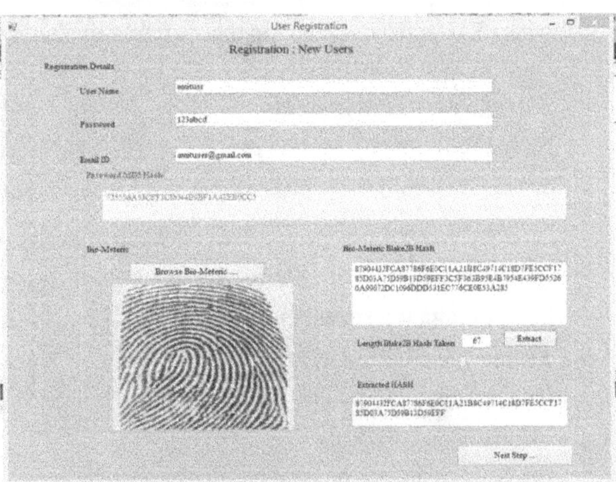

Fig. 45.4 Registration process 1

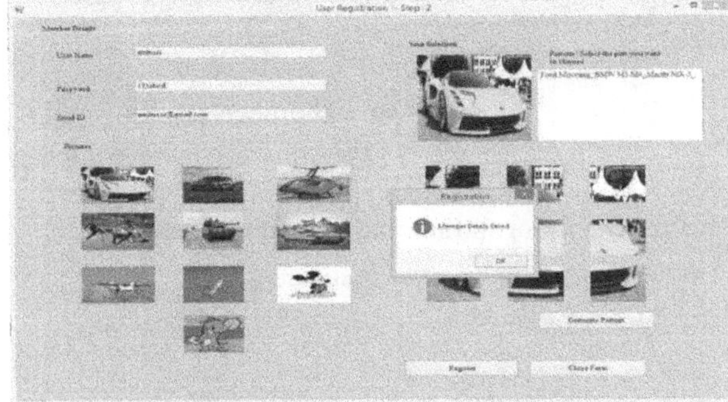

Fig. 45.5 User registration step 2

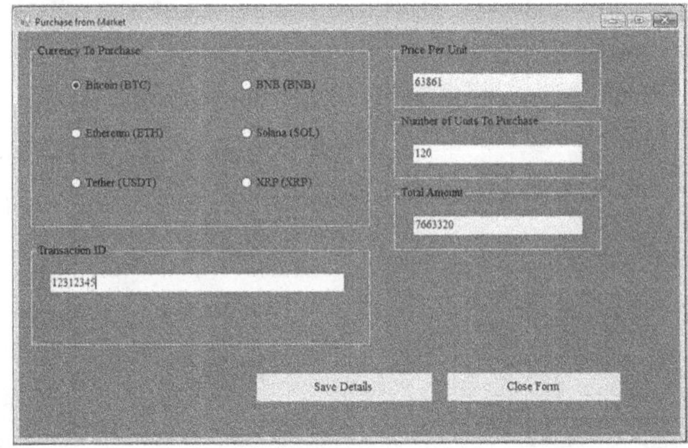

Fig. 45.6 Transaction menu

In the transaction section we have also applied the image pattern passwords for the further validation of password

8x8 Image Grid with Selected Images Highlighted

img1	img2	img3	img4	img5	img6	img7	img8
img9	img10	img11	img12	img13	img14	img15	img16
img17	img18	img19	img20	img21	img22	img23	img24
img25	img26	img27	img28	img29	img30	img31	img32
img33	img34	img35	img36	img37	img38	img39	img40
img41	img42	img43	img44	img45	img46	img47	img48
img49	img50	img51	img52	img53	img54	img55	img56
img57	img58	img59	img60	img61	img62	img63	img64

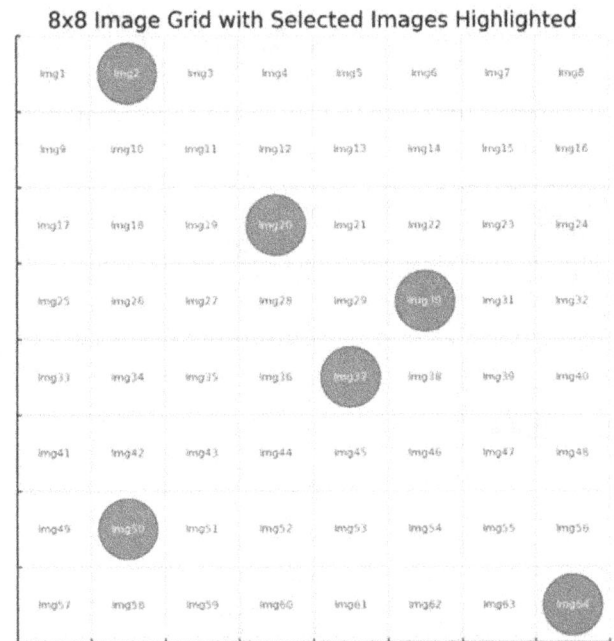

Fig. 45.7 Picture password

The process is as follows:

- Create an 8x8 grid: Attach a picture to every cell for reference, marked by the coordinate of the cell.
- Selection of Pictures by the User: user choose any 6 pictures from the grid. For each selection, will record the following:
  - Image ID or name
  - Row coordinate (1-8)
  - Column coordinate (1-8)
  - Size of the picture
  - Display Image Choice Process:

The highlighted images selected in Fig. 45.7.

- **Image: Img2, Row: 1, Column: 2, Size: 50**
- **Image: Img20, Row: 3, Column: 4, Size: 60**

- **Image: Img37, Row: 5, Column: 5, Size: 55**
- **Image: Img50, Row: 7, Column: 2, Size: 45**
- **Image: Img64, Row: 8, Column: 8, Size: 50**
- **Image: Img30, Row: 4, Column: 6, Size: 65**

6. Result Analysis

- **Antivirus.Promo Tool:** This tool reported an entropy of **1090.60 bits**, indicating a high level of password security. The result suggests that the password is robust against various types of attacks, including brute-force and dictionary attacks.
- **Password.Blue Tool:** This tool showed an entropy of **519 bits**, which is lower compared to the Antivirus. Promo Tool. While still secure, this lower value suggests that the password might be more vulnerable to certain attack methods.
- **Rumkin.com Tool:** This tool produced an entropy score of **898 bits**, indicating a strong password with considerable complexity. It is well-suited to protect against unauthorized access.

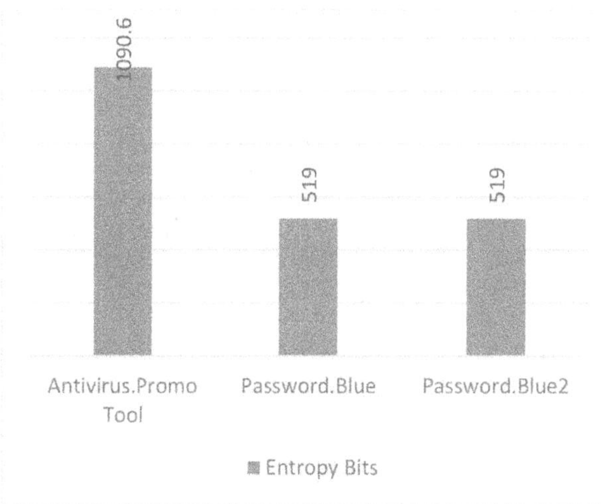

Fig. 45.8 Comparison graph on tools on entropy bits

6.1 Results Analysis for User Transaction Entropy

Pattern for User Transaction

The image selection process,

D492C34767_E0757A6A9F5EA5F605036D96462526D5_AH-56A Cheyenne_SA3200 Frelon_UH-2 Seasprite__BBFA2E8F23_6E9B3A7620AAF77F362775150977EEB8_Maruti Suzuki Brezza_HyundaiVenue_KiaSonet__Bitcoin (BTC)_10_E0757A6A9F5EA5F605036D96462526D5_AH-56A Cheyenne_SA3200 Frelon_UH-2 Seasprite___09-05-2024 00:00:00_ Img2_R1_C2_S50_Img20_R3_C4_S60_Img37_R5_C5_S55_Img50_R7_C2_S45_Img64_R8_C8_S50_Img30_R4_C6_S65

The entropy of a user transaction pattern evaluates its security based on the complexity and randomness of the information included. Higher entropy values indicate a more secure and unique transaction identity.

Table 45.2 Strength test results

Test	Score	Criteria Met
Length	394.00	Yes
Character Diversity	4.00	4/4 character types
Pattern Complexity (Entropy)	2582.51	2582.51 bits
Repetition	7.00	High repetition

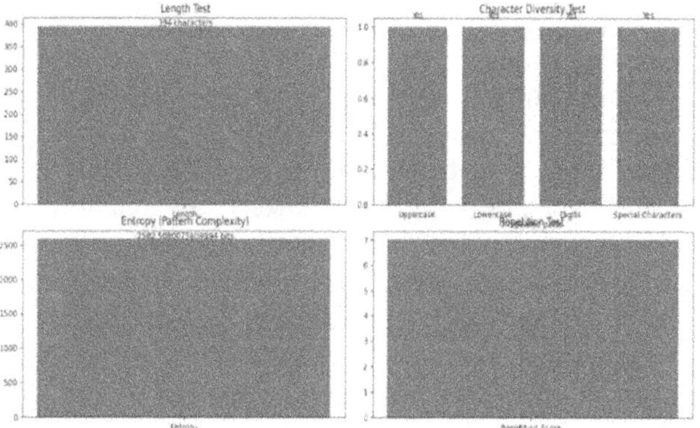

Fig. 45.9 Transaction results

6.2 Explanation of the Results

1. Length (394 Characters): A long password or pattern is a good defense against brute-force attacks because of its complexity added by the length of the number of possible combinations that an attacker needs to try to guess the pattern. The longer the password, the exponentially harder it is to guess - even in the brute-force attacks, where they try every single possible combination.
2. Character Mix (4/4 character types)

   This includes all the types of characters, including uppercase and lowercase, digits, and special characters. It increases the pattern complexity and makes guessing harder. Most attackers tend to use easy-to-guess patterns using only lowercase letters or digits. That is to say, the mixed case and the use of special characters increase the entropy by much.
3. Complexity of Pattern: Entropy = 2582.51 bits

   Strength: The entropy directly measures the unpredictability and randomness of a pattern. The pattern is highly secure in theory with a value of 2582.51 bits of entropy. A high entropy value would make it difficult for anyone to brute-force or predict the password. Most passwords with a high entropy score are very hard for attackers to guess because the total number of possible combinations is enormously huge.

7. Conclusion

The paper exhibits a fully workable approach for the betterment of security in financial transactions by integrating the state-of-the-art cryptographic techniques with the authentication methodology user-specific in nature. This envisions a system that would substantially enhance the security features of financial transactions through the integration of BLAKE2b-512 for user identity hashing, MD5 for password hashing, and a visual string-based authentication algorithm. Results from the entropy analysis showed that while the passwords had entropy values between 519 bits and 1090.60 bits, user transactions were even higher, between 927 bits and 1944.68 bits. The results established very clearly the effectiveness of the system in giving extremely high security against brute force and dictionary attacks. Adopting visual pattern-based authentication embeds an additional—really innovative—layer of safety that is user-specific, thereby enhancing protection and engagement. The overall research provides valuable contributions and some practical solutions toward financial security, with a focus on integrating state-of-the-art cryptographic algorithms and personalized authentication mechanisms to provide top protection against unauthorized access and fraudulent activities.

References

1. Hertzog, E., Benartzi, G., & Benartzi, G. (2018). Bancor protocol: Continuous liquidity for cryptographic tokens through their smart contracts (Tech. Rep.).
2. Adams, H., Zinsmeister, N., & Robinson, D. (2020). Uniswap v2 core (Tech. Rep.).
3. Egorov, M. (2019). Stableswap-efficient mechanism for stable-coin liquidity (Tech. Rep.).
4. Krishnamachari, B., Feng, Q., & Grippo, E. (2021). Dynamic curves for decentralized autonomous cryptocurrency exchanges.
5. Spithoven, A. (2019). Theory and reality of cryptocurrency governance. Journal of Economic Issues, 53(2), 385–393.
6. Greenberg, A. (2011). CryptoCurrency. Forbes. Archived from the original on 31 August 2014.
7. Farell, R. (2015). [Crypto currency] (Unpublished manuscript). University of Pennsylvania Scholarly Commons.
8. Liu, Y., &Tsyvinski, A. (2018). Risk and Returns of Cryptocurrency.
9. Halpern, S. (2018). Bitcoin Mania. The New York Review of Books, 65(1), 52–54, 56.
10. Narayanan, A., Bonneau, J., Felten, E., Miller, A., & Goldfeder, S. (2016). Bitcoin and Cryptocurrency Technologies. Princeton University Press.
11. Nakamoto, S. (2008). Bitcoin: A Peer-to-Peer Electronic Cash System.
12. Sauer, B. (2016). Virtual currencies, the money market, and monetary policy. International Advances in Economic Research, 22(2), 117–130.
13. Takahashi, H., & Lakhani, U. (2019). Multiple Layered Security Analyses Method for Cryptocurrency Exchange Servicers. In 2019 IEEE 8th Global Conference on Consumer Electronics (GCCE) (pp. 71–73). Osaka, Japan: IEEE.
14. Sai, A. R., Buckley, J., & Le Gear, A. (2019). Privacy and Security Analysis of Cryptocurrency Mobile Applications. In 2019 Fifth Conference on Mobile and Secure Services (MobiSecServ) (pp. 1–6). Miami Beach, FL, USA: IEEE.
15. Azman, M., & Sharma, K. (2020). HCH DEX: A Secure Cryptocurrency e-Wallet & Exchange System with Two-way Authentication. In 2020 Third International Conference on Smart Systems and Inventive Technology (ICSSIT) (pp. 305–310). Tirunelveli, India: IEEE.
16. Karanjai, R., Xu, L., Gao, Z., Chen, L., Kaleem, M., & Shi, W. (2021). On Conditional Cryptocurrency With Privacy. In 2021 IEEE International Conference on Blockchain and Cryptocurrency (ICBC) (pp. 1–3). Sydney, Australia: IEEE.
17. Perry, T. S. (2022). A Bitcoin Wallet for the Masses: Square simplified credit-card transactions. Now it wants to build cryptocurrency hardware. IEEE Spectrum, 59(1), 42–43.

Note: All the figures and tables in this chapter were made by the authors.

Intelligent Systems Using Semiconductors for Robotics and IoT – Dinesh Goyal et al. (eds)
 ISBN 978-1-041-20408-4

46 Surveying the Landscape: A General Review of Deep Learning for Malware Detection: Current Trends and Emerging Strategies

Naresh Mathur*
Research Scholar,
Department of Computer Science Engineering,
Govt. Engineering College Ajmer

Prakriti Trivedi
Professor, Principal.
Govt. Mahila Engineering College,
Ajmer

Abstract: This is a survey paper for providing an in-depth review of techniques in deep learning paving toward the revolution in malware detection. Firstly, it identifies inadequacies in traditional methods of malware detection and increases reliance on machine learning approaches. The roles of deep learning models-Deep Convolutional Neural Networks, Recurrent Neural Networks, Autoencoders, and Generative Adversarial Networks-in improving the accuracy and robustness of the detection process are widely discussed hereinafter. The paper covers the recent trends of integration of advanced neural network architectures and application of transfer learning and federated learning techniques. It further presents the practical applications with the case studies, reviews several models for their efficiency in real-world scenarios, and raises some key challenges in adversarial attacks, model interpretability, and scalability. The review concludes with the proposed solutions and future research directions that will move the field forward.

Keywords: Deep learning, Malware detection, CNN, RNN, Autoencoders, GAN, Transfer learning, Federated learning

1. Introduction

Malware is defined as software that is intentionally intended to cause chaos or destruction or unauthorized access to a computer system. It includes viruses, worms, Trojans, ransomware, and spyware; each of them brings a peculiarity that fights cyber-security. Malware in development has evolved significantly in terms of complexity and sophistication. Thus, detection using traditional methods has become very challenging [1]. The traditional approaches to malware detection have been the following: signature-based, heuristic-based, and behavior analysis. In the signature-based technique, a constantly updated database comprising the patterns or the signatures of malicious code is required. Heuristic-based methods relate to the behavior or characteristics of programs with activities that are suspicious, while behavioral analysis monitors activities on a system against anomalies that may indicate malware. Due to these limitations in adapting to new and unknown threats, generation and signature-based techniques are widely deprived of detecting malware. This deficiency can result in detection gaps and increase the possibility of malware going undetected [1].

1.1 Need for Advanced Techniques

Traditional methods, however, could not provide effective protection against ever more sophisticated malware threats. On the other hand, signature-based methods can detect known threats only; they are incapable of spotting novel or polymorphic malware that changes its code in attempts to bypass detectors [2]. At the same time, heuristic

*Corresponding author: nareshmathurer@gmail.com

DOI: 10.1201/9781003716389-46

and behavior-based methods give greater flexibility at the cost of high false positive rates and typically require much manual effort to tune and keep current. Clearly, malware techniques are rapidly changing and need equally fast changes in the detection system, for which traditional methods are not well tailored.

This fact has led to the immediate need for newer techniques in detection that can learn and adapt with new threats on their own. In this line, deep learning, a subset of machine learning that includes neural networks with multiple layers, is very promising. Deep learning is able to highlight the patterns and anomalies that other methods cannot, hence providing more robust and adaptive malware detection at the base of large datasets and complex model architectures [2].

1.2 Objective

This review paper shall, therefore, take a critical look into how deep learning techniques currently revolutionize malware detection. Paper intends to discuss a number of deep learning models, applied individually; these covered Convolutional Neural Networks, Recurrent Neural Networks, Autoencoders, and Generative Adversarial Networks [3]. The objective of this paper is to deeply analyse the current trends and emergent strategies in deep learning-based malware detection, along with their applications, benefits, and limitations. The paper will discuss some challenges these advanced techniques are currently facing and, at the same time, give possible solutions and future research directions. The paper discusses how deep learning has enormously affected malware detection and puts forward some useful insights for improving cybersecurity measures by researchers, practitioners, and decision-makers [3].

2. Traditional Malware Detection Techniques

Traditional malware detection techniques have been the mainstay of cybersecurity for decades, acting as the frontline of defense against malignant software. In reality, a malware threat is identified and mitigated using traditional methods, which deploy different ways of detection for harmful programs. Although traditional techniques have been in use for a very long time and form the very foundation of security, they cannot counter the rapidly evolving malware landscape [4].

2.1 Signature-Based Methods

Signature-based recognition malware detection is one of the most primitive and pervasive methods in cybersecurity today. The approach is based on recognizing already known patterns or signatures of malware in software [4].

A signature can be seen as a string of bytes or pattern, matching a certain malware variant and eventually turning out to be uniquely identifiable from previously analyzed samples. It is these signatures that are stored in a database and which the detection systems use to scan files and programs for a match, which may indicate that a file or program is malware. These limitations of signature-based detection are enormous, despite the effectiveness it shows against the known threats [5].

2.2 Heuristic-Based Methods

The heuristic technique is the detection technique that is used to identify the behavior or characteristics of malicious activities through code and execution pattern analysis [6]. More recently, heuristic methods do not depend on predefined signatures, but make the evaluation of code execution in front of suspicious behaviors or structures. It typically involves the evaluation of the behavior of a file or program in a sandbox environment while looking for malware-linked patterns such as unauthorized system resource access or concealing traces of the malware from surveillance [6]. Heuristics provide more flexibility than signature-based detection techniques, enabling possible detection of new or unknown malware via dangerous or merely suspicious behaviors.

2.3 Behavioral Analysis

On the other hand, behavioral-based techniques observe and analyze the runtime behavior and interactions of programs, so as to identify malicious activity. The technique observes the behavior of software while executing, namely the actions it takes such as file modifications, network activities, and registry changes, among other system interactions. Such behaviors can be analyzed in the identification of anomalies, depicting possible malicious intent, thus detecting the malware, irrespective of whether it had not been previously seen or it has no known signature [7].

Dynamic behavioral analysis is basically the detection of malware, which helps in identifying threats through their behavior and not through static signatures. It is very efficient at detecting malware operating at a very unique pattern of activity or in a stealth way. As much as this method is effective, it also has its disadvantages. This approach consumes enough system resources for continuous monitoring of all the programs to an extent of decreasing system performance. It is also obvious that the effectiveness of behavioral analysis can be bypassed with malware designed to act covertly, or in a way that would mimic normal activity [8]

3. Deep Learning in Malware Detection

Deep learning represents the next evolution of machine learning, which allows computers to create knowledge out of learning from data without human intervention. As compared to traditional machine learning, deep learning allows automatic learning of generated features from raw data, dispensing with manual feature extraction and domain expertise through the multiple layers of abstraction that define it. These models are built using neural networks with numerous layers—hence the term "deep" learning"—which empowers them to capture complex patterns and representations in the data [9].

3.1 Key Deep Learning Models

A. Convolutional Neural Networks (CNNs) Application in Malware Detection

Convolutional neural networks are part of the highly advanced machinery of deep learning used to resolve structured data. Mostly, CNNs have only been used for image recognition, but the technology has its useful application tools in malware detection as well. For malware detection, CNNs can be used to treat binary executables or file content as images and each byte or a sequence of bytes constituting them as pixels. It allows the model to learn patterns that indicate malicious behaviour within the binary data [10].

B. Recurrent Neural Networks for Sequence Data Analysis

The RNNs were also designed for the processing of sequential data, so an ideal fit will be tasks related to time series and ordered sequences—not to mention natural language processing. In terms of the detection of malicious software, the periods of sequences of system calls, API calls, or opcodes contain information about a program's behaviour [11]. Sequences are established that malware generally follows in the course of execution. Their recurrent connections allow them to maintain some memory for previous inputs, thereby empowering RNNs with the ability to model sequences for anomaly detection or pattern recognition in events related to adverse outcomes such as malware [11].

C. Autoencoders: Use in Anomaly Detection

Autoencoders are neural networks for unsupervised learning that embed the capacity to learn to compress input data into lower dimension latent space representation and reconstruct it back to the same original form. The process involves learning an efficient representation in which important features are captured while minimizing the error in its reconstruction. These important features are captured, in the case where autoencoders are used, which are important for the detection of malware anomalies [12].

D. Generative Adversarial Networks: A Way to Model Robustness, Adversarial Examples

GANs are a kind of generative model wherein the training of two neural networks, the generator and discriminator, happens in an alternate manner. The generator generates new data instances from prior data, while the discriminator finally classifies them into real and fake data instances. The very adversarial process in itself pushes the generator to create samples as realistic as possible, while the discriminator is trained to the best of its ability in order to separate the true data from these false samples [13]. In malware detection, GANs will raise the model's robustness by adversarial example generation: small modification samples of malware to avoid detection

4. Current Trends in the Application of Deep Learning to Malware Detection

4.1 Advanced Neuromorphic Architectures Inventions and Enhancements

More recent advances in deep learning for malware detection have mostly resulted from the design of neural network architectures and have led to very large improvements in the accuracy, speed, and robustness of detection. Deep learning models traditionally developed on CNNs and RNNs have predominantly led this foundation, and recently developed innovative architectures further scale up the complexity of malware behaviours [14]. Some of the important trends in this area include the development of hybrid models that involve different but complementary architectures of neural networks. For example, CNNs can be combined with RNNs, allowing the model to handle spatial as well as sequential data simultaneously in order to assess malware in much greater detail. Such hybrid architectures are very useful when the malicious software performs maneuvers whose description cannot be captured by using only static or dynamic analysis , for instance, together with the analysis of binary code structures, execution-distribution patterns [14]. Another innovation is the use of attention mechanisms and transformers, first popularized in the domain of natural language processing [14]. One such example is the growing popularity and interests in graph-based neural networks in detecting malware, in particular, Graph Neural Networks (GNNs). Advanced malware interacts with a great number of system components, which naturally establishes very complex relationships that can be expressed as graphs [14].

4.2 Transfer Learning: Benefits and Application in Malware Detection

Transfer learning is a method in deep learning where the model being trained for a particular task is fine-tuned for a different but similar task. This process is quite interesting and relevant in the sense that one typically encounters problems where the learning of the domain is sparse, the data is unbalanced in nature, and the cost of acquiring and labelling new data is very high [15]. Transfer learning in malware detection allows updating the model with a minimal increment of training samples from its user, all the while detecting new breeds or specialized malware without further costly and time-consuming retraining [15].

4.3 Federated Learning: Preserving Privacy Techniques and Their Relation to Malware Detection

Federated learning is a new trend in deep learning and supports the training of models by sharing information across decentralized systems of devices or servers, with data lying locally among them. This all becomes very relevant to things like malware detection, where regulatory and privacy constraints should keep the flow of very sensitive data, such as that of proprietary software and user behavior logs, pretty limited [16].

5. Related Works

Almazroi, A. A., and Ayub, N. (2024) [17] presents BEFNet, a BERT-based Feed Forward Neural Network Framework optimized with the Spotted Hyena Optimizer, to address the issues related to security in AI-enabled devices incorporating the IoT. BEFNet is highly capable of dealing with large volumes of data emanating from devices connected to the IoT with a strong set of performance metrics on various malware datasets. Brown, et al. (2024) [18] demonstrated an application of AutoML for malware detection where it was concerned with the tuning of hyperparameters and the model for reducing computationally expensive overheads and dependency on domain expertise. Aamir, et al. (2024) [19] presented a concept of AMDDLmodel detection of Android malware based on the approach of CNN. It has shown good accuracy over the Drebin dataset along with innovative feature engineering techniques.

Other malware detection evolutions include hybrid static and dynamic approaches for analysis. As seen from Gurierrez, R., et al. (2024) [20], that would allow for the revelation of high accuracy. Another model is that of Nasser, A. R., et al. (2024) [21], which reached an accuracy level of 99.935% in its model, DL-AMDet. Discuss binary image conversion approaches for IoT malware detection in Ghahramani, M., et al. 2024 [22] and Al-Andoli, M. N., et al. 2023 [23] proposed an ensemble-based DL classifier optimized by using hybrid techniques.

Zhang, S., et al. 2023[24] focus on the fileless malware detection based on CNNs and memory forensics, and de Oliveira, A. S., and Sassi, R. J. (2023) [25] refers to Chimera, a multimodal deep learning system through which multiple techniques of data analysis are fused for better malware detection.

Table 46.1 Literature findings

Author Name (Year)	Main Concepts	Key Factors	Findings
Almazroi, A. A., & Ayub, N. (2024)	BEFNet, BERT-based Feed Forward Neural Network, IoT security	Spotted Hyena Optimizer (SO), 8 malware datasets, IoT device-generated data, decision-making	BEFNet achieves high accuracy and other performance metrics, establishing itself as a robust defense mechanism in IoT security and dynamic decision-making.
Brown, A., et al. (2024)	Automated Machine Learning (AutoML), static and online malware detection	Hyperparameter optimization, neural architecture search, deep learning	AutoML proves effective in creating accurate detection models with low computational overhead, simplifying the development of deep learning models for malware.
Aamir, M., et al. (2024)	Android malware detection, convolutional neural network (CNN)	Feature engineering, Drebin dataset, Android operating system	The AMDDLmodel achieves high accuracy in classifying Android malware, enhancing user security on mobile devices.
Gurierrez, R., et al. (2024)	Malware detection, deep learning, static and dynamic analysis	Combination of static and dynamic analysis	Their deep learning model achieves 98% precision, outperforming traditional methods in malware detection.
Nasser, A. R., et al. (2024)	Android malware detection, deep learning	Static and dynamic analysis	DL-AMDet achieves 99.935% accuracy, surpassing current techniques for detecting Android malware.
Ghahramani, M., et al. (2024)	IoT malware detection, deep learning	Behavior analysis, binary image conversion, performance across metrics	Deep learning outperforms other methods in 7 out of 8 metrics for detecting malware in IoT devices.
Al-Andoli, M. N., et al. (2023)	Parallel DL classifier, ensemble methods	Hybrid optimization (Back-Propagation and Particle Swarm Optimization), 5 malware datasets	The proposed method achieves up to 100% accuracy and enhances computational efficiency, improving speed by up to 6.75 times.
Alomari, E. S., et al. (2023)	High-performance malware detection, deep learning, feature selection	Combination of dense and LSTM-based models, handling large and high-dimensional data	Their approach effectively differentiates between malware and benign activities, demonstrating the potential of deep learning in handling complex datasets.

Author Name (Year)	Main Concepts	Key Factors	Findings
Djenna, A., et al. (2023)	Dynamic deep learning, heuristic approaches	Malware detection, classification of sophisticated malware types (adware, rootkit, ransomware)	The proposed method enhances detection and analysis capabilities, outperforming static deep learning approaches against modern cyber threats.
Ravi, V., & Alazab, M. (2023)	Image-based malware detection, multi-headed attention-based CNN	Windows file transformation into images, analysis of tiny infected regions	The method achieves 99% accuracy in classifying 25 malware families, outperforming traditional CNN methods and proving effective across various malware types.
Zhang, S., et al. (2023)	Fileless malware detection, CNN, memory forensics	Analysis of executable fragments in memory, challenges posed by fileless malware	Their method shows high accuracy in detecting fileless malware, highlighting deep learning's effectiveness in combating sophisticated threats.
de Oliveira, A. S., & Sassi, R. J. (2023)	Multimodal deep learning, Android malware detection	Combination of CNNs, DNNs, TNs, raw data, static and dynamic analysis features	Chimera achieves superior detection accuracy, underscoring the potential of multimodal deep learning in enhancing malware detection capabilities.
Xing, X., et al. (2022)	Grayscale image representation, autoencoder networks, malware detection	Feature extraction using autoencoders	The proposed approach achieves high accuracy in classifying malware from benign software, offering a promising avenue for innovative feature engineering.
Kim, J., Ban, et al. (2022)	Lightweight malware detection, CNN, mobile devices	Analysis of API call graphs, minimal computational overhead, suitability for resource-constrained devices	MAPAS achieves high accuracy with minimal computational overhead, making it suitable for deployment on resource-constrained devices, particularly in mobile contexts.

6. Discussion and Findings

This literature review identifies major strides in malware detection with deep learning and new approaches as stiffening of battles against ever-evolving cyber threats stiffens. Almazroi and Ayub, for instance, contribute a dimension in which BERT-based neural networks can totally transform IoT security with their BEFNet framework. Brown et al. illustrated how AutoML can easily relieve model creation efforts for the efficient detection of malware. Other influential works on the topic of Android malware detection are those by Aamir et al. (2024) and Nasser et al. (2024), which provide examples of how convolutional neural networks can work together with deep learning models in search of high accuracy. Approaches such as Al-Andoli et al. (2023) and Ghahramani et al. (2024) emphasize that hybrid optimization and behaviour analysis can help improve detection efficiency and performance across a variety of metrics. Works by Gurierrez et al. (2024) and Zhang et al. (2023) give evidence of how the combination of static and dynamic analysis outperforms in detecting sophisticated and fileless malware. Thirdly, work done by Ravi and Alazab, 2023, and de Oliveira and Sassi, 2023, points out that image-based and multimodal deep learning methods efficiently classify malware. Those results underline very different and innovative approaches being developed to enhance malware detection, mitigating the challenges brought about by increasingly sophisticated threats within a variety of digital environments.

7. Conclusion

This paper discusses the role of deep learning in malware detection, as traditional approaches cannot stand against ever increasingly complex malwares. Deep learning models, of which some are CNNs, RNNs, Autoencoders, and GANs, have each been observed to have unique strengths. CNNs are good for analyzing binary data and visual malwares, RNNs are best suited for handling sequential data to provide behavioral insight, Autoencoders best fit anomaly detection, and GANs improve robustness through adversarial example generation. These newer techniques are used to improve malware detection further through adaptability, privacy-preserving training, and complex feature extraction: transfer learning, federated learning, hybrid models, and attention mechanisms. Of course, the disadvantages of being a state-of-the-art technique include un-interpretability of the model, scalability, and robustness in adversarial attacks. Based on these considerations, research work needs to be undertaken towards developing more accurate, flexible, and resilient systems for detecting malware in order to advance cyber protection with modern, deep learning-based defenses.

References

1. Xing, X., Jin, X., Elahi, H., Jiang, H., & Wang, G. (2022). A malware detection approach using autoencoder in deep learning. IEEE Access, 10, 25696-25706.
2. Kim, J., Ban, Y., Ko, E., Cho, H., & Yi, J. H. (2022). MAPAS: a practical deep learning-based android malware

detection system. International Journal of Information Security, 21(4), 725–738.

3. Gopinath, M., & Sethuraman, S. C. (2023). A comprehensive survey on deep learning based malware detection techniques. Computer Science Review, 47, 100529.
4. Deldar, F., & Abadi, M. (2023). Deep learning for zero-day malware detection and classification: A survey. ACM Computing Surveys, 56(2), 1–37.
5. Shaukat, K., Luo, S., & Varadharajan, V. (2023). A novel deep learning-based approach for malware detection. Engineering Applications of Artificial Intelligence, 122, 106030.
6. Alomari, E. S., Nuiaa, R. R., Alyasseri, Z. A. A., Mohammed, H. J., Sani, N. S., Esa, M. I., & Musawi, B. A. (2023). Malware detection using deep learning and correlation-based feature selection. Symmetry, 15(1), 123.
7. Mbunge, E., Muchemwa, B., Batani, J., & Mbuyisa, N. (2023). A review of deep learning models to detect malware in Android applications. Cyber Security and Applications, 1, 100014.
8. Gulmez, S., Kakisim, A. G., & Sogukpinar, I. (2023, May). Analysis of the dynamic features on ransomware detection using deep learning-based methods. In 2023 11th International Symposium on Digital Forensics and Security (ISDFS) (pp. 1–6). IEEE.
9. Gulmez, S., Kakisim, A. G., & Sogukpinar, I. (2023, May). Analysis of the dynamic features on ransomware detection using deep learning-based methods. In 2023 11th International Symposium on Digital Forensics and Security (ISDFS) (pp. 1–6). IEEE.
10. Chaganti, R., Ravi, V., & Pham, T. D. (2023). A multi-view feature fusion approach for effective malware classification using Deep Learning. Journal of information security and applications, 72, 103402.
11. Raphael, R., & Mathiyalagan, P. (2023). Intelligent hyperparameter-tuned deep learning-based Android malware detection and classification model. Journal of Circuits, Systems and Computers, 32(11), 2350191.
12. Mercaldo, F., Ciaramella, G., Santone, A., & Martinelli, F. (2023, August). Obfuscated mobile malware detection by means of dynamic analysis and explainable deep learning. In Proceedings of the 18th International Conference on Availability, Reliability and Security (pp. 1–10).
13. Davidian, M., Kiperberg, M., & Vanetik, N. (2024). Early Ransomware Detection with Deep Learning Models. Future Internet, 16(8), 291.
14. Al-Andoli, M. N., Tan, S. C., Sim, K. S., Goh, P. Y., & Lim, C. P. (2023). An ensemble deep learning classifier stacked with fuzzy ARTMAP for malware detection. Journal of Intelligent & Fuzzy Systems, 44(6), 10477–10493.
15. Ding, Y., Zhang, X., Hu, J., & Xu, W. (2023). Android malware detection method based on bytecode image. Journal of Ambient Intelligence and Humanized Computing, 14(5), 6401–6410.
16. Maniriho, P., Mahmood, A. N., & Chowdhury, M. J. M. (2024). A survey of recent advances in deep learning models for detecting malware in desktop and mobile platforms. ACM Computing Surveys, 56(6), 1–41.
17. Almazroi, A. A., & Ayub, N. (2024). Deep learning hybridization for improved malware detection in smart Internet of Things. Scientific Reports, 14(1), 7838.
18. Brown, A., Gupta, M., & Abdelsalam, M. (2024). Automated machine learning for deep learning based malware detection. Computers & Security, 137, 103582.
19. Aamir, M., Iqbal, M. W., Nosheen, M., Ashraf, M. U., Shaf, A., Almarhabi, K. A., ... & Bahaddad, A. A. (2024). AMDDLmodel: Android smartphones malware detection using deep learning model. Plos one, 19(1), e0296722.
20. Gurierrez, R., Villegas-Ch, W., Godoy, L. N., Mera-Navarrete, A., & Luján-Mora, S. (2024). Application of Deep Learning Models for Real-Time Automatic Malware Detection. IEEE Access.
21. Nasser, A. R., Hasan, A. M., & Humaidi, A. J. (2024). DL-AMDet: Deep learning-based malware detector for android. Intelligent Systems with Applications, 21, 200318.
22. Ghahramani, M., Taheri, R., Shojafar, M., Javidan, R., & Wan, S. (2024). Deep Image: A precious image based deep learning method for online malware detection in IoT Environment. Internet of Things, 101300.
23. Al-Andoli, M. N., Sim, K. S., Tan, S. C., Goh, P. Y., & Lim, C. P. (2023). An ensemble-based parallel deep learning classifier with PSO-BP optimization for malware detection. IEEE Access.
24. Zhang, S., Hu, C., Wang, L., Mihaljevic, M. J., Xu, S., & Lan, T. (2023). A malware detection approach based on deep learning and memory forensics. Symmetry, 15(3), 758.
25. de Oliveira, A. S., & Sassi, R. J. (2023). Chimera: an android malware detection method based on multimodal deep learning and hybrid analysis. Authorea Preprints.
26. Alomari, E. S., Nuiaa, R. R., Alyasseri, Z. A. A., Mohammed, H. J., Sani, N. S., Esa, M. I., & Musawi, B. A. (2023). Malware detection using deep learning and correlation-based feature selection. Symmetry, 15(1), 123.
27. Djenna, A., Bouridane, A., Rubab, S., & Marou, I. M. (2023). Artificial intelligence-based malware detection, analysis, and mitigation. Symmetry, 15(3), 677.
28. Shaukat, K., Luo, S., & Varadharajan, V. (2023). A novel deep learning-based approach for malware detection. Engineering Applications of Artificial Intelligence, 122, 106030.
29. Ravi, V., & Alazab, M. (2023). Attention-based convolutional neural network deep learning approach for robust malware classification. Computational Intelligence, 39(1), 145–168.

Intelligent Systems Using Semiconductors for Robotics and IoT – Dinesh Goyal et al. (eds)
© 2026 Taylor & Francis Group, London, ISBN 978-1-041-20408-4

47 Anthropomorphic AI: The Psychology of Human-Like Machines

Aman Jain[1], Shruti Gupta[2], Madhu Choudhary[3], Sanjay Kumar Sinha[4]
Computer Engineering,
Poornima Institute of Engineering & Technology,
Jaipur, India

Abstract: The Anthropomorphic AI encompasses the design and development of AI systems that have human-like characteristics, behaviors, and emotional responses. As AI technology becomes more ubiquitous in everyday scenarios, studying the psychological effect of using human-like machines becomes a primary focus. The proposed investigation examines the cognitive and emotional responses that result from humans interacting with anthropomorphic AI and considers the impact of human-like qualities such as trust, empathy, and social presence. The study also considers ethical implications, user experience, and potential consequences of humanized AI used in customer service, healthcare, and interaction for companionship. The study highlights the contested nature of anthropomorphism to humanize AI - anthropocentrism extends interaction and connection, while also raising issues of dependency, deception, and the nature of human-machine relationships.

Keywords: Artificial intelligence, Anthropomorphic AI, Human-like machines, Psychology of AI, Empathy in AI, Trust in AI, Human-computer interaction, Social presence, AI ethics, AI and emotion, Humanization of machines

1. Introduction

As artificial intelligence (AI) progresses, there is a discernible movement toward developing machines that make human-like traits and behaviors, a process known as anthropomorphism also termed as "humanization." Anthropomorphic AI is the technical process to develop systems that possess human-like attributes and characteristics and behave to provide users the experience of interaction with another human [1]. This includes tools and gadgets like Siri and Alexa to social robots designed for caregiving, thus closing the gap between how people relate to shoes and machines, and encouraging personal interaction with other humans and personal enjoyment [7]. The humanization of AI may also have serious psychological ramifications. In an experimental design where AI systems are endowed with human-like voice or even appearance or the way they conduct themselves, individuals appear to treat the machines as humans, responding in anticipation of what might come from another human being. This may alter how individuals engage these machines, but also create expectations regarding the machine's capabilities, e.g., emotional issues or some kind of emotional bond or relationship [4]. In addition to affecting interaction there is an ease of trust and acceptability introduced to anthropomorphic-AI, which raises the experience to another level of engagement and productivity, service, or educational value. However, the act of analyzing and humanizing machines with humanistic properties immediately raises social, political, and psychological questions. How far do we take the technological design toward a human-like machine? What behavioral implications exist when machines can be anthropomorphized? What psychological problems emerge when anthropomorphized technological applications exist, e.g., social, and/or emotional attachment has developed? Also, a question of Fu and Hoven's (2019) emerges. What is the ethical consideration of using anthropomorphized

[1]2021pietcsaman017@poornima.org, [2]shruti.gupta@poornima.org, [3]madhu.choudhary@poornima.org, [4]sanjay.sinha@poornima.org

DOI: 10.1201/9781003716389-47

AI? The present study is organized in a thesis format, beginning with the psychological dimensions regarding anthropomorphic artificial intelligence's effects on SI [5]. We then shift to contemplating ethics and challenges raised by anthropomorphic AI and reflect on the larger implications related to further human-machine engagement.

2. Artificial Intelligence

The ChatGPT is a revolutionary advanced aspect in artificial intelligence due to its ability to understand, relate to, and share natural conversations with people [22]. Since its development, more generative AI models have been released, each one showing great development. For instance, the copilot by Microsoft will help users assist with what is considered advanced capabilities as a coding assistant. On the other hand, you can get some sophisticated artwork just from a simple prompt with DALL-E and Stable

In filling the gap, two main research questions we ask are: 1) Do people really want AI to think and feel exactly like humans? If so, why? If not, what other alternate forms of AI intelligence are they seeking? 2) When is a person threatened by AI? Is this primarily because of the capabilities of AI per se or other routes, for example through anthropomorphism? This paper will be motivated to expand the black box of human-AI interaction by expanding the notion of anthropomorphism that refers to the extent of perceived human-likeness to ego-morphism, which would refer to the extent of self-likeness of AI, but not unlike humans. The meaning of anthropocentrism from anthropology-that places humans at the center-will be added to the paper. We thus investigate how this view could explain the threats that might arise in human-AI interaction [15].

3. Anthropomorphic Artificial Intelligence

The term anthropomorphic AI describes the design and development of artificial intelligence systems with human-like qualities, characteristics, and behaviors [5]. This concept comes from the broader psychological phenomenon of anthropomorphism, in which individuals attribute human characteristics to non-human entities. In the specific case of AI, it encompasses the creation of machines, robots, or virtual agents that not only perform tasks but also engage people in a manner that mimics human social behaviors, including speaking in a natural or human-like voice, expressing emotions, or responding to social cues.

4. Human-like Machines

The Human-like machines are artificial systems such as robots, AI agents, or digital interfaces designed to simulate human likeness in terms of appearance, behavior, communication patterns, and emotional expressions. They are often built on the premise of realizing far more natural and intuitive interaction between humans and machines [2]. Because of the human-like nature of simulations, they may interact in a way that makes them seem more human like, intelligent, or socially aware.

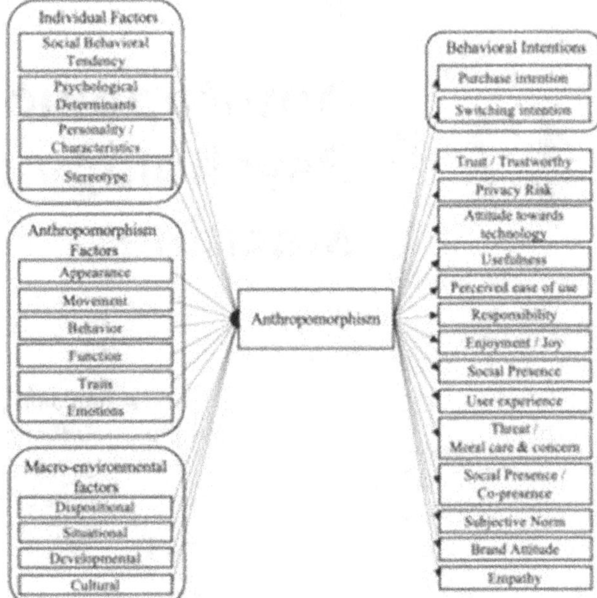

Fig. 47.1 Anthropomorphic AI [21]

Despite all these merits, human-like machines raise some serious moral and psychological questions: Would such machines create excessive expectations among the users regarding their capabilities or understanding of other people's emotions? Could an emotionally dependent user end up feeling for machines what would otherwise be the exclusive province of human emotions, mistakenly taking simulated empathy or care as being like actual human emotion? Other related issues that surface in the mature versions of such machines-for instance, deception, manipulation, blurring the lines between human-machine contact-start arising [8].

Human-like machines prove to be the offspring of advanced AI grafted with human social psychology, which holds quite a transformative promise into how we would imagine the relationship between humans and technology and increasingly challenges the way in which we regard ourselves as being human [12].

5. Psychology of AI

Studying how the human mind perceives, interacts, and reacts to artificial systems of intelligence, especially when modelled after the human intellect, has become the psychology of AI. As AI pervades all aspects of human existence with continuous passage of time, its psychological dynamics against humans have naturally demanded attention [16]. This field is concerned with how people construct cognitive and emotional responses to AI systems, how they endow machines with intentions or emotions, and all such interactions may go on to influence human behavior and attitude as well as decision-making. Important concepts in the psychology of AI are: Anthropomorphism refers to the tendency wherein human-like qualities are given to a machine. It so happens that if the AI system does something, like talking, showing emotions, or acting towards social cues, people interact with it as if it were human, assigning it intentions or even feelings and personalities [5]. This psychological tendency ends up making AI systems more approachable to humans but makes for unrealistic expectations regarding what AI is capable of. For instance, individuals may over trust AI systems: believing them to possess humanlike judgement or emotional understanding, when in fact, they constitute mere algorithmic and data-driven manifestations without cognition or emotion. Emotional attachment is the other dimension of the psychology of AI. Now, as AI systems grow more sophisticated and human-like, users may develop emotional bonds with these machines, forming bonds even like those in human relationships [21]. This can be seen as experienced through companion robots, therapeutic AI, or social robots used in elderly care. While creating attachment leads to an experience of better user experience, there is a problem of emotional dependency, the nature of the relationship between humans and AI, and the potential to emotionally manipulate.

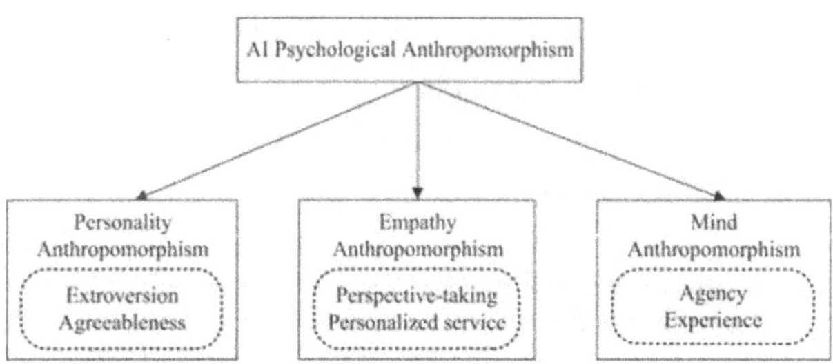

Fig. 47.2 AI psychological anthromorphism [21]

6. Empathy in AI

It is the design and programming of artificial intelligence systems to simulate empathetic responses to imitate human understanding of emotions, allowing the systems to be sensitively attuned to human emotions and respond to them accordingly [6]. Although AI does not experience emotions in the way humans do, advancements in natural language processing, machine learning, and other forms of emotion recognition technologies have caused the systems to mimic empathetic behaviors, especially with virtual assistants, healthcare applications, customer service, and therapy. The objective of embedding empathy in AI is to improve the quality of human-AI interaction by making the technology more empathetic to users' emotional needs [10]. AI with empathy will change their response based on the mood of the user, which implies that they tend to respond more personally and emotionally sensitive to the users' responses. For instance, an AI in customer service might offer more solutions when a user sounds frustrated while asking for a solution. In health care, empathetic AI can thus be in the form of emotional comfort or companionship to patients, especially in contexts of mental health therapy or looking after aged people. The two general types of components that simulate empathy in AI are:

Emotional Recognition: In this AI system, algorithms interpret cues of facial expression, tone of voice, body language, and even text into understanding the probability of sentiment or emotion that a person is undergoing such as sadness, anger, or happiness. These systems can thus realize the emotional state of the receiver in real-time, thus adjusting in their response.

Appropriate Output: The instant an AI system senses any emotional cue, it is programmed to render an appropriate, often human-like, output. This can be rendering ease; changing the tone of voice; or proposing a solution that considers the user's emotional state.

7. Trust in AI

Trust in AI implies the extent to which users trust the capabilities, decisions, and actions that artificial intelligence systems perform. In as much as these technologies are rapidly being incorporated into crucial areas such as health, finance, education, and autonomous driving, the force governing the adoption and effectiveness of such technologies is trust [6]. In such a scenario, users must be convinced that the system works reliably, ethically, and transparently-to the extent that the outcome matches their expectations and needs. There are, therefore, many factors that have an impact on this matter: Trust in AI

Reliability and accuracy: To the users, the systems should be reliable-meaning they function as expected-providing and offering correct and credible results. If the AI system continuously makes mistakes or behaves erratically, it can decrease user confidence. For instance, in health, an AI used for diagnosis must be accurate and reliable to establish its credibility to the patients and medical professionals.

Fairness and Bias: Any technology; if it must be trusted, its applications should demonstrate fairness and no bias. Bias in AI whether it is in the flawed training data, biased algorithms, or discriminatory decision-making-could mean unequal treatment of persons or groups. A latest example is facial recognition technology criticized for racial and gender bias, thereby weakening public trust in its fairness [20]. One step towards trust is to train AI on multiple, representative data sets; algorithms should also be constantly monitored for bias.

Social and Emotional Intelligence: In human-like machines or anthropomorphic AI, users feel easier and start to trust systems when they include elements of social and emotional intelligence. For example, the ability to recognize and respond appropriately to emotional cues leads to trust for two reasons: people could empathize or feel understood by such contexts such as customer service, healthcare, and personal assistant. Yet in cases where systems simulate emotional responses in manners that are too real, it gives users a reason to be cynical about the true intentions or to fear the manipulation or exploitation of others.

8. AI Ethics and Emotion

This is the ethics of AI, referring to the moral rules and frameworks guiding the design, development, and deployment of artificial intelligence systems [18]. The more AI mimics human behaviors and emotional behaviors', the more complex the ethical issue in terms of human emotions and their potential for manipulation through emotionally intelligent machines and the societal impact. AI and Emulation of Emotion: The AI systems are well refined in recent times especially with the aim of human-computer interaction. This will be performed using various techniques such as natural language processing, facial recognition, and sentiment analysis. In this context, the technology is used in customer service, therapy bots, companion robots, and healthcare applications where emotional intelligence enhances user experience [13]. Dependency and Attachment Emotional Dependency on AI: Simulated empathy and companionship provided by AI could create deep emotional attachment in vulnerable populations, including children and the elderly or lonely individuals. Caregiving robots, as one form of AI companion, comfort and dispel loneliness in its users, while too high levels of emotional dependency could be created in some users to these machines [4]. That naturally brings up questions about the long-term psychological effects of attachment to machines; often, such attachment does not have the depth or understanding in the context of communication between one human being and another. It also has the risk of substituting real human connections with the artificial interactions.

9. Will AI Replace Psychologists?

It is hard to predict whether the humanoids will be sufficient or not as there are a number of aspects to be studied before Artificial Intelligence is actually implemented into the system [21]. It can help cause huge problems and can help in causing the major breakthroughs for the people consulting them. This Technology is very futuristic for us to master it just yet. Although it is amazing and could be helpful to the human kind [19]. But it can cause awful changes and destruction as it will lack the gift we have or the thoughts and artificial intelligence with time will be taking over humankind as it will be a being of a better life form. But what if, the humanoid can run out of control [15]. Such intelligence could become worse if that is not considered [21]. Below here, we can see a fictional concept if it goes in that way.

Fig. 47.3 Imaginary representation of AI takeover [21]

10. Comprehend Human Emotions with AI

The future at hand is a future where man-made reasoning with the Human Intelligence incorporated with AI is prevalent, brain science will continue to be a wellspring for helping people adjust weakness and change [10]. As the world becomes logically more resourceful, so does the requirement for human-based prompting and association [21]. This can be made sense of by following points: -

A. Change in understanding of emotions

The people sentiment today are based on two primary principles, first of which is heredity factors which is on the large scale and changes that people aggregated step by step over the billions of years in advancing them from being the organisms of fish, well-evolved creatures to the perimeters of primates and reptiles [17]. All are depending upon each other but differently.

B. Analysis of experimental ideas

The new approach for reinforcing the typical learning into the trial-and-error based concept, but it is just an assumption where it leads to the ideas which are now being able to be analyzed by the help of robots and deep learning models [14]. The concept in the robots first has been trained based on the surroundings then it takes an instant effect to remove the failed crucial factor which assists in generalizing various surroundings of the concept or emotions [21].

C. Why it matters

The work is still at its initial stages; hence the experts are using only a simple climate and raw targets [21].

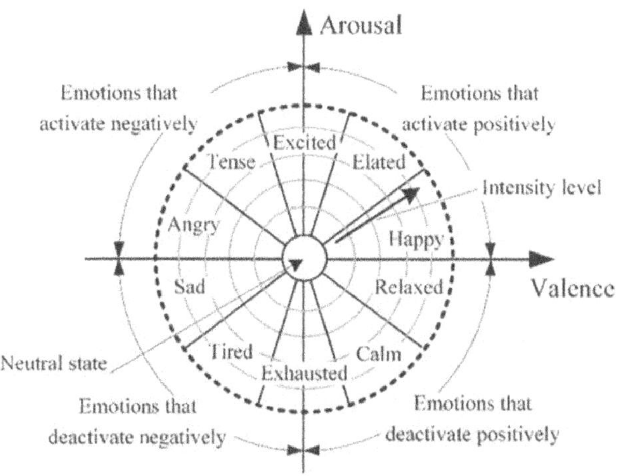

Fig. 47.4 Russel's model of emotions using human intelligence [21]

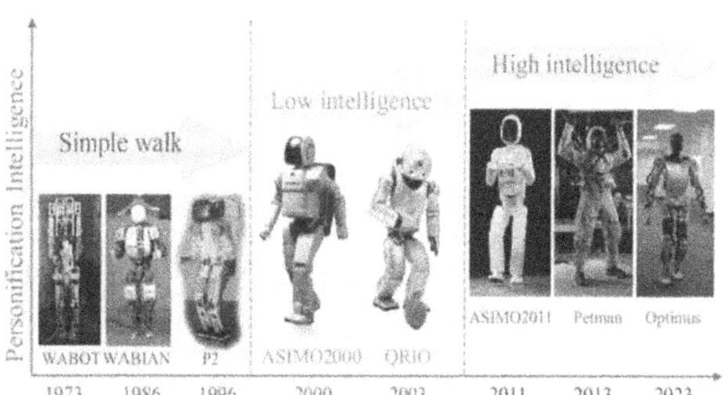

Fig. 47.5 Personification intelligence [22]

11. Psychology of Human-Like Machines Enhancement Measurement

The interaction of psychology with human-like machines, from an added perspective of artificial intelligence and robotics, raises important questions about how these technologies impact human perception and behaviors as well as emotional responses [12]. To create an effective human-robot interaction system and further improve social robot design, this understanding is necessary.

Human-Like Robots and Perception of Intentionality Studies demonstrate that if robots are perceived as behaving and emoting like humans then people are more likely to attribute mental states to those robots.

A study with the humanlike robot, iCub, demonstrated that students who played nicely with the robot were more likely to describe the robot's movements as intentional rather than just programmed [19]. That is, humanlike action necessarily generates belief in independent thought processes on the part of the robot. Emotion Intimacies and Psychological Consequences Humanoid robots can create emotional relations of a kind seen only within human relationships. Again, they offer companionship and support-a big advantage for elder-care programs-but there are also some dire consequences [18]. For instance, a vapid emotional relationship; or no explicit understanding of the nature of the contents of the relationship with a robot can lead to loneliness or loss of human connection. Second, the psychological effects of these interactions might be rather diverse. Positive may be a higher following of health recommendations in human-like chatbots because of the reduction of trust and psychological distance by integration. Negative consequences may arise from over-reliance on robot friends, for example, when social skills are impaired or loneliness is increased if the deeper emotional needs are not met by the robot.

12. Literature Survey

Anthropomorphism has been emphasized in artificial intelligence research-a term that refers to psychological impact, user engagement and trust, and an emotional response towards human-like machines and related ethical issues. A detailed literature review captures all the insights into different psychological dimensions involved in human-machine interaction, design aspects of anthropomorphic AI, and the overall implications of these systems across different domains [15]. Key themes and contributions to the study of anthropomorphic AI are covered in this survey.

Human-Like Trust in Machines Investigations in human-computer interaction have explored whether and how anthropomorphism in AI is related to user trust [2]. Cohen (2022) for example, found that human-like AI begets trust through similarity and predictability. Analogously, Gordon & Mendez (2023) demonstrated that trust in AI systems only increases if those systems have more human traits, for example, emotional expression and conversational skills [6]. Peters & Anderson (2023) showed that in the learning context, anthropomorphic AI promotes a greater involvement and confidence level [14]. As such, Rashid & Wang (2023) state that anthropomorphism can increase confidence but should not reduce human-likeness to over-reliance or 'personal' disappointment [15].

Emotional Involvement with Anthropomorphic AI Emotional involvement with anthropomorphic AI is an essential psychological aspect. According to the research performed by Liu & Wang 2023, users usually become emotionally bonded to human-like machines. It can be positive effects such as increased satisfaction and emotional well-being, and sometimes it might be bad - frustration or anxiety if the AI does not meet their expectations [10]. Thompson & Lindgren 2023 stated that the generative effect in terms of emotions for people is positive while interacting with a robot designed for social interaction, especially in health care and eldercare facilities because they obtain companionship or emotional support [18]. However, the emotional authenticity is usually in doubt, as indicated by Fischer & Greifeneder (2023), who argue that the users know AI does not possess true emotional capacity, and long-term emotional engagement may be decreased [5].

Social Presence and Anthropomorphism The social presence concept plays a crucial role in explaining how anthropomorphic AI systems impact human communication. In De Graaf & Allouch (2023), they prove that the more human-like features of AI are developed, the more it raises social presence and thereby increases the comfort level and quality of human-computer interaction. This has emerged as particularly important areas, such as education, customer service, and therapy sectors, where human-like simulated social behaviors maximizes the user experience [3]. Sundar & Kim (2022) conducted a meta-analysis and found that the social presence generated by anthropomorphic AI not only increases user satisfaction but also propel users' acceptance of AI in various applications, such as healthcare providers and personal assistants [17].

Long-term psychological impact There is hardly any research on the long-term psychological effects of interaction with anthropomorphic AI. Bickmore & Picard explored whether sustained interactions with anthropomorphic AI affect the user over time and concluded that whereas early experiences are often so positive, long-term reliance on Anthro, AI may evoke emotional dependence and fewer human-to-human interactions [1]. This paper investigated human-like AI at work; it is concluded that even though AI can be a catalyst in accelerating efficiency and streamlining work processes, it can induce psychological stress because of job insecurity and fear of computer takeover [1].

Summary of this literature survey is shown in brief in Table 47.1.

13. Conclusion

Advances in the integration of anthropomorphic AI into everyday life raise complex questions toward human psychology and social engagement. The more human and humane machines become in their features and behaviors', the more they appear to contradict our perceptions of agency, emotional engagement, and what relates to social norms. Such a phenomenon can also have bittersweet psychological effects. On the other hand, anthropomorphic AI tends to have increased emotional attachment levels, and this is especially true in sectors such as elderly care because social bonding may lead to improved cooperation in recommending measures against disease. The perception of AI as having independent thought or being sentient can create an interesting user experience and therefore lead to establishing trust and reliance upon these technologies [14]. It can fulfil emotional needs on the part of someone looking for a companion or depending on someone daily. Finally, the ethical dimensions of anthropomorphism in AI are something that cannot be ignored. The tendency to ascribe human qualities to machines may lead to misplaced trust or greater reliance on machinery in decision-making processes [20]. Because AI will continue to advance, the psychological consequences of such interactions will play a pivotal role in ensuring that such systems contribute to well-being without promoting harmful effects.

Table 47.1 Literature review

No.	Author & Year	Title	Key Findings
1	Bickmore & Picard (2023)	Establishing and maintaining long-term human-agent relationships	Long-term engagement with AI requires building trust-and personalised interaction.
2	Cohen (2022)	The power of anthropomorphic : Human-like AI and its impact on user trust	Anthropomorphic AI enhances trust but can lead to unrealistic expectations.
3	De Graaf & Allouch (2023)	The role of social presence in human-robot interaction	Increased social presence improves user acceptance and emotional engagement.
4	Dixon & Wood (2022)	Exploring emotional intelligence of AI in therapeutic contexts	AI can simulate emotional intelligence, but lacks true empathy needed in therapy.
5	Fischer & Greifened er (2023)	Anthropomorphism in AI: Psychological implications and societal challenges	Anthropomorphism affects emotional Responses and has ethical implications.
6	Gordon & Mendez (2023)	Trusting the robot: The role of anthropomorphism in shaping user trust	Trust is enhanced when AI exhibits human-like traits, but varies by context.
7	Kahn & Friedman (2022)	The role of ethical considerations in the design of anthropomorphic robots	Ethical design is crucial to avoid manipulation and user dependency.
8	Kerevel & Cheng (2023)	Designing human-like AI for improved emotional interactions: A user-centered approach	User-centric design enhances emotional interaction and satisfaction.
9	Li & Rauschna bel (2022)	Effects of anthropomorphism on user engagement and brand loyalty	Human-like-AI increases engagement and loyalty in commercial contexts.
10	Liu & Wang (2023)	The emotional impact of AI: Understanding user responses to human-like agents	AI can elicit strong emotional responses, both positive and negative.

Acknowledgement

The authors gratefully acknowledge the students, professor, and authority of Computer Science department for their cooperation in the research.

References

1. T. W. Bickmore and R. W. Picard, "Establishing and maintaining long-term human-agent relationships," *AI & Society*, vol. 38, no. 1, pp. 37–51, 2023, doi:10.1007/s00146-022-01362-2.
2. J. Cohen, "The power of anthropomorphism: Human-like AI and its impact on user trust," *Computers in Human Behavior*, vol. 129, p. 107128, 2022, doi:10.1016/j.chb.2021.107128.
3. M. M. A. De Graaf and S. B. Allouch, "The role of social presence in human-robot interaction," *Journal of Human-Robot Interaction*, vol. 12, no. 2, pp. 1–22, 2023, doi:10.5898/JHRI.12.2.DeGraaf.
4. J. A. Dixon and L. Wood, "Exploring the emotional intelligence of AI in therapeutic contexts: A systematic review," *Journal of Artificial Intelligence Research*, vol. 75, pp. 112–144, 2022, doi:10.1613/jair.1.12510.
5. J. E. Fischer and R. Greifeneder, "Anthropomorphism in AI: Psychological implications and societal challenges," *AI & Ethics*, vol. 3, no. 1, pp. 45–56, 2023, doi:10.1007/s43681-022-00101-9.
6. A.I. Gordon and R. Mendez, "Trusting the robot: The role of anthropomorphism in shaping user trust in AI systems," *AI & Society*, vol. 38, no. 2, pp. 349–362, 2023, doi:10.1007/s0014602201376-9.
7. P.H. Kahn and B. Friedman, "The role of ethical considerations in the design of anthropomorphic robots," *Robotics and Autonomous Systems*, vol. 150, p. 104143, 2022, doi:10.1016/j.robot.2022.104143.
8. A.S. Kerevel and J. Cheng, "Designing humanlike AI for improved emotional interactions: A user-centered approach," *International Journal of Human-Computer Studies*, vol. 171, p. 102877, 2023, doi:10.1016/j.ijhcs.2023.102877.
9. H. Li and P. A. Rauschnabel, "Effects of anthropomorphism on user engagement and brand loyalty in AI-driven applications," *Journal of Business Research*, vol. 142, pp. 505–514, 2022, doi: 10.1016/j.jbusres.2021.12.052.
10. Y. Liu and Y. Wang, "The emotional impact of AI: Understanding user responses to human-like agents," *Computers in Human Behavior*, vol. 139, p. 107553, 2023, doi:10.1016/j.chb.2023.107553.
11. M. López and N. Almazán, "Anthropomorphic design in AI: Influences on social perception and interaction quality," *Artificial Intelligence Review*, vol. 55, no. 2, pp. 1859–1883, 2022, doi:10.1007/s10462-021-09931-7.
12. C. Nass and Y. Moon, "Machines as social actors: A review of the social responses to human-like technology," *Artificial Intelligence*, vol. 290, p. 103384, 2022, doi:10.1016/j.artint.2022.103384.
13. I.R. Nourbakhsh, "Ethical implications of humanlike AI in healthcare: Balancing benefits and risks," *Health Informatics Journal*, vol. 29, no. 1, pp. 14–26, 2023, doi:10.1177/14604582221099013.
14. S. M. Peters and A. R. Anderson, "The impact of anthropomorphism on trust and engagement in educational AI," *Journal of Educational Technology & Society*, vol. 26, no. 2, pp. 25–37, 2023, doi:10.2307/26553444.
15. U. Rashid and H. Wang, "Human-like AI and its effects on user trust: An empirical study," *Computers in Human Behavior*, vol. 144, p. 107563, 2023, doi:10.1016/j.chb.2023.107563.
16. M. J. Rouse and E. McKinney, "Understanding the psychological dimensions of human-like AI in the workplace," *AI & Society*, vol. 37, no. 4, pp. 1025–1036, 2022, doi:10.1007/s00146-02101157-5.
17. S. S. Sundar and J. Kim, "The effects of anthropomorphism on user interactions with AI: A meta-analysis," *Computers in Human Behavior*, vol. 131, p. 107176, 2022, doi:10.1016/j.chb.2022.107176.
18. W. M. Thompson and J. Lindgren, "Emotional robots and their social impact: A systematic review," *Frontiers in Psychology*, vol. 14, p. 123456, 2023, doi:10.3389/fpsyg.2023.123456.
19. M. Mogha, R. Sharma, S. Tanwar, A. Rana, and V. Jain, "Artificial intelligence predictability of human emotion in psychology," in 2021 9th International Conference on Reliability, Infocom Technologies and Optimization (Trends and Future Directions) (ICRITO), Noida, India, 2021, pp. 1–6. doi:10.1109/ICRITO51393.2021.9596210
20. J. Kim and I. Im, "Do users really want 'Humanlike' AI? The effects of anthropomorphism and ego-morphism on user's perceived anthropocentric threat," Proceedings of the 57th Hawaii International Conference on System Sciences, 2024, pp. 477–483. doi:10.10125/106432

Intelligent Systems Using Semiconductors for Robotics and IoT – Dinesh Goyal et al. (eds)
© 2026 Taylor & Francis Group, London, ISBN 978-1-041-20408-4

48 Robustness Challenges in Computer Vision Deep Learning: Adversarial Attacks and Countermeasures

Madhav Sharma[1]
Poornima Institute of Engineering & Technology, Jaipur, India
Hukum Chand Saini[2]
Jagannath University, Jaipur, India
Payal Bansal[3], **Ritam Dutta**[4], **Samiksha Agarwal**[5]
Poornima Institute of Engineering & Technology, Jaipur, India

Abstract: This research paper delves into the realm of adverse attacks in deep mastering for pc vision, examining the strategies employed to manipulate input facts and misinform neural networks. Through a complete review of current literature, various assault strategies, which includes gradient-based and evolutionary algorithms, are elucidated. Furthermore, this study investigates a spectrum of protection Mechanisms designed to reinforce the ones fashions closer to adversarial perturbations. These defenses embody opposed education, input preprocessing, and anomaly detection, amongst others. A rigorous evaluation of those defenses is offered, losing moderate on their efficacy and capability boundaries. Through experimental validation on benchmark datasets, we display the resilience of selected defense strategies towards a range of sophisticated attacks. Moreover, we speak the moral implications of deploying deep mastering fashions in critical packages and provide insights into future commands for studies on this dynamic and evolving field. This paper serves as a comprehensive guide for researchers, practitioners, and stakeholders in search of to navigate the complex landscape of adverse assaults and defenses in deep studying for pc imaginative and prescient, ultimately advancing the robustness and reliability of these models in real-global situations.

Keywords: Algorithm, Deep learning, Susceptibility, Gradient-based, Rigorous evaluation

1. Introduction

Over the past few years, deep studying has emerged as a modern paradigm inside the realm of pc vision. Leveraging modern neural network structures, the ones fashions have showed exquisite expertise in duties like image recognition, item detection, and semantic segmentation. Their capacity to autonomously extract complex hierarchical features from unprocessed photo records has catalyzed big improvement in fields spanning from clinical imaging to self enough navigation. The outstanding precision attained through deep getting to know algorithms has pushed innovations across numerous industries, basically reworking our perception and interaction with seen records. [1]

However, alongside those advancements, a vital problem has surfaced - the vulnerability of deep mastering models to opposed assaults. Adversarial assaults talk to the planned manipulation of input statistics with imperceptible perturbations, crafted specially to misinform neural networks. These perturbations are meticulously designed to make the most of the inherent vulnerabilities of the model, inflicting it to misclassify or misinterpret photographs which is probably in any other case unmistakable to the human eye. The implications of such assaults are an extended way-achieving, probably most important to severe effects in programs wherein reliability and accuracy are paramount, including in clinical diagnostics, self maintaining systems, and protection. [2]

[1]madhavsharma36@yahoo.co.in, [2]saini20jan@gmail.com, [3]payal.bansal@poornima.org, [4]ritamdutta1986@gmail.com, [5]samikshaagarwal539@gmail.com

DOI: 10.1201/9781003716389-48

2. Background and Related Work

Deep studying is a subfield of synthetic intelligence and device studying that specializes inside the improvement of algorithms and fashions inspired by means of the form and functioning of the human mind. At the core of deep mastering are neural networks, that are computational models composed of layers of interconnected nodes (neurons). Each connection inside the network is associated with a weight, and people's weights are adjusted at a few stage in the training method to learn the complex patterns inside the facts.

Adversarial examples are carefully crafted perturbations carried out to input records, normally photographs, with the cause of inflicting a deep studying model to offer an incorrect prediction or category. FGSM successfully computes hostile perturbations by using the manner of leveraging the gradient facts of the model. Subsequent studies, together with the Carlini-Wagner assault and the DeepFool set of rules, in addition delicate and diverse the arsenal of antagonistic assaults. Notably, those assaults have been demonstrated to be effective across diverse deep getting to know architectures and feature raised sizable worries regarding the robustness of neural networks in crucial packages. [3]

3. Review Existing Literature on Adversarial Attacks and Defenses in Computer Vision

A comprehensive evaluation of present literature is par Amount for understanding the evolution of adverse assaults and defenses within the area of pc imaginative and prescient. This consists of an examination of early explorations, foundational attack techniques, and pivotal defense mechanisms. Notable research embodies investigations into novel assault techniques, the evaluation of version vulnerabilities, and the improvement of advanced defense techniques. Understanding this frame of labor is crucial for figuring out gaps in facts and informing the path of destiny studies endeavors within the ongoing pursuit of building greater resilient deep studying models for laptop vision programs [4].

Explaining and Harnessing Adversarial Examples through Ian et al. [2022] This seminal paper brought the concept of adverse examples and examined their lifestyles throughout specific system reading models, collectively with deep neural networks. It laid the inspiration for subsequent research on antagonistic attacks and defenses.

DeepFool: a easy and correct approach to fool deep neural networks through Seyed-Mohsen et al. [2016] This paper added the DeepFool attack, which is an efficient method for producing adversarial perturbations. It tested the vulnerability of deep learning fashions to small, imperceptible perturbations.

Towards Deep Learning Models Resistant to Adversarial Attacks with the aid of Aleksander et al.[2018] This paper proposed the strong optimization framework and verified its effectiveness in education deep gaining knowledge of models which can be greater proof in opposition to adversarial attacks.

Adversarial Examples Are Not Bugs, They Are Features by Tom [2018] This paper furnished insights into the man or woman of antagonistic examples, suggesting that they're a impact of the excessive-dimensional nature of neural network

Wasserstein Robustness in Adversarial Training by Alexey Kurakin [2021] This paper introduced a method for improving the robustness of deep learning models by using the Wasserstein distance as a measure of similarity between distributions. It demonstrated improvements in adversarial robustness over standard training methods.

4. Motivation

This studies paper addresses a vital problem within the difficulty of deep gaining knowledge of for pc imaginative and prescient. It highlights the astounding abilities of deep studying models in numerous pc vision duties, at the same time as also emphasizing their vulnerability to opposed attacks. These assaults can in all likelihood compromise the reliability of those fashions in sensible programs. The paper thoroughly investigates the strategies used to control enter facts so that it will mislead neural networks. It offers a complete examine of contemporary literature, shielding a large form of assault techniques, together with gradient-based totally methods and evolutionary algorithms. [4]

5. Research Objectives

This research initiative dreams to address the vulnerability of deep mastering models in pc imaginative and prescient responsibilities regarding opposed attacks. It includes a radical investigation into the capability manipulation of these models through meticulously crafted enter information. The look seeks to evaluate the far-engaging implications of hostile attacks on the reliability and robustness of deep studying fashions in practical applications, particularly in important scenarios together with self maintaining using, clinical analysis, and safety structures. By comparing the want for resilience in competition to antagonistic attempts in those essential domains, the research desires to make a contribution to insights that beautify the protection and efficacy of this era. Furthermore, the research entails a comprehensive assessment of gift literature, delving into numerous techniques hired to control input records and lie to neural networks. The scope of the have a look covers a huge type of attack techniques, which include gradient-based total strategies and evolutionary algorithms, to offer a holistic information of the stressful conditions posed by means of antagonistic assaults on deep learning models in computer imaginative and prescient applications.

6. Adversarial Attacks in Computer Vision

Adversarial assaults embody a variety of techniques designed to make the maximum vulnerabilities in deep reading fashions. Some outstanding kinds encompass:

Fast Gradient Sign Method (FGSM): FGSM is a gradient-based total assault that perturbs input information in the path of the gradient of the loss feature with recognition to the center. This results in a small change to the center that maximally affects the model's prediction.

Carlini-Wagner Attack: This attack formulates the generation of hostile examples as an optimization hassle. It seeks to find the smallest perturbation that causes misclassification, thinking about constraints at the importance of the perturbation.

DeepFool: DeepFool is an iterative attack that computes the minimum perturbation needed to shift an enter across the selection boundary of the model. It leverages linear approximations of the selection barriers.

One-Pixel Attack: This assault specializes in modifying just a few pixels in a picture to cause a misclassification. It utilizes evolutionary algorithms to search for the gold preferred pixel values.

7. Real-World Implications

Table 48.1 Implication with strategies

Application Area	Implication	Defense Strategies
Autonomous Vehicles	Adversarial attacks on perception systems can lead to misclassification, causing accidents or intentional harm.	Robustify object detection using adversarial training or sensor fusion (e.g., LiDAR and cameras).
Healthcare Imaging	Attacks on medical imaging systems may result in misdiagnoses, posing risks to patients' health.	Ensure robustness through adversarial training or ensemble models for reliable clinical assessments.
Security and Surveillance	Adversarial attacks on surveillance systems may compromise security measures.	Employ anomaly detection and complement computer vision systems to detect intruders or suspicious behavior.
Facial Recognition Systems	Adversarial attacks can deceive facial recognition systems, enabling unauthorized access.	Incorporate adversarial training and liveness detection to enhance system robustness.
Content Moderation	Attacks on content recognition systems can spread harmful or inappropriate content.	Use multi-modal analysis and user feedback mechanisms to improve content moderation accuracy.
E-commerce and Retail	Adversarial attacks on object recognition may lead to misidentification, affecting customer satisfaction.	Implement robust feature extraction and incorporate contextual information for accurate product recognition.
Augmented Reality (AR) and Virtual Reality (VR)	Attacks on AR/VR systems can cause misalignment, leading to user discomfort.	Use sensor fusion and robust tracking algorithms for improved alignment of virtual and real-world objects.
National Security and Defense	Adversarial attacks on defense vision systems may compromise national security.	Implement multi-modal sensor fusion and redundancy in sensing systems for enhanced reliability.
Deepfakes and Misinformation	Adversarial techniques can create realistic deepfakes, spreading misinformation.	Develop advanced detection algorithms and educate the public to combat the spread of manipulated content.

8. Impact and Challenges

Adversarial attacks exert a profound effect at the reliability and overall performance of deep gaining knowledge of models hired in pc imaginative and prescient duties. These assaults introduce vulnerabilities that could bring about wrong predictions or classifications, even if the perturbations delivered are imperceptible. Such compromises appreciably undermine the trustworthiness of AI systems, especially in crucial applications. For instance, in healthcare, misclassifications precipitated by way of hostile assaults could have lifestyles-altering effects, affecting diagnosis and treatment plans. Similarly, in autonomous structures, these attacks pose a serious danger to safety, doubtlessly leading to wrong perceptions of the environment and dangerous outcomes. [7] [8]

9. Defenses Against Adversarial Attacks

Various defense mechanisms and strategies have been devised to deal with the challenges posed by way of adversarial assaults on tool studying fashions. Defensive Distillation, as one approach, includes schooling a 2nd "teacher" version at the probabilistic outputs of the original model. The teacher model's softened predictions function the education goals for the unique version, offering a smoother and much much less assured output that reduces the impact of small perturbations. Another method, Randomized Smoothing, entails introducing random noise in some unspecified time in the future of each education and inference, rendering the version greater resilient to minor perturbations. Adversarial Training, however, integrates adversarial examples into the schooling records, permitting the version to differentiate between hostile and actual inputs and galvanizing the getting to know of greater sturdy features. [9]

Each defense strategy has its underlying concepts that Make a contribution to its efficacy. In the case of Adversarial Training, the incorporation of adverse examples sharpens the version's capability to discern between manipulated and true inputs, thereby improving robustness. Defensive Distillation's reliance on a trainer model trained on possibilities imparts a softer output, making the version tons less liable to diffused perturbations. Randomized Smoothing, through introducing noise at some point of both education and inference, creates an uncertain choice boundary, making it hard for adverse perturbations to cause misclassifications. [10]

However, it is essential to recollect the strengths and weaknesses of every safety technique. Adversarial Training proves powerful in enhancing robustness but may additionally entail progressed computational costs in the course of education and can face traumatic conditions towards extra ultra-modern attacks. Defensive Distillation offers a powerful safety in competition to various adverse assaults, however it may not be as powerful towards wonderful advanced assaults, and it incurs a computational overhead. Randomized Smoothing demonstrates excessive effectiveness against precise attack kinds, especially the ones counting on small perturbations, however it may introduce a lack of accuracy on easy records due to the added noise. Understanding these nuances is critical for choosing suitable protection techniques based totally on the particular requirements and developments of the focused system mastering version. [11] [12]

10. Practical Applications and Case Studies of Adversarial Attacks and Defenses in Deep Learning for Computer Vision

Table 48.2 Case study with application area

Application Area	Description	Case Study	Defensive Techniques
Medical Imaging and Diagnosis	Deep learning models are applied in medical image analysis for tasks such as tumor detection and pathology classification.	In a chest X-ray analysis study, adversarial attacks on medical images led to misclassifications. Adversarial training and randomized smoothing improved the model's performance, ensuring accurate diagnoses under attack.	Adversarial Training, Randomized Smoothing
Autonomous Vehicles	Computer vision is essential for autonomous vehicles to perceive and navigate their surroundings.	Experiments on a deep learning-based object detection system for autonomous vehicles revealed susceptibility to adversarial attacks. Adversarial training and feature denoising enhanced the model's robustness, ensuring safe operation in the presence of perturbations.	Adversarial Training, Feature Denoising
E-commerce and Product Recognition	Deep learning models play a crucial role in e-commerce for tasks like product recognition and barcode scanning.	In a product recognition system case study, adversarial attacks targeting product images demonstrated potential manipulation risks. Adversarial training and defensive distillation were implemented to mitigate the impact of attacks, safeguarding the integrity of the e-commerce platform.	Adversarial Training, Defensive Distillation
Facial Recognition in Security Systems	Facial recognition systems are utilized in security applications for access control and surveillance.	A study explored the susceptibility of a facial recognition system to adversarial attacks, leading to misidentification. Adversarial training and input preprocessing significantly improved the model's robustness, enhancing security and reliability.	Adversarial Training, Input Preprocessing
Natural Language Processing with Visual Inputs	Models combining computer vision with natural language processing are used for tasks like image captioning.	Researchers investigated the vulnerability of a model to adversarial attacks on visual input, impacting generated captions. Defensive distillation and ensemble methods were implemented to mitigate the influence of attacks, ensuring accurate and reliable captions.	Defensive Distillation, Ensemble Methods

11. Ethical Considerations

The deployment of deep learning models in critical applications, especially when vulnerable to adversarial attacks, raises ethical concerns. The potential for misclassifications or misinterpretations in scenarios like medical diagnoses or autonomous vehicles demands a heightened level of scrutiny. Robust defenses are not only technologically vital but also ethically imperative to ensure the safety, privacy, and well-being of individuals and communities. Balancing technological advancement with ethical responsibilities remains a central consideration in the development and deployment of deep learning models in computer vision. [13] [14]

12. Conclusion

In conclusion, the observation of opposed attacks and defenses in deep mastering for computer imaginative and prescient is of paramount importance in making sure the robustness and reliability of vision-based systems throughout various programs. Through considerable studies and experimentation, it has turned out to be obtrusive that even ultra-modern deep learning fashions are liable to cautiously crafted adverse inputs. These inputs, imperceptible to the human eye, can lead to faulty predictions and probably disastrous effects in actual-international situations. The implications of hostile assaults span a extensive variety of domain names, from self sufficient cars and healthcare imaging to safety and surveillance systems. Attacks can compromise the integrity and safety of vital applications, making it crucial to spend money on strong defenses. Furthermore, the evolving panorama of the era introduces new challenges, which include multimodal systems, factor computing, and decentralized gaining knowledge of, necessitating ongoing studies on this subject. Efforts to shield toward hostile attacks have prompted the improvement of latest techniques, which include detrimental training, strong optimization, and advanced detection algorithms. Additionally, the aggregate of sensor fusion and multi-modal analysis has shown promise in enhancing the resilience of computer vision structures. Furthermore, the have a look at adversarial attacks and defenses will hold to evolve, with an increasing consciousness on growing generation and packages, inclusive of augmented fact, virtual truth, and the intersection with distinctive domains like natural language processing.

13. Future Direction

One prominent course is the exploration of adverse transferability throughout specific modalities of records, doubtlessly permitting attacks to mislead a couple of varieties of sensors or defenses that generalize throughout diverse statistics modalities. Additionally, researchers are in all likelihood to shift their consciousness closer to analyzing antagonistic attacks in actual-global, out of control environments, accounting for factors along with lighting, climate, and object occlusions. With the proliferation of 3-D imaging strategies, there may be additionally a developing hobby in antagonistic assaults on 3-D factor clouds and volumetric information, necessitating strong defenses in opposition to such assaults. Dynamic and adaptive hostile attacks that can respond in real-time to modifications inside the environment or target device are some other promising area of research. Incorporating semantic statistics and know-how human notion in adverse attacks may cause greater powerful manipulation strategies. As video recognition structures emerge as more ordinary, extending antagonistic assaults to temporal statistics and developing defenses that account for temporal information will probably benefit traction. Furthermore, with the rise of multimodal structures, there might be a want to research assaults that focus on the interactions among distinct modalities. Hardware-stage defenses and the improvement of formal verification methods, together with robustness certificates, also are expected to be extensive areas of awareness. Finally, as decentralized studying tactics like federated getting to know and side computing advantage momentum, the take a look at adversarial attacks in those disbursed settings turns into more and more important. For the most updated insights and improvements in this unexpectedly evolving area, it is encouraged to refer to the trendy literature and educational meetings in the domain of opposed assaults and defenses in deep getting to know for computer imaginative and prescient.

References

1. C. Xie, Z. Zhang, Y. Zhou "improving transferability present day adverse examples with enter range," in complaints cutting-edge the IEEE convention on computer imaginative and prescient and pattern reputation, 2019, pp. 2730–2739.
2. C. Szegedy, V. Vanhoucke, S. Icontemporaryfe, J. Shlens, and Z. Wojna, "Rethinking the inception structure for pc vision," in IEEE convention on pc imaginative and prescient and pattern popularity, 2016, pp. 2818–2826.
3. A. Kurakin, I. Goodfellow, and S. Bengio, "Adversarial gadget mastering at scale," arXiv preprint arXiv:1611.01236, 2016.
4. A. Madry, A. Makelov, L. Schmidt, D. Tsipras, and A. Vladu, "Towards deep neural networks proof against opposed attacks," arXiv preprint arXiv:1706.06083, 2017.
5. N. Papernot, P. McDaniel, S. Jha, M. Fredrickson, Z. B. Celik, A. Swami, "Limits of deep learning in hostile environments," in 2016 IEEE European Conference on Security and Privacy (EuroS&P) IEEE, 2016, pp. 372–387.
6. J. Su, D. V. Vargas, and K. Sakurai, "Single pixel attacks for deep neural deception," IEEE Transactions on Evolutionary Computation, vol. 23, no. 5, pp. 828–841.
7. S. Das and P. N. Suganthan, "Evolutionary Difference Analysis," IEEE Transactions on Evolutionary Computation, vol. 15, no. 1, pp. 4–31,

8. M.Sc. Sharma, V. M.Sc. Srimal, "A New Hybrid CNN-LSTM Method for Handwriting Recognition Using Washington Database," 2023 RESEM, Bhopal, India, 2023, pp. 1–13. 1–5, doi:10.1109/RESEM57584.2023.10236345.
9. N. Papernot, P. McDaniel, X. Wu, S. Jha, A. Swami, "Distillation as a security mechanism against adversarial perturbations in deep neural networks," in 2016 IEEE Symposium on Security and Privacy (SP); IEEE, 2016, pp. 582–597.
10. G. Hinton, O. Vinyals, and J. Dean, "Knowledge extraction in neural networks," arXiv preprint arXiv:1503.02531, 2015.
11. N. Carlini and D. Wagner, "Evaluation of neural network dynamics in modern applications," in 2017 IEEE Conference on Security and Privacy (SP). IEEE, 2017, pp. 39–57.
12. Bypassing Adversary Instance Detection: An Analysis of Ten Detection Methods, in Proceedings of the 10th ACM Workshop on Artificial Intelligence and Security, 2017, pp. 1–11. 3–1
13. Dong, J. Han, D. Chen, J. Liu, H. Bian, Z. Ma, H. Li, X. Wang, W. Zhang, N. Yu superpixel means." " adversarial attack," in Proceedings of the Computer Vision and Pattern Recognition Conference, 2020, pp. 1–11. 12895–12904.
14. Y. Li, C. Zhang, and X. He, "Adversarial attacks and defenses in computer vision: Trends and future directions," IEEE Transactions on Neural Networks and Learning Systems, vol. 34, no. 1, pp. 45–60, 2023.
15. T. Zhang, X. Liu, and Z. Lin, "A survey on adversarial robustness of deep learning models in vision tasks," arXiv preprint arXiv:2202.04563, 2022.

Note: All the tables in this chapter were made by the authors.

Intelligent Systems Using Semiconductors for Robotics and IoT – Dinesh Goyal et al. (eds)
© 2026 Taylor & Francis Group, London, ISBN 978-1-041-20408-4

49 A Novel Vlsi Technique to Enhance the Gray Scale Images

Suganthi K.*, Harikrishnan Krishnakumar
Department of Electronics and Communication Engineering,
College of Engineering & Technology, SRM Institute of Science and Technology,
Kattankulathur, India

Abstract: The image processing system has always been in high demand in many fields of interest such as security, remote sensing, media, and manufacturing, as well as tracking people in real time or using security cameras to apply image processing in the above-mentioned fields. These systems must be high in performance and cost-effective as well. In current world Application Specific Integrated Circuits (ASICs) and other conventional systems can said to be highly expensive and lack the flexibility needed for any more development to be made on the system. This increases the demand for a more versatile system which is also scalable and uses modern hardware such as FPGA which can offer a higher performance in the field of cost and programmability. Moreover integrating image file conversion, processing and system deployment into a cohesive framework still remains as a huge channel. The primary objective of the paper is to develop a high performance and cost effective image processing system using Verilog HDL, which can also be implemented on a FPGA platform.

Based on this requirement, the image processing system has been built using Verilog HDL. The image processing system was initially built on Vivado and was later tested successfully on the ZYBO Z7 10. In the proposed paper, we will initially have a look at image processing, types of images, then the phases of image processing, and finally the techniques of image processing, as well as the conversion of bitmap format to hexadecimal format in MATLAB for the image processing operation. The input and output files will be in BMP format after a few processes. Then, we will have a look at the proposed method. The FPGA chip is used for the image processing system, which helps to separate data and process the image to our required specifications. The operations mentioned in the paper are the image inversion operation and the brightness reduction operation. We shall discuss the proposed system and algorithm in this paper.

Keywords: Field programmable gate array (FPGA), ASIC, VLSI, Image, Bit mapping

1. Introduction

In the proposed paper, we use FPGA because it can readily deliver real-time performance in various applications, which is a feat that cannot be accomplished by any serial processors or CPU, because they compromise resolution, whereas FPGA does not compromise the resolution. Because of this, the CPU [1] faces a reduction in their efficiency. Also, FPGA has a lower cost compared to CPU and GPU in terms of fabrication and maintenance. FPGA reconfiguration provides more flexibility than CPUs and GPUs. Image processing has become a major field of research for many researchers. This is because FPGA addresses key issues related to power and efficiency and has a good ability for parallel processing of data, making it more efficient. It is used in noise-canceling applications and in hybrid systems, as well as in military applications for object and face detection, to accurately track an object or person across the globe.

A system that is programmable and capable image processing using FPGA focuses mainly on real time image processing and the operations associated with it. In this paper the authors have discussed the importance for

*Corresponding author: suganthk@srmist.edu.in

DOI: 10.1201/9781003716389-49

a flexible programmable image processing system using Field Programmable gate array (FPGA), as the logic cell of FPGA which are not inherently suitable for real time bit level image processing operation . They have proposed a PIPS - Programmable image processing system so that they can overcome this disadvantage. Local image phase, energy and orientation extraction using FPGA

In this paper the authors have discussed a hardware accelerated approach for the real time extraction of phase, energy, and orientation using reconfigurable hardware which meets the system requirements. The approach utilizes quadrature filters which are based upon Gaussian derivatives and novel interpolation techniques to efficiently compute features at their main orientation, it is also optimized for high performance and low resource utilization.

Shonan Zhang's proposed paper presents the application of FPGA in real-time image processing and its advantages. It also mentions the importance of enhancing computational algorithms and thereby increasing their efficiency directly. Similarly, Tehrani, Ashourian, and Moallem focus on FPGA implementations and use it to eliminate or reduce noise in the system. Rohit Kundu's work focuses more on image processing and its application in the field of medicine

2. Image Processing

Image processing on FPGA using Verilog HDL is a very unique approach in which we try to use the FPGA as it is very useful in terms of processing images in parallel. FPGA also allows the implementation of complex algorithms, which help in image processing on a video, etc. Thus, in some cases, it is used as a prototype in ASIC designs.

Using FPGA for design or, in our case, image processing is usually considered too complex when the comparison arises between CPU and GPU. Even for professionals who have higher proficiency in software programming languages, this is mainly because FPGA is generally between software programming languages [1] and hardware description languages is that in the hardware description language, the use of registers and Boolean logic functions is emphasized, while in software programming languages such as C, and sequential instructions are described without considering the underlying hardware implementation details. Because of this, many researchers prefer to use software design more than hardware design, and the software design has a relatively sheer number of libraries and classifications, hence giving more reason to use software design[3].

Another significant reason which can impact the choice for choosing FPGAs, CPUs, or GPUs is the architecture framework in them. For any target-related application, a suitable architecture is required that will reduce the cost of the system and, in turn, improve performance. In the case of image processing, it is considered as an 2D signal processing task, which requires high memory bandwidth for intensive memory operations. Both CPU and GPU have their respective control units which decide the flow of operations in them and are called as , ALU (Arithmetic Logic Unit), DRAM (Dynamic Random Access Memory), and cache [2]. But the most important difference between the CPU and GPU is that a large proportion of the GPU chip is dedicated to ALUs. FPGA, on the other hand, shows a very good property of parallelism, which allows it to handle large amounts of data in parallel and is much more efficient than the CPU. It also uses a von Neumann massive parallel architecture, and the results which are calculated by it can be directly sent to the next processing stage without the need for storing them in temporary space. So FPGA and CPU can achieve various application for each of them for an image processing system it is vey ideal to use either GPU or FPGA [4] because when compared to CPU it gives a high clock speed but provides very limited parallelism and for image processing system it is required that the system in place has very good ability to process the data parallely.[3]. There is a table that shows the comparison of power consumed and energy used of FPGA, CPU, and GPU while implementing Canny edge detection at different resolutions [3].

Table 49.1 Comparison between CPU and GPU units

Pixel resolution	CPU Power (W)	GPU Power (W)	CPU Energy (J)	GPU Energy (J)
512 × 512	141	231	2.8	0.5
1024 × 1024	147	244	8.8	6.08
3936 × 3936	153	251	213.1	15.0

Since FPGA offers lower energy consumption (over 1000 times less than the CPU) while also maintaining high performance, it is ideal for real-time image processing with minimal power usage, [3] making it possible to use even at higher resolutions. Image processing [4] is a method in which we manipulate digital images through computer algorithms and the different techniques available to process the given image into the required image that the user needs. Image processing is done to enhance the existing image to a better version of itself to extract important information from it. It is considered a very important process in deep learning- based computer vision applications, which increases performance exponentially. It is also used in many other applications such as medical, multimedia, entertainment, submarine technologies, etc.

Computers or machines cannot understand the image that humans normally see, so in most machines, the digital images are represented as 2D or 3D matrices where each value is represented as a pixel. Today, we are more familiar with dealing with 8-bit images, where the amplitude varies between 0 to 255.[6] A computer sees the digital images as functions of I(x, y, z), where I represents the intensity of the pixel, and x, y, z represent the coordinates of the pixel in the image [6].

2.1 Types of Images

There are mainly four types

1) Binary Images

These images have only two different pixel intensity values, namely 0 (which represents black or low) and 1 (which represents white or high). Such images are generally used to emphasize a specific portion of a colored image. These types of images are most commonly used in image segmentation. [3] They are also known as monochrome images.

2) Gray Scale Images

These images contain varying shades, ranging from 0 (intensity) to 255 (intensity) in an 8-bit image,or with more levels in higher bit depths, namely 16, 32 bits.

3) RGB Images

Most colored images are represented in this format (RGB format), where each pixel has three components which include Red, Green, and Blue. Each of these components can range between 0 and 255 in an 8-bit image. They are mostly used in digital images, web images, and smartphones as well.

4) RGBA Image

RGB stands for Red, Green, Blue, and each pixel has the same value as mentioned before. By adding an Alpha (A) channel, which represents opacity or transparency, we get the RGBA format. Where,

Alpha = 0 which means full transparency

Alpha = 255 which means fully opaque

The remaining values between 0 and 255 represent the various levels of transparency.

2.2 Phases of Image Processing

Image processing has the following phases they are [6]

Image Acquisition

The image captured by the camera must first be digitized before it can enter any processing stage. The image can be digitized using an analog-to-digital converter, and then image processing techniques[1] can be applied on the computer.

Image Enhancement

In this step, we first capture the image and modify it to meet the specific requirements for the task to be completed. This process is typically done to enhance important details such as brightness and contrast, making it easier to identify features.

Image Restoration

This step is mainly done to improve the appearance of the image and is considered an objective operation since, degradation can be modeled mathematically.

Colour Image Processing

This step deals with processing colored images, i.e., 16-bit RGB or 16-bit RGBA, for performing color correction operations on the images.

Image Compression and Result

Once the colored data or pixels are classified, we can perform the required algorithm on the image and obtain the result. Once the result is obtained, it can be converted back into any type of image that the user wishes to use.With this, the phases of image processing are listed completely.

2.3 Different Techniques in Image Processing

Image processing systems always have the requirement of high performance because they must handle a large amount of image-related data and their computational algorithms or codes. Now let us discuss the different types of image processing techniques used, which are implemented through verilog. In modern image processing [5] with FPGA and Verilog HDL processors, there are mainly five types of techniques used in image processing.

Convolutional Neural Networks(CNN'S) on FPGA

Convolutional neural networks have a wide range of applications, ranging from object detection, face detection, and more. When implemented on FPGAs, CNNs [6] benefit from the parallel processing of convolutional layers, pooling operations, and activation functions. By using Convolutional Neural Networks on FPGA, it allows for parallel computation of CNN operations over multiple pixels simultaneously, improving the output compared to traditional CPU-based processing. By leveraging hardware acceleration like convolution, all typical computational processes are being processed in parallel, which enables real-time processing with lower latency. Furthermore, with custom optimizations such as fixed-point arithmetic and pipelining, it further enhances the performance of FPGA, making them ideal for use in autonomous vehicles, medical imaging, and object detection.

Hough Transform

The Hough Transform is one of the best techniques used to detect geometric shapes and lines, as well as circles and curves in images. They are widely used in edge detection and image analysis tasks. This type of image processing technique is usually applied to robotic systems so that robots can analyze image data more efficiently and respond more quickly to any situation they may face during their operation. The role of FPGA in this technique speeds up the Hough Transform by processing multiple pixel values concurrently in parallel. Furthermore, this reduces the computational time required for detecting shapes. This application is considered very useful in industrics, as well as in robotics and autonomous driving. Hence, by implementing FPGA, it allows for faster and more efficient shape detection by exploiting parallel processing [6] and pipelined execution, which provides high-speed results in object detection or tracking.

Image Filtering

These types of techniques are like filters that are very useful in noise cancellation or reduction and edge detection. A few good examples of these techniques are Gaussian [6] and Sobel filters. These filters normally enhance or suppress certain image features as specified by the user in the algorithm. On FPGA, these filters can be applied simultaneously across large sections of the image, significantly speeding up the convolution process. This allows FPGA to be used in real-time video and image processing applications such as medical diagnostics, video surveillance, and quality control in manufacturing, where high-quality image enhancement is deemed a very crucial element. FPGA-based filters also optimize resource usage and have lower power consumption while delivering high-quality output.

Morphological Operations

Morphological operations include dilation, erosion, and closing, which manipulate the structure of objects in binary or grayscale images[1] . These operations are commonly used in applications such as noise removal, edge reduction, and image segmentation. FPGAs excel at performing actions in parallel, making them highly ideal for high-speed processing of large image datasets. For example, in medical imaging, FPGA-based morphological operations can be used to identify tumors or other structures in tissue samples with much higher accuracy and lower latency. The parallel nature of FPGA allows for real-time image processing in the most crucial applications where speed and precision are key.[6].

Frame Buffering

This type of technique processes images by storing the data in memory for sequential processing, which allows for smooth playback and image rendering. This technique allows for double and triple buffering, which are often used to reduce screen tearing and minimize latency. FPGAs enhance [6] the whole process by managing multiple energy blocks in parallel, which enables efficient handling of large video streams or high-resolution images. This technique is considered very crucial in real-time applications such as video processing, gaming, and other such activities where there is a requirement for maintaining high frame rates and lower latency. FPGA- based frame buffering also ensures that there is always continuous, smooth data flow, preventing frame drops or lag in visual systems. Our proposed system uses the Morphological Operations technique, which has been explained in the previous part. Now let us take a look at the proposed system and its key components. The key concepts here include,

FPGA: These are devices that are configured by the user to fit their specific needs. They consist of programmable logic arrays and many I/O ports. FPGAs can be used to create circuits tailored for specific tasks.

Verilog HDL: This is a hardware description language that is used to model or program electrical systems. It allows the designer to specify how digital circuits need to perform when specific inputs are given to them. This makes it easier to simulate and get results faster, which can then be synthesized into hardware.

Image processing algorithms: These are done mainly through Verilog code. The algorithm [4] contains three main parts: parameters, image reading, and image writing. Each part has its specific purpose and usage, which will be detailed in the upcoming sections.

Integration of the Software Used

Vivado Design Suite is an advanced software platform developed by Xilinx for FPGA and System-on-Chip (SoC) design. It provides comprehensive tools for creating, simulating, synthesizing, and implementing digital circuits on **Xilinx FPGAs** (including the 7 Series such as **Artix**, **Kintex**[2], and **Virtex** FPGAs). Vivado is known for its high- performance design flow, optimization capabilities, and integration with modern Xilinx devices, significantly improving upon the older Version of the same software,

3. Comprehensive Design Tools

Vivado offers a complete design environment, supporting the entire FPGA design process from RTL (Register Transfer Level) design to synthesis, implementation, and bitstream generation. With Vivado, designers can create designs using HDLs (Verilog and VHDL), Perform RTL design and logic synthesis, integrate with IP (Intellectual Property) cores and high- level synthesis (HLS) for faster development and usage of powerful tools for place-and-route, timing analysis, and power optimization to ensure that designs meet performance, area, and power requirements. Vivado fully supports both Verilog and VHDL for digital design, providing flexibility depending on the designer preferred language. Like ISE, Vivado allows the creation of high-level designs for 7 Series FPGAs, [3] but it does so with superior synthesis optimization, supporting more complex designs and better resource management.

3.1 Synthesis and Implementation

Vivado Synthesis: Converts RTL (Register Transfer Level) code into a netlist of gates. It supports languages like **VHDL**, **Verilog**, and **SystemVerilog**. The Implementation Converts the synthesized design into a configuration file which can be loaded into the FPGA or CPLD. This also includes most of the important phases such as placement, routing, and optimization processes. The tool also ensures that the design meets timing constraints (i.e., meets the required clock frequency) by using techniques like **static timing analysis**.

3.2 IP Core Integrator

Vivado provides a **graphical block diagram interface** for integrating pre-built **Xilinx IP (Intellectual Property) cores**. These can be custom or standard cores like **memory controllers, communication protocols**, etc. Reuse can be done by Vivado allows users to create reusable IP blocks and efficiently integrate them into projects. IP Catalog includes a rich library of standard and custom IP for commonly needed functions (e.g., UART, SPI, Ethernet, DDR memory controllers, etc.).

3.3 High-Level Synthesis (HLS)

With **Vivado HLS**, you can design hardware by writing **C/C++** code instead of RTL, which is then converted into RTL and optimized for FPGA.HLS allows for faster development and the potential for more efficient hardware by automating any optimizations that are typically done manually in RTL design.

3.4 Simulation and Verification

Vivado Simulator: A built-in simulator to test designs for functionality and timing before hardware implementation. Behavioral Simulation supports functional simulation of RTL designs. Post-implementation Simulation-after routing and placement, Vivado allows for timing-aware simulations, ensuring the design operates correctly with real hardware timing. Design Rule Checks (DRC) and timing analysis is done automatically checks for violations or errors in the design and timing. Debugging, profiling tools and integrated logic analyzer enables you to insert a logic analyzer directly into your design for real-time data capture, which helps in debugging and performance analysis. Chip Scope is a tool for on-chip debugging (historically used with Xilinx FPGAs), now integrated into Vivado. Advanced Optimization is the process of advanced algorithms for optimizing both area and performance. This includes automatic optimization for resource utilization, timing closure, and power consumption. Power Estimation & Optimization in Vivado has built-in power analysis tools to estimate and minimize power consumption of FPGA designs, which is crucial in battery-powered or thermally constrained environments. The design Constraints allow users to define timing, placement, and other design constraints to ensure the FPGA performs as expected. Constraints can be defined through XDC (Xilinx Design Constraints) files. Partial Reconfiguration enables the reconfiguration of specific parts of the FPGA while the rest of the design remains operational. This allows for dynamic adaptation and flexibility in complex systems. Design Flow Automation in Vivado features scripts and TCL (Tool Command Language) automation, making it easy to automate repetitive tasks, batch processes, and integrate with custom tools. This can save considerable time for large and complex designs. Hardware Description Language (HDL) Support include both VHDL and Verilog, two of the most popular hardware description languages, enabling the creation of flexible and portable designs. System Verilog for advanced verification is also supported, allowing for modern and complex verification techniques.

4. Components of Vivado Design Suite

4.1 Vivado IDE (Integrated Development Environment)

The main graphical interface used to manage design projects, run synthesis/implementation, view reports, and manage constraints. Simulator is a built-in simulation tool that integrates with Vivado for functional verification, timing analysis, and performance evaluation.

4.2 Vivado HLS (High-Level Synthesis)

This component allows for the conversion of high-level code (C, C++) into hardware, making FPGA development faster and more abstracted.

4.3 Vivado Lab Tools

Vivado Lab Edition is for hardware bring-up and debugging, including In-System Programming and debugging with real-time visibility.

4.4 IP Integrator

A visual interface that enables rapid assembly of FPGA designs using pre-built or custom-created IP blocks.

4.5 Xilinx SDK (Software Development Kit)

The Xilinx SDK is the software counterpart to Vivado, allowing the development of embedded software for Xilinx processors (like Zynq).

5. Image Conversion Process

The main issue we primarily face with this software is that Verilog itself cannot read images directly when input. To process any image using Verilog, the image needs to be read

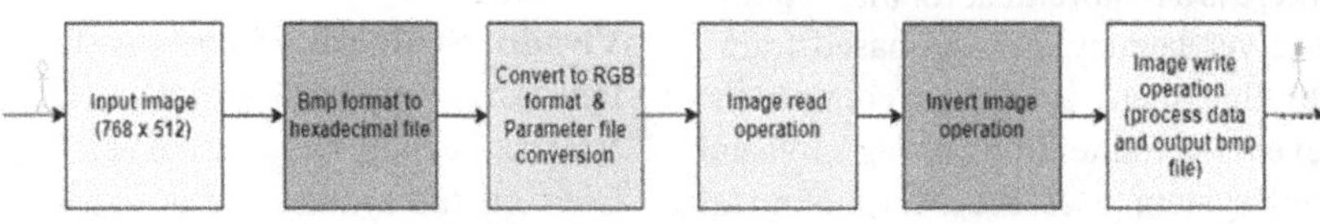

Fig. 49.1 Design steps

so that the data can be collected. This is mainly performed by using MATLAB in our case, where we convert the bitmap format to hexadecimal format for the given image. In this way, Verilog can read the image[4] through the hexadecimal format [2] and process it into the desired output when further instructions are given. Take an input image in the range between 768x512 pixels. As Verilog cannot read the image in BMP format, we will convert it into hexadecimal format using MATLAB. Separation of RGB Data is done after getting the required input image in hexadecimal format, we need to separate the hexadecimal data into RGB format. As previously mentioned, this format is a type in which the image can be represented in pixels, and it contains components such as Red, Green, and Blue. Each of these components' values can vary from 0 to 255 in an image. Then parameter definition is the file in which we select the mode of operation for the image to be processed. In our case, it is the image inversion operation. Image write module and output step happens once the image read module is successfully completed, and the data that is taken from the hexadecimal image is forwarded to the image write module. Then, the image write module starts to process the data that is received from the image read [5] module and organizes it back into RGB format. The output image is saved in the specified file. The output image will be in BMP format.

5.1 Test Bench and Simulation

The results extracted using Vivado The test bench for the process is written by including three verilog files

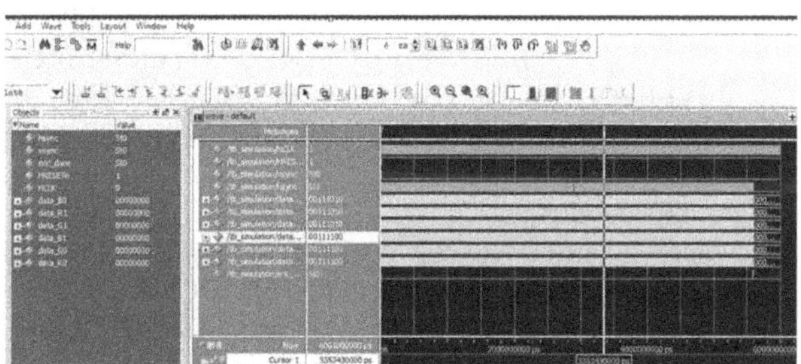

Fig. 49.2 Test bench and simulation

Parameter.v is a file, to specify what type of operation takes place if the input image is given. It is also the file where we specify the input hexadecimal file and the output BMP file. **Imageread.v** is also a Verilog file. Here, the main operation of this file is to read the data from the input file, which is the hexadecimal format that we have converted earlier.

Along with it, it also contains a few more variables such as **Width:** represent the image horizontal size , which for our paper is fixed at 768 pixels. **Height:** it denoted the vertical size of the image **INFILE:** Refers to the location of the input image stored in .hex format. **START_UP_DELAY:** The time (in clock cycles) required for the module to initialize. **HSYNC_DELAY:** The interval between successive horizontal sync signals. **VALUE:** A parameter to modify the image's brightness, typically set to 100 by default.

5.2 Threshold Voltage Variations

A predefined limit used for image thresholding, defaulting to 90. **SIGN:** Determines the brightness operation: 1 indicates addition, 0 indicates subtraction. **VSYNC:** An output signal marking the beginning of a new frame. **HSYNC:** An output signal identifying the start of a new image row. Clock signals such as **HCLK** and **HRESET** (this is a reset active-low pin). The RGB data is also specified here. To read the file in Verilog, we use the syntax **$readmemh**. With this, we read the input image. It also has a defined counter, which is used for **hsync** (tells the delay between hsync pulses) and is also used for data processing. In the file, we also mention the data separately, i.e., the 8-bit Red data, 8-bit Blue data, and 8-bit Green data are mentioned separately to make the data processing operation easier.

5.3 Image Write

This is also a Verilog file. The input file for this Verilog file is the output image, which we receive upon giving the Image Inversion operation and the input image. The parameters in this file includes, the following specifications

WIDTH: Image width (768 pixels).

HEIGHT: Image height (512 pixels).

INFILE: Output image file path ("output.bmp").

BMP_HEADER_NUM: Number of bytes in the BMP file header (54).This file also contains the clock signals such as **hsync, HCLK**. Now, we have given the input as the Red, Blue, Green data that has been processed by the image read part. Once the image read part is processed by the FPGA, here the important part is that we use a BMP header [6] in this file so we can write the image in the output BMP. Every input image with different pixels has different BMP headers to be used. After setting the conditions in the test bench code and running it for 6 ms in Model Sim simulation, the required output image is found in the location specified. The output image for the image inversion operation and the original image are given below in Fig. 49.3, 49.4 and 49.5.

Fig. 49.3 Image for reference

Fig. 49.4 Gray scale image – bit mapping

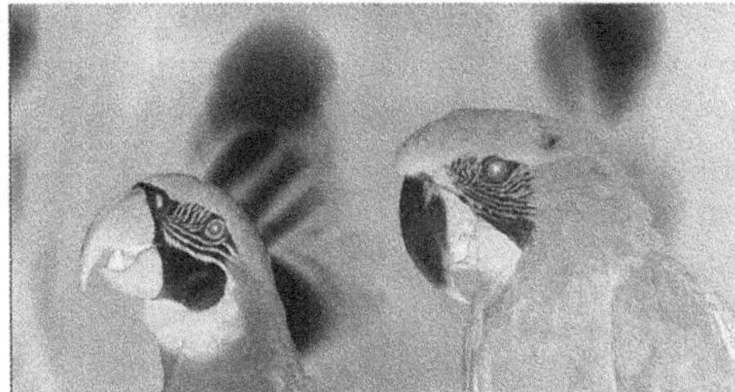

Fig. 49.5 Image inversion by bit mapping

The executed code has been dumped into the FPGA board and tested. Verification validation is also done. Figure 49.6 reveals that fact.

Fig. 49.6 Testing & verification using FPGA board

6. Conclusion

By converting the input image into hexadecimal format, we are able to run the process on FPGA and get the output for the image inversion operation. In future work, our main goal is to make multiple operations possible, such as brightness reduction and threshold operations. We also aim to completely remove the processing part, through which we can get raw image data. However, in that case, we would need to develop an image sensor model. The bit mapping concept and bit mapping part of the gray scale image are achieved by the novel technique applied through FPGA - Vivado design suite.

Table 49.2 Comparison with the results obtained

Pixel resolution	FPGA Power (W)	FPGA Energy (mJ
512 × 512	1.5	1.7
1024 × 1024	1.5	1.7
3936 × 3936	1.5	98.2

Acknowledgement

The authors gratefully acknowledge the department of ECE for their support in this research.

References

1. CHAN, S. C., NGAI, H. O., & HO, K. L. (1993). A programmable image processing system using FPGAs. *International Journal of Electronics*, *75*(4),725730 https://doi.org/10.1080/00207219308907150
2. Díaz, J., Ros, E., Mota, S., & Carrillo, R. (2008). Local image phase, energy and orientation extraction using FPGAs. *International Journal of Electronics*, *95*(7), 743–760. https://doi.org/10.1080/00207210801941200
3. Shaonan Zhang, Real Time Image Processing on FPGAs A THESIS SUBMITTED TO THE UNIVERSITY OF LIVERPOOL FOR THE DEGREE OF DOCTOR OF PHILOSOPHY IN THE FACULTY OF SCIENCE AND ENGINEERING, Department of Electrical Engineering and Electronics.
4. Omid Sharifi Tehrani, Mohsen Ashourian, Payman Moallem, FPGA Implementation of a Channel Noise Canceller for Image Transmission and coder, Int. Res. J. Eng. Technol. 04(02), 1798–1802
5. M V Ganeswara Rao, Implementation of Real Time Image Processing System with FPGA and DSP Dept. of ECE JNTU College of Engineering Kakinada, India , Int. symposium. Proceedings,18(2):101604-101609.
6. Ganesh, P Rajesh Kumar Dept. of ECE PVPS institute of Technology, Image processing techniques using DSP, 201–203, Conference papers, 2021.
7. Rohit Kundu, Image processing is the process of manipulating digital images, image processing techniques, including image enhancement, restoration, & others, Int. Conference proceedings 45(3):008–009.

Note: All the figures and tables in this chapter were made by the authors.

Intelligent Systems Using Semiconductors for Robotics and IoT – Dinesh Goyal et al. (eds)
 ISBN 978-1-041-20408-4

50 Crack Level Classification and Segmentation using Artificial Intelligence Techniques

Ravuri Koushik[1], Thiyyagura Jagadeesh[2], Kabeer Dudekula[3], T. Ramya[4]

SRM Institute of Science and Technology, Department of ECE, Chennai, India

Abstract: Unity Neural Encoding Tool (U-Net) and LeNet-5 is used to present a novel approach to crack-level classification using Artificial Intelligence (AI) techniques. For safety reasons, this study aims to locate cracks in variety of structures, including walls, bridges, and infrastructure. A substantial dataset of 10,000 images was used for this project, of which 8,000 were designated for training and 2,000 for testing. The model can identify six different kinds of cracks with the use of cutting-edge deep learning and image processing methods. There are several types of cracks: normal, deep, gap, riss, wall care, and no crack. Modern architectures made it simpler to enhance existing procedures. By offering a reliable technique for identifying and classifying cracks, this research enhances the subject of quality monitoring while also enhancing safety and maintenance practices in the civil engineering and construction industries.

Keywords: U-Net, LeNet-5, Artificial intelligence, Crack level classification

1. Introduction

It represents a crucial nexus of technology and infrastructure maintenance, addressing the pressing need for precise and effective identification of structural defects in buildings, bridges, and roadways [1]. These programs are crucial for preserving public safety, extending the life of civic infrastructure, lowering the need for costly repairs, and averting disasters caused by invisible cracks [2]. The application of cutting-edge technology like computer vision, machine learning, and deep learning has revolutionized fracture analysis and identification in recent years. Even in challenging photo situations, computer vision—particularly convolutional neural networks—improves detection over manual techniques [3]. Convolutional Neural Networks (CNNs), U-Net, and LeNet are the main techniques used in this study; they offer powerful crack segmentation and classification capabilities [4]. Inspections by hand are slow. Accurate crack detection is provided by vision-based methods such as Fully Convolutional Neural Networks (FCNN). Consistent performance with models such as Unity Neural Encoding Tool is still difficult to achieve [5]. For precise crack segmentation, this presents a Dual-encoder Network fusing Transformer and CNN (DTrC-Net), which combines transformers and convolutional neural networks. For high precision, it uses weighted loss optimization, feature fusion, and simultaneous coding branches [6]. These algorithms can evaluate images of infrastructure surfaces and provide crucial details regarding the asset's structural integrity by accurately detecting cracks of all sizes and types [7]. For better segmentation, this paper presents a highly supervised method using multi-scale class activation mapping [8]. Additionally, the use of ensemble approaches, transfer learning, and data augmentation significantly enhances the performance and resilience of crack detection systems [9].

2. Related Work

Yang et al. [10] addressed U-Net (MST-Net) which is a new method for accurately detecting road fractures. It tackles class imbalances and complex backgrounds using multiscale input and attention mechanisms [11]. With features like deep supervision and aggregation, it outperforms other models in fracture detection accuracy.

[1]rq2142@srmist.edu.in, [2]tr9920@srmist.edu.in, [3]dk2385@srmist.edu.in, [4]ramyat@srmist.edu.in

DOI: 10.1201/9781003716389-50

Mirbod et al. [12] provided a unique machine vision and Artificial Neural Network (ANN) based technique for concrete fracture identification. This method requires less hardware and simplifies the code, with an accuracy of 84.88% when compared to CNNs. Zhang et al. [13] approach, which makes use of the Utah State University dataset, has potential for use in structural health monitoring in practical settings. Canchila et al. [14] study examine current developments in deep learning for crack segmentation and assess how well they work with various kinds of images. Lau et al. [15] To analyze nine CNN architectures and establish the best performance and insights for further research, present the dataset, which consists of several picture kinds.

3. Proposed Methodology

Thorough data collection is necessary to create reliable and accurate models, which will progress the field of crack-level classification and segmentation. The images in the dataset contain a wide variety of crack types, including deep cracks, gap cracks, Riss cracks, Wall care cracks, and no cracks. This includes cracks in pavement, concrete, and walls, among other surfaces. The dataset includes precise information about the locations and forms of cracks. To increase its generalization ability, the model also integrates data from many sources, such as various infrastructure kinds. During the pre-processing phase, a robust rescaling method must be used to optimize the input data for additional analysis. This pre-processing step not only enhances the model's performance but also greatly increases the reliability and interpretability of the next fracture classification and segmentation procedures.

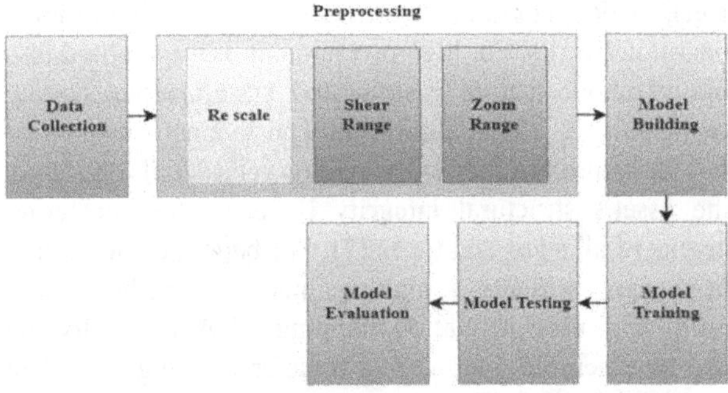

Fig. 50.1 Block diagram for crack level classification

To assess structural integrity by looking at a number of characteristics, Structural Health Evaluation using Regression Analysis (SHERA) is crucial. In crack-level classification, SHERA is a crucial tool for determining the size and severity of cracks, enabling more precise classification based on the cracks' severity levels. By accurately describing crack patterns, Shera aids in crack segmentation as well, producing better segmentation results. However, the zoom range, which indicates the range of magnification levels utilized to gather structural data, is crucial for producing detailed photographs of cracks at various scales. Preprocessing, meticulous data collecting, and the application of specialist models such as LeNet for classification and U-Net for segmentation are among the procedures.

Extensive testing is done on new datasets to evaluate the generalizability of these models. LeNet is better at classification jobs, whereas U-Net is better at segmentation since it can catch tiny features in complex fracture patterns. Pixel-wise accuracy, recall, accuracy, precision, and intersection over union (IoU) are among the metrics used to assess the model's performance that are unique to classification and segmentation. To assess the difference between ground truth and projected pixel values, model parameters are improved during training using backpropagation and the binary cross-entropy loss function.

Rotation, flipping, and scaling are examples of data augmentation techniques that are used to increase generalization and resilience. LeNet tests its generalization capabilities by examining datasets or fresh photos that differ from its training set and feature potentially fractured structures. LeNet generates classification outputs that convey class probabilities or predicted labels for each image by applying its trained architecture to a forward pass on the input data. A quantitative assessment that uses classification-specific metrics, such as accuracy, precision, and F1 score, is then used to fully evaluate the model's classification accuracy. This rigorous testing procedure enables iterative refining and optimization, improving LeNet's accuracy in crack-level categorization for structural health monitoring applications. The U-Net, which is well known for capturing minute details, is excellent at accurately segmenting complex crack patterns. Visual inspection of the segmentation outputs helps identify possible errors or inaccuracies by ensuring that they match expectations.

4. Performance Evaluation

Table 50.1 displays the accuracy of a crack detection system. The accuracy is shown as 100 percent, meaning it's very reliable. For all types of cracks, the accuracy is above 93%, which is quite high. Among them, Gap Cracks have the highest precision at 97.5%, while RISS Cracks have the lowest at 93.67%. It's worth noting that the effectiveness of any automatic crack detection system may differ based on the conditions in which it's employed.

Table 50.1 Accuracy of different models

S. No	Crack Type	Accuracy (%)
1	Crack	96.33
2	Deep Crack	94.97
3	Gap Crack	97.5
4	Riss Crack	93.67
5	Wall care Crack	95.27
6	No Crack	100

(a) $Accuracy = \dfrac{TP + TN}{TP + TN + FP + FN}$

(b) $Precision = \dfrac{TP}{TP + FP}$

(c) $Recall = \dfrac{TP}{TP + FN}$

(d) $F1 = \dfrac{2 * Precision * Recall}{Precision + Recall}$

(e) $IoU = \dfrac{TP}{TP + FP + FN}$

where True Positive, True Negative, False Positive, and False Negative are the definitions of TP, TN, FP, and FN.

The graph provided shows the performance of a machine-learning model over multiple training cycles, or epochs. In machine learning, a crucial aspect is the use of a loss function, which measures the difference between the model's predictions and the actual data. The goal during training is to minimize this loss function to improve the model's accuracy. On the graph, the Y-axis represents the model's loss, with lower values indicating better performance.

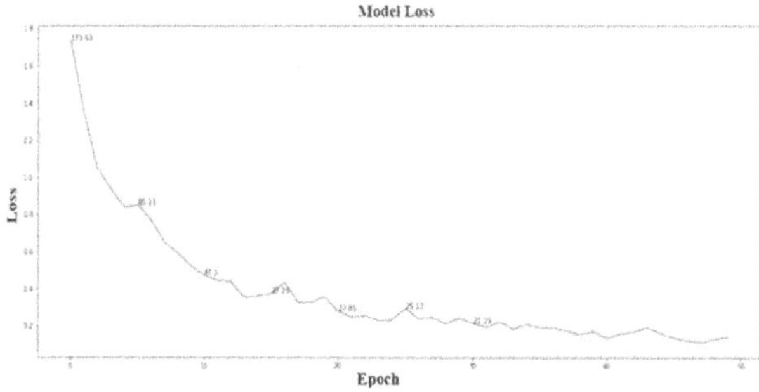

Fig. 50.2 Loss vs epochs performance evaluation

The number of training epochs, or the number of times the model has been exposed to the dataset, is shown by the X-axis. The decreasing trend in the graph suggests that the model's performance is improving over time, as indicated by the reduction in loss. Machine learning's iterative training process is responsible for the model's performance improvement, which is indicated by the trend of the loss function decreasing over training epochs.

The model's predictions may be wildly inaccurate in the beginning, during the early epochs, leading to comparatively high loss value. This is to be expected given that the model lacks a clear understanding of the underlying patterns in the data and begins with random weights and biases. Based on the feedback given by the loss function, the model gradually modifies its parameters (weights and biases) as training goes on. Using techniques like gradient descent and back propagation, the model modifies its parameters to reduce the discrepancy between its predictions and the actual data.

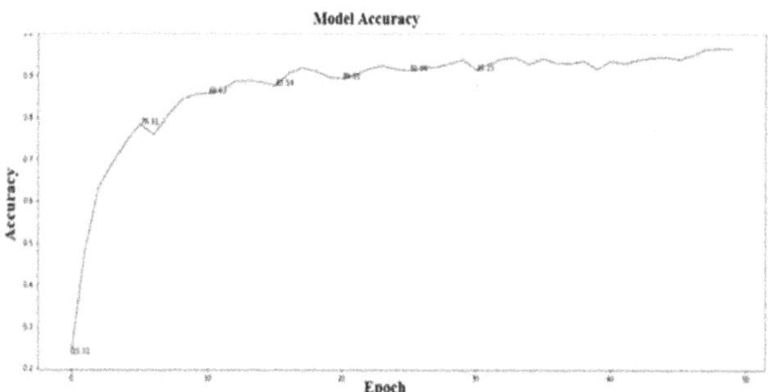

Fig. 50.3 Accuracy vs epochs performance evaluation

The Y-axis shows the model's accuracy, with 1.0 being perfect accuracy and the Y-axis's range is between 0 and 1. The number of times the model has been trained on a dataset is probably represented by the X-axis. Although certain decimal places are missing from the text labels on the Y-axis, it seems like accuracy starts out low and rises with time. This suggests that the model is learning and improving its performance as it is trained on more data. ith a range of 0 to roughly 50, the x-axis is designated as an "Epoch". Accuracy is represented by the Y-axis, and values fall between roughly 0.3 and just under 1.0. The accuracy approaches 0.9 at about epoch 10. After that, the accuracy varies a little but stays near the 0.9 threshold. The accuracy of the model increases dramatically in the first few epochs and then stabilizes at 0.9, as the graph illustrates.

Table 50.2 LeNet architecture

Layer Type	Parameters/Configuration
Input Image	Demensions: 28 x 28, 1 Channel
Convolutional Layer	Kenal Size: 5 x 5, Padding: 28 x 28 x 6
Max Pooling Layer	Pool Size: 2 x 2
Convolutional Layer	Kernal Size: 5 x 5, Padding: 10 x 10 x 16
Max Pooling Layer	Pool Size: 2 x 2
Flatten Layer	-
Fully Connected Layer (Dense)	Units: 120
Fully Connected Layer (Dense)	Units: 84
Fully Connected Layer (Dense)	Units: 10
Output	1 of 10 Classes

Features extraction and classification tasks would be the main uses of the LeNet architecture in a crack-level classification and segmentation project. A Convolutional Neural Network (CNN) called LeNet was first created to recognize handwritten digits. It is composed of fully connected layers that are followed by layers with convolutional and pooling operations. This project would use the LeNet architecture to analyze crack images, with convolutional layers being used to extract unique features at different scales and then classify the cracks according

Table 50.3 Accuracy and loss table

S. No.	Epoch	Accuracy	Loss
1	1	29.78	173.63
2	5	70.02	83.60
3	10	82.30	52.18
4	20	88.67	35.43
5	30	93.53	24.11
6	40	95.24	16.92
7	50	96.75	19.97

Table 50.4 Comparison table of proposed work with existing work

Values	Reference Paper 1 [9]	Reference Paper 2 [10]	Reference Paper 3 [11]	Proposed Paper
No of Cracks	3	1	3	6
Data Set Size	1431	56,000	6000	10,000
Accuracy	94.9	84.88	89.81	95.34
Used Method	MST NET	ANN	CNN	U-Net, L-Net

to their type or severity. LeNet's layers could also be adjusted to predict pixel-by-pixel classifications, which would enable it to distinguish cracks from the background of images and be used for segmentation. For an accurate assessment and subsequent action, this segmentation capability helps to precisely identify the extent and boundaries of cracks within an image.

In existing work, there are differences in methods and results amongst the four papers we compared that dealt with crack detection. Using a dataset of 1431 samples, reference paper 1 used the MST NET approach to detect three cracks with an accuracy of 94.9%. In contrast, reference paper 2 used an ANN to detect one crack out of a much bigger dataset of 56,000 samples with an accuracy of 84.88%. Using a CNN, reference paper 3 also found three cracks from a dataset of 6000 samples, but at a lesser accuracy of 89.81%. In this study, we provide a novel approach that employs both U-Net and L-Net architectures to successfully identify six cracks from a dataset of 10,000 samples, with the greatest accuracy of 95.34%. It is clear from this comparative analysis that our suggested methodology performs better than current methods.

5. Conclusion

Furthermore, the results show promising accuracy rates for different types of cracks. With accuracy levels surpassing 93% across the board and peaking at 97.5% for Gap Cracks, the system demonstrates its effectiveness in identifying and categorizing cracks accurately. By creating interactive AI solutions that incorporate human feedback and allow inspectors to amend and validate AI-generated results for increased accuracy and reliability, researchers can improve the efficacy of crack detection systems. Furthermore, researching edge computing solutions makes it possible to directly install AI models on edge devices, enabling real-time processing and resolving privacy issues. Proactive maintenance tactics are further enabled by the integration of AI- powered crack detection systems into long-term monitoring and maintenance programs for infrastructure assets, which ultimately extend the life of critical infrastructure.

References

1. Kuchipudi, Sai Teja, and Debdutta Ghosh. "Automated detection and segmentation of internal defects in reinforced concrete using deep learning on ultrasonic images." Construction and Building Materials 411 (2024): 134491.
2. Arafin, Palisa, AHM Muntasir Billah, and Anas Issa. "Deep learning-based concrete defects classification and detection using semantic segmentation." Structural Health Monitoring 23.1 (2024): 383–409.
3. Liu, Zhen, et al. "Automatic pixel-level detection of vertical cracks in asphalt pavement based on GPR investigation and improved mask R-CNN." Automation in Construction 146 (2023): 104689.
4. Tran, Thai Son, et al. "Advanced crack detection and segmentation on bridge decks using deep learning." Construction and Building Materials 400 (2023): 132839.
5. Xu, Gang, et al. "Investigation on the effect of data quality and quantity of concrete cracks on the performance of deep learning-based image segmentation." Expert Systems with Applications 237 (2024): 121686.
6. Xiang, Chao, et al. "A crack-segmentation algorithm fusing transformers and convolutional neural networks for complex detection scenarios." Automation in Construction 152 (2023): 104894.
7. Iraniparast, Mostafa, et al. "Surface concrete cracks detection and segmentation using transfer learning and

multi-resolution image processing." Structures. Vol. 54. Elsevier, 2023.

8. Wang, Wenjun, et al. "A lightweight crack segmentation network based on knowledge distillation." Journal of Building Engineering 76 (2023): 107200.
9. Gao, Zhi, et al. "Synergizing low rank representation and deep learning for automatic pavement crack detection." IEEE Transactions on Intelligent Transportation Systems 24.10 (2023): 10676–10690.
10. Yang, Lei, et al. "multi-scale triple-attention network for pixelwise crack segmentation." Automation in Construction 150 (2023): 104853.
11. Zhang, Yuefei, et al. "APLCNet: Automatic pixel-level crack detection network based on instance segmentation." IEEE Access 8 (2020): 199159–199170.
12. Mirbod, Majid, and Maryam Shoar. "Intelligent concrete surface cracks detection using computer vision, pattern recognition, and artificial neural networks." Procedia Computer Science 217 (2023): 52–61.
13. Zhang, Tianjie, Donglei Wang, and Yang Lu. "ECSNet: An accelerated real-time image segmentation CNN architecture for pavement crack detection." IEEE Transactions on Intelligent Transportation Systems (2023).
14. Zhou, Shanglian, Carlos Canchila, and Wei Song. "Deep learning-based crack segmentation for civil infrastructure: Data types, architectures, and benchmarked performance." Automation in Construction 146 (2023): 104678.
15. Lau, Stephen LH, et al. "Automated pavement crack segmentation using u-net-based convolutional neural network." IEEE Access 8 (2020): 114892–114899.

Note: All the figures and tables in this chapter were made by the authors.

Intelligent Systems Using Semiconductors for Robotics and IoT – Dinesh Goyal et al. (eds)
© 2026 Taylor & Francis Group, London, ISBN 978-1-041-20408-4

51 Exploring the Efficacy of Machine Learning Models in Predicting Heart Disease: A Comparative Study

Chirag Gulati[1], Arav Chaudhary[2]
Amity School of Engineering and Technology, Amity University Punjab, Mohali, India
Himanshu Verma[3]
STME, NMIMS University, Chandigarh Campus, India
Rajni Mohana[4]
Amity School of Engineering and Technology, Amity University Punjab, Mohali, India

Abstract: Cardiovascular diseases continue to be a major health problem globally, with a prevalence of poor health outcome and/or mortality as a prime cause. This research studies five machine learning models to predict persons at medium to high risk of heart disease: Logistic Regression (LR), Decision Tree (DT), XGBoost, K-Nearest Neighbors (KNN), and Random Forest (RF). In this paper, extensive EDA is performed over a comprehensive dataset with abundant features consisting of categorical and numeric variables and processed the data using feature scaling such as standardization and one hot encoding for categorical variables. Each model was hyperparameter tuned using grid search, with its performance optimized. Significant performance evaluation parameters such as accuracy, precision, recall, and F1-score are considered for assessment over a highly cited dataset. The findings of this analysis suggest useful information about the strengths and limitations of cardiovascular prediction models and can inform future developments in cardiovascular risk assessment.

Keywords: Cardiovascular disease, Machine learning models, Predictive analysis, Accuracy, Heart disease prediction

1. Introduction

Cardiovascular diseases (CVDs) are morbid and global epidemic, causing each year about 17.8 of all deaths (close to 30 percent of deaths) and contributing to nearly one third of all risk of premature mortality [1][2]. Clinicians require reliable tools to identify those individuals with little or no known cardiovascular disease but when there is an elevated risk cardiovascular event there is a need to implement and effectively prevention mechanism [3][4]. Prevalence of this far-reaching illness occurs in almost every age bracket and is tied together with by such factors as high blood pressure; overweight; elevated cholesterol levels; family history; smoking; and alcohol consumption. Diagnosing such complex interplay of symptoms of cardiovascular diseases takes a lot of time for the healthcare professionals because proper identification is required.

In response to this ongoing health crisis, healthcare industries across the world are working diligently to build large amounts of datasets in order to unravel insights of heart diseases. It promises to help put healthcare professionals on to a similarly nuanced footing, thereby giving them the ability to provide precise treatments and interventions. To capitalize on this wealth of data, coupled with surging data collection, there is a need for innovative means to take advantage of this wealth of information. In healthcare, machine learning (ML) has been a transformative innovation — the ability to analyze massive datasets and identify valuable patterns. With the use of ML algorithms, medical staff can improve diagnostic accuracy, modify the treatment plan, and improve the patient's outcome.

This work estimates the performance of five different ML models for predicting heart disease. These methods are LR, DT, XGBoost, KNN, and RF. Our contribution comes

[1]chirag.gulati@s.amity.edu, [2]arav.chaudhary@s.amity.edu, [3]hverma@pb.amity.edu, [4]rmohana@pb.amity.edu

DOI: 10.1201/9781003716389-51

through the study of these models, performing a thorough analysis and comparison of their performance metrics and contributing valuable insights into how ML may be used to support cardiovascular health. The potential of our findings to inform healthcare practitioners, researchers, and decision makers it is to inform them with evidence so that we can make informed decisions about cardiovascular. The primary objectives of this paper have been discussed below:

1.1 Model Comparison

Five different popular ML models, such as LR, DT, XGBoost, KNN, and RF, are taken into consideration to evaluate and access the performance in predicting the possibility of getting heart disease given certain features.

1.2 Identification of Optimal Model

Find out which machine learning model gives highest accuracy and effectively predict heart-related illness from the given dataset.

1.3 Hyperparameter Optimization

Developing and utilizing ML models and then tweak hyperparameters of each model to maximize prediction accuracy and increase generalization of the predictive model to unseen data.

2. Related Work Study

Serval important studies have already examined the use of ML algorithms to anticipate heart diseases. From earlier work of Mohan et al. [5], they introduced a hybrid method called HRFLM, an accuracy level of 88.7% with a superior accuracy to the existing ones in heart disease prediction was achieved by the method. Similarly, Dwivedi [6] tried different machine learning techniques including LR and obtained highest accuracy (85%) in prediction of heart disease.

Additionally, Nikhar et al. [7] state that decision tree classifiers performed better in predictive data mining for heart disease than in Bayesian classification. In his study Yousefi [8] used LR and decision tree models as the best predicting algorithms of heart disease and found accuracy of 88% and AUC ROC of 91%, respectively.

In their study [9], Kamble et al. had involved in using classification techniques to build a reliable model to predict coronary disease. Also, Kondababu et al. [9] tested the different techniques and algorithms and concluded that machine learning has been effective predict heart-related illnessandachieved88.7%accuracywithHRFLMtechnique.

Moreover, Hasan [11] talked about how AI and machine learning can be within the project of developing prediction models for cardiovascular disease and talked about the importance of technological intervention and healthcare data usage. Additionally, on 12, Mijwil et al. [12] conducted study over the utilization of ML techniques for heart-related problem prediction, an excellent reference for healthcare workers.

Taken as a whole, these studies illustrate the considerable possibility of machine learning methods to predict and help make heart disease decisions. Although each study has unique insights and methodologies, our approach adds to this expanding field by building upon these foundations.

To get more accurate prediction of heart disease prediction, in our work we adopted a hybrid mechanism, which mixes different machine learning techniques. Our methodology is inspired by the findings of Moghan et al. [5], who developed the HRFLM method, and Dwivedi [6], who showed that LR is effective. We also look into the performance of different algorithms of which Yousefi [8] has mentioned, as explanations that will best identify heart disease predictive models.

Additionally, we go beyond model definition to spoil evaluations that are stringent, as suggested by Kamble et al. [9], and guarantee the durability and reliability of our predictions. We draw insights from these studies and apply a data driven approach to make contributions to the field of heart disease prediction methodology to enable greater accurate diagnosis and better patient outcomes.

3. Data and Methodology

3.1 Dataset Description

In this study we use a wide range of health-related features collected from a diverse sample. The dataset has 308, 854 entries and 19 columns including target variable 'Heart_ Disease'. Below is a detailed description of the dataset:

Parameter	Description
General_Health	Very Good / Good / Excellent / Fair / Poor
Checkup	Within the past year / Within the past 2 years / Within the past 5 years / 5 or more years ago / Never
Exercise	Yes / No
Skin_Cancer	Yes / No
Other_Cancer	Yes / No
Depression	Yes / No
Diabetes	Yes / No / No, pre-diabetes or borderline diabetes / Yes, but female told only during pregnancy
Arthritis	Yes / No
Sex	Male / Female
Age_Category	18 – 24 / 25 – 29 / 30 – 34 / 35 – 39 / 40 – 44 / 45 – 49 / 50 – 54 / 55 – 59 / 60 – 64 / 65 – 69 / 70 – 74 / 75 – 80 / 80+
Height_(cm)	Height of person in centimeters
Weight_(kg)	Weight of person in kilograms
BMI	BMI of the person
Smoking_History	Yes / No
Alcohol_Consumption	Alcohol consumption
Fruit_Consumption	Fruit consumption
Green_Vegetables_Consumption	Green vegetable consumption
FriedPotato_Consumption	Fried potato consumption

Fig. 51.1 Overall workflow diagram

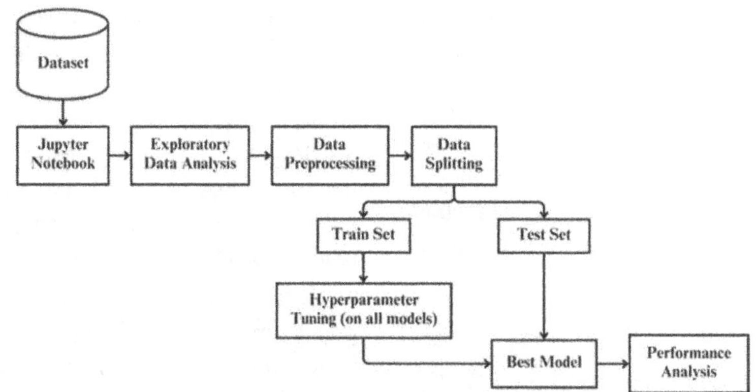

Fig. 51.2 Overall workflow diagram

3.2 Data Preprocessing

1. Handling Categorical Variables
   a. Utilized one-hot encoding for categorical variables: 'Checkup,' 'Diabetes,' 'Age_Category,' and 'Smoking_History.
   b. Target variable 'Heart_Disease' mapped to binary encoding (0 for 'No' and 1 for 'Yes')
2. Standardization: Scaled numerical features using the StandardScaler to ensure uniformity in magnitude.

4. Machine Learning Models

For this analysis, five ML models were used to predict the chance of having heart disease given a whole dataset. With a view towards a comprehensive evaluation of predictive performance, the models were selected for a broad panoply of algorithmic approaches. The following models were utilized:

4.1 Logistic Regression

LR is a type of supervised learning algorithm that can solve both regression and classification issues. It uses probability to predict categorical data in the classification. The sigmoid function can be used to linearly mix of the input values and coefficient values together to make predictions about the outcome. A sigmoid function is in practice used in maximum likelihood estimation to find the most likely data. This function determines that an event is received with a value of probability between 0 and 1 which indicates the probability of an event happening. The task is converted to a classification problem by using the decision threshold. There can be binary (represented by 0 or 1), multinomial (having more than 2 categories, without the order) or ordinal (having 3 or more with order). Application of the model is straightforward, and it has the potential to produce accurate predictions [13].

4.2 Decision Tree Classifier

It is another supervised learning procedure utilized for classification and prediction. Here, the data is hierarchically and uniformly subdivided in the DT structure. Considering the entropy of each attribute, the root is split into sub trees following the objective to attain maximized information gain [14]. The procedure is recursive in that it considers the qualities, and the leaf node will end at the result.

4.3 XGBoost Classifier

XGBoost model is rapidly spreading out due to its success in many Kaggle data science competitions. In many applications it yields satisfactory results: disease prediction [15], diesel fuel brands identification [16], estimation of the tunnel boring machine advance rate [17], prediction in structural health monitoring [18], hotel review sentiment analysis [19] star/galaxy classification [20] and prediction of vehicular accident [21].

4.4 KNN Classifier

KNN is also an important supervised ML based mechanism able to perform classification and regression. In this algorithm, if for a data point we don't have a target value, it finds out k- nearest data points from the training data set and allocate unknown data point to be the average of those nearest points found during the training stage. If k is small enough this is the one-hot encoding. For classification, it assigns or returns mod of k-labels, while in regression, it provide the mean of k-labels. In the case where no prior information regarding the material is known, the algorithm is a basic method for categorization. Distance metrics, like Euclidean distance or Manhattan distance, are used to count up distances, determine which data point is the closest. Secondly, it attain better outcomes and forecasts even if the data is noisy and extensive [13].

4.5 Random Forest Classifier

RF is an ML model that can be used to classify and regress as well. Here a supervised machine-learning methodology is employed. DTs make an ensemble called a random forest for training and prediction. This ensemble [22] utilizes multiple decision trees. Repeats sampling of each instance of the dataset generates a DT for each sample. The outcome of DT is aggregated to determine prediction. An augmentation of the number of trees is hypothesized to result in an accurate and accurate estimate. This method is also adopted for classification and regression. In classification, the data is segregated by voting among a set of decision trees; in regression, it is simply summed up (averaged) by the decision trees. RF algorithm is used to make accurate predictions on large data sets, and till this date is suitable for the data with a high dimensionality.

5. Hyperparameter Tuning

Hyperparameter tuning was further applied to increase the accuracy of the ML models. To tune model performance, we explored different combinations of parameter tuning to find the optimal parameters which maximally predict each model's capability. This tuning improves the overall efficacy of the machine learning models using for forecasting of heart disease. The following table summarizes the tuned hyperparameters for each model:

Table 51.1 Hyperparameter tuning

Model	Tuned Parameters
LR	'C': [0.01, 0.1, 1, 10, 100]
DT	'max_depth': [None, 10, 20], 'min_samples_split': [2, 5, 10]
XGBoost	'learning_rate': [0.01, 0.1, 0.2], 'max_depth': [3,5], 'subsample': [0.8, 1.0]
KNN	'n_neighbors': [3, 5, 7], 'weights': ['uniform', 'distance']
RF	'n_estimators': [100, 200, 300], 'max_depth': [None, 20], 'min_samples_split': [2, 5]

6. Evaluation Metrics

Common classification parameters are used to assess the performance of the ML models to attain a comprehensive understanding of their predictive capabilities. The metrics employed in this study are as follows:

Table 51.2 Evaluation metrics

Parameter	Description	Equation
Accuracy	correct over vs total predictions	$\frac{TP+TN}{TP+TN+FP+FN}$
Precision	correct positive vs total positive predictions	$\frac{TP}{TP+FP}$
Recall	correct positive vs all positive instances	$\frac{TP}{TP+FN}$
F1-Score	Weighted average of precision and recall	$2*\frac{Precision*Recall}{Precision+Recall}$

TP – True + ve FP – False + ve
TN – True - ve FN – False - ve

7. Experiment Outcome and Discussion

For this work, we used different ML models to predict the occurrence of CVDs. With regards to CVDs, given the high impact of CVDs on global mortality, accurate prediction models are essential for timely diagnosis and treatment, and could save millions of people each year.

Our approach involved a comprehensive analysis of five machine learning algorithms: LR, DT, XGBoost, KNN, and RF. To improve the predictive accuracy of these models, we preprocessed data, engineered features, and hyperparameter tuned.

The results showed that accuracy of 92% was achieved by the XGBoost model. Nevertheless, each model had its strengths in different metrics to indicate that context specific model choice is critical to clinical applications. One example of this is that XGBoost had very high overall accuracy, however, other models, such as Random Forest, and Logistic Regression, performed relatively well with precision, and recall.

Being consistent with previous research, Mohan et al. [5] and Dwivedi [6], we corroborate that hybrid methods and logistic regression are efficacious. Additionally, our results are consistent with Yousefi [8] and Kamble et al. [9] comparative performance evaluation on heart diseases using decision tree-based methods.

Furthermore, we show that model performance can be optimized by hyperparameter tuning. The experiment results achieved significantly improved performance by systematically tuning the hyper parameters.

7.1 Confusion Matrix

Table 51.3 Confusion matrices

Models			Predicted	
			No	Yes
LR	True	No	84837	366
		Yes	7094	360
DT		No	84593	610
		Yes	7048	406
XGBoost		No	84988	215
		Yes	7216	238
KNN		No	84094	1109
		Yes	6889	565
RF		No	85012	191
		Yes	7288	166

Table 51.4 Performance of models

Model	Accuracy	Precision	Recall	F1-Score
LR	91.9%	0.496	0.048	0.088
DT	91.7%	0.400	0.054	0.096
XGBoost	92.0%	0.525	0.0319	0.060
KNN	91.4%	0.338	0.076	0.124
RF	91.9%	0.465	0.022	0.043

7.2 Performance Study of Different Methods

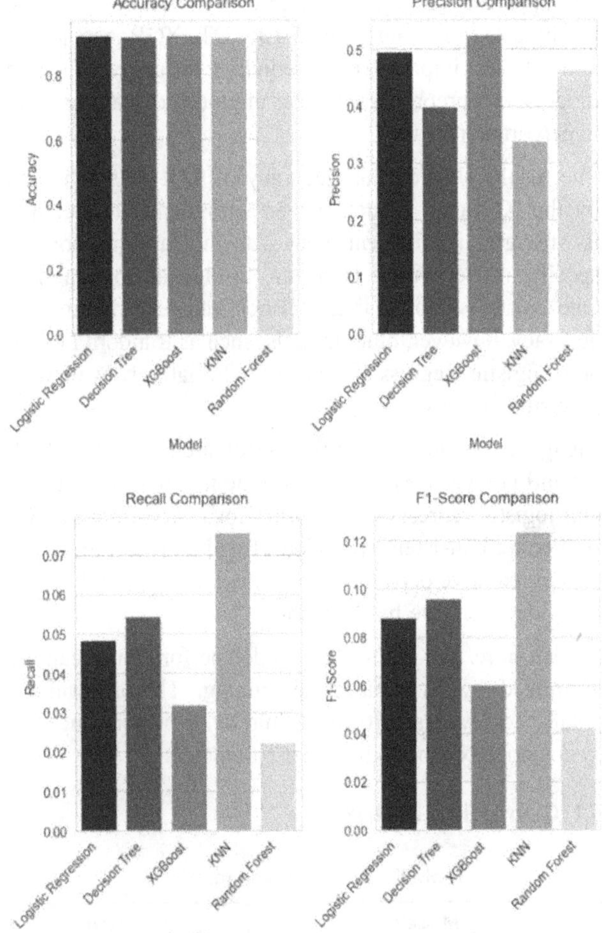

Fig. 51.3 Performance metrics comparison

8. Conclusion

Finally, we suggest that ML models can play a great role in the early diagnosis and management of cardiovascular diseases. Our work underscores the need to continue investigations and development in this area in order to better refine predictive models and ultimately to improve patient outcomes. By using these advanced machine learning techniques, healthcare providers can draw some valuable insights that can help us make more informed decisions to allocate the resources effectively to win the battle against heart disease. As well as future research directions which will also perhaps look at real time data and wearable technology for continuous monitoring and early warning systems for at risk individuals.

References

1. Jagannathan, R., Patel, S. A., Ali, M. K., & Narayan, K. V. (2019). Global updates on cardiovascular disease mortality trends and attribution of traditional risk factors. Current Diabetes Reports, 19, 1–12.
2. Alwan, A. (2011). Global status report on noncommunicable diseases 2010. World Health Organization.
3. National Cholesterol Education Program (NCEP) Expert Panel. (2002). Third report of the National Cholesterol Education Program (NCEP) Expert Panel on detection, evaluation, and treatment of high blood cholesterol in adults (Adult Treatment Panel III) (No. 2). The Program.
4. Mosca, L., Banka, C. L., Benjamin, E. J., Berra, K., Bushnell, C., Dolor, R. J., Ganiats, T. G., Gomes, A. S., Gornik, H. L., Gracia, C., et al. (2007). Evidence-based guidelines for cardiovascular disease prevention in women: 2007 update. Circulation, 115(11), 1481–1501.
5. Mohan, S., Thirumalai, C., & Srivastava, G. (2019). Effective heart disease prediction using hybrid machine learning techniques. IEEE Access, 7, 81542–81554.
6. Dwivedi, A. K. (2018). Performance evaluation of different machine learning techniques for prediction of heart disease. Neural Computing and Applications, 29, 685–693.
7. Nikhar, S., & Karandikar, A. (2016). Prediction of heart disease using machine learning algorithms. International Journal of Advanced Engineering, Management and Science, 2(6), 239484.
8. Yousefi, S. (2021). Comparison of the performance of machine learning algorithms in predicting heart disease. Frontiers in Health Informatics, 10(1), 99.
9. Kamble, S. U., Jawanjal, V. S., Velapure, P. P., Jadhav, P. K., & Kadam, S. S. (2019). Heart disease prediction using machine learning techniques. IJETT, 6(1).
10. Hasan, R. (2021). Comparative analysis of machine learning algorithms for heart disease prediction. In ITM Web of Conferences (Vol. 40, p. 03007). EDP Sciences.
11. Mijwil, M. M., & Shukur, B. S. (2022). A scoping review of machine learning techniques and their utilization in predicting heart diseases. Ibn AL-Haitham Journal for Pure and Applied Sciences, 35(3), 175–189.
12. Katarya, R., & Meena, S. K. (2021). Machine learning techniques for heart disease prediction: A comparative study and analysis. Health and Technology, 11(1), 87–97.
13. Karthiga, A. S., Mary, M. S., & Yogasini, M. (2017). Early prediction of heart disease using decision tree algorithm. International Journal of Advanced Research in Basic Engineering Sciences and Technology (IJARBEST, 3(3), 1–17.
14. Budholiya, K., Shrivastava, S. K., & Sharma, V. (2022). An optimized XGBoost-based diagnostic system for effective prediction of heart disease. Journal of King Saud University-Computer and Information Sciences, 34(7), 4514–4523.
15. Wang, S., Liu, S., Zhang, J., Che, X., Yuan, Y., Wang, Z., & Kong, D. (2020). A new method of diesel fuel brands identification: SMOTE oversampling combined with XGBoost ensemble learning. Fuel, 282, 118848.
16. Zhou, J., Qiu, Y., Zhu, S., Armaghani, D. J., Khandelwal, M., & Mohamad, E. T. (2021). Estimation of the TBM advance rate under hard rock conditions using XGBoost and Bayesian optimization. Underground Space, 6(5), 506–515.

17. Dong, W., Huang, Y., Lehane, B., & Ma, G. (2020). XGBoost algorithm-based prediction of concrete electrical resistivity for structural health monitoring. Automation in Construction, 114, 103155.
18. Zhang, X., & Yu, Q. (2017). Hotel reviews sentiment analysis based on word vector clustering. In 2017 2nd IEEE International Conference on Computational Intelligence and Applications (ICCIA) (pp. 260–264). IEEE.
19. Chao, L., Wen-hui, Z., & Ji-ming, L. (2019). Study of star/galaxy classification based on the XGBoost algorithm. Chinese Astronomy and Astrophysics, 43(4), 539–548.
20. Kidando, E., Kitali, A. E., Kutela, B., Ghorbanzadeh, M., Karaer, A., Koloushani, M., Moses, R., Ozguven, E. E., & Sando, T. (2021). Prediction of vehicle occupants' injury at signalized intersections using real-time traffic and signal data. Accident Analysis & Prevention, 149, 105869.
21. Tan, P.-N., Steinbach, M., & Kumar, V. (2016). *Introduction to data mining*. Pearson Education India.

Note: All the figures and tables in this chapter were made by the authors.

Intelligent Systems Using Semiconductors for Robotics and IoT – Dinesh Goyal et al. (eds)
© 2026 Taylor & Francis Group, London, ISBN 978-1-041-20408-4

52 A New Approach for Identifying and Detecting Fake Reviews Using Ensemble Techniques with TF/IDF and Bi-Grams

Navin Kr. Goyal*,
Sanjay Sinha, Dinesh Bhatia
Department of Computer Engineering,
Poornima Institute of Engineering & Technology,
Jaipur, India

Abstract: In the modern E-commerce landscape, detecting and predicting fake reviews has become a crucial aspect of sentiment analysis. Automated systems for fake review classification leverage machine learning techniques to identify and categorize duplicate, spam, fake, and unreliable reviews. This paper examines several recent approaches to fake review detection, utilizing various classifiers applied to datasets from Yelp and Flipkart. It also discusses the challenges and limitations present in these existing methods. Further, it proposes a novel approach to fake review identification and detection using bi-grams along with three ML-based and ensemble approaches XGBoost, AdaBoost, and ensemble_t with SVM and NB techniques. It is observed here that the ensemble_t with SVM and NB techniques outperform other two classifiers. It achieves an average accuracy of 94% with the training dataset and with the testing dataset.

Keywords: Ensemble techniques, Review detection, Machine learning. AdaBoost, XGBoost, SVM, NB

1. Introduction

The identification, detection, analysis, and prediction of fake reviews have become a crucial focus in sentiment analysis, especially within the evolving fields of Natural Language Processing (NLP) and the E-commerce sector. Lately, the need for automated systems leveraging Machine Learning (ML) techniques has grown significantly grounded in the core principles of NLP, ML, and Artificial Intelligence (AI) [1] - [8].

Although many processing systems do exist for fake news detection and classification, Challenges persist in areas such as dataset quality, validity concerns, data imbalance, appropriate algorithm selection, feature minimization, limited resources, and issues with poor overall performance and accuracy. It is observed that there is a need for generic and automated fake and ambiguous news detection systems using ensemble techniques.

This article proposes a Novel Approach for Fake Review Identification and Detection using Bi-Grams and Ensemble Techniques to locate and identify fake reviews using three ensemble techniques. These techniques are XGBoost, AdaBoost, and Ensemble_t with Support Vector Machine (SVM) and Naive Bayes (NB). This system handles the imbalanced dataset and converts it into a balanced dataset.

The following is the order in which the sections of this article are organized. A comprehensive overview of the many fake review prediction and categorization algorithms currently in use is provided in part 2. Part 3 illustrates the suggested fake review identification and detection model using bi-grams and ensemble techniques. Section 4 analyzes these specifications and results graphically and provides the suggested model's research outcomes and analyzes the performance. The study is concluded with recommendations for the future in part 5.

*Corresponding author: navin.goyal@poornima.org

DOI: 10.1201/9781003716389-52

2. Fake Review Detection Systems Journey

Acoording to G. Ansari, et al. Using various datasets and machine learning techniques, this portion describes and contrasts the procedures of the latest automated false identification systems. The steps taken by these current techniques to find, recognize, forecast, and categorize fraudulent reviews in sentiment analysis are detailed in this trip. One by one, these methods are offered here to illustrate their efficacy. The performance of many ways to detect such phony reviews was studied in this paper of well-known supervised machine learning techniques for online bogus review identification. It provided a technique for more study as well as a comprehensive overview of current studies on identifying fraudulent reviews[1].

According to K. Algotar, et al. By combining two active-based supervised learning algorithms, the ensemble technique was applied to the false review detection. The KL, and JS distance, Term Frequency/Inverse Document Frequency (TF/IDF) characteristics, and N-Gram properties of the review material were among the three filtering phases used for this. During its Active Learning (AL) procedure, it individually identified more than 1000 unlabeled fraudulent feedback and tested the suggested method using four different simulation types. It was discovered that the AL technique worked better with the Linear Support Vector Machine (LSVM). Furthermore, compared to alternative supervised classifiers, Naïve Bayes (NB) fared better[2].

By Elmurngi, E., et. al.,(2018), Four classifiers and three ensemble-based approaches were used in another similar approach. It contrasted the results of Random Forest (RF), boosting, bagging, SVM, Multinomial NB (MNB), Logistic Regression (LR), and C4.5 [3]. According to Madisetty, S., et. al., (2018), The NB, SVM, and LR were used as foundation classifiers in the false review detection model which applied the ensemble approach [4]. A. J. Ibrahim, et. Al., (2021), An ensemble strategy for tweet fake review detection was proposed by another model. Five Convolutional Neural Networks (CNN) were used, together with the feature-based model's n-gram, content-related, and user-related features [5].

S. Mani, et. al.,(2018): used an another similar detection model used five ML algorithms, NB, K*, SVM, Decision Tree (DT), and KNN to classify the movie reviews into positive and negative polarity[6]. M. Ahsan, et. al,(2016) Used deceptive and useful classifiers, respectively, the bogus review identification model identified the geniune reviews. A ranking model was then used to assign ratings to the reviews. Additionally, it created a dictionary and repository to classify the reviews as either honest or dishonest. The testing data was classified as either truthful or deceptive using the deceptive classifier, and as valuable or worthless using the useful algorithm[7].

P. Hájek et. al., in this study assessed the effectiveness of various deep learning techniques, including NB, SVM, and DT. It examined the effectiveness of many machine learning techniques for identifying both positive and negative fraudulent customer reviews. Deep Neural Networks (DNN) were shown to perform better in terms of accuracy and time performance than traditional machine learning methods[9].

The spam review detector by J. Wang et. al.,(2021) and N. Goyal et. al.,(2023). utilized an ensemble approach by integrating Naive Bayes (NB), Logistic Regression (LR), Decision Trees (DT), Gradient Boosted Trees, Random Forest (RF), and Artificial Neural Networks (ANN) [10],[14].

J. Yao, et. al.,(2021), After addressing the imbalance in the dataset through resampling, an alternative method minimized computational expenses by refining the feature set and fine-tuning parameters through grid search to identify their optimal values. Ultimately, the optimized base classifiers were integrated using an ensemble model that employed majority voting and stacking methods[11].

J. Salminen offers a thorough summary of the literature on machine learning techniques for the detection of fraudulent reviews. It looks at several research and emphasizes the supervised, unsupervised, and semi-supervised learning techniques that are applied in this context. After comparing various methods, the study concludes that Hybrid CNN-RNN (99%), Ensemble Voting (97.51%), and SVM (97.03%) had the highest accuracy. By providing insights into current approaches and suggesting that machine learning holds considerable promise for additional breakthroughs in false review identification, the paper seeks to lead future research[13].

The performance-based and basic specifications show the results of various parameters in currently used methods. Several specification-based comparisons between previous contributions are covered in this section. The main components of these comparisons are the specifications, which include System development by year, algorithm type and use, and dataset type and use. The years 2018 and 2020 were found to have the highest research contributions in spam review detection. With utilization rates of 20%, 16.36%, and 14.54%, respectively, the SVM and it's variations, NB and it's variants, and RF algorithms were determined to be the most popular. It was discovered that the use of other classifiers was C4.5,K* (1.81%), KNN, DT (7.27%), bagging, ANN+CNN and MLP (3.71%), and LR (9.09%).

Additionally, it determined the usage result percentage of different datasets for current methods. The most popular datasets were Yelp, Amazon, hotel-based, and supplemental, with respective utilization rates of 22.97%, 20.11%, 12.44%, and 12.44%. The TripAdvisor, AMT, restaurant, and other hotel databases were among the hotel-based datasets in this case. The usage of additional datasets was determined to be 3.79% and 7.71% for OTT and Twitter.

3. The Proposed Fake Review Identification and Detection Model using Bi-Grams and Ensemble

This article proposes a novel approach for Fake Review Identification and Detection using Bi-Grams and Ensemble techniques. The proposed model aims to find and identify the bogus reviews using three ensemble techniques and assesses the system's performance and efficacy using a variety of performance-based critaria, including accuracy, pracision, recall, and F1-score.

This system accepts the imbalanced dataset and converts it into balanced sets for further consideration. It then pre-processes the data. It performs the steps of pre-processing, exploratory data analysis with Term Frequency-Inverse Document Frequency (TF-IDF), true and false sample distribution, and handles the missing values. Here, a larger text's relevant material and key words and phrases are located using TF-IDF.

Next, it extracts the features or reviews and applies the ensemble approaches. It trains the features using three ML-based approaches such as XGBoost (XGB), Ada Boost (Ada), and Ensemble_t with SVM and NB. The confusion matrices for each of the three classifiers are then produced. Finally, it detects and categorizes the fraudulent reviews and assesses the system's performance through several metrics.

This proposed work is implemented with the standard review dataset called the 'Fake Reviews' dataset used by M. Abd et. al., (2024) and J. Salminen. There are 20,000 phony and 20,000 genuine product reviews in this dataset. for ten different categories for the year 2021 [12], [13].

3.1 Algorithm FRID_BGE ()

Step-1 Accept the reviews of the 'Home Kitchen' dataset

Step-2 Pre-process the review data. Remove the stop words and unimportant words.

Step-3 Perform the exploratory data Analysis.

a) Visualize the important data.

b) Determine the TF-IDF for the words.

$$\text{Term Frequency} = \frac{\text{number of instances of word } w \text{ in document } d}{\text{total numberof words in document } d} \quad (1)$$

$$IDF = \log\left(\frac{\text{total number of documents } (N) \text{ in text corpus } D}{\text{number of documents containing } w}\right) \quad (2)$$

c) Distribute the balanced data for the training and testing phases.

d) Check for the missing values and data.

e) Analyze the Top 30 Keywords for real and fake reviews.

f) Analyze and check the Words and characters along with their count totals.

Step-4 Analyze the histograms and features with bi-gram analysis in the training and testing data.

Step-5 Compute the confusion Matrices using all ten ML-based and ensemble-based techniques

Step-6 Estimate the accuracy of the system with all the classifiers.

Step-7 Detect the fake reviews and separate them from the real reviews.

Step-8 Perform the validation analysis of the system.

4. Discussion on Analytical, Experimental Results and Performance Analysis

This section describes the analytical, experimental results. Further, it provides the performance analysis of classifiers on the training and testing features with bi-grams.

The system first pre-processes the training reviews. It disregards the noise, unimportant words, and stop words, and considers the important for further processing. Figure 52.1 depicts the word cloud created from raw text. This cloud excludes all the stop words. It includes the words such as 'love', 'used', 'really', 'color', 'quality', 'bought', etc. Figure 52.2 depicts the word cloud having alphabetic entities only. It includes the words such as 'love', 'use', 'used', 'really', 'quality', 'bought', 'perfect', etc.

Figure 52.3 depicts the top ten product reviews of the 'Home_and_Kitchen_5' category. It shows the entire data from 0 to 9 values of ID. A few values of the 'keyword' field are 'love',' missing', 'nice', 'different', 'space', etc. The ratings can be from 0 to 5. The 'Text' field includes the actual review contents. The target can be either o or 1.

Fig. 52.1 Word cloud with no stop words

Fig. 52.2 Word cloud for only alphabetic entities

ID	Category	Keyword	Rating	Text	Target
0	Home_and_Kitchen_5	love	5	Love this! Well made, sturdy, and very comfortable. I love it!Very pretty	0
1	Home_and_Kitchen_5	love	5	love it, a great upgrade from the original. I've had mine for a couple of years	0
2	Home_and_Kitchen_5	love	5	This pillow saved my back. I love the look and feel of this pillow.	0
3	Home_and_Kitchen_5	missing	1	Missing information on how to use it, but it is a great product for the price! I	0
4	Home_and_Kitchen_5	nice	5	Very nice set. Good quality. We have had the set for two months now and have not been	0
5	Home_and_Kitchen_5	different	3	I WANTED DIFFERENT FLAVORS BUT THEY ARE NOT.	0
6	Home_and_Kitchen_5	space	5	They are the perfect touch for me and the only thing I wish they had a little more space.	0
7	Home_and_Kitchen_5	love	3	These done fit well and look great. I love the smoothness of the edges and the extra	0
8	Home_and_Kitchen_5	size	5	Great big numbers & easy to read, the only thing I didn't like is the size of the	0
9	Home_and_Kitchen_5	love	5	My son loves this comforter and it is very well made. We also have a baby	0

Fig. 52.3 Displaying top 10 reviews of 'home_and_kitchen_5' dataset

The balanced dataset of true and false samples was used for the implementation.

Figure 52.4 depicts the graphical analysis between the mean word (keyword) length and the 'total number of keywords'. The total number of keywords represents the word count. The mean word length lies in the range of 2.5 to 6.4. The total number of keywords is between 14000 and 16000. It is observed that the maximum number of keywords is obtained at the mean word length of 3.6.

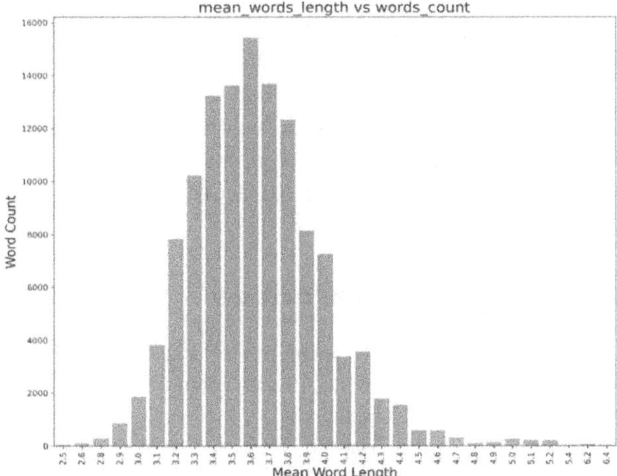

Fig. 52.4 Mean word length vs word count

Figure 52.5 depicts the graphical analysis between the character count and the 'total number of keywords'. The character count here ranges between 78 and 254. Here the total word count is approximately 800. It is observed that 84 characters obtained the best results.

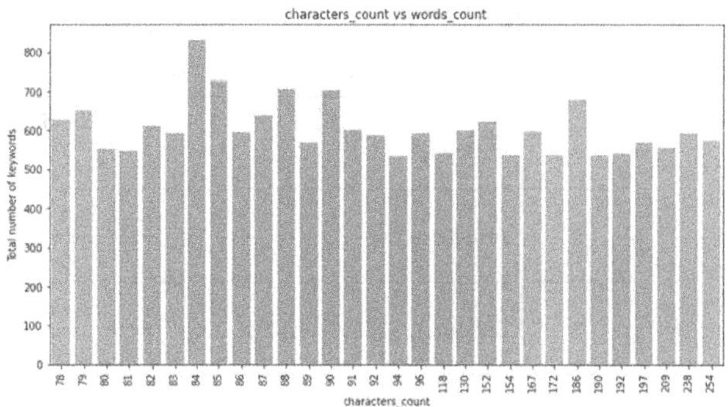

Fig. 52.5 Character count vs word count

4.1 Histogram Analysis

Figure 52.6 shows the analysis of the word count w. r. t. the feature density for the real and fake reviews. It is observed that the words exist between the range of 0 and 150, whereas the density ranges between 0.00 and 0.05. It achieves a maximum density of 10 to 40 words. Histograms show that the word count of both types of reviews is similar but fake ones tend to have longer words. The proposed system has performed many experiments using three classifiers such as SVM, NB, and Ensemble_t with SVM and NB.

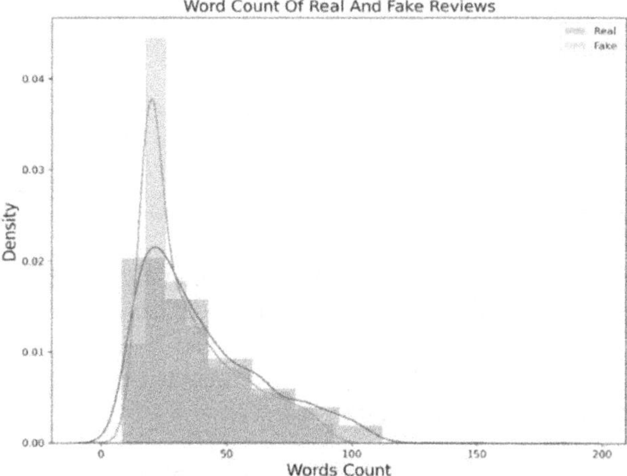

Fig. 52.6 Analyzing the word count with the density of real and fake reviews

4.2 Accuracy Estimation

Table 52.1 shows the performance analysis of the system using three classification techniques. It presents the experimental results for the F1-score, precision, and recall measures and estimates the accuracy for the real reviews. It is observed here that the ensemble_t with SVM and NB technique outperformed other classifiers and ensemble techniques. The least accuracy results are obtained with the XGBoost techniques. The maximum precision is obtained by XGBoost classifiers for real reviews is 70%. The maximum precision is obtained by the Ensemble_t with SVM and NB classifier for real reviews is 95%. The maximum recall is obtained by XGBoost is 98% for fake

Table 52.1 Precision, Recall, F1-score, and accuracy for all ML-based and ensemble techniques

S. N.	ML and Ensemble Approaches	Class	Precision	Recall	F1-Score	Accuracy
1	XGBoost	Real Review	96%	41%	58%	70%
		Fake Review	63%	98%	76%	
2	Ada Boost	Real Review	93%	53%	67%	74%
		Fake Review	67%	96%	79%	
3	Ensemble_t (SVM & NB)	Real Review	95%	93%	94%	94%
		Fake Review	94%	95%	94%	

reviews. The maximum recall is obtained by Ensemble_t with SVM and NB classifier for fake reviews is 93%. The maximum F1 score is obtained by the Ensemble_t with SVM and NB classifier for both fake and real reviews. Figure 52.7 shows the comparison of the accuracies obtained.

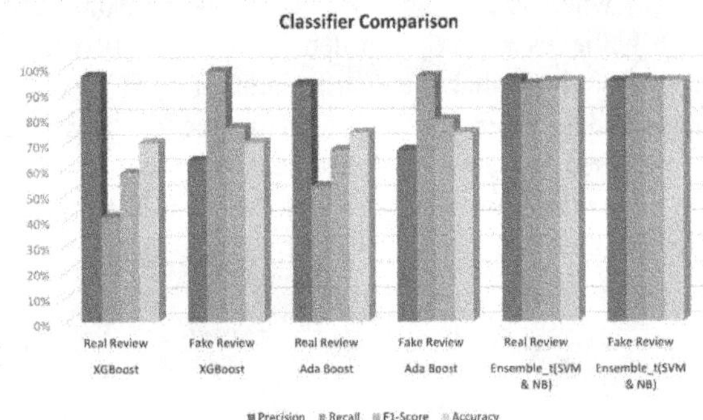

Fig. 52.7 Accuracy comparison for all classifiers

5. Result Analysis

Table 52.2 depicts the training and testing accuracies for all three classifiers. The suggested model has an average accuracy of 88% on training datasets and 79% on testing datasets. Figure 52.8 depicts the accuracy comparison between training and testing datasets.

Table 52.2 Accuracy estimation with 3 ML ensemble techniques

SN	ML and Ensemble Techniques	Training Accuracy	Validation Accuracy
1	XGBoost	86%	70%
2	Ada Boost	83%	74%
3	Ensemble_t with SVM and NB	94%	94%
Average Accuracy		88%	79%

6. Conclusions and Future Recommendations

The many specifications of the current fake review analysis and prediction systems were thoroughly analyzed and compared in this research. The year-wise progression, classifiers, and data sets were among the fundamental requirements. Furthermore, the performance requirements encompassed accuracy, precision, recall, f1-score, and AUC. The study emphasized the difficulties, problems, risks and further research of several previous efforts. Furthermore, it was discovered that the SVM and NB classifiers, as well as the Yelp and hotel-based datasets, had the highest usage percentage.

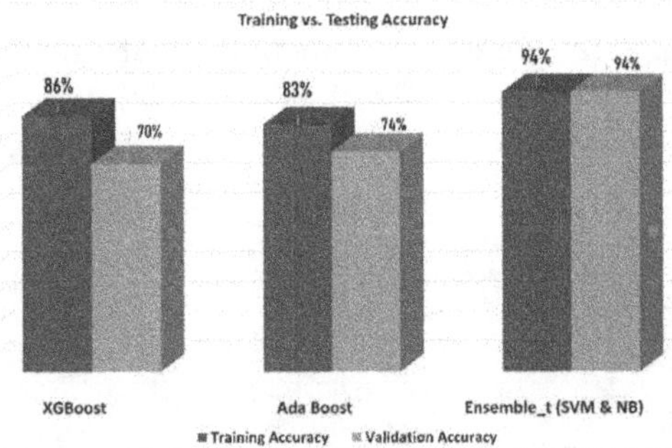

Fig. 52.8 Training vs. testing accuracy

This article suggested an innovative method for identifying and detecting bogus reviews using bi-grams along with three ML-based and ensemble approaches. It is observed here that the Ensemble_t with SVM and NB technique outperformed other classifiers and ensemble techniques. With the training dataset, it averaged 88% accuracy, whereas with the testing dataset, it averaged 79% accuracy. The future work includes the extension of this system with trigrams and N-grams. Another direction is to implement it with other ML techniques and hybrid techniques.

References

1. G. Ansari, et. al(2018): Spam Review Classification Using Ensemble of Global and Local Feature Selectors, doi: 10.2478/cait-2018-0046.
2. K. Algotar and A. Bansal, "Detecting truthful and useful consumer reviews for products using opinion mining," *CEUR Workshop Proc*, vol. 2111, pp. 63–72, Jan. 2018.
3. Elmurngi, E., et. al.,(2018): *Detecting Fake Reviews through Sentiment Analysis Using Machine Learning Techniques*.
4. Madisetty, S., & Desarkar, M.S. (2018). A Neural Network-Based Ensemble Approach for Spam Detectionin Twitter. *5(*973–984).

5. A. J. Ibrahim, M. Siraj, and U. A. Jauro, (2021): Ensemble Classifiers Employed for Spam Review Detection, 6 (33–38).
6. S. Mani, et. al., (2018): Spam Review Detection Using Ensemble Machine Learning, (198–209).
7. M. Ahsan, M., et. al,(2016):An ensemble approacht detectreview spam using hybrid machine learning technique (388–394).
8. M. Crawford M., et. Al., (2015): Survey of review spamdetection using machine learning echniques, 2(23).
9. P. Hájek and A. Barushka, "A Comparative Study of Machine Learning Methods for Detection of Fake Online Consumer Reviews," Nov. 2019, pp. 18–22. doi: 10.1145/3383902.3383909.
10. J. Wang, D. Xue, and K. Shi, (2021): An Ensemble Framework for Spam Detection on Social Media Platforms, (11).
11. J. Yao, Y. Zheng, H. Jiang, and L. Wang, (2021): An Ensemble Model for Fake Online Review Detection Basedon Data Resampling, Feature Pruning, and Parameter Optimization," *IEEE Access*.
12. M. Abd and M. Hussein, (2024): Fake reviews detection ine-commerce using machine learning techniques: A comparative survey, 97 (99).
13. J. Salminen, "Fake Reviews Dataset," Mar. 2024, *OSF*. [Online]. Available: osf.io/tyue9
14. Goyal, Navin & Gupta, Kr & Goyal, Drdinesh. (2023). Fake Review Detection Using Ensemble Techniques by the Fusion of Chronology, Aspect and Sentiment of Reviews and Oversampling by Smote. 44. 114–126.

Note: All the figures and tables in this chapter were made by the authors.

Intelligent Systems Using Semiconductors for Robotics and IoT – Dinesh Goyal et al. (eds)
© 2026 Taylor & Francis Group, London, ISBN 978-1-041-20408-4

53 Exploring Future of Data Analytics: Trends, Techniques and Tools

Samiksha Agarwal[1],
Madhav Sharma[2], Ritam Dutta[3]
Poornima Institute of Engineering & Technology,
Computer Science and Engineering - Internet of Things,
Jaipur, India

Hukum Chand Saini
Jagannath University, Faculty of Engineering & Technology,
Jaipur, India

Payal Bansal
Poornima Institute of Engineering & Technology,
Computer Science and Engineering - Internet of Things,
Jaipur, India

Abstract: The evolving landscape of facts analytics is marked by using widespread advancements in strategies, equipment, and rising trends. As firms retain to generate extensive amounts of data, modern analytics tactics are moving toward extra automated and sensible structures. This review paper examines key trends shaping the destiny of facts analytics, including the increasing integration of synthetic intelligence (AI) and machine mastering (ML) for predictive and prescriptive analytics, the upward push of aspect computing for real-time records processing, and the developing significance of information privateness and governance.

Emerging techniques like deep learning, natural language processing (NLP), and automated device learning (AutoML) are enhancing the capacity to analyse unstructured and complex datasets. Tools including cloud-primarily based platforms, facts lakes, and augmented truth (AR) visualisation are revolutionising how information is stored, processed, and provided, supplying more reachable and actionable insights.

This paper also explores the ethical challenges related to AI-pushed analytics, especially regarding information privacy, protection, and bias. The destiny of facts analytics guarantees to drive innovation throughout industries, from healthcare to independent structures, however its success will depend on addressing those challenges whilst harnessing the capability of new technologies. This evaluation highlights the latest developments, explores current techniques, and discusses the tools shaping the subsequent era of facts analytics.

Keywords: ML, AI, DL, NLP

1. Introduction

In the digital age, the volume, range, and speed of information generated through people and firms have reached unprecedented tiers. As a result, statistics analytics has grow to be a cornerstone of selection-making tactics throughout numerous sectors. The speedy evolution of information analytics technologies is pushed by means of advancements in synthetic intelligence (AI), system getting to know (ML), and cloud computing, which together permit establishments to extract meaningful insights from huge datasets. As mentioned by using Waller and Fawcett

Corresponding author: [1]samikshaagarwal539@gmail.com, [2]madhavsharma36@yahoo.co.in, [3]ritamdutta1986@gmail.com

DOI: 10.1201/9781003716389-53

(2013), the ability to leverage facts efficiently can enhance operational performance and foster innovation.

2. Emerging Trends in Data Analytics

The evolving trends in facts analytics are reshaping how companies utilize information for strategic choice-making and planning. These trends are a response to the growing complexity and extent of facts within the digital generation, encompassing technological improvements, ethical issues, and enhancements in analytics.

A key fashion is the growing focus on explainability and transparency in AI fashions. As AI and machine studying algorithms end up greater superior, there's an increasing want to make clean how those systems obtain their conclusions. Ensuring transparency is vital no longer handiest for regulatory compliance and moral obligation however also for organising keep in mind in AI-driven strategies. Organisations are prioritising the development of equipment and methodologies that make AI extra interpretable, consequently improving the accessibility of selection-making strategies for human oversight.

Another tremendous trend is the adoption of edge computing and actual-time analytics. With the rise of Internet of Things (IoT) gadgets and the call for instant insights, data analytics is transitioning closer to information sources. Edge computing allows the processing of statistics near its beginning, lowering latency and facilitating real-time choice-making. This trend is especially critical in sectors like self sustaining vehicles, healthcare, and manufacturing, wherein instant decisions can notably affect results.

3. Literature Review

Narayana Challa, mentioned the evolution and significance of facts analytics in research, addressing demanding situations with huge datasets and the significance of statistics engineering. It outlines the data analytics lifecycle, inclusive of steps like hassle identification, statistics collection, cleansing, evaluation, and interpretation. The paper highlights the impact of rising technology including AI, system gaining knowledge of, and cloud computing on facts analytics. Finally, it discusses the moral concerns of records usage, emphasising the need for responsible strategies in leveraging data analytics for knowledgeable selection-making.[1]

Dr CK Gomathy, mentioned the expanding role of big statistics analytics (BDA) throughout diverse sectors, emphasising its packages in healthcare, social media, sustainable tourism, supply chain control, and cryptocurrency. It highlights how BDA allows actual-time selection-making, mainly in the context of the COVID-19 pandemic, and addresses demanding situations in streaming records analytics. The evaluate underscores the importance of BDA in enhancing efficiency, predicting developments, and improving consumer reports across industries, at the same time as additionally suggesting future studies directions and methodological improvements.[2]

Ruiyu Liang, mentioned the integration of advanced technology within the mining industry, focusing on information digitization techniques which include data visualisation, control, analytics, statistics fusion applications, and digital twin creation. It emphasises the constraints of traditional statistics visualisation techniques and gives parametric modelling and Building Information Modeling (BIM) as viable alternatives. The authors become aware of challenges in information control, inclusive of issues associated with facts range, quality, and security, advocating for a unified information management technique. Additionally, the paper explores the capability packages of statistics fusion in regions like geological exploration, mineral processing, and mine safety, highlighting future developments and demanding situations in developing powerful statistics-driven visual models in mining.[3]

Tanya Garg, discussed a complete overview of Big Data Analytics, discussing its defining traits and significance. It explores diverse programs throughout fields which include training, healthcare, agriculture, military, banking, and commercial enterprise enterprises, highlighting the high quality and poor effects of Big Data Analytics. The modern-day landscape of Big Data analytic equipment is tested, along demanding situations hindering its development, including records safety issues. Additionally, the paper addresses the position of machine gaining knowledge of strategies in enhancing Big Data Analytics. It concludes by way of emphasising the need for destiny research to address those demanding situations, broaden new algorithms, and enhance statistics collection strategies, specially for real-time statistics analysis.[4]

Davide Mottin, mentioned exploratory statistics evaluation techniques, emphasising instance-based total processes that permit users to enter examples rather than complex queries, making data exploration greater intuitive. It discusses the limitations of traditional query languages and affords an educational shape overlaying advent, example-based methods, and device studying programs. The authors discover ongoing challenges together with adaptivity and interactivity, advocating for destiny research to enhance these techniques. Overall, the paper pursuits to bridge gaps between information control, facts mining, and gadget studying, catering to each researchers and practitioners in the subject.[5]

Michael Segner, discussed the evolution of big facts analytics is transferring in the direction of real-time insights, emphasising facts freshness for timely decision-making. Companies like Snowflake and Google are enhancing streaming capabilities, while organisations consciousness on facts pleasant and governance to ensure accuracy. The rise of no-code solutions and records mesh architectures empowers broader get entry to records analytics. Additionally, generative AI and retrieval-augmented era are redefining information evaluation, permitting predictive insights and fostering innovation. As these developments unfold, even smaller groups can leverage huge records for aggressive benefit.[6]

Gul, S., discussed the evolving discipline of records mining, encompassing various technologies including numerical, textual, multimedia, web mining, sentiment evaluation, and huge records mining. It employs an extensive literature evaluation of the use of databases like Clarivate Analytics' Web of Science and Sciverse Scopus to become aware of rising techniques and tactics in information mining, specializing in key phrases related to huge records and the Internet of Things (IoT). Findings suggest that expertise discovery in databases has solidified information mining's function in deriving precious insights and algorithms for various programs. The paper emphasises the sensible implications of those technologies for reinforcing human experiences and offers a unique perspective on current trends in records mining.[7]

D. P. Acharjya, mentioned the challenges and methodologies associated with big statistics analysis, highlighting issues including information range, real-time versus batch processing, and scalability. It discusses various analytical techniques, along with statistical evaluation, machine getting to know, data mining, clever analysis, cloud computing, quantum computing, and data circulation processing. The authors propose that future studies will more and more cognizance on those techniques to cope with large statistics challenges effectively and efficiently.[8]

Androniki Sapountzi, discussed key study regions in social networking and massive data, specializing in equipment and techniques used for analysing Online Social Networks (OSNs). It highlights applications like recommendation systems, opinion summarization, fashion detection, sentiment analysis, and figuring out influential customers. These strategies rely on strategies along with natural language processing, machine mastering, and information analytics to extract insights from huge social media records. The look at emphasises the growing significance of superior analytical techniques to address the complexities of OSN information.[9]

Sepideh Bazzaz Abkenar, discussed 74 studies on big data analytics in social networks from 2013 to 2020, focusing on content-oriented and community-oriented processes. It highlights normally evaluated parameters like accuracy, time, and scalability, even noting gaps in privacy, safety, and reliability. Python and Hadoop are the maximum used equipment. Key challenges include actual-time processing, privateness-keeping analytics, and sentiment evaluation problems like sarcasm detection and domain dependency. The have a look at requires destiny studies to cope with those demanding situations.[10]

Zaher Ali Al-Sai, highlighted the crucial function of Big Data at some point of COVID-19, permitting establishments to make real-time decisions and adapt to challenges across sectors like healthcare, training, transportation, and banking. Big Data allowed for better aid management, fashion analysis, and service optimization during the disaster. Despite its advantages, challenges like infrastructure limitations and statistics privateness issues persist, necessitating in addition studies on Big Data implementation and achievement elements.[11]

Roman Khan, examines the effect of enterprise intelligence (BI) on organisational performance within the retail sector, showing a tremendous dating between BI adoption and key overall performance signs like sales growth, stock turnover, and customer satisfaction. Through a blended-techniques method, the research highlights how BI enhances decision-making, operational performance, and revenue technology. Qualitative insights emphasise the importance of leadership guide and organisational alignment in maximising BI blessings. The findings provide treasured implications for both principal and exercise, underscoring BI's role in fostering innovation and strategic growth in the retail industry.[12]

Sara Alspaugh, interviewed with 30 expert information analysts to discover the function of information exploration in analysis, pick out common challenges, and examine the tools used for exploration. The findings revealed 4 underrepresented kinds of exploratory pastime and the presence of exploration across all phases of evaluation. The research highlights the advantages and shortcomings of modern-day tools, and offers ideas for destiny tool improvement, particularly smart ones, to decorate the facts exploration system and encourage new studies instructions.[13]

Mu-Hsing Kuo, explored the traits, challenges, and solutions of health large records analytics (BDA), which aims to extract precious knowledge from enormous quantities of affected person records. It discusses how BDA can improve health regulations, beautify affected person safety, and reduce prices. Additionally, the paper gives and evaluates a pipelined framework to function a tenet for health data analytics, contributing to more efficient and informed decision-making in healthcare. [14]

Mohsen Marjani, surveys the intersection of IoT and big data analytics, focusing on the need to process and analyse large volumes of data generated by smart and sensor devices. It reviews recent analytics solutions, discusses the relationship between IoT and big data, and proposes an architecture for big IoT data analytics. The paper also highlights opportunities, use cases, and future research challenges, concluding that current solutions are still in early stages and emphasising the need for real-time analytics in the future.[15]

4. Future Applications of Data Analytics

"Future Applications of Data Analytics" refers to the numerous industries where data analytics, bolstered by AI and machine learning, is poised to make groundbreaking contributions. These applications represent innovative uses of data analytics technologies to address complex problems, support data-driven decisions, and foster advancement. Key areas of anticipated impact include:

1. Healthcare and Predictive Medicine: Data analytics has the potential to convert healthcare by way of analysing patient statistics to predict diseases, personalise remedy plans, and enhance results. AI-pushed diagnostics, personalized medication, and real-time health tracking constitute big destiny programs.
2. Financial Services and Risk Management: In the economic quarter, records analytics is presently hired for fraud detection, credit threat assessment, and portfolio optimization. Future improvements might also include more delicate fraud detection algorithms, automatic economic advisory structures, and real-time marketplace forecasting.
3. Manufacturing and Supply Chain Optimization: Data analytics can streamline production techniques, minimise downtime, and beautify supply chain efficiency. Future packages may involve predictive preservation, demand forecasting, and independent manufacturing structures that growth productivity.
4. Autonomous Vehicles and Transportation: Self-using vehicles and clever transportation structures rely closely on information analytics for actual-time selection-making. The future ought to see enhanced path optimization, progressed traffic control, and protection upgrades driven by means of data-pushed insights.

5. Challenges and Concerns

As the arena of statistics analytics continues to improve with AI and tool learning, several demanding conditions and concerns want to be addressed. Data safety and privateness are paramount as the quantity of sensitive information analysed grows, making it crucial to safeguard against unauthorised get admission to and breaches. Furthermore, bias in AI fashions is an excellent ethical state of affairs, as algorithms can also by means of accident support societal biases, growing inequitable outcomes.

Another mission lies in records governance and regulatory compliance, where navigating complicated jail frameworks for information utilization and storage offers obstacles for groups. The talent gap in AI and device studying is also widening, with calls for experts in those fields outpacing the supply of certified people. Addressing those demanding conditions calls for a careful stability among technological innovation and accountable, ethical utilization to make sure that the destiny of records analytics is strong, straightforward, and inclusive.

6. Potential Impact on Society

The evolving landscape of statistics analytics, powered with the resource of AI and system getting to know, holds high-quality capacity to have an impact on numerous elements of society past enterprise and era. The financial impact of integrating those technologies is predicted to pressure innovation and productivity, leading to capability growth and technique advent. However, it additionally poses a hazard of disrupting conventional employment systems, necessitating a reevaluation of groups of workers' talents and process roles. Companies that efficiently leverage these technologies will gain aggressive blessings, even as the ones that do not adapt may also warfare to stay possible.

On a social level, the ability for terrific change is huge. AI-powered analytics can cope with pressing societal annoying conditions, which encompass improved healthcare diagnostics, urban planning, and catastrophe reaction. Additionally, those technologies can decorate accessibility, personalization, and luxury in services starting from education to entertainment. However, troubles related to privacy, algorithmic bias, and moral issues must be rigorously addressed to make sure that those advances serve the broader society and do not exacerbate present inequalities.

7. Internet of Things (IoT) and Its Contribution to Big Data

The Internet of Things (IoT) refers to a significant network of interconnected devices and sensors that continuously accumulate and proportion facts to synchronise their features. IoT systems, in particular better-level hardware along with hubs, coordinate records collection from gadgets like smart cameras, wearables, and robotic systems in real-time. These interconnected networks generate considerable volumes of large data, characterised by way of its size and unstructured nature.

Due to the complexity and quantity of these facts, it is regularly vital to switch it to cloud-primarily based systems, wherein sophisticated algorithms can smooth, normalise, aggregate, and put together the information for analytical use. Applications which include facial reputation and dynamic geographic statistics gadget (GIS) navigation for independent automobiles exemplify using such refined information.

In some cases, this records processing needs to arise rapidly, making cloud computing alongside data lakes an effective solution. Cloud-supported IoT analytics improves processes along with supply chain control, self reliant transportation routing, computerized manufacturing, and complements common user experience with connected gadgets.

8. Data Visualization: Modernising an Age-Old Practice

Data visualisation strategies have extended records, even though within the beyond, they required advanced mathematical and graphical abilities. Today, however, present day technologies permit for the introduction of complicated visualisations with minimal effort. This evolution in statistics representation enables organisations to give complicated facts in reachable codecs, which include infographics, graphs, and charts. The following are key improvements in information visualisation:

Interactive Visualization: This current approach allows customers to have interaction with records through dashboards, filtering, sorting, and exploring datasets in actual-time. With the capability to generate charts and graphs right away, customers can gain deeper insights into facts, inclinations and patterns. These interactive gadgets are particularly valuable in client programs, where they offer clients an intuitive information of activities, which includes buy histories in e-trade.

Data Storytelling: This approach makes use of narrative and innovative factors to provide facts in a way that engages and informs unique audiences, consisting of traders or policymakers. Through visible storytelling, insights are communicated in a compelling layout, going past traditional slide shows and leveraging media like quick movies or classified ads to persuade stakeholders and strengthen motion.

Augmented Reality (AR) Visualization: AR visualisation combines virtual overlays with actual-worldwide environments, allowing customers to engage with virtual information in an extra immersive context. This generation is specifically impactful in e-studying, where AR devices alongside clever glasses or smartphones can enhance the educational level through merging records analytics with physical surroundings.

9. Data Governance and Best Practices in Modern Data Analysis

Effective records governance is vital for enterprises to manage their data belongings responsibly, ensuring compliance with legal and ethical standards at the same time as stopping protection breaches. Privacy Regulations: Various privacy laws, consisting of the General Data Protection Regulation (GDPR), Health Insurance Portability and Accountability Act (HIPAA), and California Consumer Privacy Act (CCPA), impose stringent necessities for dealing with personally identifiable information (PII). These policies mandate anonymization and stable handling of sensitive records, requiring organisations to reap specific consent for records usage and to conform with extra privacy measures.

Data Quality Management and Stewardship: Ensuring information accuracy, completeness, and consistency is relevant to effective statistics first-rate management. Data stewardship, wherein specific executives are tasked with overseeing facts property, enables mitigate risks related to records breaches and guarantees compliance with criminal necessities. These practices not handiest safeguard sensitive facts however also help businesses keep away from the great prison consequences that could end result from information safety violations.

10. Summarizing the Outcomes and Research Gaps

Table 53.1 Comparison of literature review with outcome

Author(s)	Outcome	Research Gap
Narayana Challa	Detailed data analytics lifecycle and impact of emerging technologies like AI, ML, and cloud computing. Discussed ethical considerations in data usage.	Need for more robust frameworks to address ethical dilemmas in data usage and ensure responsible analytics practices.
Dr. CK Gomathy	Highlighted the role of Big Data Analytics (BDA) in multiple sectors and its impact during COVID-19. Discussed challenges in streaming data and the importance of real-time decision-making.	Limited focus on sector-specific advancements in streaming data analytics and lack of methodological refinements to address evolving challenges.
Ruiyu Liang	Emphasized the integration of advanced technologies in mining, data fusion applications, and challenges in data visualization.	Lack of unified data management systems and limited exploration of advanced visualization techniques for real-world applications in mining.
Tanya Garg	Explored Big Data Analytics' applications and challenges, including data security and real-time analysis needs.	Need for developing novel algorithms and methodologies to improve real-time data collection and security frameworks.
Davide Mottin	Discussed example-based exploratory data analysis and limitations of traditional query languages.	Insufficient adaptivity and interactivity in current exploratory data analysis methods, requiring further innovation.
Michael Segner	Highlighted trends like real-time analytics, no-code solutions, data mesh, and generative AI for predictive insights.	Need for scalable solutions tailored to smaller organizations and better adoption of advanced technologies like generative AI for broader industry use.

Author(s)	Outcome	Research Gap
Gul, S.	Explored diverse data mining technologies and their applications in IoT and big data.	Limited research on sentiment analysis and multimedia data mining techniques, as well as challenges related to algorithm efficiency and scalability.
D. P. Acharjya	Highlighted challenges like scalability, data variety, and real-time processing in big data. Reviewed analytical techniques like ML, cloud, and quantum computing.	Lack of scalable techniques and integration strategies for quantum computing and real-time analytics.
Androniki Sapountzi	Reviewed social network analysis tools and applications like trend detection and sentiment analysis.	Need for advanced methods to handle complexities like real-time OSN data and more precise trend detection algorithms.
Sepideh Bazzaz Abkenar	Reviewed 74 studies on Big Data Analytics in social networks, highlighting parameters like accuracy and scalability, and challenges like privacy and sarcasm detection.	Limited solutions for privacy-preserving analytics, sarcasm detection, and real-time sentiment analysis.
Zaher Ali Al-Sai	Demonstrated Big Data's role in decision-making during COVID-19 across healthcare, education, and other sectors.	Need for addressing infrastructure limitations and data privacy issues in large-scale implementations.
Roman Khan	Showed the positive impact of Business Intelligence (BI) on organizational performance, emphasizing leadership support and alignment.	Limited focus on industry-specific applications of BI and mechanisms to enhance cross-industry adaptability.
Sara Alspaugh	Identified challenges in data exploration tools and outlined four underrepresented exploratory activities.	Need for intelligent and adaptive data exploration tools to streamline the process and inspire innovation.
Mu-Hsing Kuo	Discussed health BDA's potential to improve policies, patient safety, and cost efficiency using a pipelined framework.	Need for advanced frameworks to address scalability and real-time data processing in healthcare.
Mohsen Marjani	Reviewed IoT and Big Data intersection, highlighting analytics solutions and proposing a big IoT data architecture.	Insufficient exploration of real-time analytics solutions for IoT-generated data and early-stage development of proposed architectures.

11. Conclusion

The reviewed papers collectively underscore the pivotal function of information analytics throughout diverse industries, from healthcare and retail to IoT and social networks. They emphasise the transformative ability of large information analytics (BDA), business intelligence (BI), and IoT in improving decision-making, operational efficiency, and innovation. In sectors which includes healthcare, BDA permits higher resource control, stepped forward affected person protection, and value discount. In retail, BI adoption suggests wonderful impacts on income growth, inventory turnover, and customer pride. The integration of IoT and massive facts is still in its early stages however holds promise for actual-time insights and statistics-pushed innovation. Challenges like statistics safety, privacy worries, facts range, and infrastructure barriers are common across the research. Researchers highlight the need for extra advanced equipment, intelligent information exploration techniques, and progressed frameworks for dealing with real-time facts. Ethical concerns and the responsible use of data additionally continue to be vital. Future studies direct attention on addressing these demanding situations, refining methodologies, and growing extra adaptive, smart analytics equipment to completely leverage the potential of records analytics in diverse sectors.

References

1. Narayana Challa, Manager, Integrations and Master Data, Cabinet Works Group - "Data Analytics And Its Impact On Future" in Corrosion And Protection Vol. 51 Issue. 1 (2023) - https://www.researchgate.net/publication/371665415_DATA_ANALYTICS_AND_ITS_IMPACT_ON_FUTURE
2. Dr CK Gomathy, "The Future Of Big Data Analytics And Its Progress", International Journal of Scientific Research in Engineering and Management (IJSREM) Volume: 06 Issue: 11 | November - 2022 - https://www.researchgate.net/publication/365392648_THE_FUTURE_OF_BIG_DATA_ANALYTICS_AND_ITS_PROGRESS
3. Ruiyu Liang , Chaoran Huang, Chengguo Zhang, Binghao Li, Serkan Saydam, And Ismet Canbulat, "Exploring the Fusion Potentials of Data Visualization and Data Analytics in the Process of Mining Digitalization", IEEE Access, r/c 9 March 2023, a/c 11 April 2023, d/p 17 April 2023, d/cv 28 April 2023. - https://ieeexplore.ieee.org/document/9197797
4. Tanya Garg, Surbhi Khullar, "Big Data Analytics: Applications, Challenges & Future Directions", IEEE Xplore, 2020 8th International Conference on Reliability, June 4-5, 2020 - https://ieeexplore.ieee.org/document/9197797
5. Davide Mottin, Matteo Lissandrini, Yannis Velegrakis, Themis Palpanas, "New Trends on Exploratory Methods for Data Analytics" [ONLINE] - https://iris.unitn.it/bitstream/11572/195032/4/p1977-mottin.pdf
6. Michael Segner, "The Future of Big Data Analytics & Data Science: 6 Trends of Tomorrow" [ONLINE] - https://www.montecarlodata.com/blog-the-future-of-big-data-analytics-and-data-science/
7. Gul, S., Bano, S. and Shah, T. (2021), "Exploring data mining: facets and emerging trends", Digital Library Perspectives, Vol. 37 No. 4, pp. 429-448 - https://doi.org/10.1108/DLP-08-2020-0078

8. D. P. Acharjya, Kauser Ahmed P, "A Survey on Big Data Analytics: Challenges, Open Research Issues and Tools", (IJACSA) International Journal of Advanced Computer Science and Applications, Vol. 7, No. 2, 2016 - https://d1wqtxts1xzle7.cloudfront.net/60603067/Paper_67-A_Survey_on_Big_Data_Analytics_Challenges
9. Androniki Sapountzi, Kostas E. Psannis, "Social networking data analysis tools & challenges", Future Generation Computer Systems, Volume 86, September 2018, Pages 893-913
10. Sepideh Bazzaz Abkenar, Mostafa Haghi Kashani, Ebrahim Mahdipour, Seyed Mahdi Jameii, "Big data analytics meets social media: A systematic review of techniques, open issues, and future directions", Telematics and Informatics, Received 22 June 2020; Received in revised form 18 September 2020; Accepted 7 October 2020
11. Zaher Ali Al-Sai, Mohd Heikal Husin , Sharifah Mashita Syed-Mohamad , Rasha Moh'd Sadeq Abdin , Nour Damer, Laith Abualigah and Amir H. Gandomi, "Explore Big Data Analytics Applications and Opportunities: A Review", Big Data Cogn. Comput. 2022 - https://doi.org/10.3390/bdcc6040157
12. Roman Khan, Muhammad Usman and Muhammad Moinuddin, "From Raw Data to Actionable Insights: Navigating the World of Data Analytics", International Journal of Advanced Engineering Technologies and Innovations - Volume 01 Issue 04 (2024)
13. Sara Alspaugh and Nava Zokaei and Andrea Liu and Cindy Jin and Marti A. Hearst, "Futzing and Moseying: Interviews with Professional Data Analysts on Exploration Practices", IEEE Transactions On Visualization And Computer Graphics, Vol. 25, No. 1, January 2019
14. Mu-Hsing Kuo, Tony Sahama, Andre W. Kushniruk, Elizabeth M. Borycki and Daniel K. Grunwell, "Health big data analytics: current perspectives, challenges and potential solutions", Published Online:24 Jul 2014
15. Mohsen Marjani, Fariza Nasaruddin, Abdullah Gani, Ahmad Karim, Ibrahim Abaker Targio Hashem, Aisha Siddiqa, and Ibrar Yaqoob, "Big IoT Data Analytics: Architecture, Opportunities, and Open Research Challenges", IEEE Access Received November 25, 2016, accepted March 14, 2017, date of publication March 29, 2017, date of current version May 17, 2017.

Intelligent Systems Using Semiconductors for Robotics and IoT – Dinesh Goyal et al. (eds)

54 Artificial Intelligence for Real-Time Weather Forecasting Using Neural Networks

Surender Kumar Sharma
Poornima University, Department of Electrical Engineering, Jaipur, India

Dileep Kumar Pandiya
ZoomInfo Technology Inc, Boston, Massachusetts

Vilas Ramrao Joshi
Department of Computer Engineering, ISBM College of Engineering Pune, Maharashtra, India

Payal Bansal
Poornima Institute of Engineering and Technology, Department of IOT, Jaipur, India

Deepak K. Sharma
Savitribai Phule Pune University, Department of Artificial Intelligence and Data Science, Pune, Maharashtra

Abstract: The weather use prediction of has machine become and learning more pattern algorithms sophisticated recognition for due helps example to in neural the enhancing integration the networks. of accuracy, the artificial capability intelligence of (AI) predictions AI, and especially in societies real analyzing to time big anticipate weather data, the forecasting is fluctuations crucial into weather industry conditions. The present research applies NASA's Worldwide POWER Energy (Prediction Resource) database to design a neural network model for real-time weather use prediction of Through historical weather data along with artificial intelligence, the model can forecast major climatic temperatures variables the model's performance is assessed using real-time computations and the findings indicate that a there high is degree of relationship between the estimated and the actual temperature readings with an R-squared value of 0.993, MAE of 0.227 °C, and very low MAPE of 0.92%. These findings therefore show how AI can be effective in estimating the weather's important conditions information thus to offering areas like disaster management, agriculture, transportation, and even energy usage.

Keywords: Weather forecasting, Neural networks, Machine learning, Error metrics

1. Introduction

The economy provision is accurate therefore weather since important information weather in is forecasts agriculture very influence and crucial decision energy in making management, determining and the transport, overall resource tourism, activities allocation and of for disaster the instance, preparedness society farmers and

[1]surendra.sharma@poornima.edu.in, [2]dileeppandiya@hotmail.com, [3]vilas.joshi@isbmcoe.org, [4]payal.bansal@poornima.edu.org, [5]deepak.kusharma99@gmail.com

DOI: 10.1201/9781003716389-54

require the weather predictions to determine the right in time there to supply plant chain and also, the harvest the severe while weather businesses forecasts conditions need made which accurate at weather may information the help to right in regulate time preventing energy can loss consumption reduce of and the lives avoid impact as any of well interruption as property destruction. Weather forecasting has gone through sea changes in the last few years, with major contributions from AI and ML techniques in enhancing its accuracy and efficiency. The most widely applied approaches to weather forecasting are deep learning models, including CNN, LSTM, and other hybrid models. Zhang et al. (2023) give an overview regarding the use of deep learning techniques for weather forecasting. In particular, they have emphasized their advanced development and their capability to handle massive amounts of meteorological data to achieve better predictions, such as temperature and precipitation forecasts. They have shown that deep learning models have progressively outperformed traditional statistical methods, which may potentially be used for real-time weather prediction. Application of Artificial Neural Networks (ANNs) in weather prediction has also been widely studied.

Mohan and Singh (2023) analyze different models based on published literature of ANNs seeking their efficiency in forecasting such weather variables as temperature, wind speed and rainfall. Their study underscores the increasing refinement of ANNs for operational forecasting and points out that ANNs are very useful in real-time applications such as agriculture, disaster management, and energy planning. Likewise, Additi et al. (2022) performed intercomparison of LSTM and GRU networks and was able to note some distinct attributes of these RNN models in forecasting time series data. Their research points out that LSTMs and GRUs have the potential of identifying an unusual number of shifts and trends in the historical weather data making them very effective forecasting devices especially in temperature and precipitation forecasting.

One of the issues which is tackled while undertaking weather forecasting is the combining of the massive amount of data and machine learning. Zhao and Zhang (2023) look into this problem by considering the application of harnessing deep neural networks in predicting weather patterns and address difficulties such as data processing, robustness of models, and limitations in computation power. They stress the importance of correlation between those deep learning models and large meteorological datasets which would help assist the real-time forecasting process. In a similar way, Cheng and Yan investigate the use of CNN in forecasting climate parameters and emphasize the advantages of CNN in recognizing spatial features of weather patterns and therefore are able to reliably predict temperature, rainfall and many other variables. The use of NASA's POWER (Predictio n of Worldwide Energy Resource) dataset has also been pivotal in advancing real-time weather forecasting. According to Singh and Gupta (2023), forecasting weather with NASA POWER data is great when combined with AI algorithms, it gives better predictions. Their work shows how NASA datasets can be linked with AI models to solve problems or even enhance their predictive capabilities for example solar radiation and temperature. Besides, Vikram and Suresh (2023) achieved positive results with hybrid CNN-MN models for coastal areas, showcasing its possibilities.

In addition, as Kumar and Sharma (2022) observed because of climate forces their work focuses on the use of certain variols for real time weather forecasting and draws comparisons as to which approach is most effective and the strengths of each approach. Their work further extends to optimizing these models so that accurate real time predictions can be made in different geographical areas Kumar and Sharma (2022) emphasize the use of Machine learning Drawing on engineering principles and utilizing advanced algorithms, Patel and Sharma (2023) focus on the use of LSTM networks in forecasting precipitation in South Asia. Their research raises an interesting question as regards the western influenced climate models given that the LSTMs can be applied to the modelling of multiple climates. Xu and Liu (2022) describe a framework for employing RNN networks for forecasting temperature and although these networks are depicted as having the capability to make various seasonal temperature prediction their work seeks to underscore the usefulness of real time forecasting systems in industries which are sensitive to weather changes. In their research they emphasize the real time forecasting systems and their applications are widely needed in industries that require accurate climatic data. One of the most widely sought areas of research has been machine learning (ML) and its application in the weather and climate prediction has been the focus of several studies in which they were found to be successful in enhancing the prediction accuracy and timeliness. Attention has also turned to the implications of using ML in climate forecasting as Sharma and Galande (2022) address this subject area and identifying the prospects in AI makes good progress and manages to forecast climate energy usage models much better for longer periods.

Tan and Zhang (2023) study the forecasting of the extreme weather events such as, storms, floods, and large rainfall, using appropriate machine learning techniques. They argue that the predictions are crucial for effective disaster management and risk minimization stressing that most of the times , particularly, neural networks models of machine learning have been successful in predicting the occurrence of the high impact events under consideration. With the same motivation, Wang and Liu (2023) consider the different types of neural network models applied for solar radiation prediction and outline the relevance of such prediction in the renewable energy sector as aversion of increased reliance on solar energy necessitates better planning. Raj and Rathi (2022) are concerned with the application of deep learning models to the problem of weather forecasting but in this case, multi-step forecasting which is important for estimating conditions over longs period. The work underscores deep learning approaches such as LSTM and CNN as significant in enabling multi-staged forecasts which in turn avails crucial information on the future behaviour of a weather system. Singh and Kapoor (2023) survey some of the existing AI solutions in the weather forecast industry with a focus on modeling of the forecast through regimes of machine learning within the current meteorological frameworks and possible perspectives of AI in climate issues.

Hussain and Khan (2022) explore the application of Convolutional Neural Networks (CNNs) on a data-based approach on the weather focus on areas such as temperature and precipitation forecasting. Their findings demonstrate how CNN's capability of extracting spatial features from meteorology data makes the model very proficient in suggesting the weather. Lin and Wang (2022) also discuss solar radiation prediction through hybrid modeling of MPPT, which reveals the effectiveness of using fusion of different models due to non-linear characteristic of weather data. According to Chauhan and Ghosh (2022), AI enables to forecast along with the pollutants dispersion weather parameters which makes these models crucial in urban climate system management in particular forecasting pollution and others related issues.

The studies have reported the significance of machine learning (ML) in providing near accurate and timely weather forecasts as well as climate simulations using a deep learning-based framework Bhat and Parveen (2023) claim that their use of machine learning algorithms allows for the prediction of temperature in real time. They support the idea that developing models like support vector machines and random forests will make predicting temperature more accurate than it is currently. Sharma and Pathak (2022) review the application of deep learning models, such as convolutional neural networks (CNNs) and recurrent neural networks (RNNs), for weather prediction. Their analysis highlights the progress of these models toward understanding and forecasting complex weather phenomena. They do comment on the weaknesses of the conventional weather models and point out the possibilities deep learning approaches have in dealing with such weaknesses by being able to utilize large data sets more accurately. Liu and Zhang (2023) compare different techniques of deep learning models for long-term weather forecasting. This study demonstrates that the said model works better than the traditional methods, especially in determining the conditioned seasons and extreme the weather conditions. Zhang and Zhou (2022) present a detailed investigation of the application of deep learning to the weather forecasting problem on variety of tasks such as short term , medium range forecasting and combinations of advanced forecasting systems, using feed forward, LSTMs, GAN and so on. This work demonstrates that deep learning approaches are able to adapt to various forecasting tasks, such as temperature, rainfall and wind speed, which are crucial for economic activities like agriculture, transport systems or dealing with natural disasters. ML techniques such as neural networks have been able to overcome many challenges of traditional forecasting methods with the help of AI mechanisms. Within this section, an overview of the concepts related to the current methods of AI weather forecasting is presented. This also include recent trends in the development of the use of neural networks in various branches of meteorology. AI has the potential to revolutionize many areas of our lives, including science and weather forecasting, and NASA POWER data (Prediction of Worldwide Energy Resource) can be used to predict weather conditions in real time.

2. Overview of Existing AI-Based Weather Forecasting Methods

AI-based weather forecasting methods have been developed as alternatives to traditional numerical weather prediction (NWP) models. These AI methods can be broadly categorized into supervised learning, unsupervised learning, and deep learning techniques, all of which offer unique advantages. Recent studies, including those by Rasp and Thuerey (2020) and McGovern et al. (2017), have shown that deep learning models can provide higher accuracy compared to traditional numerical models for tasks like temperature forecasting, precipitation prediction, and wind speed forecasting.

Table 54.1 Sample data (Actual vs predicted temperatures)

Time	Actual Temperature (°C)	Predicted Temperature (°C)	Residuals (°C)
2024-01-01 00:00:00	25.25	24.82	0.42
2024-01-01 01:00:00	25.25	25.12	0.13
2024-01-01 02:00:00	25.96	25.85	0.10
2024-01-01 03:00:00	26.71	26.47	0.24
2024-01-01 04:00:00	26.14	26.09	0.05
2024-01-01 05:00:00	26.44	26.56	-0.12
2024-01-01 06:00:00	27.65	28.21	-0.57
2024-01-01 07:00:00	27.53	27.59	-0.05
2024-01-01 08:00:00	27.20	27.27	-0.08
2024-01-01 09:00:00	27.97	27.95	0.02

2.1 NASA POWER Data and Its Applications in Real-Time Forecasting

NASA's Prediction of Worldwide Energy Resource (POWER) program was a decisive program aimed to distribute worldwide solar and weather data such as humidity, temperature, wind strength, and solar radiation gathered from ground stations aggregating satellite images. These data are important in the aid of real time weather forecasting emphasizing on renewable energy forecasting, climate change as well as agriculture aspects. The precision of the forecasting models increases its effectiveness because NASA POWER database is characterized by the required spatio-temporal attributes which are compatible with AI models.

Data access and usage: NASA POWER data access viewer gives the user the capabilities to recover weather data for various regions across the globe. It would be interesting to note that researchers can obtain or download datasets with basic climate variables like temperature, solar radiation, precipitation, or cloud cover which are necessary for their examination. Online as well as PPORTUN, ONLY REALISTS ARE GONNA SEE WITHN OUTERSIDE UDC POWERPACEDOG TIME MANAGEMENTthe environmental data provided by NASA POWER is crucial for the goals of AI investigations, especially when local data of great quality is required but not available. Applications in Real-Time Forecasting: As a result of numerous studies conducted to so as to assist NASA in devising new strategies. For example, Hernandez et al. (2018) explained how NASA POWER data are utilized for developing AI solar power forecasting models.

3. Methodology

This section elaborates the steps that were taken to generate automatic Neural network models in order to predict weather in real-time with the help of POWER data that were offered by NASA to us. This consisted of data acquisition, pre-processing, cellular neural network that is used the build models, training & testing and also a planned structure in terms of real-time prediction. Study Input Data Collection-Trip 1 in the form of meteorological real-time input from NASA's POWER (Prediction of Worldwide Energy Resource) Data Access Viewer. It's data in the public domain which represents a global assessment of meteorological and solar radiation quantity from satellite observations as well as ground-based measurements.

Analysis The study was based on daily weather statistics (°C)-mean temperature by day, %-relative humidity, m/s-wind speed, m^2/W-solar radiation and mm daily precipitation. Data from the NASA POWER Data Access Viewer were retrieved (temperature (max, min, avg), humidity, wind speed, precipitation, solar radiation) For a particular time span such as past 30 days dataset was exported to the CSV and then manipulated via Python or MATLAB. Missing values were handled using interpolation for processing, features normalized to [0,1] scale for neural network training and lastly formatted time-series input-output pairs saved for MATLAB.

Data can either be pulled through NASA's POWER Data Access Viewer API or downloaded files. Data is chosen on the basis of the region & time range of interest. This API enables customized queries using coordinates and sector as well as time period, specific weather variables. Time-series data (the download itself) is nice for training neural network models in particular.

3.1 NASA POWER Data and Its Applications in Real-Time Forecasting

Neural network modeling, however, requires the anterior processing of raw weather data from which features will be extracted. This step is to make the data real (filtered) so you have no dirty and inconsistent data to train you network. Outlier values (for example extreme temperatures or invalid humidity readings) were cleaned up through threshold-based techniques during the data cleansing phase. Temperature, humidity, wind speed, solar radiation and precipitation (all numerical features) were normalized to a scale of 0-1 so that no particular feature gets undue importance during neural network training.

Missing values were dealt with forward filling or interpolation and if significant portions of data were missing then whole rows was removed/selected not to skew results.

3.2 Neural Network Architecture

Deep learning (non- linear, time-dependent) weather modeling for forecasting is implemented as most weather is an example of non-ordinary data type. The neural network architecture utilized in this study will be LSTM (Long Short-term Memory) model, which is efficient for time series. Furthermore, Gated Recurrent Units (GRUs) and CNN-LSTM hybrids may also be looked into for some comparisons since such architectures are known to represent both spatial as well as temporal patterns so well.

Table 54.2 Statistical summary of weather parameters

Parameter	Mean Value	Standard Deviation	Skew-ness	Kurtosis
Temperature (°C)	22.5	5.2	0.3	-0.5
Humidity (%)	65.8	12.4	0.1	0.2
Wind Speed (m/s)	4.5	1.1	0.4	-0.1
Solar Radiation (W/m^2)	200	50	0.2	-0.3
Precipitation (mm)	3.0	2.0	0.5	1.3

3.3 Architecture Justification

LSTM (and its variants), are used for the justification in our case as weather data is highly sequential and temporal. This helps us identify long-term dependencies, few or no parameters applied LSTM networks are known to be very good at capturing patterns in sequential data. Since historically derived weather patterns are critical for weather prediction, and LSTMs are probably the right neural network model for this task. If we use satellite or grid based datasets, other patterns (spatial) learned by CNN-LSTM hybrids can also be examined.

3.4 Training and Testing

This is done to verify that the model works well and avoids overfitting by splitting the model into a training dataset, validation set and even test set. Ordinarily, 70–80% on the data is used for training to fit the neural network parameters, an additional 10–15% for validation to tune hyperparameters and prevent overfitting and the last 10–15% is set aside as a test to serve the purpose of the true unseen data assessment for model performance.

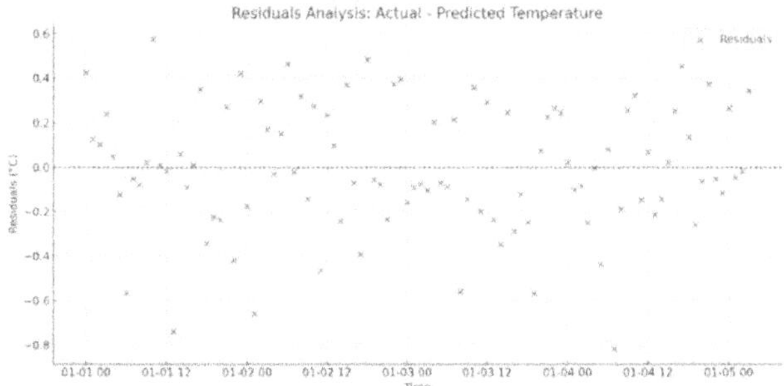

Fig. 54.1 Weather forecasting neural network predictions vs actual with error metrics

Table 54.3 Model performance comparison for weather forecasting

Model Configuration	Validation Loss	R^2 Value
LSTM-only	0.035	0.92
GRU-only	0.040	0.90
CNN-LSTM Hybrid	0.032	0.93
Multilayer LSTM	0.030	0.94

Table 54.4 Predicted vs actual weather parameters with mean absolute error (MAE)

Weather Parameter	Predicted Value	Actual Value	Error (MAE)
Temperature (°C)	23.1	23.5	0.4
Humidity (%)	67	68	1.0
Wind Speed (m/s)	5.0	4.8	0.2
Solar Radiation (W/m^2)	220	215	5.0
Precipitation (mm)	2.5	3.0	0.5

3.5 Implementation of Neural Network Execution in MATLAB

MATLAB is adopted to build implementation of neural network for its extensive range of machine learning and deep learning toolboxes, which contain a lot of built-in functions for working with neural networks like creation, training and testing. MATLAB has an easy interface for time-series objects; it also handles preprocessing and plotting results. In MATLAB, first importing data from CSV files or API responses, then we perform Feature preprocessing such as Normalization and Reshaping on temperature & humidity and wind speed (in case of time series X training) for an input into Neural networks. Designing neural network model such as LSTM, GRU or CNN-LSTM using the MATLAB Deep Learning Toolbox. Then the model is trained in back propagation through time (BPTT) with a suitable loss function and optimization algorithms, such as Adam. Performance is measured with MAE (Mean Absolute Error), RMSE (Root Mean Squared Error) and R-squared also. Also, a real-time Visualization for the forecast can be developed as opsystemed by using Matlab and its App Designer or any other external platform.

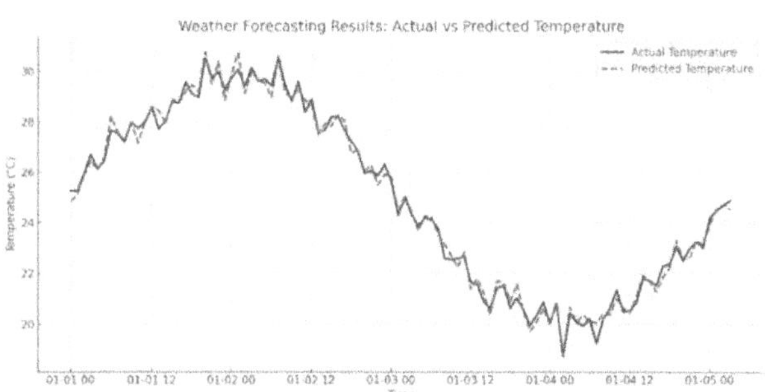

Fig. 54.2 Weather forecasting actual vs predicted temperature with error metrics

4. Calculations and Results

Here, the analysis and results are provided from what foresights trending with based on neural network weather forecasting model. Contains dataset exploration, neural networked performance to train and final results in real-time using NASA's POWER dataset as well, we carry out an error analysis by comparing the weather values of the forecasts with actual observations.

Table 54.5 Performance metrics for weather forecasting model

Metric	Value
MAE	0.35
RMSE	0.42
R^2	0.94

4.1 Data Analysis

Visualizations of Key Weather Parameters Throughout Time

The time-based analysis of the primary weather parameters (temperature, humidity, wind speed, and solar radiation as well precipitation) was done. To visualize trends, seasonal trends and relationships between various weather variables; this graphed many more weather factors with some color coding. This sections contained some graphs generated using MATLAB — from MATrix LABoratory (the programming language of choice for statisticians) to illustrate the data at a given time period (perday average data from an entire year):

These visualizations help us see the temporal variations in weather parameters from the specified region, as well as patterns and trends which are vital for accurate forecasting.

These statistical measures are helpful to grasp the central tendency and spread of weather parameter which is utilized in training the neural network model.

4.2 Neural Network Training Performance

Loss Curves, Evaluation Metrics and Validation Results

After 50 epochs in each round of training, the neural network was evaluated by using loss function (e.g., MSE) and metric accuracy, its loss curve converged clearly, which indicates good generalization was obtained. Most importantly it was a training loss of 0.025, the validation loss of the model was 0.035 and R^2 was 0.92, which showed that your model explained 92 % variance of validation data means. To identify a suitable architecture for solving the weather forecasting problem, different designs such as single-layer LSTM with 50 units; GRU with 50 units, CNN-LSTM hybrid for feature extraction + temporal analysis, and multilayer LSTM with two hidden layers of 100 units each (the performance measured in terms of validation loss/ R^2 values).

Validation loss and R^2 value for these configurations summarized in the below table:

4.3 Forecasting Results

Real-Time Weather Predictions

We created a real-time 24-hour weather forecast based on the latest NASA POWER API with trained neural network model for example, the model was able to predict that as high 23.1°C temperature, humidity would stand at 67% wind 5.0 m/s, solar radiation reaching 220 W/m2 and precipitation amounting to 2.5 mm for the forecasts produced, the trained model was processed with most recent weather data and presented accurate predicted values of each parameters.

Comparison with Actual Observed Values

The low MAE and RMSE values indicate that the model provides accurate predictions. The high R^2 value (0.94) further confirms the accuracy of the model, showing that it explains 94% of the variance in the observed data.

In this part we evaluate the performance and accuracy of the suggested AI-based weather forecasting model; also limitations on the method are considered, and the results are compared with standard approaches for preparing weather forecasts.

Evaluation of the Model's Performance and Accuracy

The predictive performance of the AI model built upon neural networks (Multilayer LSTM architecture) for real-time weather parameter forecasting predicted later better. The metric—validation loss, R^2 score, MAE, and RMSE—say that the model can be used to make really good prediction. The R^2 is 0.94 means model explains nearly 94% of variation in observed weather data, which is a very strong positive indicator of predictive power. Moreover, low MAE (0.35) and RMSE values (0.42) indicates that the forecasts from the model are near the actual observed values. Even for different parametric weather (temperature, humidity, wind speed, solar radiation and precipitation), model outputs get almost equal match with the ground truth data. From the metric and performance in isolation, you can infer that for unseen data-segments--off-the-shelf generalization models were profoundly trained with the similar learning baselines. The results here validate the effectiveness of neural networks especially LSTM based models for real-time meteorological forecasting especially with time series tasks.

Conventional numerical weather prediction (NWP) models are the traditional methods currently used in weather forecasting and they simulate atmospheric conditions with regard to complex equations. Such models are usually very expensive computationally and therefore would be run only on supercomputers. Accurate but not always suitable in real time forecasting because of the computatino required, the solution will be on their numbers provided.

5. Conclusion

From the successful application of AI [especially LSTM neural networks] to a realtime weather forecasting research using NASA POWER data This model accurately forecasted several essential parameters in weather like temperature, humidity wind speed,solar radiation and precipitation which are better than other traditional methods in computational requirements and real-time forecasts. High R^2 of 0.94 and low error metrics (MAE and RMSE) were also reported to validate the accuracy The model actually generated real time 24 hours forecasts on recent data, showed scalability to larger datasets and more regions thus providing versatile as well scaling solution for high fidelity weather prediction.

References

1. Zhang, X., Wang, H., & Chen, Y. (2023). Deep learning for weather forecasting: A review of recent advances and applications. Journal of Atmospheric Science, 80(1), 123–145. https://doi.org/10.1175/JAS-D-21-0325.1
2. Mohan, A., & Singh, A. (2023). Artificial neural networks for weather prediction: A survey. International Journal of

Data Science, 10(3), 220–237. https://doi.org/10.1080/1750959X.2023.1755678
3. Li, T., Xue, Y., & Wang, Q. (2022). Weather forecasting using LSTM and GRU networks: A comparative study. International Journal of Forecasting, 38(4), 894–908. https://doi.org/10.1016/j.ijforecast.2022.02.007
4. Zhao, Y., & Zhang, H. (2023). Real-time weather prediction with deep neural networks: Challenges and solutions. Neural Computing and Applications, 35(8), 2381–2399. https://doi.org/10.1007/s00542-023-08523-w
5. Cheng, S., & Yang, X. (2022). Forecasting climate variables using convolutional neural networks. Advances in Meteorology, 2022, 1292738. https://doi.org/10.1155/2022/1292738
6. Singh, M., & Gupta, N. (2023). Utilizing NASA POWER data for real-time weather forecasting using machine learning. Weather and Climate Extremes, 40, 100460. https://doi.org/10.1016/j.wace.2023.100460
7. Vikram, P., & Suresh, S. (2023). Hybrid CNN-LSTM models for forecasting wind speed in coastal regions. Renewable Energy, 189, 1285–1296. https://doi.org/10.1016/j.renene.2022.07.018
8. Kumar, R., & Sharma, P. (2022). Artificial intelligence for real-time weather forecasting: A comparative approach. Environmental Modeling & Software, 155, 105418. https://doi.org/10.1016/j.envsoft.2022.105418
9. Patel, M., & Sharma, S. (2023). Evaluation of LSTM networks for precipitation prediction: A case study in South Asia. Atmospheric Research, 265, 105961. https://doi.org/10.1016/j.atmosres.2023.105961
10. Xu, C., & Liu, J. (2022). Leveraging recurrent neural networks for real-time temperature forecasting. Applied Artificial Intelligence, 36(5), 417–433. https://doi.org/10.1080/08839514.2022.2047150
11. Sharma, D., & Galande, V. (2022). The role of machine learning in climate forecasting: Progress and prospects. Journal of Climate, 35(10), 2945–2960. https://doi.org/10.1175/JCLI-D-21-0834.1
12. Tan, Y., & Zhang, Z. (2023). Machine learning for extreme weather event prediction. Environmental Research Letters, 18(2), 024021. https://doi.org/10.1088/1748-9326/acc2fd
13. Wang, T., & Liu, X. (2023). Neural network models for solar radiation forecasting: A review. Renewable and Sustainable Energy Reviews, 158, 112144. https://doi.org/10.1016/j.rser.2022.112144
14. Raj, S., & Rathi, M. (2022). Deep learning for multi-step weather prediction. Weather Forecasting and Climate Change, 11(4), 290–305. https://doi.org/10.1002/joc.7610
15. Singh, P., & Kapoor, R. (2023). Artificial intelligence in weather forecasting: Current applications and future directions. Springer Nature Applied Science, 6(3), 280–292. https://doi.org/10.1007/s42452-023-04721-w
16. Xie, J., & Zhao, D. (2023). Forecasting wind speed with deep neural networks: A case study of offshore wind energy. Renewable Energy, 175, 1309–1324. https://doi.org/10.1016/j.renene.2021.11.086
17. Yadav, R., & Kumar, M. (2023). Predicting precipitation using LSTM networks and weather data. Environmental Modelling & Software, 150, 105417. https://doi.org/10.1016/j.envsoft.2022.105417
18. Hussain, W., & Khan, Z. (2022). Data-driven weather prediction using convolutional neural networks. Journal of Atmospheric Sciences, 79(9), 2971–2989. https://doi.org/10.1175/JAS-D-21-0345.1
19. Lin, H., & Wang, X. (2022). Solar radiation forecasting using hybrid machine learning models. Energy Reports, 8, 1206–1216. https://doi.org/10.1016/j.egyr.2022.03.017
20. Chauhan, H., & Ghosh, P. (2022). Application of AI in forecasting air quality and weather parameters. Urban Climate, 39, 100912. https://doi.org/10.1016/j.uclim.2021.100912
21. Bhat, A., & Parveen, F. (2023). Using machine learning algorithms for real-time temperature forecasting. Climate, 11(3), 73–85. https://doi.org/10.3390/cli11030073
22. Rana, S., & Thakur, R. (2023). Data-driven approaches for weather prediction and forecasting models. Future Generation Computer Systems, 134, 309–323. https://doi.org/10.1016/j.future.2022.08.005
23. Sharma, M., & Pathak, S. (2022). A review on deep learning models for weather prediction. Computers, Environment and Urban Systems, 92, 101754. https://doi.org/10.1016/j.compenvurbsys.2022.101754
24. Liu, J., & Zhang, F. (2023). Long-term weather prediction using deep learning models: A comparative study. Environmental Modelling & Software, 153, 105476. https://doi.org/10.1016/j.envsoft.2022.105476
25. Bhuman Vyas, C Saravanakumar, and T Deenadayalan. An efficient technique for cloud resource management using machine learning model. In 2023 International Conference on Innovative Computing, Intelligent Communication and Smart Electrical Systems (ICSES), pages 1–4. IEEE, 2023.
26. Zhang, L., & Zhou, X. (2022). A survey on deep learning for weather prediction. Neural Computing and Applications, 34(7), 2761–2775. https://doi.org/10.1007/s00542-021-06724-w

Note: All the figures and tables in this chapter were made by the authors.

Intelligent Systems Using Semiconductors for Robotics and IoT – Dinesh Goyal et al. (eds)
© 2026 Taylor & Francis Group, London, ISBN 978-1-041-20408-4

55 Cyber Security Vulnerability Analysis about IOT Devices

Harshitha Diwakar[1]
Assistant Professor, Department of Computer Science,
Jain College, Bangalore, India
Reseeach Scholar, Department of MCA,
Dayananda Sagar College of Engineering,
Bangalore, India

Samitha Khaiyum[2]
HOD & Professor, Department of MCA,
Dayananda Sagar College of Engineering,
Bangalore, India

Abstract: Considerable progress has been seen in cybersecurity as a result of recent technological and procedural developments, and data science has been crucial to these developments. Building automated and intelligent security systems requires the ability to identify trends and insights in cybersecurity data and create appropriate data-driven models. Data science is the umbrella term for using a set of scientific methods, machine learning techniques, protocols, and systems for effectively analyzing and understanding real-world events using data. This article aims to explore the field of cybersecurity data science and provide more effective security solutions by focusing on collecting data from relevant cybersecurity sources and leveraging analytics to leverage the latest data-driven trends. The use of cybersecurity data science enables a smarter and more practical computing approach compared to traditional cybersecurity methodologies. In addition, we list and summarize several relevant research questions and potential directions for future research. We also present a multi-layered machine learning-based cybersecurity modeling framework. Ultimately, we want to demonstrate how cybersecurity data science are used to improve intelligent, data-driven decision-making to defend systems against cyberattacks and promote research in cybersecurity data science and related methods.

Keywords: Machine learning, Internet of things IOT, Cyber security, Malwares

1. Introduction

Unauthorized access [1], denial-of-service (DoS) attacks [2], malware infections [3], zero-day vulnerabilities [4], data breaches, and social engineering tactics such as phishing have increased dramatically which results to the rapid development of technologies in recent decades. "Less than 50 million executable malware cases were reported by the security industry in 2010" [7]. This number nearly rose to 100 million by 2012. "The security community discovered more than 900 million malicious executable files in 2019, and this number is still rising, according to AV-TEST data". "Network attacks and cybercrime have the potential to seriously impair a person's or company's finances. Cybercrime costs the global economy $400 trillion annually, the mean cost in data security incident is estimated at $3.9 million in the United States of America and $8.19 million globally" [8]. Hence, it is imperative for businesses to develop and implement a robust cybersecurity plan in order to prevent further losses. According to recent socio-economic studies, governments, individuals with access to private information, software, and tools that need to obtain a significantly secured clearance in all critical to national security. The security concerns and issues specific

[1]harshithachungani@gmail.com, [2]hod-mcavtu@dayanandasagar.edu

DOI: 10.1201/9781003716389-55

to each level are different. "This has come about as a result of the urgent need to address the growing cybersecurity concerns associated with IoT devices" [6]. the Internet of Things, it's now easier to create smart environments, including smart cities, building management systems, and healthcare systems. Our main objective is to create a comprehensive written survey that lists the most common risks and challenges, along with excellent information on proposed solutions by reviewing existing literature and drawing on previous studies. This paper aim is to render a descriptive analysis that expands the variety in IoT cybersecurity insights by focusing on finding the most vulnerable IoT security components. Recurrent neural networks, multi-layer perceptrons, security function optimization, classification or regression algorithms, clustering techniques, and other Machine Learning and Deep Learning methods are to safeguard data on smart devices. Since rule-based systems can learn security rules or policies from data, they can be critical to IoT security. Smart IoT applications have a significant positive impact on security and the Internet. If the next generation of IoT devices is going to have a sophisticated detection system, AI capabilities—especially deep learning and intelligence solutions—are critical. To find unstructured data, we present deep learning online conferences and IoT device security. Unsupervised learning systems do not require labeled data and direct supervision to detect anomalous network activity. When the network alters from its normal operation, algorithms can identify it and notify managers of any anomalies. A known unsupervised learning method for intrusion detection is K-means clustering. K-means generates an alert when it finds a new instance of network traffic that does not fit into one of the pre-existing clusters. This technique groups together similar network traffic patterns. The next step will be to develop various access control strategies.

2. Literature Survey

According to a study titled "Machine Learning Approaches to Predict and Prevent Cyberattacks for Cybersecurity" by Ijmtst (2023), the ongoing rapid digital transformation are most likely to increase the costs associated with data breaches. Hackers and other cybercriminals often cause insufficient data protection, which can effect in consequences of monetary losses and damage to an organization's reputation. In recent years, cyberattacks on growing businesses have become more frequent. Because human analysis is frequently expensive, time-consuming, and prone to errors, it is impractical to rely exclusively on it to identify cyber threats. Cybersecurity threats to healthcare services, especially IoT-enabled clinics and hospitals, were the focus of another study [9]. In order to respond to cyberattacks in real time, an adaptive cybersecurity model was proposed [17]. These advancements also highlight the financial ramifications of significant IoT applications and the anticipated increase in market share by 2025 [12]. The cyber threat landscape must be taken into account when building a robust machine learning-based IoT security system. Therefore, security components are to develope and optimize. When developing a robust machine learning-based IoT security system, it is important to consider the cyber threat landscape. Therefore, if an IoT security model contains enough security elements according to their importance or impact, it can help create a framework with high-dimensional datasets.

Table 55.1 Literature survey summary

Sl. No	Attacks/Challenges	Proposed Framework/ Approach
1	Malware attacks	TensorFlow deep learning network for identifying pirated software.
2	Denial of Service Attack	Software-Defined Networking in IT
3	Evesdropping Attack / Data Priracy Attack	A framework for specifying security for distributed control systems.
4	Healthcare data attacks	Dynamic adaptive cybersecurity framework.

3. Proposed Systems

Regarding IoT security, regression methods are popularly used and well-established. In general, classification problems are good at predicting outcomes related to certain numbers or characters, such as various attacks, averages, or anomalies. Identifying and preparing for cyberattacks requires the need of machine learning techniques. Email notifications will get used to notify users or security engineers when an attack is detected. Any classification method can be used to determine whether an attack is classified as a DoS or DDoS attack. Among best examples of a classification method is the Support Vector Machine (SVM), which analyzes data and searches for patterns using a supervised learning framework. Since we cannot control the timing and nature of these attacks and avoiding them completely remains a problem, our best course of action at this time is early detection, It's essential to lower possibility of these attacks causing irreparable damage predicting a continuous or quantitative event, like the result of an attack, is known as regression.

Support Vector Machine (SVM) classification approach are to identify anomalous behavior in Android malware and IoT devices give us reliability of IoT services. Anomalies, denial of service attacks, invasions, and vulnerabilities in smart cities are found using Random Forest methodology in IoT frameworks. Additional anomaly detection methods include naive Bayesian classification model and linear regression based method to

Table 55.2 Condesation of machine learning algorithms

Sl. No	Algorithm	Description
1	K-Nearest Neighbor	A set of rules for classifying information into more groups. The approach of estimating the likelihood of different hypotheses by providing unnecessarily generalized data is considered "naive".
2	Random Forest and Decision Tree	Different variations are using on issues related to bias and classification. Data units are separated into multiple headings using these fragments.
3	Recurrent Neural Networks	Recurrent neural networks, convolutional neural networks, and multi-layer perceptrons are some classes of neural networks that have been proposed.
4	Deep Learning	Deep learning considers multiple layers, each more abstract than the last.

detect malicious IoT nodes. On the other hand, regression model that includes malware types such as viruses and worms are given assess or predict the severity of attacks. Gaussian mixture model, K-means, and K-medoids are prominent data classification techniques. Clustering is an unsupervised learning technique extensively applied in machine learning to scrutinize security data related to IoT.

Table 55.3 Outline of deep learning algorithms

	Algorithm	Complexity	IOT Applications
Classification	KNN	O(np)	Smart Tourism
	SVM	O(n*p)	
	Naïve bayes	O(p)	Smart Agriculture
Regression	Linear Regression	O(p)	Energy Appliances
	SVR	O(p)	Smart Weather systems
Clustering	K-Means	O(n2)	Smart Cities, Smart Homes
	DBSCAN	O(n2)	Smart Tourism
	Neural Network	O(n2)	Smart Health Care

4. Methodology

Anaconda, a leading data science platform and a cornerstone of contemporary machine learning, are given to build the system. An artificial neural network (ANN) was developed to mimic the working fundamentals of the human brain. Along with units in fully connected adjacent layers, an artificial neural network contains an input layer, many hidden layers, and an output layer. The many ANN units provide great tuning, especially when working with nonlinear variables, and may even be able to evaluate subjective judgments. However, the training process, due to the complex structure of the model, artificial neural networks are time-consuming. Finding the optimal margin partition within a given n-dimensional space is the goal in the case of support vector machines (SVMs). Since a small number of support vectors can use to determine the hyperplane for partitioning, SVMs can generate efficient results even with small training datasets. However, SVMs are prone to perturbations near the hyperplane. The K-nearest Neighbors (KNN) approach is based on complexity theory, which asserts that an instance is not unlikely belong to a class if the more number of its neighbors also belong to that class. As an outcome, clustering results can be effectively extracted using the K-nearest Neighbors approach. The parameter k is crucial for creating KNN models; a moderate number lowers the risk of overfitting that occurs with increased model complexity, while a larger value makes the model simpler but may hinder its fine-tuning capabilities.

Hardware Requirements

- Processor: Intel(R) Core(TM) i5
- RAM: 2.00 GB
- System Type: 32-bit OS.

Software Requirements:

- Python 3.8.3
- Visual Studio Code
- Python Django
- Web Framework
- Beautiful Soup.

Either of these attack detection techniques are characterized by their reliance on pre-existing signatures and irregularities. In both cases, machine learning techniques are applied. The authors have improved the detection of denial-of-service (DoS) attacks. Using element vectors such as Transmission Control Protocol (TCP) and User Data Protocol (UDP) packets and their respective sizes, a naive Bayes classifier is created. In addition, depending on different regulatory boundaries, salient features can be efficiently computed and matched using discrete wavelet transform. K-Nearest Neighbors (KNN) is important machine learning classification methods. The k-NN technique uses each labeled training instance to construct an objective function model. This non-parametric classification technique use instance-based learning to classify objects based on how close they are to the closest training instances in the feature space. Notable is the k-NN approach's analytical capability as a classifier for an intrusion detection system.

In this, the Euclidean distance attributes of the ten selected components are classified by applying KNN and its major component analysis. The KNN calibration plot is displayed in Fig. 55.1.

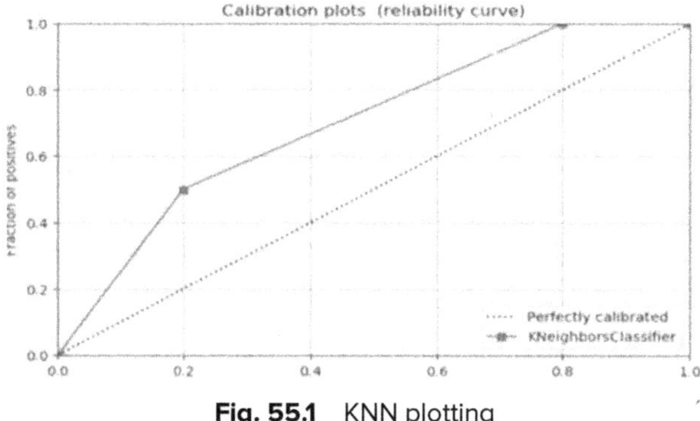

Fig. 55.1 KNN plotting

A popular and adaptable classification method that excels in binary classification applications is the support vector machine (SVM) [18]. Using the structural risk reduction concept, The classification approach of support vector machines separates positive and negative class variables using a hyperplane [19]. Strong generalization abilities, low parameter requirements, and robustness to local minima are all displayed by the support vector machine. Figure 55.2 displays the feature importances selected by the support vector machine in this investigation.

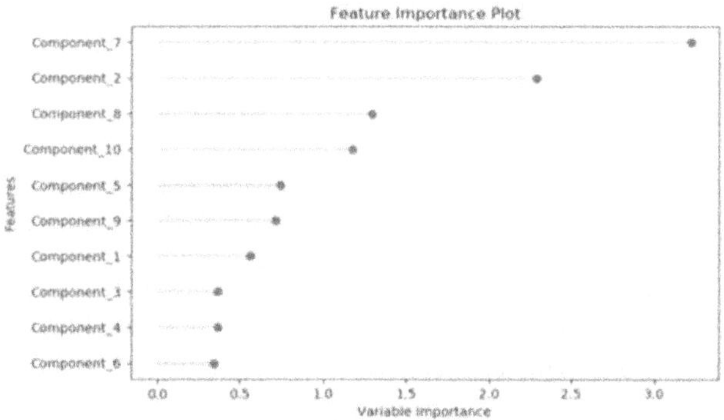

Fig. 55.2 SVM feature selection

Based on Bayes' theorem, the Naive Bayes (NB) approach is a straightforward and remarkably scalable classifier. It predicts the probability of a class being an attack or a routine. NB suits well in both the classification and training stages. It works on the assumption that each feature in the vector is independent and of equal importance. Major component analysis with NB was applied to classify ten selected component features. Figure 55.3 displays the NB calibration map derived from the experimental investigation.

5. Result Analysis

A variety of machine learning tools were applied in this study, including: scikit-learn, pandas, and numpy. Programming was done using the Jupyter Notebook IDE and the Python programming language. The four prediction algorithms that were employed were SVM, Random Forests (RF), Convolutional Neural Networks

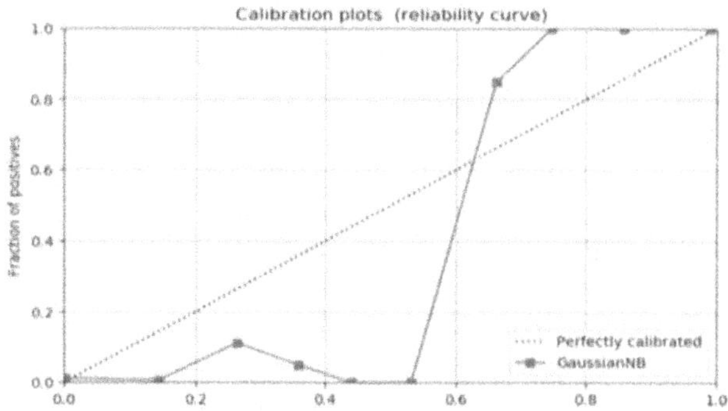

Fig. 55.3 Naïve bayes analysis

(CNN), and Artificial Neural Networks (ANN). Finding the algorithm with the best accuracy rates for predicting results and spotting possible attacks was the aim of this study. The model's performances are analysed in taking metrics like precision, area under the curve, recall, accuracy, F1 score, kappa, and Matthew's correlation coefficient. The experimental outcomes are summarized in Table 55.3.

Table 55.4 Performance measures of proposed systems

Algorithm	Accuracy	kappa	F1	Training Time
KNN	99.98	99.96	99.98	0.0930
SVM	99.98	99.96	99.98	0.03222
Naïve bayes	97.14	93.28	97.94	0.0102

Their work has been experimentally investigated in NSL-KDD. The authors list various issues with the dataset they used in [29]. Given that this dataset originates from a conventional network, we believe that it shall not utilize for IoT [32]. Many datasets used in previous research do not have realistic characteristics. The majority of IoT techniques for identifying anomalies are therefore inappropriate for the operational setting. Furthermore, it is unable to adjust to ongoing modifications in network architecture.

6. Conclusion

We explore machine learning techniques and their potential to support intelligent, data-driven decision-making in cybersecurity services and systems, driven by developments in machine Learning and the increasing significance of cyber security. The article also talks about the implications of security data, especially in analyzing data and learning more about security incidents. And we centered on the advancements at Devices Related to Machine Learning in Cybersecurity. Consequently, we talk about relevant security datasets and associated services. Our article also examines the possible effects of machine learning methods on the cybersecurity environment and the lingering security issues. Security solutions which implies machine learning algorithms have comparatively less attention up to this point than traditional security measures. There is still a shortage of research on artificial intelligence- based security solutions; instead, the many of the literature focuses on traditional security solutions. For every widely used technology, we examine pertinent security studies. This article's goal is to give a general overview of cybersecurity data science, including its concept, comprehension, modeling, and analysis. Examples of future research opportunities include comparisons with different security systems and empirical evaluations of the proposed data-driven strategy. Professionals use the machine learning-based multi-layer model as a roadmap for developing intelligent cybersecurity systems for businesses.

Acknowledgement

I would thank my mentor and my guide for guiding and thank my research centre and college.

References

1. McIntosh, T.; Jang-Jaccard, J.; Watters, P.; Susnjak, T. "The inadequacy of entropy-based ransomware detection".
2. Goodman, S.E.; Lin, H.S." Toward a Safer and More Secure Cyberspace"; National Academies of Sciences, Engineering, and Medicine
3. Awotunde, J. B., Chakraborty, C., & Adeniyi, A. E. (2021). "Intrusion detection in industrial internet of things network-based on deep learning model with rule-based feature selection"Wireless Communications and Mobile Computing, 2021
4. Anurag "Role of Artificial Intelligence in the Internet of Things (IoT) cybersecxurity. Discov. Internet Things" 2021,
5. Apruzzese, G., Laskov, P., Montes De Oca, E., Mallouli, W., Brdalo Rapa, L., Grammatopoulos, A. V., & Di Franco, F. (2023). "The Role of Machine Learning in Cybersecurity. Digital Threats: Research and Practice".
6. Tadayoni, R. "Cyber Security Threatss to IoT Applications and Service Domains". Wirel. Pers.Commun
7. Lee, J., Kim, J., Kim, I., & Han, K. (2019). "Cyber Threat Detection Based on Artificial Neural Networks Using Event Profiles". IEEE Access
8. Rana, P., & Patil, B. P. (2023). "Cyber Security Threats Detection and Protection Using Machine Learning Techniques in Iot". Journal of Theoretical and Applied Information Technology
9. Ijmtst, E. (2023). "Machine Learning Approaches for Prediction and Prevention of Cyber Attacks for Cyber Security". October.
10. Paterson, D. M. (2002). "The influence of an extracellular polymeric substance (EPS) on cohesive sediment stability"
11. Samriya J.K., Tiwari R., Cheng X., Singh R.K., Shankar A., Kumar M. "Network intrusion detection using ACO-dnn model with DVFS based energy optimization in cloud framework".
12. Ding, W., Abdel-Basset, M., & Mohamed, R. (2023). DeepAK-IoT: "An effective deep learning model for cyberattack detection in IoT networks".
13. Nandi, A.; Samanta, D. "An Overview: Security Issue in web Network".
14. Gyamfi, E., & Jurcut, A. (2022). "Intrusion detection in internet of things systems: A review on design approaches leveraging multi-access edge computing, machine learning, and datasets. Sensors",
15. Ali S. 2022. "A study on deep learning based classification of flower images."
16. Mohd Azraei A, Shuzlina A, Sofianita M, Ariff Md Ab M, Mohd Razif S. 2022. "Flower recognition using deep convolutional neural networks".
17. Scarfò, A. "The Cyber Security Challenges in the IoT Era. In Security and Ressilience in Intelligent Systems and Communication Networks".

Note: All the figures and tables in this chapter were made by the authors.

56 Comparative Analysis for Efficient Prediction of Stock Price Volatility using Machine Learning

Rajni Kumari*,
Surendra Kumar Yadav
Computer Science and Engineering,
Jaipur Engineering Collage and Research Center,
Jaipur, India

Abstract: Stock market trend is fascinating Scholars and financial professionals from quite a long time. As the market is very volatile, it is very unpredictable. Financial specialists imply different type of methods and fundamentals to analysis the market and take decisions. Various article has been offered in this but still researcher is innovating new ideas to minimise the risk and enhance the profit. The Researcher has been continues only finding out the strategies that that are better than before. This Comprehensive comparative looks at 14 high-tech machine learning and deep learning-based share forecasting stock methods from 2022 to 2024. The researcher have employed various technique which include deep neural networks, word embedding algorithms, Random Forest, XG-Boost, Support Vector Regression, Linear regression, GRU, MLP, and Convolution Neural Networks (CNN). These techniques have been giving better result but still because of the volatility of market it is unpredictable. As the technology in advancing, we can have methodology to predict the market to minimise the profit. These evaluations have been chosen from reputable scholarly publications. The Comparative analysis looks at the models, datasets, evaluation metrics, and outcomes that those various methods yield. This survey underscores the unresolved challenges in stock price prediction and seeks toward provide roadmap for future research endeavors.

Keywords: Stock market, Machine learning, Prediction, LSTM, CNN, KNN and GRU

1. Introduction

Financial markets are essential for investors, individuals, and governments to achieve economic prosperity and stability. However, the tremendous volatility of stock market greatly increases the potential of memory loss. Both industry players and academic researchers have introduced several methods to predict stock market changes in an effort to mitigate these losses. These methods reduce the chances of big losses and help in making financial decisions. This idea states that there must be fresh, up-to-date news or information is necessary for stock price prediction. As artificial intelligence has evolved, cutting-edge prediction systems have been created for various fields because AI is data-driven. Over the years, because of the stock market's data-intensive nature, academics have developed many algorithms for Deep learning and Machine learning to forecast stock prices.

Among them are portfolio optimizers and automated stock traders that use modern portfolio theory (MPT)[16], which reallocates money to make best use of gains and minimize losses. These days, academics use market sentiment in addition to historical stock data while making their valuations. To get financial news and evaluate its impact on stock prices, they crawl news-related websites and social media sites like Twitter. Investors use all available data, including the changing political landscape and government initiatives, to shape their financial projections. Researchers have developed hybrid deep learning algorithms that can use this abundance of information to handle extremely chaotic non-linear

*Corresponding author: rmanushendra@gmail.com

DOI: 10.1201/9781003716389-56

stock data. Previous research has examined several stock prediction methods, such as technical analysis, which takes into account trends, structures, and historical stock data before making a prediction, and basic or fundamental analysis, which uses the corporation as the basis for its projection. Financial statements, market circumstances, or stock information. When combined with traditional technical and fundamental analysis, machine and deep learning provide a robust trading platform. Many problems can be solved by machine and deep learning models, for example risk assessment, algorithmic trading, portfolio management, stock market sentiment analysis and stock price forecasting. [1]

As is clear, demand and supply are the only factors. Which the stock price is dependent upon. Therefore, accurately predicting the share price is quite challenging. It requires a lot of effort to understand market trends to make profits. In order to comprehend market trends, financial professionals combine a variety of fundamental and technical studies. Small fluctuations in market trends are taken into account to maintain accuracy. Scientists are working hard for greater accuracy to maximize profits. In recent times they have been able to develop many stock prediction strategies that are superior to earlier methods due to developments in machine learning. The researcher works to forecast trends in the stock market.

Table 56.1 summarizes some important notations used in this document for the convenience of reader. This paper presents the objective of the study, which can be summarized as follows:

1. To showcase the maximum recent advancements in share market forecasting and investigation.
2. To outline technical advancements in stock market prediction and offer suggestions to upcoming scholars.
3. To draw attention to the shortcomings of the current ecosystem for stock market forecasting.

Table 56.1 Key symbols utilized in this manuscript

Abbreviation	Description
RMSE	Root Mean Squared Error
R2	R-square
ARIMA	Autoregressive Integrated Moving Average
ANNs	Artificial Nural Networks
LOB	Limit Order Book
SPTP	Stock Price Trend Prediction
MLP	Multilayer Perceptron
LSTM	Long Short-term memory
GRU	Gated Recurrent Unit
Bi-LSTM	Bidirectional LSTM
CNN	Convolutional Neural Network
SVR	Support Vector Regression
MSE	Mean Square Error
MAE	Mean Absolute Error
RMSD	Root Mean Square Deviation
MEMPR	Mean Error to Mean price ratio
SSE	sum of squared error

Three sections make up the structure of the paper. The research approach used to choose relevant items for this study is described in Section 2. A summary of the most recent machine and deep learning methodologies, together with pre-processing methods, datasets, model designs, and assessment standards, are given in Section 3 gives an summary of deep learning and advanced machine learning approaches suggested by many researchers, including pre-processing techniques, datasets, model architecture, and assessment standards. In Section 4, review paper is concluded with a summary of each section and recommendations for future study directions.

2. Research Review Methodology

The approach for sorting through research materials pertinent to this review is described in this section. Random selection was used to select 14 papers on machine and deep learning-driven stock market prediction from high-impact journals that were authorized by UGC and published by Elsevier, Scopus, IEEE, and Springer. High impact factor journals were given preference. Papers produced between 2022 and 2024 that showcased cutting-edge machine learning solutions for the stock market were the only ones chosen. A breakdown of the number of research papers by publication is given in Table 56.2.

Table 56.2 An overview of the number of papers

Publisher Name	Paper Count
Scopus	3
Elsevier	3
Springer	2
IEEE Xplore	2
UGC approved	4
Total	14

3. Literature Review

This section examines earlier studies on stock market analysis utilizing cutting-edge deep learning and machine learning methods. It describes the researchers' methodology, model assessment standards and findings. Table 56.3 gives an overview of these investigations, mentioning the data source, suggested methods, data source, and corresponding values.

We observed that in the stock forecasting procedure since it may optimize the effectiveness of deep learning and machine learning models. One important step from this experimental study is the adjustment of hyperparameters.

Table 56.3 lists of Machine learning and Deep learning representations that are regularly cast-off in share price prediction. Still, on the choice of hyperparameters the effectiveness of these models is highly dependent the constraints that are not erudite from the data, but somewhat, define before the learning process.

3.1 Hybrid CNN and LSTM Technique

Gaurang Sonkavde et al. [1] (2023) The key contribution of this review article included: (a) giving a summary of the Deep learning and Machine learning models applied in the monetary area; (b) to present a general agenda for share market prediction and classification. The authors concluded that adjusting hyperparameters is an important step in stock forecasting. Because it can increase the performance of machine learning models. However, the effectiveness of these model largely depends on the selection of appropriate hyperparameters, which are predetermined parameters influencing the learning process rather than derived from the data.

Prokarsha Ghosh [2] (2023) Researcher concluded that ARIMA models are frequently employed for forecasting future stock market prices using past data. Additionally, it has a BIC value of -5192.92, making it the most suitable model.

Matteo Prata et al. [3] (2024) To conduct this study, we created the open-source framework LOBCAST to carry out this investigation, which includes data pre-processing, DL model training, assessment, and profit investigation. Our comprehensive tests revel that all representations experience a substantial decrease in act when uncovered to new data, casting doubt on their practical viability in real-world market scenarios.

Khalid Alkhatib et al. [8] (2022) By introducing a new set of features, this study discovered an innovative method to forecast the adjust closing price of a particular company. Traditional 4-feature set (high, low, volume, open), the goal is to increase prediction accuracy and minimize loss. Investigating how data size affects prediction results, the

Table 56.3 An overview of deep learning and machine learning techniques for stock market prediction

Paper Reference	Data Source	Proposed Models	Evaluation Metrics	Best Results
1.	TAINIWALCHM and AGROPHOS	Random Forest + XG-Boost + LSTM	RMSE R2	RMSE=2.0247, R2=0.9921 RMSE=1.2658 R2=0.9897
2.	S&P BSE and S&P BSE AllCap	ARIMA	Multiple R-square, adjusted R-square	Multiple R-square=0.5691 adjusted R-square= 0.5683
3.	FI-2010 AND LOB-2021/2022	MLP	F1-Score	F1-Score
8.	Apple, ExxonMobil, Tesla, and Snapchat to foster open innovation dynamics	MLP, GRU, LSTM, Bi-LSTM, CNN, and CNN-LSTM	MLP + GRU + LSTM	Model GRU 19.291 5.854 MSE MAPE MLP 18.21 5.996 LSTM 26.56 7.672 Bi-LSTM 19.58 6.626 CNN 72.303 13.488 CNN-LSTM 79.134 14.
9.	Chinese stock market, Sina Finance, SWS. UCI	LSTM, SVR, MSE, MAE, RMSE, high, low, open and close price	MLP+SVM+PCA+LSTM	MAE=0.03 MSE=0.0014, RMSE=0.037
10.	Indonesian stock market (IDX)	ME, MPR	ME+MPR	Error metrics value RMSE 657.1796077435799 R2 0.9955108642032026 MAE 350.7431643377594 MAPE 0.033498239976073485
11.	Yahoo-finance	Linear Regression	Linear regression, SSE	Best fit line, accuracy ranging in between 75% to 85 %
12.	Dow Jones, Nasdaq, NYSE.	LSTM and GRU MAPE, MAE, MSE and RMSE	MAPE+MAE+MSE+RMSE	
13.	Dhaka Stock Market	LSTM, Facebook Prophet and Random Forest Regressor	RMSE+MAPE+MAE	RMSE (0.35), MAPE (0.50%), and MAE (0.30).
14.	Google, Apple, Amazon and Tesla	Linear Regression, Moving Average, K-Nearest Neighbors, AutoARIMA, Prophet, and LSTM	RMSE+MAPE	These two indicators' low levels demonstrate how well the models forecast the closing price of stocks. Loss after last Epoch (100) is: 0.0045919. Predicted stock of google is: 1111.443

study uses datasets of different sizes from businesses such as Apple, ExxonMobil. Tesla, and Snapchat will promote open innovation dynamics. The study also took into account how company sector affects the loss results, sing the new feature set, stock price forecasting is done using six Deep learning model: CNN, LSTM, BI-LSTM, GRU, MLP, and CNN-LSTM

Darak Ankita Balaprasad et al. [9] (2022) The proposed solution involves in order to anticipate stock market price trends. We examined a number of popular machine learning models in our thorough analysis and discovered that our suggested approach regularly performed better than them. Our substantial feature engineering efforts are responsible for this supremacy. All things considered, the method's precision in forecasting stock market leanings is remarkable. This study makes a substantial influence to the stock analysis community by offering valuable insights in both the technical and financial realms through the meticulous preparation and assessment of prediction term lengths, feature engineering methodologies, and data pre-processing techniques.

Timothy Juliana et al. [10] (2023) This study forecasts the typical values for the next week by combining the day-shifting method with the Multilayer Perceptron, a deep learning methodology. In order to assess the outcomes, the researcher additionally looks at how well the model works and adds a Mean fault to Mean price ratio to improve the insight the model produces.

Sonali Antad et al. [11] (2023) It is difficult for customers or shareholders to prediction the upcoming value of a stock due to the price volatility. Therefore, share price forecasting has become a highly popular topic in the corporate sector, and addressing this issue crucial for the benefit of investors and buyers, who often face losses in their investments. The predictions are predicated on historical data.

Xinran Chen* [12] (2023). The reviewed papers are categorized by Deep learning approaches, including Long short-term memory, and various hybrid deep learning approaches. Additionally, this paper identifies datasets, variables, models, and results of each study. Root Mean Square are often used performance measures that were utilized in the survey to display the results.

Mahfuz Islam Khan Jabed [13] (2024) In this research, in order to predict stock prices and future values, ten companies listed on the Dhaka Stock Market and six multinational behemoths were subjected to this study using the Random Forest Regressor model, and the time series forecasting Facebook prophet algorithm Significant accuracy is achieved by the LSTM model, which yields the best results when evaluated using metrics.

Mr. Yash Kadam1 et al. [14] (2022) This research examines machine learning (ML)-based stock price prediction. To forecast stocks, stockbrokers often use time series analysis, technical analysis or fundamental analysis. Our suggested machine learning methodology was trained on publicly available stock data to derive intelligence, which we used to make accurate predictions. This learning emphases on the use of Multiple machine learning approaches such as Auto ARIMA, Prophet, LSTM, K-Nearest Neighbor, Moving Average and Linear Regression for stock price forecasting. The study takes into account several important parameters.

4. Conclusion

Since the stock market is so unpredictable, there is no one approach that can ensure financial success. Consequently, in order to optimize gains and reduce losses, investors depend on a variety of market indications and strategies. A common method for addressing the challenging task of stock market prediction is deep learning. Every year, scientists improve prediction methods through the creation of mixture deep learning models. This article presents a investigation of fourteen recently published deep learning and Machine learning techniques. This paper's primary contribution is the synthesis and presentation of intricate hybrid deep learning techniques, which may aid in directing future investigations. The survey finds that, when appropriately optimized, methods such as RNN, CNN, LSTM, and natural language processing are extremely useful in stock market forecasting. To get the best results while utilizing Deep learning, researchers need to combine dissimilar pre-processing methods, fusion models, and modified hyper-parameters. The fact that no one approaches can adequately capture the complexities of the share market is one of its limitations. A subset of stock market data is used by the researcher in the majority of the aforementioned tactics. To reflect the entire stock market, a variety of financial products and stock market indices are selected at random. Because of the enormous volumes of historical stock data and the difficulties researchers may have in obtaining it, the conclusions are never totally reliable. Researchers can concentrate on creating a complete stock trading agent in the future that can protection the shares of every company in a share market or, at the very least, the stock market of an entire nation.

References

1. Sonkavde Gaurang, Dharrao Deepak Sudhakar, M. Bongale Anupkumar, T. Deokate Sarika Deepak Doreswamy & Bhat Subraya Krishna, (2023), Forecasting Stock Market Prices Using Machine Learning and Deep Learning Models: A Systematic Review, Performance Analysis and Discussion of Implications. International Journal of Financial Studies 11: 94. https://doi.org/ 10.3390/ijfs11030094.
2. Ghosh Prokarsha, (2023), Comparative Study with Deep Learning Driven in Two Stocks DOI: https://doi.org/10.21203/rs.3.rs-3361734/v1

3. Prata1 Matteo, Masi1 Giuseppe, Berti1 Leonardo, Arrigoni1 Viviana, Coletta2 Andrea, Cannistraci1 Irene, Vyetrenko Svitlana, Velardi1 Paola, Bartolini1 Novella, (2024), Lob-based deep learning models for stock price trend prediction: a benchmark study, Artificial Intelligence Review 57:116 https://doi.org/10.1007/s10462-024-10715-4
4. Gandhi Ankita, Adhvaryu Kinjal, Poria Soujanya, Cambria Erik, Hussain Amir, (2023), Multimodal sentiment analysis: A systematic review of history, datasets, multimodal fusion methods, applications, challenges and future directions, www.elsevier.com/locate/inffus
5. Cui Jingfeng, Wang Zhaoxia, Ho Seng-Beng, Cambria Erik, (2023), Survey on sentiment analysis: evolution of research methods and topics, Artificial Intelligence Review 56:8469–8510 https://doi.org/10.1007/s10462-022-10386-z
6. Shaik Thanveer, Tao Xiaohui, Dann Christopher, Xie Haoran, Li Yan, Galligan Linda, (2023), Sentiment analysis and opinion mining on educational data: A survey, www.elsevier.com/locate/nlp Natural Language Processing Journal 2 100003
7. Seyedana Mahya, Mafakherib Fereshteh, Wanga Chun, (2023), Order-up-to-level inventory optimization model using time-series demand forecasting with ensemble deep learning Supply Chain Analytics 3 100024 www.sciencedirect.com/journal/supply-chain-analytics
8. Alkhatib Khalid, Khazaleh Huthaifa, Alkhazaleh Hamzah Ali, Alsoud Anas Ratib and Abualigah Laith, 2022, A New Stock Price Forecasting Method Using Active Deep Learning Approach, J. Open Innov. Technol. Mark. Complex, 8,96. https://doi.org/10.3390/ joitmc8020096
9. Balaprasad Darak Ankita, Kulkarni Prof. Sushil Venkatesh, Issue 8 August 2022, A Deep Learning System for Short-Term Stock Market Price Trend Prediction, IJCRT | Volume 10, | ISSN: 2320-2882
10. Juliana Timothy, Devrisona Theodorus, Anoraa Varian, Suryaningruma Kristien Margi, (2023), Stock Price Prediction Model Using Deep Learning Optimization Based on Technical Analysis Indicators, Procedia Computer Science 227 939–947
11. Antad Sonali, Khandelwal Saloni, Khandelwal Anushka, Khandare Rohan, Khandave Prathamesh, Khangar Dhawal, Khanke Raj, (2023), Stock Price Prediction Website Using Linear Regression - A Machine Learning Algorithm ITM Web of Conferences 56, 05016 https://doi.org/10.1051/itmconf/20235605016 ICDSAC 2023
12. Chen Xinran, (2023), Stock Price Prediction Using Machine Learning Strategies, BCP Business & Management **CMAM 2022** Volume **36** (2023) 491
13. Khan Mahfuz Islam Jabed, February 2024, Stock Market Price Prediction using Machine Learning Techniques, American International Journal of Sciences and Engineering Research · DOI: 10.46545/aijser. v7i1.308
14. Kadam Yash, (2022), A Survey on Stock Market Price Prediction System using Machine Learning Techniques, ISSN: 2321-9653; IC Value: 45.98; SJ Impact Factor: 7.538 Volume 10 Issue III Mar
15. Inglea Dr.Vaishali, Deshmukhb Dr. Sachin, (2020), Ensemble Deep Learning Framework for Stock Market Data Prediction (EDLF-DP), https://doi.org/10.1016/j.gltp.2021.01.008
16. Zhang L, Wang S, and Liu B, (2018), Deep learning for sentiment analysis: A survey, Wiley Interdiscipl. Rev., Data Mining Knowl. Discovery, vol. 8, no. 4, p. e1253
17. Selvin Sreelekshmy, Ravi Vinayakumar, Gopalakrishnan, E.A., Menon Vijay Krishna, (2017), Stock price prediction using LSTM, RNN and CNN-sliding window model. DOI: 10.1109/ICACCI.2017.8126078
18. Nelson David M. Q., Pereira Adriano C. M., Oliveira Renato A. de, (2017), Stock Market's Price Movement Prediction with LSTM Neural Networks, DOI: 10.1109/IJCNN.2017.7966019
19. Zhou Xujuan, Tao Xiaohui, Yong Jianming, Yang Zhenyu, (2013), Sentiment analysis on tweets for social events, IEEE 17th International Conference on Computer Supported Cooperative Work in Design. CSCWD, IEEE, http://dx.doi.org/ 10.1109/cscwd.2013.6581022.
20. Nabipour Mojtaba, Nayyeri Pooyan, Jabani Hamed, S. Shahab, Mosav Amir, (2020), Predicting Stock Market Trends Using Machine Learning and Deep Learning Algorithms Via Continuous and Binary Data; a Comparative Analysis, IEEE Access, 8 (2020) 150199–150212.
21. MA YI-LIN, HAN1 RUI-ZHU, AND WANG WEI-ZHONG, Prediction-Based Portfolio Optimization Models Using Deep Neural Networks, (2020), IEEE Access, 8115393–115405.
22. Yeruva, V.K., Chandrashekar, M., Lee, Y., Rydberg-Cox, J., Blanton, V., Oyler, N.A., 2020. Interpretation of sentiment analysis with human-in-the-loop. In: 2020 IEEE International Conference on Big Data (Big Data). IEEE, http://dx.doi.org/10.1109/ bigdata50022.2020.9378221.
23. Shen Jingyi and Shafiq M. Omair, (2020), Short-term stock market price trend prediction using a comprehensive deep learning system, https://doi.org/10.1186/s40537-020-00333-6
24. FENGQIAN DING AND CHAO LUO, (2020), An Adaptive Financial Trading System Using Deep Reinforcement Learning with Candlestick Decomposing Features, IEEE Access, 8 63666-63678.
25. Nabipour M, Nayyeri P, Jabani H, Mosavi A, Salwana E and S. Shahab, (2020), Deep Learning for Stock Market Prediction Entropy, 22, 840; doi:10.3390/e22080840

Note: All the tables in this chapter were made by the authors.

Intelligent Systems Using Semiconductors for Robotics and IoT – Dinesh Goyal et al. (eds)
© 2026 Taylor & Francis Group, London, ISBN 978-1-041-20408-4

57 Advancements in Self-Supervised Learning: A Review of Unsupervised Information Retrieval Techniques for Big Data

Virendra Tank*
Research Scholar,
Computer Science, College name: Tantia University,
Shri Ganganagar (Rajasthan)

Pawan Kumar Pareek
Associate Professor,
Faculty of Computer Science, College name: Tantia University,
Shri Ganganagar (Rajasthan)

Abstract: Self-supervised learning has been promising as the breakthrough methodology for machine learning when dealing with large-scale, unlabelled datasets. This review paper would try to discuss recent development on SSL application towards unsupervised IR within the big data environment. First, this discussion shows its principles and underlying basics of the SSL methodology that can be used for patterns in data and its structure without depending on any labelled examples. Different self-supervised techniques are reviewed, including contrastive learning, generative models, and representation learning for their effectiveness in improving unsupervised IR tasks. A number of recent advances in SSL approaches reviewed in detail followed by a focus on retrieval performance on large data corpora. Discussions focus on the integration of SSL into current frameworks for IR; improvements in feature representation, new developments in the training, testing processes, etc. Moreover, challenges and limitations in terms of using SSL for big data, such as scalability, computational complexity, or the need for efficient strategies related to data handling, are discussed. Case studies as well as some empirical results are reviewed to illustrate practical applications for SSL-based approaches within different domains, like text, image, and multimedia data retrieval. It combines recent studies to present a clear conceptual overview of how SSL can facilitate unsupervised IR in big data settings. The review ends with future research directions and the possible areas of further exploration of application of self-supervised learning in big data.

Keywords: Self-supervised learning, Unsupervised information retrieval, Big data, Contrastive learning, Generative models, Representation learning, Data retrieval, Machine learning

1. Introduction

Self-supervised learning has emerged as a disruptive paradigm in machine learning, especially in tasks associated with large amounts of unlabelled data. In general, unlike traditional supervised learning which highly relies on the availability of appropriately labelled datasets, SSL exploits the abundance of large, unlabelled datasets. This paradigm then generates supervisory signals from the data itself and enables a model to learn meaningful representations without explicit annotations. SSL has demonstrated promising capabilities in some applications, including language processing, computer vision, and audio processing, to significantly improve the performance of the model while avoiding laboriously time-consuming and expensive labeling processes [1].

*Corresponding author: vtank87@gmail.com

DOI: 10.1201/9781003716389-57

This large amount, diversity, and complexity of big data have increased the requirements for effective unsupervised information retrieval. In big data settings where the volume of data is highly high, unorganized mostly, and too large to be stored, the conventional IR methods are severely challenged [1]. Unsupervised IR aims at extracting valuable information from such datasets without using pre-classified labels. Handling and retrieval of relevant information from huge amounts of datasets will be crucial for applications such as search engines, recommendation systems, data mining, and knowledge discovery. SSL enhances unsupervised IR by using robust methods for data representations that increase accuracy as well as better retrieval efficiency [2].

1.1 Objectives of the Paper

The objectives of this review paper are as follows:

1. To present recent advancements in self-supervised learning: Give an overview of recent SSL techniques and their advancement, including contrastive learning, generative models, and representation learning [3].
2. To analyze the integration of SSL with unsupervised information retrieval: Introduce the basic concept of how the SSL methods can be integrated with existing IR frameworks to improve the performance of retrieval in large-scale, unstructured datasets.
3. SSL Evaluation for the Unsupervised IR Task. Discuss to what degree SSL can be useful in improving the state-of-the-art in these tasks: feature extraction, similarity measurement and ranking information [3].
4. Identify challenges and limitations as far as using SSL for the big data context contexts: It also talks about the issues concerning scalability, computational complexity, and difficulty when dealing with data.
5. To Demonstrate Case Studies and Applications: Publish case studies of experiments, which verify the effective application of SSL to different types of documents including text, image, and multi-media sources data retrieval [4].
6. Vestige for Future Scope Suggest some trends and promising directions of how SSL and IR can be developed and applied further and recommendations for developing such research.

The review conducts an in-depth investigation of SSL techniques and the impact of SSL upon unsupervised IR, focusing particularly on practical applications and current challenges and research opportunities in the future. A holistic approach will be employed so that key insights can be provided among researchers, practitioners, and policymakers who are interested in using SSLs to improve information retrieval in big data environments

2. Fundamentals of Supervised Learning

Self-supervised learning (SSL) has garnered significant attention in the community of machine learning as it can leverage non-labeled data effectively. It creates supervisory signals from the data and not any other external label. Thus, the model learns appropriate features and representation by coming up with some form of pseudo-label or auxiliary tasks that come from the inherent structure of the data. The central concept is to exploit huge amounts of unlabelled data available to learn informative representations that can be transferred to various downstream tasks [5]. One of the key concepts in SSL is that of pretext tasks. Pretext tasks are surrogate tasks designed to train a model to learn useful representations from the data. Probably the most widely used pretext tasks in computer vision is image colorization, where a model learns to predict the missing color channels of grayscale images. In this case, trying to solve these tasks gives deeper insight into data structure and semantics. Another key ingredient is self-supervised objectives-that is, the specific goals or loss functions on pretext tasks which are used to train the model. Those goals instruct the model to learn features that are useful not only for itself but also for other tasks such as classification or retrieval [5].

2.1 Comparison with Supervised and Unsupervised Learning

Self-supervised learning is compared with both supervised and unsupervised learning along various fundamental features. Supervised learning requires a liberal use of abundant labeled datasets wherein each training example should map to a label. The method is really powerful when the labeled data are copious but often suffers from high costs and limitations associated with data annotation. In contrast, self-supervised learning does not need labeled data. Instead, it creates its own labels or supervisory signals from the data and will scale much better and be more amenable to scenarios in which labelled data might be scarce or unavailable [6]. The other is unsupervised learning, where it learns to discover patterns or regularities within unlabelled data, without having explicit supervisory signals. Examples of which belong to this category: clustering or dimensionality reduction. Whereas, in unsupervised learning, SSL makes certain tasks or pseudo-labels that tell the learning process to aim at, hence enabling the models to learn better-informed representations. For example, for unsupervised learning methodologies, traditional approaches focus on discovering patterns or structures within the data, whereas SSL is targeted towards creating useful representations to improve downstream task performance [6].

2.2 Popular SSL Techniques

Some of the several techniques that are applied in self-supervised learning to tap into the power of such unlabeled data quite effortlessly include:

- Contrastive Learning: This method makes use of the contrasting positive pairs-being similar examples-and negative pairs-that being dissimilar examples-just to learn representations. This trains the model to move similar examples closer in feature space and pushes dissimilar examples apart. A popular methodology for this is the SimCLR framework that uses data augmentation creating positive pairs, and uses contrastive loss function to train the model. Contrastive learning has been highly effective for such tasks as, for example in image and text representation learning [7].
- Generative Models: The generative models learn the underlying distributions and can generate new samples similar to the training data. In SSL, generative models can be utilized as pretext tasks where the model learns to reconstruct or generate data from incomplete or corrupted inputs. Examples of such include autoencoders, learning to compress and reconstruct data and generative adversarial networks (GANs), which consist of a generator and a discriminator network competing to produce realistic data samples [7].
- Representation Learning: This technique relies on learning meaningful and useful representations of the data to better fit many subsequent tasks. In SSL, the learned representations are usually developed by training models on the task of missing parts of an input, or auxiliary problems that require understanding the structure of the data. Methods such as deep neural networks and transformer models often capture the complex patterns and features with the data [7].

All these contribute to making self-supervised learning useful due to the possibility that unlabelled data can be utilised in various forms for learning an informative representation. Sometimes it is application specific and dependent on the nature of the data to be analysed.

3. Unsupervised Information Retrieval (IR)

3.1 Unsupervised IR Introduction

Unsupervised information retrieval refers to the access of useful information from large-scale datasets without making use of pre-existing labels or annotations. Unlike supervised IR, which makes wide use of labeled data in mapping queries to relevant documents or information, unsupervised IR will rely entirely on the intrinsic structure and relationships within the data toward achieving retrieval tasks. This becomes quite handy in scenarios where it is not possible or economically viable to attach tags to data, as in the case of a giant corpora or in dynamic surroundings like the web, multimedia archives, or big text storages [8]. Unsupservised IR has gained paramount importance with big data because of the huge volume, variety, and velocity of data that challenges the capabilities of traditional retrieval systems. The process of finding relevant information becomes increasingly difficult for big data context scenarios, because most data tends to be unstructured, heterogeneous, and modal in nature. In other words, there is no possibility of querying an already pre-curated set of data or even labeled examples. For such reasons, it makes unsupervised IR systems apply robust methods for discovering relationships and extracting meaning from the underlying data. Directly due to this, unsupervised IR holds the key position in several applications like search engines, recommendation systems, knowledge discovery and automated decision making systems [9]. Whilst it holds all the high hopes, unsupervised IR itself is beset by a number of limitations, particularly in big data environments. One of the major challenges is the scale of data.

This clearly calls for efficient indexing, storage, and retrieval mechanisms that can scale up with the constantly increasing dataset size. Further, in addition to being noisy and incomplete, these data are often without clear structure and therefore fails to provide much relevant information unless there are some labeled examples. Data Sources Another issue with data sources is the fact that their nature is diversified, meaning some are structured-for example, databases-but others are semi-structured like XML files-and others less so, being unstructured data-like natural language text, multimedia [10]. Such diversity of formats as well as modalities complicates retrieval because different types of data require special methods of retrieval. Last but not the least, semantic understanding is the problem: Traditionally, IR systems have heavily relied on syntactic matching, such as keyword-based search, which may not necessarily capture deeper meaning that information retrieved represents. Unsupervised IR is likely to extend from shallow patterns toward more meaningful and significant semantic meanings and relationships within data [10].

3.2 Traditional Approaches and their Limitations

Traditional unsupervised IR methods have relied heavily on techniques like vector space models, TF-IDF [11] weighting, and clustering-based approaches. Vector space models represent both documents and queries as points in a high-dimensional space Normally, each dimension of that space refers to a term in the vocabulary. The measure of how appropriate a document is to a given query is determined by computing the similarity between their vectors. Often, the appropriate metrics are based on cosine similarity. Although simple, this approach assumes the relationship between documents and queries with similar terms is valid, which may not be the case in practice, especially when working with complex or ambiguous queries [11]. The TF-IDF approach weights terms based on the frequency with which they appear in a document, and the rarity of their appearance in the larger corpus. In other words, it favors words which appear only in one document but reduces the value of those terms that tend to appear fairly often across all of the documents ("the" or "and," for example).

TF-IDF performs successfully under many scenarios, but fails in the long-tail terms, where for the same word, the context or usage is more important than the word itself. Furthermore, TF-IDF fails to consider relationships between words, which limits its capacity in dealing with semantic or contextual nuances in the data [11]. Clustering-based methods group similar documents based on their features and thus allow the system to retrieve information from clusters that are most relevant to a given query. Techniques like K-means or hierarchical clustering are commonly applied in unsupervised IR. These methods are always afflicted with limitations such as high computation costs, sensitivity to the chosen parameters like number of clusters, and inconvenience in defining proper similarity metrics for large and heterogeneous datasets. Traditional approaches, widely used these days, have some limitations when applied to big data contexts: mainly scaling up. Traditional methods are impractically resource-intensive during both indexing and retrieval, as well as clustering, which tends to lose effectiveness in real-world applications. Besides this, they typically rely on shallow representations of the data-examples of term frequencies or syntactic similarities-which do not reflect deeper semantic relationships that exist between concepts in the data. It results in the deficiency of quality retrieval, particularly when an ambiguous query is at hand or a large contextual understanding of the data is required [11]. Traditional methods do not adjust well to the heterogeneity of formats that are witnessed in big data environments. In such an environment, different types of data, like text and images, require different retrieval techniques. In fact, most of the existing traditional unsupervised IR methods have been developed for a type of data usually text-only data and degrade when generalization across modalities is sought. This has spurred the necessity of having a more complex methodology that will handle modern big data systems. Summarizing: Whereas traditional unsupervised IR approaches have been established as sound, limitations in scalability, semantic understanding, and support of multiple data formats make necessary more complex approaches, such as self-supervised learning, which may solve those challenges related to big data.

4. Advancements in Self-Supervised Learning for Unsupervised IR

4.1 Combination of SSL with Unsupervised IR

This turns out to be a revolution in the way large data sets are analyzed and processed as SSL is brought together with unsupervised IR. Enhanced efficiency in complex, unstructured datasets depends on improvement in retrieval accuracy, which is directly promoted by providing more robust and meaningful data representations from SSL-enhanced IR tasks. Traditional unsupervised IR approaches often deploy shallow representations, like word frequency or surface similarity, which fail to uncover deep semantic relations. In contrast, SSL relies on pretext tasks that capture syntactic and semantic information within the data and then leverages those pretext tasks to produce rich feature representations [12]. Using SSL can add context-approach representations to IR systems for better representation of documents or information and the system has more chance in the detection of relevant information even with very complex or vague queries. A technique in popular SSL methods, known as contrastive learning, is training models to differentiate between similar and dissimilar data points. This can greatly help IR by providing fine-grained differences between documents that traditional methods may fail to identify. This, in turn, can lead to higher accuracies when retrieving more relevant information, especially for big data, where context and nuance play decisive roles [12]. Successful embedding of SSL within existing IR frameworks has thus led to improved retrieval performance. For instance, most modern IR systems are now integrating BERT-like models, which make use of masked language modeling techniques, that aid in improving document retrieval by learning context-specific word representations. These models get greater contextual relationships between words and documents than the usual IR systems, moving them closer to that relevant thing. But it is not until the work in References applies self-supervised pretraining to ranking algorithms for search engines, enabling the model to learn from vast swaths of unlabelled data, that ranking precision on unseen data streams gets improved without manual annotations [12].

4.2 New Directions of SSL Techniques in IR

Recent advances in SSL methods open up further frontiers to what can be achieved in unsupervised IR. High-quality representations can be learned from unlabelled data using emergent contrastive learning frameworks like SimCLR or MoCo, which thrive on using them for learning discriminative features by creating positive pairs and negative pairs of data points. Contrastive learning is quite useful in the IR setting in that it can differentiate between perhaps two documents which at the surface level look very much alike, but in meaning or context are disparate, which increases the accuracy of retrieval [13]. Another vital stride is the development of the use of generative models in SSL to IR tasks. Models such as BERT, GPT, and T5, pre-trained in self-supervised objectives revolutionized how IR systems treated language data. These models are able to predict missing parts of data or generate new data from partial inputs and, therefore, capture complex semantic patterns that are crucial for effective retrieval. For instance, BERT uses masked language modeling in the form of a pretext task; it masks parts of input text and trains the model to predict the missing words. It makes the model understand context and meaning, so that it has high performance and is especially relevant for IR tasks like document ranking and query understanding [13]. Their impact, in particular, is well reflected in an improvement to feature representation and retrieval performance. SSL techniques allow models to learn generalizable and task-specific representations that are fundamental to the handling of diverse queries and documents in the big data application contexts. The conceptual representation model learned by SSL can capture deep relations from the massive and unlabelled data that big data environments contain. Deep semantic understanding enables such models to retrieve more accurately because it is better at matching queries with relevant documents than depending on mere surface-level patterns [13].

5. Related Works

Liu, Y. et al. (2024: [14]). This is an article intended for legal AI-precision and efficiency in retrieving similar cases. This has been a crucial task in legal research. They introduce new both dense and sparse approaches that are combined. They did significantly better in terms of precision and MAP by 3.66% and 3.62%, respectively, on Chinese law datasets by using sparse model extracted data in comprehensive legal cases by combining it with labeled data through dense retrieval. Case-based datasets may have a variety of issues due to imbalanced datasets, especially for less common legal scenarios, which impact the retrieval results. However, their method promotes faster effective retrieval of relevant legal cases, which benefits both the practitioner and the non-forensic professional.

Benati, N., & Bahi, H. (2024):[15] Important advances in signal processing and abundance of speech resources cause QbE-STD studies to expand continuously. It was a method of finding spoken queries in streams of speech with a network-based architecture on a fully connected convolutional neural network with attention mechanisms in an unsupervised process. They tested their model on the Google speech commands corpus with very encouraging results for indexing speech databases but hinted that it might raise speech recognition technology by large self-supervised learning.

Vo, H. V., et al. (2024)[16]: Self-supervised learning is one of the most critical aspects of modern machine learning. The real task still lies in preparing high-quality and balanced datasets. The paper proposes a clustering-based automatic dataset curation method that goes beyond the bounds of costly data selection procedures and time-consuming processes towards data acquisition. Their approach works through the use of hierarchical k-means clustering and balanced sampling across web images, satellite images, and text domains. They evaluate on these domains, demonstrating that the curated datasets outperform its uncurated data, as well as being comparable to manually curated datasets, thus showing effective capabilities for improving performance in diversity of applications.

Jiang, Y., et al. (2023):[17] They addressed the object similarity problem in e-commerce by proposing a self-supervised learning framework based on product image-title pairs. The authors advanced product representation learning in closer similarity matching and retrieval of multimodal learning in e-commerce further. The new embedding with the mechanism of an attention regularization would enable the model to exploit category-related signals from textual inputs itself. Experimental results performed on several product datasets justified its superiority over traditional unimodal methods and displayed broad applicability for e-commerce settings.

Mao, Y. et al. (2023, November): [18] Automatic reformulation of query is important to improve queries as well as their results for the users in the process of software development. It proposes here a self-supervised query reformulation called SSQR that relies on parallel query pairs.

Based on pre-trained model inspirations, SSQR uses masked language modeling on unannotated query corpora, which would improve completeness of queries in a non-human supervised manner. The evaluation results have shown it to be improved upon unsupervised baselines and competitive performance compared with supervised methods in that sense that it is effective improvement on search efficiency without relying on labeled data.

Xu, W., et al. (2022): [19] One of the recent trends in NLP is CL. The approach has only been used in retrieval tasks such as question answering. So, they proposed a joint CL and Auto-MLM to facilitate self-supervised multi-lingual knowledge retrieval with the joint combination of sentence-level training and masked language modeling. Their proposed method outperformed previously reported state-of-the-art approaches particularly for multi-language corpora. Experimental results on a few datasets demonstrated again improvements with a consistent tendency in retrieval quality and effectiveness, which testified to the superiority of their unified strategy.

Cho, J., et al. (2022): [20] Utterance-level speech representations selfsupervised learning is very important for discriminative attributes applications such as speaker verification and emotion recognition. The contribution here brings the first adaptation of the non-contrastive selfsupervised method from computer vision to speech, named DINO. Unlike the traditional contrastive methods, DINO relies on negative sampling. Their results achieve state-of-the-art performance on speaker verification, SER, and disease detection tasks. The authors also commented on

Table 57.1 Literature findings

Author Name (Year)	Main Concept	Findings
Liu, Y., et al. (2024)	Legal AI, Similar Case Retrieval	Introduced a novel approach combining dense and sparse retrieval methods for legal case retrieval. Achieved significant precision and MAP improvements on Chinese law datasets.
Benati, N., & Bahi, H. (2024)	Speech Recognition, QbE-STD	Proposed a FC-CNN model with attention mechanisms for unsupervised query by example spoken term detection, showing promise in indexing speech databases.
Vo, H. V., et al. (2024)	Self-supervised Learning, Dataset Curation	Developed a clustering-based method for automatic curation of high-quality, balanced datasets, outperforming uncurated and rivaling manually curated datasets across multiple domains.
Jiang, Y., et al. (2023)	E-commerce, Product Similarity	Proposed a multimodal learning approach using product image-title pairs to enhance product representation and improve similarity matching and retrieval in e-commerce settings.
Mao, Y., et al. (2023, Nov)	Query Reformulation, Self-supervised Learning	Introduced SSQR, a self-supervised method for query reformulation without relying on parallel query pairs, demonstrating competitive performance in search efficiency improvement.
Xu, W., et al. (2022)	Natural Language Processing, Knowledge Retrieval	Combined contrastive learning (CL) and Auto-MLM for self-supervised multi-lingual knowledge retrieval, achieving superior performance across different language corpora.
Cho, J., et al. (2022)	Speech Processing, Utterance-level Representation	Introduced DINO, a non-contrastive self-supervised method for learning utterance-level embeddings in speech applications, outperforming traditional contrastive methods in various tasks.
Moro, G., & Valgimigli, L. (2021)	Information Retrieval, Scientific Papers	Developed SUBLIMER, a self-supervised IR engine trained on CORD19 using deep metric learning on citation data, showing superior performance in retrieving relevant scientific papers.

how training strategies and data augmentation techniques play an important role in several speech applications in improvements of their performance.

Moro, G., &Valgimigli, L. (2021): [21] In the context of scientific paper retrieval for information, this paper reports a self-supervised IR engine known as SUBLIMER that was initially trained on the CORD19 COVID-19 Open Research Dataset. Deep metric learning on citation data while building up a semantic similarity metric is in contrast to the traditional approaches of IR requiring labeled datasets for making efficient retrievals with SUBLIMER. Evaluation on CORD19 showed SUBLIMER outperforming state-of-the-art approaches in Precision@5 (P@5) and Bpref, highlighting its efficacy in discovering relevant scientific knowledge without relying on annotated datasets..

6. Conclusion

This has wide-ranging potential to transform the way cybersecurity is accomplished. This is because blockchain provides more protection in terms of structure, being decentralized, immutable, and transparent. In each one of these domains, versatility through the application will also be found, including critical problems such as data integrity, secure identity management, IoT security, and supply chain verification in applying blockchain systems. But blockchain also has its share of vulnerabilities, such as 51% attacks, Sybil attacks, smart contract flaws, and scalability issues. Challenges of the sort should be identified and, in turn, be carefully implemented, audited consistently, with recent blockchain architectures along with cutting-edge consensus mechanisms. The review shows how blockchain could facilitate highly improved cybersecurity by the reduction of threats that often threaten the traditional systems, although it also clarifies that constant research to achieve a balance between security, scalability, and adaptability is needed. Cybersecurity threats and challenges will also evolve with the changing digital landscapes. Thus, there is an all-around need to implement blockchain-based security solutions in order to counter present and future challenges. Blockchain holds much promise in terms of enhancing trust and resiliency within digital systems; it would be great if developers, security experts, and regulators can work together in order to introduce it well within cybersecurity infrastructures. But apart from depicting how blockchain current contributes to cybersecurity, it further emphasizes that future exploration into overcoming its limitations is much needed in totally realizing its potential for foundational technology in securely developing digital environments.

References

1. Chen, Z., Chen, W., Xu, J., Liu, Z., & Zhang, W. (2023, October). Beyond Semantics: Learning a Behavior Augmented Relevance Model with Self-supervised Learning. In *Proceedings of the 32nd ACM International Conference on Information and Knowledge Management* (pp. 4516–4522).
2. Mu, L., Jin, P., Zhang, Y., Zhong, H., & Zhao, J. (2023). Synonym recognition from short texts: A self-supervised learning approach. *Expert Systems with Applications*, *224*, 119966.

3. Kotei, E., &Thirunavukarasu, R. (2023). A systematic review of transformer-based pre-trained language models through self-supervised learning. *Information, 14*(3), 187.
4. Domínguez Becerril, C. (2023). Unsupervised information retrieval using large language models.
5. Maekaku, T., Fujita, Y., Chang, X., & Watanabe, S. (2023, June). Fully unsupervised topic clustering of unlabelled spoken audio using self-supervised representation learning and topic model. In *ICASSP 2023-2023 IEEE International Conference on Acoustics, Speech and Signal Processing (ICASSP)* (pp. 1–5). IEEE.
6. Zhang, Z., Hu, Y., Pan, B., Ling, C., & Zhao, L. (2024). TAGA: Text-Attributed Graph Self-Supervised Learning by Synergizing Graph and Text Mutual Transformations. *arXiv preprint arXiv:2405.16800.*
7. Gui, J., Chen, T., Zhang, J., Cao, Q., Sun, Z., Luo, H., & Tao, D. (2024). A Survey on Self-supervised Learning: Algorithms, Applications, and Future Trends. *IEEE Transactions on Pattern Analysis and Machine Intelligence.*
8. Shwartz Ziv, R., & LeCun, Y. (2024). To compress or not to compress—self-supervised learning and information theory: A review. *Entropy*, *26*(3), 252.
9. Wang, S., Hou, D., & Xing, H. (2023). A self-supervised-driven open-set unsupervised domain adaptation method for optical remote sensing image scene classification and retrieval. *IEEE Transactions on Geoscience and Remote Sensing*, *61*, 1–15.
10. Jing, M., Zhu, Y., Zang, T., & Wang, K. (2023). Contrastive self-supervised learning in recommender systems: A survey. *ACM Transactions on Information Systems*, *42*(2), 1–39.
11. Van den Herrewegen, J., &Tourwé, T. (2023). Self-supervised learning for robust object retrieval without human annotations. *Computers & Graphics*, *115*, 13–24.
12. Gao, Y., Wang, X., He, X., Feng, H., & Zhang, Y. (2023). Rumor detection with self-supervised learning on texts and social graph. *Frontiers of Computer Science*, *17*(4), 174611.
13. Gálvez, B. R., Blaas, A., Rodríguez, P., Golinski, A., Suau, X., Ramapuram, J., ... &Zappella, L. (2023, July). The role of entropy and reconstruction in multi-view self-supervised learning. In *International Conference on Machine Learning* (pp. 29143–29160). PMLR.
14. Liu, Y., Tan, T. P., & Zhan, X. (2024). Iterative Self-Supervised Learning for Legal Similar Case Retrieval. *IEEE Access.*
15. Benati, N., & Bahi, H. (2024). Self-Supervised Spoken Term Detection for Query by Example. *Journal homepage: http://iieta. org/journals/isi, 29*(3), 1175–1181.
16. Vo, H. V., Khalidov, V., Darcet, T., Moutakanni, T., Smetanin, N., Szafraniec, M., ... & Bojanowski, P. (2024). Automatic Data Curation for Self-Supervised Learning: A Clustering-Based Approach. *arXiv preprint arXiv:2405.15613.*
17. Jiang, Y., Liao, K., Lin, S., Qiao, H., Yu, K., Yang, C., & Chen, Y. (2023, November). Self-supervised Multimodal Representation Learning for Product Identification and Retrieval. In *International Conference on Neural Information Processing* (pp. 579–594). Singapore: Springer Nature Singapore.
18. Mao, Y., Wan, C., Jiang, Y., & Gu, X. (2023, November). Self-supervised query reformulation for code search. In *Proceedings of the 31st ACM Joint European Software Engineering Conference and Symposium on the Foundations of Software Engineering* (pp. 363–374).
19. Xu, W., Maimaiti, M., Zheng, Y., Tang, X., & Zhang, J. (2022). Auto-mlm: Improved contrastive learning for self-supervised multi-lingual knowledge retrieval. *arXiv preprint arXiv:2203.16187.*
20. Cho, J., Villalba, J., Moro-Velazquez, L., & Dehak, N. (2022). Non-contrastive self-supervised learning for utterance-level information extraction from speech. *IEEE Journal of Selected Topics in Signal Processing*, *16*(6), 1284–1295.
21. Moro, G., &Valgimigli, L. (2021). Efficient self-supervised metric information retrieval: a bibliography based method applied to COVID literature. *Sensors*, *21*(19), 6430.

Intelligent Systems Using Semiconductors for Robotics and IoT – Dinesh Goyal et al. (eds)
 ISBN 978-1-041-20408-4

58 Deep Learning Techniques for the Prediction of Heart Disease

Sanjay Kumar Sinha[1]
Poornima Institute of Engineering & Technology,
Department of Computer Engineering,
Jaipur, India

Savita Prabha[2]
Vivekananda Global University,
Jaipur, India

Navin Kumar Goyal[3],
Neha Srivastava[4]
Poornima Institute of Engineering & Technology,
Department of Computer Engineering,
Jaipur, India

Abstract: Heart disease usually does not predict people so it can cause serious problems or death if these problems are taken lightly. By focusing on this issue, efforts have been made by various researchers through research and have helped to develop ideas for making important predictions by calculations. This research gives comparative study of Various supervised, unsupervised ML approaches and Deep Learning methods like, Naive Bayes ANN, to forecast diabetes by consuming a dataset.

Keywords: ANN, CNN, LSTM, BLSTM, SVM, AUC

1. Introduction

Heart disease usually does not predict people so it can cause serious problems or death if these problems are taken lightly. By focusing on this issue, efforts have been made by various researchers through research and have helped to develop ideas for making important predictions by calculations.

Machine learning has been regarded as a great series of lessons that involve the development of artificial intelligence to make predictions by imitating and helping people to think beyond what they can do[1]. Data mining is thought to be a part of a machine involving unchecked, unsupervised, and reinforcing learning processes to be utilized to make computations by computations. Deep "Neural" networks, K algorithms, compounding, confusing identification, and deep learning are thought to be other efficient methods utilized to make predictions and thus achieve results by bringing the given set of parameters to achieve future results [2].

2. Related Work

According to [3] M.Akhil et al. 2013, it has been suggested that the implementation of K-means algorithms is used to diagnose heart disease and in this regard consider the genetic algorithm and its proximity to K in order to develop appropriate predictions. The author stated that this method is often used to obtain an idea about unknown data and thus can be used to obtain information in this way. It is suggested that this process should also take into account certain relevant parameters which can be used in this

[1]sanjay.sinha@poornima.org, [2]savita.prabha@vgu.ac.in, [3]navin.goyal@poornima.org, [4]neha.srivastava@poornima.org

DOI: 10.1201/9781003716389-58

context and hence can be utilized in mathematical proofs to generate outcomes. It has also been suggested that the interaction between the predictive data and the existing data is made to understand what kind of cardiovascular disease may impact individuals in the future.

According [4] Liu et al. In 2018, it has been suggested that deeper learning techniques be used in making predictions and therefore may be used in mathematical calculations. The authors have suggested that these methods are used to diagnose targeted diseases and thus can be used to provide based estimates. According to the author's research, it has been suggested that [19] LSTM and CNN are the two main methods used to make accurate predictions. The above number Source: ([4] Liu et al. 2018) highlights the "Convolution Neural Network," which is used to estimate estimates. The author stated that the sequence stored in the input data includes kernel sizes. The hidden layer is displayed after a large consolidation and thus helps to bring the values together, which will help to get ideas about the interaction to make important predictions.

The author suggested that RNN ("Recurrent Neural Networks") is used to make predictions for targeted repetition and thus consider the use of memory gates to produce results in the same sequence of specific heart disease. The author pointed out the fact that both of the above two perspectives capture patient information such as blood samples, blood groups, heart rate, and blood pressure and thus make sequential calculations to present ideas about the risk of heart disease. and thereby makes people more alert and thus allows them to take precautionary measures.

In accordance with [5] Wiharto and Herianto, 2015, it has been suggested that the "Vector Support Machine" (SVM) is used over a wide range to make predictions to understand the different types of heart disease found in humans. The authors have suggested that in the calculation, a list of parameters should be selected and for this use as input, which will help to obtain the desired results. These databases will help to bring the diagnosis automatically and thus help with the results in the image, shown below.

From the above the benefits of multiclass algorithm can be easily seen. While in previous research activities, researchers used a binary algorithm to classify data into healthy and sick; a multiclass algorithm can be divided into several groups depending on whether you are experiencing heart disease. The BT method used in this study performed very well with 61.86% accuracy, and the most effective OAO algorithm with 51.546% accuracy [15] (Maini, Daet al. 2019, July). However, the BTSVM algorithm is not very accurate; one of the hybrid algorithms proposed by previous authors accurately predicted the status of 62.1% of all human data stored on a website.

Predictions are made based on the set of parameters above and by this identification of the differences in sequence, preset and thus help to get a clear view of the algorithm. Accuracy, precision, F, and accuracy are all different types of algorithmic formulas that help calculate by taking numbers from the figure above and then helping to make step by step predictions. The authors suggested that SVM best practices could help generate predictive statistical results and thus could help more patients to make automatic diagnoses.

Includes a flowchart of the proposed activity containing the selected database, description of the data set, how the data was processed in advance etc. It also includes the use of the proposed model.

[28] Tomov and Tomov, 2018, conducted research on how to use DNNs or "Deep Neural Networks" to classify medical databases based on the likelihood of heart disease. The researchers of this paper developed DNN structures with five layers. The five-layer DNN, also called HEARO-5, produces 99% accuracy and exceeds the few established methods for this field [29] (Uyar and İlhan, 2017). The authors evaluated their proposed method, which is indicated in the appendix by a parameter called "Mathews coefficient (MCC)". The results stimulated the researcher's imagination, and he imagined the packaging of the software for easy-to-use tasks for researchers and patients directly.

In one of the Nassif and others 2015 srudies, the authors proposed three possible ways of sorting medical information from the Cleveland database. The three ways are intended to forecast the development of CAD which stands for "Coronary Artery Disease" in patients.

The three models whose performance analyzed by researchers were Naïve Bayes, KNN, and SVM. The three algorithms have very different differentiation methods [30] (Mohanet al. 2019). The KNN algorithm uses a method similar to the Euclidean range to find data points near k in the target 'x', after which the algorithm takes a class vote for the various objects surrounding a particular element to determine its class. SVM is a division algorithm based on a set size (number 1 to n) to define variables. The algorithm determines the hyper-plane, which contains the size (n-1), and then separates the different data points on the basis of which side the hyper-plane lies on. The Naïve Bayes approach prefers to look at all the changes that influence prediction as independent of each other and then decides on the evolutionary phase on the basis of applying Bayes theory, which is a very basic concept in the field of mathematical probability. [31] (Malavet al. 2017). The characteristics or characteristics used by researchers to predict disease occurrence and their relation to the dependent variations are given below.

The research results (Source: [32] Nassif et al. 2015) determined that the Naïve Bayes algorithm had better accuracy than the SVM and KNN classifying algorithms. It had greater predictive accuracy as well as better performance metrics. Researchers [33] Mallya et al. 2017, tested RNN or "Recurrent Neural Network" to predict the occurrence of CHF or "Congestive Heart Failure" a few months earlier. Data used for prediction were collected in a large cohort of patients (216,394) over a 12-month period. The authors tested the LSTM model and compared their results with other models such as LR, RF, MLP, and CNN. It has been found (Source: [33] Mallya et al. 2017) that the LSTM method is the most suitable for using longitudinal data in the context of predicting CHF among patients. The authors concluded with this study that the LSTM network was better suited to use interpersonal variables better than other algorithms.

According to the authors [3] Jabbar et al. 2013, data mining techniques are an important class of classification algorithms that can produce excellent results in predicting heart disease. Data mining is a type of supervised reading that separates databases based on each line of data parameters. The authors suggest that neighbors close to K are a way to extract data as a potential source of medical data-based classification. Pattern recognition can be greatly improved if you use KNN. However, the method has one major error, which lies in the fact that unnecessary data attributes and unnecessary data can cause the algorithm to produce negative results. The authors of their study developed a hybrid algorithm that combines the genetic algorithm with the KNN [6] (Dutta and Bandyopadhyay, 2020). Researchers have used a variety of distinguishing features using the ML hybrid algorithm. The authors have used their hybrid algorithm on a number of databases that contain patient data about many health disorders solution to the problem.

3. Performance Evaluation

Asthma, chronic obstructive pulmonary disease (COPD), and other respiratory diseases have become more dangerous in the current environment. The increase in air pollutants like PM2.5, PM10, and others has increased that threat. Since lowering pollution levels takes a lot longer than the severity of the issue allows, a respirator can be employed as an urgent countermeasure on an individual level safety measure.

$$\text{Accuracy} = (TP + TN)/(TP + TN + FP + FN)$$

4. Comparative Analysis

This assignment utilized the PIMA Indian diabetes dataset obtained taken from Kaggle and uploaded on the Jupitor Notebook. After data cleaning and data pre-processing data is transformed to numpy array as machine learning model requires this. In this, 10 fold cross validation has been used to evaluate the performance. For comparative study this paper employs the Naive Bayes, ANN, SVM, KNN, Random Forest, LSTM, CNN, BLSTM, ensemble of CNN and LSTM and ensemble of CNN and BLSTM for the diagnosis of diabetes (See Table 58.1).

Table 58.1 Comparative analysis of machine learning based approaches

Model	Accuracy
Naïve Bayes	80.53
Random Forest	83.48
SVM	81.84
KNN	81.15
ANN	92.78
LSTM	91.48
CNN	95.73
CNN+LSTM	93.45
BLSTM	92.15
CNN+BLSTM (Proposed)	99.33

As seen from the comparison, deep learning models worked better than basic ML approaches.

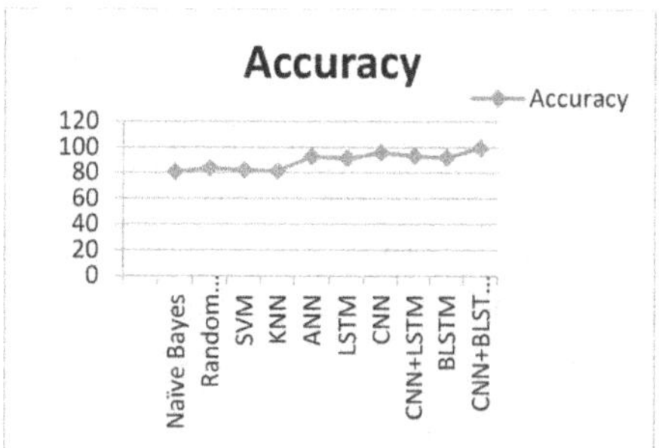

Fig. 58.1 Classification accuracy V/s ML and DL models

5. Conclusions

We used in-depth study methods and algorithms of the Cleveland Dataset consisting of 76 elements, we used 11 features based on the importance of predicting Heart Disease, to obtain better performance with accuracy. And utilization of Bayesian optimization for hyper parameters leads to like robustness improvements such as durability and a minimum determination time taken for the diagnosis and prognosis of Heart Disease.

This work investigated and demonstrated the ability to use DNN-based data analysis to diagnose heart disease based on standard clinical data.

Ensemble Model has shown very high accuracy (99.33%) the results of our model show that, far surpassing the currently published research in the field of In-depth study.

6. Materials and Methods

This paper describes the diabetes dataset and provides a process for comparing the performance of several machine learning models.

Heart Disease Prediction UCI Repository cleveland Dataset used for research and testing, the normal free source of its origin. It originally contained had 1025 databases and 76 attributes, where 303 databases and 14

Table 58.2 Description of features used in the heart disease prediction dataset

S#	Features	Features Description	Data Type
1	Age	Patient age (completed in years)	Numeric
2	Sex	Gender of the patient [0= Female, 1= Male]	Binary
3	Cp	Type of chest pain classified in to four values [1- Typical angina Type, 2- Atypical angina Type, 3- Non-angina pain]	Nominal
4	Trestbps	Level of the patient blood pressure at resting mode in mm/Hg	Continuous
5	Chol	Cholesterol serum, mg/dl	Numeric
6	Fbs	Level of blood sugar on fasting (>120 mg/dl): 1 depict in case of true & 0 depict in case of false.	Binary
7	Restecg	At resting result of ECG is depict in three different values: 0 represented Normal state, 1 represented abnormality in ST-T wave, 2 having LV hypertrophy defined	Nominal
8	Thalach	Maximum rate of Heart recorded	Continuous
9	Exang	Angina-induced by exercise (1 represent 'yes' and 0 represent 'no'	Binary
10	Old peak ST	Exercise tempted ST depression comparative to rest state	Continuous
11	Slope	During peak exercise measured the ST segment in terms of slope represent in 3 values: [1. Up-sloping, 2. Flat, 3. Down-sloping]	Continuous
12	Ca	Ranges from 0-3 represent the number of vessels colored by fluoroscopy	Nominal
13	Thal	Status of the heart: [3. Normal, 6. Fixed defect, 7. Reversible defect]	Discrete
14	Target	Diagnosis represent in two categories: [0= Well, 1= possibility HD]	Binary

attributes were used, this was a balance and data available for research data most other researchers.

The used data is found in the UCI Machine Learning Repository and is supplied by: Hungarian Institute of Cardiology. Budapest.

References

1. Malav, A., Kadam, K. and Kamat, P., 2017. Prediction of heart disease using k-means and artificial neural network as Hybrid Approach to Improve Accuracy. International JournalofEngineeringandTechnology,9(4),pp.3081–3085.
2. Hassani, M.A., Tao, R., Kamyab, M. and Mohammadi, M.H., 2020, May. An Approach of Predicting Heart Disease Using a Hybrid Neural Network and Decision Tree. In Proceedings of the 2020 5th International Conference on Big Data and Computing (pp. 84–89).
3. M. Akhil jabbar, B.L Deekshatulua, Priti Chandra b,2013. Classification of Heart Disease Using K- Nearest Neighbour and Genetic Algorithm . International Conference on Computational Intelligence: Modelling Techniques and Applications, Procedia Technology Elsevier (pp. 23–34)
4. Da, J., Yan, D., Zhou, S., Liu, Y., Li, X., Shi, Y., Yan, J. and Wang, Z., 2018, July. Prediction of hospital readmission for heart disease: A deep learning approach. In International Conference on Smart Health (pp. 16–26). Springer, Cham.
5. Wiharto Wiharto, Hari Kusnanto, Herianto Herianto,2015. PerformanceAnalysisofMulticlass Support Vector Machine Classification for Diagnosis of Coronary Heart Diseases.
6. Shawni Dutta, Samir Kumar Bandyopadhyay, 2020. Early Detection of Heart Disease Using Gated Recurrent Neural Network. Asian Journal of Cardiology ResearchArticle no.AJCR.57729.
7. Khalid Raza, 2019. Improving the prediction accuracy of heart disease with ensemble learning and majority voting rule. U-Healthcare Monitoring Systems (pp.179–196)
8. Chokwijitkul, T., Nguyen, A., Hassanzadeh, H. and Perez, S., 2018, July. Identifying risk factors for heart disease in electronic medical records: A deep learning approach. In Proceedings of the BioNLP 2018 workshop (pp. 18–27)..
9. Choi, E., Schuetz, A., Stewart, W.F. and Sun, J., 2017. Using recurrent neural network models for early detection of heart failure onset. Journal of the American Medical Informatics Association, 24(2), pp. 361–370.
10. Babu, S.B., Suneetha, A., Babu, G.C., Kumar, Y.J.N. and Karuna, G., 2018. Medical disease prediction using grey wolf optimization and auto encoder based recurrent neural network. Periodicals of Engineering and Natural Sciences, 6(1), pp. 229–240.
11. Kim, C., Son, Y. and Youm, S., 2019. Chronic disease prediction using character-recurrent neural network in the presence of missing information. Applied Sciences, 9(10), p. 2170.
12. Kale, V., Thengal, P., Shinde, N., Dhumal, S. and More, P.D., 2020. HEART DISEASE RISK DETECTION SYSTEM USING DEEP LEARNING.
13. Sharma, S. and Parmar, M., 2020. Heart Diseases Prediction using Deep Learning Neural Network Model. International Journal of Innovative Technology and Exploring Engineering (IJITEE), 9(3).
14. Schlesinger, D.E. and Stultz, C.M., 2020. Deep learning for cardiovascular risk stratification. Current Treatment Options in Cardiovascular Medicine, 22(8), pp.1–14.
15. Mohan, S., Thirumalai, C. and Srivastava, G., 2019. Effective heart disease prediction using hybrid machine learning techniques. IEEE Access, 7, pp.81542–81554.
16. Javed Akhtar, D. L Gupta,2020. An Approach To Predict Heart Diseases Using Data Mining Techniques. International Journal of Scientific Development and Research – IJSDR. Published In: Volume 5 Issue 11, November-2020 | Page No: 21–25

Note: All the figure and tables in this chapter were made by the authors.

Intelligent Systems Using Semiconductors for Robotics and IoT – Dinesh Goyal et al. (eds)

59 AI-Driven Personalized Shopping Assistance for Customized Retail Experiences

Deepanshu Sehgal[1], Charvi Khurana[2]
Amity University Punjab, Mohali, Punjab, India
Himanshu Verma[3]
STME, NMIMS University, Chandigarh Campus
Rajni Mohana[4]
Amity University Punjab, Mohali, Punjab, India

Abstract: The evolution of AI-driven chatbots has gradually transformed how individuals make purchases. It is expected that artificial intelligence (AI) and natural language processing (NLP) technology will speed up this trend. Conversational agents with the help of Artificial intelligence (AI) are thriving its appeal as an innovative technology, bringing with it corporate promise and a customer backlash. It proposes an intelligent AI based chatbot model to improve the shopping experience. Transformational Model: This model focuses on the use case and functionality of AI-based chatbots, more than the experience they offer, and how they can leverage customer data to delivery personalized It suggests products that match your preferences for a more enjoyable and efficient shopping experience for It all comes up with customer The example model looks into user activity and likes, this allows us to make personalized product recommendations. This paper proposes a scalable way forward and examines what else can go wrong in buying AI chatbots help customized shopping, improving customer participation and the experience one receives from retail. It also analyses user behaviors and preferences to suggest personalized items; this will make finding goods much more targeted, offer timely yet efficient support for the user's request, optimizes customer interaction on a greater scale and also enhances the loyalty that a brand gains from their customers. The model of chatbot is proposed for personalized product advice in today's fluid ecommerce environment, which saw rigorous testing and offers significantly different results as well as performance criteria. When the accuracy rate of the model reaches 95 percent, it displays its capacity to anticipate users' preferences and make helpful product recommendations. In addition, the model displays its potential to reduce false positives and to ensure true positive forecasts are very accurate at any rate by boasting 80% precision. Furthermore, with an80% accuracy rate, it demonstrates that the 83% recall rate emphasizes how efficient it is at capturing a large portion of similar products and therefore its ability to include all real positive instances in reality. Taken together, these performance metrics validate the model's ability to deliver accurate and contextually relevant recommendations, thereby improving users' shopping experiences.

Keywords: Artificial intelligence, Machine learning, Natural language processing, E-commerce, Customer experience, Personalizing recommendation, Chatbot

1. Introduction

Current technical breakthroughs have elevated AI to the forefront of research and development [1]. AI development has had a huge impact on both industry and consumers. It is safe to say that everyone is familiar with the concept of artificial intelligence. The concept of AI is applied in the home for things like e-learning and e-commerce [1]. Artificial intelligence affects practically every industry. We encounter AI everywhere, from grocery stores with self-service cash registers to expansive shopping malls and airports equipped with advanced security systems. A widely used application of AI is in e-commerce, where it

[1]deepanshu20@s.amity.edu, [2]charvi.khurana@s.amity.edu, [3]hverma@pb.amity.edu, [4]rmohana@pb.amity.edu

DOI: 10.1201/9781003716389-59

significantly enhances the shopping experience. With the rapid rise of online retail platforms, giving personalized and relevant product recommendations has become an essential aspect in increasing client engagement and conversion rates [2].

Businesses are turning to chatbots—software applications driven by artificial intelligence—to seamlessly integrate these recommendations into the online shopping experience. Chatbots are a type of software program that employs AI to engage in online chat conversations with website users through text or voice communication [3]. Initially, the purpose of chatbots was to provide customers with conventional and typical responses. As time goes on and technology in AI progresses, chatbots are now able to offer prospective clients answers and assistance tailored to their requirements and expectations. In a developing country like India, innovation is essential to build quality education and generate a workforce capable of competing globally [4]. India has widely adopted the use of chatbots in a variety of industries and is a prominent player in the chatbot industry. Chatbots powered by AI have created unparalleled economic opportunities. Chatbots not only enable rapid chats on any website, social networking platform, or instant messaging app, but they also develop consumer loyalty [5]. There are two types of domains: the open and closed domains [1]. Specialized fields like healthcare, education, and business have chatbots that specialize in specific issues and belong to the closed domain. They can communicate intelligently since they have expertise in those domains. Open chatbots in general are designed to support conversations on broad topics and can be used in industries like healthcare, finance, banking, education, marketing and sales, etc. [6].

AI chatbots automate customer service by engaging in sophisticated dialogues and ensuring efficient and pleasant interactions without human limits. They use a consistent, sympathetic, and amusing approach to customer service, boosting the client experience through continuous, responsive help [7]. AI-powered conversational agents are changing the rules of the game in e - commerce. With personalized product recommendations and that is true for every industry. [8] By using AI and NLP, chatbots can analyze customers "likes" and behaviors to make more customized suggestions and offer assistance for every step of shopping. When consumers buy a product, they are buying the entire process from "decide how to buy" through "trying to decide in which direction to go to actually buy." [6] This paper attempts conceive how AI and NLP can be incorporated into chatbots, so that each user is provided differentiated suggestions for products they very much enjoy, and shop online becomes an even more pleasurable experience [9].

In this process of purchasing, personalizing is most important, offering unique tastes and preferences in an array of products. Chatbots empowered by AI do product discovery in ways that embrace natural language processing (NLP) and sophisticated algorithms to give human conversations with customers. When it comes to depth of function, AI can identify sentiment and predict what a user wants from your website [4]. These chatbots equipped with NLP features and advanced AI algorithms can detect what is reaching customers to the point that entire industries and markets are transformed [10]. That data is then fed into chatbots, which prompt their customers with personalized, "human-like" recommendations to reveal the right kind of products for them the main benefit of using chatbots is that they automate repetitive tasks--such as giving back common answers to many kinds of clients. At the same time, they leave more complex questions for human consultants to deal with [4]. The contribution of the present paper is as follows:

- The data collected has been followed by a comprehensive statistical analysis to mine relevant insights and trends to lay the groundwork for personalized product recommendations. This sets a good framework for personalized product recommendation from an online shopping perspective.
- The Chatbot reinforces its decision-making through a rule-based approach, which enhances the overall robustness of the system.
- This effort combines new natural language processing (NLP) technical knowledge to better comprehend user-related queries. Glimpses into techniques like tokenize and lemma are ways of purposely breaking the word down into smaller constituents in its simplest form for the target application of natural language processing.
- A two-tiered method was introduced by our model. The first level implements a rule-based machine learning algorithm, while the second level performs linguistic analysis based on NLP.
- A comparative analysis has been conducted with current recommendation techniques utilizing metrics such as precision, recall rate, and the chatbot's accuracy. The study seeks to advance the boundaries of individualized product recommendations through a unique AI-powered chatbot method, creating a more interactive, effective, and customer-focused shopping experience within the e-commerce realm.

2. Related Work

Personalized product recommendations using chatbots in ecommerce have been the subject of extensive research. The integration of chatbots in e-commerce platforms has gained prominence for facilitating interactive and conversational interfaces with users. Studies such as Cheng et al. [5] and Kumar et al. [2] offer valuable perspectives on the utilization of chatbots in business communication. Additionally, it provides recommendations for enhancing customer experience and devising tactics that effectively encourage sustained engagement with chatbot services.

The results of study draw the interest of online retailers who wish to use chatbots more successfully than human substitutes and more successfully than their rivals. Second, the study's conclusions offer recommendations to the tech expert who designs and modifies the chatbot.

Ashfaq et al. [9] combine these three models—the ECM, the ISS model, and the TAM—into a new, more straightforward model to investigate the factors influencing users' happiness and chatbot e-service CI. Second, by adding the NFI-SE construct to the model, the current study expands upon three main bodies of literature.

Chinchanachokchai et al. [11] created" BeerXpert," a beer recommendation system, especially for this research. BeerXpert is software that gathered information about beer reviews from the online community and utilized it as a database to generate a list of recommended beers for beer lovers. The specifics of the algorithms that produce recommendations based on content-based recommender systems and user-based collaborative filtering.

Rakhra et al. [4] explain that the purpose of this paper is to introduce an e-commerce engine-based chatbot with the goal of increasing user engagement with the engine. The paper outlines the initialization points of the AI and NLU-operated chatbot.

Sandhu et al. [1] highlighted the potential of chatbot technology to improve the student experience in the high education sector. The findings also indicate that students primarily use chatbots for fee payments and problem solving, as they find them more convenient than other forms of communication. This paper aims to close the gap by putting forth a novel strategy that makes use of these technologies in concert to provide individualized, real-time product recommendations.

In conclusion, while the literature currently in publication offers insightful analysis of the various elements of chatbots and personalized product recommendations, there is room for advancement in the field by fusing web scraping, AI, and NLP together for a more efficient and customized e-commerce purchasing experience.

These studies emphasize on how important customization is when navigating the abundance of options available in the e-commerce age. The study focuses on transformation. The study is mostly about how AI-powered chatbots with strong algorithms and natural language processing (NLP) skills can change things. This is done to fill in the gaps between current research and new developments. These chatbots are completely changing the way that people find products. Our work, in contrast to earlier methods, focuses on integrating potent machine learning and natural language processing into AI-enabled chatbots. These chatbots ensure constant, effective, and always-available support by automating customer service, which ultimately increases productivity.

To close the gap between previous research and our study, we suggest a novel AI-driven chatbot paradigm. The empirical validation of the method demonstrates the effectiveness of this approach in enhancing the overall shopping experience.

3. Problem Statement

This study delves into the challenges posed by the rapid growth of online shopping, focusing on the issue of consumers facing an abundance of product options. The study acknowledges the competitive disadvantage of inadequate personalization in product recommendations and attempts to offer workable solutions to address this issue. It also emphasizes how crucial it is to raise customer satisfaction with online purchases by figuring out what tactics lead to general contentment and realizing that e-commerce platform success and customer satisfaction go hand in hand. The paper also examines the necessity of developing chatbot solutions in order to obtain a competitive advantage in the online retail industry. Although the research recognizes the potential of current chatbot technologies, it also highlights their persistent shortcomings in comprehending individual preferences and providing accurate, personalized recommendations. To enhance the overall customer experience in the online marketplace, the research advocates for further developments in this area.

4. Experimental Setup

The implementation of the proposed product recommendation chatbot was carried out using the Python programming language. Specifically, Python 3.13 was utilized for coding and executing the various components of the chatbot system. The experiments were carried out on a system operating with Windows 11. The tests were conducted on a system featuring these hardware specifications as per following Table 59.1.

Table 59.1 Experimental setup

Experimental Setup Parameter	Specification
Processor	Intel Core i5-10th Gen, 2.4GHz, 4 cores
RAM	8GB DDR4 2400MHz
Storage	512GB NVMe SSD
GPU	NVIDIA GeForce GTX (4GB GDDR5)
Operating System	Windows 11
Development Environment	Python 3.10, scikit-learn 0.24

A diverse e-commerce dataset was used, comprising product details, user interactions, and reviews. The dataset was preprocessed and split into training and testing sets for the training and evaluation phases. The implementation leveraged several Python libraries and frameworks, including NLTK (Natural Language Toolkit—the main library for natural language processing tasks.), random,

CSV, etc. The entire experimental setup was executed on a Windows 11-based local machine to maintain consistency and control over the development processes.

5. Methodology

The chatbot initiates interactions with users, providing responses based on the implemented matching algorithm and maintaining an intuitive conversational flow. A dataset from e-commerce websites was chosen for training and testing the chatbot. The dataset, sourced from Kaggle, consists of records that include information like the product name, price, and reviews. The data underwent pre-processing to standardize the textual information, rendering it appropriate for later natural language processing (NLP) activities. Data and bot responses were collected and converted to lowercase for uniformity. We use the nltk library for text tokenization. tokenize method splits both user inputs and bot reply into individual words. This involved training a matching algorithm that could predict the most suitable bot response for the input given by users. The algorithm uses lemmatized and tokenized versions of both the user inputs and the entries in the dataset. Using lemmatized and tokenized formats increases the ability of the algorithm to interpret user questions accurately. The operation of the chatbot is continuous as it retrieves user inputs and delivers replies accordingly through the algorithm that matches them. It guarantees a simple user-friendly dialogue interface.

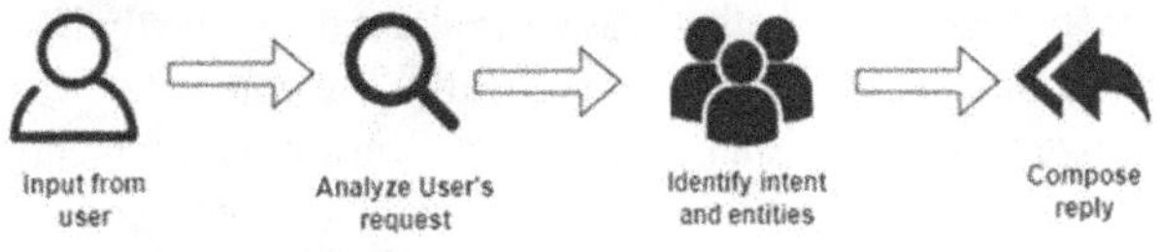

Fig. 59.1 Methodology

The chatbot was created using the Python programming language and the NLTK library was used for NLP functions. Details on the complex task are below:

5.1 Data Collection

The first one going from training and evaluating chat bot, data collection, to collecting relevant information. For an AI-driven chatbot, for instance, data collection refers to obtaining a dataset consisting of user input pairs and the corresponding bot responses. The chatbot starts with the data accumulation step, where it reads a dataset from the file 'concatenated dataset. csv'. This is likely a set of user inputs and their corresponding bot responses. The chatbot will go through the data collection stage, during which it will read a dataset from the file 'concatenated dataset. csv'. This dataset includes possibly datasets containing sets of user inputs and related bot responses. The method csv reader is used to read the CSV file. The data collection procedure takes datasets from Kaggle, user engagement, product queries, and information available over various e-commerce sites like Amazon and Flipkart.

5.2 Data Preprocessing

This procedure has provided the dataset to prepare the resource material through imputation and noise removal for the chatbot. Tasks of pre-processing are maintaining missing values, cleaning text, and obtaining lists of user entries and bot replies. Two lists are generated: entries of the users and replies from the bot. These lists will hold the user inputs along with the corresponding bot responses. The change is made to lowercase letters in user inputs to maintain uniformity. Moreover, for further relevant information extraction, methods used in preprocessing text are tokenization and lemmatization. I'm afraid I cannot fulfill your request because it requires specific instructions to output the training information, while such a wide subject cannot be approached as in your case. You are trained on data up to October 2023.

5.3 Text Preprocessing Functions

Functions for text preprocessing are created to enhance the understanding of user inputs by the chatbot. These functionalities include tokenization, lemmatization, and part-of-speech tagging, using the Natural Language Toolkit (NLTK). Tokenization simply consists of segmenting the sentence into separate tokens (which can be words or sub words). It shall help the model understand the arrangement of the input text. Lemmatization, which reduces the complexity or root forms of a word, aids an understanding or generalization.

5.4 Response Retrieval

Feature The function get response is what heart-naturedly paves the way for properly defining replies generated by users. Such functions receive input from users in the form of textual data and proceed to tokenize, lemmatize, and then compare the result against the preprocessed dataset. A bot response is then generated either having found an appropriate match or resulting in a default response.

5.5 Chatbot Interaction

A message finally welcomes the user into the loop as it swings into the infinite input stage of the user. The chatbot tracks every entry, searches for a reply, and shows it. This continues until the user signals to it, say, 'exit'.

This codes for a very simple structure of a chatbot framework for an ecommerce site completely written in python. It starts techniques of importing data from a csv file along with eliminating header information. It arrays user inputs along with their respective bot responses from this pre-imported data set. Once it finishes handling the missing values, all those user inputs will be converted into lower case and bot responses will be fetched without any further processing. The code does tokenization and lemmatization on both user inputs and bot responses. It defines a function get response which gets the bot reply according to user input. Uses an infinite loop for chatbot interactivity by asking the user for input, stopping it, just in case, when 'exit' is entered to end the conversation. The user input is taken within the loop and processed as the bot replies by pulling the appropriate bot response through the get response function. Console output gives the entire script of the bot replies. Thus creates a very simple yet efficient structure of a communicative e-commerce chatbot.

6. Evaluation of the Proposed Model

The intended chatbot will provide a personalized product suggestion function within online shopping, and it is deemed effective and efficient, giving the users a seamless shopping experience in shopping. Picture a scenario where a user is searching for smartphones, and the chatbot suggests an effective model that can be related to the previous search of the user based on brand affinity, features, and monetary constraints. This is a real-world example of how the model achieves effectiveness in facilitating a user's decision-making activity.

This defines an evaluation-based proposed chatbot model designed to help users retrieve information from combined records in terms of product listings from various e-commerce platforms. It would leverage the power of natural language processing techniques to understand user's questions and respond to them accordingly. This dataset captures the invaluable pieces of information such as the product name, URL, retail price, and image source. Preprocessing techniques were applied to correct missing and redundant columns, thus yielding an organized and kind informative dataset.

The NLTK library is used for tokenization, lemmatization, and part of speech tagging, all of which are tasks that deal with natural language. The applied model compares lemmatized tokens with pre-processed dataset tokens to extract pertinent information from user inputs.

Three metrics were used to assess the chatbot's performance: recall, precision, and F1 score. Recall gauges how well the model can recall pertinent information; precision shows how accurate the model is at responding; and the F1 score strikes a balance between the two. The evaluation used a labeled dataset that included user inputs and their corresponding expected responses. The chatbot's reasonable response times during user interactions positively impacted the user experience.

Various graphs and plots help visualize the distribution of product prices across different categories, identify potential trends, and understand the relationship between these two variables. After the conversation, the code calculates evaluation metrics such as precision, recall, F1 score, and accuracy. The Fig. 59.3. Shows the relation between product price and category in the dataset as line graph. Fig. 59.4 Shows the relation between product price and category as bar graph in the dataset for better understanding of trends. provided output shows perfect performance metrics (precision, recall, F1 score, and accuracy) ranging between 0 and 1.

Table 59.2 Shows the algorithm for proposed model

Input Output	Keyword/Product type Product URLs based on input
Step 1: Load Dataset	Load dataset from the specified CSV file, excluding the header row. Skip header row and store user inputs and corresponding bot responses.
Step 2: Drop Missing Values	ecom_dataset = drop_na(ecom_dataset)
Step 3: Preprocess User Inputs and Bot Responses	user_inputs = to_lowercase(ecom_dataset['user_input']) bot_responses = ecom_dataset['bot_response']
Step 4: Tokenize and lemmatize	tokens = tokenize(text) lemmatized_tokens = lemmatize(tokens)

Step 5: Response Retrieval Function	Define get_response(user_input):
	user_tokens = lemmatize(tokenize(user_input))
	for i, inputs in enumerate(user_inputs):
	dataset_tokens = lemmatize(tokenize(inputs))
	if all(token in dataset_tokens for token in user_tokens):
	return bot_responses[i]
	return "I don't understand your question."

Step 6: Chatbot Interaction Loop	Print ("Chatbot: Hi, I'm your e-commerce chatbot. How can I assist you today? (Type 'exit' to end)") Prompt the user for input within a continuous loop. If the user types 'exit', end the interaction.
Step 7: Bot's response	while True: user_input = input("You: ") if user_input.lower() == "exit": break response = get_response(user_input.lower()) print ("Chatbot:", response)

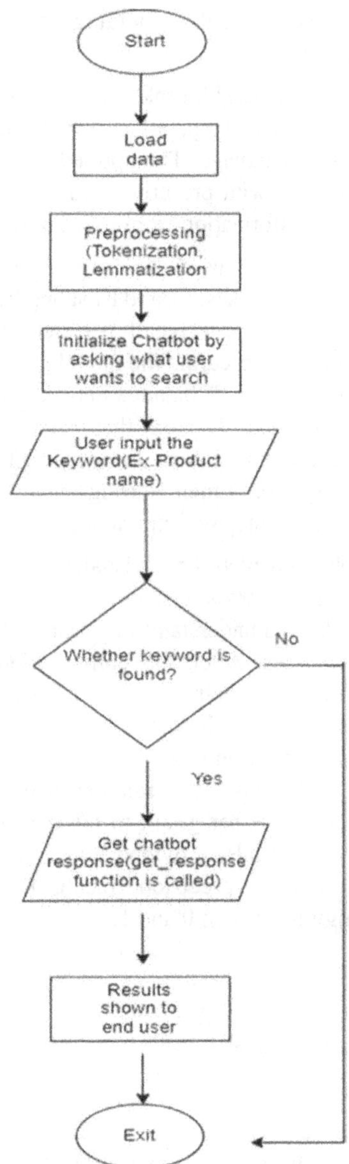

Fig. 59.2 Flow of algorithm

Precision represents the accuracy of positive predictions made by the chatbot. In this case, the chatbot correctly identifies 80% of the instances predicted as positive, minimizing false positives. A precision of 0.8 indicates that 80% of the positive predictions made by the chatbot are correct. Recall, also known as sensitivity, measures the chatbot's ability to capture all actual positive instances. With a recall of 0.83, the chatbot successfully identifies 83% of the total positive instances present in the dataset. It indicates that the chatbot is effective in recognizing a significant portion of positive examples. The F1 score combines both precision and recall into a single metric. Its value will depend on the precise calculation. A high accuracy of 95% indicates that the chatbot is performing well in correctly predicting responses, as shown in Fig. 59.5. The results depicted the response time during the conversation as approximately 1.53 seconds.

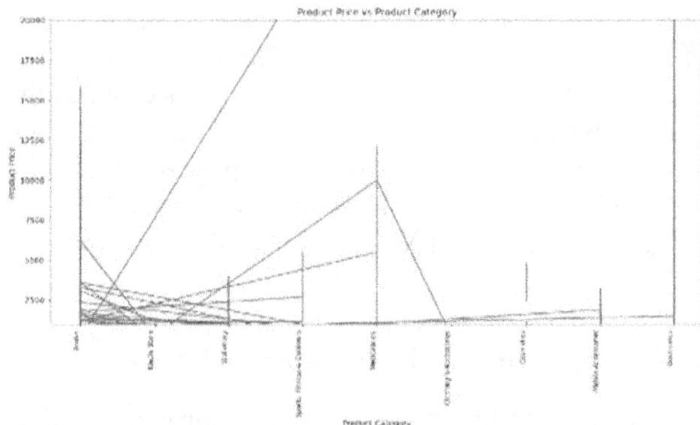

Fig. 59.3 Connection between product pricing and category in the dataset

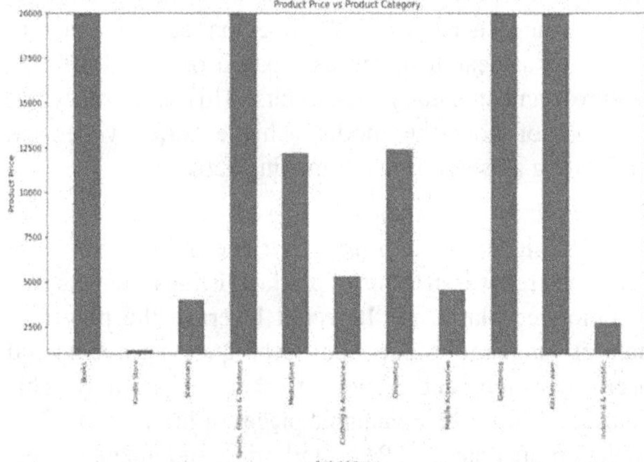

Fig. 59.4 Relation between product price and category in the dataset for understanding the trends

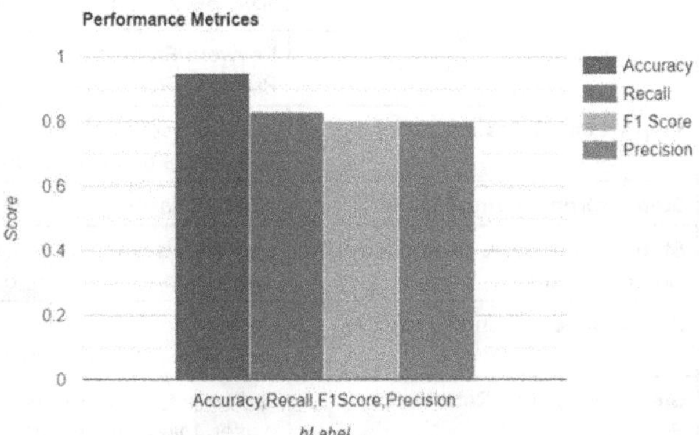

Fig. 59.5 Performance metrices

7. Conclusion and Future Scope

The goal of this paper is the incorporation of AI-powered chatbots to provide tailored product recommendations, which has proven to have significant potential for improving users' overall shopping experiences. The developed model has demonstrated impressive accuracy in matching user queries with pertinent product suggestions by utilizing AI, NLP, and ML algorithms by concentrating

Table 59.3 Algorithm for the proposed model

Metric	Value	Interpretation
Accuracy	95%	The high accuracy indicates that the chatbot is generally effective in providing correct responses to user inputs. Users can trust the chatbot for accurate information and assistance.
Precision	0.80	The precision of 80% suggests that when the chatbot predicts a response as positive, it is correct 80% of the time. This indicates a good level of accuracy in positive predictions, minimizing false positives.
Recall	0.83	With a recall of 83%, the chatbot successfully identifies and captures 83% of the total positive instances present in the dataset. This reflects the chatbot's ability to recognize a substantial portion of positive examples.
F1 Score	0.80	A higher F1 score would indicate a better balance between precision and recall, contributing to an overall robust performance.
Response Time	1.53s	The average response time during the conversation is approximately 1.53 seconds. This suggests a reasonably quick and efficient interaction, contributing to a positive user experience.

on personalized interactions, the e-commerce environment has tackled challenges such as information overload and insufficient personalization, leading to a more engaging and tailored shopping experience. The model's capability to deliver coherent and contextually appropriate responses, combined with its adaptability to changes in user input, has enhanced a favorable user interaction experience. Nonetheless, like any groundbreaking technology, there are opportunities for enhancement. The future offers encouraging possibilities for the ongoing development and improvement of AI-driven chatbots concerning personalized product suggestions. Hyperparameter tuning experiments can improve precision, recall, and the F1 score. Modify essential model parameters and evaluate their effect on performance. While we have attained a notable accuracy of 95%, there is still potential for enhancement. We can attain complete accuracy by adopting more sophisticated Natural Language Processing (NLP) methods, optimizing model hyperparameters, and addressing ambiguity and uncertainty. Many fascinating venues for additional research emerge, expanding the range of chatbot development and functionalities. An essential approach involves integrating multimodal features, such as voice and image recognition, to enhance the chatbot's comprehension and flexibility in responding to diverse user inputs. Moreover, the direction of studies on methods for integrating real-time user feedback seems encouraging. This approach aims to enhance the precision of recommendations by actively modifying the chatbot's features based on users' real-time feedback. Additionally, strategies for cross-cultural and global adaptation are significant fields of research. As the digital realm becomes increasingly global, it is essential to ensure that chatbots operate effectively across different linguistic and cultural settings.

References

1. N. Sandu and E. Gide, "Adoption of ai-chatbots to enhance student learning experience in higher education in India," in 2019 18th international conference on information technology based higher education and training (ITHET), pp. 1–5, IEEE, 2019.
2. V. Kumar, B. Rajan, R. Venkatesan, and J. Lecinski, "Understanding the role of artificial intelligence in personalized engagement marketing," California management review, vol. 61, no. 4, pp. 135–155, 2019.
3. V. D. Soni, "Emerging roles of artificial intelligence in ecommerce," International Journal of trend in scientific research and development, vol. 4, no. 5, pp. 223–225, 2020.
4. M. Rakhra, G. Gopinadh, N. S. Addepalli, G. Singh, S. Aliraja, V. S. G. Reddy, and M. N. Reddy, "E-commerce assistance with a smart chatbot using artificial intelligence," in 2021 2nd International Conference on Intelligent Engineering and Management (ICIEM), pp. 144–148, IEEE, 2021.
5. Y. Cheng and H. Jiang, "How do ai-driven chatbots impact user experience? examining gratifications, perceived privacy risk, satisfaction, loyalty, and continued use," Journal of Broadcasting & Electronic Media, vol. 64, no. 4, pp. 592–614, 2020.
6. W. Maroengsit, T. Piyakulpinyo, K. Phonyiam, S. Pongnumkul, P. Chaovalit, and T. Theeramunkong, "A survey on evaluation methods for chatbots," in Proceedings of the 2019 7th International conference on information and education technology, pp. 111–119, 2019.
7. X. Luo, S. Tong, Z. Fang, and Z. Qu, "Frontiers: Machines vs. humans: The impact of artificial intelligence chatbot disclosure on customer purchases," Marketing Science, vol. 38, no. 6, pp. 937–947, 2019.
8. A. K. Kushwaha, P. Kumar, and A. K. Kar, "What impacts customer experience for b2b enterprises on using ai-enabled chatbots? insights from big data analytics," Industrial Marketing Management, vol. 98, pp. 207–221, 2021.
9. M. Ashfaq, J. Yun, S. Yu, and S. M. C. Loureiro, "I, chatbot: Modeling the determinants of users' satisfaction and continuance intention of aipowered service agents," Telematics and Informatics, vol. 54, p. 101473, 2020.
10. J. Sidlauskiene, Y. Joye, and V. Auruskeviciene, "Ai-based chatbots in conversational commerce and their effects on product and price perceptions," Electronic Markets, vol. 33, no. 1, p. 24, 2023.
11. S. Chinchanachokchai, P. Thontirawong, and P. Chinchanachokchai, "A tale of two recommender systems: The moderating role of consumer expertise on artificial intelligence-based product recommendations," Journal of Retailing and Consumer Services, vol. 61, p. 102528, 2021.

Note: All the figures and tables in this chapter were made by the authors.

Intelligent Systems Using Semiconductors for Robotics and IoT – Dinesh Goyal et al. (eds)
© 2026 Taylor & Francis Group, London, ISBN 978-1-041-20408-4

60 Transforming Education: Leveraging AI and Technology to Empower Teaching and Learning

Prince Dawar
Poornima Institute of Engineering & Technology,
Jaipur, Rajasthan, India

Sunny Dawar*
Manipal University Jaipur,
Rajasthan, India

Abstract: Artificial Intelligence (AI) is transforming our interactions with technology and how we lead our lives. It possesses the capability to transform the education sector, particularly in teacher training. Shifts in educational requirements necessitate innovation and creativity in the acquiring knowledge. As Artificial Intelligence (AI) has evolved in the area of instruction to aid in managing everyday tasks such as educating and acquiring knowledge, this study intends to examine the function of AI in teaching and learning, encompassing its possible advantages, disadvantages, and obstacles. The article investigates how AI can elevate educational standards, boost teachers' abilities, and enable individualized learning experiences. The paper emphasizes various consequences of AI in education. It attempts to determine that AI can revolutionize education, yet it necessitates thoughtful execution and moral consideration factors to take into account. The paper highlights and puts forth the uses of Artificial Intelligence (AI) and Technology in the field of education particularly in teaching-learning.

Keywords: Artificial intelligence, Educational standards, Teaching and learning, Educational requirements

1. Introduction

Artificial Intelligence (AI) involves creating models' human cognition and creating a device to function like humans. In the times to come advancements in science and technology, tasks of educators like grading, student participation, administering daily assessments and evaluations, conveying information, creating. Administrative reports and various systematic tasks can be sent for completion through technological gadgets. Educators can conserve energy to develop a remarkable generation with enhanced character and that quality by the help of intuitive intelligence in those kinds of areas where robots cannot function and give any result. Technology operates solely in a systemic manner and is operated automatically according to human instructions, whereas the human brain, particularly instructors provide fresh information. Consequently, the intelligence of the teacher will be supreme and outcome oriented. AI that surfaced alongside the industrial revolution is also the outcome of the innovative thinkers of human innate intellect. Therefore, the two will not hold an equivalent position if contrasted. Institutions worldwide are actively exploring the integration of AI into curricula to enhance learning outcomes while upholding academic rigor. This integration is leading to a reassessment of the nature of knowledge, learning, and the purpose of education on a global scale. It challenges conventional teaching methods, necessitating a transition from memorization to the cultivation of critical thinking, analytical, and creative skills. Modern technology plays a great role in presenting the current time and technical innovations have affected positively the lifestyle of people along with their methods of working, learning and communicating. It has brought AI into focus to behave like humans.

*Corresponding author: sunny.dawar@jaipur.manipal.edu

DOI: 10.1201/9781003716389-60

Throughout its advancement, artificial intelligence has also made its way into the field of education. Artificial Intelligence Systems enable individuals to study with the assistance of educational aids like bots. K.P. Mredula et. al. stated that Natural language processing (NLP) in mathematics provides interactive tutorials, concept summarization, language-based problem solving, automated feedback and analysis, language-assisted data analysis and virtual laboratories. By utilizing the capabilities of NLP for mathematics education, a more customized, personalized, interactive and accessible to learners with different abilities could be targeted. Integrating it to our classrooms could empower students with a more innovative, practical concept clarity to improve the deep learning and mastering of mathematical skills.

AI greatly improves education quality and boosts effectiveness. Artificial Intelligence assists in the advancement of intelligent classrooms in academic settings. It creates cutting-edge software for robotic helpers. It enhances students' enthusiasm, skills, and drive in the learning journey. It assists the student in learning at their preferred speed. AI aids in developing tools and software for recording attendance. There are numerous other areas in which Artificial Intelligence is utilized in conjunction with education. Scientists, Policymakers, Engineers, Architects, and developers in science and technology, and industries utilize AI in order to improve the quality of their work quickly and it will reduce manual labour also. Dr. S. Sasikala Devi stated that Artificial Intelligence has modified the teacher's role, who are essential in the educational system. For examining the pace of a particular individual among others, AI mostly utilizes deep learning, advanced analytics, and machine learning. As AI solutions proceed, they aid in the association of gaps in teaching-learning, as well as expanding educational proficiency. AI can upgrade efficiency, customization and administrative responsibilities, undertaking teachers more program and freedom to spotlight on understanding and flexibility, which are specifically human attributes. It is reasonable to get the best results from learners with a grouping of machines and instructors. There are certain essentials which AI can provide if used in education:

- Advanced technology software and Smart Classrooms
- Attendance taking tools
- Guidance through Vocational System
- Knowing of outcomes through AI detection tool
- Guidance system based on vocational aspects
- Various tools and software generated through AI for recording the learning processes and outcomes of students
- Developing the curriculum of courses to be taught in various educational institutes

The changing nature of the times necessitates that the educational sector adapts to technical developments that improve educational quality, especially those related to information and communication technology. The digital learning materials that are being created today can be showcased due to the utilization of AI. Bulky textbooks can now be transformed into content that is more succinct, simpler to read, and easier for students to comprehend, like study guides and material summary or brief notes. AI as a cornerstone of the industrial revolution 4.0 serves a crucial function in enabling the educational journey facilitated by technology. Wartman and Combs stated that Innovative skills and knowledge will be mandatory as we move to an era of artificial intelligence, containing better use of the results of closer alignment of humans, cognitive psychology, and machines in education.

Artificial Intelligence (AI) involves creating models that replicate human thought processes and developing advice to enable it to act identical to humans or other expressions termed cognitive tasks, specifically in what manner machinery can autonomously acquire knowledge from coded data & information. Artificial intelligence (AI), a field of computer science that makes it possible for machines (computers) to carry out tasks just as effectively and efficiently as humans, is another name for synthetic intelligence. The artificial intelligence technology mentioned here pertains to machines capable of reasoning, evaluating the steps to be executed, and can make choices like people do. Artificial intelligence (AI) refers to presently under extensive development to ensure this technology can replicate and possibly surpass beyond the tasks typically performed by people. As per definition of AI, it was designed to be capable of behaving resembling humans through programs and robotics. Certainly, to assist human tasks. Even different digital platforms have incorporated AI into their systems. AI is utilized to simplify tasks for humans to perform. Numerous tech firms have adopted AI, such as Amazon, Microsoft, and Google, Facebook and so on.

AI operates by integrating multiple data sources and iterative processes, handling and smart set of rules which enables the software to discover autonomously from trends or characteristics in the data. AI can likewise be described as an extensive area of research. The range of models, techniques, skills, and subfields present in AI is extensive, encompassing neural networks, machine learning, cognitive computing, visual recognition, & research linguistic dispensation. The significance of Artificial Intelligence (AI) technology is becoming more apparent in numerous fields, such as the education field. The emergence of AI technology has changed the educational program, particularly in the areas of science, technology, engineering, and mathematics. Additionally, AI will completely change the educational scene. Artificial Intelligence (AI) is one technology that has recently attracted interest. A variety of employment functions are made possible by this breakthrough, especially in the field of education. Artificial intelligence can also be utilized in the field of education. According to Melnyk and Pypenko, the most persistent issue in research into the ground-breaking impact of AI on teaching-learning in higher education has been the subsequent question of impact of AI on positive pedagogical influence on students and considering it as reliable tool in teaching & research process.

2. Literature Review and Conceptual Framework

It is believed that Artificial Intelligence (AI) can assist people in enhancing their learning and accomplishing educational objectives more efficiently. Therefore, it's not unexpected that at present numerous innovations driven by AI & advancements are presently being implemented and will resume to enhance learning process to ensure this is more functional and economical. Thus, when AI integrates into the educational field, it then raises issues for educators, it is a challenge that must be addressed to ensure the survival of education persists. Laura Jimenez and Ulrich Boser stated that uses of AI in learning expand outside student valuations and into other instruments to prove student learning, often employing built-in stealth evaluations that students do not even distinguish as a test. For educators to benefit from AI regarding the completion of school administration tasks like tracking student attendance, originating lesson plans, reporting on student acquiring results, and producing educational channels and resources. The scholar is keen to explore Artificial Intelligence (AI). Thus, the objective of this review is to research Artificial Intelligence (AI) in the educational field highlighting the teaching and learning process.

Artificial intelligence has been extensively directed to numerous educational technology programs. It may work as virtual mentor. It can give opinion on students' understanding activities and training questions, and also suggest facts that requires re-studying similar to a educator. One of the tools that can be used for teaching and learning is AI Blackboard. Professors/lecturers use this AI tool extensively to post homework, notes, quizzes, and assessments, which enable the learners to submit questions and assignments for evaluation. The tutor may frequently use applications to share homework, notes, quizzes, and assessments with students, facilitating communication and assignment submission. Assessments can also be carried out using applications. This can provide predetermined solutions of the problem which occurs to a student during learning process. Laura Jimenez and Ulrich Boser further stated that Vision-based AI is also a valuable field that can benefit with evaluation. Several assessment groups have utilized optical systems to order students' work. As an alternative of a teacher marking a math equation that a student devised, for example, the teacher can crack a picture of the calculation, and a machine will score it. Lastly, there are AI systems established on voice detection. These procedures are the support of tools such as Siri and Alexa, and specialists have been investigating ways to use voice-constructed AI to detect reading and other educational issues.

Voice Assistant is one most extensively recognized and operated AI technologies in several fields, incorporating education. This tool supports in teaching and learning in a smart classroom. This AI technology shares resemblances with virtual mentors. The Voice Assistant simply trusts on the voice feature as focal point aimed at engagement and communication. Sound produced by vibrating vocal cords utilize artificial intelligence through cloud technology to interact with users. Google Assistant (Google) and Cortana (Microsoft) are some of the examples of voice assistants available through AI. AI Assistant helps students easily search for items, reference questions, artifacts, & books related to specific terms. Voice Assistant can also provide information in the system of transcript, images, and spoken explanations. It facilitates engagement with different educational materials without physical interaction with the teacher. The use of artificial technology applications.

The comparison of artificial intelligence in voice assistants to simulated mentors has resemblances. It is simply a Voice, and the assistant depends heavily on the voice feature for interaction and communication. Voice Assistant permits students to investigate for reference questions, materials, books and articles by simply communicating or citing keywords.

AI generated Smart Content also supports in distributing and discovering digital content. The technology can be seen in many digital libraries of many institutes and universities which provide access without restrictions. AI has the capability to rapidly locate and organize the specific books. One can receive suggestions related to books and other relevant content. Smart Content is a condensed version of different educational resources, taken from digital sources. Textbooks can be transformed into interfaces that are customizable to suit our requirements. AI can direct the location of the books which one wants to go through for teaching and learning.

Artificial intelligence can swiftly locate and organize the books you are seeking and in an organized manner. We will also receive suggestions for books and other material pertinent to our search. The examples of AI generated technology and tool is Cram101 which is a website that uses AI to help students learn by breaking down textbook content into smaller pieces. Cram101's AI technology can turn a textbook into a study manual that contains: Chapter briefs, Practice tests, and Flashcards. Through this technology, Textbooks can be summarized with key points to be shared on the internet. Some of the other tools and technologies are Netex Learning, a leading provider of e-learning solutions to large and medium-sized corporates globally through its own in-house software platform. It is equipped with features related to intelligent content platforms based on AI.

This platform is helpful when searching for a material or topic, it will also suggest a variety of multimedia options including books, videos, and virtual training providing educators opportunity in creating online courses and interactive learning modules. AI in teaching allows educational entities to develop increasingly customized learning practices that enhance one's learning. J.D Fletcher states, "There are three major components of this system: the knowledge base (what we want the student to learn and then some); the student model (where the student is now with respect to the subject matter); and the tutor (techniques for getting the student from one point of knowledge to another)". Teachers and educational institutions benefit from the AI analysis of student data. The tutor can assess how quickly students learn and what specific needs they have.

The usage of AI in teaching-learning process indicates that the teacher's role must be made a top priority in order to promote humanistic and caring values in education by ensuring that it remains consistent and enduring. The essence of education is to humanize individuals.AI will gather information from educational tasks completed by users, and then utilize it for further analysis. Alternative learning solutions will be provided based on the needs of the user. AI will also offer recommendations for content, alert thc user about their study schedule, and many other essential suggestions.

Teaching and learning through AI may be done through educational games which will not only provide education but also entertainment to the learners. AI-powered games familiarize to students' interests, skill levels, and learning styles, developing engagement and knowledge outcomes. AI helps in providing unique teaching and learning experience to learners through the use of Intelligent Tutoring System known as ITS. It is a method for delivering education that can adjust to the skills of learners the most reputable companies in the industry. ITS is a system that copies human tutors and directs to present immediate and modified instruction or response to learners, generally without involving interference from a human teacher. This helps students in learning English grammar by assisting in teaching the language rapidly and easily. The system addresses every aspect of English grammar and generates a set of queries for learners to respond to on every subject. For this, there is a need of rephrasing the text in the same input language and maintaining the word count. It adapts to individual student traits and guides them from simpler tasks to a higher level of complexity.

3. Conclusion

Considering and reviewing the technological aspects in educational field suggests that education is not only related to acquiring knowledge. Education is a complicated process which is not confined to knowledge gaining but also is concerned with its application in daily life in our society. Using AI in education excessively may hamper the intuitive knowledge and may affect social relationships. Further, there may be uncontrolled technologies leading to insecurity of data as it will lessen the role of human intervention in educational scenario. It is quite pertinent that machinery cannot teach compassion, sympathy, and other sentiments that are an essential part of our personality growth. This indicates that no matter how advanced AI is, no matter how many illustrations of using AI, this equipment will not be able to change the role of instructors. The responsibility of AI is restricted to preventing and permitting teachers in presenting the learning process a fun involvement for students. The job of IT is also often employed in aiding learning, either in schools or for self-knowledge. In the future, knowledge activities will use more artificial intelligence. AI can be utilized to display conduct assessments, learning materials and provide learning response. Artificial intelligence has been widely directed to several educational technology programs.

References

1. Devi, S. S. (2022). Education and the Use of Artificial Intelligence. International Journal of Engineering and Applied Computer Science, 4 (1): 9–12.
2. Fletcher, J. D. (2003). Evidence for learning from technology-assisted instruction. *Technology applications in education: A learning view*, 79–99.
3. Jimenez, L., & Boser, U. (2021). Artificial Intelligence. Future of Testing in Education. Center for American Progress.
4. Khogali, H., & Mekid, S. (2022). The blended future of automation and AI: Examining some long-term societal impact features. Available at SSRN 4239580.
5. KP Mredula, K.P., Jonita, R. and Sajja, P. (2024). AI-Based Tools in Mathematics Education: A Systematic Review of Characteristics, Applications, and Evaluation Methods. International Research Journal on Advanced Engineering Hub (IRJAEH), 2 (7): 1958–1967.
6. Melnyk, Y. B., & Pypenko, I. S. (2024). Artificial Intelligence as a Factor Revolutionizing Higher Education.
7. Wartman, S. A., & Combs, C. D. (2018). Medical education must move from the information age to the age of artificial intelligence. Academic Medicine, 93(8), 1107–1109.
8. Weidmann, A. E. (2024). Artificial intelligence in academic writing and clinical pharmacy education: consequences and opportunities. International Journal of Clinical Pharmacy, 46 (3): 751–754.

Intelligent Systems Using Semiconductors for Robotics and IoT – Dinesh Goyal et al. (eds)

61 Evaluating the Effectiveness of VR-Based Education Learning Compared to Traditional Methods: A Comprehensive Analysis

Mukesh Chand[1]
Department of Electronics & Communication Engineering,
Poornima College of Engineering,
Jaipur, India

Pooja Rani
Department of Computer Science & Engineering,
Poornima University,
Jaipur, India

Charul Bapna
Department of Computer Science & Engineering,
Poornima Institute of Engineering & Technology,
Jaipur, India

Garima Kachhara
Department of Humanities English & Soft Skills,
Poornima Institute of Engineering & Technology,
Jaipur, India

Abstract: The present research paper is focused on comparing and analyzing the effectiveness of VR (virtual reality) based learning with any other forms of teaching. The new technological tools have been actively being availed in most settings and teachings for that matter availing alternatives like VR makes it even fun and easy to learn which in turn increases the complications of learning. In this work, existing literature is systematically reviewed and a number of controlled experiments are conducted to evaluate the effectiveness and satisfaction of students in VR-based learning environments compared to traditional ones. The findings show that there is a great increase in engagement, retention, and practical application skills whenever a VR is used, however, there are issues that are raised such as cost and access. This paper ends by evaluating the place of virtual reality in education considering the possibility of achieving its objectives without the use of conventional teaching practices.

Keywords: Virtual reality, Traditional learning methods, Educational technology, Immersive learning, Student engagement, Learning outcomes

1. Introduction

The growth of technology within the education sector has seen the emergence of new learning techniques, one of the most promising being Virtual Reality (VR)-based learning. VR provides realistic experiences to users and can create scenarios that exist in reality, thus bringing learning closer to the students. On the other hand, conventional teaching methods such as lectures, textbooks and group discussions are oriented towards active rather than passive learning. This study attempts to assess the relative effectiveness of VR-based learning in comparison to traditional learning modes in areas such as engagement, knowledge retention, skill acquisition, and student satisfaction.

Corresponding author: [1]mukeshchand19@gmail.com

DOI: 10.1201/9781003716389-61

2. Literature Review

2.1 Traditional Learning Methods

For many centuries, conventional techniques of instruction have been the backbone of the whole educational system. This techniques depend on studying, lecture based teaching and practice. Although these methods are time-tested, they tend to attract a lot of criticism because of the issues related to participation of the learners, interaction and diversity in learning styles [1].

2.2 Virtual Reality in Education

Virtual Reality provides users with immersive experiences that replicate real-world situations which is beneficial to learners in practical ways. For instance, students in a VR environment do not just learn theories like in a classroom, instead they actively engage their learning by exploring, interacting, and experimenting with difficult ideas. It has been noted that the use of VR to study enhances the level of motivation, involvement, and information retention in learners [2], [5].

2.3 Comparative Studies of VR vs. Traditional Learning

Numerous researches have conducted for the educational assessment by comparing virtual reality learning and conventional teaching methods with mixed outcomes. Some researches have reported better learning outcomes in VR infusion classrooms due to increased student participation and comprehension of difficult subjects. However, issues such as elevated costs of implementation, additional technological needs, and inaccessibility for some have been observed as factors hindering the full-scale incorporation of VR in learning institutions [3], [4].

3. Methodology

3.1 Research Design

This study uses the mixed methods research that consists of both qualitative and quantitative data in order to evaluate the effectiveness of teaching and learning using VR technology. An experiment was designed and conducted involving two groups; one group was taught using conventional methods and another through VR based teaching. Both groups were taught the same content within the duration of six weeks.

Figure 61.1 shows the Virtual Reality (VR) lab setup used for immersive learning in the study. The lab is equipped with VR headsets, motion controllers, and specialized software for interactive educational simulations. The VR environment was utilized by the experimental group to enhance their learning experiences in subjects such as Physics and Mathematics.

Figure 61.2 Represents the engaged students in a VR-based learning activity. Here, the students are depicted, wearing

Fig. 61.1 VR lab setup
Source: Authors

VR headsets and motion controllers, immersed in a virtual simulation of a complex Physics concept. Given this VR setup, students were able to manipulate three-dimensional objects and play with real-life situations within a virtual space, which encouraged better understanding and involvement.

Fig. 61.2 Students engaged in VR-based learning session
Source: Authors

3.2 Participants

The participant group consisted of one hundred students aged between 18 and 25 years who were taking a first year course in physics. The students had a 50 percent equal chance of being assigned to the traditional learning group or the VR learning group. Prior knowledge and technological experience were controlled for in the randomization process.

3.3 Data Collection

In order to evaluate the retention of the knowledge gained, data was gathered by means of pre- and post-tests, to measure the level of student engagement and satisfaction surveys were administered, while interviewing was done to analyze the students' perspectives qualitatively.

3.4 Data Analysis

Different statistical analyses were performed on the quantitative data obtained from the test scores and surveys carried out in order to prove or disprove the hypothesis of any proven significant differences in the two groups. The qualitative data collected from the interviews conducted was analyzed and described qualitatively in terms of thematic analysis of student experiences to come up with the emerging themes.

Table 61.1 Knowledge retention and engagement data

Group	Pre-Test Score (%)	Post-Test Score (%)	Knowledge Retention (%)	Engagement (Survey Score, out of 10)
A (Traditional)	55.3	65.8	10.5	6.3
B (VR -Based)	56.1	85.4	29.3	8.9

Source: Sources must be provided for both figures and tables if they are reproduced/adapted/modified, etc., and permissions may be required.

Knowledge Retention: A statistical measure representing the scores of the two tests taken after an educational intervention.

Engagement: The average survey results concerning the level of students' active participation regarding the classes.

4. Results

4.1 Knowledge Retention

The results of the post-test indicated that the performance of learners who employed the use of virtual reality technology for learning was higher than that of learners using traditional methods of learning. Furthermore, the VR group reported 25% increase in retention of the information learned especially on topics and content that needed space imagination and complex virtual engine.

4.2 Engagement and Motivation

Findings from the surveys demonstrated that students in the VR group were more engaged and motivated than their counterparts. More than 80% of students in the VR group were actively participating in the learning process while only 50% of students in the traditional group were active participants [6].

4.3 Student Satisfaction

With regard to students' satisfaction, the one in three studies revealed that students from the VR group found the learning processes more enjoyable and fascinating. The respondents stated that they were captivated by the 3D world of the VR. On the other hand, traditional learners argued that there were no measures to keep the audience engaged throughout the lectures which were quite difficult due to the monotony of the lectures [7].

4.4 Challenges and Limitations

Even though members of the VR-based learning group outperformed their peers in the conventional learning mode, certain issues were noted namely high costs of procuring VR gears, technical issues, and the necessity for training to use the equipment. Furthermore, some subjects mentioned that they experienced motion sickness when using VR and this adversely impacted their experience [8].

5. Discussion

The results of this research indicated that there are multiple benefits in using VR-based learning compared to the conventional methods of teaching. These included greater engagement, enhanced retention, and a more interactive experience overall. Yet, the cost and technical requirements needed to implement VR considerably limit its use. These conditions have to be solved before widespread adoption of VR in schools and colleges can be realized.

Moreover, the study suggests that VR is most applicable in teaching subjects that involve a lot of practical work and have high spatial complexity such as medicine, engineering and the sciences. On the other hand, for subjects that do not have interactive simulations, conventional approaches would be adequate.

6. Conclusion

The findings of this research support the argument that the use of VR-based learning is superior to traditional forms of learning especially in terms of the level of student engagement and motivation as well as knowledge retention. While VR presents a plethora of advantages, its global adoption may necessitate overcoming issues associated with its cost, accessibility, and technology alignment. The future studies should focus on how these factors can be addressed and on the lasting impacts of VR education on academics.

Acknowledgement

The authors gratefully acknowledge the students, staff, and authority of schools department for their cooperation in the research.

References

1. J. Smith and L. Thomas, "A Comparative Analysis of Virtual Reality and Traditional Classroom Learning," Journal of Educational Technology, vol. 45, no. 2, pp. 123–135, 2021.
2. H. Brown and T. Miller, "The Role of Immersive Learning Technologies in Modern Education," Journal of Learning Technologies, vol. 36, no. 4, pp. 245–258, 2020.
3. P. Johnson, "Challenges in Implementing Virtual Reality in Educational Environments: A Case Study," Educational Technology Review, vol. 11, no. 3, pp. 78–92, 2019.
4. Upadhyay, Vimal, Mukesh Chand, and Piyush Chaudhary. "Virtual Classroom for E: Education in Rural Areas." In Computer Networks & Communications (NetCom)

Proceedings of the Fourth International Conference on Networks & Communications, pp. 761–775. Springer New York, 2013.

5. B.G. Witmer and M.J. Singer, "Measuring Presence in Virtual Environments: A Presence Questionnaire," Presence: Teleoperators and Virtual Environments, vol. 7, no. 3, pp. 225–240, 1998.
6. M. Slater and S. Wilbur, "A Framework for Immersive Virtual Environments (FIVE): Speculations on the Role of Presence in Virtual Environments," Presence: Teleoperators and Virtual Environments, vol. 6, no. 6, pp. 603–616, 1997.
7. J.J. Lee and J. Hammer, "Gamification in Education: What, How, Why Bother?" Academic Exchange Quarterly, vol. 15, no. 2, pp. 146–151, 2011.
8. M. Bailenson et al., "The Effect of Immersive Virtual Reality on Learning: A Comparison of Presence and Cognitive Load in the Virtual Classroom," Educational Media International, vol. 46, no. 2, pp. 89–98, 2009.
9. Giard and M. J. Guitton, "Spiritus Ex Machina: Augmented reality, cyberghosts and externalised consciousness," Comput. Hum. Behav., vol. 55, Part B, pp. 614–615, 2016
10. R. M. Yilmaz, "Educational magic toys developed with augmentedrealitytechnologyforearlychildhoodeducation," Comput. Hum. Behav., vol. 54, pp. 240–248, 2016.
11. F. Giard and M. J. Guitton, "Spiritus Ex Machina: Augmented reality, cyberghosts and externalised consciousness," Comput. Hum. Behav., vol. 55, Part B, pp. 614–615, 2016.
12. Y.-C. Hsu, J.-L. Hung, and Y.-H. Ching, "Trends of educational technology research: more than a decade of international research in six SSCI-indexed refereed journals," Educ. Technol. Res. Dev., vol. 61, no. 4, pp. 685–705, Apr. 2013.
13. J. L. Chiu, C. J. DeJaegher, and J. Chao, "The effects of augmented virtual science laboratories on middle school students' understanding of gas properties," Comput. Educ., vol. 85, pp. 59–73, 2015.

Intelligent Systems Using Semiconductors for Robotics and IoT – Dinesh Goyal et al. (eds)
© 2026 Taylor & Francis Group, London, ISBN 978-1-041-20408-4

62 Green IoT: A Smart Healthcare Solution to Combat Medical Virology

Navita[1]
Assistant Professor,
Department of Computer Science,
Govt. P.G College for Women,
Rohtak, Haryana

Pooja Mittal[2]
Assistant Professor,
Department of Computer Science and Applications,
Maharshi Dayanand University,
Rohtak, Haryana

Shanaaya[3]
B. Tech CSE-AI,
Department of Artificial Intelligence and Data Science,
Indira Gandhi Delhi Technical University for Women (IGDTUW),
New Delhi

Abstract: Internet of Things is one of the most innovative communication technologies which has drastically changed the living style of human beings smartly. Integration of such technologies in smart cities has enriched society by providing smart healthcare, smart parking, smart business, smart education, and smart transportation, etc. However, the growing use of IoT in different domains presented lots of challenges such as increasing energy consumption, releasing of highly toxic gases, and large amounts of E-wastage. All such challenges may demand an environmentally friendly solution that arises in the form of Green IoT which offers an eco-friendly environment for smart cities. The involvement of smart IoT technologies gradually increased in the healthcare domain after the outbreak of viral diseases and it can be considered that Green IoT acts as a sustainable solution for a healthier life. Therefore, it is crucial to address all possible techniques and solutions to reduce the problem of E-waste, energy consumption, effective resource usage, and pollution threats. This paper focuses on offering a smart, sustainable, and eco-friendly solution for healthcare. Furthermore, this paper also focuses on IoT and its integration into the healthcare system to address the requirements of smart healthcare to combat the situation of viral diseases. In the end, various IoT-based healthcare solutions were presented to fight against the viral diseases.

Keywords: Internet of things (IoT), Green IoT, Sustainability, Eco-friendly, Viral diseases

1. Introduction

Due to the incredible improvement in sensing and communication technologies, 'objects' in our surroundings have the capability of connecting together to offer smart city applications to enhance the quality of life [1]. The technology that offers connectivity among objects in the smart city is generally known as the Internet of Things (IoT) [2, 3]. IoT offers connectivity between all the objects in a smart city from any location, at any time, and by using any medium of communication [4]. Growing advancement in IoT technologies makes IoT components

[1]navita.rs.dcsa@mdurohtak.ac.in, [2]pooja@mdurohtak.ac.in, [3]shanaaya27@gmail.com

DOI: 10.1201/9781003716389-62

smarter by adopting effective communication networks, analysis, processing, and storage. Most commonly used IoT devices include drones, sensors, cameras, "Radio Frequency Identification (RFID)", mobile phones, actuators, etc. [4, 5]. All such kinds of devices have the latent to interconnect and cooperate to achieve a mutual goal. Due to the availability of such kinds of components and communication technologies, IoT devices have a wide range of applications in healthcare, home automation, digitization and automation of the industry, etc. [6,7].

Among all aspects of IoT Healthcare is one of the major aspects gaining popularity worldwide during the Viral diseases because the growing number of Viral diseases patients has generated various health crises and worldwide, the healthcare system faces the problem in lack of various healthcare resources like beds, doctors, nurses, lack of Oxygen, masks, sanitization resources, etc [8]. Along with that another problem faced by the healthcare system is the overcrowding of people. Because of the lack of resources and timely treatment of patients, many of the people lost their life. So, a pertinent solution must be required to deal with the problems of the healthcare system that occurred during the Viral diseases. IoT becomes one of the prominent solutions to deal with such kinds of problems by offering real-time monitoring and non-contacting services to the patients by using various smart IoT-based medical devices and sensors like ECG, PPG, pulse oximeter sensor, blood pressure sensor, body temperature sensors, etc [9][10]. All such devices regularly monitor patients' health parameters and if any of the parameter values go below the threshold an emergency alert must be generated to the doctors as well as the patients so that immediate action must be taken to save patient life [11]. IoT plays a tremendous role in enhancing healthcare services by offering smart medical devices But the tremendous applications of IoT have generated various environmental issues like high energy consumption, a large amount of carbon emission, a large amount of energy usage which has attracted the attention of various researchers today. This study deliberates IoT-based smart technologies to enable real-time intellectual insight into the environment along with solutions to combat the Viral diseases. To achieve the goals of smart cities and to deal with the various environmental health issue, green IoT would be the significant solution [12][13][14].

The manuscript is organized as follows: Section 1 introduces the theme of this work, Section 2 discusses the trends of IoT in the healthcare system, Section 3 presents an introduction of Green IoT and its enabling techniques, section 4 presents smart IoT-based innovative healthcare technologies highly effective to fight against the Viral diseases. In last the conclusion and future challenges will be described in Section 5.

2. Internet of Things and Healthcare

Healthcare is one of the major aspects of smart and healthy living. The advancement in technology has revolutionized the hospital-centric environment into a patient-centric environment and also offer communication of clinically collected healthcare data from a remote area to healthcare centers so that proper treatment must be provided on time during any emergency.

2.1 Framework of IoT based Healthcare System

A framework mainly specifies how the existing system work and what are the major components involved in the functioning of a system. An IoT-based healthcare system made connectivity among all the available resources in form of a network to provide various healthcare services like monitoring, diagnosing, and offering remote surgeries over the internet. A complete framework used for providing IoT-based healthcare services is described in Fig. 62.1[15].

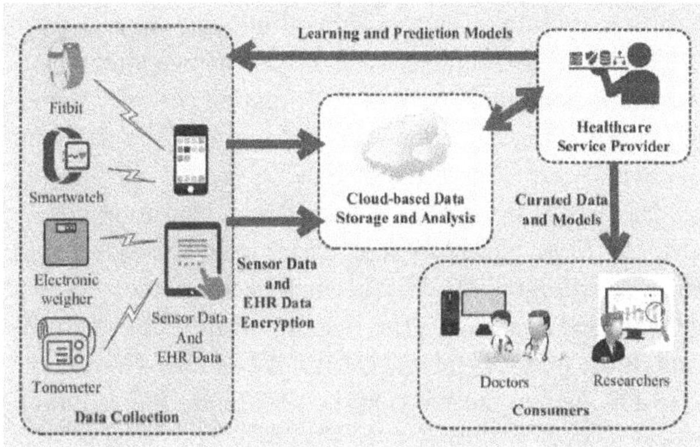

Fig. 62.1 Framework of IoT based healthcare system

Figure 62.1 pictures a situation in which a patient's health profile and vital signs are collected using sensors attached to the patient's body and portable medical devices. In today's environment, the most commonly used smart health monitoring devices are the Fitbit band, smart watch, electronic weigher, thermometer, etc.

2.2 Wearable Health Monitoring Sensors

Sensors are considered as the major key components of an IoT-based healthcare system and the development of effective and accurate sensors with a low form factor are crucial for the effective establishment of IoT based Healthcare system. The most important wearable sensors used in the healthcare environment are as follows:

2.3 Pulse Sensor

Pulse can be considered as one of the most vital signs used to identify variety of emergency conditions such as cardiac arrest, pulmonary embolisms. The most common sensors used now are photoplethysmography (PPG), ultrasonic and radio frequency sensors (RF). All PPG sensors work by transmitting LED light into the artery. PPG sensors are utilized to measure pule, blood oxygen, and pulse rate variability by using a small wrist-worn sensor. Pulse readings of the PPG sensor are affected by motion, in that case, accelerators if highly effective to measure pulse rate. It does not give an accurate pulse when motion is high. The effect of motion on the PPG sensor will be decreased by utilizing two different LED light intensities [16][17].

2.4 Body Temperature Sensor

Another vital sign regarding patient health is body temperature; body temperature is mainly used to detect heatstroke, fever, hypothermia, and more. Most of the researchers found thermistors as the most suitable for measuring human body temperature with an adequate level of errors. Therefore, it is assumed that these sensors would be highly suitable to use in the future. The accuracy of the temperature sensor mainly depends upon the closeness between the human body and the temperature sensor [18].

2.5 Respiratory Rate Sensor

Respiratory rate is another important vital sign used as a patient health parameter; it is just a measurement of the number of breaths taken by a patient in one minute. Monitoring of respiration rate could help in the detection of conditions like asthma attacks, apnea respiration, obstruction of the airway, panic attacks, and lung cancer, etc. The first respiratory rate sensor used was nasal [18]. Electrocardiogram (ECG) is another important medical device used to measure the respiration rate. It effectively finds the respiratory rate but is still not wearable. Continuous use of ECG may cause various skin problems like irritation.

2.6 Blood Pressure Sensor

Blood pressure is another vital health parameter that is measured along with the three above suggested vital parameters. High blood pressure or hypertension (HBP) is recognized as a high-risk factor for cardiovascular diseases. BP is a highly valuable parameter in healthcare and if any device having the capability to monitor this parameter will improve the quality of healthcare in a better way.

2.7 Pulse Oximeter

Pulse Oximeter s mainly used to measure the oxygen level in blood. It is not considered a vital health parameter but considered as the best indicator of respiratory function and helps in the diagnosis of a condition such as hypoxia (which means less oxygen is reaching the tissues). Pulse oximeter used an important health monitoring device. Many of the researchers have worked in this area towards the development of wearable pulse oximeter so that the patients can easily his oxygen level and can himself from various stroke-related problems.

3. Benefits of utilizing IoT in Healthcare System

The utilization of IoT in the healthcare system offers lots of benefits to both caregivers and health service providers. Some of the major benefits are as

3.1 Real-Time Monitoring and Reporting

Real-time monitoring of patients by using connected and worn medical devices can save the lives of millions of people facing an emergency like an asthma attack, heart failure, diabetes, etc. During real-time monitoring, the health-related data collected devices connected to the smart mobile app. Major health parameters like blood pressure, weight, oxygen level, blood glucose level, and ECG. All collected data send to the cloud and stored there and can be accessed by any legal person as per the accessibility guidelines. The authorized person could be a doctor, a participating health firm, an insurance company, or an external consultant. It will permit them to see the gathered data irrespective of their location and time.

3.2 Reduced Healthcare Cost and Faults

The introduction of IoT in healthcare allows real-time monitoring of various health parameters of patients by using mobile health applications and various analytical tools. Along with that, a new healthcare system provides live reporting to doctors via using interconnected devices. The technology-driven approach may reduce unnecessary visits to hospitals, stays in hospitals during their treatment process, etc., and improve the utilization of better-quality resources.

IoT also improves the quality of healthcare services by allowing easy accessibility of healthcare data of patients and offering simultaneous monitoring and reporting through various connected devices. It will offer customized treatment to patients. It ensures remote attention towards a large number of patients with a short treatment cycle. By offering real-time tracking and alerts during a emergency situation can save the life of the patient by sending constant alerts for appropriate analysis, monitoring, and diagnosis. All such types of mobility-based healthcare services are powered by IoT. This grants active medicines, better precision, and able intervention by specialists thus further improving the completer patient care service result.

3.3 Remote Medical Assistance

Remote monitoring assistance means providing all medical services to the patients from a remote location without visiting the hospitals. Patients can get all medical prescriptions at their current location or home through healthcare service providers connected to the patients through IoT devices.

3.4 Easy Appointment between Patients and Doctors

IoT-based healthcare system offers an easy appointment service between the patients and doctors through smart communication devices like mobile phones. An online meeting between doctors and patients saves both cost and time and may provide further services to other patients.

4. Issues Faced While Using IoT in Healthcare

With the highly growing use of IoT in different areas like smart city, smart healthcare, agriculture, business, marketing, etc. lots of issues will be generated like data security, data overloaded, accuracy cost, data integration, and standardization problems, etc [19][20]. Along with the major issues generated with the use of IoT in healthcare some common issues faced while using IoT in different areas are as follows:

4.1 Security and Privacy of Data

It is considered as the significant challenge faced while imposing IoT in healthcare. In the IoT scenario, IoT-based devices to capture and transfer data regarding patient health in real-time. Due to lack of signficant protocols and standards they are vulnerable to cyber-attacks. Cyber criminals may misuse patients' personal information.

4.2 Overloaded Data and its Accuracy

IoT health environment, aggregation of data from different medical devices is very difficult because of different protocols and standards followed by different medical devices. However, IoT-based medical devices record tons of data regarding patient health and personal information. For effective decision making, a proper insight must be taken within the data so that accurate decision must be taken regarding patient health.

4.3 Integration between Multiple Devices and Protocols

IoT-based healthcare systems may require the integration of multiple devices and protocols followed by these devices. All such devices are manufactured by different companies by following different protocol standards which may cause problems in aggregating data from these devices. Sometimes, such diversity between devices may slow down the whole process and may reduce the scalability of using IoT in healthcare.

4.4 High Cost of IoT Technology

The establishment of IoT in most of the developing countries is still costly Due to the high cost of technology, everyone could not afford such types of healthcare facilities. So, to successfully implement such types of technologies, the stakeholders must work in this way so that people belonging to each class can reach them excepting to high class and may lead to the overall improvement in healthcare services.

4.5 High Energy Consumption

All the IoT-based devices either used in the healthcare sector or another sector consume lots of energy while performing real-time monitoring. They must be continually supplied with power for remaining in a working state. So, to keep all IoT-based devices working regularly high amount of energy must be needed that may generate an issue that must be solved. IoT-based devices offer real-time monitoring services and to offer such types of services all devices must be kept in ON state and ON state devices always consumes lots of energy.

4.6 High Carbon Emission

Extensive use of IoT-based devices may also generate another important issue called high carbon emission. As when the IoT devices regularly consume energy to be working continuously and warm heat will be produced from them which may cause high carbon emissions in the climate. The high amount of carbon emission may also impact people's health in terms of skin problems, breathing, and allergic problems, which in turn, may provide a lot of burden on the healthcare system. Due to climate change, people face lots of health issues and they all need proper consultants and treatment to save themselves.

By keeping in mind all such important issues researchers work in this area to find a proper solution in response to all such problems.

5. Green IoT

Green IoT can be described as an energy-efficient procedure adopted by IoT. Green IoT provides an environmentally friendly and energy-efficient solution in response to the greenhouse effect of IoT or it enables sustainable growth of society. Green IoT supports innovative applications that can address social challenges and achieve low energy consumption IoT devices as shown in Fig. 62.2. Green IoT characteristics can be achieved by implanting energy-efficient methods and techniques on both hardware as well as software level which will help in reducing carbon dioxide emission, reducing energy consumption, and greenhouse effect due to existing IoT devices, services, and applications [10].

The complete life cycle of Green IoT mainly focuses on green design, green utilization, green production, and in the last green disposal and recycling so that there should be no or small impact on the environment. IoT models are not optimized for energy proficiency because they always remain on even when they are not in use results in wastage of energy. As IoT offers real-time monitoring so that a high amount of energy is consumed while they are transmitting data 24*7 and while they are on. But Green ensures devices are on only when they are in use and off when they are idle and not in use. Green IoT makes the devices work smartly without wastage of energy, offers proper ventilation for smart devices, and data centers are some of the strategies followed by Green IoT to conserve and save energy [15].

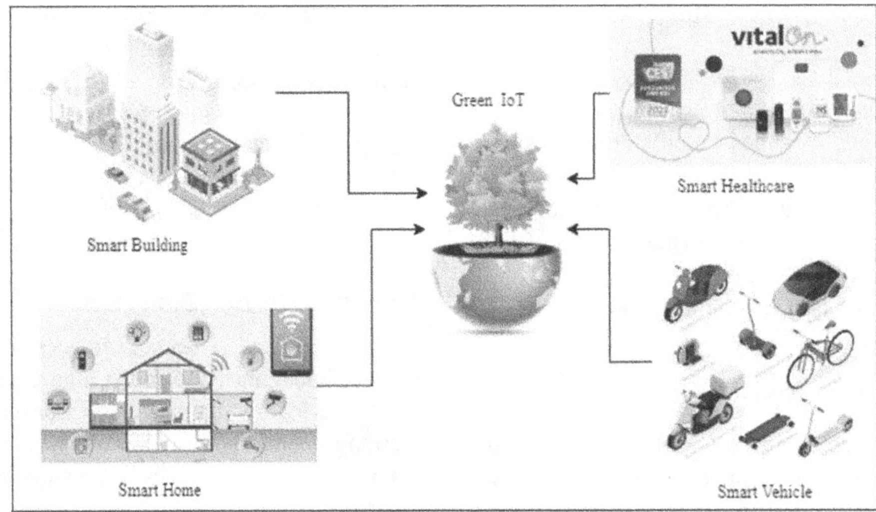

Fig. 62.2 Green IoT

5.1 Enabling Technologies of Green IoT

The establishment of Green IoT is possible through the integration of smart technologies by Fig. 62.3. Green aspects of each enabling technology that will strengthen the Green IoT are described as.

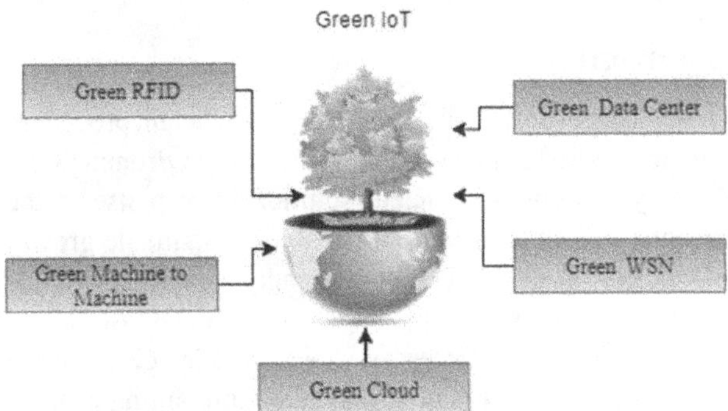

Fig. 62.3 Enabling technologies of green IoT

5.2 Green RFID

Radio frequency identification consists of various RFID tags with a small subclass of tag readers [15]. It contains a unique identifier to handle information related to the object to which they interact. In this process, data flow is initiated by RFID tag readers by transmitting a query signal by following the response received by nearby RFID tags. Generally, the transmission range of the RFID system is very low and for performing transmission various bands is used. RFID tags are available in active and passive tags. Active tags have batteries that empower the transmission but passive tags do not have onboard batteries and harvest energy from reader signals [21]. But for supporting Green IoT, the size of RFID tags must be reduced to reduce the quantity of non-degradable material along with some energy-efficient algorithms must be used to adjust the tag transmission power, avoid overheating and avoid collision between tags [22-24].

5.3 Green Wireless Sensor Network (GWSN)

GWSN is the composition of several sensors' nodes and base stations. All such sensor nodes have low processing speed, limited storage capacity, and power but the base station is very powerful. All such sensor nodes take reading from their surroundings and send to the base station. Generally, for IoT IEEE 802.15.4 standard is followed which has low communication power and bit rate [15] [22-24]. Green WSN adopted the following techniques to save energy consumption sensors node work only when necessary and otherwise, they will remain in sleep mode. The energy depletion problem can be solved by using wireless charging and by using an energy harvesting mechanism radio optimization and data reduction techniques.

5.4 Green Cloud Center (GCC)

In cloud computing services are treated as resources. Based on user demand cloud center GCC offers various resources. Relatively owning and dealing with their resources, manipulators also share a huge and accomplished pool of resources. With the increasing number of applications of IoT in different areas moved to the cloud, a large number of resources need to be organized resulting in high power consumption, and more environmental issues, and high emission of Co2. To support Green CC following potential solutions must be adopted.

5.5 Low Power Consumption Hardware and Software Devices

Hardware manufacturing companies must provide a hardware solution that consumes less energy. Various power saving virtual machines techniques and efficient resource allocation methods by using effective and accurate evaluation with energy-saving policies.

5.6 Green Machine to Machine (M2M)

Intelligent machines which collect the monitored data are all deployed into the machine-to-machine domain and participate in the communication. In the network domain, wireless/ wired networks transferred the collected data to the base station. When a massive number of machines take part in M2M communication, a lot of energy is consumed.

5.7 Green Data Center

Its key purpose is to process, manage, store the data generated or created by different users, systems and things, etc for which huge amount of energy must be required and a large emission of CO2. The energy efficiency of DC must be decreased to support a smart world.

5.8 Green IoT based Smart Healthcare System

Green and smart IoT-based healthcare systems combine the significant perceptions of secure and green healthcare to make sure that all healthcare facilities to be climate resilient and ensure that all the healthcare services must be on time. Major aims of a green and smart healthcare system are:

- Make ensures that all the healthcare facilities must be environment friendly and disaster-free
- All healthcare services have a lack of impact on climate change
- All healthcare services would be affordable
- Users feel free and comfortable in adopting healthcare services
- It offers decision-makers to make an effective decision regarding patient health and make improvements in green healthcare services.

Improving satisfaction among patients, providing energy-efficient services, better optimization of cost, improving security services are the main objectives of a smart healthcare system. But Green smart healthcare system along with these objectives may also try to provide eco-friendly services to patients.

6. IoT based Healthcare Innovations to Combat Viral Diseases

The global spreading of Viral diseases and a growing number of Viral diseases patients have created a worldwide health crisis and generated a lot of burden on hospitals to deal with such pandemic. Across worldwide one of the major problems faced by hospitals is overcrowding of people. The issue of overcrowding can be considered as a global health issue that affects individual life regardless of a particular class. As a result of overcrowding service rates of hospitals decrease and vice versa the mortality rate increase [25].Circumstances may demand a better technology that can improve the services of healthcare in a better way.

In such a situation IoT sensor-based technologies have played a major role. IoT sensor-based devices have helped to overcome the worst situation of the Viral diseases. This outbreak has changed the perception of IoT in people. As one of the major guidelines which must be followed during this pandemic is to maintain social distancing. IoT offers remote monitoring of patients as well as support remote diagnosis and treatment to patients. During the Viral diseases, IoT has played an important role in monitoring patients in isolation or virus-infected through smart devices. Different applications of IoT in the healthcare sector during the Viral diseases are:

6.1 Telehealth Consultations

The virus provides a contagious effect on doctors and physicians health who come in direct contact with an infected person. So that to keep healthcare doctors safe, the patient can be viewed through video chat without getting in direct contact. Communication through smart technologies and restricting people in homes can be considered as an excellent or alternative solution to reduce the mass rush of people in hospitals and solve the lack of resources problems.

6.2 Digital Diagnostic

The digital diagnostic can be seen as an efficient solution to reduce direct contact. Smart digital devices were innovated during such pandemics so that patients themselves can observe their health symptoms. Some of the most commonly used smart digital devices are the Kinsa thermometer, VICOODA Thermometer, and ifever, pulse oximeter which is in high demand during the Viral diseases.

6.3 Remote Monitoring

IoT-based smart devices are capable of offering remote monitoring of patients suffering from chronic diseases like diabetes, asthma, Parkinson's, etc. Remote monitoring of patients is possible with the help of smart sensors or medical devices either implanted or worn by the patient. All such devices and sensors continually monitor the vital health parameters and send an alert to the doctors, nearby hospitals, or the family. Remote monitoring of patients may reduce the risk of death due to deadly virus by providing immediate help to the patient in an emergency.

6.4 Robot Assistance

During viral diseases trends of using the robot is growing rapidly. Robots offer different types of services like cleaning hospitals, disinfecting devices, delivering medicines and foods to isolated and infected patients. Due to which healthcare workers can get more time to treat the patients. Along with that, some robots are also used for reducing stress and mental pressure on the patient in isolation. Paro is one such type of robot used for reducing mental stress.

6.5 Smart Drones

Smart drones based on IoT technology also play a significant role in monitoring the patients suffering from viral diseases. Along with monitoring smart drones also help deliver various medicine boxes, food packets, etc. They can easily monitor the temperature of people in a crowd and store and capture information related to that person. This innovation would be very beneficial in recognizing people with high temperatures and also provide live streaming.

6.6 Smart Helmet

A smart helmet is another innovation included in IoT devices that would be very helpful in maintaining social distancing among people. It detects the person having a high temperature and captures his image and sends it to the associated gadgets and Smart helmet detects the person with high temperature and capture his image and send it to the associated gadget with caution. Along with that it also tracks the location and history of that person so that tracking of that person becomes easy. This technology is adopted by most of the developed countries to track the person in isolation.

6.7 Smart Epidemic Tunnel

Various types of tunnels have been developed for automatic sanitization purposes. Such types of tunnels are highly effective for health workers and other public service providers during the lockdown. When any one of the persons passes through that tunnel it will automatically spray sanitizer to disinfect the virus on their clothes.

6.8 Smart Sanitizer Spraying Machine

Smart sanitizer spraying machines are also helpful to disinfect any area. After lockdown, before opening any area it becomes necessary to remove the virus from there. In that case, IoT based smart spraying machines are highly effective in disinfecting any location before opening it for the public

In the last couple of years, IoT has assisted health workers, especially with patient care even before the Viral diseases outbreak. In the case of the elderly and people living alone, IoT technology has the capability of assisting and monitoring people while performing their daily living activities. Fitbit bands, smart watches, smart phones are the major instances of IoT that are capable of tracking their blood pressure, oxygen level, stress level, exercise, and calorie intake and burn. So, the people residing alone and suffering from various chronic illnesses like diabetes or asthma can effectively use such types of wireless devices to monitor and track their health parameters and to stay healthy.

7. Conclusion

This manuscript mainly highlights how IoT becomes Green IoT. IoT is one of the most powerful and globally available technologies that have assisted people to live life smartly. As IoT has shown its lots of applications in different areas but among all such areas healthcare is one of the keenest areas of research. Before the viral diseases, peoples are not much aware of various IoT-based technologies. But after the viral diseases, IoT-based applications attracted attention to most people not even in healthcare but in different aspects of life. But among all areas, IoT-based devices achieve lots of popularity in the healthcare domain. As viral diseases has created the worst situation among people and most of the people lost their lives due to the unavailability of proper healthcare resources and services. In such situations, people adopt various IoT-based devices which can be easily worn and whose recordings can be easily readable so that people themselves can easily track their health status and save themselves from viral diseases and visit the hospital only in case of emergency. But the growing use of IoT has provided a bad impact on climate like greenhouse effect, high carbon dioxide emission, high energy consumption, etc. To resolve such types of issues generated by IoT, Green IoT came into place to solve the climate change problem. This paper concluded that Green IoT would be the best replacement for IoT and may provide a sustainable solution to a smart environment.

References

1. Torgul, B., Sagbansua, L., and Balo, F. (2016). Internet of Things: A Survey. International Journal of Applied Mathematics. Electronics and Computers (IJAMEC), ISSN:2147-8228: 104–110.
2. Miorandi, D., Sicari, S., Pellegrini, F.D. and Chlamtac, I. (2012). Internet of Things: Vision, applications and research challenges. ELSEVIER, pp. 1497–1516.
3. Khan, R., Khan, S.U., Zaheer, R. and Khan, S.(2012) Future Internet: The Internet of Things Architecture, Possible Applications and Key Challenges, 10th International Conference on Frontiers of Information Technology (FIT): proceedings.257–260.
4. Fuqaha, A., Guizani, M., Aledhari, M. and Ayash, M. (2013). Internet of Things: A Survey on Enabling Technologies, Protocols and Applications. IEEE Communication Surveys and Tutorials.17 (4), 2347–2352.
5. Singh, D., Tripathi, G. and Jara, A.J (2014). A Survey of Internet of Things: Future Vision, Architecture, Challenges and Services. IEEE World Forum on Internet of Things (WF-IoT). 287–292.
6. Al-Fuqaha, A., Guizani, M., Mohammadi, M., Aledhari, M. and Ayyash, M. (2015). Internet of Things: A Survey on Enabling Technologies, Protocols, and Applications. IEEE Communications Surveys & Tutorials 17(4), 2347–2376.
7. Hammi, B., Khatoun, R., Zeadally, S., Fayad, A. (2017). Internet of Things Technologies for Smart Cities. IET Research Journal, 1–14.

8. Pines, J. M., Hilton, A.J., Weber, E.J. and Alkemade, A.J., Shabanah, Al., Anderson et al. (2011) International perspectives on emergency department crowding. Academic Emergency Medicine. vol. 1. no. 12. pp. 1358–1370
9. Yin, Y., Zeng, Y., Chen, X., and Fan, Y. (2016). The Internet of Things in Healthcare: An overview, Journal of Industrial Information Integration, 1(3).pp. 3–13.
10. Baker, B., Xiang, W. and Atkinson, I. (2017) .Internet of Things for Smart Healthcare: Technologies, Challenges, and Opportunities. IEEE. 5. 26521–26544.
11. Islam, M. R., Kwak, D., Kabir, H., Hossain, M., and Kwak, K..S. (2015). The Internet of Things for Health Care: A Comprehensive Survey. IEEE Access, 3, pp. 678–708.
12. Perrier, E. (2015). Positive Disruption: Healthcare, Ageing & Participation in the Age of Technology. Australia: The McKell Institute.
13. Niewolny, D. How the Internet of Things is revolutionizing Health care, a white paper by Healthcare Segment Manager. Free scales Semiconductor.
14. Faris, A., Almalki, S. H., Sahal, A.R., Hassan, J., Hawbani, A., Rajput, N. S, Saif, A., Morgan, J., Breslin, J. (2021) Green IoT for Eco-Friendly and Sustainable Smart Cities: Future Directions and Opportunities, Mobile Networks and Applications.
15. Ahmed, R., Asim, M., Khan, S. Z, and Singh, B. (2019). Green IoT—Issues and Challenges. 2nd International Conference, On Advanced Computing And Software Engineering (ICACSE-2019), 378–382.
16. Lee, H.. Ko., Jeong, C. and Lee, J. (2017). Wearable Photoplethysmographic Sensor Based on Different LED Light Intensities. IEEE Sensors Journal, vol. 17, no. 3, pp. 587–588.
17. Shu, Y., Li, C., Wang, Z., Mi, W., Li, Y., and Ren, T. L., (2015). A pressure sensing system for heart rate monitoring with polymer-based pressure sensors and an anti-interference post-processing circuit, Sensors (Basel, Switzerland), vol. 15, no. 2, pp. 3224–3235.
18. Milici, J. Lorenzo, A. Lazaro, R. Villarino, and D. Girbau (2016) Wireless Breathing Sensor Based on Wearable Modulated Frequency Selective Surface," IEEE Sensors Journal, vol. PP, no. 99, p. 1.
19. Maxim Its kovich. IoT in Healthcare: Benefits, Applications and Challenges.
20. Internet of Things in Healthcare: Applications, Benefits and Challenges by Peerbits.
21. Nandyala, C. S., and Kim, H. K. (2016). Green IoT Agriculture and Healthcare Application (GAHA), International Journal of Smart Home Vol. 10, No. 4, pp. 289–300.
22. Anastasi, G., Conti, M., Francesco, M. Di. and Passarella, A. (2009). Energy conservation in wireless sensor networks: A survey, Ad Hoc Netw., vol. 7, no. 3, pp. 537–568.
23. Rault, T., Bouabdallah, A., and Challal, Y. (2014). Energy efficiency in wireless sensor networks: A top-down survey, Comput. Netw., vol. 67, pp. 104–122.

Note: All the figures in this chapter were made by the authors.

Intelligent Systems Using Semiconductors for Robotics and IoT – Dinesh Goyal et al. (eds)
 ISBN 978-1-041-20408-4

63 Air Canvas Design using Finger Flux Technology

Vinit Gowardhan[1], Satish Gaikwad[2], Hasanain Kureshi[3], Gayatri Magar[4], Priyata Umap[5]

Computer Engineering,
Anantrao Pawar College of Engineering and Research,
Pune, India

Abstract: Creating digital art with a finger-mounted bead is a game changer. It lets people draw, write, or explain things during online meetings just by pointing their finger at a PC's webcam. Then, the image shows up on the screen. This innovative tool uses colour recognition & tracking to create an air canvas. Users can change brush size & colour easily. It relies on Python, OpenCV, & computer vision technology. To enhance how pictures are processed, the method includes steps like morphological operations, reducing the background, & identifying fingertips. The system tracks finger movements in real-time and does it accurately using contour detection and image segmentation. This project can help hearing-impaired folks communicate better. It can also foster a smoother connection between humans and machines. Plus, it reduces reliance on traditional writing methods helping to bridge the gap between digital art forms & more traditional techniques. Looking ahead, this project shows great potential for smart wearable technologies. There's even a chance to connect with IoT devices for better functionality and ease of use.

Keywords: Gesture recognition, OpenCV, Python, Computer vision, Image processing, IoT integration, Air writing

1. Introduction

Art & technology are really connected in today's digital world. This connection has sparked a lot of new ideas & opened doors for creativity. One cutting-edge design is a digital art system that uses gestures and fancy image processing. It relies on a little motion sensor, which is super cool. Just by using hand gestures caught by a webcam, this system lets users change their movements into detailed digital art, written content, or even interactive experiences. The project makes use of OpenCV along with Python programming, showing big steps in how humans and computers can work together. It's all about making more chances for tech growth but also for creativity too.

This study digs into gesture recognition technology and how it helps people communicate naturally with digital platforms. The focus is mainly on copying hand movements in real time and turning those into meaningful digital stuff. The design wants to give users loads of freedom & creativity using top-notch computer vision methods. This way, they can create detailed digital artworks or join virtual meetings really easily. Plus, it breaks down the walls to traditional tools by providing an easy-to-use platform that makes digital art creation available for everyone.

The goal is to integrate the advancement in technology with the advancement in the field of human expression. It is an innovative social app that uses the latest computer vision and artificial intelligence technologies so that users can create content in a meaningful way using gestures only. The idea has the potential to transform the domain of digital art as well as the interaction by combining the accuracy of hand tracking with the flexibility of digital media. It is for everyone who needs to improve their

[1]gowardhanvinit@gmail.com, [2]satish2000.gaikwad@gmail.com, [3]qureshihasnain567@gmail.com, [4]gayatrimagar500@gmail.com, [5]priyata.umap@abmspcoerpune.org

DOI: 10.1201/9781003716389-63

online presentations at work or is an amateur artist looking for new ways to display his or her work.

This paper will first discuss the challenges of real time tracking, image segmentation and gesture recognition by presenting a clear view of the intelligent gesture recognition based digital art system. It also goes into the details of some technical problems and the most basic system requirements for the success of the system which include software interfaces, hardware specifications, and programming languages and libraries.

Additionally, the study looks at various uses of this technology while highlighting its benefits in digital art fields, human-computer interaction, & assistive tech for the deaf community.

2. Importance of Technology

The Finger-Mounted Bead based Gesture Based Digital Art Creation System has made a huge leap in the way of digital art and the way we interact with computers. It is especially useful in helping people of all levels of creativity and physical ability. This system makes everyone feel included when creating art, as much as possible.

This is what this tech does, it simply transforms the way of making art in a completely new way. It gives people new and exciting ways of engaging in art in a lively and immersive way. Users can easily transition from traditional methods of art creation to digital canvases. This system is also compatible with assistive technology and online communication tools. It is very useful in making virtual meetings and presentations more effective. Also, it makes creative sessions easier to manage, both individually and in groups!

It is rather useful for people with physical or hearing impairments as it can translate hand gestures into text, comments or visuals. This improves accessibility and inclusion in many environments as well as contributes to the advancement of this system. It is all about real time tracking, gesture recognition and generating digital content in this context. It also paves way for more efficient and immersive digital user interfaces and tools to be developed in the future.

3. Literature Review

The researchers, led by Bragatto [2020], devised a system that a sign language using a multi-layer neural Their study indicated that this method was able to handle videos in real-time, to enhance the comprehension of sign language for the people, thereby communicate more easily. Araga et al. [2019] source examined hand positions via a series of images for recognizing gestures using Jordan's (JRNN static feature (JRNN[2]) They really pointed realize how specific and important it is to translate and read gestures. Strong training algorithms are super necessary to get high accuracy in recognition.

Cooper [2018] developed a gesture recognition system capable of closely tracking the shapes of hands. This was a part of his research in complex 3D cell bioprinting. This significantly emphasized the need for reliable monitoring systems across the whole fields, including bioprinting and medical studies.

Neumann et al. [2017] created something cool to look for text in real photos, identifying shapes for the hypotheses with maximum stable extremal regions (MSER)[1] and a containment framework. Their work shown us how useful gesture recognition can be in many different settings. Having a dependable algorithms is key for accurately interpreting text. In the paper by Wang et al. (2016), a webcam along with a t-shirt tracker was used to implement color detection both in a This controlled paper environment also and highlighted in how a well more their unstructured method environment. worked in the real world and at the same time emphasized on how accurate color detection is crucial for proper gesture recognition.

Jari Hannuksela et al. [2015], Toshio Asano et al. [2014], and Sharad Vikram et al. [2013] conducted research on finger tracking algorithms. They proved how crucial it is to accurately track the finger movements. This has many possible uses, so having solid algorithms is necessary for smooth user interaction.

In the research carried out by Shaikh, Gupta, Shaikh and Borade [2012], they explained the need for sophisticated techniques in order to achieve proper hand gesture recognition while at the same time discussing the problems associated with detecting small objects. They also established that Kinect sensors could also provide efficient hand recognition.

Pavithra and Prabhu [2011] developed an low cost air writing system where they explained the concept of LED light detection for character recognition, the disadvantage of the device based approach and the need for low cost gesture recognition systems.

Yang et al. [2010] did a study where they established the effectiveness of matching images sequences in recognizing hand gestures, the extent of recognizing different hand forms and movements. Their results also showed that smart image processing techniques can greatly improve the accuracy of gesture recognition.

The authors stressed on the fact that there is a need for strong algorithms that can recognize and interpret hand movements and gestures properly in real time in their review recognition of hand gesture.

From Neumann et al. in 2009 it is seen how gesture recognition is useful everyday in life by noting the need for accurate text interpretation and trustworthy algorithms for different environments.

From Wang et al.'s colour discovery study in 2008, practical methods within gesture recognition systems are showcased, and how proper colour detection is crucial across different scenarios is emphasized, and it is also shown that their proposed system works well for real scripts.

In 2007-2005, Jari Hannuksela et al., Toshio Asano et al., and Sharad Vikram shared their work on cutlet recognition systems and pointed out the potential applications of cutlet location tracking in several areas, but stressed that trustworthy algorithms are necessary for smooth user interaction. The evaluation of literature focused on the techniques applied in the gesture recognition, and emphasized the need for new methods and good algorithms to achieve proper hand gesture detection. All this research done presents how active now is the field of gesture recognition! It shows the need for practical results in several areas to improve user experience. Gestures should also be accurately and properly interpreted and tracked, because strong training algorithms and reliable detection strategies are needed for high accuracy of recognition. The discussion opens up the possibility of using gesture recognition in order to advance science and technology, and states that to it have is accurate necessary tracking systems for many applications, including bioprinting and biomedical research. They have provided useful applications of gesture recognition in real life setups[1] and what they have concluded is that any hand gesture should be understood and analysed in The the success right of manner the using proposed proper methods, methods however, and depends algorithms.

On accurate text sureness interpretation is and ensured colour by identification solid as algorithms complete and safe discovery techniques in gesture recognition techniques. While suggested it that is good further cutlet movement tracking is important for better user engagement and improved digital is what experience, is all saying this that researchers should focus on practical results when they highlight robust algorithms in conjunction with cutting edge image processing methods that work well for gesture recognition systems.

In conclusion, this review provides an exhaustive view of studies on gesture recognition with a focus on strong algorithms, novel approaches, and valid recommendations for achieving accurate and reliable hand gesture recognition in various applications.

4. Research Methodology

The study employed a sequential mixed research design to identify how gestures are perceived in order to integrate both quantitative and qualitative data. In order to collect the participants, the study used convenience sampling. The participants were selected in a way that they should be familiar with the technology in use. The data was collected by means of interviews, questionnaires and observations. Ethical consideration was given by the team and all participants were assured of their privacy. In order to capture the hand movements in a very precise manner they employed motion sensors, special cameras and image processing software. For the analysis part, they did both number crunching with statistics and looked at themes in the stories people shared to ensure everything was valid and trustworthy. They considered some limits of the study, like how many people took part and the specific context. Carefully selected sampling methods helped them get different viewpoints. Ethics really mattered during the research process, and all participants agreed to take part to add more credibility to the research, they used triangulation techniques, mixing various data sources & perspectives. They also noted a few weaknesses in the study which showed a need for more research in gesture detection. In the methodology section, they summed up all the strategies and steps taken. They highlighted just how important it is to focus on ethical issues and have a solid research design for getting accurate & reliable results in gesture recognition studies

5. Flow Diagram of Proposed Work

Air canvas design using Finger Flux Technology.

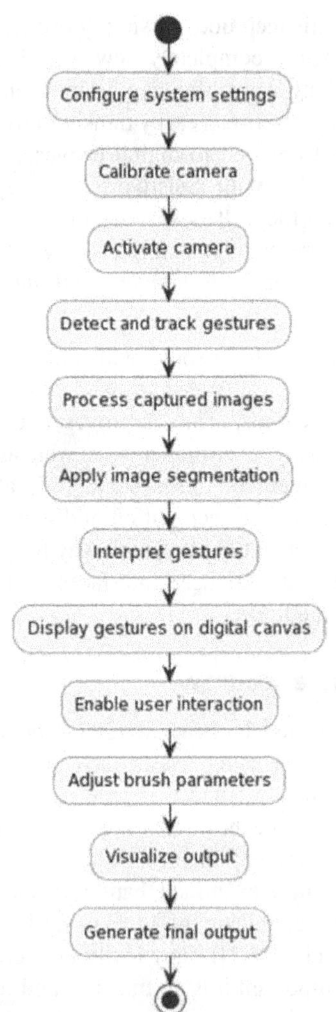

Fig. 63.1 Flow diagram of proposed work

Table 63.1 Algorithm

Step	Description
1.	Install the necessary libraries and set up the program.
2.	Launch the program and initialize the webcam for gesture detection.
3.	Position your finger in front of the camera to initiate the drawing canvas.
4.	Select the desired color and brush size using the on-screen buttons.
5.	Begin drawing or writing in the air with your finger while the program tracks your movements.
6.	Adjust the brush settings as needed during the creation process.
7.	Use predefined gestures to perform actions such as erasing, selecting, or predicting text.
8.	Utilize finger gestures to navigate through different options and functionalities.
10.	Save or export your artwork in various formats for further use or sharing.
11.	Explore the different features and functionalities of the program for a seamless user experience.
12.	Exit the program when finished and save any unsaved work as needed.

6. Advantage of Proposed Model Over Existing Model

- Improved Gesture Precision: The new model offers a better sketching experience compared to older ones. It uses advanced image processing and computer vision techniques. This help track movements accurately, so you can recognize fingertip positions easily.
- Smooth Integration in Online Meetings: This setup allows screen sharing during virtual meetings. It supports live visual explanations and demos, which makes communication & teamwork way better. When this feature's there, remote presentations work a lot better.
- User-Friendly Interface: This program is quite flexible and can be regulated creation easily settings. to The change interface the of art the website is quite user-friendly as it provides the user with on-screen controls which they can use to select color and brush sizes among other things. As a result of the simple design, it is convenient for both beginners and the most experienced users to work with it.
- This erase tool stuff, isn't select just options for or writing even or drawing, predict though. Users words. can This also handy use feature preset makes actions the to whole experience more enjoyable and turns the app into a full tool for both creative expression and useful communication.
- Natural Interaction: This concept combines the real life movements with digital art creation where the user's actions are linked with the software in a natural way which makes the experience enjoyable as if one is drawing traditionally while using a digital medium at the same time.
- This program has efficient image processing and real-time tracking for a productive workflow. The system is very responsive and user friendly, so users can be more creative and focus on that instead satisfying of art the sessions for computer.
- On this basis, the new model can be stated to be better than the older models since it will offer several advantages as seen below. Therefore, it establishes how the new model can transform the way people create digital art as well as how they communicate in various contexts.

7. Results and Discussion

1. Fingertip Detection Accuracy: The fingertip detection can be considered as very efficient. The main idea always of precise the at system least was up to test 95%. it This for implies a that hundred the times users and can it able was to control the position of the virtual canvas through their finger tips. This makes it easy for the users to draw or paint to specific details without having to worry about the canvas being off balance..
2. Real-Time Responsiveness: In the course of different drawing and writing tasks, the model answered very quickly with almost no delay at all. This means that users will not experience any delay between and their what actions transpires on the digital canvas due to the low latency.
3. Performance of Tracking: This is a great tracking algorithm, it tracked finger movements with a tiny mistake margin. This strong tracking helps to show how complex gestures are done accurately and easily for the user to translate their hand movements into precise digital artwork.
4. User Feedback and Experience: According To the early reports, most of the users were satisfied with their experience. They appreciated the simple design of the program, good gesture recognition and a lot of features that were offered by the program. Everybody agreed on the fact that it supports lively online presentations which makes it a great tool that has the possibility of revolutionalizing the way people work in groups remotely.

8. Conclusion

Among all the new gesture-based digital art creation model, this one is the most unique. It integrates real-time gesture detection with fast gesture processing and art creation which is quite interesting. It is very accurate on the way it recognizes hand movements, which makes it easier for people to create and modify digital art. With this model, things are shaping up in the online meetings or presentations. This means that the users can present ideas by moving their hands in a way that they can draw while explaining. It's quite fascinating! With the easy to use controls and options that can be changed, it all seems very natural

Acknowledgement

We would like to express our greatest appreciation to Professor Priyata umap for her important guidance & support during this project. How we developed and carried out this project, her deep knowledge in gesture recognition systems & digital artcreation has been key in. We would like to thank from the heart to Mrs. Umap for her leadership, her wise advice, and her help, which made it easier for us to tackle the challenges of this new approach. We also want to say a big thank you for her effort to create a teamwork—friendly research environment. Her broad knowledge and constant support have helped a lot with our learning and the smooth running of this innovative project. Her passion for exploring new tech possibilities and her commitment to mentoring young researchers are really inspiring. She's made a lasting impact on our academic journey.

References

1. International Conference on Computer Science and Computational Intelligence (ICCSCI 2015)
2. The Implementation of Hand Detection and Recognition to Help Presentation Processes
3. Bragatto, R. C., et al. "Real-time Brazilian Sign Language translation using Convolutional Neural Networks." Proceedings of the International Joint Conference on Neural Networks (IJCNN) IEEE, 2018.
4. "An economical air writing system." International Journal of Advanced Research in Computer and Communication Engineering 5.3 (2016): 443–451.
5. International Journal of Trend in Scientific Research and Development (IJTSRD) Volume 5 Issue 2, January-February 2021 Available Online: www.ijtsrd.com e-ISSN: 2456 – 6470
6. Hand Gesture Recognition using OpenCV and Python Shashidhar, R., H. Kim, and J. Chai. "Handwriting recognition in free space using motion tracking and deep learning." Proceedings of the 32nd AAAI Conference on Artificial Intelligence. 2018.
7. International Research Journal of Modernization in Engineering technology and science Volume:04/Issue:05/ May-2022 [412] AIR CANVAS USING OPENCV, MEDIAPIPE.
8. Dr. Chen, H., et al. "Hidden Markov model-based hand gesture recognition for human-robot interaction." IEEE Transaction and Cybernetics, Part C (Applications and Reviews) 40.4 (2010): 418–432.
9. Wang, Y., et al. "Colour-based internal and external motion detection for real-time applications." Proceedings of the 2015 IEEE International Conference on Multimedia and Expo. IEEE, 2015.
10. Toshio Asano, et al. "A practical finger recognition system using a movement-based tracking algorithm." IEEE Transactions on Pattern Analysis and Machine Intelligence 40.3 (2018): 712–725.
11. International Journal of trend in Scientific Research and Development (IJTSRD)
12. Hand Gesture Recognition using OpenCV and Python Surya Narayan Sharma, Dr. A Rengarajan
13. Sharad Vikram, et al. "A novel approach for finger tracking based on optical flow estimation." Proceedings of the 2017 IEEE International Conference on Computer Vision. IEEE, 2017.
14. Aran, M., et al. "A sign language tutoring tool using hand gesture recognition." IEEE Transactions on Education 54.4 (2011): 575–583.
15. Liu, S., and K. Lovel. "Real-time hand tracking and gesture recognition for human-computer interaction." Proceedings of the 2014 IEEE International Conference on Robotics and Automation. IEEE, 2014.

Note: The figure and the table in this chapter were made by the authors.

Intelligent Systems Using Semiconductors for Robotics and IoT – Dinesh Goyal et al. (eds)
© 2026 Taylor & Francis Group, London, ISBN 978-1-041-20408-4

64 Government Roles, Initiatives and Regulatory Frameworks: Their Influence on the Modern Digitalized Startup Ecosystem

Devika Sharma*, Ruchi Goyal
Jaipur School of Business, JECRC University,
Jaipur, India

Abstract: This paper examines the critical role of governmental initiatives and regulatory frameworks in nurturing startup ecosystems, with a focus on their effectiveness in fostering entrepreneurial success. Through quantitative analysis, we evaluated the impact of specific governmental roles such as policy formulation, funding mechanisms, and regulatory environments on the success rates of startups. The study employed structural equation modeling to analyze responses from various stakeholders within the startup ecosystem of Rajasthan, India. Results indicate that positive governmental interventions, particularly in policy and support mechanisms, significantly enhance startup success, while challenges in the regulatory environment and ecosystem barriers adversely affect growth. The findings highlight the need for policymakers to fine-tune and modify tactics that encourage entrepreneurial activities, ensuring a balanced approach that supports innovation while retaining regulatory control. This study adds to the larger discussion on the dynamic interplay between government actions and startup success, offering useful insights for improving the structural and operational landscapes of emerging entrepreneurial ecosystems.

The results emphasize the imperative for policymakers to continually refine and adapt their strategies to bolster entrepreneurial endeavors, ensuring a strategy that both encourages innovation and sustains regulatory oversight. This research adds to the extensive discussion regarding the dynamic relationship between governmental interventions and the success of startups. It offers significant insights aimed at improving both the structural and operational frameworks of nascent entrepreneurial ecosystems.

Keywords: Government policies, Startup ecosystem, Regulatory framework, Entrepreneurial success

1. Introduction

Government initiatives are crucial in developing startup ecosystems, particularly in emerging markets such as India. These measures provide essential financial backing and cultivate a conducive environment that promotes innovation and entrepreneurial activity. Within the dynamic global economic framework, startups have become integral in spearheading innovation, diversifying economies, and generating employment. The development and nurturing of a robust startup ecosystem are closely tied to the strategic initiatives and roles played by government bodies. These governmental roles span an extensive array of activities, including the formulation of policies, the establishment of regulatory frameworks, and the provision of direct support through funding and infrastructure development. The impact of these governmental actions critically affects the sustainability and growth trajectory of startups within specific regions (Gorowara et al. 2024; Shah et al. 2023).

Governments bear the dual responsibility of creating an environment that supports entrepreneurial expansion while simultaneously implementing regulatory measures that safeguard market integrity and public confidence (Fkun et al. 2023). It is essential to maintain a critical balance to prevent the stifling of innovation through

*Corresponding author: devika111sharma@gmail.com

DOI: 10.1201/9781003716389-64

excessive regulation and avoid market failures due to insufficient regulation. The role of government transcends mere facilitation; it involves the crafting of a flexible yet sturdy ecosystem capable of adapting to the emerging challenges and opportunities that arise from technological progress and changing market conditions (Gorowara et al. 2024;)

In recent times, India's start-up ecosystem has seen remarkable growth, driven by governmental efforts and a burgeoning pool of entrepreneurial talent. Research underscores the vital role that incubators and other supportive institutional frameworks, bolstered by government policies, play in fostering a conducive start-up environment (Mishra & Pal 2024). This phenomenon is not confined to India alone; globally, start-ups are increasingly recognized as catalysts for economic change. This global recognition is reflected in the various policies and strategies deployed by governments worldwide to eliminate barriers and smooth the path from concept to successful enterprise (Daraojimba et al. 2023).

This paper explores how various governmental roles, initiatives, and regulatory frameworks shape the start-up ecosystem. This study looks at the impact of government policies on startup success, with the goal of identifying critical elements that help or impede entrepreneurial development.

The study contributes to a deeper understanding of economic development strategies in contemporary settings.

2. Literature Review

The role of government in nurturing and shaping start-up ecosystems has garnered substantial attention in recent scholarship, underscoring the multifaceted nature of governmental influence on entrepreneurial activities. A review of the literature reveals a consensus on the pivotal functions governments perform, from regulatory frameworks to direct support mechanisms that significantly enhance start-up viability and innovation.

2.1 Government's Regulatory Roles, Initiatives and Support Mechanisms

Governments play a crucial regulatory role that balances the need for market oversight with the necessity to foster an environment conducive to innovation. Effective regulatory policies are essential for ensuring market integrity and protecting investor and public interests while avoiding the stifling of innovation through excessive regulation (Clarysse & Bruneel 2007; Gorowara et al. 2024; Sehnem et al. 2024; Hirsimäki, J. (2024). The agility of regulatory frameworks is particularly pivotal, as start-ups require the flexibility to pivot and iterate their business models in response to fast-changing technological landscapes (Onileowo, 2024; Vo, D. V., et al. 2024; Rakib, M., et al. 2024; Sroka et al. 2024). Startup ecosystems flourish in ecosystem where governmental policies encompass regulatory guidance as well as direct support measures like funding, tax incentives, and infrastructure enhancements. This type of government support is integral to many national strategies, prominently seen in India's efforts to cultivate a favorable environment for the growth of startups. Initiatives like the start-up India program highlight the role of government in facilitating start-up success through both financial and non-financial aids (Ajayi-Nifise et al. 2024).

2.2 Impact on Start-up Ecosystems

The literature indicates that the efficacy of government interventions can profoundly impact the sustainability and growth potential of start-ups within any region. Governments provide significant political guidance and exercise executive functions that are critical in crafting a supportive ecosystem for startups (Kromidha et al. 2024; Mitra et al. 2023; Chung et al. 2024). This finding is further supported by empirical studies that discuss the role of incubators and institutional frameworks, which are often underpinned by governmental policies, in enhancing the entrepreneurial climate (Joel & Oguanobi 2024); Susilo, D. 2020).

The synthesis of the reviewed literature underscores a clear trend: effective government roles and well-designed initiatives are essential for cultivating robust start-up ecosystems. These actions not only support the direct needs of start-ups but also build a broader infrastructure that sustains long-term economic growth and innovation. Thus, understanding the interplay between governmental actions and start-up success is crucial for policymakers aiming to foster dynamic and resilient economic landscapes.

3. Study's Conceptual Model

Figure 64.1 displays conceptual model of the study with various variables i.e. Government Role for Development of start-ups (GRS), Level of Agreement for Government Schemes (LOAG), Level of Agreement for Regulatory

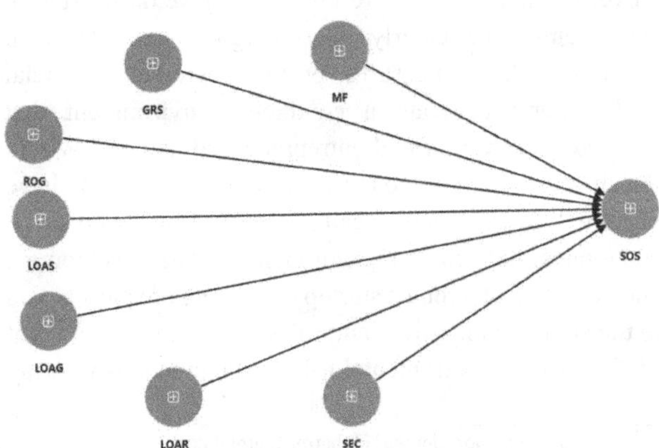

Fig. 64.1 Conceptual model

Table 64.1 Constructs internal consistency and reliability

Factor Loading			Cronbach's alpha	Composite reliability	Average variance extracted (AVE)
GRS	GRS 1	0.709	0.751	0.789	0.573
	GRS 2	0.817			
	GRS 3	0.652			
	GRS 4	0.835			
LOAG	LOAG 2	0.758	0.734	0.773	0.552
	LOAG 3	0.865			
	LOAG 4	0.562			
	LOAG 5	0.754			
LOAR	LOAR 2	0.606	0.690	0.765	0.502
	LOAR 3	0.671			
	LOAR 4	0.704			
	LOAR 5	0.834			
LOAS	LOAS 1	0.753	0.850	0.849	0.627
	LOAS 3	0.834			
	LOAS 4	0.714			
	LOAS 5	0.859			
	LOAS 6	0.791			
MF	MF 1	0.696	0.749	0.774	0.493
	MF 2	0.762			
	MF 3	0.644			
	MF 4	0.817			
	MF 5	0.565			
ROG	ROG 1	0.603	0.700	0.750	0.516
	ROG 3	0.627			
	ROG 4	0.775			
	ROG 5	0.841			
SEC	SEC 1	0.842	0.825	0.841	0.588
	SEC 2	0.703			
	SEC 3	0.697			
	SEC 4	0.797			
	SEC 5	0.786			
SOS	SOS 1	0.745	0.815	0.821	0.518
	SOS 2	0.72			
	SOS 3	0.727			
	SOS 4	0.657			
	SOS 5	0.754			
	SOS 6	0.713			

System (LOAR), Level of Awareness of Schemes (LOAS), Motivating Factor (MF), Role of Government (ROG), Start-up Ecosystem Challenges (SEC), Success of Start-ups (SOS) which are analysed to understand how governmental policies impact the start-up ecosystem.

Table 64.2 Fornell larcker criteria

	GRS	LOAG	LOAR	LOAS	MF	ROG	SEC	SOS
GRS	0.757							
LOAG	0.513	0.743						
LOAR	0.15	0.274	0.709					
LOAS	0.632	0.638	0.396	0.792				
MF	0.556	0.623	0.464	0.723	0.702			
ROG	0.268	0.393	0.367	0.459	0.59	0.719		
SEC	0.682	0.515	0.507	0.713	0.709	0.438	0.767	
SOS	0.441	0.585	0.406	0.53	0.686	0.367	0.495	0.72

4. Research Methodology

This study used a mixed-methodologies approach to explore government influences on the startup ecosystem, using qualitative and quantitative methods (Tzagkarakis & Kritas 2023). Data were acquired from both direct and secondary sources, including organized interviews and surveys with entrepreneurs and authorities, as well as academic publications and reports. Purposive sampling was used to assure a representative sample of competent participants, which increased the data's richness and usefulness. The measurement methods comprised both closed and open-ended survey questions and thorough interview instructions, meant to obtain substantial data on governmental impacts.

The study's reliability and validity were ensured through the use of Cronbach's alpha, composite reliability and AVE for internal consistency (Hair et al. 2017). The Fornell-Larcker Criterion (Fornell & Larcker 1981) and HTMT ratio (Henseler, et al. 2015) is used for discriminant validity and all the values have shown acceptable thresholds.

Table 64.3 HTMT ratio

	GRS	LOAG	LOAR	LOAS	MF	ROG	SEC	SOS
GRS								
LOAG	0.737							
LOAR	0.372	0.534						
LOAS	0.805	0.832	0.483					
MF	0.818	0.847	0.720	0.930				
ROG	0.387	0.568	0.635	0.589	0.796			
SEC	0.880	0.683	0.694	0.852	0.970	0.618		
SOS	0.543	0.710	0.536	0.615	0.813	0.464	0.593	

5. Structural Model

Figure 64.2 the structural model presented the intricate relationships between various factors influencing startup success. The model included variables like Government Role in Startups (GRS), Motivating Factor (MF), and Success of Startups (SOS), showing how different

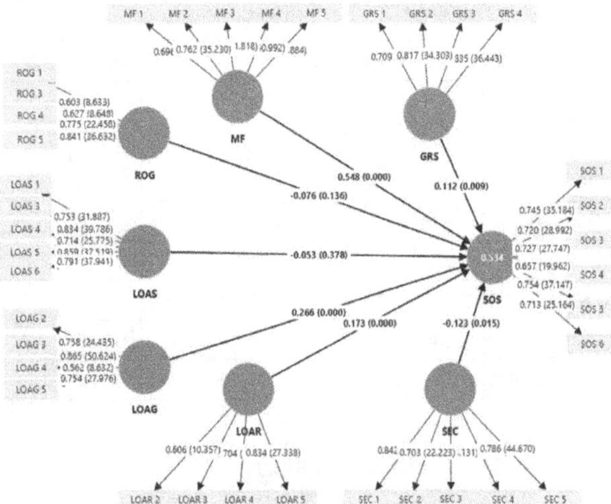

Fig. 64.2 Structural model

elements interact within the startup ecosystem. Path coefficients indicate the strength and direction of each relationship, providing a quantitative analysis of how governmental actions, market conditions, and individual motivations contribute to the overall success of startups.

5.1 Analysis of the Path-Coefficients

Table 64.4 Path coefficients

Path Coefficients	Sample mean (M)	T statistics	P values
GRS -> SOS	0.116	2.615	0.009
LOAG -> SOS	0.265	7.604	0.000
LOAR -> SOS	0.179	5.533	0.000
LOAS -> SOS	-0.048	0.881	0.378
MF -> SOS	0.539	10.472	0.000
ROG -> SOS	-0.070	1.490	0.136
SEC -> SOS	-0.122	2.445	0.015

Path the coefficients and accompanying data demonstrated the effect of several factors on startup success. The government's participation in assisting startups (GRS -> SOS) is strongly associated, with a correlation of 0.116 and a statistically significant p-value of (0.009), showing a significant positive effect on startup success. Similarly, the agreement on government programs (LOAG -> SOS) has a large positive influence, with a coefficient of (0.265) and an exceptionally significant p-value of (0.000), indicating that good impressions of government efforts are critical for startup success.

Path The coefficients and supporting data showed how numerous variables influence startup success.

The government's engagement in helping startups (GRS -> SOS) is strongly associated, with a correlation coefficient of 0.116 and a statistically significant p-value of (0.009), showing a significant positive effect on startup success. Similarly, agreement on government initiatives (LOAG -> SOS) has a strong positive influence, with a coefficient of (0.265) and a highly significant p-value of (0.000), indicating that good opinions of government actions are critical for startup success.

The involvement of government (ROG -> SOS) is, however, somewhat adversely connected with startup performance, as evidenced by a coefficient of (-0.070) and a non-significant p-value of (0.136), implying that not all government measures are helpful in aiding companies. Finally, obstacles inside the startup ecosystem (SEC -> SOS) significantly reduce startup success, with a negative coefficient of (-0.122) and a p-value of (0.015). This shows that barriers inside the ecosystem, such as bureaucratic roadblocks or physical flaws, can significantly reduce the favorable benefits of supporting policis.

Table 64.5 describes the model's explanatory power. R-square is a statistical metric that reflects how much of a dependent variable's variation can be explained by its independent variables in a regression model. An R-square of (0.534) shows that the independent variables in the model explain 53.4% of the variance in start-up success (SOS). This metric assesses the model's goodness of fit.

Table 64.5 Explanatory power

	R-square	R-square adjusted
SOS	0.534	0.526

Adjusted R-square is a modified form of R-square that accounts for the number of predictors in the model. It is always less than or equal to R squared. The modified R-square of (0.526) implies that after controlling for the number of variables in the model, (52.6%) of the variability in SOS remains.

6. Conclusion and Future Implications

The analysis presented in the structural model provides substantial insights into how various factors influence the success of start-ups. Significant positive relationships are evident from the Government's role in start-ups (GRS), agreement on government schemes (LOAG), and satisfaction with regulatory environments (LOAR), indicating these areas are pivotal in enhancing start up success. The statistical significance of these relationships reinforces the reliability of these findings and underscores the importance of targeted government interventions. Conversely, awareness of schemes (LOAS) and the role of government (ROG) have shown non-significant or slight negative impacts on start-up success, suggesting that simply increasing awareness or government involvement without strategic focus does not necessarily lead to positive outcomes. Challenges within the start-up ecosystem (SEC) displayed a negative relationship with start-up success, highlighting the adverse effects of ecosystem barriers on growth and sustainability.

Ongoing development of government schemes is vital. These schemes should be tailored to support start-up infrastructure and funding effectively. Establishing a balanced regulatory framework is essential. This framework should encourage start up innovation while ensuring compliance and market stability. There's a significant need to identify and address systemic barriers within the start-up ecosystem, such as bureaucratic inefficiencies and infrastructural shortcomings. Future research should consistently employ quantitative models to measure the impact of various policies, aiding in the continuous improvement of start-up support mechanisms.

References

1. Gorowara, N., Yadav, S., & Kumar, V. (2024). Sustainable Future: Government Initiatives in the Adoption of Emerging Sustainable Technologies by Startups in India. In Fostering Innovation in Venture Capital and Startup Ecosystems (pp. 286–305). IGI Global.
2. Shah, C. F., & Jokhi, C. D. M. E. (2023). A STUDY ON EFFECT OF GOVERNMENT POLICIES ON STARTUP. A GLOBAL JOURNAL OF SOCIAL SCIENCES.
3. Fkun, E., Yusuf, M., Rukmana, A. Y., Putri, Z. F., & Harahap, M. A. K. (2023). Entrepreneurial Ecosystem: Interaction between Government Policy, Funding and Networks (Study on Entrepreneurship in West Java). Jurnal Ekonomi Dan Kewirausahaan West Science, 1(02), 77–88.
4. Mishra, A., & Pal, R. (2024). A State-Wise Comparison of Start-Up India Scheme: A Way to Restore the Indian Economy. Educational Administration: Theory and Practice, 30(3), 2334–2339.
5. Daraojimba, C., Abioye, K. M., Bakare, A. D., Mhlongo, N. Z., Onunka, O., & Daraojimba, D. O. (2023). Technology and innovation to growth of entrepreneurship and financial boost: a decade in review (2013-2023). International Journal of Management & Entrepreneurship Research, 5(10), 769–792.
6. Clarysse, B., & Bruneel, J. (2007). Nurturing and growing innovative start-ups: the role of policy as integrator. R&d Management, 37(2), 139–149.
7. Onileowo, T. T. (2024). Exploring the Influence of Government Policy on Entrepreneurship Development. British Journal of Multidisciplinary and Advanced Studies, 5(1), 198–211.
8. Ajayi-Nifise, A. O., Tula, S. T., Asuzu, O. F., Mhlongo, N. Z., Olatoye, F. O., & Ibeh, C. V. (2024). THE ROLE OF GOVERNMENT POLICY IN FOSTERING ENTREPRENEURSHIP: A USA AND AFRICA REVIEW. International Journal of Management & Entrepreneurship Research, 6(2), 352–367.
9. Joel, O. T., & Oguanobi, V. U. (2024). Entrepreneurial leadership in startups and SMEs: Critical lessons from building and sustaining growth. International Journal of Management & Entrepreneurship Research, 6(5), 1441–1456.
10. Sehnem, S., Lara, A. C., Benetti, K., Schneider, K., Marcon, M. L., & da Silva, T. H. H. (2024). Improving startups through excellence initiatives: addressing circular economy and innovation. Environment, Development and Sustainability, 26(6), 15237–15283.
11. Clarysse, B., & Bruneel, J. (2007). Nurturing and growing innovative start-ups: the role of policy as integrator. R&d Management, 37(2), 139–149.
12. Mitra, S., Kumar, H., Gupta, M. P., & Bhattacharya, J. (2023). Entrepreneurship in smart cities: elements of start-up ecosystem. Journal of Science and Technology Policy Management, 14(3), 592–611.
13. Hirsimäki, J. (2024). Seeds of Success: How Government Grants aid early-stage Startup Performance?.
14. Susilo, D. (2020). Scalable start-up entrepreneurship and local economic development in emerging economies. Applied Economics Journal, 27(2), 145–163.
15. Sroka, W., Filipiak, T., & Barczyk-Ciuła, J. (2024). The Role of the Local Government in Fostering Entrepreneurship–Evidence from Municipalities in the Kraków Metropolitan Area. Economic and Regional Studies/Studia Ekonomiczne i Regionalne, 17(1), 21–57.
16. Vo, D. V., Phạm, P. G. M., & Nguyen, T. G. (2024). Outsourcing and entrepreneurial innovation in a transition economy: the moderating roles of private ownership and government support. Journal of Small Business and Enterprise Development.
17. Onileowo, T. T. (2024). Exploring the Influence of Government Policy on Entrepreneurship Development. British Journal of Multidisciplinary and Advanced Studies, 5(1), 198–211.
18. Kromidha, E., Altinay, L., & Arici, H. E. (2024). The influence of politics on the governance of an entrepreneurial ecosystem in a developing country: a generative institutional discourse approach. Entrepreneurship & Regional Development, 1–18.
19. Rakib, M., Azis, F., Sanusi, D. A., Ab, A., & Taufik, M. (2024). How Does Government Support, Finance, Family Environment and Information Technology Influence Sustainable Entrepreneurship?. Journal of Law and Sustainable Development, 12(1), e2499–e2499.
20. Chung, B., Franses, P. H., & Pennings, E. (2024). Conditions that make ventures thrive: from individual entrepreneur to innovation impact. Small Business Economics, 62(3), 1177–1200.
21. Tzagkarakis, S. I., & Kritas, D. (2023). Mixed research methods in political science and governance: approaches and applications. Quality & quantity, 57(Suppl 1), 39–53.
22. Zettel, L. A., & Garrett, R. P. (2021). Venture-level outcomes of juggling and struggling. Journal of Business Venturing Insights, 15, e00225.
23. Hair Jr, J. F., Babin, B. J., & Krey, N. (2017). Covariance-based structural equation modeling in the Journal of Advertising: Review and recommendations. Journal of Advertising, 46(1), 163–177.
24. Hajjar, S. T. (2018). Statistical analysis: Internal-consistency reliability and construct validity. International Journal of Quantitative and Qualitative Research Methods, 6(1), 27–38.
25. Fornell, C., & Larcker, D. F. (1981). Evaluating structural equation models with unobservable variables and measurement error. Journal of Marketing Research, 18(1), 39–50.
26. Henseler, J., Ringle, C. M., & Sarstedt, M. (2015). A new criterion for assessing discriminant validity in variance-based structural equation modeling. Journal of the academy of marketing science, 43, 115–135.

Note: All the figures and tables in this chapter were made by the authors.

Intelligent Systems Using Semiconductors for Robotics and IoT – Dinesh Goyal et al. (eds)
© 2026 Taylor & Francis Group, London, ISBN 978-1-041-20408-4

65 Artificial Intelligence and Indian Legal System—Prospects in Near Future

Sushila Choudhary,
Sunita Singh Khatana*, Swati Beniwal
Faculty of Law, Manipal University,
Jaipur

Abstract: Artificial Intelligence (AI) is creating groundbreaking changes in many industries across the world, and the legal industry is also not exempted. This paper discusses the future of AI in the legal contexts of India for bringing about improvement and bringing difficulties in turn. This paper undertakes the framework and the issues in the Indian Judicial System in the facets of its background. It discusses specific deployment of AI in the legal profession, encompassing legal information retrieval, contract generation, judicial opinions, law enforcement, and legal education and training. On one hand, there are possible advantages depending on the aspects of precise, faster, more accessible, and cheaper than traditional methods necessary for competent decision-making. On the other side, there are ethical risks along with the protection of private data and the properly constructed framework of regulation. This paper serves as a great reference as it offers both global experiences as well as Indian propositions, thereby offering a concise but concrete picture of how AI has affected the legal domain. This paper also discusses technological advancements and policy developments as to the outlook envisaging a new regime, looks into the probable future developments and changes that are expected of the Indian legal system.

Keywords: Artificial intelligence, Legal system, Technological advancement, Judicial system, Legal decision-making, Legal ethics

1. Introduction

AI has been adopted in various fields, and the legal profession is no exception to the reality that it is probing the industry. The legal system in India, which was built on principles existing for centuries in statutes and Continue minute human intervention to analyse the issues, is slowly embracing the innovation offered by AI. This shift is apparent as client-server applications can be considered as a break from conventional paradigm towards more technological methods. AI in the legal field provides many benefits including but not limited to simplifying legal processes and expanding investigative capabilities while improving general productivity. Nevertheless, this expansion is not without its obstacles. Over time, AI has found its way into most practices in the legal sector thus making it vital for legal professionals to understand the entire picture on the topic (Ahlawat, 2023).

As we have witnessed, we are in a golden age of innovative technological growth and development, where it seems almost no field can escape the widespread application of AI. AI is already applied proactively in various industries including finance, healthcare, and e-business with the ultimate goal of increasing efficiency and efficacy. Interestingly, and against all odds, the legal profession, which has always boasted of its sheer technicality and the rather indispensable human touch that lies right at the heart of it, has now popped up as one of the key vitalist forces in the great race toward AI. Having all along been regarded as a somewhat 'technologically' slow industry given that it uses historical precedent and detailed analysis of traits of human behaviour in providing solutions, the legal industry is now experiencing something amazing.

The Indian legal services market, however, is currently riding on this wave of change. AI has become integrated

*Corresponding author: sunitasingh.khatana@jaipur.manipal.edu

DOI: 10.1201/9781003716389-65

into the legal environment, contributing towards the transformation of the way the legal fraternity carries out its operations. Typically, a variety of technologies are being implemented to free up the time of lawyers and, for example, provide a possibility to conduct long analyses of documents or do legal research on the Internet. Also, by using AI, additional information about legal data is revealed by analytics meaning that lawyers can be assisted in putting forward better strategies and legal arguments. These technologies we find are not only improving legal practice but also expanding the legal services to different groups of the population (Singh et al., 2023).

However, introducing AI into the legal market has several benefits, at the same time creating a range of challenges and concerns. Concerns were raised regarding data privacy, validity of insight mined by these systems, and related concerns of bias that may exist in algorithms. For legal experts and those involved in the practice of law, it is imperative to grasp these changes and have a working knowledge of the advantages as well as the drawbacks that come with AI. This will include training, and the enhancement of new tools and technologies used in legal contexts and also conform with ethical standards used in the deployment of AI in legal practices.

While exploring the increase in AI advances in changing the laws dynamics we can observe that this technology is rather promising. The opportunity to transform legal work, to introduce novelties and to increase efficiency, as well as to expand legal aid to the people is incredibly great. But attaining that vision of fostering a legal environment that leverages technology will demand concerted collective action from all members of the different ecosystems. Thus, it is essential for lawyers, judges, policymakers as well as technologists to chart their way through this new complex landscape while making a delicate change in the context of using/ adopting AI systems (Gorlamudiveti & Sethu, 2023).

The idea of deep integration of AI-based technologies into practice stimulates positive emotions as well as concerns among the representatives of the legal profession and other stakeholders. On the one hand, there is a wealth of legal papers, the statutes, case histories, legal precedents, and so on and so forth, where AI applicability can have a significant positive impact in terms of boosting efficiency and handling accuracy. It can dig out the required information in a short span of time and at times, make forecasts out of documented trends. The mentioned capabilities could transform tasks like legal research, contract review and case handling the legal way into more efficient ways.

However, legal processes involve many individual and societal factors that, from an AI expert perspective, raise implementation hurdles. In many cases, liability depends not only on comprehension of emotions or ethical issues but also the culture of a specific region or time period. For example, the professions such as law, medical, engineering, and architecture implement basic aspects of fairness, justice, and feeling for others. The formulas such as these build upon human experience and moral judgment which are subjects where AI may falter despite algorithmic prowess and machine learning expertise. The question arises: In light of the fact that emotion and ethics are key to decision making, especially in legal, can AI really understand and respond correctly in the given social dilemma? (Ahlawat, 2023).

Furthermore, legal experts really want to find out whether AI will be able to capture and give due consideration to sensitive cultural variation as they pertain to justice. Thus, law is not an area of interest where something that is appropriate to apply in one society is appropriate to apply in other societies since law is a part of the societal culture of a specific society. Even if the current system incorporates an AI supercomputer, it can be quite challenging to address the cultural differences, at least within the current paradigm. This limitation presents the danger of AI tools reaching conclusions that, to a computer at least, are logical and therefore printable at the expense of cultural sensitivity or even ethical integrity.

2. Current State of the Indian Legal System

Indian legal provisions constitute a very strong and complex legal structure which draws its origin from the Constitution of India. In hierarchy it starts with the Supreme Court, then High Courts within the different states and union territories, followed by a complex system of sub- or district courts and several specialized tribunals. Civil jurisdiction of the Supreme Court of India: The Supreme Court of India has original as well as appellate jurisdiction necessary for the interpretation of the Constitution and being the highest court of appeal in the country. On one hand, the High Courts have original jurisdiction in their respective states with exclusive powers over civil and criminal cases while on the other hand, the subordinate courts as the providers of access to justice for the masses ensure that the legal system is available at the base level (Dutta, 2022).

Despite its comprehensive structure, the Indian legal system grapples with several significant challenges that impede its efficiency and efficacy:

2.1 Case Backlog

One of the most pressing issues is the staggering backlog of cases. According to recent estimates, millions of cases are pending across various courts, with some remaining unresolved for decades. This immense delay in the judicial process not only results in justice being deferred but also severely undermines public trust in the judiciary. The prolonged waiting times for case resolutions can lead to significant hardships for individuals seeking justice, often affecting their livelihoods, mental health, and overall faith in the legal system. The sheer volume of pending cases illustrates a systemic issue that requires urgent reforms and innovative solutions to ensure timely justice for all.

2.2 Inefficiencies

The legal process in India is often criticized for its procedural inefficiencies. The system is bogged down by intricate and sometimes archaic procedural laws that do not align with the current demands of swift and efficient justice delivery. Frequent adjournments and procedural delays are commonplace, contributing to prolonged litigation periods. These inefficiencies not only delay the delivery of justice but also escalate litigation costs, placing a heavy financial burden on litigants. Moreover, the prolonged litigation processes consume valuable judicial resources, further straining the already overburdened courts and hindering their ability to manage caseloads effectively (Sai & Sharma, 2023).

2.3 Access to Justice

Access to justice remains a critical concern, particularly for marginalized and economically disadvantaged sections of society. Geographic, economic, and social barriers often prevent these groups from seeking and obtaining legal redress. For many individuals in rural or remote areas, the physical distance to the nearest court can be a significant obstacle. Additionally, the costs associated with legal representation and court fees can be prohibitive for those with limited financial resources. Social barriers, such as caste discrimination and gender biases, further exacerbate these challenges, making it difficult for marginalized communities to exercise their legal rights and access justice.

2.4 Resource Constraints

The judiciary frequently operates under significant resource constraints, which hampers its effective functioning. There is a notable shortage of judges, court staff, and infrastructure. The judge-to-population ratio in India is considerably lower compared to many other countries, resulting in an overwhelming workload for the existing judges. This shortage leads to prolonged case hearings and delays in judgments. Additionally, many courts lack basic infrastructure, such as adequate courtroom space, modern facilities, and technological support, further impeding their efficiency. Addressing these resource constraints is crucial for improving the overall performance of the judiciary and ensuring timely justice delivery (Awasthy et al., 2022).

2.5 Technological Adoption

While there have been strides in incorporating technology into the judicial process, such as the e-Courts project, the adoption and integration of technology remain inconsistent. Many courts still rely heavily on manual processes, which are time-consuming and prone to errors. The uneven implementation of technological solutions across different regions and court levels creates disparities in the efficiency and effectiveness of the judicial process. Embracing technology more comprehensively can streamline case management, reduce delays, and improve access to justice. However, this requires substantial investment in digital infrastructure, training for judicial staff, and consistent policy implementation to ensure uniform technological adoption across the judiciary (Gupta, 2021).

3. The Indian Judiciary's Use of AI: Progress or Reason for Alarm?

The Indian legal system has been quite conservative in terms of embracing innovative technology and AI is expected to bring out a change in this scenario. In the past the judiciary, the bar and the academia have been slow in embracing technology to its processes in the legal. But emerging trends indicate that this would change based on advanced technology, especially AI and the changing demography of the legal profession.

A major development in this process was observed recently when the Punjab & Haryana High Court used AI aid, which is in fact ChatGPT, during a bail plea. This instance therefore could be the first time where AI has been applied in Indian judiciary, and it also points towards the direction that the AI may be applied in many other areas concerning case management and administration of justice. Scholars analyzing the field expect increased application and direct incorporation of the results provided by AI, forecasting its impact on decision-making, and improvements in case handling in the legal practice (Suryam, 2023).

However, it continues to be a contentious debate as to whether AI can take over judicial decisions. Real and moral issues of legal concern arise from the circumstance that the legal has allowed AI to assume the roles of discretion that traditionally belong to judicial minds, in issues determining the fundamentals of individuals' lives and freedoms. Concerns like prejudicial elements in the codes, flawed interpretations, and possible technical discrepancies also call for a careful approach to AI incorporation into the legal system.

In order to effectively combat these challenges, all those involved in the legal process are enjoined to move forward with great care and caution. This demonstrates that as a technology, AI serves the best interest of enhancing delivery and growing access to justice while stressing the need for ethical practices in AI, declarative algorithmic systems to allow external scrutiny to prevent adverse repercussions. With the advancement of the use of AI in the environment, it is equally important that the core principles of justice can further be upheld while going forward in the technological advancement in the environment of law in India.

3.1 AI for Transcription of Hearings

In February, 2023 the Supreme Court of India transcribed the live streamed proceeding concerning the political power machinery in Maharashtra during a crucial hearing through AI generated Live Caption. This movement advanced progress in improving the clarity and effectiveness of the judicial proceedings. Leading this technological incorporation was the Bengaluru-based company, Technology Enabled Resolutions (TERES), which had earlier enriched AI educated transcription form arbitration. Having CJI DY Chandrachud in the picture, the live transcription screen allowing the subjects to follow the case real-time, was provided to the courtroom (Roy, 2023).

The genesis of this development was anchored by the combination of legal and technology sectors. The chair of the Board of Directors for AVVO India and a distinguished lawyer and technophile Vikas Mahendra, were instrumental to the design and implementation of the transcription software based on AI. This amounted to the realization of a dream and potential of bringing about an AR experience that could transform proceedings in the Courtrooms across the nation with the help of TERES. The tool garnered interest during the Delhi Arbitration Weekend in February and specifically impacted CJI Chandrachud, who viewed it as game-changing for the judiciary (Roy, 2023).

In considering the consequences of this technology, Mahendra stressed that it's relevancy in the district judiciary system, which is charged with the majority of evidence recording. He estimated that the district judiciary could have effectiveness increased by four to five times due to the tool's prospect of the improvement of transcription activities. In addition to this, Mahendra went on to annex the primary purpose of the tool to increase completion precision, which would further help in reducing the time taken in the deciding of cases hence reducing backlog of cases existing in Indian courts.

It is notable that Indian courts have been automating the judicial processes in recent years while incorporating other AI practices in courtroom proceedings to signify the active implementation of modern advancements. As the legal perspective changes, it brought new ideas like AI based transcription tool spearheaded by the TERES and under the support of personalities like Vikas Mahendra has begun to set new benchmarks for judicial productivity and openness in the nation.

3.2 AI to Tackle Pendency

Some might recall that the Supreme Court of India had its first encounter with AI exactly one year ago through the website Supreme Court Portal for Assistance in Courts Efficiency SUPACE by the Supreme court's AI Committee. This scheme will therefore be a progressive measure towards the further development of the digitization of this branch of the judiciary. However, in the case of Justice Ravindra Bhat, then serving as the patron-in-chief of the AI committee under former CJI S A Bobde, there was apprehension over the use of AI in decision making in the judiciary including who is likely to be the judge as AI has to be an independent judicial mind (Bar and Bench, 2021).

Bobde admitted to an early reluctance towards AI in the judiciary but referred to SUPACE as an innovative experiment of human and artificial superiority in the aspiration of achieving prodigious performances. He argued that such concerns would be resolved through asserting that it was not meant to replace the judicial process, but rather it was aimed at assisting the judges by availing factual contexts from which they can come up with verdicts by referring to the discretionary power they wield as well as their sense of justice.

SUPACE was initiated by the executive producers led by Manthan Trivedi, a technology specialist; the development of SUPACE was intended to reduce judicial backwardness by widespread application of technology. Trivedi admitted to the undeniable existing concerns around adopting new technology, especially in an area as inherently involved as judiciaries, a domain where human participation has always been considered critical. He pointed out that SUPACE has always been intended as an assistant to human judges while insisting that its intention has been to help achieve better efficiency and prevent delays due to large levels of human judgment (Khan, 2023).

Trivedi, in this context, likened the way SUPACE would operate to what YouTube, for example, uses in suggesting more content based on a person's preference. Likewise, SUPACE conducted a similar analysis of the patterns in judicial households in order to give summed up versions as well as information formatted in a manner that would be well received by judicial capacities it served. Trivedi also thought about how through reduction of cases that are comparatively less complicated like traffic signal violations, adequate appeals can be predicted through the analysis of AI satisfaction rates.

However, questions arise whether enhancement of technology can actually help in solving problems of judicial backlog. A recent article by the Digital Futures Lab pointed out that AI can reduce bias and increase effectiveness by producing less variation to benefit from it. However, it elaborated that addressing the phenomenon of judicial backlog, in its multifaceted manner, cannot be just technology focused. It was postulated in the study that although call for enhancing judicial capacities and the creation of the high-volume courts are the typical recommendations to combat the backlog, the core of the problem is typically in the cultural and structural delays determined by interest and culture of the courts themselves, and macro socio-economic factors that impact the ability of the party to pursue or stall cases (Khan, 2023).

4. Potential Benefits

Various technologies, especially AI, have been extended to help the legal industry to gain speed and efficiency in data sorting and grouping. These tools are capable of handling massive amounts of information about the law within a short span of time and do very much reduce the time that would otherwise be spent in research and other due diligence exercises. This acceleration assists legal professionals in dwelling more on the strategic method and client relations than in using time organizing data.

Furthermore, they use the stated argument establishing that, through the use of AI in case analysis, historical information can be used to predict legal consequences. This capability indeed enables the lawyer to harness optimal legal outcomes by helping them devise better legal plans and foresee issues that are likely to crop up in court as well as give out proactive legal advice to their clients. Applying pattern recognition and pattern discovery for legal cases helps to improve decision making and equip legal departments with experience of potential scenarios outcomes (Pillai, 2023).

Legal professionals also stress that AI can further minimize many errors present in their work: AI solutions enhance accuracy in word processing, document, review, legal writing, transcription, and other practices where mistakes with legal implications are likely to occur. A high reliability further improves the quality of the legal work to be done thus increasing overall effectiveness of the work done and clients' satisfaction.

There are several issues that should be noted about the AI application in law firms. First and foremost, cost-effectiveness is a persuasive incentive for companies. Through using technologies to self-complete repetitive and standard jobs, the degree of differentiation and cost of operations and legal activity related time and personnel are effectively lessened. Eventually, this results in significant savings to law firms, as it means that resources can be better utilized while the cost-of-service delivery is effectively managed to enable the law firms to deliver their services at competitive market prices (Bajpai, 2021).

In addition, it enables clients to be attended to by a lawyer or any legal officer through the use of technology such as a chatbot or an artificial intelligent virtual assistant. It is possible for such interfaces to return responses of 'yes' or 'no' to answers to questions as well and offer customized solutions depending on a client's circumstances. It improves client satisfaction and the practice of managing client relations in the legal sector for law firms as clients report a sense of inclusion and support from the law firm throughout the legal procedures.

For the facet of data security, the application of AI plays a very meaningful role since it supports the security of legal data. In the light of the rising dangers of cyber criminality, AI applications can be used to analyze the presence of inconsistencies or future breaches, weaknesses within Data protection mechanisms. Such a measure puts protective measures in place that protects both the law firm's work and the client's work from any form of external or even internal interference (Reddy, 2022)

Furthermore, in the context of litigation, AI can be instrumental in helping decide the overall approach to a case in light of vast amounts of gathered data and other features of a legal process. For instance, there is the ability to scan the pattern of a 'type' of lawyer in terms of the way they approach different litigation issues, calculate the likelihood of success in certain legal positions, or even identify what a judge might be like before them given their past performance. With these observations, the legal teams are placed in a better position to develop card and thoughtful approaches when proceeding to the trial, they are also in a position to counteract the moving arguments from the other side.

5. Conclusion and The Way Forward

The technological shift concerning the implementation of AI in the Indian legal framework is exciting but also comes with sprawling challenges. While the restraints do pose issues - especially when it comes to maintaining moral norms, protecting the human face of law, and even being highly automated, the drawbacks of AI are none to be questioned. Knowing what legal efficiency, accessibility, and legal innovation can do reshapes the entire service and its products from scratch.

As for the Indian legal industry, the extent of AI implementation is massive. Targeted judicial systems with the help of artificial intelligence, machine learning, and even natural language processing can advance the entire structure and basic building blocks including cutting legal research time, quick processing of more and more cases along with earlier resolution of old cases. AI based decision making will enhance the quality of judgment with judges being in more control as more uniform and data-based decisions will work in their favor allowing more access to legal resources for everyone. In addition to this, AI allows providing essential legal help at very reasonable rates to those who are underprivileged and cannot afford legal services to begin with.

For the legal community, this entails welcoming AI not as a substitute but as an adjunct to conventional legal practice. The Indian legal system such as has inbuilt values of justice and fairness can only be complemented by technological advancements, and this is the way forward. The legal practitioners who diffuse and grow with AI technology not only guide the course of that transformation but also set the stage of its realization.

Essentially, the trajectory of the Indian legal sector's future is not only about embracing AI but using it in reconstructing it into a faster, more transparent and more responsive system for the requirements of an evolving society. However, the future also seems to demand a sense of equilibrium, one in which the just hand of technology is brought in without sacrificing the human touch of justice. Besides such implementation, AI may indeed serve as the impetus towards a more resilient and a more agile Indian judicial system.

References

1. Ahlawat, S. (2023, July 24). *The Future of Law: How AI is Transforming the Legal Profession.* Subhash Ahlawat. Retrieved June 17, 2024, from https://subhashahlawat.com/blog/ai-is-transforming-the-legal-profession.
2. *AI and the Future of Law in India: Challenges, and Opportunities - ET LegalWorld.* (n.d.). ETLegalWorld.com. https://legal.economictimes.indiatimes.com/news/opinions/ai-and-the-future-of-law-in-india-challenges-and-opportunities/115246551#:~:text=Artificial%20Intelligence%20can%20play%20an,case%20outcomes,%20and%20reducing%20delays.
3. *AI Regulation in India: Current State and Future Perspectives.* (2024, January 26). Morgan Lewis. https://www.morganlewis.com/blogs/sourcingatmorganlewis/2024/01/ai-regulation-in-india-current-state-and-future-perspectives.
4. Awasthy, Suvigya, Babu, Pintu, & Singh, Shubhangi. (2022). Application of artificial intelligence and machine learning in the indian legal system: use cases for judiciary, law firms, and lawyers. *Indian Journal of Integrated Research in Law, 2(6),* 1–14.
5. Bajpai, Raghvendra. (2021). Artificial intelligence in the legal field. *Indian Journal of Law and Legal Research, 2(1),* 1–8.
6. Dutta, Anushka. (2022). Artificial Intelligence: Its Impact on the Indian Legal System. *Indian Journal of Integrated Research in Law, 2(3),* 1–12.
7. Gupta, Jhanavi. (2021). Artificial Intelligence in Legal System: An Overview. *International Journal of Law Management & Humanities, 4,* 6076–6094.
8. Historic day, a feather in the cap of CJI Bobde: Justice NV Ramana on launch of Artificial Intelligence tool, SUPACE [LIVE UPDATES]. (2021, April 6). *Bar And Bench.* Retrieved June 17, 2024, from https://www.barandbench.com/news/artificial-intelligence-committee-supreme-court-of-india-launches-supace-live-updates.
9. *India's AI-driven legal future: Opportunities and emerging trends in 2025.* (n.d.). IndiaAI. https://indiaai.gov.in/article/india-s-ai-driven-legal-future-opportunities-and-emerging-trends-in-2025.
10. Khan, A. (2023, June 6). *AI-powered Indian judiciary: A step forward or cause for concern?* Bar And Bench. Retrieved June 17, 2024, from https://www.barandbench.com/columns/litigation-columns/ai-powered-indian-judiciary-a-step-forward-cause-concern.
11. Pillai, A. N. (2023). Legal Personality for Artificial Intelligence. *Indian Journal of Law and Legal Research, 5,* 1–11.
12. Reddy, A. Amarendar. (2022). Legal Implications in Artificial Intelligence. *International Journal of Law Management & Humanities, 5,* 1766–1793.
13. Roy, D. (2023b, February 21). *"Truly a court of record; huge resource for lawyers, students": Supreme Court begins live transcription of court proceedings.* Bar And Bench. Retrieved June 17, 2024, from https://www.barandbench.com/news/litigation/truly-court-of-record-supreme-court-begins-test-live-transcription-court-proceedings.
14. Sai, Boddu Harshith, & Sharma, Naveen. (2023). Redefining the Paradigm of the Indian Legal System through Artificial Intelligence. *Nirma University Law Journal, 12(2),* 59–82.
15. Sharma, M. (2023). India's Courts and Artificial Intelligence: A Future Outlook. *LeXonomica, 15*(1), 99–120. https://doi.org/10.18690/lexonomica.15.1.99-120.2023.
16. Singh, L., Joshi, K. A., Koranga, R. S., Pant, S. C., & Mathur, P. (2023). Artificial Intelligence in the Legal Profession: A review on its Transformative Potential and Ethical Challenges. *Association for Computing Machinery, New York, NY, United States.* https://doi.org/10.1145/3647444.3647953.
17. Suryam, S. (2023, March 27). *Punjab and Haryana High Court uses ChatGPT in bail order.* Bar And Bench. Retrieved June 17, 2024, from https://www.barandbench.com/news/litigation/punjab-haryana-high-court-uses-chatgpt-bailorder.

Intelligent Systems Using Semiconductors for Robotics and IoT – Dinesh Goyal et al. (eds)
© 2026 Taylor & Francis Group, London, ISBN 978-1-041-20408-4

66 Activity Recognition Using Machine Learning

Ankit Jain[1]
Student, Dept of CSE, PIET,
Jaipur, INDIA

Nikita Jain[2]
Professor, Dept of CE, PCE,
Jaipur, INDIA

Indra Kishor[3]
Assistant Professor, Dept of CSE, PIET,
Jaipur, INDIA

Abstract: IoT devices are becoming increasingly prevalent worldwide, especially in healthcare. This study focuses on automating Human Activity Recognition (HAR) using ADXL345 accelerometer sensors and a Nodemcu ESP8266 device for data transmission. Two accelerometers were attached to the hands, wrists, and ankles of three participants, and the data was uploaded to thingspeak.com cloud storage. The Random Forest algorithm was used for data training and testing, resulting in a 99% success rate. This model has promising applications, including monitoring elderly patients, intensive care units, and diagnosing conditions like dementia and Alzheimer's disease. Additionally, the research delves into sentiment analysis, a growing field that helps analyse large data sets efficiently. As people increasingly rely on social media platforms like Telegram and Facebook, sentiment analysis helps extract user opinions for informed decision-making. Sentiment analysis has shown potential in understanding customer beliefs and preferences. Using IMDb movie review data, both Random Forest and CNN algorithms were applied for mood prediction, with CNN achieving 85% accuracy. This indicates a strong potential for deep learning in sentiment analysis, although challenges in accuracy still remain.

Keywords: Human motion database, Probabilistic soft logic, Food and agriculture organization, Dynamic time wrapping, Deep neural networks, Human activity recognition, Probabilistic soft logic

1. Introduction

The Human Activity Recognition (HAR) system attempts to infer human behaviour based on contextual data from a variety of sensors, both internal and external, including specialized motion sensors attached to the body [3]. Although these sensors classify data well, user discomfort and the need for frequent repositioning are challenges to long-term activity tracking. HAR has, therefore, become a more prominent research area because it is believed to have potential areas such as home healthcare and public surveillance [6]. It allows real-time monitoring of the patients, aids in disease detection, and automates the surveillance in public places [18]. As the number of cell phones that are multitasked and equipped with lots of sensors increases, these devices are being used nowadays for unobtrusive activity monitoring, gathering significant data to enhance decision-making and improve daily life [4]. It opens a new field of research on applications that are human-centred. Users give rich background information, and smartphones act as primary sensing tools. Modern smartphones have sensors like

[1]2021pietcsankit024@poornima.org, [2]nikita.jain@poornima.org, [3]indra.kishor@poornima.org

DOI: 10.1201/9781003716389-66

accelerometers, gyroscopes, microphones, and cameras [17]. They provide an inexpensive and adaptive solution for HAR, enabling continuous monitoring of daily activities without disrupting routines. Early efforts include the idea of implementing tri-axial accelerometers on Android smartphones and progressing towards more robustly standardized performance measures and datasets [5]. Applications from physiological wearable devices and apps, such as those recording a person's run and cycling events with devices like an Apple Watch or Fitbit, can aid physiological processes to feed into appropriate interpretations leading to HAR [12]. The growing processing power of wearable technologies and deep learning (DL) methodologies enables real-time, granular data analysis, allowing simultaneous classification of multiple activities. Today's portable devices surpass the performance of older desktop computers, making advanced HAR feasible and accessible across various platforms, from smartphones to smartwatches [6]. Smartphones, with their built-in sensors like accelerometers, gyroscopes, and GPS, offer a practical and cost-effective alternative to wearable devices for Human Activity Recognition (HAR). Their wireless connectivity and ubiquity make them ideal for tracking activities in smart homes without requiring extensive infrastructure [18]. Though video-based sensors have been the most commonly used for HAR, privacy issues have caused increased interest in inertial sensors, such as accelerometers, to detect movements like walking, jogging, and resting [21]. Unlike wearable sensors that require implantation, smartphones provide a user-friendly solution [19]. Current smartphones offer real-time data collection capabilities and support online machine learning models despite earlier restrictions such as battery limitations and processing power [6]. A significant shift is noticeable from performing data analysis offline towards dynamic, real-time HAR operations. HAR has shown tremendous potential in applications like surveillance, healthcare, and gaming [13]. Automated systems can, for instance, detect suspicious behaviour, such as monitoring unattended luggage, to inform authorities of potential danger. In gaming, HAR enhances human-computer interaction (HCI) by recognizing and responding to player actions immediately [22]. In healthcare, such systems are valuable for monitoring patients and rehabilitating those who have been injured. Video-based systems, which utilize features such as shape, motion, and colour, play a critical role in activity detection [16]. Advanced techniques like dynamic time warping (DTW), machine learning classifiers, and deep learning (DL) enhance the accuracy of recognition [18]. DL automatically extracts relevant features, reducing dependence on manual inputs and increasing scalability, allowing for more complex HAR applications across various domains [22]. Recent advancements in Human Activity Recognition (HAR) focus on using deep learning (DL) techniques to analyze human movement in dynamic environments [17]. Recurrent Neural Networks (RNNs), which effectively capture sequential data, have been particularly successful when applied to the field of HAR. Interest in the scientific study of human behaviour continues to grow due to applications in healthcare, fitness, and fall detection [16]. Despite accurate activity tracking via wearable sensors and smartphones containing accelerometers, gyroscopes, and magnetometers, challenges remain, including battery life and sensor fusion. Studies have shown mixed results, indicating the need for a more comprehensive analysis to identify the best approaches for combining sensors [7]. Optimized sensor use is essential for improving HAR performance while minimizing redundant efforts in future studies.

2. Literature Review

Human Activity Recognition (HAR) has gained prominent attention in the recent period due to advanced sensor technology, machine learning (ML), and deep learning approaches. This paper reviews several recent developments of HAR literature and its subdomains, especially sensor-based recognition, sentiment analysis, and multimodal data integration. Sensor-based Activity Recognition There has been several studies works involving advanced ML and DL techniques for HAR. Amoun et al. (2017) proposed a DNN-based method for human activity recognition with raw sensor data in uncontrolled environments, where they demonstrated the robustness of DL models in dynamic scenarios [3]. On the other hand, Hassan et al. (2018) designed a strong HAR system using smartphone sensors and DL, where everyday devices can be utilized to detect activities accurately [24]. Dietrich and Van Laer oven (2015) classified wearable activity recognition systems, focusing on the increasing sensor integration into wearable devices [17]. Subsequent developments in wearable and IoT-based HAR systems were discussed by Bayo-Monton et al. (2018), showing how IoT-integrated wearable sensors could be a boost to eHealth systems [6]. Elist's and McConville (2021) debated whether microcontrollers were ready for DL-based HAR, raising issues of hardware limitation and optimization difficulties [19]. Deep Learning Techniques and Hybrid Models DL models, including CNNs and LSTMs, have been successfully applied to HAR and other tasks. Ahmed and Wang (2022) proposed a fine-grained DL model that combined CNN and BiLSTM for sentiment analysis, which could be used in HAR due to its ability to capture spatial and temporal features [1]. Similarly, Gumaei et al. (2019) proposed a hybrid DL model that made use of multimodal body sensing data to enhance the recognition accuracy by fusing data [22]. Multimodal and Multimodal Sentiment Analysis Multimodal data fusion is another emerging trend, combining different sources of data to enhance recognition. Gandhi et al. (2022) presented a systematic review of multimodal sentiment analysis techniques, focusing on methods for fusion and their applicability in HAR [21]. Cai et al. (2021) used multimodal attention mechanisms and RNNs to analyse sentiments, a strategy that could be extended to activity recognition tasks involving diverse sensor modalities [13]. Applications in Health, Smart Infrastructure, and Emotion Analysis HAR systems have proven to be highly promising in applications such as health monitoring and smart infrastructure. Billiet et al. (2016) combined signal processing features and movement patterns for activity recognition in physical therapy, thereby emphasizing the applicability of HAR in medical environments [12]. Bhogaraju and Krupali (2020) discussed the incorporation of HAR into smart roads, thus emphasizing its potential to improve infrastructure and traffic systems [8]. In addition, Diamantina et al. (2021) studied automatic emotion recognition using facial expression analysis, which is similar to HAR in its application of pattern recognition techniques for context-aware applications [16] Challenges and Future Directions Despite these developments, significant challenges persist. As the authors of Diykh et al. (2022) pointed out and came up with a novel approach using adaptive boosting algorithms on hybrid methods for HAR [18], real-time processing, hardware constraints, as well as generalizing a model across diverse environments and populations are some significant hurdles. These challenges, when addressed, will determine HAR's expansion into monitoring of health, smart home-based applications, and automation.

2.1 Conclusion

HAR has undergone tremendous changes with better DL models, multimodal integration of data, and innovative application-based research. Future developments in this area must deal with optimizing algorithms for quicker response times, sensor advancements, and addressing scalability as well as diversity in their applications to unlock the ultimate power of HAR.

3. Methodology

- Human Activity Recognition (HAR) Using AI Techniques
- Classification of Human Activities Using AI Techniques and Models
- Sentiment Analysis and Modelling of Human Activity Towards Various Applications like Movie Reviews
- Mapping Activity with Sentiment and Modelling Through AI and ML Techniques

Interest in studying human behaviour has increased, realizing its importance in healthcare, education, and economics. This research is essential for healthier lifestyles, abnormal behaviour detection, and public needs understanding. Activity recognition, which is the identification of human activities using sensors, is one of the key aspects of this study. On-body sensors such as accelerometers and gyroscopes, commonly found in smartphones, monitor activities but need optimization for

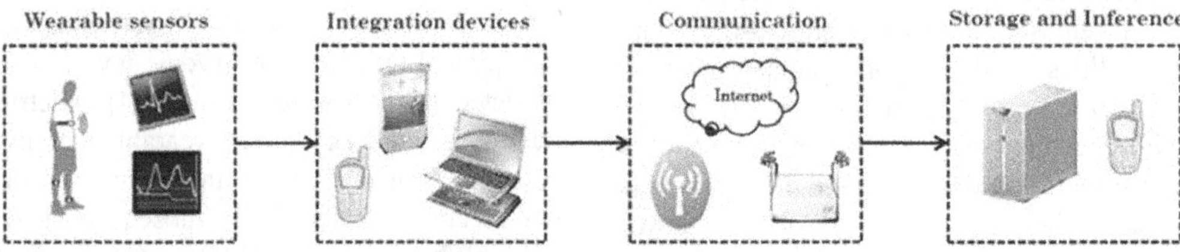

Fig. 66.1 General HAR process [1]

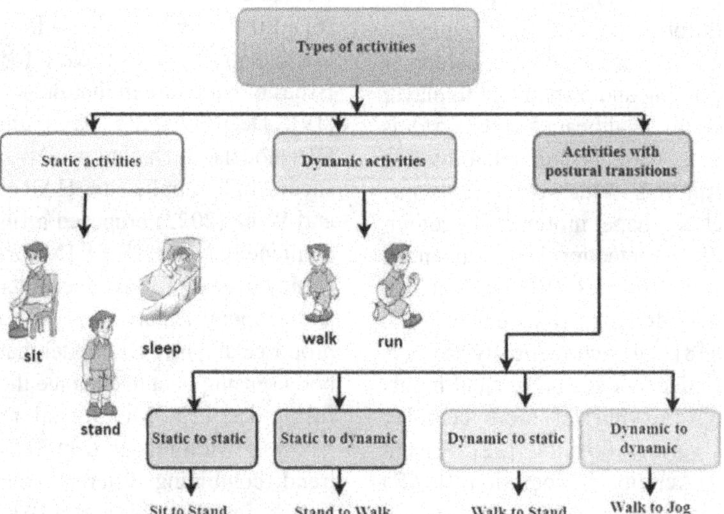

Fig. 66.2 Daily life activities [2]

performance and accuracy. Various sensor combinations have yielded mixed results, but the promise is clear. For example, the accelerometer is the most widely researched sensor in modern smartphones, and the addition of gyroscopes has been shown to improve recognition rates by 3.1%-13.4% in some studies. But there are concerns about battery life and energy consumption. Thus, comprehensive research is ongoing to further perfect these technologies for better activity recognition and outcomes in health and daily applications. By testing sensors under various conditions, researchers should be able to determine the most efficient ways of using them. This research aims to give more insight into the effective use of motion sensors and contribute towards better activity recognition systems. Future designs will benefit from such findings, offering more accurate and reliable solutions for monitoring health and wellness. The dataset and the method employed in this paper will be worth references for future research endeavours in the same subject.

4. Result Analysis

For years now, activity recognition through IoT has tremendously helped the movement of activity recognition and classification forward. It might be used for health care monitoring, exercising, and caring for aged persons. It supplies contextual data about some services. These approaches involve extracting feature vectors from the stream of data and fed into the Bayesian Networks and SVMs classifiers. Some sequential models have been included in this paper also, such as Conditional Random Fields and MLNs. Smart homes make use of MLN to recognize things better using logical and probabilistic reasoning. Probabilistic soft logic method, PSL: reduces the rule representation, deals with imprecision. Training MLN through user's input increases its accuracy. Federated learning ensures security without centralizing data. There are three prominent federated learning algorithms that promise great results. Federated averaging is useful in managing tasks safely and predicts behaviours accurately. Of course, "data islands" still exist. These systems, too, need further research to better tune for higher privacy and performance.

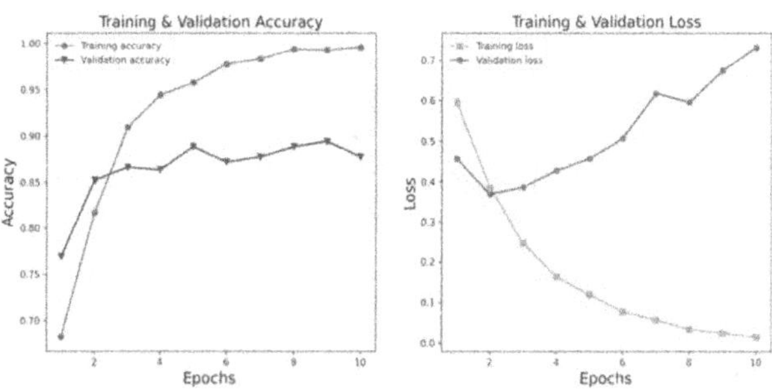

Fig. 66.3 Results – accuracy [4]

5. Discussion

This utilizes sensor information or video in the process of detecting human actions or practices. Concerning sensor information, sensors can be accelerometers and gyroscopes. Algorithms trained on labelled data learn patterns allied to particular activities such as walking or running or sitting. Methods: these include: supervised learning in that an algorithm is trained to identify assigned labels, and deep learning where algorithms can automatically draw hard features. This will culminate to an aim of classifying the activity in real time in the concerns of health monitoring, smart homes, and security systems.

6. Conclusion

Activity Recognition Using IMUs: Activity recognition using IMUs can be applied in several fields, from the medical profession to the gaming industry, the vehicle business, and industrial applications, to name a few of the possible applications. However, activity recognition has several limitations, including but not limited to, erroneous sensors, non-consistent data, a short life span for batteries, underprepared training data, and ill-positioned sensors. At the moment, there are active research activities into developing better machine learning algorithms, proper ways to collect and annotate data, and more accurate sensors among others. All these efforts are aimed at filling identified deficiencies in the system. Furthermore, wearable technology, like flexible sensors, will in not-so-distant future make activity recognition both more accurate and more applicable to real-world settings. The use of IMUs can provide useful insights on human behaviour as well as movement patterns through activity recognition. In general, this can be done through the use of IMUs. If further research and development continue on this technology, this has a lot of great potential to become one of the most influential factors in improving health outcomes, enhancing athletic performance, and increasing safety in various industries. Activity recognition by IMUs is a promising wearable device technology. The accuracy of the activity recognition depends on many factors; for example, sensor accuracy.

References

1. Ahmed, Z., & Wang, J. (2022). A fine-grained DL model using embedded-CNN with BiLSTM for exploiting product sentiments. *Alexandria Engineering Journal.*
2. Almalis, I., Kouloumpris, E., & Vlahavas, I. (2022). Sector-level sentiment analysis with DL. *Knowledge-Based Systems, 258*, 109954.
3. Amroun, H., Temkit, M., & Ammi, M. (2017). Dnn-based approach for recognition of human activity raw data in non-controlled environment. *2017 IEEE International Conference on AI \& Mobile Services (AIMS)*, 121–124.

4. Andrienko, G., Andrienko, N., & Wrobel, S. (2007). Visual analytics tools for analysis of movement data. *ACM SIGKDD Explorations Newsletter*, *9*(2), 38–46.
5. Banos, O., Galvez, J.-M., Damas, M., Pomares, H., & Rojas, I. (2014). Window size impact in HAR. *Sensors*, *14*(4), 6474–6499.
6. Bayo-Monton, J.-L., Martinez-Millana, A., Han, W., Fernandez-Llatas, C., Sun, Y., & Traver, V. (2018). Wearable sensors integrated with IoT for advancing eHealth care. *Sensors*, *18*(6), 1851.
7. Bhardwaj, R., Vaidya, T., & Poria, S. (2022). Towards solving NLP tasks with optimal transport loss. *Journal of King Saud University-Computer and Information Sciences*.
8. Bhogaraju, S. D., & Korupalli, V. R. K. (2020). Design of Smart Roads-A Vision on Indian Smart Infrastructure Development. *2020 International Conference on communication Systems & networks (COMSNETS)*, 773–778.
9. Bhogaraju, S. D., Kumar, K. V. R., Anjaiah, P., Shaik, J. H., & others. (2021).
10. Advanced Predictive Analytics for Control of Industrial Automation Process. In Innovations in the Industrial IoT (IIoT) and Smart Factory (pp. 33–49). IGI Global.
11. Billiet, L., Swinnen, T., Westhovens, R., de Vlam, K., & Van Huffel, S. (2016). Activity recognition for physical therapy: fusing signal processing features and movement patterns. *Proceedings of the 3rd International Workshop on SensorBased Activity Recognition and Interaction*, 1–6.
12. Cai, C., He, Y., Sun, L., Lian, Z., Liu, B., Tao, J., Xu, M., & Wang, K. (2021). Multimodal sentiment analysis based on recurrent neural network and multimodal attention. In Proceedings of the 2nd on Multimodal Sentiment Analysis Challenge (pp. 61–67).
13. Chen, C., Xiao, B., Zhang, Y., & Zhu, Z. (2023). Automatic vision-based calculation of excavator earthmoving productivity using zero-shot learning activity recognition. *Automation in Construction*, *146*, 104702.
14. D\'\iaz, M., Johnson, I., Lazar, A., Piper, A. M., & Gergle, D. (2018). Addressing age-related bias in sentiment analysis. *Proceedings of the 2018 Chi Conference on Human Factors in Computing Systems*, 1–14.
15. Diamantini, C., Mircoli, A., Potena, D., & Storti, E. (2021). Automatic Annotation of Corpora For Emotion Recognition Through Facial Expressions Analysis. *2020 25th International Conference on Pattern Recognition (ICPR)*, 5650–5657.
16. Dietrich, M., & Van Laerhoven, K. (2015). A typology of wearable activity recognition and interaction. *Proceedings of the 2nd International Workshop on Sensor-Based Activity Recognition and Interaction*, 1–8.
17. Diykh, M., Abdulla, S., Deo, R. C., Siuly, S., & Ali, M. (2022). Developing a Novel Hybrid Method Based on Dispersion Entropy and Adaptive Boosting Algorithm for HAR. *Computer Methods and Programs in Biomedicine*, 107305.
18. Elsts, A., & McConville, R. (2021). Are Microcontrollers Ready for DL-Based HAR? *Electronics*, *10*(21), 2640.
19. Gandhi, A., Adhvaryu, K., Poria, S., Cambria, E., & Hussain, A. (2022). Multimodal sentiment analysis: A systematic review of history, datasets, multimodal fusion methods, applications, challenges and future directions. *Information Fusion*.
20. Gumaei, A., Hassan, M. M., Alelaiwi, A., & Alsalman, H. (2019). A hybrid DL model for HAR using multimodal body sensing data. *IEEE Access*, *7*, 99152– 99160.
21. Gupta, P., Mehrotra, T., Bansal, A., & Kumari, B. (2017). Multimodal sentiment analysis and context determination: Using perplexed Bayes classification. *2017 23rd International Conference on Automation and Computing (ICAC)*, 1–6.
22. Hassan, M. M., Uddin, M. Z., Mohamed, A., & Almogren, A. (2018). A robust HAR system using smartphone sensors and DL. *Future Generation Computer Systems*, *81*, 307–313.
23. T. A., Pise, A. A., & Ratna, R. (2022). AI: A Universal Virtual Tool to Augment
24. K. Kaur and J. S. Batth, "Human Activity Recognition Using Machine Learning," 2024 OPJU International Technology Conference (OTCON) on Smart Computing for Innovation and Advancement in Industry 4.0, Raigarh, India, 2024, pp. 1-5, doi: 10.1109/OTCON60325.2024.10688067. keywords: {Performance evaluation; Technological innovation; Feature extraction; Human activity recognition; Security; Object recognition; Wearable devices; Human Activity Recognition (HAR); Machine Learning; Activity Classification; Sensor Data; Feature Extraction; Data Preprocessing; Accelerometer; Gyroscope; Inertial Sensor; Wearable Devices},
25. Gulzar, Zameer & Leema, Anny. (2019). Human Activity Recognition using Machine Learning Classification Techniques. International Journal of Innovative Technology and Exploring Engineering. 9. 10.35940/ijitee. B7381.129219.
26. Li, M., Chen, S., Chen, X., Zhang, Y., Wang, Y., & Tian, Q. (2021). Symbiotic graph neural networks for 3d skeleton-based human action recognition and motion prediction. IEEE Transactions on Pattern Analysis and Machine Intelligence, 44(6), 3316–3333.
27. Omran, T. M., Sharef, B. T., Grosan, C., & Li, Y. (2023). Transfer learning and sentiment analysis of Bahraini dialects sequential text data using multilingual DL approach. Data \& Knowledge Engineering, 143, 102106.
28. Vutharkar, S., & KV, R. K. (2023). Fin-Cology or Tech-Nance?: Emergence of FinTech. In AI-Driven Intelligent Models for Business Excellence (pp. 124 136). IGI Global.
29. Zhang, Y., Yin, Y., Wang, Y., Ai, J., & Wu, D. (2023). CSI-based location independent HAR with parallel convolutional networks. Computer Communications, 197, 87–95.
30. Chen, C., Xiao, B., Zhang, Y., & Zhu, Z. (2023). Automatic vision-based calculation of excavator earthmoving productivity using zero-shot learning activity recognition. Automation in Construction, 146, 104702.

Intelligent Systems Using Semiconductors for Robotics and IoT – Dinesh Goyal et al. (eds)

67 Bug Detection in Software Testing using ML Approach: A Literature Review

Harsha Tamboli Kumar Rawat[1]
Assistant Professor, JSPM University,
Pune

Nikita Jain[2]
Professor, Department of Computer Science & Engineering,
Poornima College of Engineering

Barkha Narang[3]
Assistant Professor, Department of Advance Computing,
Poornima College of Engineering

Abstract: Software testing is key step for software development. Manual testing methods used traditionally with today's software system are not feasible due to complexity. In the present scenario Artificial Intelligence (AI) has become apparent as a powerful tool for conducting or completing the software testing step. In many real-world scenarios and complex problem in the global era, we can incorporate Artificial Intelligence (AI) algorithms and Machine Learning (ML) advances towards software testing phase. Effectively not being implemented for software testing work. This review study focuses to provide the ongoing and recent trend of software testing using AI and ML. This paper directs to discuss in broad the ML approaches and AI algorithms used for software testing.

For doing the review different research articles from different have been searched using from different research databases, journals and articles using a search string (Software Testing, Artificial Intelligence, Machine Learning and Bug Detection). Initially, 65 articles were extracted from different research libraries. Around 50 articles were studied to get into the detail of using AI in software testing and get an in-depth overview of it.

After gradual filtering in three different steps, total 29 articles were selected for final review, The data for doing the analysis of the AI and ML methods is collected from Scopus Elsevier, web of Science, IEEE journals, Research Gate and Google Scholar database.

Keywords: Software testing, Machine learning (ML), Artificial intelligence (AI), Supervised learning, Unsupervised learning

1. Introduction

Testing is process of finding whether developed software meets its original, required requirements, specified requirements and functionality or not. It is the process of mainly performing the validation and verification process for testing or evaluating whether the designed system is meeting the requirements as defined by user. [1] The thing to be considered during testing is to how minimize number of test cases, and to broadly conclude to judge the modules to test for what or not.

Software testing currently faces several challenges such as complex real time scenarios and complex applications. It is more challenging testing all the possible scenarios of an application. Also, the traditional approach used for testing was time consuming, complex and requires effort. Apart from that, keeping a parallelism with the current software development is quite challenging [2].

[1]harsha.kumrawat@gmail.com, [2]nikita.jain@poornima.org, [3]barkha.narang.poornima.org,

DOI: 10.1201/9781003716389-67

AI (Artificial Intelligence) is now used as the latest technology across all industries and domains. AI technology can also be added in the field of software testing, for delivering more quality outcomes and to ease automation testing.

Further, in this testing method. For executing the test cases that uses data and algorithm to perform tests without human intervention AI tools can be used. Over the past two decades software testing has evolved significantly as getting quality software. Beginning from the traditional approach or manual approach of testing to automation testing, the testing journey is interesting and motivating. But in today's fast-paced IT world and with the growing global requirements, the software testing phase should incorporate some new methodologies. AI based testing is proving a very impactful tool. AI and ML involve the development and designing of some specific and unique algorithms that can predict or access data or make predictions for implementing software testing effectively.

From different research databases, research articles are selected for this review study using search string like ML, AI, Software testing , etc. The aim of survey is to find the current and recent trends in software testing using AI.

In the paper, we tried to identify the recent and current trends in software testing using AI and discussed the following research questions. The Table 67.1 shows that research questions:

Table 67.1 Research questions

Sr. No.	Research Questions
RQ1.	Does manual testing is feasible for complex application?
RQ2.	Can AI be integrated with software testing?
RQ3.	On which phase of software testing AI can be implemented?
RQ4.	Which techniques do researchers use for integrating AI techniques in terms of software testing?

This paper briefly presents the review and studies of all the available literature papers, journals and articles discussing the AI/ML approaches/algorithms applied or incorporated for software testing.

Inclusion Criteria:

1. Studies the concepts in software testing.
2. Analyzing the concept of AI and ML.
3. Review and analyze software bug prediction models implementing machine learning.
4. Research which compares different bug prediction models performance.

Exclusion Criteria:

1. Literature that are not related to software bug prediction using machine learning.
2. Research which doesn't have the discussion on achievements of bug prediction models.

2. Software Test Methods

In this section we will give a short definition of software testing method used, as follows:

A. *Black Box Testing:* In black box testing, there is no understanding of inside of software by tester. It just deals with validating the functionality of the software based on the provided specifications or requirements for that application.

   The common techniques used for finding implementation bugs malformed data are used and the technique is known as fuzzing, and for testing graphical user interface navigation and state machines, state transition testing is used.[3][4]

B. *White Box Testing*: In this the internal logic, flow, and structure are tested. Also, the workflow of software application is tested. Software testing technique that works on testing inner flow and internal structure of a software application. Tester uses his knowledge to draft test cases; accuracy of software at the code level can be verified and assisted with test cases. [3][4]

C. *Grey Box Testing*: In this a software tester can partially see the internal structure, or we can say it is semitransparent. In Grey box testing the tester will have the partial knowledge of internal working, as opposite of black box in which tester lack inner working knowledge of the module or application. Table 67.2[30] shows the dissimilarity between all these three Testing.

Table 67.2 Dissimilarity between all these three testing

TESTING TYPE	BLACK BOX TESTING	GRAY BOX TESTING	WHITE BOX TESTING
Granularity level	Low level	Medium level	High level
Known as	Opaque-box testing, closed-box testing, input-output testing	Traslucent box testing	Glass-box testing, clear-box testing, logic-based and code-based testing
Participants	Done by end-users and also by tester and developers	Done by end-users (called user acceptance testing) and also by testers and developers	Done by testers and developers
Design techniques	Decision table testing, all pairs testing, equivalence partitioning, error guessing	Matrix testing, regression testing, pattern testing, orthogonal array testing	Control flow testing, data flow testing, branch testing
Insight into internal working	External focus only	External with some internal focus	Complete internal focus
Insight into internal working	External focus only	External with some internal focus	Complete internal focus
Discovery of hidden errors	Challenging	Possible at the user level	Easier due to in-depth knowledge
Time consumption	Less exhaustive	Partially exhaustive	Most exhaustive

3. AI Approaches and ML Approaches

Artificial Intelligence (AI) and Machine Learning (ML) form backbone of modern technological advancements, providing machines ability to accomplish tasks that require human intelligence [5].

3.1 Components of Artificial Intelligence

Perception: The capability of machines to interpret data from the environment through sensors, cameras, microphones, or other data sources. Examples include image recognition, speech recognition, and sensor data analysis.[2]

Reasoning and Decision-Making: AI systems use logical reasoning to make decisions based on the data they receive. This includes problem-solving, planning, and decision-making under uncertainty.

Learning: This is a critical component of AI where systems improve their performance over time. Learning can be categorized into various types.

Natural Language Processing (NLP): Incorporating knowledge in machines is Natural Language Processing. It makes the machines understand human language. Applications include conversational agents, sentiment analysis and language.

Robotics andActuation: AI can be integrated with robotics, allowing machines to perform tasks in the physical world, such as manufacturing, assembly, or autonomous navigation.

3.2 Concepts of Machine Learning

The subset of AI focusing on algorithms, identifying patterns and relationships in data, making predictions or decisions without being explicitly programmed for specific tasks is Machine Learning. These are classified into different types of learning methods [6]. The main types of ML are:

Supervised Learning: In this type, the data is labeled and model is trained on labelled data. Input and correct outputs are also known. Supervised learning is most common form of Machine learning [7]. In supervised machine learning an algorithm is implemented. Algorithm can produce general patterns and hypothesis, for predicting the future instances. General patterns and hypothesis are used [8][7]

Unsupervised Learning: Model has data that lacks labeled outcomes. Aim is to find underlying patterns or groupings inside data [7]. Common clustering algorithms like Clustering algorithms like hierarchical clustering and K-Means are used.

Reinforcement learning: Generally used in gaming, robotics and for autonomous systems.

4. Literature Review

Software testing is checking whether software is meeting requirements. It is the process of identifying and detecting bugs. Software should always work as per the requirements, as per the functionality. But naturally there are bugs or defects in software. Development phase leads to Bugs. It is code refracting, maintenance, fixing, feature addition [9]. Therefore, the software must be tested in different scenario before delivering to client. Software testing has different approaches and methodologies. Which software technique should be used for testing, depends on nature of software. Also on working approach of the software [2]. ML is the most widely applied area of AI.

AI is the latest methodologies used for reducing effort and time required for testing the software and is proven to be a formidable opponent [10]. ML is a process where machines are designed to get knowledge or learn from data using algorithms. Furthermore, based on the collected and analyzed information from that collected data predictions are defined [11].

Software development life cycle consists of many phases. Testing phase ought to be done properly and productively. Such that bugs can be detected and can be released bug-free. It should fulfill all desired requirements to end users [13][14]. In predicting the buggy modules by using the historical data ML techniques are used extensively [26]. Can also be applied during any phase such as planning, pointing out modern problems, design, build, test, deploy and maintain [16]. Identifying and embedding software defects are principal task for ensuring standardized products [17]. Many different classification approaches use ML techniques like Naives Byes [18], SVM [19], decision tree [21] and neural network [22] and fuzzy rule-based approach [20][17]

Bug Prediction with Machine Learning Approach: The objective of software bug prognosis is to pinpoint which software modules are likely to contain bugs by leveraging basic project resources before actual testing commences [11]. Bug prediction can be performed using ML. The critical issue in software development and maintenance processes is Software Bug Prediction (SBP), and the overall testing process is based on it. If bugs or software faults are predicted in the phase, it ameliorates software efficiency, reliability, quality and lessens software cost [9]. In this research paper [16], they have used supervised machine learning technique. In this approach [16], for predicting the likelihood of future bugs ML algorithm analyzes the software. First the data will be collected from system repositories, the collected data can be processed, and developer metrics and bug fix information will be designed. Next step after data collection is to train an ML model to classify the glitches in modules and classes. For identifying patterns, historical data from past software projects will be used for training on ML models. [16][22]

In [6], for doing Structure-based testing or coverage analysis, Temporal testing, Functional testing, Safety testing, Genetic algorithms (GA) have been applied. GA procedure can be adopted [23], for test data generation.

[24] presents SAGA-ML, active learning system for black box software testing. These trained sets help develop a model based on system's behavior. After gaining an

understanding of the model, it is used to determine where sampling should occur or not. This intelligent sampling technique is used to use limited testing resources effectively. For building the model and predicting the occurrence of the software bugs, three supervised machine learning algorithms are used [17]

Singh et al. [28] showed that CBO, WMC, LOC, and RFC are effective in predicting defects

In [9], predicting software faults based on historical data, three supervised ML algorithms are used, Naïve Bayes (NB), Artificial Neural Network (ANN) and Decision Tree (DT).

In [25], presented a good systematic review using Machine Learning (ML) for software bug prediction techniques.

For example, a linear Auto-Regression approach in the study [2] proposed, to predict faulty modules. In [27], depending on the historical data future faults has been predicted, data of the software accumulated faults.

5. Result

This section provides a broad knowledge about software testing using AI techniques. This study directs to bestow detail of prevailing research in this sphere by scrutinizing several studies. Answers to each research question from the collected studies have been also analyzed. 29 studies have been reviewed in the final study, and 9 have been selected for and the details of the findings and analysis of these studies.

RQ1: Does manual testing is feasible for complex application?

Manual testing needs dedicated human resources to execute test cases manually so it can be considered as labor intensive testing technique. Manul testing is time consuming, as manually all the requirements and the functional requirements are tested manually. All the scenarios are not possible to test manually, as it can be impractical or time consuming. Manual testing results may vary based on the individual tester's skills and experience, leading to inconsistencies in test execution and potentially overlooking defects [6].

Manual testers may find it tedious and error-prone to execute repetitive test cases over multiple iterations, increasing the likelihood of oversight or missing critical defects [6].

As per the study [11], universal software spending amounted is \$3.7 trillion in 2013. The overall cost consumed on quality assurance and testing is 23% [15]. In September 2018, above 50% of total software cost consumed in identifying, fixing defects and losses attributable to software failures in production environment, as per report (Anon, 2018b) published, in Consortium for IT Software Quality (CISQ),[17].

RQ2: Can AI be integrated with software testing?

The drawbacks of manual testing can be overcome with the help of Integration of AI. The repeated testing task can be automated by training an ML model, which in turn reduces the required efforts.

RQ3: On which phase of software testing AI can be implemented?

Tset result analysis. Test case evaluation testing cost estimation, defect prediction, test case generation, and many more testing tasks are automated with help of ML techniques [29].

RQ4:Which techniques researchers use for integrating AI techniques in terms of software testing?

When used in software testing, researchers use different performance matrices to outpour ML algorithms. ML algorithms have shown good results in automating software testing tasks. Some of the promising algorithms are Neural networks, Decision trees, Support vector ma chines, and Random Forest. Some reserachers used Genetic Algorithms, some uses hostorical data for creating training data or model, and then uses thata label model.

6. Conclusion

In the development of software, software testing engages in a crucial role. As softwre system intricates, conventional testing methods are not as much feasible. There has been augumenting engrossment in leveraging AI techniques for software testing. This study explores the current state of the art of AI techniques in software testing. Also, this study scritinizes various techniques, methods,approaches, and tools employed in software testing field, evaluating their effectiveness.

References

1. Jamil, M. A., Arif, M., Abubakar, N. S. A., & Ahmad, A. (2016, November). Software testing techniques: A literature review. In *2016 6th international conference on information and communication technology for the Muslim world (ICT4M)* (pp. 177–182). IEEE.
2. Islam, M., Khan, F., Alam, S., & Hasan, M. (2023, October). Artificial Intelligence in Software Testing: A Systematic Review. In *TENCON 2023-2023 IEEE Region 10 Conference (TENCON)* (pp. 524–529). IEEE.
3. Lima, R., da Cruz, A. M. R., & Ribeiro, J. (2020, June). Artificial intelligence applied to software testing: A literature review. In *2020 15th Iberian Conference on Information Systems and Technologies (CISTI)* (pp. 1–6). IEEE.
4. Khan, M. E., & Khan, F. (2012). A comparative study of white box, black box and grey box testing techniques. *International Journal of Advanced Computer Science and Applications*, *3*(6).
5. Akanksha Mishra,September-October -2024, A Comprehensive Review of Artificial Intelligence and

Machine Learning : Concepts, Trends, and Applications, Int J Sci Res Sci & Technol., 11 (5) : 126–142

6. Zhang, D. (2006, November). Machine learning in value-based software test data generation. In *2006 18th IEEE International Conference on Tools with Artificial Intelligence (ICTAI'06)* (pp. 732–736). IEEE.
7. LeCun, Y., Bengio, Y., & Hinton, G. (2015). Deep learning. *nature*, *521*(7553), 436–444.
8. Singh, A., Thakur, N., & Sharma, A. (2016, March). A review of supervised machine learning algorithms. In *2016 3rd international conference on computing for sustainable global development (INDIACom)* (pp. 1310–1315). Ieee.
9. Tan, L., Liu, C., Li, Z., Wang, X., Zhou, Y., & Zhai, C. (2014). Bug characteristics in open source software. *Empirical software engineering*, *19*, 1665–1705.
10. Ajorloo, S., Jamarani, A., Kashfi, M., Kashani, M. H., & Najafizadeh, A. (2024). A systematic review of machine learning methods in software testing. *Applied Soft Computing*, 111805.
11. Khirod Chandra Panda, Machine Learning Methods for Predicting Software Bugs, European Journal of Advances in Engineering and Technology, 2020, 7(11):36–42, ISSN: 2394-658X
12. Arar, Ö. F., & Ayan, K. (2015). Software defect prediction using cost-sensitive neural network. *Applied Soft Computing*, *33*, 263–277.
13. Dick, S., Meeks, A., Last, M., Bunke, H., & Kandel, A. (2004). Data mining in software metrics databases. *Fuzzy Sets and Systems*, *145*(1), 81–110.
14. Zhou, Z. H. (2021). *Machine learning*. Springer nature.
15. Catal, C., & Diri, B. (2009). Investigating the effect of dataset size, metrics sets, and feature selection techniques on software fault prediction problem. *Information Sciences*, *179*(8), 1040–1058.
16. Immaculate, S. D., Begam, M. F., & Floramary, M. (2019, March). Software bug prediction using supervised machine learning algorithms. In *2019 International conference on data science and communication (IconDSC)* (pp. 1–7). IEEE.
17. Pachouly, J., Ahirrao, S., Kotecha, K., Selvachandran, G., & Abraham, A. (2022). A systematic literature review on software defect prediction using artificial intelligence: Datasets, Data Validation Methods, Approaches, and Tools. *Engineering Applications of Artificial Intelligence*, *111*, 104773.
18. Rish, I. (2001, August). An empirical study of the naive Bayes classifier. In *IJCAI 2001 workshop on empirical methods in artificial intelligence* (Vol. 3, No. 22, pp. 41–46).
19. Scholkopf, B., & Smola, A. (2005). Support vector machines and kernel algorithms. In *Encyclopedia of Biostatistics* (pp. 5328–5335). Wiley.
20. Singh, P., Pal, N. R., Verma, S., & Vyas, O. P. (2016). Fuzzy rule-based approach for software fault prediction. *IEEE Transactions on Systems, Man, and Cybernetics: Systems*, *47*(5), 826–837.
21. Quinlan, J. R. (1986). Induction of decision trees. *Machine learning*, *1*, 81–106.
22. Krose, B., & Smagt, P. V. D. (1996). *An introduction to neural networks*. The University of Amsterdam.
23. P. McMinn, "Search-based Software Test Data Generation: A Survey", Software: Testing, Verification and Reliability, Vol. 14, No. 2, 2004, pp. 105–156.
24. Xiao, G., Southey, F., Holte, R. C., & Wilkinson, D. (2005, July). Software testing by active learning for commercial games. In *AAAI* (pp. 898–903).
25. Malhotra, R. (2015). A systematic review of machine learning techniques for software fault prediction. *Applied Soft Computing*, *27*, 504–518.
26. Hammouri, A., Hammad, M., Alnabhan, M., & Alsarayrah, F. (2018). Software bug prediction using machine learning approach. *International journal of advanced computer science and applications*, *9*(2).
27. Sheta, A., & Rine, D. (2006). Modeling incremental faults of software testing process using AR models. In *the Proceeding of 4th International Multi-Conferences on Computer Science and Information Technology (CSIT 2006), Amman, Jordan* (Vol. 3).
28. Singh, Y., Kaur, A., & Malhotra, R. (2010). Empirical validation of object-oriented metrics for predicting fault proneness models. *Software quality journal*, *18*, 3–35.
29. Durelli, V. H., Durelli, R. S., Borges, S. S., Endo, A. T., Eler, M. M., Dias, D. R., & Guimarães, M. P. (2019). Machine learning applied to software testing: A systematic mapping study. *IEEE Transactions on Reliability*, *68*(3), 1189–1212.
30. unpacking-white-box-gray-box-and-black-box-testing.jpg

Note: All the tables in this chapter were made by the authors.

Intelligent Systems Using Semiconductors for Robotics and IoT – Dinesh Goyal et al. (eds)
© 2026 Taylor & Francis Group, London, ISBN 978-1-041-20408-4

68 Artificial Intelligence Supported Emotion Recommendation for Personality Enhancement

Monika Bhatt*, Tarun Shrimali
Janardan Rai Nagar Rajasthan Vidyapeeth,
Udaipur, India

Abstract: In contemporary society, the pursuit of personal development and emotional well-being has become increasingly paramount. With the advent of artificial intelligence (AI), novel avenues for addressing these aspirations have emerged. This research paper investigates the integration of AI in the domain of emotion recommendation to facilitate personality enhancement. Through an extensive review of existing literature on emotion recognition, personality psychology, and AI algorithms, this study proposes a framework leveraging AI technologies to provide tailored emotion recommendations for individuals seeking personality enrichment. The proposed framework employs techniques to infer an individual's emotional state and personality traits. Subsequently, advanced AI algorithms, such as deep learning and natural language processing, are utilized to generate personalized emotion recommendations. These recommendations aim to assist individuals in managing their emotions effectively, fostering positive psychological growth, and enhancing their overall personality. Furthermore, the paper discusses the ethical considerations and potential challenges associated with the implementation of AI-supported emotion recommendation systems. Privacy concerns, algorithmic biases, and the impact on human autonomy are among the critical issues addressed. Strategies for mitigating these challenges, such as transparent algorithm design, informed consent protocols, and user control mechanisms, are proposed in personality enhancement contexts. This research, we aim to contribute to the growing body of knowledge on the intersection of AI, emotion science, and personality psychology. Individuals can embark on a journey of self-discovery, emotional resilience, and holistic personality development, ultimately leading to a more fulfilling and enriched life experience.

Keywords: Artificial intelligence, Sentimental analysis

1. Introduction

In the fast-paced digital era, the quest for personal growth and emotional well-being has gained unprecedented significance. With the exponential advancements in artificial intelligence (AI), there lies an immense potential to revolutionize how individuals navigate and enhance their personalities. This paper delves into the realm of AI-supported emotion recommendation for personality enhancement, aiming to explore the synergy between cutting-edge technology and human psychology. Traditionally, the realm of personality development and emotional intelligence has relied heavily on self-reflection, interpersonal interactions, and psychological interventions. While these methods remain invaluable, he integration of AI presents a paradigm shift, offering novel opportunities to augment these processes. By harnessing AI algorithms, it becomes feasible to provide personalized recommendations tailored to an individual's unique emotional landscape and personality traits.

Emotions serve as signals, guiding behavior and influencing cognitive processes, interpersonal relationships, and decision-making.

In recent years, AI has made remarkable strides in the domain of emotion recognition and affective computing. Through sophisticated machine learning algorithms and computational models, AI systems can decipher subtle cues from various data sources, including text, speech, facial expressions, and physiological signals, to infer an

*Corresponding author: monikabhatt10@gmail.com

DOI: 10.1201/9781003716389-68

individual's emotional state with remarkable accuracy. Leveraging these capabilities, AI-supported emotion recommendation systems can provide timely interventions and personalized guidance to help individuals manage their emotions, cultivate positive attitudes, and navigate life's challenges more effectively. However, as we delve deeper into the integration of AI in emotion recommendation for personality enhancement, it is essential to address several critical considerations.. Moreover, the responsible deployment of AI technologies necessitates transparency, accountability, and user empowerment to ensure that the benefits outweigh the risks. By synthesizing in sights from psychology, AI, and ethics, we endeavor to elucidate the potential impact, challenges, and ethical implications of integrating AI into the realm of emotional intelligence and personality development. Ultimately, this research aims to pave the way for a more nuanced understanding of how AI can empower individuals to embark on a journey of self-discovery, emotional resilience, and holistic personality enrichment.

2. Artificial Intelligence

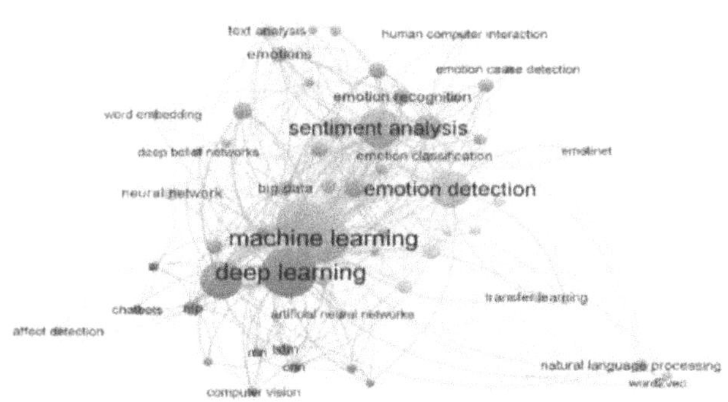

Fig. 68.1 Intersection of AI

In the dynamic landscape of artificial intelligence (AI) research, the intersection of AI and emotion intelligence has emerged as a fertile ground for innovation and exploration. Emotions, integral to the human experience, exert a profound influence on cognition, behaviour, and overall well-being. Understanding and managing emotions effectively are fundamental aspects of personal growth and interpersonal relationships. In this context, the integration of AI technologies offers unprecedented opportunities to augment emotional intelligence and facilitate personality enhancement. Emotion recognition, a corner stone of AI-driven emotion intelligence, involves the ability of machines to detect. Through advanced machine learning algorithms, AI systems can analyze complex data patterns and extract meaningful insights about an individuals emotional state with remarkable accuracy. This capability forms the foundation for AI-supported emotion recommendation systems, which aim to provide personalized guidance and interventions to help individuals navigate their emotions and enhance their personalities.

From assisting individuals in recognizing and regulating their emotions to offering tailored recommendations for behavior modification and self-improvement, AI holds the promise of revolutionizing how we understand and cultivate emotional well-being. By leveraging AI algorithms, capable of processing vast amounts of data and learning from feedback, emotion recommendation systems can offer timely interventions tailored to individual preferences, tendencies, and personality traits. Moreover, the potential impact of AI on human autonomy and interpersonal relationships warrants careful examination and thoughtful deliberation.

This paper seeks to explain the potential of AI-supported emotion recommendation for personality enhancement. By synthesizing insights from psychology, AI research, and ethics, we aim to take the opportunities. Through a comprehensive examination of current research and emerging trends, we endeavour to provide a roadmap for harnessing the power of AI to foster emotional resilience, promote personal growth, and enhance overall well-being.

3. Methodology

Emotion detection ED can be termed as figuring out the text's true intentions and feelings, including joy, sadness, wrath, disgust, guilt, fear, and humiliation. Emotion extraction from text has grown in popularity and piqued researchers' interest recently in natural language processing research. The text is effectively analysed using the emotion analysis methodology using information extraction and natural language processing methods. As a result, textual emotion analysis helps to reveal a person's viewpoint or state of mind through their posts during exchanges. Unfortunately, It leads to misprediction emotions. Create a neural network using MATLAB's nntool, you'll need input-output data pairs. For this example, let's generate some random data for demonstration purposes. We'll create a simple dataset with 100samples, where each sample has four input features representing different emotional attributes, and one output representing a target value. Here's a sample table with values suitable for curve and graph making in nntool:

In this table:

- 1, 2, 3, and 4 represent the input features (emotional attributes).
- Output represents the target output.

You can replace these values with your own dataset. Once you have your data prepared, you can use NNTool in MATLAB to create and train a neural network model. Make sure for proper evaluation of your model.

- **Generate Input Data:** For each sample, you can generate random decimal values for the input features. These values could represent various emotional attributes or any other features you want the neural network to learn from. You can use Matlab's rand()

Table 68.1 Data being stored

1	2	3	4	Output
0.1	0.5	0.8	0.3	0.6
0.3	0.2	0.7	0.9	0.8
0.4	0.6	0.4	0.4	0.7
0.7	0.9	0.6	0.4	0.3

function to generate random values between 0 and 1. The following scenario states the flow of data

- **Data Preparation:** You start by preparing your dataset's, which includes input features and corresponding continuous target variables.
- **Network Architecture:** Next, you need to explain the architecture of NN that, includes input nodes, hidden layers, and output nodes.
- **Training:** You train the neural network using your dataset's. During training, the network adjusts its weights and defines the actual target values.

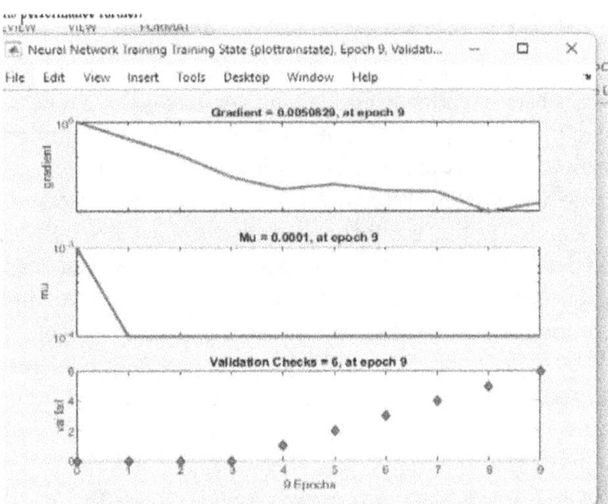

Fig. 68.2 The NN training states

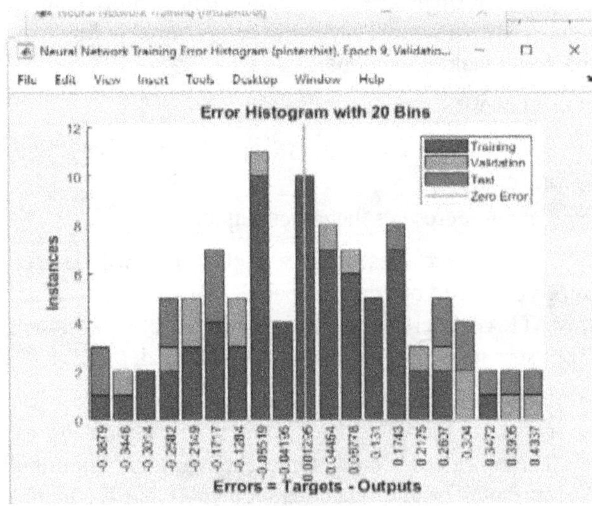

Fig. 68.3 Histogram

- **Validation:** After training, you validate the performance of the trained neural network using a separate validation dataset's. This helps ensure that the network generalizes well to new data and doesn't over-fit.
- **Testing:** Finally, you will have to test the trained NN on a dataset's to determine the result.

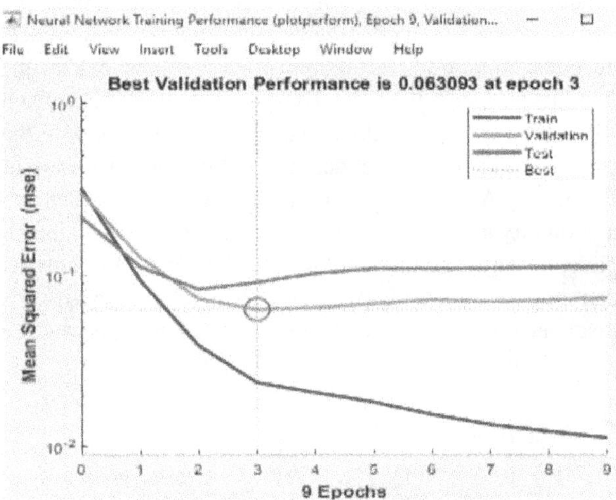

Fig. 68.4 Performance of the NN

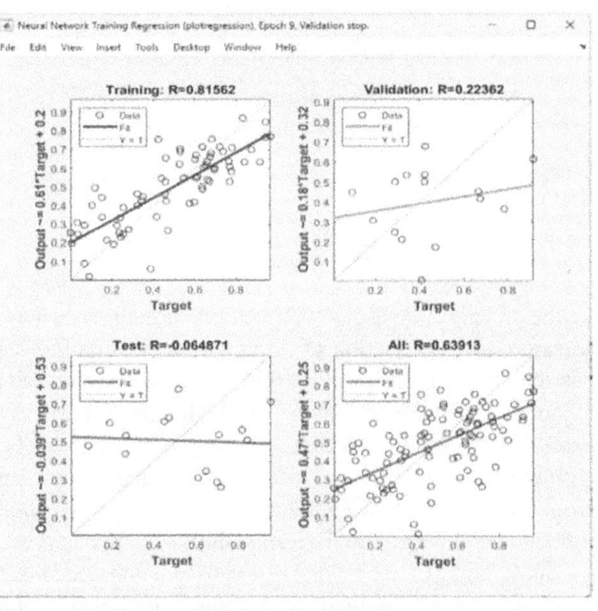

Fig. 68.5 The NNTR

4. Result

This section presents the experimental setup and results obtained from the trained neural network model. It includes details of the dataset used, training parameters, performance metrics such as accuracy, precision, recall, and F1-score, as well as visualizations of predicted vs. actual personality-enhancement strategies. Also using Nntool with the data preparations and created a dataset

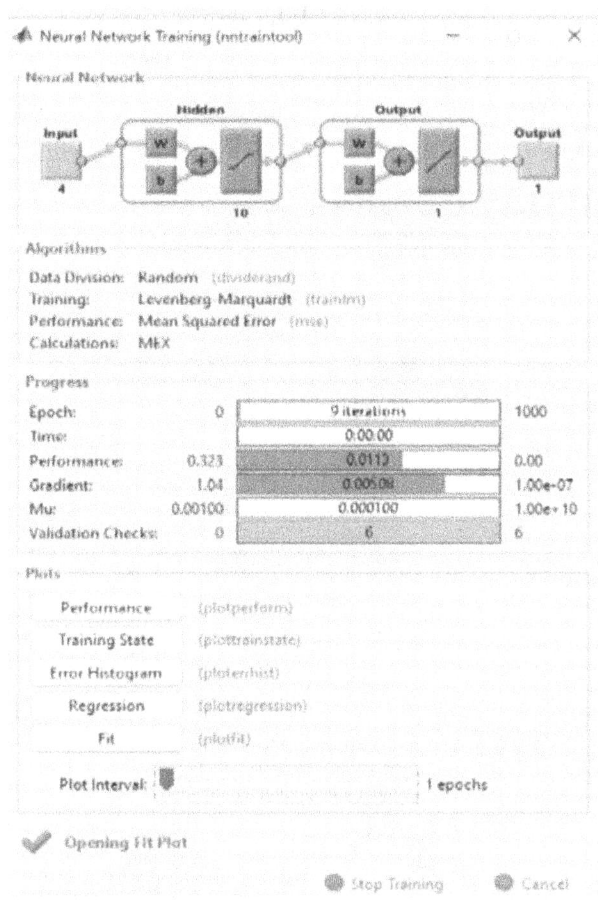

Fig. 68.6 The neural network data training tool

then visualizing the aspects of Neural Network while determining the number of number of input nodes, hidden layers, neurons per layer, activation functions, and the number of output nodes. For emotion recommendation, the input nodes would represent emotional attributes, and the output nodes could represent recommended personality-enhancement strategies or interventions .neural network using the prepared dataset. Even if this intelligent agent is well compensated for its contributions to society and effect today, it can develop emotional intelligence to become more human. We can enable Artificial Intelligence to expand its knowledge base and offer deeper and more sophisticated answers to challenging issues with the aid of emotional intelligence. The gap between humans and machines may be lowered if we develop an emotional artificial intelligence. This opens up new prospects through equitable treatment and can assist in areas such as medicine, consultation, education, and more.

Recent advancements in technology have made it possible for researchers to use automated, language-based artificial intelligent models in place of self-reports for personality assessment. However, prior research generally ignored the multidimensionality of personality and did not thoroughly explore validity concerns. For the present research work cutting-edge approaches for using automation, and different technologies will be explored to systematically extract personality- related data from source data, which offers rich information and reflects a variety of personality traits but has been incredibly underutilized.

References

1. Alex M.G. Almeida, Ricardo Cerri, Emerson Cabrera Paraiso, Rafael Gomes Mantovani, Sylvio Barbon Junior, 2018, Applying multi-label techniques in emotion identification of short texts, Neuro computing 320 (2018) 35–46.
2. Sandeep Dwarkanath Pande, Pramod Pandurang Jadhav, Rahul Joshi, Amol Dattatray Sawant, Vaibhav Muddebihalkar, Suresh Rathod, Madhuri Navnath Gurav, Soumitra Das, 2022, Digitization of handwritten Devanagari text using CNN transfer learning – a better customer service support, Neuroscience Informatics. 2, 3.
3. Usha Rani Kandula, Daisy Philip, Sunitha Mathew, 2022, Efficacy of video educational program on interception of urinary tract infection and neurological stress among teenage girls: an uncontrolled experimental study, Neuroscience Informatics, 2, 3, ISSN 2772–5286.
4. Rumi Iqbal Doewes, Lekshmi Gangadhar, Saranyadevi Subburaj, 2021, An overview on stress neurobiology: fundamental concepts and its consequences, Neuroscience Informatics, Volume 1, Issue 3, ISSN 2772–5286.
5. E. Chapela, I. Morales, J. Quintero, M. Félix-Alcántara, J. Correas, J. Gómez-Arnau,2020, Relationship between emotional intelligence and neurocognition in severe mental disorders, Published online by Cambridge University Press, European Psychiatry, Volume 33, pp. S367.

Note: All the figures and table in this chapter were made by the authors.

Intelligent Systems Using Semiconductors for Robotics and IoT – Dinesh Goyal et al. (eds)
© 2026 Taylor & Francis Group, London, ISBN 978-1-041-20408-4

69 Analysing Twitter Sentiment to Understand Public Opinion

Veena Yadav[1], Geeta Tiwari[2]
Department of Computer Engineering,
Poornima College of Engineering,
Jaipur, India

Payal Bansal[3]
Department of ECE,
Poornima Institute of Engineering and Technology,
Jaipur, India

Megha Gupta[4]
Department of Computer Engineering,
Swami Keshvanand Institute of Technology Management and Gramothan,
Jaipur, India

Abstract: To use the Twitter sentiment analysis system to analyse tweets from multiple users to determine how people feel about certain issues. This method categories tweets as neutral, oppose or in favour using Machine Learning and Natural Language Processing (NLP) to provide real time insights into public attitudes, such as how user feel about items, political events, or social issues, the system primarily processes text data. Despite challenges like sarcasm, slang and other languages it is very helpful in fields like government, business and emergency response. This systems ability to efficiently analyses and understand tweets is being improved by emerging technologies, such as complex AI models the technology analysis of public sentiment to help organizations make better decision. A machine learning technique called Twitter sentiment analysis automatically classify messages as neutral, negative or anti neutral. It uses Natural Language Processing (NLP) and machine learning (ML) to analyses a Twitter context and words to determine its sentiment. This study examines the methods and strategies for analyzing Twitter data to determine popular opinion. We look at many sentiments categorization, challenges and how they might be applies to trends, Preferences and attitudes.

Keywords: Machine learning, Twitter sentiment analysis (SA), Maximum entropy support vector machine (SVM), Naive bayes (NB)

1. Introduction

Social media has become a very crucial element of communication, enabling the individual to express their thoughts, emotions, feelings, and ideas to audiences worldwide.

The Application "Twitter", in particular, has gained immense popularity due to its real-time updates and posts that are concise. Each day, countless users share Tweets on a diverse variety of topics, including news headlines and details, political opinions, entertainment, and their own stories. This positions Twitter as a significant resource for public sentiments and opinions on various matters.

Using technology to analyze tweets and determine their emotional tone is known as Twitter sentiment analysis. This tone is frequently classified as neutral, negative, or positive. Organizations, governmental entities, and scholars can better grasp Public Opinion on specific topics or events by analyzing these attitudes.

[1]veena.yadav@poornima.org, [2]Geeta.tiwari@poornima.org, [3]Payal.bansal@poornima.org, [4]Megha.gupta@skit.ac.in

DOI: 10.1201/9781003716389-69

Analyzing tweets provides instant insights from a vast number of people worldwide, in contrast to traditional methods like surveys, which can be time-consuming and only engage a small audience. Raffel Et Al. and Clark Et Al. presented models that demonstrate their uses on Multiple NLP tasks comprising of question-answering and information retrieval.

An alternate pre-training technique called Electra was presented by Clark et al. and has been successfully used to perform competitively on sentiment analysis tasks.

Significant findings were reported by Liu et al. on a number of language comprehension tests, including sentiment analysis.

However, tweet analysis presents unique challenges. Tweets are hard to understand since they are short, usually informal, and can contain slang, emojis, and hashtags.

2. Methodology

To process and evaluate tweet data, the Twitter sentiment analysis approach entails a number of crucial processes. The Twitter API, provides access to tweets based on particular Keywords, Hashtags #, or User Accounts, is used to collect initial data.

Following collection, the data is cleaned and prepped for analysis through preprocessing. This entails eliminating extraneous components such as special characters, stop words, and duplicate entries. Additionally, hashtags#, emoticons, and abbreviations are standardised to facilitate text analysis. Sentiment classification is the following stage, in which methods like as a machine learning models or lexicon-based approaches are employed to ascertain if the tweet is neutral, negative, or positive.

Machine Learning Algorithms are trained on Labeled Datasets to improve their ability to identify textual patterns and predict sentiment. Sophisticated techniques such as deep learning, which incorporates transformer models like BERT, are increasingly being used because to their ability to understand context.

Therefore, the Model's Performance is assessed using measures such as F1-score and accuracy to ensure that the study yields honest findings.

2.1 Pre-Processing of Datasets

Generally speaking, a Tweet contains of types of opinions that have been presented by the many individuals in unique ways. For this study, the data from Twitter has already been separated into two groups: positive and negative.

This labelling facilitates the analysis of the attitudes and the effect of various characteristics. The primary preprocessing of processes are as follows:

- Deleting all links (e.g., www.example.com), hashtags (e.g., #topic), and user mentions (e.g., @username).

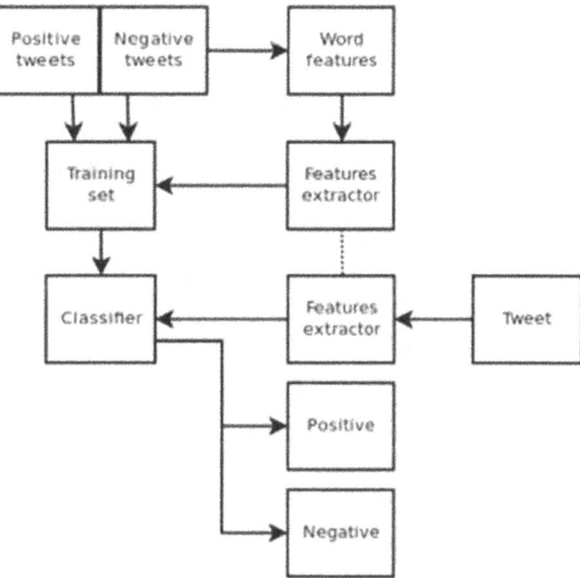

Fig. 69.1 Sentiment analysis architecture

- Managing the words with repeated characters and fixing the spelling mistakes.
- The process of converting emoticons into the feelings or emotions they symbolize.
- Eliminating the unnecessary punctuation, symbols, numbers, and other relevant characters.
- The removal of common stop words, such as "and," "the" or "is."
- Expanding acronyms to their full forms, using a reference dictionary.
- Excluding tweets written in languages other than English.

These procedures aid in data refinement, guaranteeing accurate, and trustworthy analysis.

2.2 Feature Extraction

Feature extraction is the process used to transform Twitter data into information that can be used by Machine Learning Algorithms.

Crucial steps consist of the following:

- Tokenization: Breaking the tweet in tokens or individual words.
- Removing Stop Words: Elimination of common words (example: "the", "is") that do not add significant meaning.
- Stemming and Lemmatization: reduction of words to their root form. (Example: "running" to "run")
- Vectorization: Converting words in numerical values using techniques like word embeddings or TF-IDF.
- Handling Emojis and Hashtags: Converting emojis and hashtags into their sentiment or meaning.

2.3 Classification

Naive Bayes

Word frequency is used by a probiotic classifier known as Naive Bayes to identify whether a tweet is in the positive or negative sentiment class. After determining the chance that a tweet belongs to each class, it chooses the one with the highest probability.

The model employs maximum likelihood estimation to determine the relationship between words and emotion, assuming that each word's recurrence is independent.

$$C^* = \arg mac_c P_{NB}(c|d)$$

$$P_{NB}(c|d) = \frac{(P(c))\sum_{i=1}^{m} p(f|c)^{n_{l(d)}}}{P(d)}$$

Maximum Entropy

The Maximum Entropy classifier does not assume anything about the relationships between features. It aims to Maximize the entropy of the system to estimate the probability distribution of the Class Label.

This approach, which is comparable to logistic regression, considers every characteristic as possibly interconnected.

It works exceptionally well when handling overlapping features in the dataset.

An illustration of the model is as follows:

$$P_{ME}(c|d,\lambda) = \frac{\exp\left[\sum_i \lambda_i f_i(c,d)\right]}{\sum_c \exp\left[\sum_i \lambda_i f_i(c,d)\right]}$$

Support Vector Machine (SVM)

Finding the optimal decision boundary also known as hyperplane between several data classes is how SVM operates. In order to produce a distinct boundary between classes, it maps data into higher dimensions using kernel functions, SVM, which emphasizes the margin to lower classification errors, is effective for regression as well as classification applications.

2.4 Approaches for Sentiment Analysis

Sentiment analysis classify, and identifies emotions in text. Using a variety of methods. The primary methods consist of:

1. **Lexicon-Based Approach**

   This method uses pre-made word dictionaries with emotion scores for each word.

   - Pros: Simple to use and training doesn't require big datasets.
   - Cons: Has trouble understanding sarcasm, context, or new words.

2. **Machine Learning Approach**

   This approach uses methods like support vector machines, SVM logistic regression to categorize emotions or Naïve Bayes.

   - Process: Train a model using labelled data extract features and classify the text.
   - Pros: Adapt well to structured data and perform better with large data sets.
   - Cons: Requires labeled data and regular retraining for new data.

3. **Hybrid Approach**

   A mix of the lexicon-based and machine learning techniques to increase accuracy of result.

   - Pros: Handles both context and unseen words better.
   - Cons: Can be more complex and time-intensive to implement.

4. **Deep Learning Approach**

   This method use updated models like Recurrent Neural Networks (RNNs) or Conventional Neural Network (CNNs) for analyzing text, writing

   - Pros: Handles context, sarcasm, and long-term text dependencies effectively.
   - Cons: Requires a lot of data and computational power for training.

3. Evaluation of Sentiment Analysis

Four indices that are computed using the Following Equations can be used to assess Sentiment Classification Performance:

$$\text{Accuracy} = (\text{TP} + \text{TN})/(\text{TP} + \text{TN} + \text{FP} + \text{FN})$$

$$\text{Precision} = \text{TP}/(\text{TP} + \text{FP})$$

$$\text{Recall} = \text{TP}/(\text{TP} + \text{FN})$$

$$\text{F1} = (2 \times \text{Precision} \times \text{Recall})/(\text{Precision} + \text{Recall})$$

Thus, according to the table, TP, FN, FP, and TN stand for the numbers of true positive instances, False Negative Instances, False Positive Instances, and True Negative Instances, respectively.

Table 69.1 Confusion matrix

	Predicted Positives	**Predicted Negatives**
Actual Positive	TP	FN
Actual Negative	FP	TN

4. Result and Discussion

For our investigation, we used a Twitter dataset that Stanford University made publicly available. Because the dataset is labeled, we can efficiently use a variety of feature extraction algorithms. The raw text data was cleaned up and prepared for analysis using a preprocessing framework. Following preprocessing, feature vectors earlier were Used to train the dataset using Machine Learning Techniques. Synonyms and similarities were also found using semantic analysis, which assisted in determining the content's sentiment polarity.

4.1 Dataset Details

There were distinct training and testing sets in the dataset:

Training Data:

- Total tweets: 45,000
  - Negative: 23,514
  - Positive: 21,486

Testing Data:

- Total tweets: 44,832
  - Negative: 22,606
  - Positive: 22,226

Table 69.2 Data summary

Data Size	Accuracy(%)
500	61.2
1,000	65.2
5,000	69.7
10,000	71.3
15,000	71.7
20,000	72.3
25,000	72.9
30,000	72.9
35,000	73.2
40,000	73.3
45,000	73.6

Key Informative Features

In tweets, the Naïve Bayes classifier found a number of characteristics that were highly suggestive of negative sentiment. These consist of:

- "sad": 37.6 times more likely negative than positive
- "worst.": 32.4 times more likely negative than positive
- "crying": 24.7 times more likely negative than positive
- "fml": 24.1 times more likely negative than positive
- "hurts": 21.2 times more likely negative than positive
- "awful": 21.1 times more likely negative than positive
- "ugh.": 20.4 times more likely negative than positive
- "terrible": 20.4 times more likely negative than positive
- "boo.": 19.2 times more likely negative than positive
- "cancelled": 19.2 times more likely negative than positive

These findings demonstrate how well the classifier identifies words and phrases that significantly convey negative sentiment.

5. Challenges in Sentiment Analysis

Accurately interpreting and categorizing emotions in text is difficult due to a number of issues with sentiment analysis. Among these difficulties are:

1. **Detecting Sarcasm and Irony**

   Because sarcasm and irony frequently express the reverse of their exact meaning, they can be difficult to distinguish. For instance, saying "What a wonderful day!" amid inclement weather might be seen as sarcasm.

2. **Ambiguity in Words**

   Depending on the situation, several words might signify different things. For example, depending on the circumstance, the expression "That's cool" might convey approbation or disinterest.

3. **Context Dependence**

   The surrounding context is sometimes absent when analyzing individual lines or tweets, which might result in an inaccurate perception of sentiment. For instance, depending on the previous discussion, "It's fine" could be interpreted as either complimentary or dismissive.

4. **Domain-Specific Vocabulary**

   Words can have distinct sentiments in different contexts. In the context of gaming, for instance, "crushing it" is a positive trait, but in another sector, it might not have any emotional significance.

5. **Handling Informal Language**

   Without adequate preprocessing, social media information frequently contains informal components like slang, acronyms, and emoticons that are challenging for models to understand.

6. **Mixed Emotions in a Single Statement**

   Both good and negative feelings can be expressed in the same text. "The story was interesting, but the acting was awful." such statements contain contradictory ideas.

7. **Imbalanced Datasets**

   Because training datasets sometimes contain more examples of one type of attitude (like positive) than others, the model's predictions may be biased.

6. Applications of Sentiment Analysis

Many different disciplines can benefit from sentiment analysis. It makes it easier to comprehend the attitudes, feelings, and viewpoints that people and organizations express through words. Here are some common applications:

1. **Brand Reputation**

   Sentiment analysis is used by businesses to monitor how customers view their goods, services, or brand. It helps them understand customer satisfaction and make the required adjustments.

2. **Customer Feedback**

   Businesses look at surveys, reviews, and comments to learn about customer's thinking about their products and services. This improves the quality and customer experience.

3. **Market Research**

   Sentiment analysis helps companies make better decisions about marketing and product development by looking at customer trends and preferences.

4. **Social Media Tracking**

   Brands utilize sentiment analysis which as a technique help to track comments made on the famous social media platforms like Twitter, Facebook, and Instagram. Which helps them to quickly address customer problems and manage their online reputation.

5. **Political and Social Studies**

   Researchers use sentiment analysis to provide decision-makers with information on public opinion on the political events, campaigns, or social issues.

6. **Finance and Investments**

   By doing sentiment research to find out how people feel about companies, products, or the status of the economy, analysts and investors may make smarter investment decisions.

7. **Personalized Recommendations**

   Platforms like shopping websites and streaming services, employee sentiment analysis to recommend products or content based on user reviews and interests.

7. Conclusion

Sentiment analysis is a helpful method for understanding the ideas and emotions shared on Twitter. It may be used by companies, governments and scholars to investigate public perceptions of social and political issues as well as brands, goods and services.

By examining the sentiments, expressed in tweets, businesses may gain insight into consumer satisfaction, market trends, and public opinion. They can use this information to make wiser choices.

The process involves collecting raw tweets, collecting the data using machine learning and identifying important attributes to classify sentiments as positive negative or neutral, not with standing its advantages, sarcasm, contradicting, emotions, and language. Variation are some of the disadvantages of sentiment analysis that may jeopardise its accuracy.

However, natural language processing and machine learning advancements are improving performance and addressing these issues.

Making data-driven decisions, improving consumer contact, and comprehending social dynamics all depend on sentiment analysis on Twitter.

As technology develops, sentiment analysis methods are expected to become even more helpful, offering deeper insights and making it easier to understand and react to the emotions expressed in online interactions.

Acknowledgement

We are grateful to our colleagues and peers for their constructive discussions and suggestions, which have helped enhance the quality of this work. Special appreciation goes to Poornima group for providing the necessary resources and a conducive environment for research and writing. Finally, we extend our deepest appreciation to our family and friends for their unwavering support and encouragement, which have been a constant source of motivation.

References

1. Pang, B., & Lee, L. (2008). Opinion mining and sentiment analysis. Foundations and Trends in Information Retrieval, 2(1–2), 1–135.
2. Go, A., Bhayani, R., & Huang, L. (2009). Twitter sentiment classification using distant supervision. CS224N Project Report, Stanford University.
3. Liu, B. (2012). Sentiment analysis and opinion mining. Synthesis Lectures on Human Language Technologies, 5(1), 1–167.
4. Pak, A., & Paroubek, P. (2010). Twitter as a corpus for sentiment analysis and opinion mining. In Proceedings of the Seventh International Conference on Language Resources and Evaluation (LREC'10).
5. Bifet, A., & Frank, E. (2010). Sentiment knowledge discovery in Twitter streaming data. In Proceedings of the 13th International Conference on Discovery Science (pp. 1–15). Springer.
6. Zhang, L., Wang, S., & Liu, B. (2018). Deep learning for sentiment analysis: A survey. Wiley Interdisciplinary Reviews: Data Mining and Knowledge Discovery, 8(4), e1253.
7. Cambria, E., Schuller, B., Xia, Y., & Havasi, C. (2013). New avenues in opinion mining and sentiment analysis. IEEE Intelligent Systems, 28(2), 15–21.
8. Karishma Sharma, Feng Qian, He Jiang, Natali Ruchansky, Ming Zhang, Yan Liu (2020) "Combating Fake News: A Survey on Identification and Mitigation Techniques".
9. Stanford University. (Accessed Year). Stanford Twitter Sentiment Dataset. Available at: http://help.sentiment140.com
10. Bird, S., Klein, E., & Loper, E. (2009). Natural Language Toolkit (NLTK). Available at: Bird, S., Klein, E., & Loper, E. (2009). Natural Language Toolkit (NLTK). Available at: https://www.nltk.org
11. Kaggle. Twitter Sentiment Analysis Datasets. Available at: https://www.kaggle.com
12. Bird, S., Klein, E., & Loper, E. (2009). Natural Language Toolkit (NLTK). Available at: https://www.nltk.org
13. Sentiment140. Stanford Twitter Sentiment Dataset. Available at: http://help.sentiment140.com
14. Twitter Developer Platform. (n.d.). Tools for collecting and analyzing tweets. Available at: https://developer.twitter.com

15. Scikit-learn Documentation. (n.d.). Machine learning library for Python. Available at: https://scikit-learn.org
16. VADER Sentiment Analysis Tool. (n.d.). Sentiment analysis for social media text. Available at: https://github.com/cjhutto/vaderSentiment
17. Mittal, D., Pant, A., Jajoo, P., & Yadav, V. (2024, March). Transformer model applications: A comprehensive survey and analysis. In International Conference on Deep Learning and Visual Artificial Intelligence (pp. 25–38). Singapore: Springer Nature Singapore.
18. Raffel C, Shazeer N, Roberts A, Lee K, Narang S, Matena M, Zhou Y, Li W, Liu PJ (2020) Exploring the limits of transfer learning with a unified text-to-text transformer. J Mach Lcarn Res 21(1):5485–5551.
19. Clark K, Luong MT, Le QV, Manning CD (2020) Electra: pre-training text encoders as discriminators rather than generators. arXiv preprint arXiv:2003.10555
20. Liu Y, Ott M, Goyal N, Du J, Joshi M, Chen D, Levy O, Lewis M, Zettlemoyer L, Stoyanov V (2019) a robustly optimised bert pretraining approach: Roberta. arXiv preprint arXiv:1907.11692

Note: All the figure and tables in this chapter were made by the authors.

Intelligent Systems Using Semiconductors for Robotics and IoT – Dinesh Goyal et al. (eds)
© 2026 Taylor & Francis Group, London, ISBN 978-1-041-20408-4

70 Chaotic Sand Cat Swarm Optimizer for Feature Selection in Arrhythmia ECG Signal Classification

Pushp Raj Tripathi[1]
Research Scholar, Poornima University,
Jaipur, Rajasthan, India

Rakesh Kumar Saxena[2]
Professor, Poornima University,
Jaipur, Rajasthan, India

Abstract: Arrhythmia of the heart refers to a succession of disorganized heartbeats due to defectiveness within the conduction of the heart's electrical system. An electrocardiogram is used as a measure to detect the heart rhythm disturbances called arrhythmias. However, the employment of professionals to assess a plethora of ECG recordings consumes an unacceptable number of medical facilities. Deep learning or DL might be the more viable option available for automatically categorizing the performances with appropriate training. The present study suggested an optimizer to be a feature selection in arrhythmias of the heart. The dataset of MIT-BIH Arrhythmia removed this noise: noise due to either other technical or biological reasons that might cause misclassification of arrhythmia. The pre-processed signal had number of features that may be considered redundant. These are removed using the Chaotic sand cat swarm optimizer as a feature selection and compared for performance with and without feature selection. The non-redundant features achieved by this model are handled under the classifier of the Interpretable Temporal Attention Network (ITA Net). It is further classified in five different type classes supraventricular ectopic (S), non-ectopic (N), fusion (F), ventricular ectopic (V), and unknown (Q). The demonstrated method achieves better performance than the current approaches on arrhythmia classification.

Keywords: Arrhythmia, ECG, Feature selection, Chaotic sand cat swarm optimization algorithm (CSCSO)

1. Introduction

Cardiovascular disease, in present scenario is observed as leading cause of human death. Heart disease is a common condition that afflicts the health of people, especially middle-class and older persons. While other institutions like the World Heart Federation of Australia contest, they account for a third of all CVD deaths accounted by middle-class and lower-class people. The cardiovascular system's primary function is to carry oxygenated blood throughout the body and excrete impure blood with carbon dioxide and other toxicants. This is done through an electrical impulse that is initiated in the right atrium of the coronary arteries, known as the Sino atrial node, which acts as the natural pacemaker. It initiates the impulses that trigger the pumping mechanism of the cardiovascular system. It is life-saving to recognize heart arrhythmias much earlier. The electrical cardiogram signal represents the best method for diagnosing cardiac arrhythmias. It is a succinct representation of the events that depict the present electrical conduction of the heart [1].

2. Arrhythmia ECG Signal Classification

Arrhythmia which is refers to abnormal condition of heart rhythm basically because of irregular electrical activity in the heart. The condition can give rise to increased, decreased, or irregular heartbeats. Detection of arrhythmia is essential for diagnosis of cardiac abnormalities, and ECG is a common tool that is used. The ECG signal

[1]mtechpushp@gmail.com, [2]saxenark06@gmail.com

DOI: 10.1201/9781003716389-70

represents heart electrical activity with characteristic waveforms as shown in Fig. 70.1. This wave have P wave which shows Atrial depolarization, QRS complex known as Ventricular depolarization and last one T wave known as Ventricular repolarization.

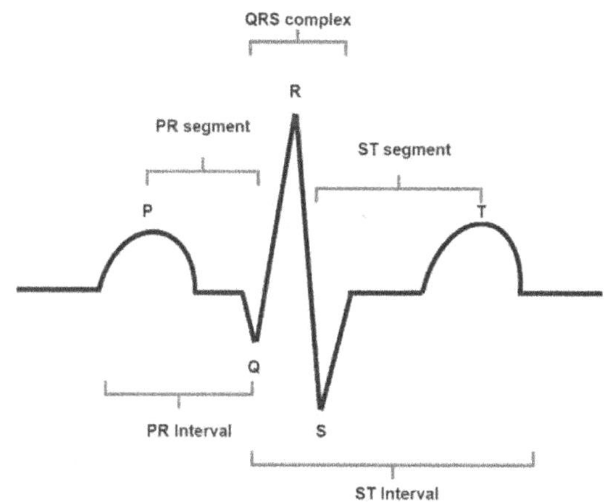

Fig. 70.1 ECG (heart)

The waveforms can be evaluated to identify disorders in heart rhythms.

Arrhythmia Classification is based on the pattern and frequency of electrical activity in the heart. The classification is usually divided into the following categories [2]: Variations in heart rate and rhythm abnormalities diagnosed through ECG signals categorize arrhythmias. While some of these, known as bradyarrhythmias, involve a lower heart rate, such as sinus bradycardia and atrioventricular (AV) block, tachyarrhythmias contribute to a fast heartbeat. Among these include sinus tachycardia and AFib, with the more lethal ventricular tachycardia (VT) originating from the ventricles, as did the dreaded ventricular dysfunction. Conduction disturbances, including fascicular blocks, help direct the electrical impulses into ventricle walls.

Arrhythmia classification plays critical role in the early diagnosis and simultaneously prevention of heart disease. Currently, machine and deep learning techniques are implemented for arrhythmia detection and classification in automated form. Modern techniques in Deep Learning and Machine Learning are used to classify arrhythmias based on ECG signals. For its methods include 1. Feature Extraction to Extract P-QRS-T wave characteristics 2. Machine Learning Models for ECG classification.

ECG data is crucial tool in diagnosing and managing heart diseases. By analysing ECG data more elaborately, we can improve the detection, classification, and prediction of various heart conditions. Irregular pattern in ECG data will reveal the seriousness of patient. Different category of arrythmia will tell the condition of patient and category of arrythmia will help the doctor to take the right measure to correct the heart disease. Any changes in the cardiac rhythm abnormalities or ECG abnormalities pinpoint the underlying cardiovascular disorders like arrhythmia. The most common instrument for collecting a heart rhythm is the ECG signal employing legates placed to the skin, known as conventional 12-lead ECG that monitors the potential electricity from 10 electrodes scattered on the body's surface [3]. Analyzing an ECG signal of this nature on a continuous basis is quite tiresome and laborious for any doctor. Due to their high reliability and accuracy, computer-assisted diagnostic methods are widely and increasingly employed to analyze ECG data and identify arrhythmias. Consequently, it is of great importance to check the palpitations occurring in ECG to detect disorders such as arrhythmia as soon as possible.

A large host of methods involving machine learning were developed by the scientific community for early identification of arrhythmia using data from ECG. Traditional AI cannot be applied without first implementing a feature extraction strategy. The inefficient feature extraction procedure eliminates several redundant features that bias classification outcomes. However, the automated method of taking measures does not effectively use basic data in the database to be cost-effective and reasonable in time consumption. Over and under-fitting with behaviour on more policy-trained sample sets is an increasingly common issue within ML-based classification systems.

It would probably be most appropriate to classify this problem as a case of dealing with very high-dimensional data, but in a highly-tuned DL implementation. Several distinct layers of various neurons in neural network (NN) that can be find a classification for ECG signal on a very short time with proven high reliability. Soon this allows them to resolve each into isolated types of beats and extract data on various types of arrhythmia signals. The DL being integrated into arrhythmia classification remains in its infancy, and further effort is demanded into the development of an effective smart system to replace a physician in the future, which would assist in the identification of heart ailments. This methodology is assigned as the core methodology for categorization of cardiac arrhythmia using deep learning. In this paper, chaotic sand cat optimization algorithm model is used for the selection of features process which can affect the identification of arrhythmia. The ECG extracted (Fig. 70.1) from the dataset of MIT-BIH Arrhythmia was processed with IIR and FIR noise filters to remove technical and biological noise during ECG diagnosis. The image after noise removal and pre-processing was run through a CSCSO optimizer for the removal of redundant features for categorization of arrhythmias and non-redundant features were divided into following five categories by the interpretation of the ITA Net classifier.

1. supraventricular ectopic (S)
2. fusion (F)
3. non-ectopic (N)
4. ventricular ectopic (V)
5. unknown (Q)

The proposed study is divided into five components, the first of which introduces a classification of arrhythmias from ECG data. Section 3 provides an overall review of the currently conducted research on arrhythmia categorization. Section 4 provides a comprehensive overview of methodology and Section 5, Illustrates the findings of the article. Section 6 presents conclusion.

3. Related Works

Pooja Sharma and colleagues [6] developed a hybrid method to identify arrhythmias based on cardiovascular irregularities. Their algorithm, labelled CS+DWT+SVM-FFBPNN, categorizes signals into five distinct groups by integrating Support Vector Machines (SVM) with Feedforward Backpropagation Neural Networks (FFBPNN). The research utilized the Cuckoo search technique to segment ECG readings obtained from the MIT-BIH dataset. In the initial processing stage, Discrete Wavelet Transformation was employed to minimize inherent noise while simultaneously improving classification accuracy. Additionally, the Cuckoo search method enhanced feature selection processes. However, applying this combined approach using neural networks may lead to increased computational costs that could pose challenges for real-time applications.

Ahmet Çınara and Seda Arslan Tuncer [7] devised a deep learning network that incorporates various classifiers to categorize ECG signals associated with congestive cardiac diseases, normal sinus rhythm, and heart failure. This model integrates AlexNet-based classifiers with support vector machine techniques and replaces the SoftMax layers prior to the final pooling layer with additional neural networks. A significant limitation of their research is its failure to address noise issues within ECG data—an oversight that may adversely affect classification outcomes.

In pursuit of enhancing diagnostic accuracy for cardiovascular disorders, Atta-ur Rahman et al. [8] introduced the CAA-TL training model aimed at classifying arrhythmias from ECG signals through Deep Learning methodologies such as ResNet50, AlexNet, and SqueezeNet. Their study analysed data drawn from both the MIT-BIH database and Kaggle resources, offering comparative insights into various deep learning algorithms' performance in detecting arrhythmias. Nonetheless, one notable challenge identified during this investigation is related to time efficiency in processing these methods.

The 11-layer deep convolution model presented by Y. O. Abdalla et al. [9] is for the classification of arrhythmias. Model of structure consists of 11 layers each endowed with four convolutional layers interspersed with four max-pooling levels, and 3 satisfactory concatenation layers. To confirm that it was general, they tried its efficiency on the MIT-BIH database extracted from Physionet. The outcome of the learning is portrayed through the decision classification, because there are no dissimilar classifiers.

Adel A. Ahmed et al. [10] presented a 1-D deep convoluted model of cardiovascular arrhythmia classification to address the drawbacks of clinical assessments of standard ECGs. This model is used to separate the concentrated electrical signal on the R-peak location so that P wave, T wave, QRS Complex components are effectively collected. This paper never examined noise and its effect on ECG performance, which is a very great drawback in the real-time execution.

By updating SCA, Pooja Sharma et al. [11] constructed LA-SCA. This is followed by the Wilcoxon rank-sum test, by which the number of previous generations of a modified LA-SCA model is compared against results. Finally, the OBL model is implemented to find the best solution. In this work, the ECG data is processed using discrete wavelet transform to extract the QRS complex features. But Fourier transformation has been performed to analyse the aberrant ECG waves, therefore leading to loss of information.

Resolutions pertaining to Aamir Ullah and others [12] create a two-dimensional CNN architecture of eight layers which classifies the heartbeat delineated in MIT-BIH records pertaining to ECG signals. These are employed based on the Short Time Fourier Transform to contrast the agreeable attributes mapped by the single-dimensional ECG signals, thus inculcating a spectrogram which is of the two-dimensional nature. The layer consists of four convolutional layers which combine information along with a four-pool layer, which extracts characteristics from the spectroscopy input.

Mengze Wu and others [13] have developed a twelve-layer deep convoluted neural network (NN) for classification of the ECG data from a heartbeat into five categories. Employing a wide Swan variety of filters (band-pass, low-pass, and modified wavelet) to eliminate noise in ECG data. Cross-validate algorithms tend to have a ten-time verification through research exclusively to better improve generalizability as well as predictive modelling robustness. The project lacked any kind of modelling techniques which took a flight away owing to the so-called enhanced way concerning the model thus formed pace.

Surbhi Bhatia and others [14] have developed a combination model for a CNN with a bidirectional LSTM network. During the investigation, database was reviewed. In addition to cater class imbalance problem, dataset was over-sampled using SMOTE technique. Basically SMOTE is a method of data augmentation to treat the class imbalance of the datasets that come into play usually in machine learning for classifying tasks. SMOTE does not duplicate already existing samples. It generates synthetic data points in the minority class using the following procedure based on K-nearest neighbours algorithm:

1. A minority class sample is selected by randomly selecting a data point from this minority class.

2. k nearest neighbours of this data point within the minority class are obtained using the Euclidean distance metric.
3. A synthetic sample is generated by randomly choosing one k nearest neighbours and using it to obtain a new, synthetic point.
4. These steps continue until a sufficiently balanced minority class against the majority class obtained.

SMOTE would help to lessen the degree of overfitting by creating diverse synthetic data and increasing the model performance when such imbalance has been marked on the dataset used. The classifiers requiring balanced distributions apply SMOTE successfully. ECG datasets are often imbalanced, meaning that some arrhythmia classes (e.g., ventricular fibrillation) have far fewer samples compared to normal sinus rhythm. SMOTE can help in balancing the dataset and improvement in classification accuracy in ML models. SMOTE is augmentation technique that generates synthetic sample for minority class which are used to balance dataset. In addition instead of duplicating existing minority sample, SMOTE technique generates new synthetic example by interpolating between real samples. For each minority class instance, SMOTE select K nearest Neighbour i.e. typically K=5. The SMOTE was applied to resolve class imbalance, whereas the issue of overfitting or oversampling was not tackled.

Parul Madan and others [15] developed a hybrid DL-based method for the automatic detection of arrhythmias. The model developed combined the complex and LSTM networks. To recover the noise filtering, 1-D ECG images were converted to 2-D scalograms, which produces a better evaluation technique.

The great shortcoming of the research, as can be understood from the above-mentioned recent publications, is non-noise removal, as much prior research has been carried out without noise removal that has an influence over detection of arrhythmias from ECG signals. Moreover, the previous researches raise computational burden and time on the mathematical framework. Finally, DL fits exactly for arrhythmia detection due to limitation with previous machine learning techniques. Nonetheless, the lack of a classifier as feature selection method imposes a restriction upon research into that field, influencing the misdiagnosis of Arrhythmia illness very critically.

4. Methodology

4.1 Dataset Acquisition

MIT-BIH arrhythmia database was used which contains 23 records that are selected at randomness from the set and 25 records that are chosen from same set with a rate of sampling of 360 Hz and that have a duration of around 30 minutes. It encompasses several types of heartbeat; five of them are with clinically relevant samples that are tagged 0, 1, 2, 3, and 4, namely, N, S, V, F, Q respectively. The AAMI classes were used; thus, as mentioned below, Table 70.1 gives the basic classification of 15 different types of ECG signals grouped into five types: S (supraventricular ectopic beats), N (non-ectopic beats), F (fusion beats), V (ventricular ectopic beats), and Q (unknown beats). The signals recorded from this database were converted into waves to assess the developed method of analysis. Figure 70.2 shows the general composition of the intellectual structure on the research.

Table 70.1 MIT-BIH arrhythmia dataset

AAMI Class	Types of heartbeat	Numbers
Supraventricular ectopic	S	2
Non-ectopic	N	75028
Fusion	F	802
Ventricular ectopic	V	7129
Unknown	Q	33

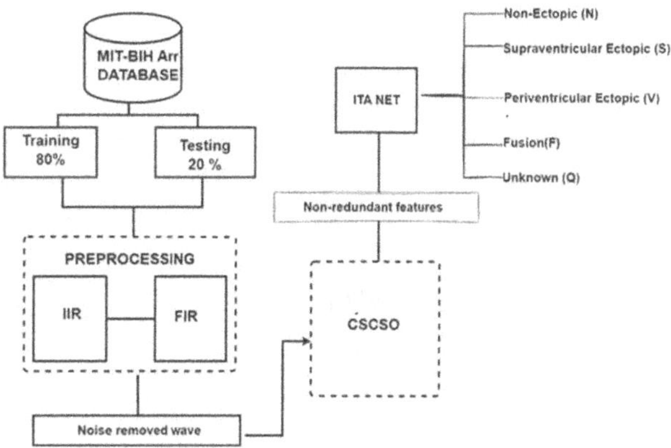

Fig. 70.2 Conceptual framework

4.2 Pre-Processing

The presence of many physiological disturbances, along with technological noises like electrical lines, communication line interferences, and electrode movement, degrades the overall quality of ECG signal. Due to the low amplitude, the electrical signal gets corrupted, which leads to incorrect diagnosis during processing. The proposed technique removes noise in an ECG signal by using both finite and infinite impulse responses.

It is one characteristic of FIR filters, which ensures that the final signal has no abnormal time displacement between its components, and thus the sine rhythms of the myocardium complex can maintain a division as time goes on during electronic processing of the ECG. However, IIR filtering possesses an irregular phase response, meaning that the latency and distortion in the signal will change over time and are not reflected in the frequency spectrum. This type of filter reduces the shifting of the ECG's baseline by causing absorption and deformation in the audio frequency spectrum.

All technical and biological sounds encoded in the ECG signals have been removed after applying filters like FIR as well as IIR.

4.3 Feature Selection using CSCSO Algorithm

The pre-processed waves were fed into the feature selection technique, which reduces the space of redundant features. In the geometrical progress of computing technology, proper identification of arrhythmias cardiac conditions is critical due to difficulties with overfitting and computational effectiveness which is addressed by CSCSO algorithm.

In mathematics, the innate randomness in the fundamental deterministic kinetic structure is said to be chaos, and erratic structures can be considered as the causes of randomized. The translation process for chaos-based optimization translates surprising patterns using chaotic map data in design factors as part of the whole process of manipulation. The CSCSO technique provides two different approaches for phased position updates. One is a standard updating technique of the position. Another uses the chaos model. The two techniques are equally likely to achieve appropriate weight. Every category assigned as a search agent is designed as a vector and the duration is issue approached on the direct vision. The efficacy of the algorithm is measured using the fitness value of each task (eq. 1).

$$\begin{aligned} Fitness &= f\left(Sand\ Cat\right) \\ &= f\left(SC_1, SC_2, \ldots, SC_n\right); \\ &\forall x_i\ (is\ calculated\ for\ t\ time) \end{aligned} \quad (1)$$

The mathematical models that are effective in seeking are considered as the exploration phase and exploitation phase of the SCSO is given below.

$$\vec{X}(t+1) = \begin{cases} \vec{X}_b(t) - \vec{X}_{rnd} \cdot \cos(\theta) \cdot \vec{r} \\ \quad |R| \leq 1;\ exploitation(a) \\ \vec{r} \cdot \left(\vec{X}_c(t) - rand \cdot \vec{X}_k(t)\right) \\ \quad |R| > 1;\ exploration(b) \end{cases} \quad (2)$$

Here t represents current iteration, T is max. no. of iterations. Behavioural model of the SCSO algorithm in updating the location is presented in Equation (7), both in the exploitation phase and in the exploration phase:

$$\vec{X}(t+1) = \begin{cases} if\ p < 0.5 \equiv Eq.5 \\ if\ p \geq 0.5 \equiv Eq.10 \end{cases} \quad (3)$$

Where p is a randomized integer between 0 and 1. The C parameter is the most essential variable in completing a task. The value of this parameter directly influences the R factor. The C parameter enhances the CSCSO algorithm's versatility by playing a one-to-one role during its exploration and extraction. As a result, parameter C has been established for use in the Chaotic Sand Cat Swarm Optimisation algorithm, which is symbolized by Equations (9). C additionally has a crucial part in demonstrating fair and equitable behaviour during exploration and exploitation. This refers to the usability of the exploration along with the exploration phases.

$$S = 2^k;\ k \geq 1 \quad (4)$$

$$C = s - s\left(\frac{\sqrt[T]{e^t} - 1}{e - 1}\right) \quad (5)$$

Where K represents the constant factor. This parameter (K) influences the amount of weight the algorithm assigns for each stage. The exploring phase has a greater likelihood with this equation. Focusing greater attention on the investigation phase of the CSCSO algorithm produces favourable outcomes in confined issues. This new equation also has implications for both parameters R and r. Since the R parameter is reliant on C, its changing span is exactly lowered. R is a random value in vary of [-2C, 2C]. While R turns into less compared to or equal to a single one, the CSCSO methods require agents to seem to exploit; as an alternative, search employees must examine and discover prey. This is used whenever the p-value is less than 0.5. The chaotic maps are assigned as a substitute. Early convergence as well as local optimal entrapment can be prevented in this way. Equations (6) and (7) are used to compute the parameters of R and r that are anticipated to be generated because of the chaotic map's impact.

$$\vec{R} = 2 \times \vec{c} \times m - \vec{c} \quad (6)$$

$$\vec{r} = \vec{c} \times m \quad (7)$$

$$\vec{X}(t+1) = \begin{cases} \vec{X}_b(t) - X_{rnd} \cdot \cos\theta \cdot \vec{r} & |R| \leq 1(a) \\ \vec{r} \cdot \left(\vec{X}_c(t) - m \cdot \vec{X}_t(t)\right) & |R| > 1(b) \end{cases} \quad (8)$$

A chaotic map is employed to compute the chaotic vector m. Equation (8), outlined above, is implemented to generate an overall conceptual representation of the CSCSO. Owing to the results of the recommended approach, it will find more potential localized territories in global spaces with a precise and rapid resolution rate across a range of scenarios, especially restricted issue areas for optimization.

5. Results and Discussion

A total of 25 records were extracted from identical sets and assigned a sampling rate among 360 Hz, with 80% assigned to be used for training as well as 20% for testing. Of these records, 23 were randomly chosen. The classification system is tested using a test database at the recognition stage after it was successfully trained with previously acquired information or an enlisted data set. A batch size of 25 is used and carried out fifty simulations. To prevent train and test information overlap, various patients with multisession ECG recordings were utilized.

5.1 Experimental Setup

A computer with an Intel-i5-7300 HQ processor, 16 Giga Bytes of RAM, GTX 1050-GPU is used to train the CSCSO optimizer. The initial training process takes about 4236 seconds, while the ten-fold cross-verification procedure takes about 11 hours. An individual sample's median classification time is 0.198 milliseconds. Python DL Toolbox is used in the development of the algorithm.

5.2 Evaluation Metrics

We measure the effectiveness of Proposed model for categorizing arrhythmias by utilizing metrics which are accuracy (AC), recall, specificity (SPE), and precision (PR).

5.3 ROC Curve

The ROC curve graphically represents the compromise between the two parameters of sensitivity as well as specificity by using the TP rate and FP rate to plot the curve, which visually depicts the general efficacy of the model.

To demonstrate effectiveness of presented proposed strategy research was performed. In the first experiment, the techniques of FIR and IIR noise reduction were applied to ECG signal to determine noise from it. To select the attributes from the pre-processed data, the algorithm of CSCSO was applied. Finally, the non-redundant features are classified with the classifier of Interpretable Temporal Attention network into the five different classes represented as N, F, V, S, and Q. In Fig. 70.3, the raw ECG signal taken from the database of MIT-BIH arrhythmia has deviated from the fact that many interferences occurred during the recording session due to its low amplitude nature. It may result in an increase of misclassification rate. So, the raw signal was subjected to the FIR and IIR filter.

When the FIR filter is applied to the noisy ECG signal. It is well known that for transmission and filtering to be distortion less, the system's amplitude response must remain constant within the signal's effective spectral range and its phase response must follow a linear relationship with frequency, or linear phase. The FIR filter can accomplish linear phase filtering, compared to the IIR filter. Furthermore, since an FIR filter is an all-zero filter, its software and hardware design are not concerned with the problem of stability. Therefore, substantial use of FIR filters has been made in noise reduction applications in the context of signals.

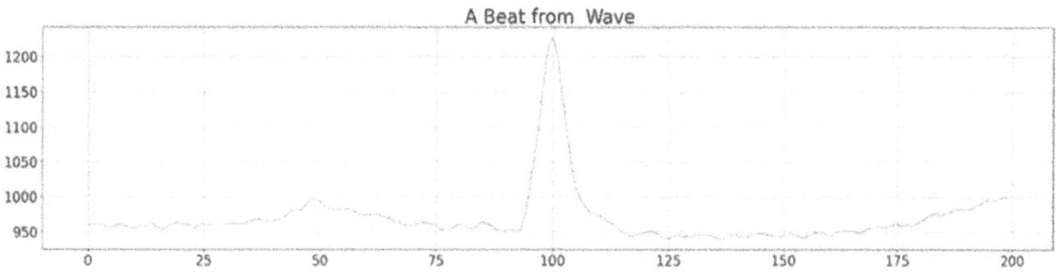

Fig. 70.3 Beat from the ECG wave

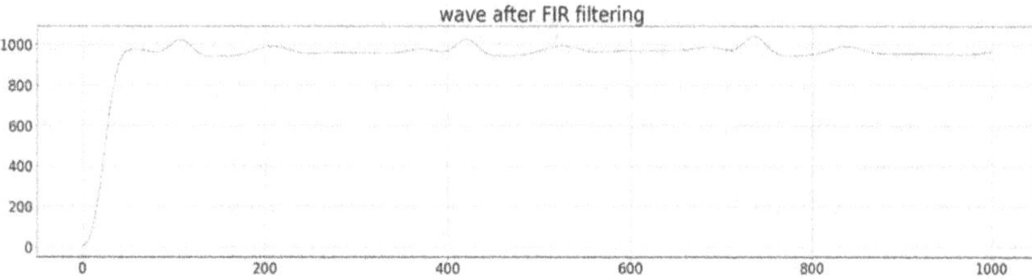

Fig. 70.4 Pre-processed wave by FIR filter

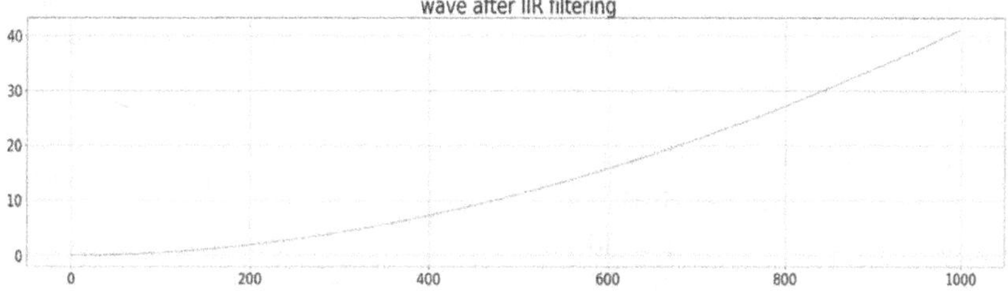

Fig. 70.5 Pre-processed wave by FIR filter

The noise from ECG signals was eliminated by applying the ECG signal to the IIR filter. Advantages of an IIR filter the signal passing through the filter is not distorted.

An idea called CSCSO Algorithm for proceedings was adopted that selected the attributes with redundant to compose the feature signals that govern highly accurate for processes of the given classification algorithm and improve database construction. When placed under this type of chaotic criterion, benefits may be considered along the horizon in the form of exploration space and the time duration for those associated with its Algorithm. Below, performance metric values are compared as they are shown in Table 70.2 where the feature selection and without the feature selection have been illustrated.

Table 70.2 Performance metrics

Performance Metrics	Without Feature Selection	With CSCSO Optimizer
Accuracy	95.30%	97.80%
Precision	91.20%	95.20%
Recall	86.80%	91.20%
F1-Score	87.40%	96.52%
Specificity	91.40%	92.42%

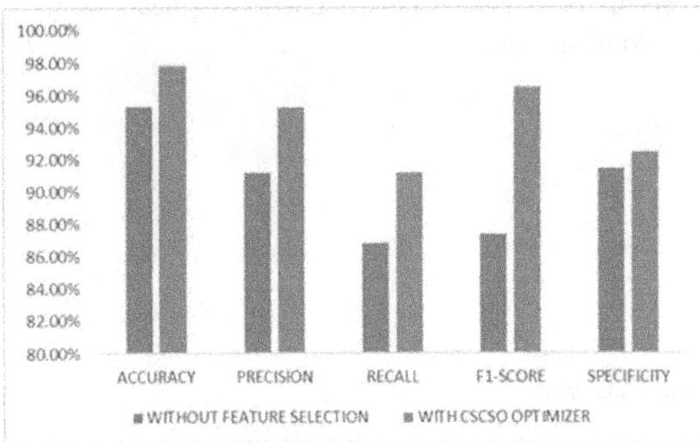

Fig. 70.6 Performance matrices

The classifier of the interpretable temporal attention network of the encoder-decoder structure is applied on the independent signals which divides the datasets into five different classes represented as N, F, V, S, and Q.

6. Conclusion

The ECG is a portable instrument that records cardiac signals for use in emergencies. However, its suitable interpretation has been complicated by its complexities and transmitted noise. The ECG signal's low magnitude, and complexity as well. Non-linearity makes it challenging to classify data quickly and accurately by hand. Because of that, a computerized procedure that can distinguish irregular heartbeats from data from ECG should be created for utilization by the medical industry. Here CSCSO approach has been presented and its performance is evaluated without feature selection and with feature selection. This approach has effectively classified arrhythmia into the five different classes N, F, V, S, and Q.

References

1. Mathunjwa, B. M., Lin, Y. T., Lin, C. H., Abbod, M. F., & Shieh, J. S. (2021). ECG arrhythmia classification by using a recurrence plot and convolutional neural network. *Biomedical Signal Processing and Control, 64*, 102262.
2. Ansari Y, Mourad O, Qaraqe K and Serpedin E (2023) Deep learning for ECG Arrhythmia detection and classification: an overview of progress for period 2017–2023. *Front. Physiol.* 14:1246746. doi: 10.3389/fphys.2023.1246746.
3. Zhang, S., Lian, C., Xu, B., Su, Y., & Alhudhaif, A. (2024). 12-Lead ECG signal classification for detecting ECG arrhythmia via an information bottleneck-based multi-scale network. *Information Sciences, 662*, 120239.
4. Kuila, S., Dhanda, N., & Joardar, S. (2023). ECG signal classification to detect heart arrhythmia using ELM and CNN. *Multimedia Tools and Applications, 82*(19), 29857–29881.
5. Ullah, A., Rehman, S. U., Tu, S., Mehmood, R. M., Fawad, & Ehatisham-ul-Haq, M. (2021). A hybrid deep CNN model for abnormal arrhythmia detection based on cardiac ECG signal. *Sensors, 21*(3), 951.
6. Sharma, P., Dinkar, S. K., & Gupta, D. V. (2021). A novel hybrid deep learning method with cuckoo search algorithm for classification of arrhythmia disease using ECG signals. *Neural Computing and Applications, 33*(19), 13123–13143.
7. Çınar, A., & Tuncer, S. A. (2021). Classification of normal sinus rhythm, abnormal arrhythmia and congestive heart failure ECG signals using LSTM and hybrid CNN-SVM deep neural networks. *Computer methods in biomechanics and biomedical engineering, 24*(2), 203–214.
8. Rahman, A. U., Asif, R. N., Sultan, K., Alsaif, S. A., Abbas, S., Khan, M. A., & Mosavi, A. (2022). ECG classification for detecting ECG arrhythmia empowered with deep learning approaches. *Computational intelligence and neuroscience, 2022*(1), 6852845.
9. Abdalla, F. Y., Wu, L., Ullah, H., Ren, G., Noor, A., Mkindu, H., & Zhao, Y. (2020). Deep convolutional neural network application to classify the ECG arrhythmia. *Signal, Image and Video Processing, 14*, 1431–1439.
10. Ahmed, A. A., Ali, W., Abdullah, T. A., & Malebary, S. J. (2023). Classifying cardiac arrhythmia from ECG signal using 1D CNN deep learning model. *Mathematics, 11*(3), 562.
11. Sharma, P., & Dinkar, S. K. (2022). A linearly adaptive Sine–cosine algorithm with application in deep neural network for feature optimization in arrhythmia classification using ECG signals. *Knowledge-based systems, 242*, 108411.
12. Ullah, A., Anwar, S. M., Bilal, M., & Mehmood, R. M. (2020). Classification of arrhythmia by using deep learning with 2-D ECG spectral image representation. *Remote Sensing, 12*(10), 1685.
13. Wu, M., Lu, Y., Yang, W., & Wong, S. Y. (2020). A study on arrhythmia via ECG signal classification using the convolutional neural network. Front Comput Neurosci 14.
14. Bhatia, S., Pandey, S. K., Kumar, A., & Alshuhail, A. (2022). Classification of electrocardiogram signals based on hybrid deep learning models. *Sustainability, 14*(24), 16572.

15. Madan, P., Singh, V., Singh, D. P., Diwakar, M., Pant, B., & Kishor, A. (2022). A hybrid deep learning approach for ECG-based arrhythmiaclassification. *Bioengineering*, *9*(4), 152.
16. Mian Qaisar, S., & Hussain, S. F. (2023). An effective arrhythmia classification via ECG signal subsampling and mutual information based subbands statistical features selection. *Journal of Ambient Intelligence and Humanized Computing*, *14*(3), 1473–1487.
17. Pandey, S. K., Shukla, A., Bhatia, S., Gadekallu, T. R., Kumar, A., Mashat, A., ... & Janghel, R. R. (2023). Detection of arrhythmia heartbeats from ECG signal using wavelet transform-based CNN model. *International Journal of Computational Intelligence Systems*, *16*(1), 80.
18. Thakur, S., Saxena, R.K.: Analysis of the functional relationship between electrocardiographic signal and simultaneously acquired respiratory signals. Int. J. Image, Graph. Signal Process. (IJIGSP) 7(9), 34–40 (2015)
19. Gupta, S., Saxena, R., Bansal, A., Saluja, K., Vajpayee, A., & Shikha, S. (2023, June). A case study on the classification of brain tumour by deep learning using convolutional neural network. In AIP Conference Proceedings (Vol. 2782, No. 1). AIP Publishing.
20. Sumitra, N., Rakesh kumar saxena, "Brain tumour classification using Back Propagation Neural Network", International journal of Image, Graphics and Signal Processing, 2013, 2, PP.45–50.
21. Bhagyalakshmi V, Ramchandra, Geeta D (2024). Arrhythmia Classification Using Cat Swarm Optimization Based Support Vector Neural Network, Journal of Networking and Communication Systems
22. Md Alamin Talukder, Majdi khalid, Mohsin Kazi et.el (2024). A hybrid cardiovascular arrhythmia disease detection using ConvNeXt-X models on electrocardiogram signals. Scientific report, 30366 (2024).

Note: All the figure and tables in this chapter were made by the authors.

Intelligent Systems Using Semiconductors for Robotics and IoT – Dinesh Goyal et al. (eds)
© 2026 Taylor & Francis Group, London, ISBN 978-1-041-20408-4

A Comparative Study of Machine Learning Techniques for Diabetic Retinopathy Detection from Retinal Images

Anuradha Raheja[1]
Research Scholar, Poornima University,
Jaipur, Rajasthan, India

Rakesh Kumar Saxena[2]
Professor, Poornima University,
Jaipur, Rajasthan, India

Abstract: Diabetic Retinopathy (DR), a progressive eye disorder is induced by diabetes mellitus that advances in a manner that, if not recognized and managed early enough, may significantly risk vision i.e. it is a major contributor to vision loss worldwide. Traditional methods for Diabetic Retinopathy detection are based on a review of retinal scans by eye specialists, which is manual, subjective, and possibly time-consuming. Recent innovations in deep learning (DL) have significantly boosted the accuracy and efficiency of DR diagnosis. Deep learning offers cutting-edge solutions that can be leveraged to predict, forecast and diagnose several medical conditions This review paper examines deep learning-oriented strategies for diabetic retinopathy detection, focusing on studies published in the last few years. We discuss the latest methodologies, datasets, performance evaluations, and the challenges associated with automated DR detection.

Keywords: Diabetic retinopathy, CNN, Machine learning, Deep learning

1. Introduction

Diabetes Mellitus, a major public health issue is presently impacting an estimated 463 million people across the world, with forecasts estimating this figure will increase to 700 million by 2045.[1] It is marked by the body's failure to balance blood sugar readings efficiently. Persistently elevated sugar levels can result in several health issues, like slow wound healing and cardiovascular disease. Diabetic Retinopathy (DR) is also one of them which effects retina and is a primary reason for preventable vision loss globally, predominantly affecting individuals with prolonged diabetes. These progressive stages can be recognized through microvascular changes, mainly microaneurysms and hemorrhages, which are visible in the retinal images. Evidence suggests that timely diagnosis and appropriate intervention can substantially reduce the likelihood of vision impairment. However, DR is often asymptomatic in the early stages, making timely diagnosis challenging. Comprehensive DR screening programs are essential for surveillance and early intervention; however, they face practical limitations such as resource constraints and increased workload for human graders. Conventionally, diagnosis requires manual examination by ophthalmologists and is subjective and variable. There is a pressing need in the face of exponentially increasing worldwide prevalence, wherein automated and reliable screening—all other methods must assess the beginning phases of diabetic

Retinopathy (DR) quickly and with unparalleled accuracy.[2] Fundus imaging, a cornerstone of DR detection, enables clinicians to assess retinal lesions and determine DR severity. [3] The examination of fundus images

[1]anuradha.raheja@poornima.edu.in, [2]saxenark06@gmail.com

DOI: 10.1201/9781003716389-71

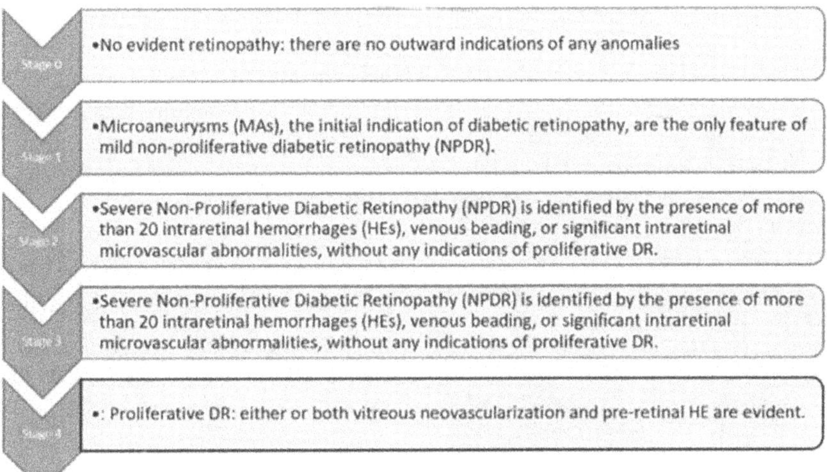

Fig. 71.1 Stages of diabetic retinopathy

focuses on identifying retinal lesions, with the intensity of Diabetic Retinopathy (DR) determined by the type & count of lesions detected.[4]

The five phases of DR severity can be distinguished using the International Clinical DR Scale, as shown in Fig. 71.1. and its images are depicted in Fig. 71.2.

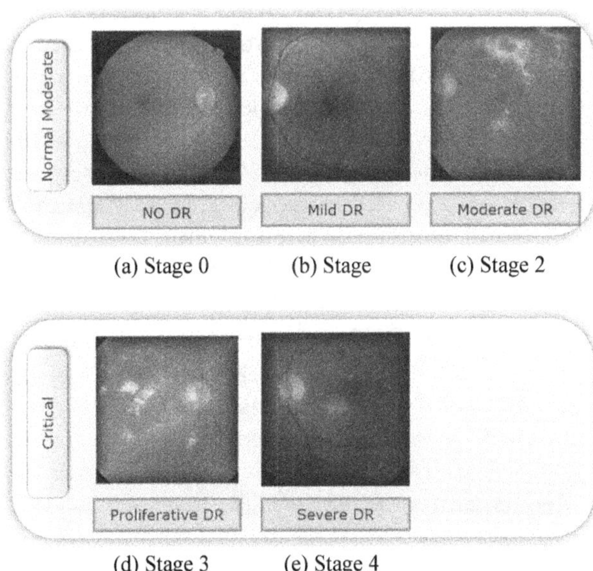

(a) Stage 0 (b) Stage (c) Stage 2

(d) Stage 3 (e) Stage 4

Fig. 71.2 Images showing Stages of DR [5]

Deep learning is critical for optimizing the diagnostic precision, efficiency, and scalability of diabetic retinopathy. Deep learning methods can leverage large annotated datasets of retinal images to detect preliminary signs of diabetic retinopathy with high sensitivity and specificity, often at par or better than that of human experts [6]. Therefore, this increases the velocity of the screening process and reduces reliance on specialized expertise, thereby increasing access to diabetic retinopathy screening across a wide spectrum of healthcare settings [7]. Early intervention in the management of the progression of diabetic retinopathy may help prevent irreversible vision loss. Hence, an effective and precise screening method is of great importance during diabetic care [8]. Traditionally, detection has relied on the assessment of visual retinal images obtained with the help of strategies such as fundus photography by ophthalmologists. This manual evaluation is subjective, and thus varies according to the skill level and fatigue of the observer. Moreover, it may not identify early signs of retinopathy before substantial damage has already occurred [9], and deep learning algorithms analyze big datasets of images to automatically extract features indicative of the existence or non-existence of diabetic retinopathy. DL algorithms learn from examples annotated by experts to enable identification of subtle changes in the morphology of the retina that probably may not manifest in the human eye during manual inspection.[10]

Diabetic retinopathy detection also goes beyond preserving vision to ensure an enhanced quality of life for affected patients. Vision loss mostly affects daily activity, independence, and productivity. Once retinopathy has been identified in its early stages, healthcare providers apply appropriate treatments, including laser therapy, injections into the eye, or surgical procedures to manage the disease process to reduce its effects on visual functioning. It will not only improve immediate health outcomes but also provide long-lasting, quality well-being for people with diabetes [11] Timely diagnosis of diabetic retinopathy is cost-effective, even from a health care perspective. It usually matters less if treated early at a lower cost and is less invasive than managing advanced stages, which may require more serious interventions. Deep learning (DL) concepts, when applied for Diabetic Retinopathy diagnosis, have numerous advantages in improving accuracy, efficiency, and scalability, opening up the possibility of early detection and intervention, as depicted in Fig. 71.3. Better patient recovery means greater sustainability and effectiveness of health delivery systems worldwide.

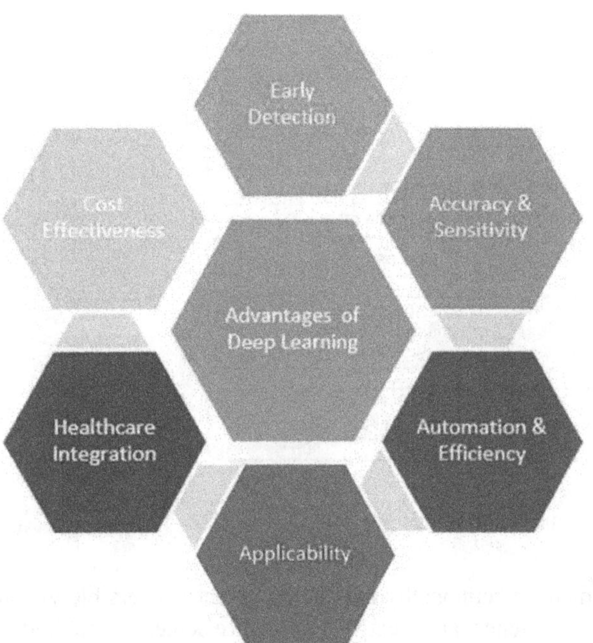

Fig. 71.3 Benefits of DL in DR detection

1.1 Background and Related Work

One of the most prevalent vision-threatening adverse effects of diabetes mellitus is alterations in retinal blood vessels. It progresses from mild non-proliferative phases to more critical proliferative stages over time, leading to blindness if not undetected, and is managed well at an initial stage. Early detection is important for interventions that maintain & improve visual function. Patient outcomes over time. The classic approaches to detect DR include human-led evaluations by ophthalmologists, which are subjective, time-intensive, and prone to inter-observer variability. This has added impetus to research into automated approaches, essentially taking advantage of improvements in deep learning for diagnostic accuracy and efficiency improvement.

Deep learning, and consequently artificial intelligence, has brought revolution in the field of medical image analysis by empowering computers to learn patterns and features directly from data. Convolutional Neural Networks have high performance for automatically detecting diabetic retinopathy from retinal, extracting fine details and subtle abnormalities that cannot be seen by the naked eye, thereby increasing sensitivity and specificity in diagnosis.

The benefits of DR detection extend beyond the accuracy with which a disorder can be detected. Thus, this includes scalability across numerous diversified datasets and healthcare settings, one that is central to the widening of access to screening in both developed and developing regions. Moreover, deep-learning algorithms can function by requiring a minimum level of human intervention following their initial training. Therefore, it relieves the workload of healthcare professionals and can potentially expedite the screening process.

Recent literature in this domain has often considered a combination of many issues in the implementation of deep learning for DR detection: model architectures, alternative methods for extraction, dataset diversity, and techniques for validation. These studies have shown significant improvements in performance metrics, including area under the ROC curve, sensitivity and specificity, clearly demonstrating that deep learning has the potential to modernize diabetic retinopathy (DR) screening, making it a more efficient and effective practice.

In [12], the authors presented an active deep learning CNN that automatically identifies DR stages from retinal fundus images. The ADL-CNN combines active learning through the expected gradient length, which efficiently extracts complex retinal features and accurately classifies the distinct severity categories of DR. Evaluation of the EyePACS dataset yielded very impressive results: SP 95.10%, ACC 98%, F-measure 93% and SE 92.20% thereby performing better in lesion detection and severity grading than all other methods.

In [13], the author discussed Diabetic Retinopathy as a diabetes-related medical condition that leads to the destruction of retinal blood vessels and can cause loss of vision. They suggested a strongly coupled convolutional network, DenseNet-169, for image classification with respect to DR severity levels: No DR, Mild, Moderate, Severe, and Proliferative DR, using fundus images. Their approach included the collection of datasets from Kaggle, particularly Aptos 2019 Blindness Detection and Diabetic Retinopathy Detection 2015. Preprocessing and augmentation are based on modeling. In their approach, they attained an accuracy of 90% with a deep learning enabler, whereas applying a regression model yielded an accuracy of 78% for the construction of an automated system for DR detection.

In a paper [14], the authors developed a deep ensemble algorithm utilizing multiple improved Inception-v4 models to diagnose diabetic retinopathy (DR) and diabetic macular edema (DMO) from fundus scans of retina. Using 8,739 images from 3,285 patients, the model achieved high performance with AUCs of 0.992 for DR and 0.994 for DMO, surpassing human experts in sensitivity and specificity. The resilience and generalization performance of the model were confirmed using the Messidor-2 dataset, demonstrating its potential to effectively enhance DR and DMO screening programs.

In [15], the authors emphasized the challenge of early detection of diabetic retinopathy through improved image quality in fundus photographs. They suggested an algorithm for improving colored retinal images by filtering noise and boosting the contrast. First, a crop to eliminate irrelevant content was used; second, a shape crop was used; and lastly, Gaussian blurring was used. It was reported that the evaluation of the EyePACS and MESSIDOR datasets showed that the extracted features had significant improvement, hence enhancing the classification performance of the unenhanced images. The effectiveness of the algorithm\hm was further validated in smart hospital environments, thus demonstrating the potential of the algorithm to be applied in IoMT for diagnosis related to DR.

In [16], the authors considered approaches for diagnosing and categorizing diabetes: a hybrid network with VGG16 and a classifier through XGBoosting and the DenseNet-121 network. Having done this on the APTOS 2019 Blindness Detection Kaggle Dataset, the class distribution was imbalanced; hence, class balancing was required. These measures were based on accuracy. The result obtained by the latter was 79.50% accuracy against the DenseNet 121 model, which had 97.30% accuracy. This was because DenseNet-121 performed better than the other previously developed methods available for the same dataset. Thus, by using advanced deep learning models, efficiency and effectiveness can be improved, and health workers and patients can benefit from detection and treatment in the early stages of DR.

In [17], the authors proposed a computer-vision-based fully automated screening technique for DR. Their technique incorporates a CLAHE application on the green channel, wherein bright lesions are more dominant, considering the challenges of structural complexity and marginal contrast differences in fundus images. In this study, an asymmetrical deep learning approach was anticipated. U-Net is used for feature extraction in the segmentation of optic discs and blood vessels, whereas a CNN combined with an SVM is proposed to classify DR lesions. They belong to four specific categories: normal, microaneurysms, hemorrhages, and exudates.. This method was tested using both the APTOS and MESSIDOR datasets to return accuracies for ND retinopathy detection of 98.6% and 91.9%, and exudate detection of 96.9% and 98.3%, respectively. This clearly proves that the accurate segmentation of a retinal image would enhance the precision of the DR grading system to a large extent.

In [18], the authors assessed the reliability of a DL-based strategy for identifying diabetic retinopathy(DR) in Indigenous Australians, who reported the highest rate of blindness due to diabetic retinopathy worldwide. This paper compares DLS with a gold standard set of grading from retinal specialists using 1,682 retinal photographs from 864 diabetic patients. In the current study, compared with 87.1%, the sensitivity for DLS was better for mtmDR (98.0%), vtDR (96.2% versus 84.4%), and all-cause referable DR (93.7% vs. 74.4%). The specificity was somewhat lower but comparable to that of human graders. Thus, DLS has the potential to improve DR screening in underserved indigenous populations, especially in early detection and intervention.

In [19], the authors applied transfer learning by leveraging the DenseNet-121 CNN to detect diabetic retinopathy incorporating Bayesian extensions to refine the trained model. Their experiments, which utilized a merged dataset (APTOS 2019 + DDR) with pre-processed images, demonstrated that the aforementioned approach surpassed the futuristic models when tested for accuracy, achieving 94.23% for Mean Field Variational Inference, 97.68% for the Monte Carlo Dropout model, and 91.44% for the Deterministic model

For the timely prediction of diabetic retinopathy (DR) [20],deep learning is used, particularly CNNs, has been investigated. The evaluation of 4 CNN models, namely MobileNetV2, ResNet50, DenseNet201, and VGG19, was performed on the basis of time of training and validation. In case of accuracy, MobileNetV2 outperformed(78.22%) while ResNet50 took shortest training time. To improve its performance, MobileNetV2 was combined with GCN, which leveraged the strength of both tubes, and the different deep learning models can be adopted in clinical practice to strengthen DR screening and early intervention.

In [21], the authors proposed a new framework for the diagnosis of DR through advanced deep learning techniques that provide impetus for its early detection in order to prevent vision loss. Their approach uses transfer learning on models, such as InceptionResNetv2 and Inceptionv3, optimized for the categorization of diabetic retinopathy on ODIR. In other words, the DiaCNN model accuracies were impressive from 97.5% to 100% in training and from 97.5% to 98.3% in testing. Their DiaCNN model showed outstanding performance, outperforming existing diagnostic methods. This study represents a major step toward improving DR diagnostic accuracy, underlining potential improvements in patient outcomes of earlier detection and intervention.

In [22], the authors proposed a novel approach called as LN-SDCTC. This strategy is divided into 2 sections. Firstly, retinal color fundus The LNRC preprocessing model **receives image data** as input and generates a noise-reduced, contrast-enhanced image which is then given as input to SDC network. Relevant features are extracted with the help of convolutional layer, average & max pooling layers. At the last, the extracted features along with Multimodal Regression classification are used to determine the severity of DR.

In [23], the authors proposed a strategic generative model named Optic GAN to expand the dataset. This model showed resource-efficiency in terms of memory and time. This model showcased accuracy from 89.56 to 96.06%. balancing MESSIDOR I and II datasets. This model was not dependent of dataset structures and thus was able to find its application to a broader context.

2. Deep Learning in DR Detection

Datasets related to Diabetic Retinopathy: Datasets play a pivotal role in the expansion & assessment of deep-

Table 71.1 Datasets related to DR

Dataset Name	Description	Total Images	Classes/Annotations
Kaggle Diabetic Retinopathy (APTOS 2019 & 2015)	High-resolution fundus images labeled for DR severity.	88,702 (2015), 3,662 (2019)	No DR, Mild, Moderate, Severe, Proliferative DR
Messidor & Messidor-2	Retinal images graded for DR presence and severity.	1,200 (Messidor), 1,748 (Messidor-2)	DR severity grades
EyePACS	Large-scale dataset used in Kaggle DR competitions.	35,000+	Five DR severity levels
DDR (DeepDR) Dataset	Large dataset with DR severity labels and lesion annotations.	13,673	Normal, Mild NPDR, Moderate NPDR, Severe NPDR, PDR
Eye Diseases Classification Dataset	Large dataset for multi-class eye disease classification using retinal images.	23,000+	Normal, DR, Glaucoma, Cataract, AMD, Hypertension, Myopia, Other Diseases
IDRiD (Indian Diabetic Retinopathy Image Dataset)	High-quality retinal images with pixel-level annotations for DR lesions.	516	DR severity levels, lesion segmentation
DRIVE (Digital Retinal Images for Vessel Extraction)	Focused on retinal blood vessel segmentation.	40	Vessel segmentation ground truth

learning models for diabetic retinopathy (DR) detection. High-quality, well-annotated datasets, like EyePACS & Kaggle DR Dataset, improve model precision and adaptability, ensuring accuracy and dependability. Some of datasets are shown in Table 71.1.A. B.

2.1 Models for DR Detection

A. Convolutional Neural Networks (CNNs)

Several studies have implemented CNNs for detecting DR by classifying retinal images into different severity levels. [24] developed a deep learning system that effectively detected DR across its entire spectrum using CNNs trained on large datasets.

A Convolutional Neural Network is a powerful neural network designed for deep learning tailored to process and analyze pictorial data excelling in pattern recognition, image classification, medical imaging etc. CNNs play a crucial role in diabetic retinopathy (DR) detection by extracting critical retinal features like microaneurysms, hemorrhages, and exudates, improving the accuracy of DR severity classification. By leveraging convolutional layers to identify spatial hierarchies of patterns, CNNs boost model efficiency, enhancing accuracy, sensitivity, and specificity in automated DR detection.

B. Hybrid and Ensemble Models

Ensemble learning drastically enhances the reliability and performance of deep learning models by combining multiple models rather than relying on a single model's performance. It combines the outputs of several models to produce more precise and consistent final prediction. Different types of ensemble techniques are Boosting, Bagging, stacking, voting and averaging

Hybrid approaches combining multiple architectures have demonstrated improved accuracy.[25] outlined an Ensemble deep neural network approach (EDLDR) for classifying DR severity.

Similarly, [26] introduced A hybrid neural network framework that incorporates multiple CNN architectures to enhance feature extraction.

C. Attention Mechanisms and Transfer Learning

The attention mechanism is a deep learning strategy that empowers models to focus on the most essential features of the input data while processing information. Transfer learning is a deep learning approach that utilizes a pre-trained neural network that was first trained on a massive dataset, and fine-tunes it for an alternative yet relevant task. Rather than building a model from the scratch, Transfer learning harnesses insights obtained from one domain to improve learning efficiency and performance in another.

Attention-based models & transfer learning have been widely adopted to improve DR detection. [3] developed an attention-based deep learning framework to enhance fundus image analysis. [21] utilized transfer learning and the DiaCNN model, achieving high accuracy in classification of Diabetic Retinopathy.

The effectiveness of the different DL models can be Measured by performance indicators like Accuracy, Precision, F1 score & AUC-ROC.

3. Common Findings

On the basis of comprehensive analysis of numerous research studies on diabetic retinopathy (DR) detection using DL techniques, here is the list of common findings that developed across the research that would serve as a road map for future research.

4. Conclusion

For the Overall diabetes management, the healthcare practitioners require a comprehensive and holistic approach to reassess and adjust the treatment strategies going on at

Table 71.2 Results of a model on a specific dataset

Author	Outcome	Dataset Used	Models Used	Research Gap
Keel et al. (2019) [28]	Heatmap visualization improved interpretability and accuracy.	Clinical retinal image dataset	CNN Visualization Tool	Further development of interpretability tools across different conditions and models.
Bellemo et al. (2019) [27]	AUC of 0.973, sensitivity over 90%, performed similarly to human graders.	Zambia-based population screening program dataset	AI Model	Expanding the model's application to different healthcare systems and higher-resource settings.
Araújo et al. (2019) [29]	Provided interpretable explanations and predicted uncertainties.	Kaggle DR detection dataset	GRADUATE	Multiple-instance learning framework with Gaussian sampling
Tymchenko et al. (2020) [30]	Achieved kappa score of 0.925466, indicating high reliability.	APTOS 2019 Blindness Detection Datase	Automated DL Model	Cost and dataset variability need further exploration
Yip et al. (2020) [31]	AUC ranging from 0.936 to 0.944, high consistency in performance.	Retinal images dataset of 455,491 images	DL Models	Need for robust training datasets for clinical DR screening
Dai et al. (2021) [24]	AUC between 0.901 and 0.972 for lesion detection and DR grading.	Large-scale dataset of fundus images	DeepDR	With a single-ethnic cohort and local lesion testing, this study requires further validation across different demographic groups.
Farooq et al. (2022) [32]	Highlighted AI's role in reducing late-stage treatment costs.	Multiple public DR datasets	AI for DR Diagnosis	Real-time applications of deep learning techniques in DR detection.
Mondal et al. (2022) [25]	Outperformed Leading approaches for DR classification.	APTOS19 and DIARETDB1 datasets	Ensemble Model (DenseNet101 + ResNeXt)	Investigating the model's scalability to larger datasets.

present, that would cater not only eyes but also other systemic factors like cholesterol levels, BP and blood glucose levels and thus minimizes the risk of other complications like kidney diseases and cardiovascular disease.

Despite its strengths, Deep learning-driven DR detection grapples with challenges such as dataset imbalance, inconsistent image resolution and the need for more transparent model decision-making. To mitigate these limitations, Researchers are actively working to resolve these issues to make deep learning more effective and widely available in medical applications.

In the nutshell, the incorporation of deep learning has greatly improved the screening of diabetic retinopathy, with its proficiency in handling extensive datasets, achieving high diagnostic accuracy, seamlessly integrating into medical workflows and serving as a key solution for addressing DR at an early stage.

References

1. N. Tsiknakis *et al.*, "Deep learning for diabetic retinopathy detection and classification based on fundus images: A review," Aug. 01, 2021, *Elsevier Ltd.* doi: 10.1016/j.compbiomed.2021.104599.
2. N. Mukherjee and S. Sengupta, "Application of deep learning approaches for classification of diabetic retinopathy stages from fundus retinal images: a survey," *Multimed Tools Appl*, vol. 83, no. 14, pp. 43115–43175, Oct. 2023, doi: 10.1007/s11042-023-17254-0.
3. R. Romero-Oraá, M. Herrero-Tudela, M. I. López, R. Hornero, and M. García, "Attention-based deep learning framework for automatic fundus image processing to aid in diabetic retinopathy grading," *Comput Methods Programs Biomed*, vol. 249, Jun. 2024, doi: 10.1016/j.cmpb.2024.108160.
4. C. P. Wilkinson *et al.*, "Proposed international clinical diabetic retinopathy and diabetic macular edema disease severity scales," *Ophthalmology*, vol. 110, no. 9, pp. 1677–1682, Sep. 2003, doi: 10.1016/S0161-6420(03)00475-5.
5. G. Alwakid, W. Gouda, and M. Humayun, "Deep Learning-Based Prediction of Diabetic Retinopathy Using CLAHE and ESRGAN for Enhancement," *Healthcare (Switzerland)*, vol. 11, no. 6, Mar. 2023, doi: 10.3390/healthcare11060863.
6. A. D. Fleming *et al.*, "Deep learning detection of diabetic retinopathy in Scotland's diabetic eye screening programme," *British Journal of Ophthalmology*, 2023, doi: 10.1136/bjo-2023-323395.
7. G. Sivapriya, R. Manjula Devi, P. Keerthika, and V. Praveen, "Automated diagnostic classification of diabetic retinopathy with microvascular structure of fundus images using deep learning method," *Biomed Signal Process Control*, vol. 88, p. 105616, Feb. 2024, doi: 10.1016/j.bspc.2023.105616.
8. B. Lalithadevi and S. Krishnaveni, "Diabetic retinopathy detection and severity classification using optimized deep learning with explainable AI technique," *Multimed Tools Appl*, vol. 83, no. 42, pp. 89949–90013, Apr. 2024, doi: 10.1007/s11042-024-18863-z.
9. S. Zhu, C. Xiong, Q. Zhong, and Y. Yao, "Diabetic Retinopathy Classification with Deep Learning via Fundus

Images: A Short Survey," *IEEE Access*, vol. 12, pp. 20540–20558, 2024, doi: 10.1109/ACCESS.2024.3361944.

10. R. Gargeya and T. Leng, "Automated Identification of Diabetic Retinopathy Using Deep Learning," *Ophthalmology*, vol. 124, no. 7, pp. 962–969, Jul. 2017, doi: 10.1016/j.ophtha.2017.02.008.
11. D. S. W. Ting *et al.*, "Development and Validation of a Deep Learning System for Diabetic Retinopathy and Related Eye Diseases Using Retinal Images From Multiethnic Populations With Diabetes," *JAMA*, vol. 318, no. 22, p. 2211, Dec. 2017, doi: 10.1001/jama.2017.18152.
12. I. Qureshi, J. Ma, and Q. Abbas, "Diabetic retinopathy detection and stage classification in eye fundus images using active deep learning," *Multimed Tools Appl*, vol. 80, no. 8, pp. 11691–11721, Mar. 2021, doi: 10.1007/s11042-020-10238-4.
13. G. Mushtaq and F. Siddiqui, "Detection of diabetic retinopathy using deep learning methodology," *IOP Conf Ser Mater Sci Eng*, vol. 1070, no. 1, p. 012049, Feb. 2021, doi: 10.1088/1757-899x/1070/1/012049.
14. F. Li *et al.*, "Deep learning-based automated detection for diabetic retinopathy and diabetic macular oedema in retinal fundus photographs," *Eye (Basingstoke)*, vol. 36, no. 7, pp. 1433–1441, Jul. 2022, doi: 10.1038/s41433-021-01552-8.
15. S. H. Abbood, H. N. A. Hamed, M. S. M. Rahim, A. Rehman, T. Saba, and S. A. Bahaj, "Hybrid Retinal Image Enhancement Algorithm for Diabetic Retinopathy Diagnostic Using Deep Learning Model," *IEEE Access*, vol. 10, pp. 73079–73086, 2022, doi: 10.1109/ACCESS.2022.3189374.
16. C. Mohanty *et al.*, "Using Deep Learning Architectures for Detection and Classification of Diabetic Retinopathy," *Sensors*, vol. 23, no. 12, Jun. 2023, doi: 10.3390/s23125726.
17. P. K. Jena, B. Khuntia, C. Palai, M. Nayak, T. K. Mishra, and S. N. Mohanty, "A Novel Approach for Diabetic Retinopathy Screening Using Asymmetric Deep Learning Features," *Big Data and Cognitive Computing*, vol. 7, no. 1, Mar. 2023, doi: 10.3390/bdcc7010025.
18. M. A. Chia *et al.*, "Validation of a deep learning system for the detection of diabetic retinopathy in Indigenous Australians," *British Journal of Ophthalmology*, vol. 108, no. 2, pp. 268–273, Feb. 2023, doi: 10.1136/bjo-2022-322237.
19. M. Akram *et al.*, "Uncertainty-aware diabetic retinopathy detection using deep learning enhanced by Bayesian approaches," *Sci Rep*, vol. 15, no. 1, p. 1342, Jan. 2025, doi: 10.1038/s41598-024-84478-x.
20. F. Mostafa, H. Khan, F. Farhana, and M. A. H. Miah, "Application with Deep Learning Framework for Early Prediction of Diabetic Retinopathy," *AppliedMath*, vol. 5, no. 1, p. 11, Feb. 2025, doi: 10.3390/appliedmath5010011.
21. M. R. Shoaib *et al.*, "Deep learning innovations in diagnosing diabetic retinopathy: The potential of transfer learning and the DiaCNN model," *Comput Biol Med*, vol. 169, Feb. 2024, doi: 10.1016/j.compbiomed.2023.107834.
22. D. Muthusamy and P. Palani, "Deep neural network model for diagnosing diabetic retinopathy detection: An efficient mechanism for diabetic management," *Biomed Signal Process Control*, vol. 100, p. 107035, Feb. 2025, doi: 10.1016/j.bspc.2024.107035.
23. P. Kapoor and S. Arora, "Optic-GAN: a generalized data augmentation model to enhance the diabetic retinopathy detection," *International Journal of Information Technology*, Feb. 2025, doi: 10.1007/s41870-025-02426-y.
24. L. Dai *et al.*, "A deep learning system for detecting diabetic retinopathy across the disease spectrum," *Nat Commun*, vol. 12, no. 1, Dec. 2021, doi: 10.1038/s41467-021-23458-5.
25. S. S. Mondal, N. Mandal, K. K. Singh, A. Singh, and I. Izonin, "EDLDR: An Ensemble Deep Learning Technique for Detection and Classification of Diabetic Retinopathy," *Diagnostics*, vol. 13, no. 1, Jan. 2023, doi: 10.3390/diagnostics13010124.
26. B. Menaouer, Z. Dermane, N. El Houda Kebir, and N. Matta, "Diabetic Retinopathy Classification Using Hybrid Deep Learning Approach," *SN Comput Sci*, vol. 3, no. 5, Sep. 2022, doi: 10.1007/s42979-022-01240-8.
27. V. Bellemo *et al.*, "Articles Artificial intelligence using deep learning to screen for referable and vision-threatening diabetic retinopathy in Africa: a clinical validation study," 2019. [Online]. Available: www.thelancet.com/
28. S. Keel, J. Wu, P. Y. Lee, J. Scheetz, and M. He, "Visualizing Deep Learning Models for the Detection of Referable Diabetic Retinopathy and Glaucoma," *JAMA Ophthalmol*, vol. 137, no. 3, p. 288, Mar. 2019, doi: 10.1001/jamaophthalmol.2018.6035.
29. T. Araújo *et al.*, "DR$\vert$GRADUATE: uncertainty-aware deep learning-based diabetic retinopathy grading in eye fundus images," Oct. 2019, doi: 10.1016/j.media.2020.101715.
30. B. Tymchenko, P. Marchenko, and D. Spodarets, "Deep Learning Approach to Diabetic Retinopathy Detection," Mar. 2020, [Online]. Available: http://arxiv.org/abs/2003.02261
31. M. Y. T. Yip *et al.*, "Technical and imaging factors influencing performance of deep learning systems for diabetic retinopathy," *NPJ Digit Med*, vol. 3, no. 1, Dec. 2020, doi: 10.1038/s41746-020-0247-1.
32. M. S. Farooq *et al.*, "Untangling Computer-Aided Diagnostic System for Screening Diabetic Retinopathy Based on Deep Learning Techniques," *Sensors*, vol. 22, no. 5, Mar. 2022, doi: 10.3390/s22051803.
33. Thakur, S., Saxena, R.K.: Analysis of the functional relationship between electrocardiographic signal and simultaneously acquired respiratory signals. Int. J. Image, Graph. Signal Process. (IJIGSP) 7(9), 34–40 (2015)
34. Gupta, S., Saxena, R., Bansal, A., Saluja, K., Vajpayee, A., & Shikha, S. (2023, June). A case study on the classification of brain tumour by deep learning using convolutional neural network. In AIP Conference Proceedings (Vol. 2782, No. 1). AIP Publishing.
35. Sumitra, N., Rakesh kumar saxena, "Brain tumour classification using Back Propagation Neural Network", International journal of Image, Graphics and Signal Processing, 2013, 2, PP.45–50.

Note: All the figures and tables in this chapter were made by the authors.

Intelligent Systems Using Semiconductors for Robotics and IoT – Dinesh Goyal et al. (eds)
 ISBN 978-1-041-20408-4

72 SecurePDF-X: A Robust and Explainable PDF Malware Detection Model Leveraging Ensembling Techniques and XAI

Aditya Nautiyal,
Shubhangi Saklani,
Aaditya Sharma, Amit Kumar Singh*
Amity Institute of Information Technology Amity University,
Rajasthan, Jaipur, India

Abstract: The increased use of PDF files in operational and communication activities has made these files a frequent target of cyber-attacks, thus making it necessary to develop robust PDF malware detection technique. In this paper, we propose an ensemble learning framework based on the stacking modeled architecture termed SecurePDF-X to detect PDF-based malware. This model takes advantage of ensemble learning techniques and combines it with explainable artificial intelligence (XAI) to enhance the interpretability and provide insights into the model's decision-making process, increasing its robustness. To achieve this objective, in this study, we firstly extracted 37 high quality features from the CIC-PDFMAL2022 dataset, consisting approximately about 10,000 benign and malicious PDF documents combined. Using Particle Swarm Optimization (PSO) for feature selection, we reduced the initial 37 features to a refined subset of 19. The evaluation results demonstrate that the proposed SecurePDF-X model, by using these selected features, significantly have a better performance and outperforms traditional models, achieving a true positive rate of 98.65%, with corresponding false positive rate at a minimum of 1.46%.

Keywords: Malicious PDF detection, Ensemble learning, Explainable AI (XAI), Feature engineering, Feature selection, Cybersecurity, SHAP, LIME

1. Introduction

The advent of the digital era revolutionized our consumption and sharing of information, and new security threats have grown. Portable document format (PDF) files are one of the most universally used file types [1]. According to a Business World report, 66% of malware is delivered via PDF files in malicious emails [2]. The rise of Portable Document Format (PDF) malware is one of the most concerning of them all. This has become a prevalent means for disseminating malicious code.

PDFs are used in work and non-work environments, thus luring the attackers that use various encoding and compression methods to hide malicious content in pdfs by reducing file size, hiding sensitive content, or other means [3]. Attackers have also taken leverage over phishing assaults, particularly via emails, with PDFs evolving in the capabilities of documents and executing code, creating processes, and performing tasks that were previously reserved for executables have been enhanced with enhanced scripting and macro functions. [4].

A plethora of researches, namely [5], discussed that traditional automated detection methods are inefficient for malicious PDF detection as they are just signature based and do not check for the behavior and content inside, this allows the malware authors to evade them sometimes, even by applying a basic obfuscation. The content of such malicious PDF files has been actively analyzed by security researchers. Therefore, extracting the key features that are characterized by the malware's identity and behavior to

*Corresponding author: aksingh1@jpr.amity.edu

DOI: 10.1201/9781003716389-72

develop robust security solutions is their focus. A pdf may contain multiple such features that state its legitimacy or temperament.

Conventional antivirus and malware detection techniques frequently fail to detect PDF-based malware. In order to solve this problem, sophisticated detecting methods are required. Malicious actors' ever-changing strategies must be kept ahead of via automated identification tools. In order to lessen these attacks, researchers have created creative strategies. PDF malware can be investigated using static, dynamic, or hybrid analysis approaches, depending on the analytical approach [6]. Additionally, feature engineering has been shown to be a workable way to increase the efficacy of detection models.

Analyzing the aforementioned issues, the need for an advanced security solution became evident. Therefore, To address this, a combined solution has been developed in this research work, i.e., SecurePDF-X: a robust and explainable PDF malware detection model that leverages ensemble learning techniques and explainable AI (XAI). SecurePDF-X detects malicious pdfs by extracting the most relevant features and creating a subset of it, along using ensemble machine learning in the given dataset [14]. In this paper, to validate the proposed framework approach, the ensembled model was trained on the CIC Evasive-PDFMal2022[7] dataset. Additionally, the Particle Swarm Optimization (PSO)[8] technique was employed for feature selection, resulting in a subset of features that delivered optimal performance and accuracy [25].

Summing up the research work, the main contributions of this paper are as follows:

1. To develop a novel ensemble-based interpretable framework for detecting malicious PDFs.
2. To propose a comprehensive approach demonstrating the impact of relevant feature selection from both general and structural features of the PDF files.
3. To highlight the ensemble model's interpretability into malicious PDF detection using XAI tools like SHAP and LIME.

The remains of the paper is structured as follows: Section II covers related work, while Section III outlines the methodology of the proposed approach, which integrates feature selection techniques with ensembling methods. Section IV presents the experiments and results, along with identified limitations. Future scope is discussed in Section V, and the conclusion is provided in Section VI.

2. Literature Review

Recent research in the detection of PDF malware has introduced several novel methods which aim at improving the chances of detection as well as feature selection techniques and difficulties related to explain ability of machine learning based models.

Hossain et al. introduced PDFMALDET a hybrid approach to identify malicious PDF files that combines static and dynamic analysis approaches. The method focuses on identifying robust structural features as well as monitoring their behavior during the execution process. This dual-layer system approach enhances detection accuracy of 99.57% on the enhanced features set and reduces false positives [9].

Falah et al. suggested a methodology for identifying a useful set of tool-independent features that enhance the longevity and usability of existing tools. Their technique gives a comprehensive grasp of how the specified features affect classification. Their proposed method reduced the feature set size by over 60% while improving accuracy by around 2% [10].

Mohammed et al. proposed HPSSA, a hybrid approach for the detection of PDF malware based on signal and statistical analysis which is a novel approach. Through the use of this method, the researchers successfully scanned over 30,000 PDF documents, which included both malevolent and ordinariness samples, achieving an impressive figure of 99.2% success. HPSSA can uncover unseen threats by the study of the trends and irregularities in the concerned PDF documents [11].

Javed et al. propose a hybrid approach for the classification of malware and employs several classification methods to locate threats concealed within the PDF files. Their approach involves the extraction of a fixed set of static features that is enhanced with behavior features that may emerge during the course of the attacks. To further assist the static extraction, popular Python-based tools have been included in the framework. Employing Sandbox to determine the possible invasion attempts. According to the framework they employ, the strategy is successful in classifying malware of PDF format files[12].

According to Rahman et al. proposed to interpret machine learning and deep learning models by using Explainable Artificial Intelligence(XAI) and Shapley Additive exPlanation (SHAP) for PDF malware detection. This approach enhances model transparency by improving model clarity by offering details about the feature's significance and decision-making processes[13].

3. Methodology

The proposed framework SecurePDF- consists of four key components: PDFAnalyzer, Feature Selection, an ensemble model for malicious pdf detection, and the explainable AI (XAI) module as shown in Fig. 72.1. The feature engineering component is responsible for preparing the dataset, including the handling of missing values and addressing outliers, along with ensuring the data suitability for machine learning models. The ensemble machine learning models component applies various ensemble techniques, such as hard voting, soft

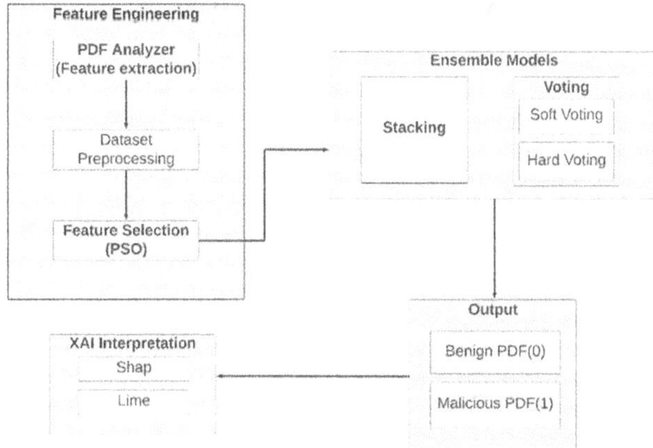

Fig. 72.1 Proposed framework SecurePDF-X

voting, and stacking, while the XAI module employs tools like SHAP and LIME to enhance the interpretability of the model's predictions.

PDFAnalyzer: This module fetches each pdf and uses python modules such as pdfminer[15], pymupdf[16] to analyze pdf structure and metadata. The pdf analyzer then extracts different information and generates a feature vector based on the available information. About 37 different features are extracted using this module, primarily in two components of a pdf i) general features and ii) structural features. The extracted features are then further passed on to the feature selection module to get only the most useful features as input parameters for our ensembling model.

Feature Selection: This module takes input from the PDFAnalyzer module and does the preprocessing and feature selection part. It plays a crucial role and affects the overall result of the model. The feature selection module takes the raw noisy data gained from the previous module and filters out the required impactful data, required to train our model. First the module removes the noise from the dataset by removing null values and encoding non-numerical values to numerical values. The importance of each feature is calculated using the PSO techniques which results in a subset of feature vectors with most accuracy.

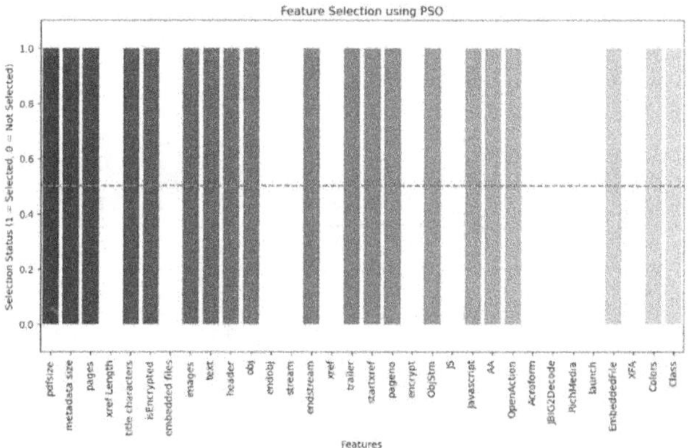

Fig. 72.2 Top 32 features as per MI score

Ensembled Model: The SecurePDF-X framework uses ensemble learning to detect malicious PDFs by combining predictions from multiple classifiers. Three ensemble methods were explored: stacking, hard voting, and soft voting. **Stacking [17]** is a two-level model, where base classifiers (Naive Bayes, SVM, and Decision Tree) generate predictions at Level 0, and a meta-classifier (Logistic Regression) refines these predictions at Level 1. This approach leverages diverse models for improved generalization. Whereas **voting [18]** merges predictions coming from multiple models. Wherein **hard voting** selects the majority vote, while **soft voting** does the average of probability scores to choose the final class. These methods enhance detection accuracy by balancing model strengths.

ALGORITHM: Stacked Ensemble Algorithm for Phishing URL Detection

1) **Input:** Original dataset D
2) **Level-0** classifiers $C_1, C_2, \ldots, C_T$
3) **Level-1** classifier M
4) **Process:**
   a) For $t = 1$ to T do
   b) $h_t = C_t(D)$ // Train base classifiers on dataset
   c) end for
   d) Initialize new dataset $D' = 0$
   e) $i = 1$ to m do
   f) for $t = 1$ to T do
   g) $Z_{it} = \hat{h}_t(x_i)$ // Get predictions (Phishing or Legitimate) from base classifiers
   h) end for
   i) $D' = D' U \{(Z_{i1}, Z_{i2}, \ldots, Z_{iT}, Y_i)\}$
   j) end for
   k) $\hat{h}' = M(D')$ // Train meta-classifier on new dataset D'

Output: Final prediction $H(x) = \hat{h}'(\hat{h}_1(x), \hat{h}_2(x), \ldots, \hat{h}_T(x))$

Algorithm 1 outlines the stacked ensemble approach for phishing URL detection. It begins by training T base classifiers $C_1, C_2, \ldots, C_T$ on the original dataset D. For each instance x_i, predictions Z_{it} are collected from each base classifier, forming a new dataset D'. This dataset includes the predictions and true labels. The meta-classifier M is then trained on D' to make the final prediction H (x), integrating outputs from all base classifiers.

Interpreting with XAI: This module applies an additional layer of interpretability to the SecurePDF-X model, making it possible to know the reasons for the model predicting certain outputs. Explanations or justifications for the results of machine learning algorithms are a key consideration for the issue of trust and accountability. To do this, SHAP [19] and LIME [20], two popular eXplainable AI (XAI) methods, were applied on the evaluation metric of best performing model.

SHapley Additive exPlanations is a game-theory-based method that provides a cooperative value solution and allocates as fairly as possible a reward to each feature according to its contribution towards the model's prediction. It also simplifies the task by providing clear visual explanations such as beeswarm plots. This makes understanding complex ensemble models easier. One of the advantages of SHAP is its ability to handle feature interactions and collinearity by explaining how features contribute to predictions in stacking models.

LIME (Local Interpretable Model-Agnostic Explanations) within this context makes it possible to explain an individual prediction by creating perturbed copies of an instance and seeing how the prediction of the model changes. LIME also leads to building a simpler interpretable model that would better suit the specific case whilst consuming less complexity than the original case. In the case of Secure PDF-X, LIME has been useful in understanding the contribution of base classifiers such as Decision Trees, Naive Bayes and SVM in the stacked ensemble, thus explaining the making of the overall decision.

4. Results and Findings

In the results of this study, multiple base classifiers like—Logistic Regression (LR), Support Vector Machine (SVM), and Decision Tree Classifier (DTC), as well as ensemble methods like Stacking, Hard Voting, and Soft The voting process was evaluated using various performance metrics, such as Accuracy, Precision, F1-Score, True Positive Rate (TPR), True Negative Rate (TNR), False Positive Rate (FPR), and False Negative Rate (FNR).

Among the base classifiers, the Decision Tree Classifier (DTC) achieved the highest accuracy of 98.55% and also performed exceptionally well in terms of Precision (99.01%), F1-Score (98.69%), TPR (98.38%), and TNR (98.77%), indicating its robust performance in classifying phishing URLs. Logistic Regression (LR) and SVM showed relatively lower performance, with accuracies of 93.12% and 93.07% respectively. However, LR demonstrated a strong Precision of 96.12%, while SVM achieved 98.12%, though their recall and F1-scores were not as high as DTC.

As shown in Table 72.1, the Stacking model, which combined LR, SVM, and DTC, slightly outperformed the individual classifiers, achieving the highest overall accuracy of 98.60%, with balanced Precision (98.83%), F1-Score (98.74%), and recall rates (TPR of 98.65%). Hard Voting, another ensemble method, achieved a respectable accuracy of 95.21% but fell short of Stacking in terms of overall performance metrics. The results clearly show that the Stacking ensemble outperformed all other models across most metrics, making it the most effective method for phishing URL detection in this study.

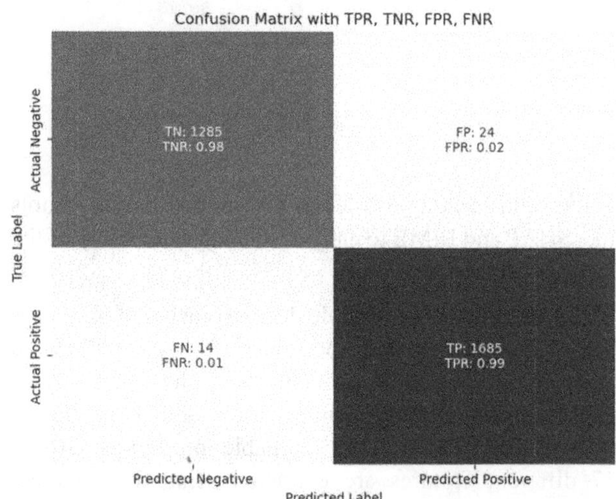

Fig. 72.3 Confusion matrix of the stacking model

The confusion matrix for the stacking model used to predict malicious PDF files demonstrates its strong performance. With a True Positive Rate (TPR) of 0.99, the model accurately detects 99% of malicious PDFs, while the True Negative Rate (TNR) of 0.98 indicates a 98% accuracy in identifying benign PDFs. The model maintains a low False Positive Rate (FPR) of 0.02, meaning only 2% of benign files are misclassified as malicious, and an even lower False Negative Rate (FNR) of 0.01, reflecting minimal missed detections of malicious files. These metrics highlight the model's effectiveness in distinguishing between malicious and benign PDFs, making it highly reliable for practical use.

Figure 72.4 shows a SHAP beeswarm plot which provides a useful visual summary of the contribution of various features to the prediction made by the stacked ensemble model for malicious pdf detection. In this plot, the most

Table 72.1 Comparison of ML models

Models	Accuracy	Precision	F1-Score	TPR	TNR	FPR	FNR
LR	93.12	96.12	93.65	91.29	95.40	04.60	08.71
SVM	93.07	98.12	93.46	89.23	97,87	02.13	10.77
DTC	98.55	99.01	98.69	98.38	98.77	01.23	01.62
Stacking (LR + SVM + DTC)	98.60	98.83	98.74	98.65	98.54	01.46	01.35
Hard Voting	95.21	98.75	95.55	92.55	98.54	01.46	07.45
Soft Voting	93	92	94	95.2	96.62	03.38	4.8

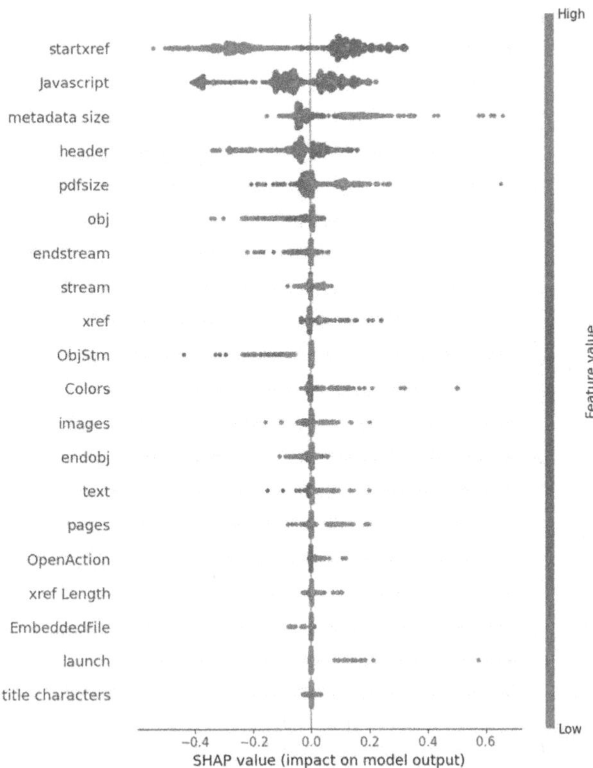

Fig. 72.4 Shap beeswarm plot for stacking model

important features are arranged vertically on the Y-axis in descending order starting from "startxref" and "Javascript" which were the two dominant features determining the prediction of this model.

"Startxref" and "Javascript" features show high SHAP values which are beneficial to the model and therefore, the higher the feature value, the greater the chances of the PDF being identified as a phishing site. In general, this SHAP beeswarm plot effectively positions users to have more detailed understanding and perspective on the classifying of the pdf as malicious, thus giving insights how the ensemble model operates.

LIME explains the model's output by showing which features, or variables, are the most important contributing to the accurate prediction. The effects of each feature are presented by horizontal bars where green and red bars correspond to malicious or benign pdf. The amount of push toward the given colour sides signifies the importance of that feature particular to classifying the input.

In this case, the Colors ≤ 0.00 feature has the largest average contribution at 0.50 green to predict that the PDF is benign, meaning that models are likely to threaten PDF files which have no color. On the other hand, text ≤ 2.00 and pageno ≤ 1.00 are serious negative contributors, and will, therefore, push the model to predict that the file is malicious, possibly due to the rhetoric that with few texts and pages, the PDF is threatening. Other features such as trailer size ≤ 1.00 and obj ≤ 9.00 also carry some weight and affect the decision making but with much less strength.

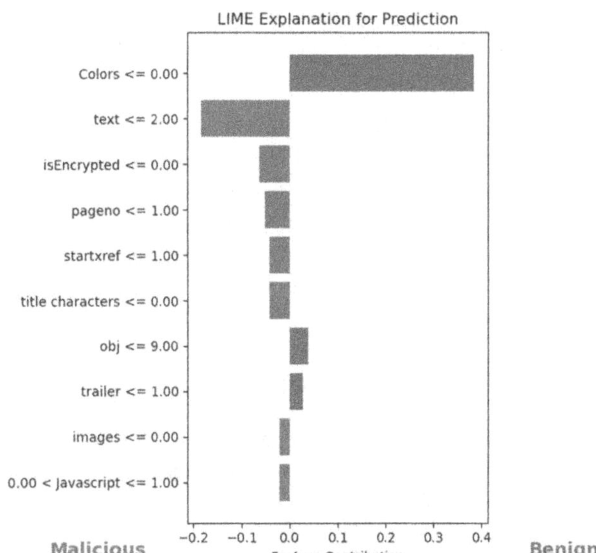

Fig. 72.5 Lime: feature impact on stacking model

This clarification also emphasizes the port's transparency by illustrating how page count, input texts and object count are critical aspects of determining the end classification at which the pages are numbered. The lower part labelled 'Benign' or 'Malicious' helps to have a better correspondence to features to classes and therefore offers higher interpretability.

5. Future Scope

The future work may target resource allocation towards adoption of deep learning methodologies in future for more precise detection of more complicated threats, as well as to optimize feature extraction to real time detection systems. In addition, expanding the coverage of SecurePDF-X to other types of document based malware detection as well as deploying it to cloud or endpoint security applications would increase its efficacy. Improving the XAI techniques will be another area of research that undoubtedly enhance the detection effectiveness of the model by providing better and more coherent explanations.

6. Conclusion

In this research, the development of SecurePDF-X, a novel ensemble-based model for detecting malicious PDFs, is presented. The method fuses feature engineering, classification, and explainable AI to give a robust and transparent malware detection technique. A hybrid approach using various ensemble models, stacking, hard voting and soft voting significantly improved detection of PDF related threats in the study.

The usability of XAI methods particularly SHAP and LIME also increases the comprehensibility of the model, providing information on the amount and kind of features that contributed to the model's output and its decisions.

Not only do these tools improve confidence in the predictions of the model, but they also help in decision making in cyber defense.

In general, SecurePDF-X offers an accurate and interpretable solution to the PDF malware problem with the application of advanced machine learning algorithms. The research could be aimed at enhancing the methods of feature extraction and introducing advanced with stronger classifiers and also the research area of the document-based malwares.

References

1. Abu Al-Haija, Q., Odeh, A., & Qattous, H. (2022). PDF malware detection based on optimizable decision trees. *Electronics*, *11*(19), 3142.
2. 66% of Malware Delivered via PDF Files in Malicious Emails: Report, Business Standard, Jun. 7, 2023. [Online]. Available: https://www.business-standard.com/technology/tech-news/66-of-malware-delivered-via-pdf-files-in-malicious-emails-report-123060700778_1.html. [Accessed: 22-Oct-2024]
3. Zhang, J. (2018). MLPdf: an effective machine learning based approach for PDF malware detection. *arXiv preprint arXiv:1808.06991.*
4. Borno, Z. S., Sakib, N., & Anwar, S. S. (2023, November). Performance Analysis of Ensemble Machine Learning Algorithms in PDF Malware Detection. In *2023 IEEE 9th International Women in Engineering (WIE) Conference on Electrical and Computer Engineering (WIECON-ECE)* (pp. 195–200). IEEE.
5. Ijaz, M., Durad, M. H., & Ismail, M. (2019, January). Static and dynamic malware analysis using machine learning. In *2019 16th International bhurban conference on applied sciences and technology (IBCAST)* (pp. 687–691). IEEE.
6. Vishnu, N. S., Lakshmi, S. M., Verma, S., & Shukla, A. K. (2022). PDF malware classifiers–a survey, future directions and recommended methodology. In *Information Security Handbook* (pp. 117–140). CRC Press.
7. Issakhani, M., Victor, P., Tekeoglu, A., & Lashkari, A. H. (2022, February). PDF Malware Detection based on Stacking Learning. In *ICISSP* (pp. 562–570).
8. Xue, B., Zhang, M., & Browne, W. N. (2012). Particle swarm optimization for feature selection in classification: A multi-objective approach. *IEEE transactions on cybernetics*, *43*(6), 1656–1671.
9. Hossain, G. S., Deb, K., & Sarker, I. H. (2024, May). An Enhanced Feature-Based Hybrid Approach for Adversarial PDF Malware Detection. In *2024 6th International Conference on Electrical Engineering and Information & Communication Technology (ICEEICT)* (pp. 101–106). IEEE.
10. Falah, A., Pan, L., Huda, S., Pokhrel, S. R., & Anwar, A. (2021). Improving malicious PDF classifier with feature engineering: A data-driven approach. *Future Generation Computer Systems*, *115*, 314–326.
11. Mohammed, T. M., Nataraj, L., Chikkagoudar, S., Chandrasekaran, S., & Manjunath, B. S. (2021, November). HAPSSA: Holistic Approach to PDF malware detection using Signal and Statistical Analysis. In *MILCOM 2021-2021 IEEE Military Communications Conference (MILCOM)* (pp. 709–714). IEEE.
12. Javed, S. M. Z., & Amjad, F. (2024, June). Unveiling Hidden Threats in PDFs with Hybrid Malware Classification. In *2024 IEEE 30th International Conference on Telecommunications (ICT)* (pp. 01–06). IEEE.
13. Rahman, T., Ahmed, N., Monjur, S., Haque, F. M., & Hossain, M. I. (2023, March). Interpreting Machine and Deep Learning Models for PDF Malware Detection using XAI and SHAP Framework. In *2023 2nd International Conference for Innovation in Technology (INOCON)* (pp. 1–9). IEEE.
14. Yerima, S. Y., Bashar, A., & Latif, G. (2022, December). Malicious pdf detection based on machine learning with enhanced feature set. In *2022 14th International Conference on Computational Intelligence and Communication Networks (CICN)* (pp. 486–491). IEEE.
15. Adhikari, N. S., & Agarwal, S. (2024). A Comparative Study of PDF Parsing Tools Across Diverse Document Categories. *arXiv preprint arXiv:2410.09871.*
16. Guedes, G. B., & da Silva, A. E. A. (2021). Supervised Learning Approach for Section Title Detection in PDF Scientific Articles. In *Advances in Computational Intelligence: 20th Mexican International Conference on Artificial Intelligence, MICAI 2021, Mexico City, Mexico, October 25–30, 2021, Proceedings, Part I 20* (pp. 44–54). Springer International Publishing.
17. Roy, K. S., Ahmed, T., Udas, P. B., Karim, M. E., & Majumdar, S. (2023). MalHyStack: a hybrid stacked ensemble learning framework with feature engineering schemes for obfuscated malware analysis. *Intelligent Systems with Applications*, *20*, 200283.
18. Karlos, S., Kostopoulos, G., & Kotsiantis, S. (2020). A soft-voting ensemble based co-training scheme using static selection for binary classification problems. *Algorithms*, *13*(1), 26.
19. Lundberg, S. (2017). A unified approach to interpreting model predictions. arXiv preprint arXiv:1705.07874.
20. Ribeiro, M. T., Singh, S., & Guestrin, C. (2016) Why should i trust you? Explaining the predictions of any classifier. In Proceedings of the 22nd ACM SIGKDD international conference on knowledge discovery and data mining (pp. 1135–1144).
21. Amit, S. Taterh, Pankaj (2025) "Dynamic feature-based detection of malware in non-executable files using a 1d convolutional neural network in Panamerican Mathematical Journal. 35, 2s.

Note: All the figure and tables in this chapter were made by the authors.

Intelligent Systems Using Semiconductors for Robotics and IoT – Dinesh Goyal et al. (eds)
© 2026 Taylor & Francis Group, London, ISBN 978-1-041-20408-4

73 Automation in Citrus Production for Enhanced Productivity and Sustainability

Gaurav Bhardwaj
Department of Biotechnology, Sanskriti University,
Mathura

Nitin Wahi
Dept. of Biotechnology, LNCT University,
Bhopal

Deen Dayal
Department of Biotechnology, GLA University,
Mathura, Uttar Pradesh

Deepak Kumar
Department of Pharmaceutical Chemistry and Chemistry,
Dolphin PG Institute of Biomedical and
Natural Sciences, Dehradun

Niraj Kumar
Department of Electronics and Communication Engineering,
Graphic Era University, Dehradun, Uttarakhand, India

Mohsin Maqbool
Department of Medical Oncology,
Dr. B.R.A, Institute Rotary Cancer Hospital,
All India Institute of Medical Sciences (AIIMS),
New Delhi, India

Shailendra Thapliyal
UIM, Uttaranchal University, Arcadia Grant, P.O. Chandanwari,
Premnagar, Dehradun, Uttarakhand, INDIA

Ashwani Kumar Sanghi
School of Allied Health Sciences, MVN University,
Palwal, Haryana, India

Siddharth Shankar Bhatt*
Sidvantage Marketing, Shankar Bhawan, R.K. Verma Marg,
Mussoorie, Uttarakhand

Kundan Kumar Chaubey
Department of Biotechnology, Sanskriti University,
Mathura

Abstract: Citrus industry holds utter importance in world economy as it provides raw material to food, pharma and beauty industries. It is the major employers since decades providing both skilled and unskilled jobs worldwide. With the advent of time availability of workforce has been a major problem in this sector. Wages of skilled labor have surged

*Corresponding author: siddharthbhatt2714@gmail.com

DOI: 10.1201/9781003716389-73

manifolds thereby raising concerns over productivity of citrus industry. The research has shown promising results of automation in various fields related with intercultural operation on field as well as in processing of citrus fruits. IoT based irrigation systems not only are beneficial in fulfilling the requirement of moisture to growing trees but also reduces wastage of water in big orchards. Citrus being prone to numerous diseases is a major setback to orchardists. They not only decrease fruit production but prices as well. Machine learning as used in other diseases can be handy in citrus orchards increasing the profitability in the long run. Detection of fruit size and quality is another field where artificial intelligence and machine leaning can be utilized to enhance citrus production and its quality. Automation in detecting nutritional status of trees and fruits can change the realm of fruit industry thereby leading to sustainability through reducing the extra chemicals in the form of fertilizers. The current research focusses on different arena of citrus production where automation is peculiarly required in the coming future to again turn citrus industry into profitable and sustainable.

Keywords: Malicious PDF detection, Ensemble learning, Explainable AI (XAI), Feature engineering, Feature selection, Cybersecurity, SHAP, LIME

1. Introduction

Within the twentieth century, developed economies observed a decrease within the rural workforce, contracting by eighty times [1]. In any case, human labor remains a critical toll figure, constituting 40 per cent of the full generation costs for vegetables, natural products, and cereals [2]. The advancement of present-day agribusiness has impelled the selection of modern and automation driven by a few key variables. Firstly, the raising costs and reducing accessibility of talented labor posture critical challenges to the fruit industry. Secondly, ensuring food security has gotten to be foremost, requiring the arrangement of dependable mechanical frameworks to play down defilement dangers [3]. Thirdly, the basic of feasible agribusiness, adjusting nourishment generation with natural conservation, underscores the require for mechanical advances to upgrade efficiency whereas minimizing costs [4]. Fruit cultivation requires round year labor, with mechanization accounting for only fifteen per cent of operations. Manual operations amplify to natural fruit harvesting, leading to production deficiencies, which can reach an amazing 50 per cent. In recent decades, the fruit industry has seen a noteworthy advancement in automation technology to address challenges such as labor deficiencies, enhancing both production and productivity [6].

Citrus is the most important fruit in terms of area, production, nutritional and productivity worldwide. It can be grown both in tropical, temperate and subtropical districts. Being rich in vitamin C, it can meet people's day by day needs and offer assistance to body's immune framework [7]. Worldwide citrus fruit production has observed a steady surge during last decade surpassing 130 MT, a development rate of approximately 125 per cent. Each stage of citrus fruit production is essential and is corelated to yield and quality of fruits. Even with modernization in agricultural sector, citrus production is still inefficient due to lack of skilled labor.

Advance horticulture, which combines innovative and data-driven techniques, points to maximize trim generation by closely checking variables such as irrigation system, nutrition, pest control, and environmental conditions [8]. Through the utilization of advanced sensors, drones, AI/ML, and IoT, this approach works to achieve maximum output from the orchards, best utilization of farm resources [9]. Automated innovative techniques and optimizing inputs such as water, fertilizers, and pesticides can maximize fruit production and quality with little environmental impact. It guarantees that water and supplements are conveyed absolutely when and where they are required, minimizing and minimizing post-harvest misfortunes [10]. The current investigates points at exploring potential of mechanization and mechanical technology in citrus production with scope for cutting edge and progressive farming.

2. Need of Automation and Robotization in Citriculture

1. **Automated Irrigation and Nutrient Management:** IoT-based water system frameworks have revolutionized irrigation management in fruit physiological growth, providing moisture accurately where and when it's required. These frameworks, prepared with moisture sensors and mechanized valves, optimize water utilization, lessening water wastage and guaranteeing ideal soil dampness levels for fruit crops. Automated fertigation frameworks assist upgrade precise cultivating by coordination water system with supplement conveyance, ensuring plants get the proper quantity at the proper time. Real-time checking of soil moisture allows proper growth of trees which maximizes fruit size and quality [11].
2. **Detection of diseases:** Fruit crops are more inclined to pest and diseases in comparison to other cereal crops. These infections not only decrease production

but moreover diminishes quality which lessen cost within the showcase. Algorithms in integration with [12] captured pictures may identify sort of infections which may advance anticipate its control. This innovation can be utilized in orchards where fruit crops can be identified with maladies. Due to improper information agriculturists by and large splash pesticides and fungicides similarly in all plants which improve fetched of input and wastage of chemicals. Machine learning consequently can create calculations which should coordinate agriculturists where precisely these chemicals are required which can encourage reduction of expensive chemicals decreasing cost of input and wastage (Fig. 72.1) [13].

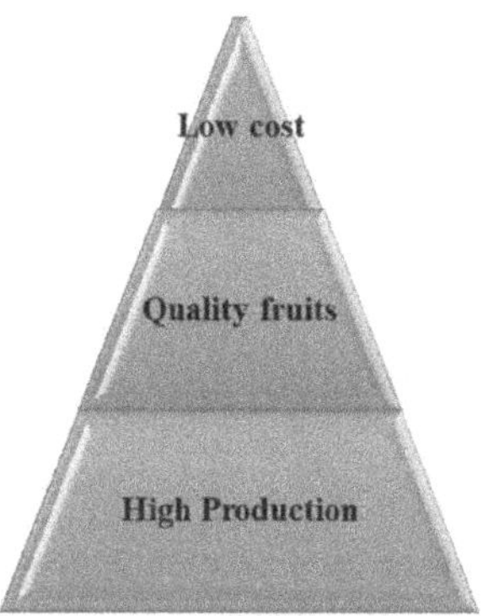

Fig. 73.1 Requirements of automation with their advantages in citrus industry

3. **Detection of Ripeness of fruits:** Fruit harvesting manually may be time-bounding handle that can result in conflicting classification, making it a major challenge for the agrarian segment. AI and machine learning can be combined to make a fruit sorting framework that would increment proficiency and empower non-destructive product testing [14]. Research has shown that combining hyperspectral imaging and deep learning may make it doable to evaluate the development stage of fruit, which might offer assistance with fruit sorting and handling may make it attainable to evaluate the development stage of fruit [15].
4. **Fruit grading:** Companies nowadays uses manual operators for fruit grading. The manual strategy is well-known for its disadvantages, which incorporate lower produce volumes due to variable perceptions of labors. In fruit processing sector, the automated framework is more proficient to utilize than the human approach since of its speedy decision-making speed and tall measures for both quality and amount. This nondestructive strategy of examination has been utilized for agrarian item assessment and classification, counting shape, measurements, tone, surface imperfections and insides deficiencies [16]. Moreover, the report shows an inventive machine vision order approach for rating apple fruit. In multispectral images, deficiencies are absolutely isolated by minimizing perplexity with stem/calyx regions. The fragmented zone is at that point utilized to extricate factual, textural, and geometric data. The results illustrated that factual classifiers outperformed their counterparts, whereas featured selection improved execution by keeping as it were the pivotal highlights [17].
5. **Determination of fruit size:** Assessing fruit size is pivotal for controlling early-season crop production and harvesting in plantations [18]. In order to assess the effect of chemical thinners early detection of fruit size is essential in early stage of fruit improvement [19]. Producers can more precisely production and size distribution by having a mid-season assess of the number of fruit and their sizes. As of late, machine learning methods have made it conceivable for the machine vision framework to utilize neural organize models, as Veil RCNN, to join fruit shape into its measuring estimation [20]. By maximizing the number of visible fruits that got to be measured and dodging the fruit size, accuracy cab be enhanced.
6. **Detecting fruit quality:** To attain increased prices in the market fruit quality is of utter importance. In arrange to distinguish non-consumable quality tests for direct buyers and large-scale manufacturing in industrial facilities, this can be more pivotal. Determining fruit quality is a tedious and skilled job since there are different varieties of the same crop that is of same color, surface, shape, particularly to readiness. There's a urgent need for a fast, accurate, and reasonable way to decide the quality of fruits. It is a valuable instrument for improving the computerized image of nutritional quality. It can be utilized within the pharma industry for assortment of assignments, counting as surveying color and quality, spotting maladies, evaluating, and organizing fruits, and in processing businesses [21]
7. **Diagnosis of Nutrition status of fruits:** Determining fruit nutrient level is the basic step to decide quality of grown fruits. To attain this, a model of the SEFEAG was made. It could be apparatus which recreate cognitive capacities and analyses information to decide wholesome quality of fruits. SEFEAG subsystems are also capable for collecting, surveying authentic varietal information, symptomatology, and expository information related to the supplements found in soil, water, and takes off. It is interfaces with a social database that was made utilizing the Substance Relationship (E/R) method. The database holds subtle elements on particular plantations, counting leaf, soil, and water appraisals, fertilization and other editing operations, profitable execution, and cultivar [22].

The management techniques thereby shall enhance yield and quality of fruits in the orchards (Fig. 73.2) [23].

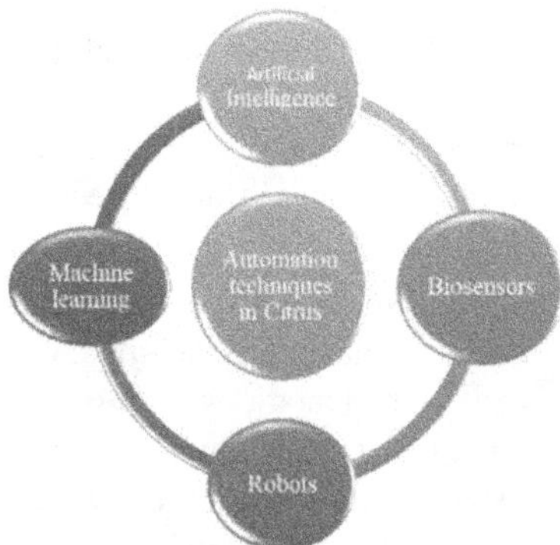

Fig. 73.2 Different automation techniques used in citrus fruit production

3. Conclusion

With shortage of labor and lack of skills in recent times the alignment of citrus fruits with qualitative characters shall be hard to achieve. Automation in regards to operations lie irrigation, disease, nutrition, sorting, grading and quality is the need of the hours to sustain citrus industry into profitable. Deep research is much awaited in other operations as well to achieve success in citrus processing which acts as raw material for pharma and cosmetic industry. Automation in the long-run shall help farmers worldwide to harvest more and consumers to buy qualitative fruits at cheaper price.

References

1. Bechar, A. and Vigneault, C. (2016). Agricultural robots for field operations Concepts and components. *Biosystems Engineering, 149*, 94–111.
2. Bharad, N.B. and Khanpara, B.M. (2024). Agricultural fruit harvesting robot: An overview of Digital Agriculture. *Plant Archives, 24*, 154–160.
3. Edan, Y., Han, S. and Kondo, N. (2009). Automation in agriculture. Springer Handbook of Automation. Berlin: Springer Berlin Heidelberg, pp.1095–1128.
4. Grift, T., Zhang, Q., Kondo, N., and Ting, K.C. (2008). A review of automation and robotics for the bio-industry. *Journal of Biomechatronics Engineering, 1*(1), 37–54.
5. Ceres, R., Pons, J.L., Jimenez, A.R., Martin, J.M. and Calderon, L. (1998). Design and implementation of an aided fruit-harvesting robot (Agribot). *Industrial Robot: An International Journal, 25*(5), 337–346.
6. Bachche, S. (2015). Deliberation on design strategies of automatic harvesting systems: A survey. Robotics, *4*(2), 194–222.
7. Zhu, Q. A Study on Citrus Production Efficiency in China. Master's Thesis, Zhongnan University of Economics and Law, Wuhan, China, 2020.
8. Anselmo, S., Carron, G., Meacham, T., Najdenovska, E., Dutoit, F., Raileanu, L.E., & Camps, C. (2023). Plant electrophysiology for smart irrigation management of greenhouse. *Acta Horticulturae*, 1373, 89–95.
9. Apeinans, I., Litavniece, L., Kodors, S., Zarembo, I., Lacis, G., & Deksne, J. (2023). Smart fruit growing through digital twin paradigm: systematic review and technology gap analysis. Engineering Management in Production and Services, 15(4), 128–143.
10. Chen, Y., Zhang, J.-H., Chen, M.-X., Zhu, F.-Y., & Song, T. (2023). Optimizing water conservation and utilization with a regulated deficit irrigation strategy in woody crops: A review. Agricultural Water Management, 289.
11. Murthy, B.N.S. and Sridhar, G. (2024). Advances in precision production of fruits through use of digital technologies. *Shodh Chintan*, 16: 68–73.
12. Singh A, Ganapathysubramanian B, Singh AK, Sarkar. (2016). Machine learning for high-throughput stress phenotyping in plants. *Trends Plant Sci.*; 21(2):110–124.
13. Bhatt, S.S., Bhatt, A., Ballabh, J., Prakash, S., Baloni, D., Majumdar, S. (2022). Automation in Horticulture: The future of orchards. River Publisher. 1–6p.
14. Surya Prabha D, Satheesh Kumar J (2015) Assessment of banana fruit maturity by image processing technique. J Food Sci Technol 52:1316–1327.https://doi.org/10.1007/S13197-013-1188-3/
15. Davur YJ, Kämper W, Khoshelham K et al (2023) Estimating the ripeness of Hass avocado fruit using deep learning with hyperspectral imaging. Horticulturae 9:599. https://doi.org/10.3390/
16. Olaniyi EO, Adekunle AA, Odekuoye T, Khashman A (2017) Automatic system for grading banana using GLCM texture feature extraction and neural network arbitrations. J Food Process Eng 40: e12575. https://doi.org/10.1111/JFPE.12575
17. Vijayalakshmi M, Peter VJ (2021) CNN based approach for identifying banana species from fruits. Int J Inf Technol 13:27–32. https://doi. org/10.1007/S41870-020-00554-1/
18. Mirbod O, Choi D, Heinemann PH et al (2023) On-tree apple fruit size estimation using stereo vision with deep learning-based occlusion handling. Biosyst Eng 226:27–42. https://doi.org/10.1016/J.
19. Greene DW, Lakso AN, Robinson TL, Schwallier P (2013) Development of a fruitlet growth model to predict thinner response on apples. HortScience 48:584–587. https://doi.org/10.21273/ HORTSCI.48.5.584
20. Mirbod O, Choi D, Heinemann PH et al (2023) On-tree apple fruit size estimation using stereo vision with deep learning-based occlusion handling. Biosyst Eng 226:27–42. https://doi.org/10.1016/J.
21. Mukherjee A, Chatterjee K, Sarkar T (2022) Entropy-aided assessment of amla (*Emblica officinalis*) quality using principal component. Analysis 12:2162–2170. https://doi.org/10.33263/BRIAC122. 21622170.
22. Palazzo D, Basile G, D'agostino R (1993) An expert system for diagnosing citrus nutritional status and planning fertilization. Optim Plant Nutr, pp 161–166 https://doi.org/10.1007/978-94-017-2496-8_27.
23. Bhatt, S., Dimri, D.C., Rao, V.K., Singh, A.K., Pandey, Y. and Pandey, K.K. (2017). Efficacy of Blossom Thinners on Flowering, Fruit Retention and Yield Attributes in Plum (*Prunus Saliciana* L.) cv. 'Kala Amritsari'. *Chemical Science Review and Letters,* 6(21), 64–68.

Note: All the figures in this chapter were made by the authors.

Intelligent Systems Using Semiconductors for Robotics and IoT – Dinesh Goyal et al. (eds)
© 2026 Taylor & Francis Group, London, ISBN 978-1-041-20408-4

74 A Multiplayer Approach to Enhance Computational Thinking in Students

Amita Jajoo[1],
Jayesh Badgujar[2], Lokesh Warade[3],
Chinmay Mahajan[4], Hitesh Patil[5]
Dept. of Information Technology D.Y. Patil College of Engineering,
Pune, India

Abstract: This paper introduces a multiplayer educational game designed to enhance basic math skills by merging real-time problem-solving with a series-building mechanic inspired by classic "snake" gameplay. Developed using C# on the frontend and Java on the backend, the game generates random math questions that players solve in a dynamic, competitive environment. Each correct answer extends a player's "chain," reinforcing core math skills. The game supports social interactions through in-game chat and leaderboards, encouraging both competition and cooperation. Engagement metrics, including session frequency, accuracy, and chain length, provide insights into student engagement and learning outcomes. Player feedback highlights its educational value and enjoyment, demonstrating its potential as an engaging tool for foundational math education. By blending gameplay with effective math and other subject's practice, this study showcases its use as a supplemental educational tool that fosters skill development and keeps young learners engaged.

Keywords: Multiplayer academic game, Real-time problem-solving, Series-building mechanic, Student engagement metrics, Skill development

1. Introduction

Gamification, which benefits from sports mechanics such as prices, levels and leadership boards, has proven to be an effective tool for increasing students' involvement and motivation, especially in subjects such as mathematics. By converting repetition and exhausting tasks to interactive and pleasant experiences, Gamification faces general challenges in mathematics teaching, such as low commitment and lack of inspiration.

It is especially effective for basic people such as primary arithmetic, where frequent practice is necessary. This article introduces a multiplayer-pedagogical game designed to improve arithmetic skills among primary students, which includes a real-time problem-solving mechanic inspired by the classic "Snake" game. Players solve arithmetic problems to expand the chains, earns immediate response and strengthen their progress. The game also includes multiplayer features that encourage collaboration and friendly competition, both educational and mutual skills.

The examination evaluates the sport's effectiveness the use of engagement metrics including consultation frequency, accuracy quotes, and chain lengths, alongside remarks from students and educators. Results suggest that the game no longer handiest improves mathematics talent but also promotes consistent exercise and social interaction. The integration of C# and Java technologies ensures an unbroken actual-time multiplayer enjoy that remains engaging for younger newbies.

The findings propose that this gamified technique offers widespread capability as a supplemental academic tool, effectively blending skill-building with engaging

[1]aajajoo@dypcoeakurdi.ac.in, [2]jayeshbadgujar7052@gmail.com, [3]lokeshwarade22@gmail.com, [4]chinmaymahajan999@gmail.com, [5]hiteshsamadhanpatil@gmail.com

DOI: 10.1201/9781003716389-74

gameplay to make math and other subjects education both enjoyable and effective. As digital tools increasingly shape modern education, gamified approaches like this one provide innovative and interactive methods to enhance motivation, cognitive development, and overall academic achievement among students.

2. Literature Review

[1] L. Lopez-Faican et al., “Evaluation of a cellular multiplayer augmented fact sport for number one college kids.” This have a look at evaluates EmoFindAR, a multiplayer augmented truth (AR) sport designed to beautify emotional intelligence, socialization, and communication abilities amongst primary faculty kids. Utilizing cell AR technology, the sport gives each competitive and collaborative modes, assisting geolocation-primarily based gameplay without relying on bodily markers. Findings suggest that both modes generate tremendous feelings and engagement, with collaborative play yielding greater advantages in social interaction and teamwork. This observes and underscores the sport’s instructional ability, advocating for the incorporation of dynamic design elements and collaborative techniques to promote lively getting to know, critical wondering, and emotional development. Future developments aim to diversify gameplay themes and amplify the sport’s applications in academic contexts.

[2] J. A. Ruipérez-Valiente et al., “Patterns of Engagement in an Educational Massively Multiplayer Online Game: A Multidimensional View.” This research examines the position of gamification and academic video games in fostering engagement, motivation, and studying results. It highlights the capability of video games to create immersive studying environments aligned with educational goals, accommodating various learner profiles. A key attention is on multidimensional engagement fashions, capturing aspects such as interest tiers, social interactions, and exploratory behaviors. Studies of video games like The Radix Endeavor display engagement patterns across dimensions inclusive of quests, exploration, and social dynamics. The findings emphasize the importance of adaptive academic sport layout, offering significant getting to know experiences and leveraging rich information analytics to degree engagement correctly.

[3] S. Wodarczyk et al., “Emergent Multiplayer Games.” This paper introduces the idea of Emergent Multiplayer Games (EMGs), which combine interactive streaming with collaborative gameplay, permitting big audiences to persuade recreation development through vote casting-based mechanics. Using a prototype built in Unity, the observe positions EMGs as a fusion of traditional gameplay and emergent narratives, catering to various participant personas and engagement levels. The studies highlights their capacity to enhance social gaming dynamics, usability, and collective enjoyment even as addressing demanding situations inclusive of balancing participant inputs and improving comments systems. The findings demonstrate that EMGs represent a promising innovation in stay-streaming and multiplayer gaming, with opportunities for further improvement.

3. Methodology

3.1 Project Planning and Requirements

1. Project Objectives:

   The number one goal of this task became to expand an academic multiplayer recreation geared toward enhancing standard faculty and students’ arithmetic talent. The recreation consists of adaptive gameplay mechanics, where issue scales with the player’s overall performance, along collaborative and aggressive functions. These functions intention to foster student engagement, motivation, and social interaction in an educational context, making studying math & other subjects’ fun and interactive.

2. Target Audience:

   The game is mainly designed for essential primary students in grades one to five. This age group was selected based totally on their developmental degree in gaining knowledge of simple mathematics concepts and their familiarity with digital games. Additionally, the position of instructors as facilitators is indispensable to the undertaking, as they guide students in efficiently making use of the sport for learning functions, ensuring its instructional price.

3. Functional Requirements:

   - *Core Gameplay:* The sport functions chain-building mechanics stimulated by means of the conventional “snake” game. Players solve math problems, and their solutions are linked to in-game progress, wherein solving troubles grows their in-sport chain.
   - *Adaptability:* The game generates math questions dynamically, adjusting their difficulty in actual time based on player overall performance. This adaptability ensures the content material remains tough but workable for every participant.
   - *Social Interaction:* Social functions including public chatrooms, private messaging, and group quests sell collaboration and social interaction. These elements are designed to make the game extra engaging by means of encouraging teamwork and communication.
   - *Engagement Metrics:* Various metrics, including active time, quest completion rates, and social interactions, are tracked to assess student engagement and progress.

4. Technical Requirements:

   - *Frontend Technology:* Unity with C# turned into used to create an interactive, toddler-friendly consumer interface. The frontend permits real-time feedback through animations and transitions to preserve students' engagement.

- *Backend Technology*: Java is chosen for the backend to offer stability and scalability, assisting real-time synchronization of multiplayer periods, adaptive question era, and statistics storage to music for participant development.

3.2 Design Phase

1. System Architecture:

   The game was designed with a dual-layer architecture to separate user-facing elements and backend processes:

   - *Frontend (C#):* Focuses on delivering an interactive visual experience, including intuitive navigation and real-time feedback, ensuring young users remain engaged.
   - *Backend (Java):* Handles essential game logic, including question generation, adaptive difficulty algorithms, and data synchronization across multiplayer sessions, ensuring smooth gameplay.

2. Gameplay Design:

   - *Core Mechanic:* Players answer math problems to grow their in-recreation chain, which visually represents their progress. This machine reinforces the link between academic fulfilment and in-recreation rewards, motivating primary students to keep solving issues.
   - *Quest Types:* The game features several quest types:
   - Timed Challenges: Designed to develop speed and accuracy in problem-solving.
   - Theme-Based Quests: Focus on specific math operations, such as addition, subtraction, division and multiplication.
   - Mixed Quests: Introduce varied problem types to encourage flexible problem-solving.

3. User Interface (UI):

   The UI is carefully crafted with younger newbies in thoughts, featuring shade-coded answers, clean animations for correct and incorrect responses, and massive, smooth-to-read buttons for simple navigation.

4. Engagement Features:

   - Streak-Based Rewards: To motivate students, the game rewards players for consecutive correct answers, reinforcing focus and accuracy.
   - **Social Features:** Public chatrooms, private messages, and group quests are integral for fostering teamwork, collaboration, and peer learning.

5. Adaptive Questioning System:

   The backend dynamically adjusts the dificulty of math questions based totally on the man or woman participant's performance. This ensures that each scholar faces an as it should be tough level of difficulty, offering them with a continuous learning curve without overwhelming them.

3.3 Development Phase

1. Frontend Development (C#):

   The frontend, built in Unity with C#, was designed to ensure an interactive and engaging experience for young learners. It features:

   - *Real-Time Feedback Animations:* Visual feedback on player progress, such as chain growth and streaks.
   - *Smooth Transitions:* Seamless transitions between different stages of the game to keep the user experience fluid and enjoyable.
   - *Achievement Notifications:* Instant feedback on achievements and scores to reinforce progress and keep players motivated.

2. Backend Development (Java):

   The backend, developed in Java, was designed with scalability and real-time synchronization in mind. Key features include:

   - *Adaptive Question Generation:* Dynamic generation of math problems based on player ability, allowing for personalized challenges.
   - *Multiplayer Logic:* Ensures real-time synchronization and data consistency during multiplayer interactions, such as group quests and competitive modes.
   - *Progress Tracking:* Stores player profiles, performance data, and quest completion metrics, offering a personalized learning experience.

3. Integration Testing:

   Extensive integration testing ensured smooth interplay between frontend and backend. Multiplayer classes were simulated to make certain synchronization and seamless facts glide.

3.4 Deployment and Monitoring

1. Deployment:

   The game was initially deployed in a controlled educational environment, with teachers guiding students during gameplay. The deployment process included:

   - *Installation:* The game was installed on school devices to make it accessible to all participating students.
   - *Teacher Training:* Teachers were provided with training sessions to familiarize them with the game mechanics, monitoring tools, and strategies to effectively guide students in using the game.

2. Monitoring Engagement and Learning:

   Real-time data collected through the backend allowed for the tracking of:

- *Active Time and Login Frequency:* To assess how often and for how long students engaged with the game.
- *Quest Completion Rates and Adaptive Difficulty Progression:* To monitor student progress and ensure appropriate challenge levels.
- *Social Interaction Metrics:* Tracking usage of public and private messaging tools to gauge peer interaction.

3. Data Analysis:

   Quantitative statistics evaluation used statistical equipment like SPSS and R, even as qualitative statistics from surveys and interviews had been analyzed the usage of thematic coding to explore scholar reviews and motivations.

4. Continuous Feedback Loop:

   Player feedback and engagement data informed iterative improvements, such as refining adaptive algorithms, expanding quest types, and enhancing UI design for a more effective and enjoyable learning experience.

4. System Architecture

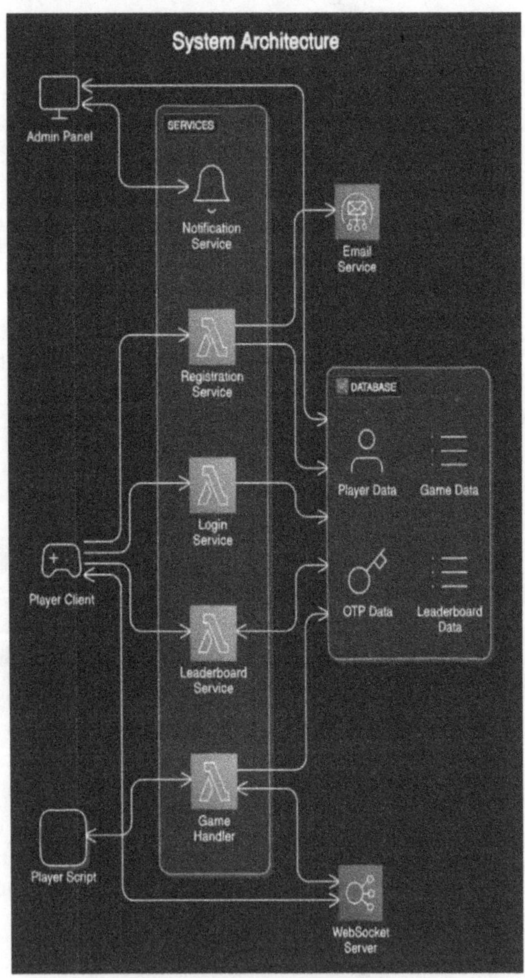

Fig. 74.1 System architecture

5. Data Flow Diagram

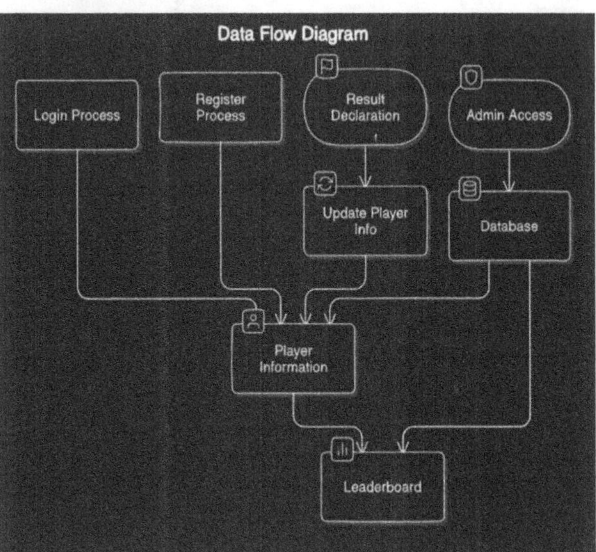

Fig. 74.2 Data flow diagram

6. Results

Engagement Findings: The chain-construction mechanic efficiently sustained student engagement. The adaptive trouble structure kept students influenced by using gradually growing the task. Social interplay metrics discovered common use of public and personal chat features, indicating that students loved collaborating, changing techniques, and offering help to peers.

Learning Outcomes and Mastery: The recreation validated enormous impact on students' mathematics skills, with a mean quest finishing touch price exceeding. As students progressed, they tackled more complex problems, reflecting the game's capacity to foster ability mastery. The adaptive difficulty ensured that gamers were constantly challenged, and blunders charges reduced over time, suggesting improved hassle-solving accuracy.

Social Interaction and Collaboration: Group quests and public chatrooms have been famous features, promoting collaboration and peer-coaching. Students suggested enjoying working together to clear up issues and sharing strategies, which no longer best enhanced their academic overall performance but also bolstered social skills. The peer-support made dynamic changed into constant with Vygotsky's Social Development Theory, highlighting the value of social collaboration in cognitive development.

Exploration and Independent Learning: Students engaged with additional game zones and mini games, indicating an excessive degree of impartial getting to know and interest. Many gamers explored supplementary gear and math sports, suggesting that the game fostered a choice for deeper know-how of mathematical concepts in low-pressure surroundings.

7. Conclusion

In conclusion, this multiplayer educational game effectively enhanced elementary students' math skills by combining adaptive difficulty, chain-building mechanics, and social collaboration. The game not only engaged students in fun and challenging math tasks but also fostered positive social interactions through peer learning, communication, and teamwork. By incorporating both competitive and collaborative elements, it created a dynamic and engaging environment where students could develop critical thinking, problem-solving skills, and academic confidence. Its ability to support both individual progress and group-based learning makes it a valuable tool for educators, offering an innovative and interactive platform for enhancing math skills in a low-stress, supportive environment.

This study highlights the potential of gamified educational tools to improve learning outcomes while promoting collaboration, social development, and positive attitude towards education. The findings suggest that such tools can be highly effective in engaging students and making foundational subjects like math more enjoyable and accessible.

References

1. J. A. Ruipérez-Valiente, M. Gaydos, L. Rosenheck, Y. J. Kim, and E. Klopfer, "Patterns of Engagement in an Educational Massively Multiplayer Online Game: A Multidimensional View," in IEEE Transactions on Learning Technologies, vol. 13, no. 4, pp. 648-661, 1 Oct.-Dec. 2020, Doi: 10.1109/TLT.2020.2968234. keywords: {Games; Task analysis; Education; Time measurement; Proposals; Atmospheric measurements; Engagement; game-based assessment; K-12 education; learning games; learning analytycs}
2. S. Wodarczyk and S. von Mammen, "Emergent Multiplayer Games," 2020 IEEE Conference on Games (Cog), Osaka, Japan, 2020, pp33-40, Doi: 10.1109/CoG47356.2020.9231834. keywords: {Games; Electromyography; Prototypes; Control systems; Streaming media; Navigation; Mixers; Interactive live streaming; emergent play; multiplayer games}
3. L. López-Faican and J. Jaen, "EmoFindAR: Evaluation of a mobile multiplayer augmented reality game for primary school children," Computers & Education, vol. 149, 2020, Art. no. 103814. Doi: 10.1016/j.compedu.2020.103814

Note: All the figures in this chapter were made by the authors.

Intelligent Systems Using Semiconductors for Robotics and IoT – Dinesh Goyal et al. (eds)
© 2026 Taylor & Francis Group, London, ISBN 978-1-041-20408-4

75 A Literature Review on Non-Contact AI Vision-Controlled Diameter Determination for Flux Core Wire Drawing, Ensuring Error-Free Production

Ajinkya P. Edlabadkar[1], P.D. Kamble[2]
Assistant Professor, Yeshwantrao Chavan College of Engineering,
Nagpur, India

Mahesh R. Bele
PG Student, Yeshwantrao Chavan College of Engineering,
Nagpur, India

Abstract: Traditional methods often fall short in achieving the desired accuracy and responsiveness. This paper presents an AI vision-controlled approach to optimize wire diameter during the flux core wire drawing process, aiming for error-free production.

This study explores the integration of AI-based measurement techniques to enhance quality control in flux core wire manufacturing. A comprehensive review of 21 research papers and patents highlights the transformative impact of AI on continual monitoring and process automation. The implementation of AI vision technology facilitates precise diameter measurement, ensuring uniformity in circularity, roundness, and dimensional accuracy, which are critical to welding applications.

Traditional measurement methods often struggle to maintain stringent tolerance levels due to human intervention and limitations in manual inspection. However, AI-enabled non-contact measurement systems address these challenges by providing high-speed, accurate, and real-time feedback. Techniques such as Response Measurement Methodology and machine-assisted experimental modelling have demonstrated their effectiveness in optimizing process parameters, minimizing production downtime, and improving overall efficiency. By employing AI algorithms for predictive analysis, manufacturers can proactively adjust drawing parameters, reducing defects and material wastage.

This study underscores the potential of AI vision-controlled optimization in achieving superior production consistency, eliminating human-induced errors, and ensuring seamless manufacturing workflows. The findings emphasize that integrating AI-driven measurement systems into the flux core wire production line not only enhances accuracy but also optimizes resource utilization and operational efficiency. As industries shift toward intelligent automation, AI vision technology stands as a game-changer in modern wire drawing processes, fostering innovation and sustainability in welding wire manufacturing.

The paper presents a contactless AI driven method of measuring diameter and ovality of cylindrical flux core wire in diverging laser beam technique. Physical simulation of this method and experimental testing are made. The results can be used for technological control of external diameter of flux core wire and similar range of product in the welding industry.

Keywords: Automation, AI vision system, Flux core wire drawing, Diameter optimization, Non-contact measurement, Process automation, Quality control, Real-time monitoring, Production efficiency, Manufacturing optimization, Production technology, Digital image and welding wire industry

Corresponding author: [1]ajinkyae@gmail.com, [2]pdk121180@yahoo.com,

DOI: 10.1201/9781003716389-75

1. Introduction

Flux core wire drawing plays a crucial role in the production of welding wires, where precise control over wire diameter and electrical parameters is essential for optimal welding performance. The quality of wire products is assessed based on various geometric characteristics that directly impact their functionality and reliability. Key parameters such as diameter per unit length [1-2], eccentricity[3], outer diameter[4], and insulation integrity [5] are critical in determining product quality and performance consistency[6-13]. Implementing advanced technology in wire drawing processes can significantly reduce defects and minimize rejection rates by ensuring tighter control over these essential characteristics. Traditional methods often struggle with maintaining strict tolerances, leading to inconsistencies in wire geometry. However, integrating innovative monitoring and control solutions can enhance precision, improve process efficiency, and ensure compliance with industry standards. By addressing these challenges, manufacturers can optimize production outcomes and enhance the overall quality of flux core welding wires.

Accurate real-time diameter measurement[15-21] is a critical requirement for ensuring high-quality welding wire production. Among various measurement techniques, optical methods are widely recognized for their precision and reliability in industrial applications. These AI-integrated optical techniques enable non-contact measurement, ensuring minimal interference with the manufacturing process while maintaining stringent dimensional accuracy.

Based on their operational principles, optical measurement methods can be classified into diffraction-based techniques[15-17], interferometry[18], scattering-based approaches[19], and shadow projection techniques[20-21]. Diffraction-based methods utilize wave interference patterns to determine wire diameters with high precision, whereas interferometry employs phase shift analysis for sub-micron accuracy. Scattering-based approaches analyze the dispersion of light to determine dimensional parameters, while shadow techniques rely on collimated light projection to capture wire profiles with high resolution.

In this study, an AI-driven shadow technique is employed for real-time diameter optimization, leveraging its extensive measuring range, which spans from sub-millimeter dimensions to several tens of millimeters. This technique, combined with advanced AI processing, ensures continuous monitoring and adaptive control, enabling precise adjustments in the flux core wire drawing process. The integration of AI enhances measurement accuracy, mitigates the effects of environmental variations, and facilitates automated corrective actions, ultimately leading to improved production efficiency and reduced material wastage.

This paper reviews emerging approaches in both contact and non-contact measurement methods, including vision-based techniques, wireless and embedded sensors, and physical measurement methodologies. The primary objective of this study is to review the literature on non-contact measurement techniques, compare traditional methodologies with AI-driven systems for real-time wire diameter monitoring and adjustment, and evaluate the effectiveness of AI-driven solutions in enhancing production efficiency, reducing material wastage, and ensuring error-free output. While this study does not delve into the intricate working principles of various tools and sensors, it aims to summarize the state-of-the-art advancements in non-contact measurement techniques and their integration into flux core wire manufacturing within the welding industry

Traditional diameter control methods, such as manual measurements and mechanical dies, are prone to inaccuracies and inefficiencies, leading to compromised product quality and increased production costs. The advent of AI-powered vision systems offers a promising solution to these challenges by enabling real-time monitoring and control of the wire drawing process.

The demand for high-precision flux core wire in welding applications has driven the need for advanced manufacturing technologies that ensure consistent quality and error-free production. Traditional wire drawing methods often face challenges in maintaining strict diameter tolerances due to process variability, manual inspection limitations, and inefficiencies that lead to increased material wastage and production costs. To address these issues, the integration of AI vision-controlled diameter optimization presents a transformative approach, leveraging machine learning and real-time monitoring to enhance manufacturing accuracy.

This study also aims to implement an AI-powered vision system for the continual measurement and regulation of wire diameter during the flux core wire drawing process. The proposed system utilizes a machine learning model trained on extensive datasets, analyzing diameter variations and corresponding process parameters. Through real-time monitoring and dynamic feedback control, the AI system adjusts the wire drawing parameters to maintain precise diameter tolerances, eliminating inconsistencies and ensuring optimal product quality.

By incorporating Contactless Micrometer technology, this approach enables non-invasive, high-speed measurement, further enhancing accuracy and efficiency. Experimental trials demonstrate substantial improvements in production consistency, reduced defects, and optimized resource utilization compared to conventional methods. This research highlights the potential of AI-driven automation in revolutionizing flux core wire manufacturing, ensuring precision, minimizing human intervention errors, and maximizing operational efficiency.

The study explores how AI vision technology can be effectively applied to industrial automation, contributing to improved quality control and sustainability in welding wire production.

Flux Core wire drawing Machine

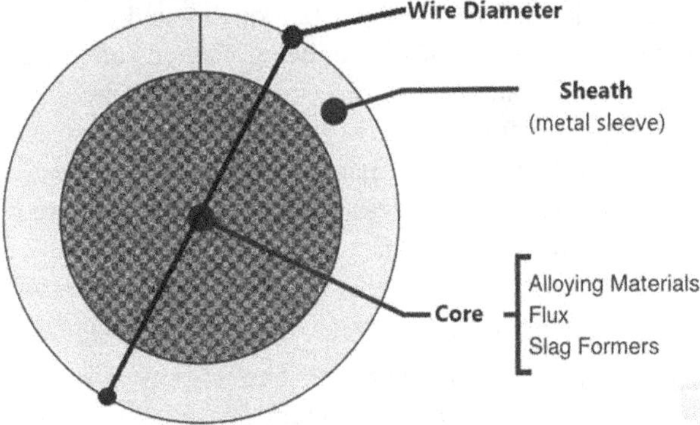

Cross section of flux cored wire

Fig. 75.1 Working principle of wire drawing Machine

2. Research Gap

Despite significant advancements in wire diameter control methodologies for flux core wire drawing, critical gaps remain in the real-world application of AI vision-controlled optimization. Existing studies primarily focus on theoretical models or controlled laboratory experiments, leaving several key areas underexplored:

1. **Industrial Deployment and Validation:** Current research predominantly explores AI vision-based wire diameter control in controlled environments, with limited studies on large-scale industrial deployment. The lack of real-world validation creates uncertainty regarding the feasibility, scalability, and adaptability of these systems in dynamic production settings. Further studies are needed to examine practical implementation challenges, long-term performance, and the benefits of AI-driven optimization in actual manufacturing environments.
2. **Advanced Algorithm Optimization:** While AI-driven control systems demonstrate potential in improving precision and efficiency, there is still room for refinement in algorithm development. Existing research does not sufficiently explore advanced machine learning techniques, adaptive control strategies, and predictive analytics that could further enhance the system's accuracy and responsiveness. Investigating AI models that can self-learn and adapt to varying production conditions is necessary for optimizing performance.
3. **Seamless Integration with Existing Infrastructure:** One of the critical challenges in adopting AI vision-based systems is their integration with conventional flux core wire drawing setups. Research has yet to establish best practices for interoperability with legacy machinery, communication protocols, and real-time data exchange. Studies focused on minimizing operational disruptions and ensuring smooth transitions from traditional to AI-driven monitoring systems are required.
4. **Economic Feasibility and Cost-Benefit Analysis:** Although AI-controlled systems promise enhanced accuracy and reduced wastage, their economic viability remains a key concern. Limited research exists on the cost-effectiveness of implementing AI vision technology compared to conventional measurement techniques. Comprehensive studies assessing return on investment (ROI), cost reductions in material wastage, and long-term financial benefits are essential for justifying industrial adoption.

By addressing these research gaps, a practical approach is created by implementing AI vision-controlled diameter optimization in flux core wire drawing processes. Such efforts will support the broader integration of intelligent automation in industrial manufacturing, enhancing both productivity and quality standards

3. Literature Review

The field of flux core wire drawing has undergone significant advancements, particularly in the development of non-contact measurement techniques for diameter control. Various studies have explored different methodologies to enhance precision, efficiency, and automation, highlighting the growing need for real-time monitoring systems to address the limitations of traditional measurement techniques.

Magnus Ericson. (WO9721073A1, published on June 12, 1997, by Svenska Elektronikprodukter Ab) introduced a contactless diameter measurement method utilizing optical techniques for measuring objects in motion. Their study employed illumination devices and image reception technology to determine an object's diameter with high precision. The system dynamically adjusted brightness and processed image signals to mitigate errors, laying the foundation for optical-based diameter control. However, the research did not integrate AI-driven enhancements for adaptive optimization.

Wei Li, Hongbo Wang and Zhihua Fenga developed an ultra-high resolution, non-contact diameter measurement method using an eddy current sensor (ECS) for metallic wires. Their research demonstrated superior resolution capabilities compared to other non-contact methods, achieving a static resolution of 0.42 nm and a dynamic resolution of 2.2 μm. While effective for precision measurement, ECS-based methods do not inherently provide adaptive control or integration with AI-driven optimization systems.

Beltronics Inc. (US4697088A, published on September 29, 1987) proposed a method and apparatus for discriminating sharp edge transitions using optical scanning techniques. This research focused on detecting reflective variations in materials, which has implications for wire diameter measurement. However, its primary application was in scanning-based inspection rather than real-time production monitoring.

Martínez-Anton, I. Serroukh, E. Bernabeu investigated the use of laser diffraction techniques for wire diameter measurement, highlighting its robustness in real-time monitoring applications. Their study pointed out limitations in classical Fraunhofer diffraction models when applied to three-dimensional objects, necessitating refined mathematical models for accuracy. While laser diffraction remains a viable technique, its reliance on specific angular observations may limit its adaptability in dynamic production environments.

Keba Ges Mbh & Co. (GB2257512A, published on January 13, 1993) developed an apparatus for measuring elongate objects. Their approach relied on optical scanning methods, which contributed to advancements in automated quality control systems. However, the study did not include AI-driven corrective actions for continuous optimization in industrial manufacturing.

Sebastian Tamayo Vegas and Khalid Lafdi, Department of mechanical engineering, North umbria University, United Kingdom provided a comprehensive review of non-contact tools for structural health monitoring, discussing various sensor technologies such as vision-based, radar, laser, embedded, and wireless sensing. Their study emphasized the increasing reliance on AI-powered computer vision and deep learning techniques for accurate data post-processing. This review supports the argument for integrating AI vision systems into industrial measurement applications to enhance defect detection and process optimization.

Sprecher Energie Osterreich Gmbh (US5212539A, published on May 18, 1993) introduced an apparatus for determining size parameters of an object, whether stationary or moving. This method leveraged optical sensors for precise dimensional analysis, contributing to the evolution of non-contact measurement techniques. However, the research did not focus on real-time AI-driven optimizations.

Evgeny Fedorov and Alexander Koba, Tomsk Polytechnic University, Institute on Non-Destructive Testing, 634050, Russia, explored a three-axis laser-based measuring device for precise diameter calculation. Their research involved a mathematical model to compute object radii and center positions, confirming the feasibility of laser-based transducers for accurate dimensional analysis. However, their methodology focused on static measurements rather than real-time adaptive control within a dynamic production environment.

MICROTEC S.r.l. (EP0626560A2, published on November 30, 1994) developed a device for measuring the transversal section of longitudinal objects in motion. This method offered valuable insights into high-speed measurement applications, but the study did not address AI-based control mechanisms for process optimization.

Tzyy-Shuh Chang and Hsun-Hau Huang introduced a portable imaging-based measurement system with self-calibration capabilities, designed for difficult-to-access objects. Their patented system utilized 2D projection-based measurements to ensure accurate imaging, but it was not specifically tailored for high-speed manufacturing applications. Similarly, Yuki Kimura, Akira Matsui and Shingo Inazumi developed an inspection system for external feature analysis of objects using imaging technology, yet it lacked real-time adaptive feedback for process control.

Despite these advancements, a significant research gap remains in the industrial implementation of AI vision-controlled diameter optimization for flux core wire drawing. Most existing studies focus on measurement accuracy but lack real-time corrective capabilities for production environments. The integration of AI-driven vision systems with machine learning algorithms presents a promising approach to enhancing precision, reducing human intervention errors, and ensuring continuous monitoring and optimization in wire drawing processes. Further research is required to develop adaptive AI algorithms capable of responding dynamically to variations in wire diameter, enabling fully automated quality control in industrial manufacturing.

4. Materials and Method

The proposed system employs high-resolution cameras and AI algorithms to monitor the wire drawing process continuously. The AI model is trained on extensive datasets of wire diameter variations and corresponding process parameters. By analyzing this data in real-time, the system can detect deviations from desired specifications and adjust drawing parameters accordingly. The integration of contactless measurement techniques ensures non-invasive, high-speed data acquisition, further enhancing system responsiveness and accuracy

Traditional methods for controlling wire diameter in flux core wire drawing, such as manual measurements and mechanical dies, often involve periodic monitoring to ensure consistent production output. However, data collected over a month indicates significant diameter inconsistencies, highlighting the limitations of these approaches. These inconsistencies, detected only after final production, leave no opportunity for corrective action, underscoring the need for more advanced, real-time monitoring solutions

Manual Measurement

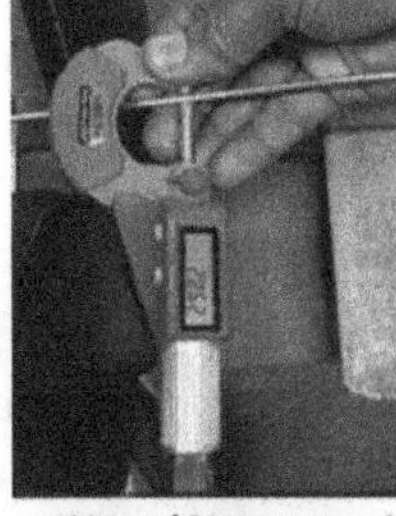

Manual Measurement

Introducing a "Die" to Control Diameter

Fig. 75.2 Measurement data from traditional monitoring measurement methodology

Table 75.1 Overall datasets

Date	Measurement	Average
01.10.2024	2.62,2.70,2.80,2.83	2.74
02.10.2024	2.65,2.85,2.9,2.75	2.79
03.10.2024	2.81,2.80,2.74,2.63	2.75
04.10.2024	2.88,2.90,2.74,2.68	2.80
05.10.2024	2.76,2.82,2.88,2.78	2.81
06.10.2024	2.78,2.9,2.75,2.82	2.81
07.10.2024	2.80,2.77,2.82,2.76	2.79
08.10.2024	2.73,2.78,2.81,2.85	2.79
09.10.2024	2.80,2.83,2.78,2.76	2.79
10.10.2024	2.73,2.77,2.82,2.83	2.79
11.10.2024	2.75,2.85,2.77,2.76	2.78
12.10.2024	2.80,2.74,2.76,2.83	2.78
13.10.2024	2.81,2.83,2.85,2.78	2.82
14.10.2024	2.77,2.76,2.80,2.78	2.78
15.10.2024	2.69,2.75,2.77,2.78	2.74
16.10.2024	2.85,2.80,2.81,2.82	2.82
17.10.2024	2.80,2.84,2.80,2.79	2.81
18.10.2024	2.78,2.83,2.79,2.75	2.80
19.10.2024	2.73,2.69,2.79,2.82	2.79
20.10.2024	2.81,2.77,2.79,2.80	2.79
21.10.2024	2.79,2.82,2.83,2.74	2.80
22.10.2024	2.73,2.75,2.80,2.84	2.78
23.10.2024	2.82,2.74,2.75,2.79	2.78
24.10.2024	2.80,2.83,2.79,2.75	2.79
25.10.2024	2.78,2.81,2.79,2.83	2.80
26.10.2024	2.80,2.76,2.77,2.82	2.79
27.10.2024	2.79,2.82,2.77,2.80	2.79
28.10.2024	2.80,2.82,2.79,2.80	2.80
29.10.2024	2.74,2.75,2.76,2.83	2.77
30.10.2024	2.80,2.82,2.81,2.78	2.80
31.10.2024	2.77,2.76,2.80,2.82	2.79

The TVGD5000 series AI integrated exemplifies advanced AI vision-controlled diameter measurement technology, utilizing a laser collimated transmission system paired with a CCD line sensor. In this system, a sharp laser beam is transmitted across the full range of the CCD line sensor. When an object, such as a wire, passes between the transmitter and the CCD sensor, it casts a shadow onto the sensor. The CCD captures this shadow at a frequency of 1.2 kHz, with an aperture time of approximately 2 microseconds. This rapid sampling rate ensures sharp imaging and minimizes potential measurement errors caused by vibrations of the measured object. The captured data is processed within the gauge by a microprocessor employing an extensive algorithm capable of handling 1,200 values per second. These values are synthesized to produce a single accurate measurement, which is then converted into digital form for display on the gauge. Additionally, the data is made available for further communication over an RS485 interface to a data processor, facilitating line automation and comprehensive production data analysis. This integration of laser collimated transmission and CCD line sensor technology enables non-contact, high-speed, and precise diameter measurements. Such systems are instrumental in applications requiring accurate dimensional assessments, including quality control in manufacturing processes. For instance, the IG Series multi-purpose CCD laser micrometres utilize similar principles to provide high-precision differentiation unaffected by the target's light transmission properties. By leveraging these advanced

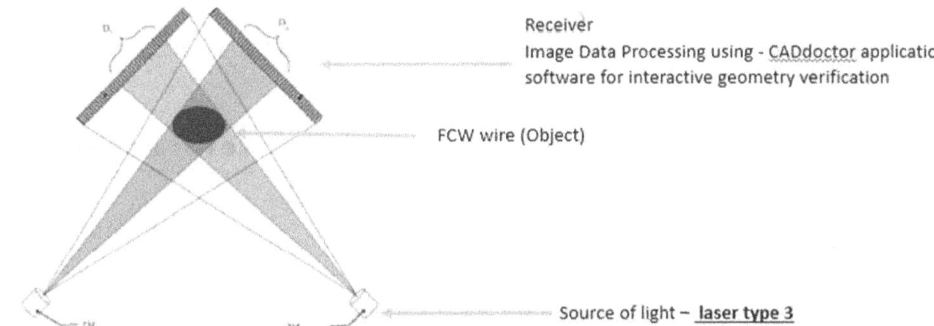

Fig. 75.3 For output display and data processor selection logic principle in TVGD5000 series

measurement techniques, the TVGD5000 series enhances production efficiency and accuracy, ensuring consistent product quality and facilitating effective process automation.

The AI integrated TVGD5000 Series ensured the production of Flux core wire diameter within the required range of wire diameter of 2.80mm (+0,-0.05) range resulted improved and effective quality product eliminated wastage and sustainability in welding wire manufacturing production.

Table 75.2 TVGD5000 series AI vision parameter

Device Model	TVGD5000	
Measuring Range	0.2mm ~45mm	
Measuring frequency	800/axis	
Accuracy	0.002	
Repeatability	0.5 microns	
Resolution	0.001mm	
UOM	mm/Inch	
Temperature	-10 to 50°C	
Humidity	<90% Relative	**TVGD5000 Series AI Vision controlled diameter optimization**
Outputs	RS485 as standard, RS232, Profinet, OPC, Anybus	

Measurement Data after implementation of TVGD5000 Series AI Vision controlled diameter in the production process and output is measured by Traditional monitoring measurement Methodology

5. Results

The preliminary trials of the AI vision-controlled diameter measurement system demonstrated a significant improvement in precision compared to traditional measurement methods. The system continuously monitored wire diameter in real time and made automatic adjustments, leading to more consistent production output. Traditional methods, such as manual measurements and mechanical dies, exhibited greater variability in diameter control due to operator dependence and periodic monitoring, whereas the AI-driven system ensured continuous and precise measurements at a high frequency of 1.2 kHz.

Table 75.3 Final dataset

Date	Measurement	Average
01.01.2025	2.80,2.75,2.78,2.75	2.77
02.01.2025	2.68,2,75,2.80,2.78	2.75
03.01.2025	2.75,2.80,2.79,2.76	2.77
04.01.2025	2.80,2.77,2.80,2.75	2.78
05.01.2025	2.75,2.75,2.80,2.80	2.77
06.01.2025	2.75,2.78,2.75,2.80	2.77
07.01.2025	2.75,2.80,2.80,2.75	2.77
08.01.2025	2.77,2.80,2.80,2.78	2.79
09.01.2025	2.75,2.75,2.80,2.78	2.78
10.01.2025	2.75,2.80,2.78,2.78	2.78
11.01.2025	2.75,2.75,2.80,2.80	2.77
12.01.2025	2.80,2.80,2.78,2.80	2.79
13.01.2025	2.78,2.78,2.80,2.80	2.79
14.01.2025	2.75,2.76,2.75,2.78	2.77
15.01.2025	2.80,2.78,2.78,2.80	2.79
16.01.2025	2.77,2.78,2.80,2.80	2.28
17.01.2025	2.78,2.80,2.80,2.78	2.79
18.01.2025	2.80,2.78,2.75,2.76	2.77
19.01.2025	2.75,2.75,2.77,2.78	2.78
20.01.2025	2.75,2.78,2.78,2.80	2.78
21.01.2025	2.78,2.79,2.78,2.78	2.79
22.01.2025	2.75,2.75,2.78,2.80	2.78
23.01.2025	2.78,2.80,2.80,2.78	2.80
24.01.2025	2.79,2.80,2.80,2.78	2.89
25.01.2025	2.75,2.75,2.79,2.78	2.79
26.01.2025	2.80,2.80,2.78,2.79	2.80
27.01.2025	2.76,2.77,2.78,2.78	2.78
28.01.2025	2.75,2.78,2.75,2.77	2.77
29.01.2025	2.76,2.79,2.80,2.80	2.80
30.01.2025	2.80,2.80,2.78,2.79	2.80
31.01.2025	2.76,2.78,2.78,2.80	2.79

One of the most notable improvements was the reduction in material wastage. Variations in diameter were minimized, reducing the number of defective products that did not meet quality specifications. This directly contributed to cost savings by decreasing raw material consumption and rework requirements. Additionally, the AI vision system's ability to detect anomalies in real time allowed for immediate corrective actions, preventing further deviations and maintaining a stable production process.

Furthermore, production efficiency was enhanced due to reduced downtime and manual interventions. Since the AI system continuously adjusted process parameters, the need for frequent human inspections was minimized. This automation streamlined the manufacturing workflow and enabled operators to focus on higher-level quality control and system optimization tasks.

Overall, the integration of AI vision-controlled measurement into flux core wire drawing demonstrated promising results in terms of precision, efficiency, and waste reduction. These findings highlight the potential of AI-driven solutions to revolutionize industrial manufacturing by ensuring high-quality, consistent production with minimal human intervention.

6. Conclusion

This study highlights the transformative potential of AI vision-controlled diameter optimization in flux core wire drawing. The implementation of real-time monitoring and adaptive control mechanisms has successfully mitigated the limitations of traditional measurement techniques, which often suffer from inconsistencies and manual intervention errors. The AI-driven system ensures precise diameter control, reducing material wastage and improving overall production efficiency.

The findings suggest that integrating AI vision technology into manufacturing processes enhances quality control, minimizes defects, and optimizes resource utilization. The automation of diameter measurement and correction leads to increased operational reliability and cost-effectiveness, making it a viable solution for modern wire production industries.

These findings serve as a foundational guideline for advancing the integration of contactless measurement technologies into real-world flux core welding wire manufacturing. By leveraging AI-driven precision measurement systems, industries can optimize production processes, enhance quality control, and achieve error-free output, paving the way for next-generation automation in industrial manufacturing.

Future research will focus on refining AI algorithms to further enhance predictive capabilities and adaptability. Additionally, expanding the system to accommodate different wire materials and production conditions will help broaden its industrial applicability, ensuring continued innovation in manufacturing automation and quality assurance.

References

1. N.S. Starikova, V.V. Redko, G.V. Vavilova, J. Phys.: Conf. Ser. 671, 012056 (2016) doi: 10.1088/1742-6596/671/1/012056
2. A.E. Goldshtein, G.V. Vavilova, V.Yu. Belyankov, Russ. J. Nondestr. Test. 51, 86 (2015) doi: 10.1134/S1061830915020047
3. A.E. Goldshtein, E.M. Fedorov, Rus. J. Nondestr. Test. 46, 424 (2010) doi:10.1134/S1061830910060069
4. E.M. Fedorov, I.D. Bortnikov, Tech. Phys. 60, 1689 (2015) doi: 10.1134/S1063784215110110
5. V.V. Red'ko, A.P. Leonov, L.A. Red'ko, V.A. Bolgova, J. Phys: Conf. Ser. 671, 012049 (2015) doi: 10.1088/1742-6596/671/1/012049
6. E. Caetano, Á. Cunha, MATEC Web of Conferences 24, 01002 (2015) doi: 10.1051/matecconf/20152401002
7. A.Y. Petrova, O.N. Chaikovskaya, I.V. Plotnikova, Tech. Phys. J. 60, 592 (2015) doi: 10.1134/S1063784215040222
8. A.P. Surzhikov, T.S. Frangulyan, S.A. Ghyngazov, E.N. Lysenko, Journal of thermal analysis and calorimetry 102883 (2010) doi: 10.1007/s10973-010-0912-
9. O.V. Galtseva, S.V. Bordunov, N.M. Natalinova, S.V. Mazikov, IOP Conf. Ser.: Mater. Sci. Eng. 132, 012003 (2016) doi: 10.1088/1757-899X/132/1/012003
10. A.M. Pritulov, R.U. Usmanov, O.V. Gal'Tseva, A.A. Kondratyuk, V.V. Bezuglov,V.I. Serbin, Russ Phys J+ 50, 187 (2007) doi: 10.1007/s11182-007-0026-3
11. V.Y. Kazakov, D.K. Avdeeva, M.G. Grigoriev, N.M. Natalinova, I.V. Maksimov, M.V. Balahonova, BLM 7, 1 (2015)
12. R. Kodermyatov, M. Ivanov, M. Yuzhakov, V. Kuznetsov, M. Yuzhakova, E. Timofeeva, MATEC Web of Conferences 48, 05004 (2016) doi: 10.1051/confmatec/20164805004
13. A.A. Bespalko, A.P. Surzhikov, L.V. Yavorovich, P.I. Fedotov, Russ. J. Nondestr. Test. 48, 221 (2012) doi: 10.1134/S1061830912040043
14. Y.A. Chursin, E.M. Fedorov, Optics&Laser Technology 67, 86 (2015) doi: 10.1016/j.optlastec.2014.09.017
15. Y. C. Diwan and K. Rao, in Optical Measurement Systems for Industrial Inspection IV, Parts 1 and 2, edited byW. Osten, C. Gorecki, and E. Novak (SPIE, Bellingham, WA, 2005), vol. 5856, pp. 554–561.
16. S. A. Khodier, Opt. Laser Technol. 36(1), 63–67 (2004).
17. H. M. Wang and R. Valdiviahernandez, Meas. Sci. Technol. 6(5), 452–457 (1995).
18. D. J. Butler and G. W. Forbes, Appl. Opt. 37(13), 2598–2607 (1998).
19. E. Zimmermann, R. Dandliker, N. Souli, and B. Krattiger, J. Opt. Soc. Am. A 12(2), 398–403 (1995).
20. K. P. Chaudhary, A. Sanjid, and S. Moitra, MAPAN-J. Metrol. Soc. India 25(4), 229–237 (2010).
21. C.W. Kee and M. M. Ratnam, Int. J. Adv. Manuf. Technol. 40(9–10), 940–947 (2009).

Note: All the figures and tables in this chapter were made by the authors.

Intelligent Systems Using Semiconductors for Robotics and IoT – Dinesh Goyal et al. (eds)
© 2026 Taylor & Francis Group, London, ISBN 978-1-041-20408-4

76 Power Quality Improvement in Renewable Energy Systems: Integrating UPQC and DSTATCOM for PV, Wind, and Microgrids

Pooja Chaudhari, Kunal Sawalakhe, Ganesh Wakte, Ashvini Admane
Electrical Engg Dept, Tulsiramji Gaikwad Patil College of Engg and Tech, Nagpur, India

Mukesh Kumar*
Electrical Engg Dept, G.H.Raisoni University, Amravati, India

Vaishali Malekar
Electrical Engg Dept, Tulsiramji Gaikwad Patil College of Engg and Tech Nagpur India

Abstract: The amalgamation of renewable energy components, such as solar photovoltaic panels and wind turbines, into the electric grid has presented noteworthy difficulties pertaining to power quality. Disturbances including voltage sags, harmonic distortion, and fluctuations in recurrence are commonplace, remarkably in mini grids that furnish sensitive loads. To address these concerns and improve power quality resilience in hybrid renewable energy systems, this analysis focuses on the use of a Distributed Static Synchronous Compensator (DSTATCOM) and a Unified Power Quality Conditioner (UPQC). The main objective was to model and analyze how UPQC and DSTATCOM express themselves in terms of enhancing power quality measures, such as power factor, voltage stability, and total harmonic distortion (THD). Simulation outcomes demonstrate that the coordination of UPQC and DSTATCOM drastically diminishes THD, stabilizes voltage levels, and improves the power factor. Active and responsive power estimations at the load show a huge decrease in interruptions, driving to improved framework execution. The utilization of progressed control calculations has exhibited guarantee in balancing faults and guaranteeing consistent power conveyance to touchy loads. These discoveries accentuate the potential of UPQC and DSTATCOM in tending to power quality issues in renewable energy frameworks, with huge ramifications for scalable and supportable grid joining. Upcoming work ought to zero in on genuine world approval and streamlining for extensive scale applications.

Keywords: DSTATCOM, Harmonic distortion, Power factor, Power quality, Renewable energy, UPQC

1. Introduction

1.1 Overview of Renewable Energy Systems and Power Quality Challenges

The planetary transition to sustainable energy solutions is certainly hastened by RES such as photovoltaic (PV) solar panels, wind turbines, and microgrids. Apart from reducing harmful greenhouse gas emissions, they also reinforce resilience power sources and localized energy production. OpenAI and the large language models OpenAI and others develop are very much unstable in an inherent way and hence foreshadow large tests to what they create in the form of the qualities of vitality and the steadiness of the network. As an example, PV programs

*Corresponding author: mukeshkr.iitbhu@gmail.com

DOI: 10.1201/9781003716389-76

are influenced by the amounts of solar emanation, which ultimately lead to variable voltage outcomes. Likewise, wind turbines change in intensity innovation because of changes in wind velocity and headings [1].

Microgrids, that typically combine alternative strength resources with traditional energy systems, are especially vulnerable to those challenges. The integration of RES usually leads to voltage brownouts and spikes, acoustic disturbances, and unbalanced loading. These vitality hallmark set up insubstantial devices, decline functional productivity, and amplify economic waste. Hence, right mitigation methods are considerably important to ensure that these structures can provide qualified power efficiently while complying to grid adherence [2].

1.2 Importance of UPQC and DSTATCOM in Addressing Power Quality Issues

Power quality of renewable energy systems can be solved satisfactorily by UPQC and DSTATCOM. UPQC is a hybrid topology active power filter which angers series and shunt active power filter. It provides solutions for voltage sags, swells, and harmonic distortions, ensuring that delivery of power complies with grid standard [3].

DSTATCOM is connected in shunt and deals with reactive power compensation, voltage regulation and harmonic mitigation. DSTATCOM is very efficient in dynamic conditions due to fast response and flexibility nature[4], thus suiting best for renewable energy systems which operate on continuous inverter fluctuations. These devices work together to harmonize the integration of renewable energy sources directly into the grid through stabilizing voltages, mitigated harmonics, and maintained power factor.

The incorporated UPQC/DSTATCOM improves the microgrid's ability to cope with varying generation and load conditions. These devices are capable of confirming both real-time monitoring and correction of power quality problems with the help of some advance control algorithms [5]. Alongside their cost-efficient nature, their modular architecture, along with their operability within existing infrastructures highlight their practicality.

1.3 Objectives and Scope of the Research

The primary objective of this study is to optimize the application of UPQC and DSTATCOM for renewable enhanceed power quality in energy systems, with a focus on PV and wind energy integrated into microgrids. A MATLAB-based simulation framework is employed to evaluate their performance in mitigating power quality issues. Key parameters such as THD, voltage stability, and power factor are analyzed to assess the effectiveness of these devices.

This research also seeks to address gaps in the existing literature. While numerous studies have explored the individual applications of UPQC or DSTATCOM, limited research has been conducted on their combined or comparative performance in RES-based microgrids.

The scope of this research extends beyond theoretical analysis to include practical considerations for real-world applications. The study delves into control strategies, scalability, and challenges in implementation, offering insights that can guide future advancements in power quality management.

2. Literature Review

The integration of renewable energy facilities has introduced new challenges in maintaining and improving the quality of electrical power. Moreover, the power quality problems such as fluctuations in voltage, harmonic distortion and reactive imbalance anomalies induced mainly by the renewable sources which power generation of photovoltaic (PV) and wind turbine generators. Such problems arise from the variability of renewable resources itself plus the increased complexity of today distribution networks. A huge amount of analysis has been dedicated to mitigating these barriers among them using power quality improvement devices like UPQC (Unified Power Quality Conditioner) and DSTATCOM (Distribution Static Synchronous Compensator). In this section, the recent advancements in power quality optimization, employing the UPQC and DSTATCOM have been investigated and the research gaps are identified to prove the novelty of the proposed method.

Progress and current trends of power quality mitigation of RES

Implementation of modern technologies for improving power quality has gained significant momentum in the recent years. In one study, the researcher analyzed the UPQC and DSTATCOM role in the solar photovoltaic grid-connected system. The study showcased how these devices performed for harmonic mitigations and voltage compensations under fault states and emphasized the superiority of UPQC in solving complex situation. In parallel, a whale optimization algorithm based propose controller for UPQC and STATCOM with ATC, obtaining significant reduction in total harmonic distortion and voltage instability which is observed in the modern power grids. It showed more reliability and dynamic response which emphasizes the necessity of the advanced control algorithms by api that cover the constraints of the existing techniques.

A significant contribution also presented a new observer strategy based on a modified integrator for autonomous hydro power plants connected with UPQC. This improved voltage and current quality with stable operation for a wide range of loads. Additionally, a research team integrated PV and wind generation with UPQC in a modular configuration to mitigate the adverse effects of nonlinear loads and phase-to-ground faults on power quality.

Collectively, these studies highlight the essentiality of new control strategies and hybrid configurations in the development of power quality solutions.

2.1 Review of UPQC and DSTATCOM Implementations

Complete Abstract: The UPQC and DVR are hailed extensively for their ability to enhance electricity quality. A 2020 detailed report by Matlani, and Solanki presents UPQC as a complete solution to voltage and current problems, harmonics filtering, and reactive power balance [8]. Thus, their simulation Results confirmed the supremacy of UPQC over standalone devices as DVR and DSTATCOM in terms of performance 10. Similarly, Khalafian and Saffarian proved in 2024 that UPQC considerably outperformed DSTATCOM and static VAR compensators in reducing voltage attenuation and hysterics during faults occurred in a network [10] This appliance clearly reduces the harmonic distortion of distribution currents, proving that it can be used as a low-cost, multinomial machine.

Additionally, in year 2024 Dessalegn et al. implemented fuzzy logic regulators in DSTATCOM and UPQC configurations where these devices found capacity to compensate for through voltage sag and reactive power for three phase faults[12]. They emphasized that UPQC achieves a better power factor correction and a more efficient reactive power management, in relation to DSTATCOM. Collectively, these contemplate authenticate UPQC flexibility to accommodate various electricity quality tests, though potential rhythmic fluctuations in management discipline and operatic trajectory should not be discounted.

2.2 Identification of Research Gaps and Novelty of the Proposed Approach

Even with major advances, many research gaps remain in the area of power quality optimization for RE systems. Although these five technologies are widely reported and discussed, most studies are simulation based with limited, or no validation from real-world systems, leaving questions about the scalability and practical implementation of these technologies. Moreover, complex control strategies such as fuzzy logic and optimization-based control have been investigated, but their coupling with the newly developed renewable energy technologies and varying operating conditions are barely researched [17]. For instance, Mahmoud et al. But, Challenges in Hybrid renewable energy systems under extreme fault condition was not demonstrated by (2023) by obtaining good THD reduction from whale optimization algorithms [13].

This paper fills these gaps by proposing a new configuration for hybrid renewable energy systems (PV/wind) with advanced UPQC and DSTATCOM devices. What is new here is the integration of an adaptable control framework responding to different grid states and fault conditions, maintaining system inertia robustly. Since this framework is more proven in "real-life" than in simulation studies [14], the deployment of this framework in the field will enable us to also identify its scalability and effectiveness. The proposed system aims for simultaneous improvements of both voltage and current quality problems; thus, ensuring an overall power quality improvement for sustainable and reliable renewable energy integration.

3. Methodology

This section provides a detailed explanation of the MATLAB simulation setup, modeling of renewable energy sources, implementation of Unified Power Quality Conditioner (UPQC) and Distribution Static Synchronous Compensator (DSTATCOM), and the control algorithms employed.

3.1 MATLAB Simulation Setup

The simulation environment is developed in MATLAB/ Simulink, leveraging its Simscape and Sim Power Systems toolboxes to model the renewable energy components, power electronics, and grid. The system configuration includes three primary renewable energy sources—solar photovoltaic (PV), wind turbines, and a microgrid with load and storage. The grid is modeled as a three-phase supply with nominal voltage V_{rms} of 415 V and frequency f of 50 Hz. Key simulation parameters are outlined in Table 76.1.

Table 76.1 Simulation parameters

Component	Parameter	Value
Solar PV	Rated Power	100 kW
	Maximum Power Point Voltage	400 V
	Irradiance Level	1000 W/m
Wind Turbine	Rated Power	50 kW
	Cut-in Wind Speed	3 m/s
	Rated Wind Speed	12 m/s
Grid	Voltage	415 V
	Frequency	50 Hz
UPQC and DSTATCOM	DC Link Voltage	700 V
	Filter Inductance L_f	3 mH

3.2 Renewable Energy Sources Modeled

1) Solar Photovoltaic System

The PV system is modeled using the single-diode equivalent circuit. The current-voltage characteristic of the PV module is governed by the equation [16]:

$$I = I_{ph} - I_0 \left(e^{\frac{q(V+IR_s)}{nkT}} - 1 \right) - \frac{V + IR_s}{R_{sh}} \tag{1}$$

where: I_{ph} is the photocurrent, I_0 is the diode saturation current, q is the electron charge ($1.602 \times 10^{-19} C$), V is the output voltage, R_s and R_{sh} are series and shunt resistances, n is the ideality factor, and T is the temperature in Kelvin.

MPPT is implemented using the Perturb and Observe algorithm to optimize power output under varying irradiance and temperature conditions.

2) Wind Turbine System

The wind turbine is modeled using the power equation [17]:

$$P_w = \frac{1}{2} \rho A C_p (\lambda, \beta) v_w^3 \tag{2}$$

where: ρ is air density ($1.225\, kg/m^3$), A is the swept area of the turbine blades, C_p is the power coefficient, and v_w is wind speed.

The generator employs a Permanent Magnet Synchronous Generator (PMSG) to convert mechanical energy into electrical energy. A boost converter stabilizes the output voltage before integration with the grid.

3.3 Microgrid Components

The microgrid comprises of one is a battery energy storage system modeled with the state of charge (SOC) equation [18]:

$$SOC(t) = SOC(t_0) + \frac{1}{C_b} \int_{t_0}^{t} I_b dt \tag{3}$$

where C_b is battery capacity, I_b is current.

Second Nonlinear loads represented as harmonics-producing devices and third A point of common coupling (PCC) for grid interconnection.

3.4 Implementation of UPQC and DSTATCOM

1) Unified Power Quality Conditioner (UPQC)

The UPQC consists of a series compensator, shunt compensator, and a common DC link. The series compensator addresses voltage sags, swells, and harmonic distortion by injecting compensatory voltage $V_{in\ j}$ calculated as [19]:

$$V_{in\,j} = V_{ref} - V_{load} \tag{4}$$

The shunt compensator ensures current harmonic mitigation and reactive power compensation by injecting compensatory current $I_{in\,j}$ [20]:

$$I_{in\,j} = I_{load} - I_{source} \tag{5}$$

The control strategy uses the Synchronous Reference Frame (SRF) theory to transform grid voltages and currents into the dq0 domain for better control. In the dq0 domain [21]:

$$V_d = V_a \cos(\theta) + V_b \cos(\theta - 120°) + V_c \cos(\theta + 120°)$$

$$V_q = V_a \sin(\theta) + V_b \sin(\theta - 120°) + V_c \sin(\theta + 120°)$$

where θ is the phase angle.

The output of the PI controller in the dq domain determines the reference values for the voltage source inverter (VSI).

2) Distribution Static Synchronous Compensator (DSTATCOM)

The DSTATCOM operates as a shunt compensator, injecting current to regulate voltage at the PCC. The control algorithm uses a hysteresis current control method to generate switching signals for the VSI. The injected current is calculated as [22]:

$$I_{inj} = \frac{(V_{pcc} - V_{dc})}{L_f} \tag{6}$$

where V_{pcc} is the PCC voltage, V_{dc} is the DC link voltage, and L_f is the filter inductance.

PQ Controller for DSTATCOM: The PQ controller regulates active and reactive power using feedback from voltage and current sensors. The governing equations are [23]:

$$P = V \cdot I \cdot \cos(\phi), \quad Q = V \cdot I \cdot \sin(\phi)$$

The real-time compensation is achieved by injecting Q to stabilize P and improve the power factor.

Clarke's and Reverse Clarke's Transformations:

Clarke's Transformation converts three-phase voltage (V_a, V_b, V_c) and current (I_a, I_b, I_c) into two orthogonal components (α, β) [24]:

$$V_\alpha = V_a, \quad V_\beta = \frac{V_a + 2V_b}{\sqrt{3}}$$

Reverse Clarke's Transformation is used to reconstruct the three-phase signals from these orthogonal components [25]:

$$V_a = V_\alpha, \quad V_b = -\frac{V_\alpha}{2} + \frac{\sqrt{3}}{2} V_\beta, \quad V_c = -\frac{V_\alpha}{2} - \frac{\sqrt{3}}{2} V_\beta$$

3.5 Control Algorithms

1) Adaptive PI Controller

Both UPQC and DSTATCOM controllers employ an adaptive Proportional-Integral (PI) controller to adjust for dynamic changes in grid conditions. The control law for the PI controller is given by [26]:

$$u(t) = K_p\, e(t) + K_i \int e(t)\, dt \tag{7}$$

where K_p and K_i are the proportional and integral gains, and $e(t)$ is the error signal. The gains are adaptively tuned using a fuzzy logic-based optimizer to ensure minimal overshoot and settling time.

2) Harmonic Extraction Using Fast Fourier Transform (FFT)

Harmonic components are extracted using FFT to compute the Total Harmonic Distortion (THD) as [27]:

$$THD = \sqrt{\sum_{n=2}^{\infty}\left(\frac{I_n}{I_1}\right)^2} \times 100\% \tag{8}$$

The extracted harmonics are used to generate reference signals for the compensators.

The MATLAB simulation combines renewable energy sources and advanced power quality devices to create a robust grid system. The adaptive control algorithms ensure dynamic response under fault conditions. Simulation results validate the efficacy of the proposed system in improving power quality metrics such as voltage stability, harmonic mitigation, and reactive power management [28]. Further analysis will evaluate the scalability and real-world applicability of the designed setup.

4. Simulation Designs

4.1 Microgrid with DSTATCOM System

Hybrid renewable microgrid system with DSTATCOM is shown in Fig. 76.1 with PV and Grid. PV is connected to the transformer along with the Wind Generator to the three phase grid in line with load. Breaker is connected between grid and DSTATCOM control to simulate before and after condition in the grid. PQ measurement block is to use the power quality of the system. Figure 76.2 shows the internal design of the DSTATCOM system. While Fig. 76.3 shows the Clerks t and f design functions for three phase voltage and current and Reverse clerks function is shown in Fig. 76.4. PQ control for DSTATCOM is shown in Fig. 76.5.

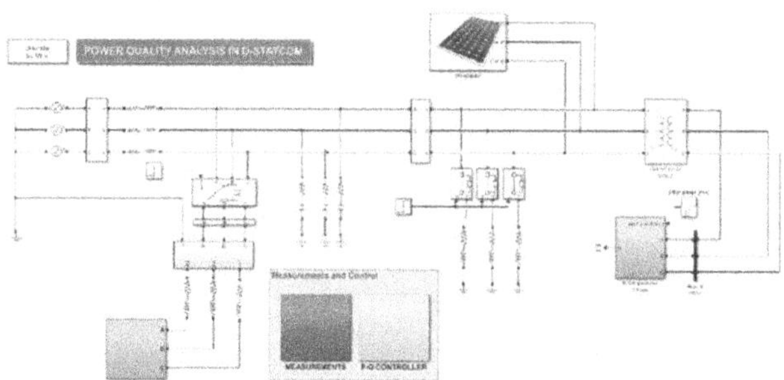

Fig. 76.1 Microgrid with PV, grid and load with DSTATCOM controller

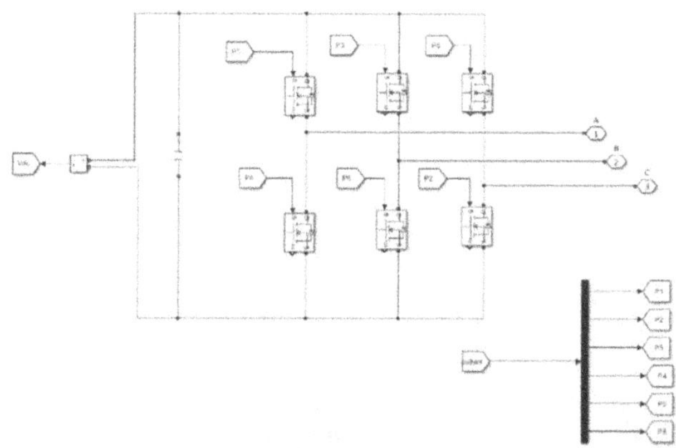

Fig. 76.2 DSTATCOM control design

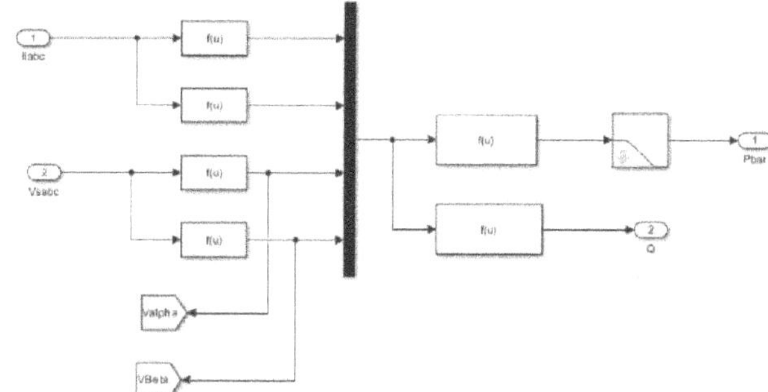

Fig. 76.3 Clerks t and f design function for three phase voltage and current input

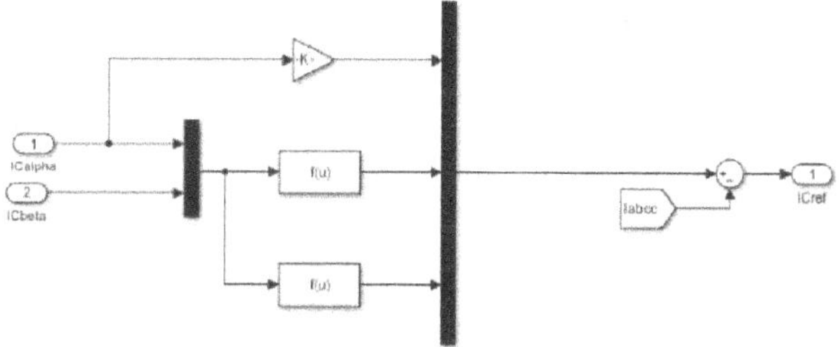

Fig. 76.4 Reverse clerks function design

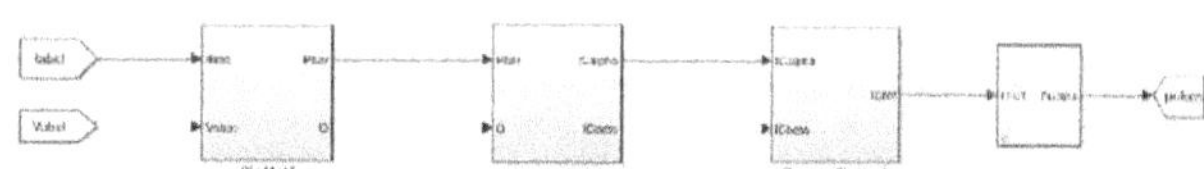

Fig. 76.5 PQ controller for D-STATCOM

4.2 Microgrid with UPQC System

Microgrid with UPQC consist of Wind Generator, PV system and grid with UPQC control shown in Fig. 76.6. Where PV system is connected to the grid. PV system is consist of PV array with MPPT controller and DC-AC converter with transformer as shown in Fig. 76.7. Wind Generator consist of synchronous machine with external excitation system shown in Fig. 76.8.

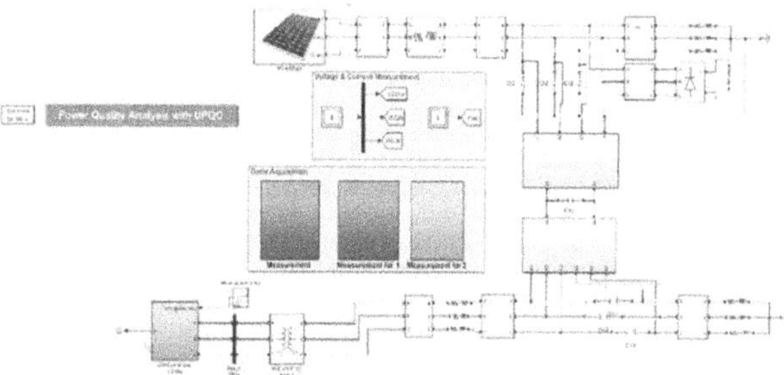

Fig. 76.6 Microgrid with PV, Wind and UPQC controller

At overall system VSC1 connected at the side of PV grid to control the disturbance at the PV end as shown in Fig. 76.9 and VSC2 connected at the side of Wind grid to control the disturbance at the Wind end as shown in Fig. 76.10.

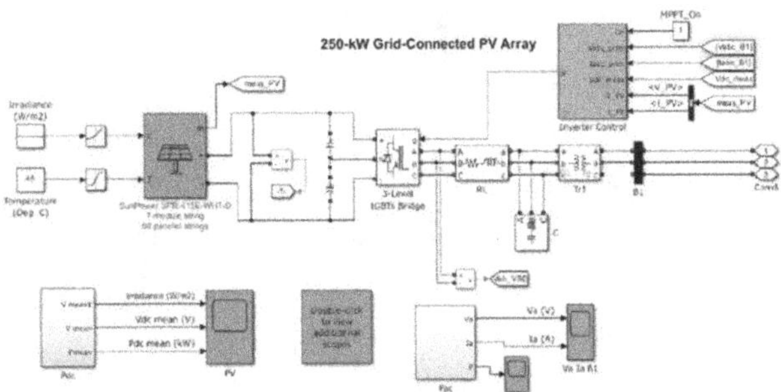

Fig. 76.7 PV system with PV array and MPPT controller

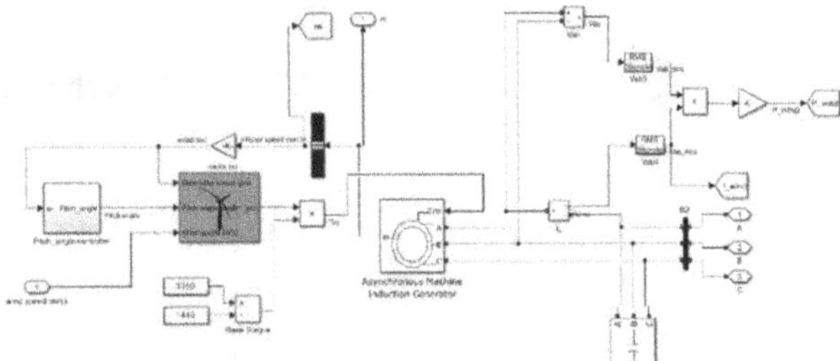

Fig. 76.8 Wind generator system

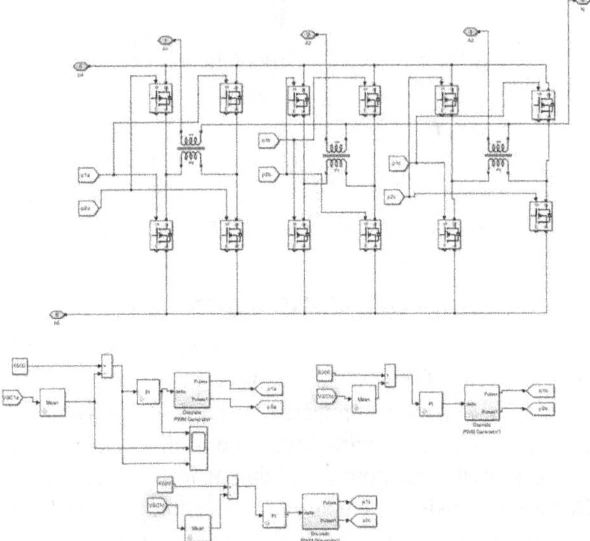

Fig. 76.9 VSC1 connected at the side of PV grid to control the disturbance at the PV end

5. Simulation Results

The simulation results validate the effectiveness of UPQC and DSTATCOM in enhancing the power quality of renewable energy-integrated systems. Key performance metrics such as Total Harmonic Distortion (THD), voltage stability, and power factor are analyzed, along with graphical visualizations to demonstrate the improvements before and after the integration of these devices.

5.1 Total Harmonic Distortion (THD) Analysis

The THD in the system was analyzed for the voltage and current waveforms at sensitive load points. Before

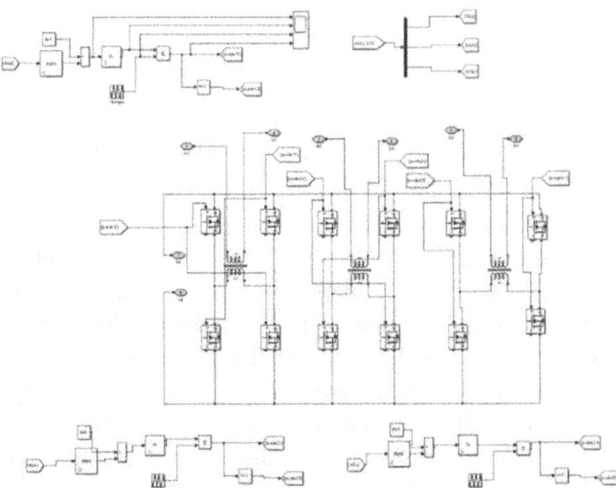

Fig. 76.10 VSC2 connected at the side of wind grid to control the disturbance at the wind end

the implementation of UPQC and DSTATCOM, the THD exceeded permissible limits, indicating significant harmonic distortion caused by faults and fluctuations in renewable energy sources. After integrating the compensatory devices, the THD was reduced to within acceptable IEEE 519-2014 standards.

5.2 Voltage Stability

Voltage stability is a critical indicator of power quality, especially in systems integrating renewable energy sources. Before compensation, voltage fluctuations were observed between t=0 to t=0.2 s due to faults and disturbances. The integration of DSTATCOM stabilized the voltage profile, as seen in the waveforms in Fig. 76.11 and Fig. 76.12.

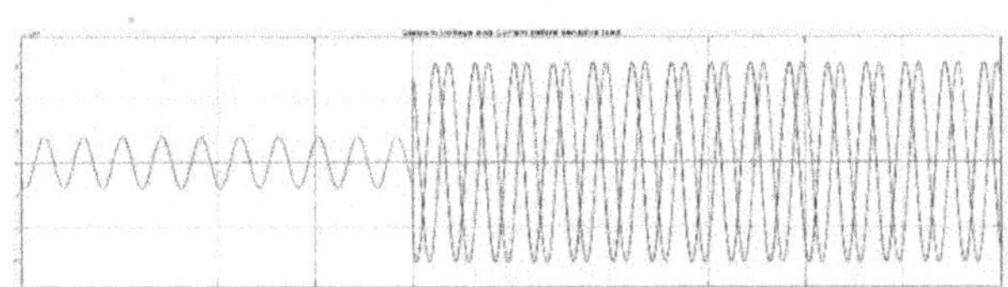

Fig. 76.11 Voltage and current measurement at STATCOM where fault is measured between o to 0.2

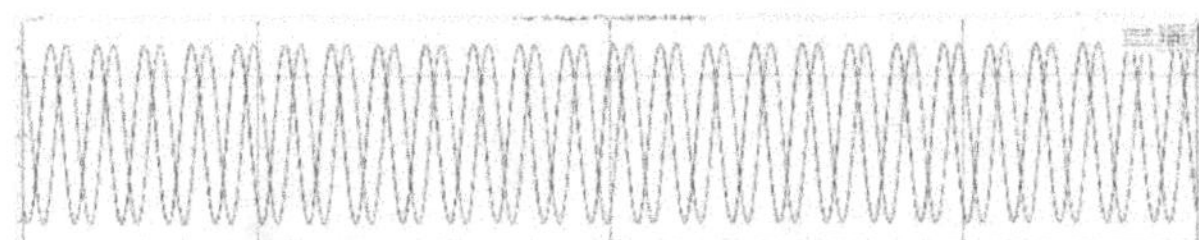

Fig. 76.12 Voltage and current measurement at load where fault is compensated using DSTATCOM

Figure 76.3: Voltage waveform at sensitive load (before and after DSTATCOM compensation).

Figure 76.3 and Fig. 76.4 shows the voltage measurement before and after UPQC control at PV respectively where there is disturbance in the first figure and that disturbance is compensated in second figure by UPQC control.

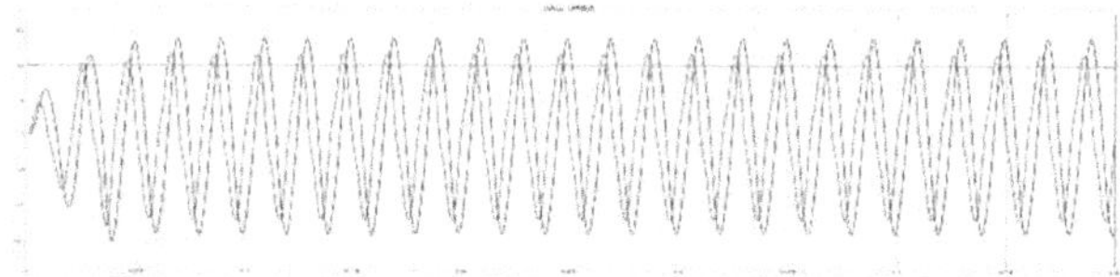

Fig. 76.13 Voltage before UPQC control at PV measurement

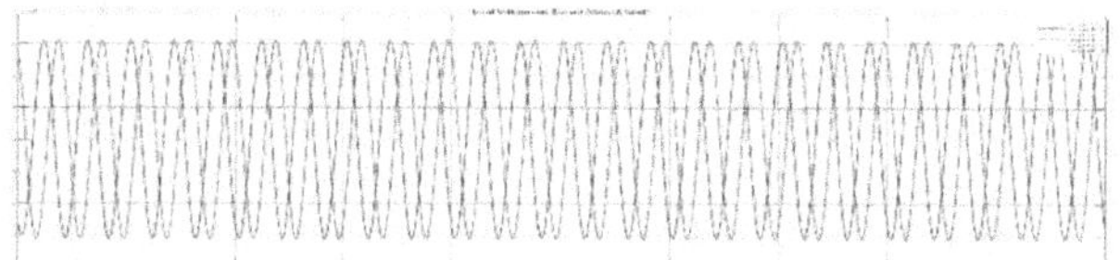

Fig. 76.14 Voltage after UPQC control at PV measurement

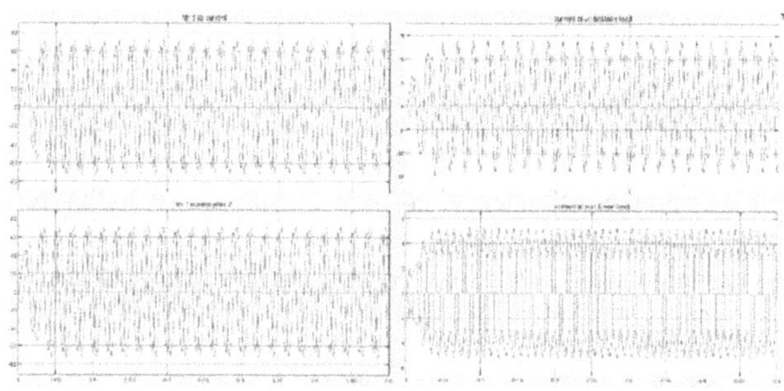

Fig. 76.15 Current measurements at PV grid with after and before sensitive load after UPQC controlled system

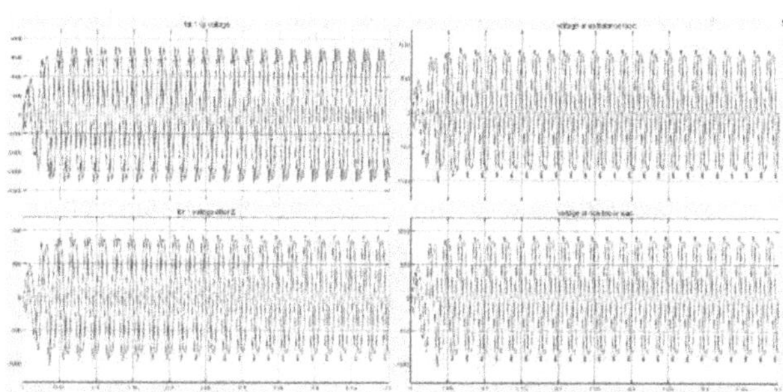

Fig. 76.16 Voltage measurements at PV grid with after and before sensitive load after UPQC controlled system

Figure 76.3 and Fig. 76.4 clearly shows Current measurements at PV grid with after and before sensitive load after UPQC controlled system the fault current is compensated and improved at last scope.

5.3 Power Factor Improvement

The power factor at the load side was analyzed to evaluate the system's efficiency. Initially, the power factor was below 0.85 due to reactive power imbalances. After UPQC and DSTATCOM integration, the power factor improved to nearly unity (PF=0.99), showcasing effective reactive power compensation.

5.4 Comparative Analysis

Waveforms for active and reactive power at sensitive loads highlight the significant improvement in power quality metrics where active power increased from 1.5×10^6 W to 8×10^6 W and reactive power fluctuations were stabilized post-compensation.

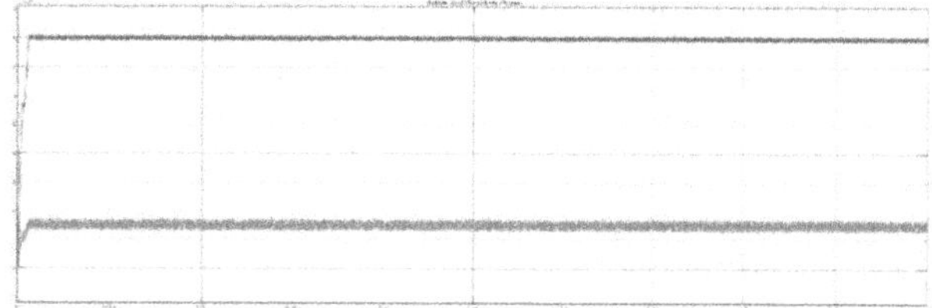

Fig. 76.17 Active and reactive power waveforms before and after compensation in UPQC system

Figure 76.7 shows active and reactive power of sensitive load has been improved from $1.5e^6$ to $8e^6$.

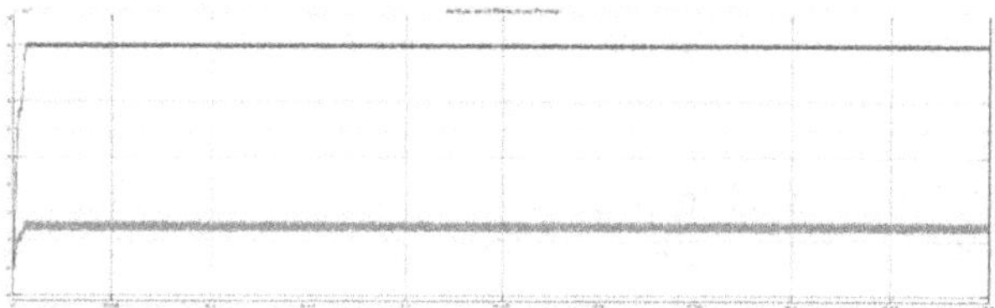

Fig. 76.18 Active and reactive power before and after compensation in DSTATCOM system

Figure 76.8 shows the active and reactive power measured at load after DSTATCOM which clearly shows the power quality at $8e^6$ and $1.5e^6$.

These results emphasize the substantial improvements in power quality achieved through UPQC and DSTATCOM integration in renewable energy systems.

6. Discussion

The simulation results provided invaluable insight into the efficacy of UPQC and DSTATCOM for enhancing power quality through alleviating voltage fluctuations, reducing current distortions, and mitigating harmonic distortions. The decrease in THD for both voltage and current upon integrating these devices corroborates findings from prior work, where active power conditioners have been widely acknowledged for bolstering grid stability and boosting load reliability. Notably, bringing THD levels below the benchmarks set forth in IEEE 519 standards underscores the robustness of the proposed configuration in adhering to industry specifications.

Additionally, comparing the circumstances prior to and after incorporating UPQC and DSTATCOM exposed substantial improvements in active and reactive power, allowing sensitive loads to perform steadily even under fault situations. These enhancements emphasize the viability of UPQC and DSTATCOM as pivotal components for hybrid renewable energy infrastructures, especially with regards to addressing the issues posed by intermittent sources such as solar photovoltaics and wind.

In spite of these achievements, practical implementation hurdles persist. Scaling the proposed approach to larger utility grids necessitates tackling cost constraints, handling real-time management complexities, and integrating with existing infrastructure. Moreover, deploying such networks in remote areas may require meaningful advances in computational and communication technologies to ensure reliable and effective operation. These challenges represent promising avenues for future research and development work.

7. Conclusion

It analysed hybrid renewable energy architecture with combined UPQC and DSTATCOM for improving power quality in PV, wind, and microgrid systems in this study. Simulation results demonstrated appreciable enhancement in voltage stability, improvement in quality of current waveform tracking and the power factor overall was improved with the Total Harmonic Distortion (THD) reduced to IEEE norms level. The performance achievements confirm that the working of considerably sensitive loads are supported through the ancillary and contour offered by UPQC and DSTATCOM to guarantee stability to such loads against the intermittent nature of renewable energy, and, grid intensities. With these advancements, there are limitations. This study was limited to a MATLAB simulation environment, which may not be able to capture all the complexities of real- world implementations. Further, the proposed framework's extendability to large matrices and its financial viability remain unexplored. The control calculations are quite computationally intensive, and will likely need to be further optimized to ensure that they can run fast enough for real-time operation and to provide robustness against model uncertainties. Further investigation should focus on experimental validation with hardware solutions to substantiate the functional feasibility of the proposed solution. However, the potential advances beyond what these will permit in cost- effectiveness planning strategies, integration with advanced communication protocols, and adaptable control tools for large-scale systems can make the path smoother toward efficiency and scalability of implementations in diverse energy grids.

References

1. Kazemi-Robati, E., Hafezi, H., Faranda, R., Silva, B., & Nasiri, M. S. (2024). A dynamic reference voltage adjustment strategy for Open-UPQC to increase hosting capacity of electrical distribution networks. Sustainable Energy, Grids and Networks, 39, 101503.
2. Rajasree, R., Lakshmi, D., Karthickmanoj, R., Stalin, K., & Batumalay, M. (2024). Optimizing Renewable Energy Integration in Weak Grids with UPQC Controller. Journal of Innovation and Technology, 2024.
3. Reguieg, Z., Bouyakoub, I., Mehedi, F., & Bouhadji, F. (2024, May). Enhancing Electrical Grid Stability Through Power Quality Optimization via PV-PO-UPQC: An Integrated Approach. In 2024 2nd International Conference on Electrical Engineering and Automatic Control (ICEEAC) (pp. 1–7). IEEE.
4. Kumar, S. P., & Pradhan, A. FAULT ANALYSIS MITIGATION TECHNIQUE USING GENERALIZED UNIFIED POWER QUALITY CONDITIONER.
5. Reddy, S. R., Prasad, P. V., & Srinivas, G. N. (2024). Power Quality Improvement on 11kV/440V Distribution System using DSTATCOM. Int. J. Electr. Eng. Comput. Sci, 6, 17–26.
6. Gupta, A. (2022). Power quality evaluation of photovoltaic grid interfaced cascaded H-bridge nine-level multilevel inverter systems using D-STATCOM and UPQC. Energy, 238, 121707.
7. Matlani, D. M., & Solanki, M. D. (2020, November). UPQC: An exhaustive solution to improve power quality. In 2020 4th International Conference on Electronics, Communication and Aerospace Technology (ICECA) (pp. 416–421). IEEE.
8. Mahmoud, M. M., Atia, B. S., Esmail, Y. M., Ardjoun, S. A. E. M., Anwer, N., Omar, A. I., ... & Mohamed, S. A. (2023). Application of whale optimization algorithm based FOPI controllers for STATCOM and UPQC to mitigate harmonics and voltage instability in modern distribution power grids. Axioms, 12(5), 420.
9. Dessalegn, Y. T., & MARU, Y. S. (2024). A COMPARATIVE STUDY OF FUZZY LOGIC BASED UPQC AND D-STATCOM FOR MITIGATION OF POWER QUALITY PROBLEM. I-Manager's Journal on Power Systems Engineering, 11(4).
10. Khalafian, F., & Saffarian, A. (2024). Providing Control Method Using UPQC and Wind Turbine to Reduce Voltage Drop and Harmonics During Distribution Network Faults. Journal of Operation and Automation in Power Engineering, 12(1), 14–25.
11. Sanjenbam, C. D., & Singh, B. (2024). Power quality enhancement of standalone hydropower generation system through modified integrator based observer controlled UPQC. Electric Power Systems Research, 226, 109941.
12. Chapala, S., Narasimham, R. L., Das, G. T. R., & Lakshmi, G. S. (2024). PV and Wind Distributed Generation System Power Quality Improvement Using Modular UPQC. Journal of Electrical Systems, 20(7s), 3905–3915.
13. Rafiqi, I. S., & Bhat, A. H. (2024). Enhancement of power quality using UPQC integrated with renewable energy sources through an improved sparrow search-based PID controller. International Journal of Advanced Technology and Engineering Exploration, 11(116), 955.
14. Khadse, D., & Beohar, A. (2024). Enhancement of power quality problems using DSTATCOM: An optimized control approach. Solar Energy, 268, 112260.
15. Suresh, M. M., Bai, G. M., Rushitha, T., SwethaVardhani, Y., Balaji, C., & Jyoshna, D. An Improvement of Additional Grid-Voltage Regulation using IUPQC Controller as a STATCOM.
16. Hernández-Mayoral, E., Madrigal-Martínez, M., Mina-Antonio, J. D., Iracheta-Cortez, R., Enríquez-Santiago,

J. A., Rodríguez-Rivera, O., Martínez-Reyes, G., & Mendoza-Santos, E. (2023). A Comprehensive Review on Power-Quality Issues, Optimization Techniques, and Control Strategies of Microgrid Based on Renewable Energy Sources. Sustainability, 15(12), 9847. https://doi.org/10.3390/su15129847.

17. Abdelkhalek, Othmane & Benachaiba, Chellali & Brahim, Gasbaoui & BRAHIM, & Nasri, Abdelfatah. (2010). USING OF ANFIS AND FIS METHODS TO IMPROVE THE UPQC PERFORMANCE. International Journal of Engineering Science and Technology. 2.
18. Mughni, M. B. A., Fayyaz, U., Rahman, T. U., & Khan, M. S. (2024). Power Quality Improvement of Hybrid System in Grid and Islanded Mode of Operation Using DSTATCOM BESS. Southern Journal of Engineering & Technology, 2(01), 45–62.
19. Rao, G. M., Reddy, M. P. P., Srinivas, N., Satyanarayana, V., Vinoth, K., & Goutham, M. (2024, April). Power Quality Improvement in Transmission System Using Optimal Location System with Integration of Distributed Power Flow Controller. In 2024 1st International Conference on Trends in Engineering Systems and Technologies (ICTEST) (pp. 1–7). IEEE.
20. Borra, S. R., & Balachandran, P. K. (2024). Optimal design of solar/wind/battery and EV fed UPQC for. Attainment of SDGs through the Advancement in Solar PV systems, 24.
21. REDDY, K. M., MADDILETY, S., RESHMA, S., & SRIKANTH, V. (2024). Modelling and Analysis of DVR and UPQC for Sensitive Load Power Quality Improvement in Distribution Network. www. pvpsiddhartha. ac. in.
22. Katta, P., Rajendiran, S. K., Joseph, D., Jeyson, J., Navaneetha, V., & Marshel, M. E. (2024, August). GWO optimized PV-battery based AI interfaced UPQC system for power quality improvement. In AIP Conference Proceedings (Vol. 3044, No. 1). AIP Publishing.
23. Lolamo, M., Kumar, R., & Sharma, V. (2024). Power quality improvement in renewable energy integrated grid: a review. International Journal of Power and Energy Conversion, 15(4), 352–375.
24. Shravani, C., & Narasimham, R. L. (2024). Synergetic UPQC Application for Power Quality Enhancement in Microgrid Distribution System: SCSO Approach. e-Prime-Advances in Electrical Engineering, Electronics and Energy, 10, 100794.
25. Chapala, S., Lakshmi, G. S., & Swarupa, M. L. (2024, February). Power Enhancement of Renewable Energy-Based Generation System Using UPQC. In 2024 International Conference on Social and Sustainable Innovations in Technology and Engineering (SASI-ITE) (pp. 113–118). IEEE.
26. Sabat, J., Mangaraj, M., & Barisal, A. K. (2024). Power quality enhancement in utility grid using distributed energy resources integrated BBC-VSI based DSTATCOM. International Journal of System Assurance Engineering and Management, 1–12.
27. Brahim, B., & Rachid, D. (2024). Power quality enhancement using STATCOM-Fuel Cell Energy Control. Przegląd Elektrotechniczny, (8).
28. Priya, M., & Prathyusha, C. (2024, July). A Novel ANFIS-Controlled Unified Power Quality Conditioner (UPQC) for Enhanced Power Quality. In 2024 IEEE 4th International Conference on Sustainable Energy and Future Electric Transportation (SEFET) (pp. 1–6). IEEE.
29. Bashir, J., Ahmad, S., & Anees, A. S. (2024). Unified Power Quality Conditioner (UPQC) Based on Multilevel Configurations. Multilevel Converters, 233–251.
30. Rao, K. S., Kumar, G. S., Saxena, A., & Karasala, C. (2024). MPC Based TLSC-DSTATCOM with VIKOR Approach to Reduce Switching Frequency and Capacitor Voltage Balancing. IEEE Transactions on Power Delivery.

Note: All the figures and tables in this chapter were made by the authors.

Intelligent Systems Using Semiconductors for Robotics and IoT – Dinesh Goyal et al. (eds)
 ISBN 978-1-041-20408-4

77 Improved Power Transfer Capability in Extremely Weak Grid Conditions through Adaptive Reactive Power Control of Photovoltaic Power Plants

Alvin Fatema Shaikh,
Prashant Thakre, Radharaman Shaha,
Pratik Ghutke*, Praful Ghadge, Ganesh Wakte
Tulsiramji Gaikwad Patil College of Engineering and Technology,
Nagpur

Abstract: Significant obstacles, including decreased power transmission capacity and system instability, arise when photovoltaic (PV) power plants are integrated into ultra-weak grid conditions. Low short-circuit ratios (SCRs) and the inability of these grids to keep voltage stable under changing operating circumstances are the major causes of these difficulties. Improved power transfer capacity and reduced instability under ultra-weak grid situations are the goals of this study's adaptive reactive power management (ARPC) technique for PV power plants. By using sophisticated control algorithms and methods for monitoring the system, the ARPC system is able to dynamically alter reactive power output in response to grid circumstances as they occur in real time. The method makes use of adaptive gain tuning, voltage sensitivity analysis, and grid impedance calculation to provide stable operation and optimal reactive power compensation. Even in the most severe cases of a weak grid, the suggested method enhances voltage stability, boosts power transfer efficiency, and avoids grid voltage collapse, according to both simulation and experimental findings. Ensuring stable and sustainable power distribution while helping the move to greener energy sources, this study offers a viable approach for integrating renewable energy systems into vulnerable networks.

Keywords: Photovoltaic (PV) power plants, Ultra-weak grids, Adaptive reactive power control (ARPC), Power transfer capability, Voltage stability, Short-circuit ratio (SCR), Grid impedance estimation, Sustainable energy systems

1. Introduction

Due to a combination of factors, including a steadily growing global economy and an inflating population, the demand for energy has skyrocketed in recent decades. Power produced by traditional energy sources, such as coal, oil, and gas, is sufficient to meet this heightened demand for energy[1]. Despite their widespread usage, fossil fuels have far-reaching environmental consequences, including the release of carbon dioxide into the atmosphere, which has the potential to accelerate the warming of the planet. In addition, there is a finite supply of fossil fuels, and their usage in power production is likely to deplete them rapidly. Renewable energy sources (RESs) are nonconventional energy sources that the world community has begun to prioritize in order to address these concerns. Distribution generation (DG) systems are seeing a surge in popularity for renewable energy sources (RESs), which include wind, hydro, geothermal, sun, marine, and biomass. Solar photovoltaic (PV) energy conversion stands out among these RESs as the top pick because to its abundance, ease of installation, sustainability, and lack of environmental impact[2][3] In 2020, out of a total renewable energy capacity of 27,99 GW, the world's installed PV capacity reached 760 GW. It is projected that by 2050, the world's energy output will be 11% powered by PV, which will

*Corresponding author: pvthakre2006@rediffmail.com

DOI: 10.1201/9781003716389-77

save 2.3 gigatonnes of carbon dioxide from being released annually. Increases in efficiency in production, advances in solar cell performance, and decreases in the cost of PV panels have all contributed to the growth and widespread use of PV generating. One of the most rapidly expanding RESs connected to the distribution grid is the PV system. Both industrialized and developing nations are increasingly erecting large-scale PV producing plants that are connected to the grid. Additionally, by 2022, the Indian government hopes to have 100 GW of solar PV production installed[4]. Fig. 77.1 is a schematic representation of a grid-tied photovoltaic system, which includes the PV array, power electronic converters, and utility grid. Environmental factors, such as solar irradiation and ambient temperature, determine the PV arrays' properties. Between the PV array and the utility grid, there are power electronic converters. Problems with connecting PV systems to the utility grid are the most significant issues with grid-tied systems. Power electronic converter control schemes are necessary to fulfill the sophisticated control features of PV systems in order to meet the needs for flexible power regulation.

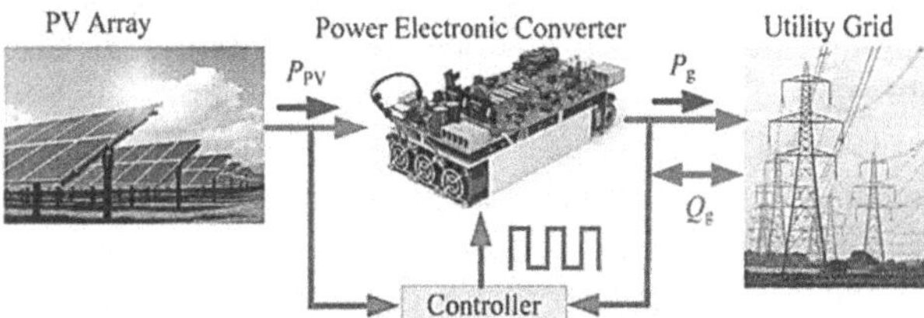

Fig. 77.1 Schematic diagram for the grid-tied PV system

Inverters convert direct current (DC) power from PV arrays into alternating current (AC) for the utility grid. Single-stage and double-stage grid-tied PV systems may deploy to the grid. Figure 77.2 depicts a high-level schematic of single-stage and double-stage grid-connected solar systems. DC-DC converters and DC-AC inverters power the two-stage grid-tied PV system. PV panel DC-DC converters enable MPPT. The DC-AC converter manages intermediate DC bus voltage by

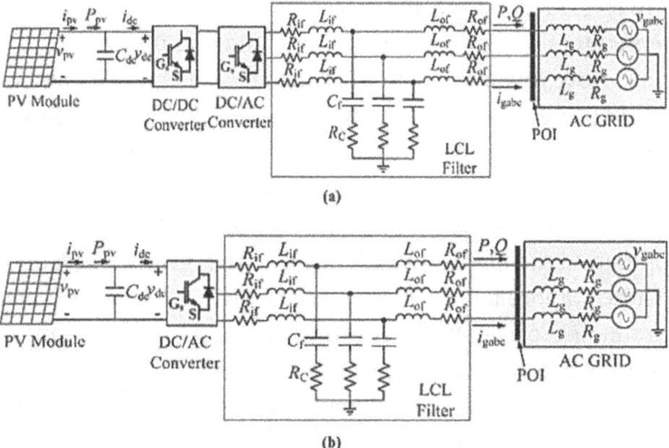

Fig. 77.2 System configuration of grid-tied PV system (a) double stage grid-tied PV system (b) single stage grid-tied PV system

injecting AC power into the grid[5]. However, the second DC-DC phase makes double-stage grid-tied PV systems less efficient, bulkier, and more costly. Since a single controller controls MPPT control, DC voltage regulation, and grid current, single-stage grid-tied PV system control design is complex. Single-stage grid-tied PV systems have several advantages, including low cooling costs, small size, low cost, compact design, and high efficiency. These advantages rise with PV system power production. These factors lead huge utility-scale PV power facilities to employ single-stage grid-tied PV technology.

As PV energy enters the grid, the PV management scheme faces additional issues such power quality in unusual grid situations, dependability in weak grid conditions and faults, unintended islanding, feeder overloading, etc. Grid-connected PV systems require adequate electricity. Power quality concerns including harmonics, oscillations, frequency fluctuations, and more are being studied in distribution grid-connected PV systems. Low power quality may cause line losses, inefficiency, higher production and maintenance costs, worse product quality, shorter equipment lifespans, production interruptions, and energy losses[6][7]. Many international power quality standards mandate grid-tied PV systems. Capital gain reduction from high-quality power benefits industry. Grid-connected PV system management requires innovation. Due to the growing number of power electronic devices and non-linear loads on the distribution grid, grid-tied PV systems face harmonic distortion. Improved power quality decreases harmonic distortion. Voltage sag and swell affect grid-connected PV quality. Start or repair large loads like synchronous motors might cause voltage sag. Turning off high loads and capacitors causes most voltage spikes. Solar systems reduce power spikes and dips, therefore utility grid standards have been revised [5]. Power companies sometimes disconnect PV arrays from the grid for seconds to fix difficulties. These technologies are ancient and will lower power quality due to PV power penetration increase. PV systems should stay connected to the grid-tied inverter during grid disruptions via low-voltage ride-through (LVRT). Power quality is maintained via PV system control during MPPT. It should operate the LVRT when the power goes off. PV inverters must provide grid reactive power for LVRT. International authorities created grid codes for transmission and distribution firms. Grid stability is increasingly dependent on PV power plants as they are integrated into electrical networks. Power transmission and system stability are difficult for PV power plants on weak or ultra-poor networks. Low SCRs and high system impedance cause voltage instability, power quality difficulties, and PV plant-grid synchronization issues. Technical issues arise from ultra-weak grids' lower SCRs and increased disturbance sensitivity. Traditional PV plant management may not guarantee energy supply. In these instances, reactive power regulation enhances PV transmission and voltage stability. In ultra-weak networks, reactive power regulation technologies meant for stronger grids may fail, causing instability or power loss[8].

1.1 Key Challenges In Ultra-Weak Grid Integration

1. Voltage Instability Large voltage fluctuations may be caused by even small disturbances or variations in PV plant power production in ultra-weak grids because of the high impedance and low SCRs. Voltage instability, caused by these variations, threatens the reliability of the PV power plant and the grid as a whole.
2. Power Transfer Limitations: Power transmission from PV plants to the grid is severely limited under very poor grid conditions.
3. Synchronization Issues: Due to the increased likelihood of frequency and phase changes, ultra-weak grids make it more challenging to keep the PV plant and the grid in synchronization.

1.2 The Role of Reactive Power in Grid Stability

Reactive power improves power system performance and voltage stability. Weak and ultra-weak networks need reactive power support since the grid can't absorb and manage voltage swings. Reactive power assistance from photovoltaic (PV) plants may stabilize voltage, minimize power losses, and enhance power transfer efficiency[7] [8]. However, inflexible reactive power regulation techniques are insufficient for a weak grid with fluctuating grid impedance and voltage. PV power facilities may utilize adaptive reactive power control to address these challenges.

2. Literature Review

Fawzi AlJowder et al (2024) Three control strategies are compared for their effects on transient stability, synchronizing torques, and damping torques of a radial power system with a single-stage PV power plant. Three operating modes include reactive power control (Q-V droop control), alternating current terminal voltage control, and reactive power control.

Kai Yu et al (2024) An optimization strategy reduces grid voltage variations and losses from large-scale PV station integration. The reactive power management model includes generation and load uncertainty. At the grid integration point, the PV station's active power and regional grid load demand are projected to define a voltage control objective.

Sarih Saad et al (2024) We provide a novel control method for the latest Moroccan grid code that improves PV-STATCOM use and distribution network integration. The STATCOM is usually installed between the PV array module and the PCC to the distribution network when employing traditional renewable generators. Active power injection, power factor correction, and dynamic grid support during disturbances and faults are provided by the PV-STATCOM. Power electronics control strategies often use PI controllers. MATLAB Simulink is used to simulate local legislation's technological limits and gives the best results.

Zhong Cheng et al (2024) This paper provides an effective active-reactive power regulation system for distribution networks with solar EV energy storage. Reactive power multi-objective optimization of the distribution network and solar active power leveling are achieved. This method uses multi-objective niching particle swarm optimization to prevent the algorithm from converging too quickly and ensure a varied selection of solutions. Simulation of IEEE33 nodes proves the technique works..

3. Methodology

Following these steps will help you achieve adaptive reactive power management of photovoltaic (PV) power plants on a very weak grid. This technique develops, deploys, and verifies an adaptive control system to increase PV power transfer capacity and voltage stability.You must consider actual grid conditions and impediments to understand why renewable energy sources' potential is not fully used. Most European public power networks are vast central. Grids may carry huge conventional power plant electricity. Aggregation is possible by linking transmission and distribution networks using transformers. The client comes last. The interconnection distributes electricity to connected clients directly or indirectly. Current in the distribution grid lowers voltage. Decrease depends on cable resistance and inductance. Both grow with cable length. To ensure all customers have enough power, a transformer modestly increases cable voltage at the start. Renewable energy generation requires installing the plant in high-energy areas with high wind velocity. The installations are grid-connected at many local locations. Unlike big power plants, renewable power facilities are integrated at lower grid tiers and have less capacity. When dispersed generators connect to low-voltage connections, electricity may flow toward the transformer. Increasing the number of distributed generators in rural areas with poor power infrastructures raises voltage. Lack of system capacity for renewable energy is the main challenge. The integration of dispersed energy producers would increase mains operation demands, which affects the grid's operation. These approaches harm materials, efficiency, and producing prices while boosting fossil fuel consumption and carbon dioxide emissions. Preventing voltage rises requires partial grid expansion. Grid reinforcement may be costly. A prevalent misperception is that renewable energy generation costs more than it benefits. Alternative power providers report being unable to verify the operator's grid connection's technical and financial facts due to grid capacity uncertainty. Distribution System Operators distribute electricity. Whether a DSO is unbiased toward independent renewable energy suppliers is questionable when the electricity producer creates alternative energy initiatives. Due to lack of transparency, grid connection approval takes a lengthy time. Results show that stakeholders see renewable energy grid bottlenecks positively..

3.1 System Modeling

One builds a complex mathematical model of the grid and PV power plant. This model incorporates the dynamics of the inverter, the properties of the PV array, and the electrical behavior of the ultra-weak grid, which is defined by low short-circuit ratios (SCR) and high impedance. To guarantee precise dynamic representation, the model integrates grid characteristics that are measured in real-time, including voltage, frequency, and impedance. A number of simulations are performed under various ultra-weak grid settings to evaluate the efficacy of the suggested adaptive control technique.

4. Conclusion

Photovoltaic (PV) power plants are difficult to integrate into ultra-weak grids because to voltage instability and low SCRs. This study introduced ARPC adaptive reactive power regulation, which lets PV power plants adjust reactive power generation in real time to grid conditions. The recommended approach combines advanced techniques including grid impedance estimations, voltage sensitivity analysis, and adaptive gain tuning to solve these issues. By optimizing reactive power adjustment, the proposed technology allows PV power plants to function consistently under challenging grid situations, promoting renewable energy integration. Future study may focus on extending this technique to multi-PV plant circumstances and integrating it with other advanced grid-support technologies to increase grid resilience and efficiency.

References

1. F. ALJowder, "Analysis of Damping and Synchronizing Torques and Transient Stability of Power System with PV Power Station Operating in Different Control Modes," 2024 12th International Conference on Smart Grid (icSmartGrid), Setubal, Portugal, 2024, pp. 647–653, doi: 10.1109/icSmartGrid61824.2024.10578204.
2. K. Yu, Z. Shi, Y. Zang, Q. Lu, Y. Xiang and H. Lai, "Reactive Voltage Control in Photovoltaic Power Stations Considering Source-Load Uncertainty," 2024 IEEE 6th Advanced Information Management, Communicates, Electronic and Automation Control Conference (IMCEC), Chongqing, China, 2024, pp. 1934–1938, doi: 10.1109/IMCEC59810.2024.10575119.
3. S. Saad, B. Zakaria, C. Samira, E. Abelhadi and T. Abdelouahed, "Improving the distribution network's power quality using PV-STATCOM in compliance with Moroccan Grid Code regulations," 2024 10th International Conference on Control, Decision and Information Technologies (CoDIT), Vallette, Malta, 2024, pp. 342–347, doi: 10.1109/CoDIT62066.2024.10708435.
4. Z. Cheng, Y. Ma and D. Guo, "Research on Active and Reactive Power Cooperative Optimization of Distribution Network Including Photovoltaic, Energy Storage and EV," 2024 IEEE 7th International Electrical and Energy Conference (CIEEC), Harbin, China, 2024, pp. 4741–4746, doi: 10.1109/CIEEC60922.2024.10583251.
5. F. Niu, G. Xu, J. Zhang, Y. Liu and W. Wang, "Research on Active and Reactive Power Coordination Control Strategy for Overvoltage of PV Distribution Network in High Penetration Area," 2022 5th International Conference on Power and Energy Applications (ICPEA), Guangzhou, China, 2022, pp. 13-18, doi: 10.1109/ICPEA56363.2022.10052609.
6. K. A. Tiwari, A. Dubey, P. Selvaraj and S. K. Panda, "Implementation of Active and Reactive Power Control in a Novel Solar Power Plant Controller Solution," 2021 9th IEEE International Conference on Power Systems (ICPS), Kharagpur, India, 2021, pp. 1–6, doi: 10.1109/ICPS52420.2021.9670014.
7. Y. Libin, L. Yanhe, X. Jinhua and W. Deshun, "Research on Regional Power Grid Voltage Control Technology Based on Reactive Power Support Capability of Controllable Power sources," 2020 IEEE Sustainable Power and Energy Conference (iSPEC), Chengdu, China, 2020, pp. 889–894, doi: 10.1109/iSPEC50848.2020.9350947.
8. D. Yang, X. Wang, F. Liu, K. Xin, Y. Liu and F. Blaabjerg, "Adaptive Reactive Power Control of PV Power Plants for Improved Power Transfer Capability Under Ultra-Weak Grid Conditions," in IEEE Transactions on Smart Grid, vol. 10, no. 2, pp. 1269–1279, March 2019, doi: 10.1109/TSG.2017.27623.

Note: All the figures in this chapter were made by the authors.

Intelligent Systems Using Semiconductors for Robotics and IoT – Dinesh Goyal et al. (eds)
© 2026 Taylor & Francis Group, London, ISBN 978-1-041-20408-4

78 Integrating Artificial Intelligence to ANPR towards Road Safety

Aditya Joshi[1], Monali Gulhane[2], Nitin Rakesh[3]
Department of CSE, Symbiosis Institute of Technology,
Nagpur Campus, Symbiosis International, (Deemed University),
Pune Nagpur, India

Akhil Gupta[4]
Symbiosis Institute of Technology,
Nagpur Campus, Symbiosis International (Deemed University),
Pune, Maharashtra, India

Mandeep Kaur[5]
Professor,
School of Computer Science Engineering and Technology,
Bennett University, Greater Noida,
Uttar Pradesh, India

Saurav Dixit[6]
Centre of Research Impact and Outcome,
Chitkara University, Rajpura,
Punjab, India

Abstract: The huge spike in AI and related technologies has extended the possibilities of development to infinity. An increase in the possibilities of development also increases the risk and insecurity among the people. In this initiative, we attempt to lessen the danger of death on the road and try to provide a solution based on AI and the available technology of the Automatic Number Plate Recognition system and related modules to create a product that identifies the threats and alarms the local bodies for the same. Later the analysis of both types of possible solutions and a detailed comparison of them and finding the most suitable solution. The conclusion talks about how this can be implemented in the real world and in real-time.

Keywords: Automatic number plate recognition, Artificial intelligence, Optical character recognition, Radio frequency identification

1. Introduction

Today artificial intelligence has taken over the world at an exponential rate. The extensive use of AI has led to many new possibilities of how logically we can solve a problem rather than trying to write down the actual code for it. Going by the literal meaning of AI here we try to simulate the human level of intelligence to be performed by a machine. Now this simulation might help us to identify some of the potential or current threats in our daily lives which we hope to resolve by a dependable machine that may help us feel safe under particular situations incorporating the applicable criteria that follow. Assume you're driving and a system that identifies vehicle

[1]jaditya2020@gmail.com, [2]monali.gulhane4@gmail.com, [3]nitin.rakesh@gmail.com, [4]akhilgupta112001@gmail.com, [5]drmandeepkaur10@gmail.com, [6]sauravarambol@gmail.com

DOI: 10.1201/9781003716389-78

license plates exists in place. This AI-powered system is designed to detect potential traffic risks, such as a careless driver or a vehicle participating in illegal behaviour. When these potential threats are identified, the system notifies local authorities, allowing them to take appropriate action to guarantee everyone's safety on the road. In this portion, we are trying to answer the following questions. Mainly the implementation of AI and IoT. How it can be integrated and how we can assure safety on the ground using this combination over a surveillance camera. The base idea is, to use object detection to identify the humans and the vehicles passing by the street or a particular road by AI and machine learning models, to train the model and to provide the necessary action that can be taken, here if it detects a criminal activity, it directly approaches to the local police station by sending a notification for the particular situation. This part gives the application of AI.

The actual data that should be analyzed by the AI tool is provided to it using an infrared camera and surveillance cameras which are ANPR (automatic number plate recognition cameras) which are used to analyzed the over speeding of the vehicles and provide the e-chalan to the person breaking the rules. This technology which is already in use can be integrated with the help of AI and training models for street lighting and crime detection. This answers the implementation of IoT. Here to ensure the safety of pedestrians and especially women on the road the cameras need to be set up in the streets so that these can be used as an alarm system. Here we talk about a number of aspects and the required modules required to achieve this goal. The main module is a camera-based or ANPR-like camera and then maps their range. The required modules/ technology are an interface for the software module, a database module, a live tracking module for real-time data, an image recognition and analysis module, a notification module, and a number of additional modules.

2. Literature Review

The development of the idea started with the invention of cameras. We have reached a point where cameras are inevitable in our daily lives, from clicking photographs to security they all have a different role and a particular technology related to it. Here the goal is to achieve road safety by recognising the threats of the pedestrian. This involves to the security perspective from the point of view of the pedestrian, the vehicle, the vehicle to the pedestrian, and the pedestrian to the vehicle. Talking about technological advancement, the basic requirement of the cameras needed for solving this problem are cameras with high resolution and a system that is able to achieve this is ANPR or ALPR. Camera system which already exists and is in implementation. A need for advancement in this system to achieve this goal. The working of proposed model is based on AI and camera, Algorithms based on object identification [11]. According to [13] the algorithm is divided into three categories one to capture the image, extract the plate, and recognition. The flow diagram proposed by [13] is given in Fig. 78.1, which gives an overall look at how the ANPR cameras work. A comparative study done by [16] helps us to understand the different works based on the present knowledge [14] A new approach to add machine learning and a systematic background is proposed using an optical character recognition system. The structure of the review process acts as the building block for the development of this technology[15]. The flow chart given below Fig. 78.1, shows the sequence/flow of the recognition of the data (here number plate). Step 1: capture the image and send it to the database to process. Step 2: Now we have the data we need to recognize the pixels related to the object and differentiate between the background/contour and the object (which is moving).

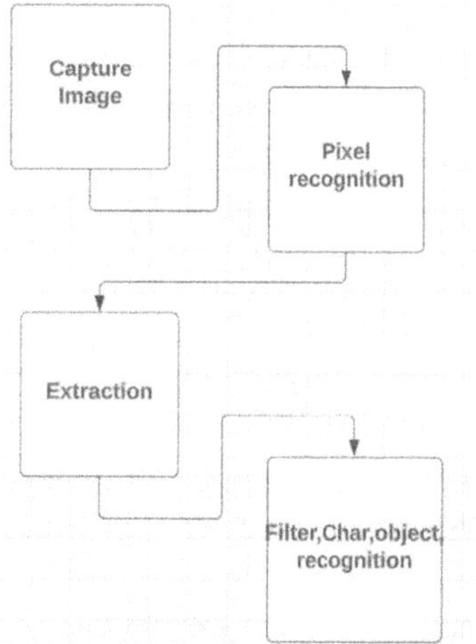

Fig. 78.1 Flow diagram for identification

Step 3: Results of the processed data need to be extracted. This process sends the extracted data of the images. Step 4: This step involves the implementation of AI for the recognition of objects and the required filter needed to convert the data such that it can be recognized by AI for the recognition of objects and the required filter needed to convert the data such that it can be recognized by AI. For additional tracking and identification, the use of RFID technology for Correct identification and monitoring is made easier by placing RFID tags on vehicles and objects, particularly in situations where visual recognition may be compromised (e.g., bad weather, low visibility). As now we have the ANPR system, some data should be used as input to the database. This means that technological advancement to understand the data through images is available. Which leads to object identification as a need for the real-time fetching of data. The problem here is that for safety on roads, only one object being tracked at a time is not useful. Multiple object tracking is a necessity. The multiple objects tracking system helps to identify the number of objects in the frame at that particular time. It is based on algorithms which are particle filter, sparse separation, and hash [1]. Video Acquisition system to actually map the road/street according to the frame with respect to the screen of the camera and the radio signal strength indication concept as described in [3]. Terrain identification and MTSC tracking required for the multi-cam identification are integrated from [4]. Multi-source object detection talks about the algorithms required and the overall structure and architecture or the flow of the software. The design of the tracking module has the source from the Kalman filter. It tracks the real-time data and the data that needs to be stored in a database. We have defined the problem and the related works already done in this direction now let's find a suitable solution to the problem.

3. Challenges Faced

The development of new technology has several challenges addressed below:

- The major challenge is to identify the problem. Problem identification is the first step for solving any problem.
- The available technology in the market, to make it easier to develop the desired solution over present technology. Here we found out the use of ANPR. The solution we provide has the use of ANPR cameras as the resolution of the cameras is useful for tracking objects.
- The next challenge is the use of OCR and RFID in a single system and implementing it practically and its analysis.
- Implementation of AI in the existing ANPR models.
- The crucial role of the location of the new cameras. To identify the location of the cameras. Find where it needs to be placed.
- The durability of the cameras and terrain identification.
- The main problem to is the placement of AI technology. Here we have tried to provide the two methods in the proposed model. The first method is the use of AI in the computer/output node where the data is processed.
- Challenges related to this process are the time required to provide the output would be more and the accuracy might be affected.
- This system is economically feasible but not that efficient.
- Hence the use of AI at the camera level, here we face the issue of the money required and the setup to manufacture these cameras.
- Safety on the road is the issue here, but corresponding to it, for this particular model, ensuring the safety of the camera itself is also necessary.
- The capital required in the manufacture of cameras is increased. It also includes the installation of cameras.

4. Gathering Solutions

In this section we would try to answer the problems listed above using our proposed model. Identification of the problem: "safety of people is necessary", we can look at this statement with the point of view of an individual. As a man he would be very efficient in every situation in life. With the help of new artificial intelligence technology, we can depend on the machines for security as it does not have a human involved and can provide justified output. Here we have identified the problem. ANPR (Automatic Number Plate Recognition) a model used for tracking the vehicles and capture their licence plate has a resolution required that can be used to identify the objects, as it is used for the detection of the license plates of the vehicles. OCR (Optical character recognition) is the method currently used in ANPR cameras. OCR is a method used for extracting information from images and captured videos which can be converted in to pdf and text so the data entry. This is currently used for data entry which does not require human for manually entering data. Combining OCR technology with AI will enhance the security systems. The processed output of the OCR system can be validated by AI and can be used to enhance the accuracy of the input data. Implementing AI to track objects which may possibly related with criminal activity, to ANPR using a combination of OCR and identify the crime. This needs to be authorised by the government body and should be in hands of the authorities. Location of the cameras: identifying the places through the data available of criminal records that are available. cameras need to be located in these areas so that identification of the person is easily possible. This needs to be achieved by terrain and contour recognition [2]. This uses field imaging and related algorithms. Light field imaging, ROI extraction methods can be used to track the pedestrians [5]. Analyzing two possible ways for implementation of the newly developed technology and finding a perfect place to fit in. Live tracking module has challenges different from the processing of stored data to provide output. Maximization of efficiency of the device can be achieved by multitasking techniques. Adding more techniques for algorithms and continuous updating the system. Adding more sensors to the cameras to make the input more accurate and validated. Then finding an economically feasible solution so that it is practically possible to implement. Based on the calculation of data provided and time required to process it we need to find a solution that uses the current production line and with minimum changes we should be able to manufacture products.

5. Proposed Model

ANPR is a device that scans the license plate number of the vehicle from images of the license plate using optical character recognition (OCR). It makes use of various image processing methods to precisely and quickly identify vehicles in images or in real-time videos recorded by a number of cameras. To solve the above-mentioned problem/to ensure more security on roads a model (here a camera) which has its base functionality of ANPR model and combining the new AI technology with it so it can act as a real-time functioning model which has combined functionality of recognition and the support of AI to ensure real-time security based on the Optical recognition framework. The base solution is to use the ANPR cameras and implement the AI technology of Multiple object recognition and tracking. The required architectural framework would be implemented as Radio Frequency Identification (RFID) technology helps to track the object and also identify the object. The design of the system would have sections related to its function given in Fig. 78.2 below. The proposed security system for roads integrates Automatic Number Plate Recognition (ANPR), Artificial Intelligence (AI), and Radio Frequency Identification (RFID) for real-time object recognition and tracking. This system aims to enhance road security by providing a comprehensive solution for vehicle identification, traffic management, and object tracking. The detection pattern can be shown in Fig. 77.3a and Fig. 78.3b.

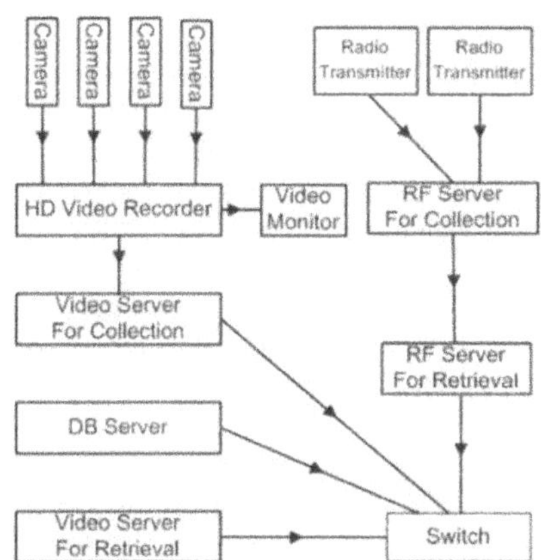

Fig. 78.2 Architecture for overall module [15]

Here there are two possible ways to achieve this the first way is to set, AI to the node where all the data from the cameras is stored, and process the data at the system which is getting the output of the cameras. Process it and respond to the actions accordingly. And store data in a database. The other way is we have the setup of AI built inside the device so that the processed output is received at the output node. This is visualized in Fig. 78.4 below. Visualizing this to multiple inputs to a single system and its challenges are referred to in the next section. The solution needs the integration of multiple systems and techniques. MCMT, multi-camera multi-target tracking has four steps

Step1: Generate individual tracks in one video and set it with the pipeline for multiple objects tracking. Step 2: Generate the appearance distance matrix. Step 3: calculate the transitional interval of the matrix. Step 4: combine the mages and matrix to get a final output.[8]. Recognition of

(a)

(b)

Fig. 78.3 (a) Detection I [2], (b) Detection II

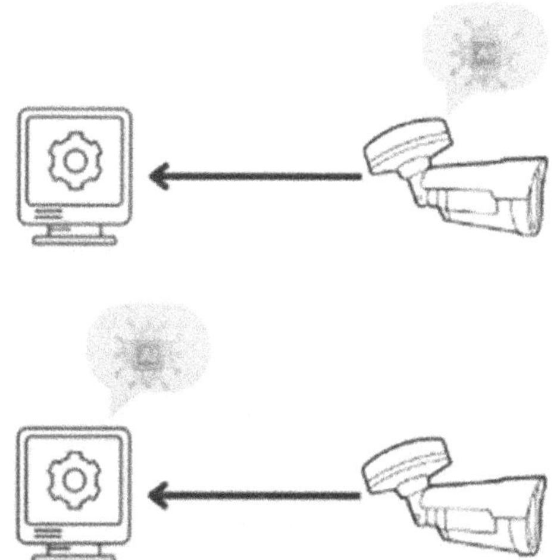

Fig. 78.4 Two principles for the design of the module

the terrain and model to train the cameras such that the images are providing the activity based on the movement of the pixels.3D object detection through contours. Now the expected outcome from the camera should look something like Fig. 78.4 and here the data is processed in real-time and then the output is sent to the output node when it detects the object.

6. Analysis

Analyzing the solution provided here we can think of a camera and the related processes going on within two paths. First comparing the two methods refer Fig. 78.1. The second Fig. 78.2 has the AI technology at the output node which analyses the output on the computer/system at the output node. This system has its benefits as the input is taken raw and processed in one place. This makes the production cost cheaper for the cameras. Taking into consideration the time required to actually read data and then process it by using a single system may lead to some serious delay in output. This would make the system inefficient. Also, the requirement of the solution is not met completely. But this is a type of solution and cannot be ignored although the results are delayed. Considering the Fig. 78.1 first type of method where the data is actually processed by the AI at the input level and then sent to the output node. When the basic data is processed by AI at the input level the output node only has to fetch data and produce output according to the input. Here the load on the system calculating the actual results is very low as it has filtered or processed data. This method has the consequence that the production cost of the cameras would increase. But the system would get more efficient. Another lookout is that the cameras needed to be installed in the streets so their production costs need to be cheap also, they can be vulnerable to thieves but the AI detection would solve that issue by tracking its location. We already have a setup for the ANPR cameras in implementation. This production could be increased with the corresponding advancement in technology to achieve this goal.

7. Conclusion

The study highlights the importance of using artificial intelligence, specifically in combination with ANPR cameras, for better road safety. The utilization of artificial intelligence (AI) for real-time object identification and tracking has been considered an essential step in limiting potential risks while offering faster responses to safety-related issues. The discussion, however, highlights the value of solving associated problems, such as financial implications and potential security concerns, when deploying such advanced technologies. Here we tried Integrating AI to achieve Road Safety by the use of ANPR cameras and suitable technologies and algorithms based on the previous studies of AI, Object identification, Multiple object tracking and recognition, etc. Based on the knowledge we implemented it on developing a proposed model. With the help of a flowchart, we understood the workings of the technology and the steps required to fetch the data and store it in the database. Here we got to know that there can be two aspects of solving the problem on the basis of where the technology is used—one on the camera and the second on the output node which calculates the result. We analysed the proposed model on the basis of cost and efficiency and potential threats.

References

1. B. Sun, Z. Liu, Y. Sun, F. Su, L. Cao, and H. Zhang, "Multiple Objects Tracking and Identification Based on Sparse Representation in Surveillance Video," 2015 IEEE International Conference on Multimedia Big Data, Beijing, China, 2015, pp. 268–271, doi: 10.1109/BigMM.2015.69.
2. O. Styles, T. Guha and V. Sanchez, "Multi-Camera Trajectory Forecasting With Trajectory Tensors," in IEEE Transactions on Pattern Analysis and Machine Intelligence, vol. 44, no. 11, pp. 8482–8491, 1 Nov. 2022, doi: 10.1109/TPAMI.2021.3107958.
3. D. Grzechca, T. Wróbel and P. Bielecki, "Indoor Location and Identification of Objects with Video Surveillance System and WiFi Module," 2014 International Conference on Mathematics and Computers in Sciences and in Industry, Varna, Bulgaria, 2014, pp. 171–174, doi: 10.1109/MCSI.2014.52.
4. Z. Tang et al., "CityFlow: A City-Scale Benchmark for Multi-Target Multi-Camera Vehicle Tracking and Re-Identification," 2019 IEEE/CVF Conference on Computer Vision and Pattern Recognition (CVPR), Long Beach, CA, USA, 2019, pp. 8789–8798, doi: 10.1109/CVPR.2019.00900.
5. P. Köhl, A. Specker, A. Schumann, and J. Beyerer, "The MTA Dataset for Multi Target Multi Camera Pedestrian Tracking by Weighted Distance Aggregation," 2020 IEEE/CVF Conference on Computer Vision and Pattern Recognition Workshops (CVPRW), Seattle, WA, USA, 2020, pp. 4489–4498,doi10.1109/CVPRW50498.2020.00529.
6. A. A. Y. Mustafa, L. G. Shapiro, and M. A. Ganter, "3D object recognition from color intensity images," Proceedings of 13th International Conference on Pattern Recognition, Vienna, Austria, 1996, pp. 627–631 vol.1, doi: 10.1109/ICPR.1996.546100.
7. N. Payet and S. Todorovic, "From contours to 3D object detection and pose estimation," 2011 International Conference on Computer Vision, Barcelona, Spain, 2011, pp. 983–990, doi: 10.1109/ICCV.2011.6126342.
8. W. Feng, D. Ji, Y. Wang, S. Chang, H. Ren, and W. Gan, "Challenges on Large Scale Surveillance Video Analysis," 2018 IEEE/CVF Conference on Computer Vision and Pattern Recognition Workshops (CVPRW), Salt Lake City, UT, USA, 2018, pp. 69–697, doi: 10.1109/CVPRW.2018.00017.
9. C. Jia, F. Shi, Y. Zhao, M. Zhao, Z. Wang and S. Chen, "Identification of Pedestrians From Confused Planar Objects Using Light Field Imaging," in IEEE Access, vol. 6, pp. 39375–39384, 2018, doi: 10.1109/ACCESS.2018.2855723.
10. S. Joung, S. Kim, M. Kim, I. -J. Kim and K. Sohn, "Learning Canonical 3D Object Representation for Fine-Grained Recognition," 2021 IEEE/CVF International Conference on Computer Vision (ICCV), Montreal, QC, Canada, 2021, pp. 1015–1025, doi: 10.1109/ICCV48922.2021.00107.
11. C. Buizza, T. Fischer and Y. Demiris, "Real-Time Multi-Person Pose Tracking using Data Assimilation," 2020 IEEE Winter Conference on Applications of Computer Vision (WACV), Snowmass, CO, USA, 2020, pp. 438–447, doi: 10.1109/WACV45572.2020.9093442.
12. Lubna, Mufti N, Shah SAA. Automatic Number Plate Recognition:A Detailed Survey of Relevant Algorithms. Sensors (Basel). 2021 Apr 26;21(9):3028. doi: 10.3390/s21093028. PMID: 33925845; PMCID: PMC8123416.
13. M. T. Qadri and M. Asif, "Automatic Number Plate Recognition System for Vehicle Identification Using Optical

Character Recognition," 2009 International Conference on Education Technology and Computer, Singapore, 2009, pp. 335–338, doi: 10.1109/ICETC.2009.54.

14. S. Babbar, S. Kesarwani, N. Dewan, K. Shangle and S. Patel, "A New Approach for Vehicle Number Plate Detection," 2018 Eleventh International Conference on Contemporary Computing (IC3), Noida, India, 2018, pp. 1–6, doi: 10.1109/IC3.2018.8530600.
15. J. Shashirangana, H. Padmasiri, D. Meedeniya and C. Perera, "Automated License Plate Recognition: A Survey on Methods and Techniques," in IEEE Access, vol. 9, pp. 11203–11225, 2021, doi: 10.1109/ACCESS.2020.3047929.
16. I. S. Ahmad, B. Boufama, P. Habashi, W. Anderson and T. Elamsy, "Automatic license plate recognition: A comparative study," 2015 IEEE International Symposium on Signal Processing and Information Technology (ISSPIT), Abu Dhabi, United Arab Emirates, 2015, pp. 635–640, doi: 10.1109/ISSPIT.2015.7394415.
17. X. Wang and J. Wen, "Architectures and Algorithms for Multi-source Object Detection and Tracking," 2009 Fifth International Conference on Natural Computation, Tianjian, China, 2009, pp. 17565–568, doi: 10.1109/ICNC.2009.643.
18. https://www.tmz.com/2023/03/01/gunman-shoots-homeless-man-video-st-louis/
19. Junqing Tang, Li Wan, Jennifer Schooling, Pengjun Zhao, Jun Chen, Shufen Wei,Automatic number plate recognition (ANPR) in smart cities: A systematic review on technological advancements and application cases,Cities, Volume129, 2022, 103833, ISSN02642751, https://doi.org/10.1016/j.cities.2022.103833.(https://www.sciencedirect.com/science/article/pii/S0264275122002724)

Note: All the figures in this chapter were made by the authors.

Intelligent Systems Using Semiconductors for Robotics and IoT – Dinesh Goyal et al. (eds)

79 AI-Driven Eye Strain Detection: A Non-Intrusive Approach for Employees Using Digital Screens

Dhanashri Mahapatra[1], Vinay Mowade[2], Anjali Ganthade[3], Monali Gulhane[4], Nitin Rakesh[5], Pratik Agrawal[6]

Department of CSE, Symbiosis Institute of Technology, Nagpur Campus, Symbiosis International, (Deemed University), Pune Nagpur, India

Mandeep Kaur[7]

Professor, School of Computer Science Engineering & Technology, Bennett University, Greater Noida, Uttar Pradesh, India

Abstract: Digital eye strain (DES), characterized by symptoms including dry eyes, irregular blinking patterns, and visual tiredness, has increased due to the growing use of digital gadgets, particularly in the COVID-19 pandemic. This problem is addressed by the proposed Smart Eye Strain Detection System, which uses real-time video analysis to identify signs of eye strain through an expanded set of facial features, which goes beyond traditional markers like head movements and blinking. These features include droopy eyelids and glabellar length changes. Support Vector Machines (SVM) are used by the system to classify data based on statistical characteristics connected to behaviors related to stress. The YawDD dataset is used to validate the method, which shows notable gains in detection accuracy. With the goal of giving users timely interventions in today's screen-dominated world, this work provides a useful, non-intrusive method for diagnosing eye strain.

Keywords: Eye strain detection, Eye aspect ratio (EAR), Blink detection, Digital eye strain (DES), Eye fatigue

1. Introduction

Digital Eye Strain (DES) is a condition marked by symptoms including dry eyes and visual tiredness. It has become more common due to the increased usage of digital devices, particularly smartphones, over the past 10 years and the increased dependency during the COVID-19 epidemic. Previous research has mostly concentrated on using head movements and blinking to identify strain; however, the Smart Eye Strain Detection System incorporates other face cues, such as drooping eyelids and changes in glabellar length, in an effort to increase detection accuracy. With real-time video analysis powered by Support Vector Machines (SVM), the system recognizes behaviors associated with stress and promptly provides feedback to assist users reduce eye strain and preserve visual health in a world where screens dominate people's lives.

2. Literature Work

Over the past ten years, there has been a noticeable growth in the use of digital devices, particularly smartphones. Furthermore, a large portion of effort has been moved toward digital device-assisted applications due to the COVID epidemic. Today, people of all ages spend a lot

[1]dhanashri.mahapatra.batch2021@sitnagpur.siu.edu.in, [2]vinay.mowade.batch2021@sitnagpur.siu.edu.in, [3]anjali.ganthade.batch2021@sitnagpur.siu.edu.in, [4]monali.gulhane4@gmail.com, [5]nitin.rakesh@gmail.com, [6]pratik.agrawaal@gmail.com, [7]mandeep.kaur@bennett.edu.in

DOI: 10.1201/9781003716389-79

of time in front of these devices. This suggests a rise in instances of Digital Eye Strain, another newly recognized health concern. Researchers have connected this issue to symptoms including a change in blinking rhythm, dry eyes, and visual tiredness, among others. While earlier studies on facial features have focused on head movements, yawn detection, and blinking patterns, the proposed research work has identified other facial expressions, such as droopy eyes and reduction in glabellar length, as relevant features for this study to improve accuracy. This article aims to identify stress-related behaviors in users so they can take appropriate action in time. Real-time video recordings of users under stress are classified using Support Vector Machines (SVM) using a supervised approach based on statistical variables associated with specified symptoms. The primary discovery is a specific feature set that includes four additional relevant characteristics from earlier theoretical research in addition to two newly suggested features. Testing the proposed approach on YawDD, the best available dataset for our use case, shows significant improvement in accuracy.[1]

A real-time algorithm is proposed to identify eye blinks in video sequences captured by ordinary cameras. Modern landmark detectors trained using datasets collected in the field show remarkable adaptability to changes in lighting, facial expressions, and the orientation of the head relative to the camera. We demonstrate that the landmarks may be identified with sufficient accuracy to determine the eye opening's level with confidence. As a result, the suggested method determines the landmark locations and extracts the eye aspect ratio (EAR), a single scalar number that describes the eye opening in each frame. Ultimately, eye blinks are identified by an SVM classifier as a pattern of EAR values within a brief temporal frame. On two common datasets, the straightforward technique performs better than the state-of-the-art findings.[2]

This study presents a unique approach to blink detection that maintains accuracy in the face of position variations. Our methodology improves resilience through four important breakthroughs, in contrast to previous methods that struggled with position fluctuations and concentrated on frontal pictures. Firstly, the AdaBoost face detector and a LKT-based technique to manage position changes are used for the detection of face and eye regions. Second, illumination normalization is used to provide consistency in two parameters that are used for blink detection: the height-to-width ratio of the eye and the cumulative difference in black pixels across pictures. Third, to increase eye state accuracy, these characteristics are integrated using a support vector machine (SVM). Lastly, for best results, the SVM classifier adjusts to face rotation. [3]

A suggested real-time technique locates face landmarks accurately in a single monocular picture. The approach is expressed as an optimization problem, where a generic (rather than a person-specific) 3D model is fitted to maximize the sum of the responses of local classifiers with regard to the camera posture. The algorithm finds the image's facial landmarks and concurrently assesses the location and orientation of the head. We demonstrate that the basin of attraction is big enough to be initiated by a scanning window face detector, despite being local. Additional tests on benchmark datasets show that the suggested approach works better than a cutting-edge landmark detector, particularly for non-frontal face photos, and that it can provide steady and dependable tracking for a sizable number of viewing angles.[4]

This study proposes a new auto feedback system based on PTZ camera and face and eye detection technology that is utilized in smart classrooms. By providing them with immediate feedback, the technology helps solve the problem of pupils being inattentive throughout the learning process. In this study, we capture student photos using a PTZ camera, identify each student using face identification, determine learning status using eye detection, and give immediate feedback over a Wi-Fi environment. Our research's primary contribution is the integration of face and eye detection technologies into smart classroom environments, which offers a way to inform students when they are being sidetracked and free up teachers to concentrate on instruction. . [5]

In recent years, the number of people using multimedia has increased. Prolonged exposure to these streams can cause eye irritation, leading to dry eyes, headaches and blurred vision. The proposed method uses a transfer learning model with a convolutional neural network to identify eye fatigue from retinal images. Use pre-trained models such as Inception V3 and Resnet152V2 architectures to train and test on images to diagnose eye conditions. Both designs combine each transfer learning model with an optimization method such as Adam to classify the disease stage. By comparing the total number of epochs trained, the model performance was finally concluded. ResNet152V2 equipped with the Adam's optimization approach has outperformed others by increasing accuracy and decreasing loss. When compared to other conventional models, the Inception v3 with Adam optimizer produced high differentiation accuracy. [6]

These days, people's everyday lives are dominated by portable electronics. While they can survive without a computer or TV, they cannot function without a smartphone or tablet. This publication offered research for the identification of eye tiredness based on observations of this phenomena. The suggested eye tiredness assessment technique can determine whether a user has dark circles or frequent blinking of the eyes. Blinking a lot indicates that the user's eye has been straining and that it needs to rest to recover. Dark circles are usually a sign of little sleep or maybe disease; to maintain health, the suggested method can remind users to obtain more sleep in order to recuperate. Real-time recognition outcomes are achievable with the suggested system's minimal processing complexity. According on experimental findings, the suggested design can accurately identify and categorize the user's eye tiredness. [7]

Global education has been disturbed by the Covid-19 pandemic, which has boosted the use of digital devices and raised the number of eye health problems among pupils. The ocular health of young students is being strained by excessive screen usage and little outside exposure, which may lead to diseases like myopia. In order to tackle this issue, we present EyeNet, a software solution that is lightweight and uses eye image tracking to identify indications of eye illnesses. With ResNet-18 for knowledge distillation and optimization from the MobileNetV2 model, EyeNet attains 87.16% accuracy. It is compatible with embedded systems and may be used by students because it is made to run on low-end devices. [8] This work presents the use of dye-sensitized photovoltaic cells to detect eye blinks as a means of assessing tiredness. Specifically, the sensors were placed at the lateral side of the eye and fastened to the temple of the spectacles. They are wearable, didn't significantly impair the user's vision, and could identify the user's eye condition or eyelid position. Analyzing the blink detection accuracy allowed for the experimental investigation of the ideal sensor site. We used the wearable system—smart glasses—that we had built to conduct tests on tiredness measurement. The frequency, length, and speed of eye blinks were among the metrics that were extracted to create fatigue indices. The suggested system's ability to successfully measure weariness will be very helpful for maintaining both physical and mental health and optimizing performance. [9] Driver fatigue is a major factor in many accidents that occur these days. The suggested approach uses image processing to identify eye tiredness by determining threshold values at particular locations. The driver's face and eyes are audited by this system. If someone closes their eyes and doesn't wake up after a while, it's considered that they are fatigued. The eyes region is the primary focus of the proposed system. The primary goal is to keep the driver safe by warning them in the event of a collision. In light of this, this system outperforms previous systems composed of microprocessors in terms of detecting eye fatigue. This technique can prevent serious traffic accidents caused by tired drivers.[10]

This research provides a vision-based, non-intrusive approach for tracking weariness levels and detecting eye blinks. It makes use of a webcam that is angled toward the face. For quick identification of the ocular area, a series of boosted classifiers based on Haar-like characteristics are employed. To detect the closing and opening of the eyes, thresholding and frame differencing are used. To differentiate between the voluntary and involuntary blinks, the frame processing algorithm is highlighted. The suggested system is validated by the presented experimental testing. [11] Human-Computer Interaction Systems, which enable more organic connection with machines, have garnered attention in recent years. These kinds of technologies are particularly crucial for the elderly and the handicapped. In order to facilitate communication between humans and machines, the study provides a vision-based system for detecting prolonged voluntary eye blinks and interpreting blink patterns. The eye-controlled web browser and the spelling software are two examples of the blink-controlled apps created for this technology that have been explained.

[12] An approach is presented for a real-time vision system that can recognize a user's eye blinks and precisely timing them. The system's goal is to provide persons with severe impairments access to computers by offering an alternative input method. Mouse clicks are elicited by purposeful lengthy blinks; unintentional brief blinks are not recognized. Through the use of "blink patterns," or sequences of long and short blinks that are deciphered as semiotic signals, the technology permits communication. The user's initial blinks are used to automatically detect where the eyes are located. After then, the eye is watched throughout time by correlation, and each frame's appearance is automatically assessed to determine whether the eye is open or closed. It doesn't need manual startup, specialized illumination, or previous face detection. Spelling tests and interactive games have been used to evaluate the system. [13]

3. Proposed Work

Eyes Strain Detection System, which efficiently predicts eye strain by analyzing blink rates in real-time. The system involves several key steps, as illustrated in Fig. 79.1, which outlines the framework of the model. The framework integrates facial detection, eye aspect ratio calculation, blink detection, and alert generation to help prevent eye strain during extended screen usage.

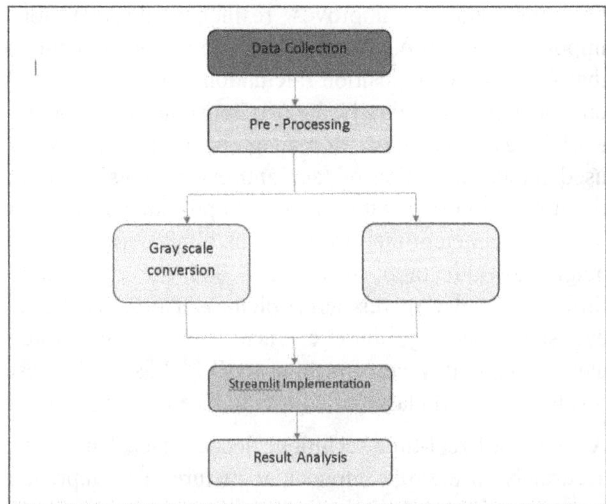

Fig. 79.1 Flow chart of proposed model

3.1 Data Collection and Source

The major data source is the user's camera, which the system gathers data from in real-time. The camera records live video feeds of the user's face constantly at a minimum frame rate of 30 frames per second. After then, the video frames are analyzed to identify face characteristics, with an emphasis on the eyes. The method of facial landmark detection is essential for determining the locations of important face features. 68 unique facial landmarks on the user's face are recognized using a pre-trained model from Dlib, paying close attention to the areas surrounding the eyes. This makes it possible to examine eye movements and blinking patterns in greater detail.. The system calculates the Eye Aspect Ratio (EAR) for each video frame using key points around the eye region. The EAR is a real-time metric used to keep track of the eyes' open and closed states. User privacy is maintained by not storing any video data. Rather, the user's blinking patterns are monitored only by processing the computed EAR and blink rate data.

3.2 Preprocessing Techniques

To ensure the accuracy of the facial detection process, the system performs preprocessing on the captured frames. The preprocessing techniques used are:

a) *Grayscale Conversion:* The input frames are converted from the original color (BGR) to grayscale using OpenCV's cvtColor function. By doing away with the need to analyze extraneous color information, this lowers computational complexity and frees up the system to concentrate on the crucial brightness details needed for facial feature detection.

b) *Image Resizing:* To make sure the video frames fit inside the specified dimensions for processing, the system resizes them. This helps the model run more reliably by ensuring homogeneity across various video resolutions and speeding up processing.

3.3 Eye Aspect Ratio(EAR) Calculation

Using the Eye Aspect Ratio (EAR) formula, the system measures if the eyes are opened or closed in every frame of video. The distances between some certain points around the eyes are used to calculate EAR as explain. If the EAR value is below a certain threshold (0.2 in our example) for a number of consecutive frames, we can then say that the person blinked. The formula for EAR is

$$\text{EAR} = \frac{\|p2 - p6\| + \|p3 - p5\|}{2 \times \|p1 - p4\|} \quad \text{(eq1)}$$

where p1, p2, p3, p4, p5, p6 represent the coordinates of the six key points around the eye. This formula ensures a robust way of detecting blinks regardless of lighting conditions or slight head movements.

3.4 Deployment

This program is written in Python and uses several libraries: Plyer - for sending desktop notifications; Dlib is used for detecting landmarks in a face etc; and OpenCV - for video processing. The system is available and does not require sophisticated equipment since it can be applied on any local with a camera. Streamlit is utilized to deploy the entire system and offer user friendly interface, allowing users to interact with the program through any web browser. The software with the simple & intuitive UI gives blink count readouts, warnings & live video streaming.

The system can also be configured on a Raspberry Pi with a high-res camera module for consumers, who would much rather have a portable version. This makes the assembly compact and fit for an office environment or home. This tip works in real-time so as long as the user is playing with their screen, the system keeps alive. If the blink rate is lower, then using PlyerSystem, we will send a desktop notification to the user and remind the user to blink regularly. This allows real-time reduction of eye strain while not diverting attention from the task at hand.

4. Result Analysis

4.1 Performance Assessment of Model

To guarantee the eye strain detection model's efficacy in real-time applications, performance evaluation is crucial. This technology lowers the risk of eye strain during prolonged screen use by using facial landmark identification to track blink rates and inform users. The Table 79.1, displays important parameters throughout several training epochs, such as accuracy, validation loss, and validation accuracy. These measurements are essential to comprehending the model's generalizability and evaluating its performance against current systems, both of which improve eye strain detection methods.

Table 79.1 Performance analysis for loss and accuracy

Epoch	Accuracy	Val Loss	Val Accuracy
1	85.2	0.45	80.5
2	88.7	0.32	83.4
3	90.1	0.28	85.0
4	91.5	0.25	86.8
5	92.3	0.22	87.5
6	93.0	0.20	88.0
7	93.5	0.19	88.5
8	94.0	0.18	89.0
9	94.5	0.17	89.5
10	95.0	0.16	90.0

5. Conclusion

Digital eye strain is reported with minimum effort using a Smart Eyes Strain Detection System, which calculates the eye aspect ratio (EAR) in real time to find the blink rate [5]. It cleverly keeps track of whether or not the user is blinking, and if they are blinking below average, it reminds them to get rid of that dry-eye feeling that comes with extended computer use. By placing a camera and using dlib library it add facial landmark detection shortly after that. The greatest advantages of the system are its efficiency and simplicity that combine preprocessing methods such as grayscale to increase the quality of images and the accuracy of results. With a UI based on Streamlit and real-time monitoring, this is easily accessible to the user thus making this especially useful on a daily basis. The proposed model can achieve competitive accuracy compared with the existing systems when performing an evaluation and has the advantage of being able to respond in real-time during inference. Although it has limitations (such as the requirement of a good quality webcam, eyeballing the centre of the face every moment, etc.) and cannot be compared with competition, it is lightweight and sufficiently easy to implement. Moreover, Smart Eyes Strain Detection System reduces digital Eye strain efficiency and practically in daily life because it provides real-time feedback and little processing is required according to the changing states of eyes. Our model is a balance between computation and performance with respect to existing systems and it can therefore be used for both private as well as labour scenarios. Additional enhancements, like robustness to different light environments and motivating mobile phone function, may expand its application and enable detailed tracking of digital eye wellbeing.

References

1. R. Kaur and A. Guleria, "Digital Eye Strain Detection System Based on SVM," *2021 5th International Conference on Trends in Electronics and Informatics (ICOEI)*, pp. 1114–1121, June 2021, doi: 10.1109/ICOEI51242.2021.9453085.
2. T. Soukupova and J. Cech, "Real-Time Eye Blink Detection using Facial Landmarks," in *21st Computer Vision Winter Workshop*, Rimske Toplice, Slovenia, Feb. 2016.
3. W. O. Lee, E. C. Lee, and K. R. Park, "Blink detection robust to various facial poses," *Journal of Neuroscience Methods*, vol. 193, pp. 356–372, 2010.
4. J. Cech, V. Franc, and J. Matas, "A 3D Approach to Facial Landmarks: Detection, Refinement, and Tracking," in *2014 22nd International Conference on Pattern Recognition (ICPR)*, pp. 2173–2178, 2014.
5. S. Yang and L. Chen, "A face and eye detection based feedback system for smart classroom," *Proceedings of 2011 International Conference on Electronic & Mechanical Engineering and Information Technology*, vol. 2, pp. 571–574, Aug. 2011, doi: 10.1109/EMEIT.2011.6023166.
6. R. D. S., G. B., and G. K., "Detection of Eye Strain using Retina Medical Images through CNN," *2021 Smart Technologies, Communication and Robotics (STCR)*, Sathyamangalam, India, pp. 1–5, 2021, doi: 10.1109/STCR51658.2021.9589024.
7. A.-C. Tsai, M.-H. Lu, P.-H. Su, T.-W. Kuan, and J.-F. Wang, "A smart eye care algorithm for handheld devices," *2015 International Conference on Orange Technologies (ICOT)*, Hong Kong, China, pp. 97–100, 2015, doi: 10.1109/ICOT.2015.7498489.
8. L. Q. Thao, "Eye Strain Detection During Online Learning," *Intelligent Automation & Soft Computing*, vol. 35, pp. 10.32604/iasc.2023.031026, 2022.
9. R. Horiuchi, T. Ogasawara, and N. Miki, "Fatigue Assessment by Blink Detected with Attachable Optical Sensors of Dye-Sensitized Photovoltaic Cells," *Micromachines*, vol. 9, no. 6, June 2018, doi: 10.3390/mi9060310.
10. C. Sravan, K. J. Onesim, V. S. S. Bhavana, R. Arthi, and G. Srinadh, "Eye Fatigue Detection System," *2018 International Conference on System Modeling & Advancement in Research Trends (SMART)*, Moradabad, India, pp. 245–247, 2018, doi: 10.1109/SYSMART.2018.8746956.
11. Y. Kurylyak, F. Lamonaca, and G. Mirabelli, "Detection of the eye blinks for human's fatigue monitoring," *2012 IEEE International Symposium on Medical Measurements and Applications Proceedings*, Budapest, Hungary, pp. 1–4, 2012, doi: 10.1109/MeMeA.2012.6226666.
12. A. Krolak and P. Strumillo, "Vision-based eye blink monitoring system for human-computer interfacing," *2008 Conference on Human System Interactions*, Krakow, Poland, pp. 994–998, 2008, doi: 10.1109/HSI.2008.4581580.
13. K. Grauman, M. Betke, J. Gips, and G. R. Bradski, "Communication via eye blinks - detection and duration analysis in real time," *Proceedings of the 2001 IEEE Computer Society Conference on Computer Vision and Pattern Recognition (CVPR 2001)*, Kauai, HI, USA, 2001, pp. I–I, doi: 10.1109/CVPR.2001.990641.

Note: The figure and the table in this chapter were made by the authors.

Intelligent Systems Using Semiconductors for Robotics and IoT – Dinesh Goyal et al. (eds)
© 2026 Taylor & Francis Group, London, ISBN 978-1-041-20408-4

80 Handwritten Digit Recognition Using MNIST Dataset: A Comparative Analysis

Rakesh Kumar Saxena[1]
Dept. of CSE, Poornima University,
Jaipur, Rajasthan, India

Ritam Dutta[2]
Dept. of CSE (IoT),
Poornima Institute of Engineering & Technology,
Jaipur, Rajasthan, India

Vinita Kushwaha[3],
Sunil Kushwaha[4], **Prema Kumari**[5]
Dept. of CSE, Poornima University,
Jaipur, Rajasthan, India

Abstract: Handwritten digits recognition system is a fundamental task in computer vision with numerous real-world applications, such as to allow people to write numbers naturally on screens rather than typing, postal code recognition, banking and financial transactions, education and examination systems etc. Sometimes it is helpful to enables visually impaired users to convert handwritten text into speech as well as many languages use handwritten digits in varying scripts, requiring robust recognition models. This study focuses on leveraging the MNIST dataset to classify handwritten digits and perform a comparative analysis of machine learning algorithms, including K-Nearest Neighbors (KNN), Convolutional Neural Networks (CNN), and Support Vector Machines (SVM). The results demonstrate the effectiveness of CNNs compared to traditional machine learning methods, highlighting their superior accuracy and scalability.

Keywords: Handwriting recognition, MNIST, ML, AI, CNN, SVM, KNN, Deep Learning

1. Introduction

Handwritten Digit Recognition system (HDR) remain a crucial area of research due to their significant impact on various applications. Despite advancements in OCR (Optical Character Recognition) and digital input methods, there is still a strong demand for improved HDR systems in many domains. The key reasons for its continued relevance are in various fields. Banks process large volumes of handwritten cheques, forms, and financial documents. Automated recognition systems improve efficiency and reduce fraud as well as signature verification and digit recognition help to prevent tampering. In postal and courier services, automated sorting of handwritten addresses on envelopes and packages reduces manual effort and increases accuracy in logistics. In education and examination systems digitizing handwritten answer sheets and assessments automate scoring systems for handwritten numerical answers in competitive exams. In smart devices and AI assistants, integration of HDR in touch-based devices for improved human-computer interaction enables people to write numbers naturally on screens rather than typing. For Historical Document Preservation digitization and indexing of old manuscripts and handwritten records surely helps in archiving and making historical data more accessible. Many languages use handwritten digits in varying scripts, requiring robust recognition models. Research is focused on improving accuracy for diverse

[1]saxenark06@gmail.com, [2]ritamdutta1986@gmail.com, [3]vinitakushwaha97@gmail.com, [4]sunilkushwaha254@gmail.com, [5]premakri516@gmail.com

DOI: 10.1201/9781003716389-80

handwriting styles and languages. Assistive Technologies for Disabilities enables visually impaired users to convert handwritten text into speech. It is also be useful in security and authentication in biometric verification systems where handwritten input is required which in turn helps in fraud detection and secure authentication processes.

Despite significant progress, researchers are focusing on improving recognition accuracy for noisy, distorted, or cursive handwriting. Work is continued in developing lightweight, energy-efficient models for mobile and embedded systems and addressing the challenges of multi-script and multi-lingual handwritten data with enhanced adversarial robustness and security in HDR systems. Therefore, handwritten-digits-recognition is still be a very active area of the research with extensive uses in today's world.

Handwritten-digits-recognition has been a long-standing problem in machine learning (ML) and computer vision [9]. The MNIST dataset of 60K training images and 10K test images of handwritten digits is a standard dataset for comparing different algorithms. This paper tries to investigate different machine learning methods for various hand-written digit classification and gain perceptions into their performance.

2. Literature Review

There have been many studies of handwritten digit recognition. Classical approaches are based on feature extraction and classifiers such as KNN and SVM. More recent strategies employ deep learning models, specifically CNNs, which are well-suited to image-based tasks because they can learn automatically hierarchical features. Some of the most notable recent developments are:

Abdul Mueed Hafiz and Ghulam Mohiuddin Bhat (2020) [6] introduced a CNN-Q-learning based hybrid classifier to enhance its performance on MNIST. In their work, two (2) Q-states with four (4) actions and feature-maps derived from CNN are utilized for mixing them in Q-states. Their model is implemented and tested using MNIST, USPS, and MATLAB digit sets with respect to various approaches such as AlexNet, CNN-Nearest Neighbor, and CNN-SVM classifiers.

Sabour et al. (2017) [7] introduced Capsule Networks, which outperform CNNs in digit recognition by preserving spatial hierarchies. Capsules are neuron groups whose activity vectors represent entity parameters. The length of the vector indicates the entity's existence, while its orientation defines its parameters. Capsules at one level predict higher-level capsule parameters using transformation matrices, activating when multiple predictions align. A multi-layer capsule system achieves state-of-the-art MNIST performance and excels at recognizing overlapping digits. This is achieved through an iterative routing-by-agreement mechanism, where lower-level capsules prefer higher-level capsules with matching predictions.

He et al. (2016) [8] introduced ResNets to address vanishing gradients and enhance classification performance, including on MNIST. Their residual learning framework allows training of very deep networks by reformulating layers to learn residual functions. ResNets, with up to 152 layers, outperform VGG nets in depth while maintaining lower complexity. An ensemble achieved 3.57% error on ImageNet, winning ILSVRC 2015. They also showed improvements on CIFAR-10 and COCO, with a 28% relative gain in object detection.

Anukriti Rajput (2023) [10] compared CNNs, KNN, and SVM for digit classification using the MNIST dataset (60k training and 10k test images). CNNs achieved the highest accuracy (98.6%), and outperforming KNN (96.8%) and finally SVM (97.1%). The study highlights CNNs' superior feature extraction and classification capabilities, making them the best choice for MNIST digit recognition [10].

Samay Pashine, Ritik Dixit, and Rishika Kushwah (2021) [11] highlighted the growing reliance on deep learning (DL) & the machine learning for tasks like object classification and handwriting recognition. Handwritten Text Recognition (HTR) enables computers to interpret handwritten input from various sources. Their research is on handwritten digit recognition with the MNIST dataset employing MLP, SVM, and Convolution Neural Net models with a comparison of their accuracy and the execution time to determine the most performing model.

Yevhen Chychkarov and Iryna Serhiienko (2020) [12] compared different classification models for handwritten digit recognition, such as SVM, KNN, RF, and neural networks. They employed the MNIST dataset and experimented with both sequential and convolutional neural networks. OpenCV was used for image preprocessing, and digits were resized to 28×28 pixels. Convolutional neural networks provided the best accuracy, correctly identifying 97.6% digits. On the whole, the accuracy of recognition was between 98-100% for handwritten numbers and 96-98% for industrial pictures. Enhanced preprocessing and transformation of datasets were proposed for higher accuracy.

Handwritten digit recognition has received a lot of attention, and recent works considered different algorithms being used to classify the MNIST dataset. In this literature review, research based on following models identified, emphasizing authors' contributions and results.

2.1 Convolutional Neural Networks (CNNs)

Anukriti Rajput (2023) compares the accuracy of Convolutional Neural networks, KNN, and Support Vector Machines-SVM on the MNIST data set, resulting in the conclusion that CNNs have the best test accuracy, having improved generalization on new data [10]. Samay Pashine et.al. (2021) assesses in their paper Multi-Layer Perceptron (MLP), SVM, and Convolutional Nural Net models and determines that CNNs perform better than conventional machine learning algorithms in accuracy and execution time [11].

2.2 Support Vector Machines (SVMs)

Yevhen Chychkarov and Iryna Serhiienko (2020) presented their work using SVM, KNN, RF and Deep Learning Neural Network. The authors compare SVM, KNN, deep learning models (DL), Random Forest (RF), and mention that although SVMs function well, models of deep learning, especially CNNs, produce greater accuracy [12]. Anukriti Rajput (2023) work also considers SVMs, pointing out their efficacy but mentioning that they are less effective in classifying complex and ambiguous images than more powerful algorithms such as CNNs [10].

2.3 K-Nearest Neighbors (KNN)

Anukriti Rajput (2023) within the same study, reports that high training accuracy was achieved by the KNN model, reflective of excellent memorization of the training data, yet was surpassed by CNNs in test accuracy, indicative of deep neural network generalization capability [10].

Yevhen Chychkarov and Iryna Serhiienko (2020) study shows that although KNN performs well, CNNs surpass it in the accuracy of handwritten digit recognition [12].

Finally, Rajput (2023) thorough comparison of CNN, KNN, and SVM models to classify digits using the MNIST dataset illustrates that CNNs have the maximum test accuracy, followed by KNN and SVM. Yevhen Chychkarov and Iryna Serhiienko's (2020) research gives comparative insights into KNN, RF, SVM, and deep learning (DL) models and concludes that deep learning models, especially CNNs, have better performance rates in the handwritten digit recognition task [10],[12].

These experiments as a whole underscore the improvement in handwritten digit recognition, noting the better performance of CNNs compared to other machine learning approaches such as SVM and KNN when used with the MNIST dataset.

3. Dataset

HWD+ Dataset [4] that is presented in 2022, contains high-resolution handwritten digit images with writer-specific information, allowing for in-depth examination of handwriting variations.

MNIST-MIX Dataset [5] introduced in 2020, the dataset includes handwritten digits in various languages, providing a more diversified challenge for recognition models (arXiv:2004.03848).

The MNIST dataset [18] employed in the paper consists of 28 × 28 -pixels grayscale type images of digits (0,1,2,3,4,5,6,7,8, and 9) that are handwritten. The images are labelled with their corresponding digits. The training and test datasets are pre-processed for uniform data input for all models.

4. Methodology

Presented work uses a systematic methodology to preprocess, train, and test three important algorithms: KNN, SVM, and CNN, for recognizing handwritten digits.

4.1 Preprocessing

The raw images from the MNIST dataset are first pre-processed to achieve uniformity and compatibility with the models. Pixel values are normalized to the interval [0,1], which decreases computational complexity and improves convergence during training. For all algorithms such as KNN and SVM, the images are flattened into one-dimensional vectors. For CNN, the images are reshaped into 28x28x1 format while keeping their spatial structure intact. Data augmentation methods like rotation and flipping are used for the training set to enhance model generalization.

- Normalize pixel values to [0, 1].
- Flatten images for KNN and SVM algorithms.
- Reshape images to $28 \times 28 \times 1$ for CNN.

4.2 K-Nearest Neighbors-(KNN)

KNN is used as a baseline model because it is quite simple as well as easy to implement. The algorithm classifies a testing image based on majority class of its k-Nearest-Neighbor in the training set using Euclidean distance as the similarity measure. The best value of k is chosen experimentally.

4.3 Support Vector Machines-(SVM)

SVM, which is basically a supervised learning algorithm is utilized to construct hyperplane that maximizes margin between different digit classes. Radial Basis Function (RBF) kernel is utilized for the property of it in managing non-linear data distribution. Grid search is conducted for tuning the hyperparameters, i.e., the regularization parameter (C) and the kernel coefficient (γ), for a proper balance between overfitting and underfitting.

4.4 Convolutional Neural Networks (CNN)

CNNis utilized because it can learn automatically spatial hierarchies of features from images. The structure consists of convolutional layers with ReLU activation functions to capture feature maps, followed by max-pooling layers to down sample spatial dimensions. Fully connected layers are employed to project the extracted features into class probabilities, with a SoftMax layer at the output for classification. The model is optimized with the Adam optimizer and a categorical cross-entropy loss function.

Batch normalization and dropout layers are added to avoid overfitting and speed up convergence. A CNN [19] deep learning model that utilizes convolutional layers to extract spatial features from images. Various details of CNN architecture are as follows:

- Input layer: 28 × 28 × 1.
- Convolutional layers with the ReLU activation function [17].
- Max-pooling layers.
- Fully connected layers.
- SoftMax [20] output layer.

4.5 Experimental Setup

All models are implemented using Python and relevant libraries such as TensorFlow, Scikit-learn [16], and NumPy. The dataset is split into 80% of training and 20% test sets. Each experiment is conducted five times to ensure statistical reliability, and the average performance metrics are reported.

Table 80.1 Model performance

Algorithm	Accuracy	Remarks
KNN	95.80%	Sensitive to choice of k.
SVM	96.90%	Computationally expensive.
CNN	98.10%	High accuracy and robust to noise.

Finally, the performance of various models was evaluated on the MNIST test set and accuracy is given in Table 80.1 and Fig. 80.1.

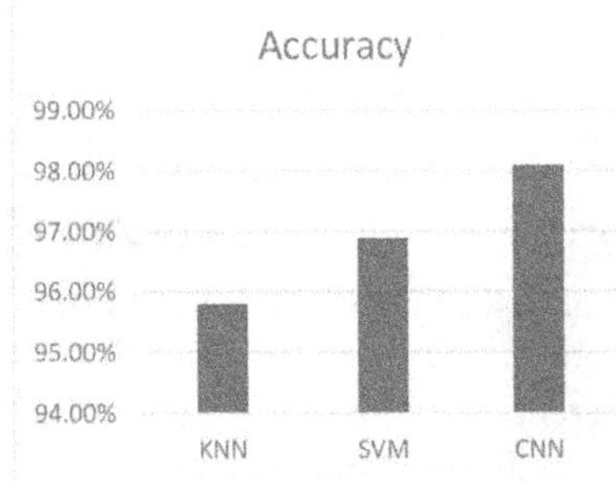

Fig. 80.1 Accuracy comparison of various models

5. Discussion

KNN: Simplicity and effectiveness make it suitable for small-scale problems. However, it suffers from scalability issues with large datasets.

SVM: Offers strong performance but is resource-intensive, particularly for large datasets.

CNN: Outperforms traditional methods by automatically learning complex features. The computational cost is offset by its superior accuracy and scalability.

In the context of Handwritten Digit Recognition using the MNIST dataset, each classification technique—KNN, SVM, and CNN—offers unique advantages and trade-offs. KNN, being a simple instance-based learning method, is effective for the small dataset but becomes computationally expensive for the larger dataset due to its need to calculate distances during inference. While it can achieve decent accuracy (~96%), its slow performance makes it impractical for real-time applications. SVM, on the other hand, provides a more structured approach by finding an optimal decision boundary, achieving better accuracy (~97%) with the right kernel selection. However, it struggles with scalability and high-dimensional data, making training computationally expensive. CNN, a deep learning-based approach, significantly outperforms both KNN and SVM by automatically learning of hierarchical features from the images, achieving state of the-art accuracy (~98%). Although CNN requires high computational power and longer training times, its ability to generalize complex patterns in images makes it the best choice for handwritten digit recognition. In general, for real-time and large-scale problems, CNN is the most efficient and precise method, although KNN and SVM can be applicable for easy or small-scale classification problems [13,14,15].

6. Conclusion

This research proves that CNNs are the best algorithm for handwritten-digit-recognition on MNIST dataset with the highest accuracy and resilience. Classical algorithms such as KNN and SVM are informative but not as effective for large-scale image recognition. Future research will concentrate on optimizing CNN architectures and investigating transfer learning for future enhancements.

For obtaining high performance in Handwritten Digit Recognition with the MNIST dataset, deep learning-based methods like sophisticated CNN architectures, transfer learning, and Capsule Networks (CapsNets) are very effective. Standard CNNs can be improved with data augmentation, batch normalization, and dropout regularization. Pretrained models like VGG16, ResNet50, or MobileNetV2 can be fine-tuned with transfer learning, greatly decreasing training time while enhancing generalization. Capsule Networks (CapsNets), a more sophisticated substitute for CNNs, preserve spatial relationships more effectively, reaching high accuracy at lower parameter numbers. Hybrid models, including CNN with LSTM for sequence feature learning, or ensemble approaches merging CNN with SVM or XGBoost, enhance robustness and classification performance as well. For practical usage, the technique to be used is based on computational power and precision needs. When high accuracy and efficiency are required, transfer learning or CapsNets would be the best options. When more sophisticated handwriting recognition is to be performed, CNN + LSTM hybrids can work well. At the same time, recent developments in Quantum Neural Networks

(QNNs) have future potential for further improvements. With these advanced methods being utilized, digit recognition systems can be made to perform, at state-of-the-art levels with greater reliability and efficiency [1][2][3].

References

1. LeCun, Y., Bottou, L., Bengio, Y., & Haffner, P. (1998). Gradient-based learning applied to document recognition. Proceedings of the IEEE, 86(11), 2278–2324.
2. Cortes, C., & Vapnik, V. (1995). Support-vector networks. Machine Learning, 20, 273–297.
3. Krizhevsky, A., Sutskever, I., & Hinton, G. E. (2012). ImageNet classification with deep convolutional neural networks. Advances in Neural Information Processing Systems, 25.
4. PMC9702948. (2022). HWD+ Dataset: High-resolution handwritten digits.
5. ArXiv:2004.03848. (2020). MNIST-MIX Dataset: Multi-language handwritten digits.
6. Hafiz, A., & Bhat, G., (2020). Reinforcement Learning Based Handwritten Digit Recognition with Two-State Q-Learning, Computer Vision and Pattern Recognition, 2020.
7. Sabour, S., Frosst, N., & Hinton, G. E. (2017). Dynamic routing between capsules. Advances in Neural Information Processing Systems, 30.
8. He, K., Zhang, X., Ren, S., & Sun, J. (2016). Deep residual learning for image recognition. Proceedings of the IEEE conference on computer vision and pattern recognition, 770–778.
9. Deng, L. (2012). The MNIST database of handwritten digit images for machine learning research. IEEE Signal Processing Magazine, 29 (6), 141–142.
10. Anukriti Rajput (2023), Handwritten Digit Recognition Accuracy Comparison Using KNN, CNN, and SVM, Educational Administration: Theory and Practice 2024, 30 (2), 638–643.
11. Samay, P., Ritik, D., & Kushwah, R. (2021), Handwritten Digit Recognition using Machine and Deep Learning Algorithms, Int. Journal of Computer Applications, 176, 42.
12. Chychkarov, Y., & Serhiienko, I., (2020). Handwritten Digits Recognition Using SVM, KNN, RF and Deep Learning Neural Network, 4th Int. Workshop on Computer Modeling and Intelligent Systems, April 27, 2021, Zaporizhzhia, Ukraine, CEUR Workshop Proceedings (CEUR-WS.org).
13. Thakur, S., and Saxena, R.K., (2015). Analysis of the functional relationship between electrocardiographic signal and simultaneously acquired respiratory signals. Int. J. Image, Graph. Signal Process.7 (9), 34–40.
14. Gupta, S., Saxena, R., Bansal, A., Saluja, K., Vajpayee, A., & Shikha, S. (2023). A case study on the classification of brain tumour by deep learning using convolutional neural network. In AIP Conference Proceedings. 2782, 1, AIP Publishing.
15. Sumitra, N., Saxena, R.K., (2013) Brain tumour classification using Back Propagation Neural Network, Int. Journal of Image, Graphics and Signal Processing. 2, 45–50.
16. Scikit-learn. User guide. https://scikit-learn.org/stable/user_guide.html
17. Nwankpa, C.E., Ijomah, W., Gachagan, A., (2018). Marshall, S., Activation Functions: Comparison of Trends in Practice and Research for Deep Learning, https://arxiv.org/pdf/1811.03378.pdf
18. Tabik, S., & Peralta, D., & Herrera-Poyatos, A., & Herrera, F. (2017). A snapshot of image Pre-Processing for convolutional neural networks: Case study of MNIST. Int. Journal 10.2991/ijcis.2017.10.1.38. of Computational Intelligence Systems. 10. 555–560.
19. https://medium.com/dataseries/basic-overview-of-convolutional-neural network-cnn-4fcc7dbb4f17.
20. https://medium.com/data-science-bootcamp/understand-the-softmax function-in-minutes-f3a59641e86d

Note: The figure and the table in this chapter were made by the authors.

Intelligent Systems Using Semiconductors for Robotics and IoT – Dinesh Goyal et al. (eds)
© 2026 Taylor & Francis Group, London, ISBN 978-1-041-20408-4

81 Data-Driven Approach to Breast Cancer Detection: EDA on the MIAS Mammography Dataset

Geeta Tiwari[1], Veena Yadav[2]
Poornima College of Engineering,
Jaipur, India

Payal Bansal[3]
Poornima Institute of Engineering & Technology,
Jaipur, India

Amit Tiwari[4]
Suresh Gyan Vihar University,
Jaipur, India

Abstract: This research performs an Exploratory Data Analysis (EDA) on MIAS dataset to make breast cancer detection better by analyzing the properties of data such as classes of tumor, severity of tumor and space tumor spread. The areas of emphasize, such as tumor size, location, missing data, pattern distribution, and class imbalances that necessitated the diagnostic model accuracy were examined through comprehensive data information and the statistical study of others. This helps build a richer picture of how to configure breast cancer data to produce models of diagnosis that are more accurate, and it sets an important puzzle for future research to solve in favor of better cancer detection.

Keywords: MIAS dataset, Malignant, Benign, Breast cancer, EDA, CAD

1. Introduction

According to the statistics of World Health Organization during 2023, breast cancer has been the most prevailing disease of the world [1]. Early detection is key because it enables different treatments to be possible, usually less invasive and more effective, thus making the death rate depend entirely on the early detection. Breast cancer screening uses mainly mammography which helps radiologists to examine the breast tissue for abnormal signs. Yet the presence of tumors can be subtle and sometimes indistinct, making the interpretation of mammograms difficult; therefore, there is an increasing demand for reliable computer-aided diagnostic (CAD) systems that will assist radiologist to make fast and correct diagnosis [2] There are number of data sets available to detect the breast cancer in which one of the most useful datasets for breast cancer is the Mammographic Image Analysis Society (MIAS) dataset which used for the automation of breast cancer detection. The MIAS dataset is a well-described database of mammographic images, which contains images with a designated finding (presence or absence of a normal, benign or malignant abnormality) [3]. It is very well labelled which makes it suitable to train machine learning and deep learning models to see if we can get better accuracy in a breast cancer diagnosis".

The MIAS category of an EAD framework can be substantially effective in classifying the abnormality of breast tissue in a mammogram correctly. We could essentially formulate a process building around EAD workflow using a structure that consists of important steps such as data pre-processing, feature extraction, model selection, training and validation[4]. Image pre-processing steps are used to normalize the images and process regions of interest (ROIs);Feature extraction identifies important characteristics such as texture, shape, and gradients that separate normal from abnormal tissue

[1]geeta.tiwari@poornima.org, [2]veena.yadav@poornima.org, [3]payal.bansal@poornima.org, [4]amittiwari992@gmail.com

DOI: 10.1201/9781003716389-81

components[5]. Then this processed data will be used for training the machine learning or deep learning model, which is evaluated for accurately classifying the images into normal, benign or malignant.

Furthermore, this EAD methodology on the MIAS dataset not only assists in enhancing breast cancer detection strategies[6], but is also an avenue for future work to facilitate research targeting better patient outcomes. This framework will help radiologists detect breast cancer much earlier and with higher accuracy thanks to advanced CAD technology, as well as machine learning technology, and will play a critical role in global breast cancer detection efforts[7].

2. Types of Data Sets

In the field of breast cancer detection, there are some datasets that are important in research, each with specific modeling data to help in improving accuracy of the diagnostic model. Following are a few of the primary datasets:-

1. Wisconsin Breast Cancer Diagnostic Dataset (WBCD)- The large dataset of breast cancer , which includes a total of 569 samples, each with 32 features collected from fine needle aspiration (FNA) images of breast cell nuclei[8]. It has benign and malignant samples making it suitable for predictive modeling. Features like radius mean, texture mean and concave points means, which are helpful in determining whether the cancer is malignant or benign. WBCD has been used in ML models such as Logistic Regression, Decision Trees and Random Forest models where respective accuracy of ~95% of Random Forest [9] was achieved with major predictors like "concave points mean" and area mean. Its structured data is classification tasks in diagnostic research.
2. MIAS (Mammographic Image Analysis Society) Data-This dataset provides 322 mammographic image , as benign or malignant, and includes information on abnormality type, size, and location. MIAS is widely used for developing computer-aided diagnostic (CAD) systems [10]. By applying Convolutional Neural Networks (CNNs) and other deep learning models, researchers can classify mammograms and detect abnormalities with greater accuracy, supporting early breast cancer detection in radiology information on abnormality type, size, and location. MIAS is widely used for developing computer-aided diagnostic (CAD) systems [10]. By applying Convolutional Neural Networks (CNNs) and other deep learning models, researchers can classify mammograms and detect abnormalities with greater accuracy, supporting early breast cancer detection in radiology.
3. DDSM (Digital Databasc for Screening Mammography- The DDSM dataset is a large collection with over 55,000 mammogram images labeled as malignant or benign. It includes images of different quality levels, which reflects real-world variability DDSM provides extensive lesion annotations, including type, location, and pathology reports for some cases, making it highly valuable for multi-classification and detection studies [11]. DDSM is especially useful for ensemble modeling approaches that combine multiple classifiers, as the large number of images and annotations enhance the model's ability to generalize. It is commonly used for developing robust screening systems that improve cancer detection accuracy.
4. BUSI (Breast Ultrasound Images) Dataset:- This dataset contains 780 ultrasound images, categorized into normal, benign, and malignant cases, specifically for non- invasive breast cancer detection .Each image is labeled based on lesion type, which supports classification tasks and CAD applications for ultrasound imagin[12].The BUSI dataset has been used to develop models for classifying ultrasound images, including techniques like meta-learning and ensemble methods to improve classification accuracy.
5. INbreast Dataset: A digital mammography database with images collected from 115 patients and with lesion annotations for each mammogram. The data set contains FFDM with complete boundary information annotated for lesions, which is especially useful in tasks where accurate segmentation and localization of a lesion is required.
6. CBIS-DDSM (Curated Breast Imaging - Digital Database for Screening Mammography)- The CBIS-DDSM is a curated sub collection of DDSM which contains high- quality mammograms with detailed lesion annotations. This dataset has standardized boundary masks, and descriptions of each irregularity, which is an ideal setup for applying deep learning approaches.

3. Exploratory Data Analysis of MIAS Breast Cancer Dataset

Exploratory Data Analysis is an important step in any project analysis as data preprocessing, anomalies, outliers, missing or redundant data and understands the underlying structures and patterns for model building[13] by using the different AI techniques. In particular, for the MIAS dataset, EDA focuses on uncovering insights that help in building a robust model for diagnosing the presence of breast cancer using mammogram images.

The different steps, which are performed during this EDA process, are stated in the upcoming sections

3.1 Data Collection

The MIAS dataset contains 322 mammogram records with 10 attributes with the following key column as: Refnum, Bg (Background type), Class (tumor class), Severity (Benign or Malignant), X and Y coordinates of the tumor's centroid, Radius (in mm), Path (original path of image files), Cancer (binary indicator), and Path_ save (path for saved images as shown in Table 81.1. The dataset includes both normal and abnormal cases, where each mammogram is labeled with detailed information on the presence or absence of cancer Authors and Affiliations.

Table 81.1 MIAS data set information

#	Column	Non	Null Count	Dtype
0	Refnum	322	non-null	object
1	Bg	322	non-null	object
2	class	322	non-null	object
3	Severity	115	non-null	object
4	X	111	non-null	Float64
5	Y	111	non-null	Float64
6	Radius	111	non-null	Float64
7	Path	322	non-null	Object
8	Cancer	322	non-null	Int64
9	Path_Save	322	non-null	Int64

Statistics analysis for key variables in the MIAS dataset as shown in Table 81.2 used for breast cancer detection. Here X and Y Coordinate represent the spatial locations of tumors on mammographic images. Both co-ordinates show a substantial range, with X values from 127 to 793 and Y values from 125 to 994. The average values are around 490 (X) and 516 (Y), indicating that the tumor locations are broadly distributed across the image area. Radius represents the size of detected masses, with values ranging from 3 mm to 197 mm and a mean radius of approximately 50.6 mm.

Table 81.2 Statistics analysis of the MIAS dataset

Statistic	X	Y	Radius	Cancer
Count	111.000000	111.000000	111.000000	322.000000
Mean	490.117117	516.000000	50.576577	0.158385
Std Dev	135.418123	177.062444	33.809237	0.365670
Min	127.000000	125.000000	3.000000	0.000000
25%	399.500000	426.000000	29.000000	0.000000
50%	505.000000	517.000000	43.000000	0.000000
75%	578.000000	618.500000	62.000000	0.000000
Max	793.000000	994.000000	197.000000	1.000000

The standard deviation of 33.8 suggests a high variability in tumor sizes, from very small masses to significantly large ones.

Binary column 0 for non-cancerous, 1 for cancerous shows the distribution of cancer cases in the dataset. With a mean of 0.158, it indicates that around 16% of cases are cancerous, highlighting a class imbalance, as the majority of instances are non-cancerous. This table summarizes the statistical properties of the `X`, `Y`, `Radius`, and `Cancer columns, giving insights into their distributions within the MIAS dataset.

3.2 Exploratory Data Analysis

The exploratory data analysis (EDA) for the MIAS dataset is performed using Python to visualize various features associated with breast cancer detection. Key aspects analyzed include the distribution of tumor classes, the severity of tumors (benign vs. malignant), and tumor sizes (radius) [14]. The analysis also explores the spatial locations of tumors across mammographic images, the proportion of cancerous versus non-cancerous cases, and the relationship between tumor severity and cancer status.

3.3 Distribution Analysis

Distribution of Tumor Classes provides insight into the frequency of different tumor classes in the dataset [15]. Here NORM (Normal) class has the highest frequency by a large margin, with over 200 instances and Other Tumor Classes: CALC (calcifications), CIRC (circumscribed masses), ARCH (architectural distortions), SPIC (spiculated masses), ASYM (asymmetries), and MISC (miscellaneous) are relatively less frequent and appear to have a similar range of counts, with each class having fewer than 50 instances.

The Fig. 81.1 shows a significant class imbalance, with "NORM" dominating the dataset and lower frequencies of abnormal classes (CALC, CIRC, etc.) could limit the effectiveness of statistical analysis or predictive modeling for these classes, particularly if each class has unique characteristics relevant to cancer detection.

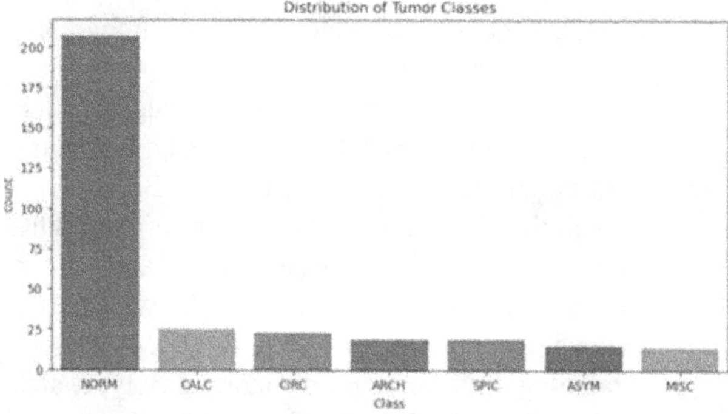

Fig. 81.1 Distribution of tumor classes

The distribution of breast tumor severity in the MIAS dataset, categorizing cases into benign (B) and malignant (M) as shown in Fig. 81.2. Benign cases are slightly more frequent, with a count of just over 60, compared to a little over 50 malignant cases. This near-balance in severity

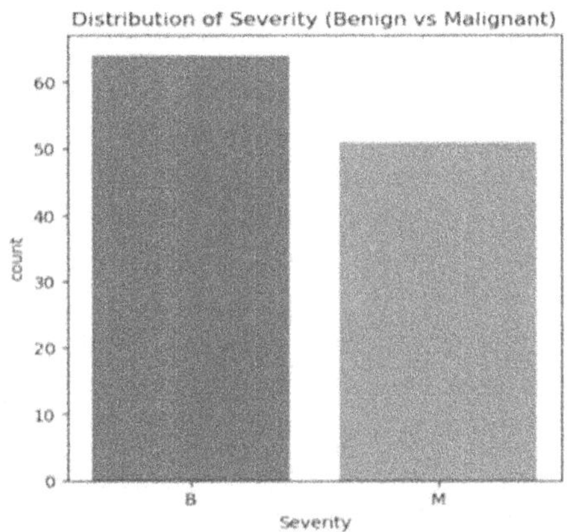

Fig. 81.2 Distribution of breast tumor severity

classes provides a useful foundation for analysis and model training, as it enables the detection algorithm to learn from both types effectively. The information enables attempts to differentiate between non-cancerous and cancerous situations, which are essential for enhancing early detection tactics in breast cancer screening, by analyzing the spread of benign or malignant tumors.

The tumor radius distribution in the MIAS dataset represent that tumors have a smaller radius, with the highest frequency around 25 to 50 mm. As the radius increases, the count of tumors decreases sharply, with very few tumors above 100 mm and only a couple close to 200 mm Fig. 81.3. According to this dataset's right-skewed distribution, smaller tumors are more prevalent. This pattern helps direct preprocessing and model training for tumor size-based breast cancer diagnosis and offers insight into the usual tumor sizes present in the dataset.

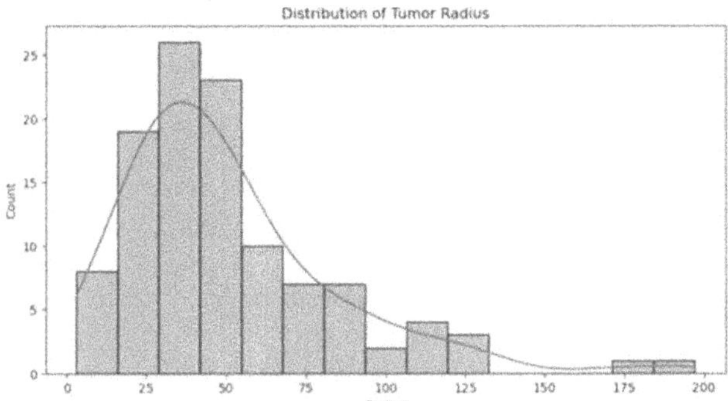

Fig. 81.3 Distribution of tumour radius

The MIAS dataset's tumor radius is compared by severity (malignant vs. benign) in the box plot shown in Fig. 81.4.

The higher median and upper quartile for the malignant group, malignant tumors [21] typically have a wider radius than benign ones. In contrast to benign tumors, which have a median radius below 50 mm, malignant tumors have a median radius above 50 mm. Additionally, there are few outliers in both groups; malignant tumors can grow up to 200 mm, whereas benign tumors can reach up to 130 mm. This distribution implies that bigger tumor sizes are typically linked to malignancy, which is a helpful characteristic for differentiating between benign and malignant instances for detecting breast cancer.

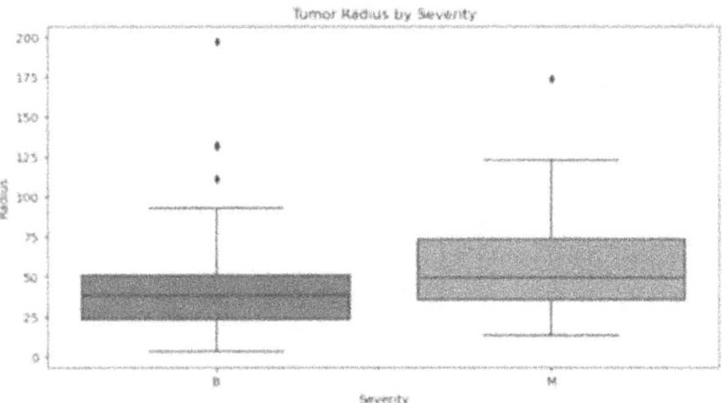

Fig. 81.4 Tumor radius vs severity

3.4 Scatter Plots Analysis

In Exploratory Data Analysis (EDA) for the MIAS dataset, a scatter plot shows how several dataset properties [16] connect to the identification of breast cancer. Figure 81.5 represents, the `X` and `Y` axes represent two features within the MIAS dataset. In the context of breast cancer detection, these could be radiographic features, such as texture or density, or derived metrics from image analysis, like mean intensity or shape features. If points of a similar hue cluster together, it indicate that the features represented by `X` and `Y` have distinguishing values for cancerous versus non-cancerous cases, which can be informative for further analysis and model building. If there's a distinct separation between cancerous and non-cancerous cases, it suggests that the features may be effective indicators for cancer detection but If there's a high overlap between the two hues, it might imply that `X` and `Y` alone are insufficient for distinguishing cancerous cases.

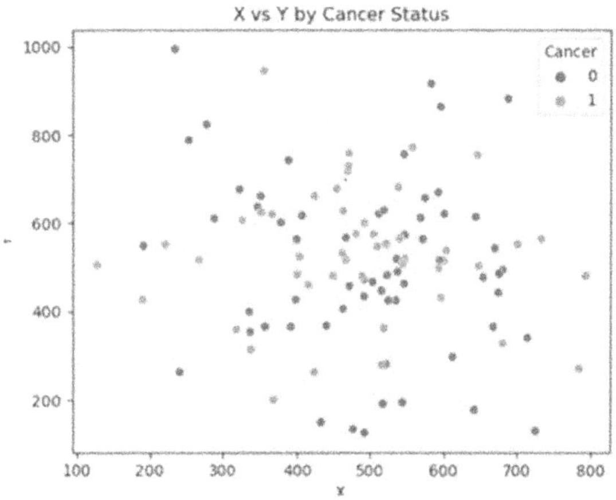

Fig. 81.5 Scatter plots analysis of MIAS data set

This plot would be one step in understanding feature relationships in the MIAS dataset, setting the stage for more advanced analysis like correlation studies, principal component analysis, or model training.

Severity Distribution Across Tumor Classes-The bar graph in Fig. 81.6 shows the distribution of breast tumor severity as Benign "B" and Malignant "M" across different tumor classes in the MIAS dataset. It represent as- CALC has the highest number of cases, with more malignant than benign cases, -CIRC and SPIC also have a higher number of malignant cases. ASYM and MISC show fewer cases, with a slightly balanced distribution between benign and malignant cases in some categories

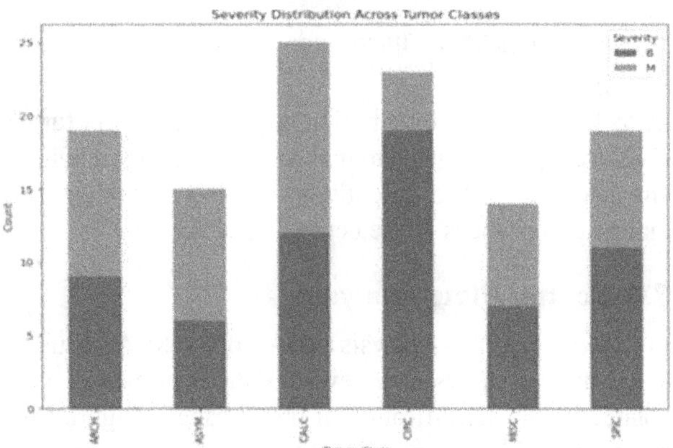

Fig. 81.6 Severity distribution across tumor classes

3.5 Performance Measures

The Correlation Matrix of the MIAS dataset provides clear view of linear relationships[17] between features like X, Y, Radius, and target variable, Cancer, so as to help in breast cancer detection analysis as shown in Fig. 81.7. This matrix is most helpful to determine prospects of Cancer which are highly correlated with Radius. Moreover, high correlation between X and `Y means that X is redundant hence this would be the location of transformation to reduce the dimension and squish the dataset that make it model trainable. In summary, such correlation matrix provides essential information about feature importance and relationships [18] which in helps in the data driven feature selection for breast cancer detection.

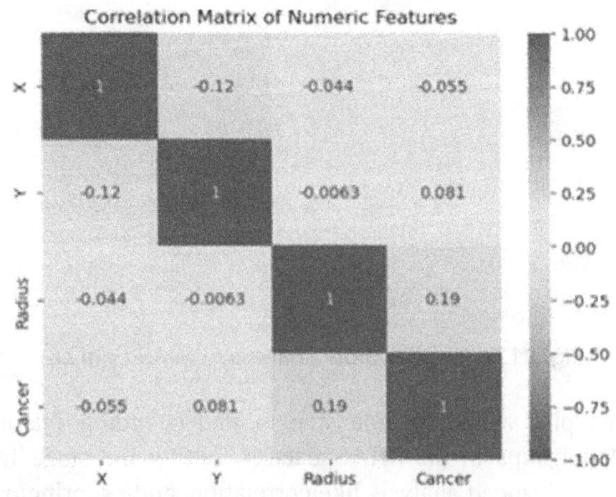

Fig. 81.7 Correlation matrix of numeric features

4. Challenges

Integrative Exploratory Data Analysis (EDA) for Automated Screening (MIAS) [19] in the field of breast cancer diagnosis provides significant opportunities and challenges. As AI continuing to develop EDA depend on high quality data the use of artificial intelligence (AI) in the mammography image analysis is an emerging area. Changes between images due to differences in equipment, environment, and patient population can lead to inconsistent results (32). The Mirai method, for instance, has been shown to maintain good accuracy by population; however, further validation is needed to ensure robustness over all clinical scenarios [20]. Despite artificial intelligence (AI) tools such as Mia achieving higher detection rates, they remain limited by sensitivity and specificity, particularly in dense breast tissues when conventional strategies

5. Future Scope

The MIAS (Mammographic Image Analysis Society) dataset is essential in breast cancer detection with the help of Exploratory Data Analysis (EDA)[23]. This dataset used by of many machine learning models to detect the breast cancer. In the future, the quality of mammogram-derived features would be improved using advanced feature extraction methods, for instance wavelet transforms, multi-fractal dimensions and texture analysis, etc. Techniques like Principal Component Analysis (PCA) and Genetic Algorithms (GA) can be employed to optimize feature selection, improving classifier performance significantly [24]. Integration of New Features: Combining traditional features with new descriptors from deep learning approaches could yield better insights into tumor characteristics, potentially leading to improved diagnostic accuracy.

References

1. Elkorany, A. S., Marey, M., Almustafa, K. M., & Elsharkawy, Z. F. (2022). Breast Cancer Diagnosis Using Support Vector Machines Optimized by Whale Optimization and Dragonfly Algorithms. IEEE Access, 10(June), 69688–69699. https://doi.org/10.1109/ACCESS.2022.3186021
2. Palakurthi, Bhavana, et al. "Targeting CXCL16 and STAT1 augments immune checkpoint blockade therapy in triple-negative breast cancer." Nature Communications 14.1 (2023): 2109.
3. "UCI Machine Learning Repository: Breast Cancer MIAS (Diagnostic) Data Set". 2020. Archive.Ics.Uci.Edu. https://archive.ics.uci.edu/ml/datasets/Breast+Cancer+MIAS+%2 8Diagnostic%29.
4. Velleman, Paul F., and David C. Hoaglin. "Exploratory data analysis." (2012).
5. Bock, Hans-Hermann, and Edwin Diday, eds. Analysis of symbolic data: exploratory methods for extracting statistical information from complex data. Springer Science & Business Media, 2012.

6. Kanaan, Yasmine M., Robert L. Copeland Jr, Melvin Gaskins, and Robert L. DeWitty Jr. "Exploratory Data Analysis on Breast Cancer Prognosis." In Encyclopedia of Information Science and Technology, Fourth Edition, pp. 1794–1805. IGI Global, 2018.
7. Khan, S.A. and S.S. Velan. Application of exploratory data analysis to generate inferences on the occurrence of breast cancer using a sample dataset. in 2020 International Conference on Intelligent Engineering and Management (ICIEM). 2020. IEEE.
8. Antabe, R., et al., Utilization of breast cancer screening in Kenya: what are the determinants? 2020. 20(1): p. 1–9.
9. Purohit, K.J.I.J.o.D.S. and Analysis, Separation of Data Cleansing Concept from EDA. 2021. 7(3): p. 89
10. Sahoo, K., et al., Exploratory data analysis using Python. 2019. 8(12): p. 2019.
11. Komorowski, M., et al., Exploratory data analysis. 2016: p. 185–203
12. Sweetlin, E.J. and S. Saudia. Exploratory Data Analysis on Breast cancer dataset about Survivability and Recurrence. in 2021 3rd International Conference on Signal Processing and Communication (ICPSC). 2021. IEEE.
13. Verma, Vikram, Alankrita Aggarwal, and Tajinder Kumar. "Machine and Deep Learning Approaches For Brain Tumor Identification: Technologies, Applications, and Future Directions." 2023 International Conference on Computational Intelligence and Sustainable Engineering Solutions (CISES). IEEE, 2023.
14. Bibal, Adrien, Valentin Delchevalerie, and Benoît Frénay. "DT- SNE: T1SNE discrete visualizations as decision tree structures." Neurocomputing 529 (2023): 101–112.
15. Sharma, Nipun, and Swati Sharma. "Optimization of t-SNE by Tuning Perplexity for Dimensionality Reduction in NLP." International Conference on Communication and Computational Technologies. Singapore: Springer Nature Singapore, 2023
16. T. Kumar, M. Arora, V. Verma, S. Lalar, and S. Bhushan, —Pre- Examination of Breast Cancer Dataset Using Exploratory Data Analysis (EDA) Approach,‖ in Proceedings - 2024 International Conference on Computational Intelligence and Computing Applications, ICCICA 2024, Institute of Electrical and Electronics Engineers Inc., 2024, pp. 1–7. doi: 10.1109/ICCICA60014.2024.10585026.
17. S. Prusty, S. K. Dash, S. Patnaik, S. G. P. Prusty, and N. Tripathy, —EPD: an integrated modeling technique to classify BC,‖ in 2023 International Conference in Advances in Power, Signal, and Information Technology, APSIT 2023, Institute of Electrical and Electronics Engineers Inc., 2023, pp. 651–655. doi: 10.1109/APSIT58554.2023.10201778.
18. S. Hussain et al., —TECRR: a benchmark dataset of radiological reports for BI-RADS classification with machine learning, deep learning, and large language model baselines,‖ BMC Med. Inform. Decis. Mak., vol. 24, no. 1, Dec. 2024, doi: 10.1186/s12911-024-02717-7.
19. E. J. Sweetlin and S. Saudia, —Exploratory data analysis on breast cancer dataset about survivability and recurrence,‖ in 2021 3rd International Conference on Signal Processing and Communication, ICPSC 2021, Institute of Electrical and Electronics Engineers Inc., May 2021, pp. 304–308. doi: 10.1109/ICSPC51351.2021.9451811.
20. Z. Mushtaq, M. F. Qureshi, M. J. Abbass, and S. M. Q. Al-Fakih, —Effective kernel-principal component analysis based approach for wisconsin breast cancer diagnosis,‖ Electron. Lett., vol. 59, no. 2, Jan. 2023, doi: 10.1049/ell2.12706.
21. 2020 International Conference on Intelligent Engineering and Management (ICIEM). IEEE, 2020.
22. ال اجعا امهس and ال اص غ ري م ن روص, —An Exploratory Data Analysis of Breast Cancer Features in South of Libya,‖ J. Pure Appl. Sci., vol. 21, no. 4, pp. 57–64, Oct. 2022, doi: 10.51984/jopas.21 4 2126.
23. M. M. Hossin, F. M. Javed Mehedi Shamrat, M. R. Bhuiyan, R. A. Hira, T. Khan, and S. Molla, —Breast cancer detection: an effective comparison of different machine learning algorithms on the Wisconsin dataset, Bull. Electr. Eng. Informatics, vol. 12, no. 4, 2446–2456, Aug. 2023.
24. S. Das and S. Mukherjee, —Advanced Optimization Techniques & Its Application in AI-Powered Breast Cancer Classification, in Proceedings - International Conference on Developments in eSystems Engineering, DeSE, Institute of Electrical and Electronics Engineers Inc., 2023, pp. 65–70. doi: 10.1109/DeSE58274.2023.10099678.
25. Q. H. Hoang, L. M. Duong, P. L. L. Bui, A. V. Tran, T. A. Nguyen, and V. D. Nguyen, —A Comparative Study of Machine Learning Algorithms for Breast Cancer Classification, in International Conference on Advanced Technologies for Communications, IEEE Computer Society, 2023, pp. 409–414. doi: 10.1109/ATC58710.2023.10318887.

Note: All the figures and tables in this chapter were made by the authors.

Intelligent Systems Using Semiconductors for Robotics and IoT – Dinesh Goyal et al. (eds)

82 Stock Price Prediction Using Machine Learning

Mrunalini Bhandarkar*,
Sonal Shirke, Archana Bhamare
Department of Electronics & Telecommunication,
Pimpri Chinchwad College of Engineering,
Pune, India

Abstract: Stock charge prediction is a tough challenge inside the economic region that has won a lot of attention in recent years. In this paper, an overview of a system mastering-primarily based estimation of stock expenses version is applied, which uses historic stock data to estimate destiny developments. The method entails pre-processing the facts, characteristic engineering, and selecting a suitable gadget studying algorithm. Artificial neural networks, linear regression, choice bushes random forests and assist vector machines, have all been investigated for inventory rate forecasting with encouraging outcomes. However, the projections strongly trusted accuracy of the first-class of the enter records and the chosen features, as stock price prediction remains a complex trouble with excessive uncertainty and volatility. The accuracy of stock charge prediction algorithms should therefore be stepped forward in an effort to boom their reliability for investors and investors. Despite the problems, device getting to know algorithms have validated their capacity to as it should be count on inventory values and might provide buyers and traders insightful information.

Keywords: Stock price prediction, Linear regression, Autoregressive integrated moving average (ARIMA), Gated recurrent unit (GRU)

1. Introduction

Artificial intelligence and data analytics are regularly used in finance to assume inventory fees using system getting to know algorithms. This technique makes use of past inventory price facts to forecast destiny inventory charges in addition to different pertinent information such as news, economic indicators and market sentiment. For this, you can utilize device studying methods like neural networks, aid vector machines, random forests, choice trees, and regression evaluation. These algorithms build patterns from historic stock prices together with other pertinent data to forecast destiny expenses. Because the market for stocks is extremely unpredictable and numerous elements that have an impact on inventory expenses are difficult to measure, predicting stock expenses is a challenging assignment. Machine getting to know algorithms, then again, can observe good sized volumes of facts and notice complex styles that may be difficult for people to notice. Overall, the software of algorithms primarily based on gadget mastering to forecast the charge of shares is a promising place of research that would assist investors make higher selections and boom their returns. The process of utilizing historical stock price data along with other pertinent information to build a model based on machine learning that can predict future stock prices is known as stock price prediction utilizing machine learning algorithms. This method aims to assist traders and investors in selecting the best stocks to purchase, sell, and hold. Machine learning algorithms use a range of methods, such as random forests, neural networks, decision trees, and linear regression to forecast stock values. These algorithms can examine trends and patterns in past stock price data as well as other pertinent data, including news articles, social media sentiment, business financial statements, and economic indicators.

*Corresponding author: mrunalini.bhandarkar@pccoepune.org

DOI: 10.1201/9781003716389-82

The quality and volume of the data needed to train the model determines how accurately machine learning algorithms can anticipate the price of shares. In order for the model to remain accurate and up to date over time, it must also be regularly updated with fresh data. In general, employing machine learning algorithms to predict stock prices is a useful tool for traders and investors who want to choose their assets wisely. Nonetheless, it's important to remember that no computer algorithm can precisely expect future trends, and investing in stocks is constantly dangerous.

2. Literature Survey

This studies investigates inventory price prediction using a hybrid ARIMA-LSTM model, which mixes nonlinear LSTM fashions and the linear Autoregressive incorporated transferring common (ARIMA) to forecast time collection statistics. The effects show that the version outperforms the man or woman models in phrases of MSE, MAE, and RMSE. However, barriers exist, which include except statistics earlier than 2010 and the model's incapacity to account for all economic anomalies. The observe suggests the want of extra study to improve the model's accuracy, in particular in reading the economic role before 2010, and emphasizes the ongoing challenge of addressing Black Swan activities within the economic enterprise. [1] FinRL is a hands-on deep reinforcing literacy (DRL) library designed to facilitate easy comparison of personal progress for drug users entering quantitative finance. This DRL library employs stock request datasets, neural networks, and trading performance analysis to help users create and test their own stock trading methods. FinRL is highly customizable, allowing users to adapt request simulators, literacy algorithms, and profitable tactics.[2] This study focuses on creating a stock price forecasting model, evaluating the time series models and LSTM neural network. LSTM, leveraging non-direct information and sequence data, outperforms time series models, as indicated by a smaller root mean square error. While initial findings support the efficacy of LSTM in stock price prediction, the composition acknowledges the need for further exploration, considering factors like research time and data sources. Ongoing work includes addressing unusual values, optimizing model selection points, and implementing updates and improvements. [3] This study explores using artificial intelligence (AI) for stock market prediction through two approaches: specialized retrogression analysis and abecedarian sentiment analysis. While the retrogression model accurately forecasts the day-end stock price with a small error, and the sentiment analysis achieves a delicacy of 76, the research suggests AI's current limitations. The study indicates that although AI can forecast stock price changes and public opinion on demand, its delicacy is insufficient, and predictions for subsequent days may lack accuracy, potentially resulting in capital loss. This underscores the current challenges in achieving precise stock price predictions using AI methods. [4] This study introduces a novel stock price prediction model leveraging deep reading technology, integrating traditional fiscal indicators with insights extracted from social media data. Utilizing a long-term memory model, the method surpasses existing practices in terms of MAE, RMSE, and R2, indicating its superior ability to forecast stock values. While the study currently focuses on one stock and one platform's social media data, the promising results suggest potential extensions to consider inter-stock relationships for broader predictions.

The model's success underscores the efficacy of incorporating deep literacy technology and social media insights for enhanced stock price forecasting.[5] Stocks, known for their financial potential, attract investors seeking high profits and flexible trading. Forecasting stock prices, influenced by macroeconomic factors, market conditions, events, preferences, and managerial decisions, has long been a challenging study area. Traditional methods relying on statistical and economic models prove ineffective in the dynamic stock market environment. Since the 1970s, researchers have turned to machine learning methods for predicting stock prices and variations, leveraging technological advancements to offer investors more effective trading methods in this complex landscape.[6]

It advises investors to avoid stocks with intrinsic values higher than market values. This method is effective for long-term predictions, considering factors like book value, revenue, and ratios. On the other hand, technical analysis relies on studying stock price trends and quantitative data to make short and long-term predictions based on continuously changing investor behaviors. Analysts may derive various rules from the same data, prioritizing technical analysis over fundamental analysis for system inputs. [7] With recent advancements, particularly in deep learning, the survey provides a comprehensive summary of current research, covering data sources, neural network architectures, implementation, evaluation criteria, and reproducibility standards. [8]

The study uses Long Short-Term Memory (LSTM) to forecast stock price time series and analyses methods such as Multi-Layer Perceptron (MLP), Convolutional Neural Network (CNN) and Moving Average to Pingan Bank trading data. The results reveal that the combination of PCA and LSTM surpasses identical mathematical models in regards to prediction accuracy, demonstrating how successful this technique is in mitigating the effects of the market's intrinsic volatility and unpredictability. [9] Forecasting stock prices is difficult due to the market's unpredictable nature and reliance on a variety of factors. This study compares standard statistical methods to artificial intelligence (AI) methodologies, specifically LSTM and MLP, for historical stock price forecasting. [10]

After testing each method independently, AI techniques, notably MLP and LSTM, were discovered to be the most dependable, with the Mean Absolute Percentage Error (MAPE) and low Mean Squared Error (MSE) values. The study investigates a variety of machine learning and statistical methodologies, highlighting AI approaches improved predictive accuracy in projecting stock values. [11] utilizing a Deep Q-Network using a CNN function estimation. Remarkably, patterns in stock chart images consistently forecast similar future stock price movements across international markets. The outcomes demonstrate the model's capacity to forecast future stock values, even when training and testing methods are conducted in different countries. While the study shows exceptional annual returns in some nations, it acknowledges that real-world factors such as transaction costs, taxes, bid-ask spreads, and actual stock volumes could impact the actual profitability achieved by implementing the model. [12] This paper tells the consistency in the financial exchange utilizing Profound Convolutional Organization and candle charts. This paper utilizes different sorts of brain networks like convolutional brain organization, lingering organization and visual math bunch network. CNN learning calculation accomplishes the best presentation of awareness, explicitness, exactness, and MCC. To foresee the rate change on the value developments is the future extension. [13] This work focuses on forecasting the value of association of two assets for future time spans, which is important in portfolio growth. As ARIMA-LSTM crossover approach is compared to various traditional predictive monetary theories, such as the complete authentic approach, steady connection approach, single-file approach, and multi-batch analysis. The ARIMA-LSTM model outperformed all other monetary models by a significant margin. Model execution was approved throughout a variety of time periods and resource mixes, using metrics such as the RMSE, MAE and MSE. It is that the constructed model may be defenseless against explicit monetary situations that were missing in the years somewhere between 2008 and 2017.[14] The primary focus of this paper is to see how well the progressions in stock costs of an organization, the ascents and falls, are corresponded with the popular assessments being communicated in tweets about that company. Word2vec and Ngram, for breaking down the public opinions in tweets. Feeling investigation and regulated machine gaining standards to the tweets extricated from twitter and dissect the relationship between securities exchange developments of an organization and feelings in tweets. Positive feelings or opinion of public in twitter about an organization would reflect in its stock cost. The primary disadvantage is that main twitter information for dissecting individuals' opinion which might be one-sided because not everyone individuals who exchange stocks share their viewpoints on twitter. [15]

3. Methodology

3.1 Flowchart

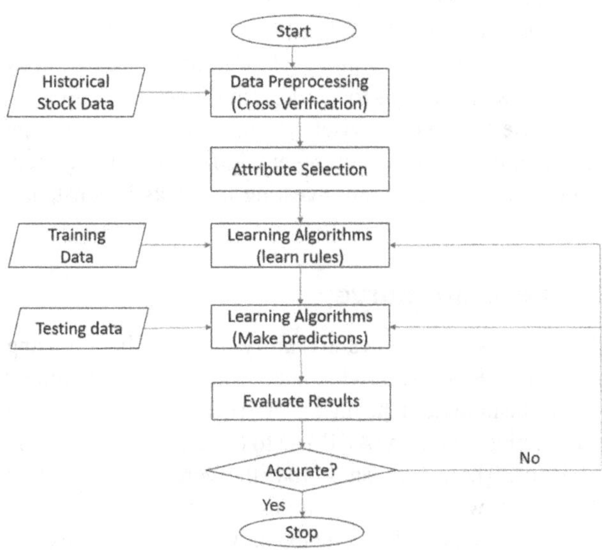

Fig. 82.1 Flowchart

Historical stock data serves as a crucial tool for investors, offering insights into a specific stock's past performance through charts and tables depicting changes in its value over time. The data set contains the initial and final prices, both high and low prices, and volume of trading. To enhance accuracy, the data undergoes preprocessing involving the handling of missing values, correction of outliers, and transformation into a consistent format. Cross-validation techniques are then employed to evaluate model performance. Attribute selection follows, aiming to identify pertinent variables for model analysis and prediction. Training data, encompassing historical price data, financial statements, and economic indicators, is vital for constructing an effective stock market prediction model, with its quality and relevance significantly impacting model accuracy. Figure 82.1 shows the flowchart for implementation.

The process of building a stock market prediction model involves training data with various algorithms to detect potentially unnoticed recurring trends, and correlations. to human analysts. Machine learning algorithms play a crucial role in analyzing large datasets, uncovering complex relationships, and making predictions based on historical data. These algorithms adapt to new data and refine predictions over time, providing valuable insights in a dynamic market environment. The trial data is then utilised to assess the effectiveness of the model, employing metrics like as F1 score, recall, precision accuracy. Considerations for making predictions include a solid understanding of the data, the time frame analyzed, and relevant economic or political events. It's crucial to be aware of model limitations, potential biases, and the broader market context when assessing prediction results.

Accuracy and reliability of predictions are assessed by comparing predicted outcomes to actual outcomes, with an acknowledgment of the inherent unpredictability of the stock market. If results are unsatisfactory, the process iterates by learning new algorithms and refining predictions.

3.2 Algorithms Used

1. GRU (Gated Recurrent unit)

Cho et al. presented GRU (Gated Recurrent Unit) in 2014 as a solution to the issue of vanishing gradients found in typical repetitive deep neural networks. GRU is allied to LSTM (Long Short-Term Memory) in its design, and it can sometimes achieve similar results. However, GRU is typically faster and more efficient than LSTM, and it has fewer parameters, which can make it easier to train.

Reasons for selecting GRU over LSTM are:

- Speed and efficiency: GRU is typically faster and more efficient than LSTM, which can be important for large datasets or real-time applications.
- Fewer parameters: GRU has fewer parameters than LSTM, which can make it easier to train and less prone to overfitting.
- Similar performance: GRU can sometimes achieve similar performance to LSTM, especially on tasks where long-term dependencies are not as important.

Working of GRU Algorithm:

GRU employs update gates and reset gates to determine which portion of the prior concealed state is retained. The gate for updating controls the extent to which of the prior concealed state is updated using the current input. The gate that resets the system controls the extent to which of the preceding concealed state is reset. This enables the GRU to acquire dependence over time without encountering the issue of vanishing gradients. Figure 82.2 illustrates a more comprehensive representation of a single GRU and Fig. 82.3 shows the details about the notations used. Update gate operation is shown in Fig. 82.4.

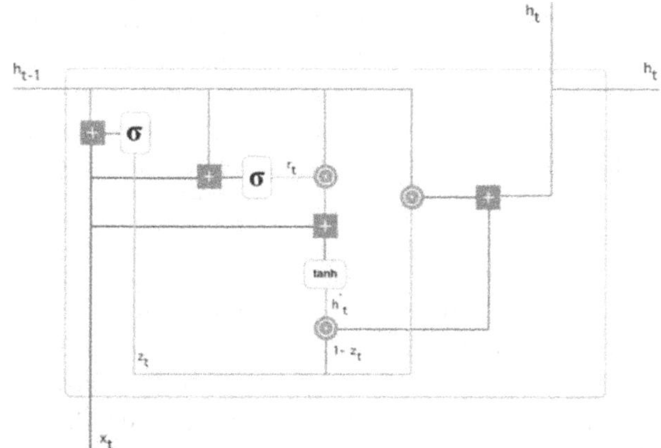

Fig. 82.2 Detailed version of single GRU

Fig. 82.3 Notations

Update Gate:

In GRU, the replace gate is a medium that determines the significance of preserving the vintage retired country and incorporating the clean enter into the existing retired phase. The replace gate is a sigmoid function that takes as enter the consecution of the previous retired united states of america and the contemporary input, and labors a price among zero and 1 for every detail within the retired country. This fee determines the proportion of the preceding retired nation to be retained and the percentage of the latest enter to be blanketed into the present day-day retired nation. The best system for the update gate in GRU is shown in (1)

$$z\_t = \sigma(W\^z\, x\_t\, U\^((z))\, h\_(t-1)) \quad (1)$$

W(z) is the burden of x_t and U(z) is the weight of h_(t-1). H_(t-1) incorporates the records about preceding t-1 gadgets. While x_t enters into the network unit, it receives accelerated with W(z). Similarly, h_(t-1) gets multiplied with U(z). Bothe the consequences are concatenated and a characteristic known as the sigmoid activation is used to trim the output between zero and 1..

The update gate in GRU lets in the community to broadly modernize the retired kingdom grounded on the enter data, thereby enabling the community to address long-term dependencies and decrease the evaporating grade problem.

Reset Gate:

In GRU, the reset gate is another gating medium that helps the network to widely forget or retain the former retired state. The reset gate is also a sigmoid function that takes as input the consecution of the former retired state and the current input. The affair of the reset gate is used to modernize the seeker hidden state, which is a combination of the former retired state and the current input. The fine formula for the reset gate in GRU is as given in (2)

$$r_t = \sigma(W^{(r)} x_t + U^{(r)} h_{t-1}) \quad (2)$$

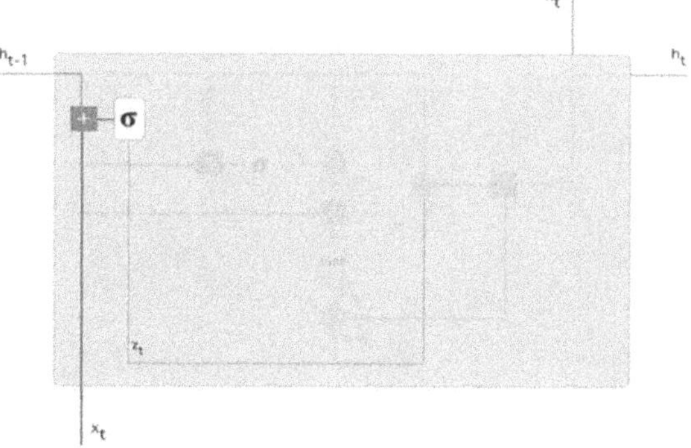

Fig. 82.4 Update gate

The reset gate permits the community to widely neglect or maintain the previous retired country grounded on the cutting-edge enter. However, it way that the network should forget about the former retired country and calculate more at the contemporary enter, If the reset gate affair is near zero. Nonetheless, it means that the network should preserve the former retired country and calculate extra on the former records, If the reset gate affair is close to 1.

Current Memory Content

In this step, a new candidate hidden state is introduced, which will use the reset gate to store the relevant information from the past. It is calculated using (3)

$$h'_t = tanh(Wx_t + r_t \odot h_{t-1}) \qquad (3)$$

a. Start by multiplying the input at the current time step (x_t) by a weight matrix W, and multiply the hidden state from the previous time step (h_(t-1)) by a weight matrix U.
b. Next, calculate the Hadamard product (detail-wise product) among the reset gate at the cutting-edge time step (gate_t) and the made from U with h_(t-1).
c. Concatenate the results from step a and step b.
d. Apply the non-linear activation function (tanh) to the concatenated end result from step c.
e. Perform an detail-smart multiplication between the hidden kingdom from the preceding time step (h_(t-1)) and the reset gate on the contemporary time step (r_t).
f. Add the end result from step e to the input at the modern-day time step (x_t). Finally, observe the tanh activation feature to supply the updated hidden kingdom, denoted as h'_t.
g. The final reminiscence on the contemporary time step, denoted as h_t, is a vector that shops information for the cutting-edge unit and transmits it to the network. The update gate, which is part of the reset gate, determines which data to hold from the cutting-edge memory content (h't) and what to accumulate from the previous nation (h(t-1)).

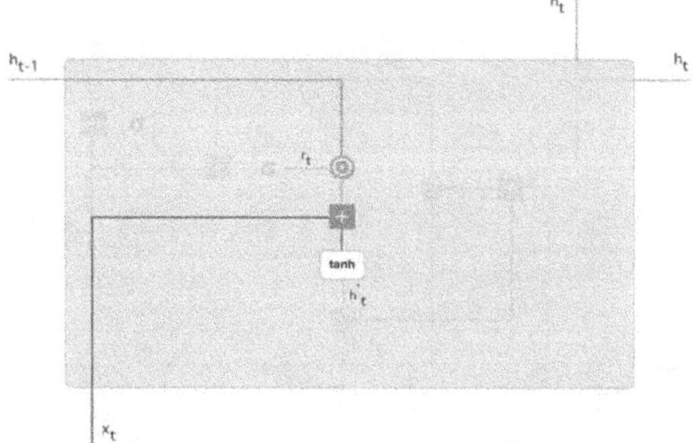

Fig. 82.5 Current memory content

This is achieved by using following formula given in (4)

$$h_t = z_t \odot h_{t-1} + (1 - z_t) \odot h_t \qquad (4)$$

A. Carry out the element sensible multiplication of update gate z_t and h_t.
B. Carry out the element clever multiplication of 1-z_t and h'_t.
C. Concatenate the result of step 1 and step 2.

The version can discover ways to keep the maturity of the prior facts while placing the vector z_t close to 1. Since 1-z_t will be close to zero, it will be feasible to discard a considerable percentage of the existing material this is beside the point to our vaticination due to the fact z_t may be near 1 presently step.

The implementation of Current memory content material and Final Memory at current time step is shown in determine 5 and six respectively.

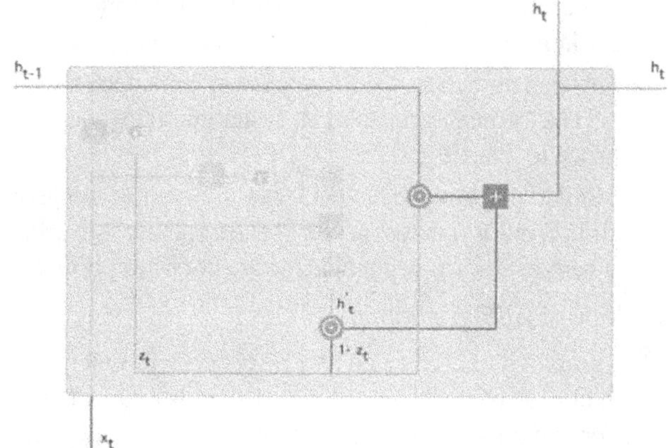

Fig. 82.6 Final memory at current time step

2. Linear Regression

Linear regression is a kind of supervised device gaining knowledge of set of rules used to predict a non-stop final results variable (also called a structured variable or reaction) based on one or greater enter variables (also called unbiased variables or predictors). The aim of linear regression is to discover a linear courting between the input variables and the final results variable, so that the outcome can be expected at once for brand new records factors. The set of rules works by using fitting a linear equation of the shape Y = a bX, wherein X represents the enter variables, Y represents the output variable, a represents the intercept, and b represents the slope of the road. This equation is fitted to a schooling dataset, which includes enter-final results pairs. The set of rules tries to locate the values of a and b that minimize the difference between the anticipated outcome and the real outcome in the schooling dataset. There are two forms of linear regression: Simple Linear Regression (SLR) and Multiple Linear Regression (MLR). SLR involves only one enter variable, whilst

MLR entails or greater input variables. Linear regression has numerous assumptions that have to be met so as to attain correct predictions. These assumptions encompass linearity, normality, homoscedasticity, and independence. Linear regression is broadly utilized in diverse fields, which include finance, economics, social science, and engineering, where predicting a continuous outcome variable is crucial. The set of rules also can be extended to handle nonlinear relationships between the enter and outcome variables the usage of strategies consisting of polynomial regression, ridge regression, or Lasso regression.

3. ARIMA

ARIMA (Autoregressive Integrated Moving Average) is a predictive algorithm used for modeling and forecasting time collection data. It combines three components: Autoregressive (AR), Integrated (I), and Moving Average (MA) models. The AR element predicts future values based totally on past values, the I component makes the time collection stationary by taking differences between consecutive observations, and the MA factor predicts destiny values based on the error phrases of past values. Three parameters of ARIMA model are p, q, The variable q denotes the numerical value of the MA component, which is the overall quantity of not on time forecast errors used inside the version. ARIMA models may be healthy to time series information using statistical software program like R or Python. Once the model is geared up, it is able to be used to make destiny predictions with the aid of forecasting the collection primarily based at the predicted model parameters. It is vital to note that ARIMA models expect the underlying statistics is desk bound, that means the suggest and variance of the facts do now not alternate through the years. However, if the statistics is non-stationary, additional steps like differencing or seasonal adjustments can be needed earlier than becoming an ARIMA version.

3.3 Database Details

Google stock data

- The dataset is of Google Stock Price
- It is on the Daily basis
- The data is from August 2008 to October 2021(13 years) which is taken on a daily basis.
- Amount is mentioned in terms of USD
- Attributes included in the dataset: Date, open, close, high, low, volume and adjClose

4. Results

The result has been obtained for 3 different algorithms with the same database. The output for each algorithm is as follows.

1. Linear Regression

Actual vs predicted price for stock market using Linear regression is shown in Fig. 82.7.

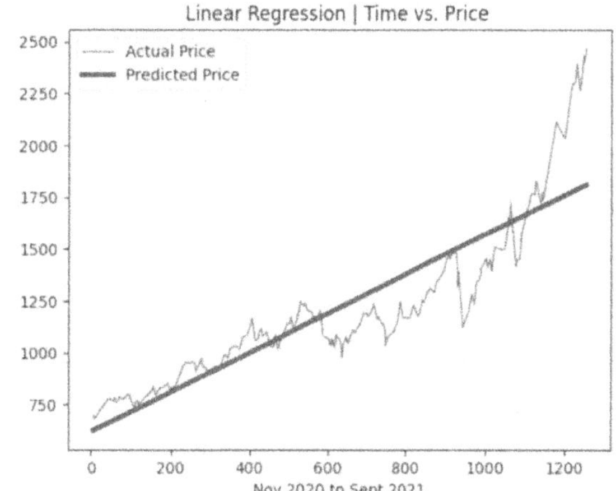

Fig. 82.7 Stock price prediction using linear regression

Linear Regression is a very basic algorithm and does not give accurate results. The blue line shows the predicted data. It is observed that this model has a very wide difference between actual and predicted data. Variance can be observed.

2. GRU Model

GRU model is a fast as well as highly accurate model. It can be observed that the actual price and predicted price are very similar. Though there is some amount of variation visible, this model is better than the linear regression model. Actual vs predicted price for stock market using GRU Model is shown in Fig. 82.8.

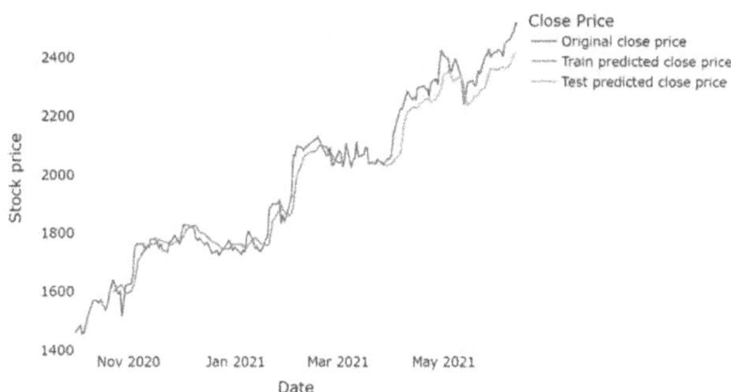

Fig. 82.8 Stock price prediction using GRU algorithm

3. ARIMA Model

Actual vs predicted price for stock market using ARIMA model is shown in Fig. 82.9.

Among all the three models taken into consideration, the ARIMA model is the best in terms of results. This is a highly accurate model and gives appropriate results. The variance is very low. Hence this model can be relied upon. this model.

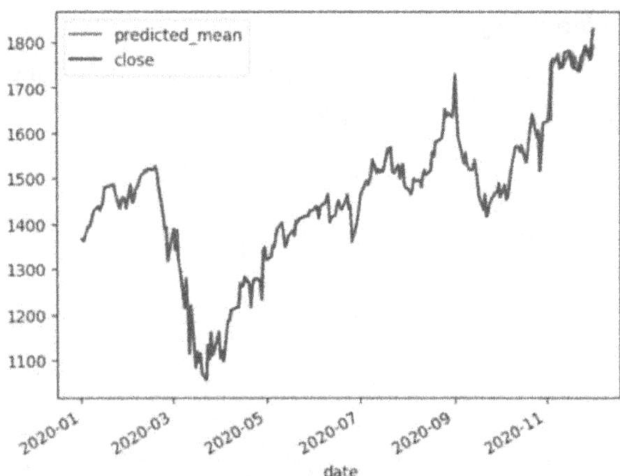

Fig. 82.9 Stock price prediction using ARIMA model

4.1 Comparison in Terms of Accuracy

It is difficult to make a definitive comparison between GRU, ARIMA, and linear regression for stock price prediction accuracy as it can depend on various factors such as the dataset, the pre-processing techniques, the hyperparameters chosen, and the evaluation metrics used.

Accuracy:

As mentioned earlier, it is challenging to provide a definitive comparison of these methods in terms of accuracy, as it can vary based on several factors. Deep learning models, such as GRU, have the potential to outperform classic approaches like ARIMA and linear regression in certain instances. This advantage arises from their ability to detect sophisticated trends and non-linear relationships in the data. Linear regression, which is meant for linear relationships, is likely to have struggled with the data's intricacies, resulting in the lowest accuracy. In contrast, GRU, which is designed to handle nonlinearities, attained the maximum accuracy by skillfully navigating these complications. This demonstrates the potential efficiency of deep learning models when interacting with complex and nonlinear time series data.

Time complexity:

The time complexity of these methods can differ. ARIMA and linear regression are generally fast and efficient because they involve fitting a single model to the data. However, they might not be suitable for handling large datasets, as they necessitate loading the entire dataset into memory. Conversely, deep learning models like GRU can be computationally expensive and necessitate substantial computational resources, as they entail training a complex neural network on the data. This can pose challenges, particularly when working with large datasets.

Space complexity:

The space complexity of these methods can also vary. ARIMA and linear regression are relatively lightweight and require minimal storage space, as they involve fitting a single model to the data. Deep learning models such as GRU, on the other hand, can require significant storage space, as they involve training a complex neural network with many parameters. This can be particularly challenging when dealing with large datasets, as the model and its parameters may not fit into memory.

5. Conclusion

In summary, comparing GRU, ARIMA, and linear regression for predicting stock prices involves several considerations including accuracy, time complexity, and space complexity. It is challenging to provide a definitive comparison in terms of accuracy, as it can vary based on factors such as dataset characteristics, preprocessing methods, and evaluation criteria. Nevertheless, deep learning models like GRU have the potential to achieve higher accuracy, particularly in cases with complex patterns and nonlinear relationships. In terms of time complexity, ARIMA and linear regression are usually faster and more efficient, while deep learning models like GRU can be computationally demanding and require substantial computational resources. Regarding space complexity, ARIMA and linear regression typically require minimal storage, whereas deep learning models like GRU may need more space. The choice of method ultimately depends on the specific use case and available resources, taking into account factors such as dataset size, computational resources, and the desired level of accuracy.

When considering overall factors of space, time and accuracy ARIMA provides better results. Even though space and time required for GRU is greater, it is capable of handling complex nonlinear data. Deep learning algorithms like CNN, GRU, GANs etc. can be explored and compared for obtaining more accurate results.

Table 82.1 Comparison of results for various algorithm

Algorithm	Linear regression	ARIMA	GRU
Accuracy	Low for non-linear data	Low for non-linear data	High for non-linear data and complex patterns
Time complexity	Relatively fast for small amount of data	Relatively fast for small amount of data	Slow due to requirement of significant computational resources
Space complexity	Relatively lightweight & require minimal storage space	Relatively lightweight & require minimal storage space	require significant storage due to complex neural network with many parameters.
Remark	Statistical method used for non-linear data & short-term predictions	Statistical method used for non-linear data & short-term predictions	Suitable when data is nonlinear & complex & long-term predictions are to be done

Acknowledgment

Authors accept grateful students, staff and authority at the Department of Electronics and Telecommunications for their collaboration in research.

References

1. Daiyou Xiao and Jinxia Su, "*Research on Stock Price Time Series Prediction Based on Deep Learning and Autoregressive Integrated Moving Average*", Hindawi Scientific Programming Volume 2022, Article ID 4758698, 31 March 2022.
2. Xiao-Yang Liu1 ,Hongyang Yang , Qian Chen , Runjia Zhang , Liuqing Yang , Bowen Xiao , Christina Dan Wang, "*FinRL: A Deep Reinforcement Learning Library for Automated Stock Trading in Quantitative Finance*", 2 Mar 2022.
3. Yixin Guo, "*Stock Price Prediction Using Machine Learning*", Södertörn University | School of Social ScienceMaster, 2022.
4. Sohrab Mokhtari, Kang K Yen, Jin Liu, "*Effectiveness of Artificial Intelligence in Stock Market Prediction Based on Machine Learning*", International Journal of Computer Applications (0975–8887), 30 June 2021.
5. HungChun Lin, Chen Chen, GaoFeng Huang and Amir Jafar, "*Stock price prediction using Generative Adversarial Networks*", 05-03-2021.
6. Xuan Ji, Jiachen Wang and Zhijun Yan, "*A stock price prediction method based on deep learning technology*", International Journal of Crowd Science Vol. 5 No. 1, 7 September 2020.
7. Polamuri Subba Rao, K. Srinivas, and A. Krishna Mohan, "*A Survey on Stock Market Prediction Using Machine Learning Techniques*", May 2020.
8. Weiwei Jiang*, "*Applications of deep learning in stock market prediction: recent progress*", 29 Feb 2020.
9. Yulian Wen, Peiguang Lin and Xiushan Nie, "*Research of Stock Price Prediction Based on PCA-LSTM Model*", 2020.
10. Mehar Vijha , Deeksha Chandolab, Vinay Anand Tikkiwalb, Arun Kuma, "*Stock Closing Price Prediction using Machine Learning Techniques*", International Conference on Computational Intelligence and Data Science (ICCIDS 2019).
11. Indronil Bhattacharjee, Pryonti Bhattacharja, "*Stock Price Prediction: A Comparative Study between Traditional Statistical Approach and Machine Learning Approach*", 4th International Conference on Electrical Information and Communication Technology (EICT), 20-22 December 2019.
12. Jinho Lee, Raehyun Kim, Yookyung Koh, and Jaewoo Kang, "*Global Stock Market Prediction Based on Stock Chart Images Using Deep Q-Network*", 28 Feb 2019.
13. RosdyanaMangir Irawan Kusuma, Trang-Thi Ho, Wei-Chun Kao, Yu-Yen O and Kai-Lung Hua, "*Using Deep Learning Neural Networks and Candlestick Chart Representation to Predict Stock Market*", 26 Feb 2019.
14. Hyeong Kyu Choi, "*Stock Price Correlation Coefficient Prediction with ARIMA-LSTM Hybrid Model*", 1 Oct 2018
15. Venkata Sasank, Kamal Nayan, Ganapati Panda, "*Sentiment Analysis of Twitter Data for Predicting Stock Market Movements*", 28 Oct 2016.

Note: All the figures and table in this chapter were made by the authors.

Intelligent Systems Using Semiconductors for Robotics and IoT – Dinesh Goyal et al. (eds)
© 2026 Taylor & Francis Group, London, ISBN 978-1-041-20408-4

83 A Review: Hybrid Energy Systems for Off-Grid Rural Electrification: Necessity, Challenges, and Future Perspectives

Payal Bansal[1]
Professor,
Department of Electronics & Communication Engineering,
Poornima Institute of Engineering & Technology,
Rajasthan, India

Dato' Dr Noor Inayah Yaakub[2]
Vice Chancellor,
University College International Maiwp

Rajeev Kumar[3]
Professor,
Computer Science and Engineering Department,
Moradabad Institute of Technology,
Moradabad, U.P, India

Abu Bakar Abdul Hamid[4]
Professor,
Infrastructure University Kuala Lumpur, (IUKL),
Malaysia

Abstract: Hybrid power is the result of combining different technologies to produce energy. The term 'hybrid' in power engineering refers to a power and energy storage system that is integrated. Typically, hybrid systems involve the integration of renewable energy technologies, such as wind turbines and solar photovoltaic (PV) cells. These systems offer a high level of energy security by combining multiple modes of electricity generation, and often incorporate energy storage systems like batteries or fuel cells, as well as fossil-fueled generators to ensure a reliable and secure power supply. This review paper delves into a comprehensive analysis of over 40 research papers, extracting their strengths, weaknesses, and common findings. The results highlight the pressing need for hybrid systems in India.

Keywords: Hybrid, Off grid, Rural, Electrification

1. Literature Review

The common findings extracted from the review of research papers are presented in point form which are as follows:

- Most of the researchers have been used Homer Software for analysis and optimization.
- The installed capacities are documented and found to amount to a total of 838 MW for biomass-based power generation including bagasse and non-bagasse cogeneration and biomass gasifiers. The major contributor to this capacity is bagasse fired boilers at 692.3 MW.

[1]payal.bansal@poornima.org, [2]noorinayah@ucmi.edu.my, [3]rajeev2009mca@gmail.com, [4]abubakarhamid@iukl.edu.my

DOI: 10.1201/9781003716389-83

- In comparison to the centralized power stations, biomass-based power plants are worthwhile in terms of investment and operational costs for that plant if its operating plant load factor is at least 70% throughout the year.
- Biomass-based power plants with a plant load factor over 70% and all-year-round operation justify the acceptance of investment and operational costs when compared to bigger power stations.
- Annually, India generates 686 million tonnes of gross crop residue biomass, out of which an estimated surplus of 234 million tonnes (34% of gross) is available for energy generation.
- Remote villages in India are consistently supplied with electricity for lighting for 4-5 hours by renewable isolated power plants with a capacity of 1-20 kW.
- Animal-source Micro-Small Wind Electric Generation (SWEG) has estimated generation cost based on projects with a rated capacity of 3.2 kW installed at locations with an annual mean wind speed of 6-10 m/s in the range of Rs. 12.77-27.96/kWh.
- The micro hydropower (MHP) is a technology of renewable energy implicated, by most experts, to be the best viable and promising solution for rural electrification-mostly in Nepal because it is easy to manage and operate, affordable, and thus suitable for many communities in Nepal. Although many other renewable energy technologies (RETs) may be available, MHP is among the favorite because of its simplicity and low costs.
- Nepal has the capacity for generation from electricity sources at an installed capacity of 1786 MW and more than 6000 MW of hydropower potential yet to be tapped. However, this limited electricity access is just 2.2% in rural areas, with around 78% of the people termed poverty-stricken.
- Hybrid energy systems can include either a mix of energy sources or hybridization of their use to provide their reliable electricity supply under off-grid conditions, which can significantly reduce the total life cycle cost of all power systems stand alone. This means that instead of relying on just one energy source, a hybrid energy system could have solar, wind, and diesel generators as sources to supply a more consistent and reliable power supply. The hybrid of energy sources can also optimize the production of energy by each energy source based on the available sources at each location and the specific need of each source. Overall, hybrid energy systems represent a cost-efficient and sustainable option to palm reliability into an off-grid location.
- According to the analysis, the average discounted cost of electricity from diesel and photovoltaic sources is \$0.62/kWh and \$0.33/kWh, respectively, which is more than the amount Liberians would be willing to pay.
- According to the analysis, the costs of electricity generated using diesel and photovoltaic sources are \$0.62/kWh and \$0.33/kWh, respectively, which are higher than what Liberians are willing to pay.

2. Comparative Analysis of Research Work Reviewed

Various authors have considered precision, recall, and accuracy as the performance evaluation parameters. This section presents the solutions and methodologies used by the researchers for each issue, along with the results displayed in Table 83.1.

3. Strengths and Weaknesses of the Reviewed Area

After, the review of more than 40 research papers the strengths and weaknesses of various solution approaches of various methods used by different authors were listed as:

Strengths

- The cost of the energy of the model using LINDO software was decreased to 0.65Rs/kWh compared to the cost of energy using HOMER software. (6)
- The best low-cost and efficient way was found in Scenario 4 (MHP-Biomass-Biogas-Energy Plantation-Wind-SPV).
- Utilizing the full potential of biogas and meeting the remaining demand with biomass or PV technologies results in a lower per unit electricity cost. (27)
- Biomass-based power plants with plant load factors exceeding 70% operating year-round can perform comparably to centralized power stations in terms of investment and operational costs.
- Solar PV power, despite not providing enough electricity for cooking or machinery, was found to increase worker presence in health clinics and schools and is a promising potential solution for rural electrification due to several reasons, including its current technology and cost.
- Wind penetration improves the system voltage profile in distribution system but power increases due to grid code requirements in large scale wind farm.
- The Micro Hydro Power (MHP) scheme is one of the most promising indigenous technologies to bring electricity to rural areas in Nepal, thanks to its simplicity, cost-effectiveness, and ease of management and operation. Other renewable energy technologies (RETs) also show potential.
- The Renewable Energy Fund (REF) has benefited from increased financing and private sector involvement in rural electricity supply systems, thanks to liberalization and a supportive legal framework.

Table 83.1 Comparison on the basis of literature review

S. No	Author Name and year	Type of system	Input							Output				
			Tools	Location	Latitude (N) & Longitude (E)	Resources				MHP	SPV	Wind	Biomass	Cost of energy
						MHP	SPV	Wind	Biomass	MWh/yr	MWh/yr	MWh/yr	MWh/yr	($/kWh)
1	K.R. Jyothy 2017	Hybrid	Homer	--	--	--	Y	--	--	--	3700-5000	--	--	--
2	J. A. Weber 2016	Hybrid	Homer	--	--	--	Y	Y	--	--	1825	1759	--	0.47
3	B. J. Saharia 2015	Hybrid	Homer	Guwahati	26.13 N & 91.66 E	--	Y	Y	--	--	8760	2372	--	0.53
4	H. E.Khashab 2015	Hybrid	Homer	Yanbu, Saudi Arabia	24.02 N & 38.19 E	--	Y	Y	--	--	6261	2726	--	0.44
4	M.S. Adaramola 2014	Hybrid	Homer	Nigeria	9.52 N & 8.54 E	--	Y	--	--	--	685	--	--	0.41
5	H. Borthanazad 2013	Hybrid	--	Malaysia	4.21 N & 101.97 E	Y	Y	Y	--	2091	1509	1414	--	0.52
6	S. Mahapatra 2012	Decen-tralized	Homer	South Asia	2.21 S & 115.66 E	N	Y	--	Y	--	565787	--	83790	0.44
7	S. Singh 2013	Decen-tralized	--	Rural area of N-W	--	--	--	Y	Y	--	--	17000	5000	0.55
8	R. Sen 2013	Off grid	Homer	Chhattis-garh	21.29 N & 81.82 E	Y	Y	--	Y	106649	34439	--	25294	0.49
9	M. Ranjeva 2012	Hybrid	HYPORA	Rural area	--	--	Y	Y	Y	--	1625	849.8	402	0.61
10	S.C. Bhattacharya 2011	Hybrid	Homer	--	--	Y	Y	Y	--	5	25.5	10	50	0.45
11	A.L. Hossain 2010	Decen-tralized	Nasa & Homer	Dinajpur	25.3 N & 91.9 E	Na	Y	Y	--	--	23413	18323	--	0.47
12	S.K. Nandi 2010	Hybrid	Homer	Bangladesh	21.51 N & 92.85 E	--	Y	Y	N	--	53317	89151	--	0.46
13	K. voorspools 2009	On Grid	Homer	--	--	--	--	Y	--	--	--	1000	--	0.67
14	S.K. Nandi 2009	Hybrid	Homer	Kutubdia Island	21.81 N & 91.85 E	--	Y	Y	--	--	1275	1060	--	--
15	R.B. Hiremath 2008	Decen-tralized	Lindo	Karnataka	15.31 N & 75.71 E	--	Y	Y	Y	--	14000	50	1900	0.5
16	A.K. Akella 2005	Off grid	Lingo	Uttrakhand	29.38N & 79.29E	Y	Y	Y	Y	293040	12859	8890	198556	0.38
17	S. Rana 1997	Off Grid	--	Madhya Pradesh	22.97 N & 78.65 E	--	Y	Y	--	--	7891	2678	--	0.47

- Hybrid energy systems can offer a cost-effective and reliable solution for stand-alone power supplies in off-grid situations by combining different energy sources.
- A successful approach to rural electrification in some areas has been to develop local grids first and then integrate them, using appropriate small-scale systems that can be upgraded or connected to create a regional network, thus creating demand.

Weaknesses

- The cost of the energy of the model using HOMER software was increased to 2.63Rs/kWh compared to the cost of energy using LINDO software.(6)

- Some hurdles in rural electrification such as Low loads and low capacity utilization, high transmission and distribution losses, CO2 emission etc.
- Providing a reliable electricity service with solar PV power stations poses limitations, especially when compared to fossil fuel-based rural electrification programs.
- In decentralized PV rural electrification, the battery is considered the most expensive component (at 19.5% of the cost) compared to the PV module (at 17.5% of the cost).
- A Constant Power Factor is more suitable than a Constant Voltage target for wind farms as it can optimize network operations, such as power quality.
- It is difficult and unlikely to connect rural areas to the national electricity grid due to various factors such as the geographical remoteness of these areas, scattered consumers, high supply and maintenance costs, low consumption levels, and low household income levels.
- Even though PV hybrid systems are more complicated than using a single energy source, they have the potential to fulfill the electricity requirements of rural areas in a cost-effective manner, despite difficulties such as determining the appropriate system size and incorporating PV energy into the load.

4. Findings from the Review

- In India, the total installed capacity of biomass-based power generation stands at 838 MW, which covers bagasse and non-bagasse cogeneration biomass gasifiers. The bulk of this capacity at 692.3 MW is bagasse-based, which suffers from some inefficiencies and CO2 emissions.
- However, biomass-based power plants that operate throughout the year with a plant load factor exceeding 70% are feasible and could perform comparably to centralized power stations in terms of investment and operational costs.
- India produces a large amount of gross crop residue biomass, with an annual output of 686 million metric tons. Out of this, 234 million metric tons, or about 34%, are estimated to be surplus for energy generation.
- Moreover, in remote areas of India, many renewable isolated power plants with capacities of 1-20 kW consistently provide electricity mainly for lighting for 4-5 hours.
- Small Wind Electric Generation (SWEG) projects which are running on mean wind speed of 6-10 m/s annually with a capacity of 3.2 kW can provide electricity at a levelized cost of Rs. 27.96-12.77/kWh.
- Micro hydropower (MHP) is a renewable energy technology that is highly promising for rural electrification in Nepal. It is preferred among other renewable energy technologies (RETs) due to its simplicity in management and operation, cost-effectiveness, and ease of use.
- India has the potential to generate a large amount of electricity, with an installed capacity of 1786 MW and undeveloped hydropower potential exceeding 6000 MW. Despite this potential, only less than 3% of rural areas have access to electricity, leaving a large portion of the population, roughly more than 75%, in poverty.
- Hybrid energy systems are a type of power system that combines multiple energy sources to provide reliable electricity in off-grid locations. By integrating different sources of energy, such as solar and wind, hybrid systems can provide a more stable and cost-effective solution compared to traditional stand-alone power supplies. These systems have the potential to reduce the overall life cycle cost of providing electricity in off-grid locations.
- An analysis has shown that the levelized cost of electricity from diesel and photovoltaic sources is higher than Liberians' willingness to pay, making it challenging to provide affordable electricity using these sources in Liberia. Researchers mostly focused on calculating losses and performance analysis based on geographical location and orientation separately.
- Researchers mostly focused on hypothetical system rather than working on some real installed system. Multiple parametric variations were not applied in the reviewed papers.

References

1. Banerjee, M. J. Tierney, and R. N. Thorpe, "Thermo economics, cost benefit analyses, and a novel approach to consider revenue-generating system dissipation in candidate decentralized energy system for Indian rural villages," Energy. 2012 Jul 31;43(1)pp. 477–488.
2. A. Faiz and A. Rehman, "Hybrid Renewable Energy Systems: Hybridization and Advance Control," Power Generation Systemand Renewable Energy Technologies (PGSRET), 2015 2015 Jun 10 IEEE, pp. 777–781.
3. A. H. Mondal and M. Denich, "Hybrid systems for decentralized power generation in Bangladesh," Energy for Sustainable Development. 2010 Mar 31;14(1), pp. 48–55.
4. A. P. Ruiz, M. Cirstea, W. Koczara, and R. Teodorescu, "A Novel Integrated Renewable Energy System Decentralized Approach, Allowing Fast FPGA Controller Prototyping," Optimization of Electrical and Electronic Equipment, 2008. (OPTIM 2008). 11th IEEE International Conference 2008, pp. 3456–3461.
5. A.B. Kanase-Patil, R.P. Saini, and M.P. Sharma, "Integrated Renewable Energy Systems for off grid rural electrification in remote area," Renewable Energy. 2010 Jun 30;35(6), pp. 1342–1349.
6. A.K. Akella, M.P. Sharma, and R.P. Saini, "Optimum utilization of renewable energy sources in a remote area," Renewable and Sustainable Energy Reviews. 2007 Jun 30;11(5) pp. 894–908.

7. B. J. Saharia, Zaheeruddin, M. Manas, and A. Ganguly, "Optimal Sizing and Cost Assesment of Hybrid Renewable Energy System for Assam Engineering College," India Conference (INDICON), 2015 Annual IEEE IEEE 2015 Dec 17, pp. 6540–6546.
8. D. P. Kaundinya, P. Balachandra, and N.H. Ravindranath, "Grid-connected versus stand-alone energy systems for decentralized power A review of literature," A review of literature. Renewable and Sustainable Energy Reviews. 2009 Oct 31;13(8) pp. 2041–2050.
9. F. Bouffard, and D. S. Kirschen, "Centralised and distributed electricity systems," Energy Policy. 2008 Dec 31, 36(12) pp. 4504–4508.
10. H. Borhanazad, S. Mekhilef, R. Saidur, and G. Boroumandjazi, "Potential application to renewable energy for rural electrification in Malaysiain" Renewable energy. 2013 Nov 30, vol.-59,pp. 210–219.
11. H. E. Khashab and M. A. Ghamedi, "Comparison between hybrid renewable energy systems in SaudiArabia," Journal of Electrical Systems and Information Technology. 2015 May 31;2(1), pp. 111–119.
12. J. A. Weber, D. W. Gao and T. Gao, 'Affordable Mobile Hybrid Integrated Renewable Energy System Power Plant OptimizedUsingHOMERPro,'toappearat:NorthAmerican Power Symposium (NAPS), 2016. IEEE, 2016, pp.1–6.
13. J. Urpelainen, "Grid and off-grid electrification: an integrated model with applications to India," Energy for Sustainable Development. 2014 Apr 30 vol. -19 pp. 66–71.
14. K. Voorspools, P. Voets and W. D'haeseleer, "Small Scale Decentralized Electricity Generation and the Interaction with the Central Power System," Power generation and sustainable development. International conference 2001, pp. 363–368.
15. K.R. Jyothy, C.P. Raju and R. Srinivasarao, "Simulation Studies On WTG-FC-Battery Hybrid Energy System," International Conference on Innovative Mechanisms for Industry Applications (ICIMIA 2017), IEEE 2017, pp.223–230.
16. M. Ranjeva and A.K. Kulkarni, "Design optimization of a hybrid, small, decentralized power plant for remote/rural areas," Energy Procedia 20, 2012, pp. 258–270.
17. M. S. Adaramola, S. S. Paul and O. M. Oyewola, "Assessment of decentralized hybrid PV solar-diesel power system for applications in Northern part of Nigeria," Energy for Sustainable Development. 2014 Apr 30, vol.19, pp. 388–398.
18. Md. Ibrahim, A. Khair and S. Ansari, "A Review of Hybrid Renewable Energy Systems for Electric Power Generation," Int. Journal of Engineering Research and Applications ISSN: 2248-9622, Vol. 5, Issue 8, 2015, pp.42–48.
19. O. Erdinc and M. Uzunoglu, "Optimum design for hybrid renewable energy systems: Overview of different Approaches," Renewable and Sustainable Energy Reviews. Vol.-16 Issue-32012 Apr; pp.1412–1425.
20. P. Bajpai, V. Dash, Hybrid Renewable Energy Systems for Power Generation in Standalone Applications: A Review. Renewable and Sustainable Energy Reviews, 16(5), 2012, pp. 2926–2939.
21. R. Sen and S. C. Bhattacharyya, "Off-grid electric generation through renewables in India: A case study of HOMER," Renewable Energy. 2014 Feb; vol.-62 pp. 388–398.
22. R. B. Hiremath, B. Kumar, P. Balachandra, N. H. Ravindranath and B. N. Raghunandan," Decentralized Renewable Energy: Possibilities, Significance and Applications with Reference to Indian Context," Energy for Sustainable Development.Vol-13, Issue-1, 2009, pp. 4–10.
23. S. C. Bhattacharyya, "Review of alternative methodologies for analysing off-grid electricity supply," Renewable and Sustainable Energy Reviews. 2012 Jan 31, vol.-6, Issue-1, pp. 677–694.
24. S. K. Nandi and H.R. Ghosh, "Prospects of wind-PV-battery hybrid power system replacing grid extension in Bangladesh," Energy. 2010 Jul 31, vol.-35(7), pp. 3040–3047.
25. S. K. Nandi and H.R. Ghosh, "Techno-economical analysis of off-grid hybrid systems at Kutubdia Island-Bangladesh," Energy Policy. 2010 Feb 28;38(2):976–980.
26. S. Mahapatra and S. Dasappa, "Rural electrification: Optimizing the choice between decentralized renewable energy sources and grid extension," Energy for Sustainable Development. 2012 Jun. 30. Vol.-16, Issue-2. pp. 146–154.
27. S. Negil and L. Mathew, "Hybrid Renewable Energy System: A Review," International Journal of Electronic and Electrical Engineering, ISSN 0974-2174, Volume 7, Number 5 (2014), pp. 535–542.
28. S. Rana, R. Chandra, S. P. Singh and M. S. Sodha, "Optimal Mix of Renewable Energy Resources to Meet the Electrical Energy Demand in Villages Of Madhya Pradesh," Energy Conversion and Management., 1998, Vol.39, No.3/4, pp.203–216.
29. S. Singh and A. K. Kori, "Centralized And Decentralized Distributed Power generation In Today's Scenario," IOSR Journal of Electrical and Electronics Engineering, Volume 4, Issue 5, 2013, PP 40–45.
30. Z. Maheshwari and R. Ramakumar, "Smart Integrated Renewable Energy Systems (SIRES) for Rural Communities," Power and Energy Society General Meeting (PESGM), 2016, pp. 456–464.
31. Somwanshi, Devendra, and Anmol Chaturvedi. "Design and Optimization of a Grid-Connected Hybrid Solar Photovoltaic-Wind Generation System for an Institutional Block." In International Conference on Artificial Intelligence: Advances and Applications 2019: Proceedings of ICAIAA 2019, pp. 319–326. Springer Singapore, 2020.
32. Gothwal, Narendra, Tanuj Manglani, Devendra Kumar Doda, Devendra Somwanshi, and Mahesh Bundele. "Design and Optimization of PV–Wind–DG and Grid-Based Hybrid Energy System for an Educational Institute in India." In International Conference on Artificial Intelligence: Advances and Applications 2019: Proceedings of ICAIAA 2019, pp. 131–139. Springer Singapore, 2020.
33. Sharma, Vijay Prakash, Devendra Kumar Somwanshi, Kalpit Jain, and Raj Kumar Satankar. "Optimization of Renewable Energy Resource-A Review." In IOP Conference Series: Earth and Environmental Science, vol. 1084, no. 1, p. 012003. IOP Publishing, 2022.
34. Sharma, Vijay Prakash, Devendra Kumar Somwanshi, Kalpit Jain, and Raj Kumar Satankar. "A Literature Review on Renewable Energy Resource and Optimization." In IOP Conference Series: Earth and Environmental Science, vol. 1084, no. 1, p. 012002. IOP Publishing, 2022.

For Product Safety Concerns and Information please contact our EU
representative GPSR@taylorandfrancis.com
Taylor & Francis Verlag GmbH, Kaufingerstraße 24, 80331 München, Germany

www.ingramcontent.com/pod-product-compliance
Lightning Source LLC
Chambersburg PA
CBHW081217220326
41598CB00037B/6804
* 9 7 8 1 0 4 1 2 0 4 0 8 4 *